T0142252

Smart Innovation, Systems and Technologies

Volume 337

The Smart Innovation, Systems and Technologies book series encompasses the topics of knowledge, intelligence, innovation and sustainability. The aim of the series is to make available a platform for the publication of books on all aspects of single and multi-disciplinary research on these themes in order to make the latest results available in a readily-accessible form. Volumes on interdisciplinary research combining two or more of these areas is particularly sought.

The series covers systems and paradigms that employ knowledge and intelligence in a broad sense. Its scope is systems having embedded knowledge and intelligence, which may be applied to the solution of world problems in industry, the environment and the community. It also focusses on the knowledge-transfer methodologies and innovation strategies employed to make this happen effectively. The combination of intelligent systems tools and a broad range of applications introduces a need for a synergy of disciplines from science, technology, business and the humanities. The series will include conference proceedings, edited collections, monographs, handbooks, reference books, and other relevant types of book in areas of science and technology where smart systems and technologies can offer innovative solutions.

High quality content is an essential feature for all book proposals accepted for the series. It is expected that editors of all accepted volumes will ensure that contributions are subjected to an appropriate level of reviewing process and adhere to KES quality principles.

Indexed by SCOPUS, EI Compendex, INSPEC, WTI Frankfurt eG, zbMATH, Japanese Science and Technology Agency (JST), SCImago, DBLP.

All books published in the series are submitted for consideration in Web of Science.

José Luís Reis · Marc K. Peter ·
José Antonio Varela González · Zorica Bogdanović
Editors

Marketing and Smart Technologies

Proceedings of ICMarkTech 2022, Volume 2

Editors
José Luís Reis
University of Maia—ISMAI
Maia, Portugal

Marc K. Peter
FHNW School of Business
University of Applied Sciences and Arts
Olten, Switzerland

José Antonio Varela González
University of Santiago de Compostela
Santiago de Compostela, Spain

Zorica Bogdanović
Faculty of Organizational Sciences
University of Belgrade
Belgrade, Serbia

ISSN 2190-3018 ISSN 2190-3026 (electronic)
Smart Innovation, Systems and Technologies
ISBN 978-981-19-9101-1 ISBN 978-981-19-9099-1 (eBook)
https://doi.org/10.1007/978-981-19-9099-1

This Springer imprint is published by the registered company Springer Nature Singapore Pte Ltd.
The registered company address is: 152 Beach Road, #21-01/04 Gateway East, Singapore 189721, Singapore

Preface

This book is composed by the papers written accepted for presentation and discussion at the 2022 International Conference on Marketing and Technologies (ICMarkTech'22). This conference had the support of the University of Santiago de Compostela. It took place at Santiago de Compostela, Spain, 1–3, 2022.

The 2022 International Conference on Marketing and Technologies (ICMarkTech'22) is an international forum for researchers and professionals to present and discuss the latest innovations, trends, results, experiences, and concerns in the various fields of marketing and technologies related to it.

The Program Committee of ICMarkTech'22 was composed of a multidisciplinary group of 312 experts and those who are intimately concerned with marketing and technologies. They have had the responsibility for evaluating, in a 'double-blind review' process, the papers received for each of the main themes proposed for the conference: (A) Artificial Intelligence Applied in Marketing; (B) Virtual and Augmented Reality in Marketing; (C) Business Intelligence Databases and Marketing; (D) Data Mining and Big Data—Marketing Data Science; (E) Web Marketing, E-Commerce and V-Commerce; (F) Social Media and Networking; (G) Omnichannel and Marketing Communication; (H) Marketing, Geomarketing, and IoT; (I) Marketing Automation and Marketing Inbound; (J) Machine Learning Applied to Marketing; (K) Customer Data Management and CRM; (L) Neuromarketing Technologies; (M) Mobile Marketing and Wearable Technologies; (N) Gamification Technologies to Marketing; (O) Blockchain Applied to Marketing; (P) Technologies Applied to Tourism Marketing; (Q) Metaverse and NFT applied to Marketing; (R) Digital Marketing and Branding; (T) Innovative Business Models and Applications for Smart Cities.

ICMarkTech'22 received about 220 contributions from 37 countries around the world. The papers accepted for presentation and discussion at the conference are published by Springer (this book, volume 1 and volume 2) and will be submitted for indexing by ISI, EI-Compendex, SCOPUS, DBLP and Google Scholar, among others.

We acknowledge all of those that contributed to the staging of ICMarkTech'22 (authors, committees, workshop organizers and sponsors). We deeply appreciate their involvement and support that was crucial for the success of ICMarkTech'22.

Santiago de Compostela, Spain
December 2022

José Luís Reis
University of Maia—ISMAI
Maia, Portugal

Marc K. Peter
FHNW School of Business
Olten, Switzerland

José Antonio Varela González
University of Santiago de Compostela
A Coruña, Spain

Zorica Bogdanović
Faculty of Organizational Sciences
University of Belgrade
Belgrade, Serbia

Contents

Part II Business Intelligence Databases and Marketing

About the Editors

José Luís Reis has Ph.D. in Technologies and Information Systems from the University of Minho and is Professor with the title of specialist in Management and Administration by IPAM—Porto. He is Professor at University of Maia—ISMAI and ISCAP.IPP and Integrated Researcher in LIACC—Laboratory of Artificial Intelligence and Informatics of the University of Porto. It carries out activities in the area of training and information systems and technologies in various organizations, coordinating various national and international projects in the area of information management, applied marketing and strategic regional planning. He is Author of scientific papers and articles in the fields of marketing automation, artificial intelligence, augmented and virtual reality, information systems modeling, multimedia, gamification, and data mining. He is Author and Co-author of several books, namely *Personalization in Marketing—Technologies and Information Systems*, *Marketing in Agri-food—Fundamentals and Case Studies*, *Gamification Model for SMEs*, *Marketing and Smart Technologies*, and *Information Systems—Diagnostics and Prospectives.*

Marc K. Peter is Professor of Digital Business and Head of the Competence Center Digital Transformation at the FHNW School of Business in Olten, Switzerland. He received his Doctorate from CSU Sydney, an Executive MBA from UAS Bern/Babson College/PKU Beijing, and a Master of Marketing from the University of Basel. He is Fellow of both the British Computer Society and the Chartered Institute of Marketing. His research and teaching areas are digital transformation, digital marketing, new work, and cyber-security.

José Antonio Varela González has been Professor at the University of Santiago de Compostela (USC) since 1976, when he began teaching marketing. He was one of the forerunners of this discipline in Spain. The quality of his teaching and research activity earned him a University Professor position at USC in 1989. He has dedicated his career to this, except for the period between 2005 and 2010—during which he served as President of the Court for the Defense of Competition of Galicia. He has led, as Principal Investigator, more than a dozen competitive projects and contracts with companies and institutions. He has directed 17 doctoral theses. He is Author of more than 60 articles in national and international journals with the greatest impact in marketing. He has presented more than 60 papers at the most important international and national marketing conferences. He has been Director of the Business Organization and Marketing Department for 8 years and Director/Coordinator of the POMARK Research Group from 2010 to 2021. He has belonged to the founding group of the *European Journal of Business Management and Economics*, of which he has been Editor for more than 20 years.

Zorica Bogdanović, Ph.D., is Professor at the Faculty of Organizational Sciences, University of Belgrade, Serbia. She teaches subjects in the areas of e-business and e-business technologies on B.Sc., M.Sc., and Ph.D. studies. Her professional and scientific interests include e-business, Internet marketing, Internet technologies, and Internet of things. Results of her research have been published in many well-known international journals and conference proceedings. She is Member of IEEE and Secretary of IEEE Computer chapter CO 16. She is in Chair of the seminar of IEEE Computer chapter CO 16. She is in the chair of the summer school "E-business technologies" at the Faculty of Organizational Sciences since 2014. Since 2016, she is in Chair of the Center for the Internet of things. She was Head of the Department of e-business at the Faculty of Organizational Sciences 2017–2021.

Part I
Blockchain Applied to Marketing

Chapter 1
The Use of Cryptocurrencies as a Tool for the Development of Marketing in Tourism

Lidia Minchenkova, Alexandra Minchenkova, Vera Vodynova, and Olga Minchenkova

Abstract Through this article, the authors discuss the problems arising in connection with the use of cryptocurrencies in payment for tourist services: cyberthreats and information security in online payments of such services. In response to this, a system will be built to the use of reducing the risks of threats arising from the payment with cryptocurrencies for travel services. In addition, it is proved that filling in the indicators of the proposed system will lead to a significant reduction in such risks and an increase in the use of cryptocurrencies when paying for travel services with the high volatility of cryptocurrency and the characteristic significant risks of irretrievable loss of funds due to fraud, hacker attacks, lack of legal protection. The paper proposes an attack threat factor system on the cryptocurrency—a factor system based on which it became possible to identify the factors in the threats of attacks on cryptocurrencies and show possible ways to neutralize the consequences in the implementation of these threats when paying for tourist services with cryptocurrency.

1.1 Introduction

By cryptocurrency, we understand a digital form of money functions with the help of different technologies and techniques, and several individuals and institutions can create them with the target of future economic decentralization being under the potential absence of central emission [1].

Currently, the tourism industry has been declining due to the COVID-19 pandemic. In response to this, the necessity for new approaches to expand the range of users, so

L. Minchenkova (✉)
Cybersecurity Center, Bernardo O'Higgins University, Santiago, Chile
e-mail: lidia.minchenkova@ubo.cl

A. Minchenkova (✉)
Department of Social Sciences, Bernardo O'Higgins University, Santiago, Chile
e-mail: aleksandra.minchenko@ubo.cl

V. Vodynova · O. Minchenkova
Plekhanov Russian University of Economics, Moscow, Russia

J. L. Reis et al. (eds.), *Marketing and Smart Technologies*, Smart Innovation, Systems and Technologies 337, https://doi.org/10.1007/978-981-19-9099-1_1

many companies and services attracting new customers decided to turn their attention to cryptocurrency as a tool for paying for their services [2].

Back in 2015, Felix Weiss proved that, having only digital coins in his wallet, "you can see almost the whole world," making an unusual experiment visiting 27 countries of the world paying only with bitcoins and showing that cryptocurrencies began to be accepted by different hotels, restaurants, air carriers, and booking services around the world. Also, the travel agency "CheapAir" became a pioneer in this area, switching to cryptocurrency payments in 2013, using a third-party processor to convert bitcoins into dollars, which can then be used for paying to airlines and hotels; this is due to the necessity that still the many travel services only accept payment in "classic" currencies. Another example of the use of "cryptos" is the Australian travel agency "Travala," created in 2017, which accepts not only payments with bitcoin but also 25 other "coins," including its cryptocurrency, called AVA.

Recently, the travel company Expedia returned to payment within cryptocurrency after refusing to accept payments whit them in 2018.

Related to the mentioned above, the ING International Survey analysts' report indicates that only 6% of the total number of travelers around the world were ready to pay for tourist trips with bitcoin [3].

Bitcoin as an example of cryptocurrency (English bit—a unit of information "bit," English coin—coin) is a virtual currency that has "no real value" and the most important characteristic of the system in which bitcoin circulates is decentralization, so it does not have a single emission center, and it does not depend on the banking system implicating that financial market regulators do not control apply control over this currency [4].

Bitcoins are "produced" around the world by their users who have installed special programs on their computers, the so-called bitcoin wallets, that are tools for storing received bitcoins and making operations with them; these virtual wallets are "tied" to the name and passport of their owners, allowing them to execute transactions on several cryptocurrencies' networks [5].

To use bitcoins, you need to purchase a virtual wallet, which is "tied" to the name and passport of its owner. After that, it becomes possible to buy bitcoins using this system. Now, there are more than 620 types of cryptocurrencies in the world.

Among the more than 620 types of cryptocurrencies in the world, bitcoin market share is approximately 35%, followed by Ethereum—22%, and Ripple—8% [6]. Bitcoin and Litecoin are widely used; they are accepted by all existing exchanges, as well as exchange offices; meanwhile, the rest of the cryptocurrencies are built based on the open code of bitcoin and are, in fact, derivatives of bitcoin [7].

During the last years and given the volatility of cryptocurrencies, the number of travel agencies and platforms that accept cryptocurrency for payment has only increased. The fundamental difference of crypto coins from other means of payment lies in the method of issue (issue) of payment and the organization of the system for their storage and payments. Cryptocurrencies have no real value and do not reflect the general state of the economy of a particular country, since they are an international currency [8].

Some disadvantages are that due to the increasing number of cryptocurrencies comes the "cryptocurrency theft," so the industry observers tend to focus on the attack against large organizations—namely hacks of cryptocurrency exchanges or ransomware attacks against critical infrastructure. On the other hand, over the last few years is noted that hackers use malware to steal smaller amounts of cryptocurrency from individual users. Malware refers to malicious software that carries out harmful activity on a victim's device, usually without their knowledge. Malware-powered crime can be as simple as stealing information or money from victims but can also be much more complex and bigger in scale. For instance, malware operators who have infected enough devices can use those devices as a botnet, having them work in concert to carry out distributed denial-of-service (DDOS) attacks, commit ad fraud, or send spam emails to spread the malware further. Some examples of malware are:

1. Cryptojackers: It makes unauthorized use of the victim device's computing power to mine cryptocurrency.
2. Trojans: It is a virus that looks like a legitimate program but infiltrates the victim's computer to disrupt operations, steal, or cause other types of harm.
3. Info stealers: It collects saved credentials, files, autocomplete history, and cryptocurrency wallets from compromised computers.
4. Clippers: It inserts new text into the victim's clipboard, replacing text the user has copied. Hackers can use clippers to replace cryptocurrency addresses copied into the clipboard with their own, allowing them to reroute planned transactions to their wallets [9] (Fig. 1.1).

The clearest example is the use of cryptocurrencies as payment method may be the Spanish travel agency "Destinia," which added a new option of paying and buying

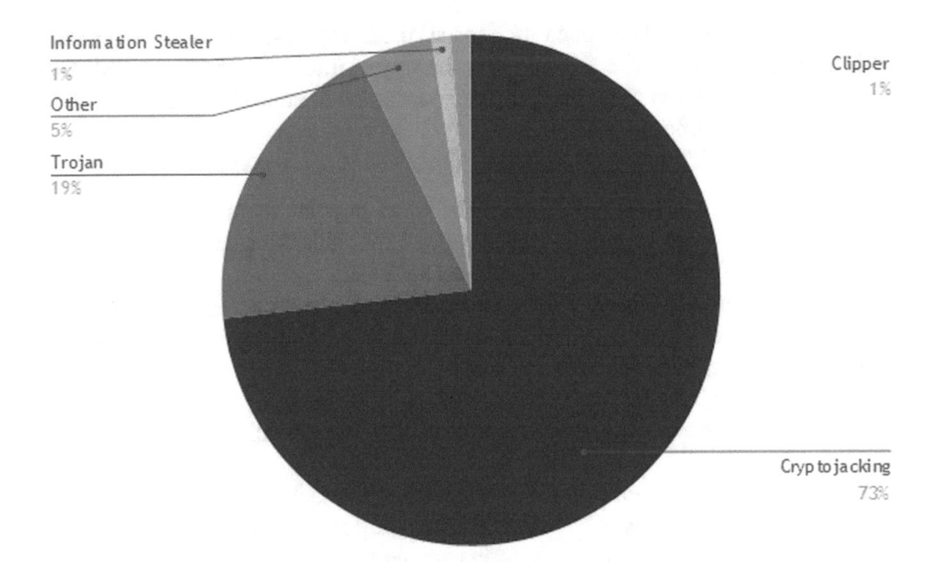

Fig. 1.1 Total value received by malware type. *Source* Chainalysis

air tickets with bitcoins, where the most active customers are from Spain, Sweden, Germany, and Argentina.

South Korea is planning to adopt a law on the legalization of cryptocurrency. Japan is the only country in the world in which there is state control over transactions with cryptocurrencies. In the UK, Coinbase has opened to its UK and Spanish customers the possibility of instantly purchasing bitcoins using credit and debit cards with 3D Secure technology. On the other hand, the Chinese authorities have recently announced a ban on the production of bitcoins, as well as the Government of India, which intends to completely ban the use of cryptocurrencies in the country.

In the next section will be view a attack threat factor system on cryptocurrencies, which can be the basis for neutralizing the risks of paying for tourist services with cryptocurrency.

1.2 Materials and Methods

The paper proposes an attack threat factor system on the cryptocurrency—a factor system based on the possible to identification of factors in the threats of attacks on cryptocurrencies and shows possible ways to neutralize the consequences in the implementation of these threats when paying for tourist services with cryptocurrency. In general, the factors of threat of an attack on cryptocurrencies can be described by the following factor system:

$$\mathbf{TACCi} = \left[\mathbf{V}; \mathbf{OI}; \mathbf{WI\,TACC}; \mathbf{MpCC}; \mathbf{NC}\right] \tag{1.1}$$

where **TACCi** is the i-th threat of an attack on cryptocurrencies; **V** is a current violator; **OI** is the object of influence; **WI TACC** is the way of implementing TACC; **MpCC** is a means of protection of cryptocurrencies, designed to neutralize the TACC; **NC** is negative consequences.

It is possible to highlight the following threats of attack on cryptocurrencies:

- Application of hacking techniques: The most popular methods of attack are malware, phishing, keyloggers, DDoS attacks, clickjacking and watering hole attacks, eavesdropping attacks, or cookies theft [10].
- By hacking hot wallets that are created on third-party Internet resources, the storage of which is carried out in the cloud, and having access to hot wallets, the attacker can move the funds to any place.
- Hacking through social engineering is carried out when receiving confidential data and subsequent access to them. This is often done by disguising yourself as a reliable source.
- Illegal operations are carried out as a result of the lack of currency control of cryptocurrencies. When converting "normal" money into cryptocurrency, it can be cashed out anywhere in the world in a matter of hours. Therefore, in this case, the problem of laundering and circumvention of currency regulation is acute.

Cryptocurrency fraud is carried out by accepting money through the site of exchange offices that exist on the market and work in the "gray" zone; in addition to them, there are clone sites that completely imitate exchange resources. Attackers collect money from users and then go into hiding.

There is also a type of fraud using sites that exchange "normal" money for cryptocurrency. It is carried out as follows: money from cards such as VISA or MasterCard is transferred in the form of payment for some goods or services, and then, the attackers cash out these monetary resources. Another method of fraud is the creation of one-day crypto exchanges.

And the most popular type of fraud since mid-2017 is the Initial Coin Offering (ICO), due to the lack of market regulation, it is quite simple to conduct an ICO, and this is used by attackers to collect cryptocurrency from users with the further disappearance of the initiators of the placement.

a. **Current violators include**:

- Individual hackers who seek to obtain financial or other material benefits.
- Criminal structures also pursue the receipt of financial or other material benefits.
- Owners of crypto exchanges and crypto-exchangers, who, by their unintentional or careless actions, can endanger the cryptocurrency.

b. **The object of influence may be**:

- E-wallets can be created by both ordinary users and crypto exchanges. "Wallets" can be "hot" and "cold." Hot ones may include those that are created on third-party resources. Cold "wallets" are created on workstations, and the storage of money is on users' machines.
- Crypto exchange is a specialized resource, trading which is carried out in cryptocurrency and is used mainly to play on the difference in rates.
- Ordinary users—persons who own and trade cryptocurrency.
- Miners—persons who produce cryptocurrency using their computer.
- The state—the spread of cryptocurrencies carries significant risks for the state for the economy and financial stability; its use as a means of payment for goods and services creates a risk for the country's monetary circulation and the loss of sovereignty of the national currency.

c. **Ways to implement TACC are**:

- Opening the infected file in the user's device is exposed to malware.
- The appearance of sites-clones of exchange offices.
- Installing software on infected machines and mining cryptocurrencies, while users will notice that their computer is working more slowly, and attackers at this time will mine cryptocurrency.

d. **Cryptocurrencies protection tool designed to neutralize TACC** [11]:

  - Two-factor authentication is mandatory in the field of cryptocurrency. Passwords are protected until they can be cracked. Hacking can be related to the incompetence of the user or the same combination for several accounts.
  - Multi-factor authentication is the next level of security for a password that is requested when you sign in or withdraw funds. Two-factor authentication can be in the form of electronic messages to a phone number or the form of an email. One way to authenticate is to use special mobile apps like Google Authenticator, which creates a unique numeric code that lasts for a certain amount of time.

An account that is tied to a specific IP address contributes to a significant reduction in the number of malicious attacks. If an account on a cryptocurrency exchange is not tied to a single IP address, it will be much more difficult to bypass it.

1.2.1 Creation of Cryptocurrencies by Travel Services, Negative Consequences

The long-term potential and application of cryptocurrencies for settlements are limited, and their rapid growth is determined by their speculative demand, which counts on further growth of the rate, which leads to the formation of a bubble. Cryptocurrencies are characterized by their similarity with a financial pyramid since their growth is supported by demand from new participants entering the market.

The spread of cryptocurrencies poses sufficient threats to the well-being of citizens:

- High volatility of the exchange rate, which has spread significantly in fraud.
- Cryptoization, like currencyization, removes the sovereignty of monetary policy, as a result of which, to contain inflation, it will be necessary to maintain a higher level of the key rate permanently. This will reduce the availability of credit for citizens and businesses.
- The spread of cryptocurrencies leads to the withdrawal of citizens' savings beyond the perimeter of the financial sector and, as a consequence, a reduction in its ability to finance the real sector and a decrease in the potential growth of the economy, which reduces the number of jobs and the potential for growth in citizens' incomes.
- Cryptocurrencies are actively used in illegal activities (laundering of donations) moves, drug trafficking, financing of terrorism, etc.). Their proliferation creates favorable conditions for criminal transactions, extortion, and bribery and is a challenge to the global system for combating money laundering and terrorist financing. It is impossible to ensure the necessary transparency of cryptocurrency circulation.

Using this tool, it is possible to increase the profitability of the tourism industry by attracting payments for tourist services.

1.3 Results

We can specify in the system the factor of the cryptocurrency protection tool designed to neutralize the TACC on the example of the online travel agency TRAVALA.com; we can conclude that the use of its crypto coins has a positive effect on increasing sales of tourist services (Table 1.1).

The results obtained are as follows:

AVA was once the most-used individual cryptocurrency in Q4, with BTC coming in second. Travelers who choose to pay with AVA receive additional benefits, including further discounts on top of the Smart Program incentives.

It was also noted that when filling the system with specific indicators, it is possible to minimize the risks of threats to the cryptocurrency and increase the attractiveness of paying for tourist services with cryptocurrency. For example, you can note the filling of various system indicators, such as: ensuring the platform that can handle the volume and the platform secure. To meet this demand hires within the engineering and security departments made up the bulk of the new hires this quarter, as well as a handful in the customer support team. As a result, the most notable improvements to the platform include:

- Significantly enhanced security, including further enhanced defense to protect from DDoS attacks, fixes, and patches for previously identified vulnerabilities.
- Infrastructure upgrades to streamline core booking functionality and improve our database for internal planning and analysis.

Table 1.1 Table payment method for hotel bookings

Hotel bookings by payment option				
Monthly	October 2021 (%)	November 2021 (%)	December 2021 (%)	Q4 2021 Total (%)
AVA	14	12	12	13
Binance pay	16	15	17	16
BTC	9	8	8	8
Other cryptocurrencies	40	47	43	44
Travel credits	11	10	13	11
Traditional currencies	10	8	7	8
Total	100	100	100	100

Source https://blog.travala.com/category/reports/

- Improved flight booking tools to inform travelers about COVID-related rules and restrictions.
- Report TRAVALA, which allowed them to increase sales of cryptocurrency and reduce the risks of threats to cryptocurrencies.

1.4 Discussion

Today, the world is becoming much easier to navigate, and more and more people are traveling to places that throughout history have not been prepared for tourists. And even more so for unprepared tourists. Sometimes, travelers may find themselves in less developed countries and be left without access to ATMs, as well as the ability to pay for goods and services with a bank card. This is where the use of cryptocurrencies and smartphones can help.

In modern literature, there is a rather small number of authors considering the use of cryptocurrencies in tourism in their works, but some authors dealing this problem [12]. So, the author considered the problem of using cryptocurrencies in the hotel sector [13]. On the other hand, the author studied the use of cryptocurrencies in medical tourism [14]. Another author raised the issue of the use of cryptocurrency in the Asian market [15].

Making an overview of the prospects for using cryptocurrencies as payment method for touristic services, we can see that their use is becoming more and more active. At the same time, different approaches of specialists to the possibilities of their use as payment method for touristic services, lead to different ways to address this issue. Nowadays, a variety of trends in the practice of using cryptocurrency payments can be observed, despite the existing disagreements, cryptocurrency continues to develop in new markets, and in several countries around the world, cryptocurrencies have been legalized.

1.5 Conclusion

According to the results of the study, it can be argued that the proposed factor methodology has drawbacks, for example, as a large amount of data is needed, the lack of documentation and automated means of determining the current TACC, as well as the need for highly qualified specialists in the field of information security. The need to track trends in cyberthreats from attackers who are aimed at hacking cryptocurrency data is noted; it is necessary to develop effective ways to protect against cryptocurrency fraud. It should also be noted that by this tool, it is possible to increase the profitability of the tourism industry by attracting payments for tourist services.

Today, in the world, travelers can, being in undeveloped countries in which there is no access to ATMs, pay for goods and services using cryptocurrencies using their phones.

It is important to add that having cash with you in developing countries is dangerous, and paying for goods and services by mobile phone is becoming more common.

One of the important points is to consider different ways of storing cryptocurrency, for example, using a hardware crypto wallet and a bank card. Experienced travelers also are advised to keep money for everyday expenses in a hardware wallet or on a card, and store savings in a multi-signature crypto wallet, and always use two-factor authentication. Also, the information about points that accept cryptocurrency can be found on the CoinMap map or other similar services.

Soon, the number of hotels, restaurants, and shops that accept payment in cryptocurrencies will increase, which will allow international payments to be made in a matter of seconds.

References

1. Monia, M.: Cryptocurrency. Ekonomika. **64**(1), 105–122 (2018)
2. SHOBHIT SETH: What Is Crypto Tourism (2022). Available at: https://www.investopedia.com/tech/what-cryptotourism/
3. Exton, J., Brosens, T.: From cash to crypto: the money revolution, pp. 11–13 (2019). https://think.ing.com/uploads/reports/IIS_New_Tech_Cryptocurrencies_report_18092019.pdf
4. Ingale, S., Bhalekar, P., Pathak, K., Mr. Dixit, S.: Bitcoin Int. J. Sci. Technol. Manage., **5**(07) (2016). Available at: https://www.researchgate.net/publication/362015640_BITCOIN
5. Čavalić, A., Brajić Ex Čabro, S., Hadzic, F.: Cryptocurrencies: a critical review of concepts and definitions (2018)
6. ElBahrawy, A., Alessandretti, L., Kendler, A., Pastor-Satorras, R., Baronchelli, A.: Royal society evolutionary dynamics of the cryptocurrency market (2017). Available at: https://doi.org/10.1098/rsos.170623
7. The Central Bank of Russia: Cryptocurrencies, trends, risks and measures, Moscow Russia (2022). http://www.cbr.ru/content/document/file/132241/consultation_paper_20012022.pdf
8. Karadeniz, E., Beyazgül, M., Günay, F., Kahiloğulları DALAK, S.: Investigation of Using Bitcoin And Other Cryptographic Currencies in The Tourism Sector Social Mentality and Researcher Thinkers J., **12**, 723–731 (2018). https://doi.org/10.31576/smryj.116
9. Chainalysis: The 2022 Crypto Crime Report (2022). Avaible at: Crypto-Crime-Report-2022.pdf
10. Liu, C., Yu, J.: Analysis of DoS attacks on Wireless LAN (2006)
11. Abdullah Alqahtani, F.T. Sheldon.: A survey of crypto ransomware attack detection methodologies: an evolving outlook. Sensors (2022). https://doi.org/10.3390/s22051837
12. Zrnić, M., Njegus, A., Brdar, I., Košutić, J.: Challenges and usage of cryptocurrencies in tourism, SCIndex Iss. **29**, 47–60 (2022). https://doi.org/10.5937/turpos0-37409
13. Blagopolychna, A.G.: Using of cryptocurrency in the sphere of the hotel and restaurant state to tourism (2021). https://doi.org/10.31395/2415-8240-2021-99-2-115-123
14. Çapar, H.: Using cryptocurrencies and transactions in medical tourism. J. Econ. Admin. Sci., **37**(4), 677–693. https://doi.org/10.1108/JEAS-07-2019-0080
15. Treiblmaier, H., Leung, D., Kwok, A.O.J., Tham Min-En, A.: Cryptocurrency adoption in travel and tourism—an exploratory study of Asia Pacific travelers. Curr. Issues Tourism **24**(22) (2020). https://doi.org/10.1080/13683500.2020.1863928
16. TRAVALA: Report Q4 2021 (2021). Avaible at: https://blog.travala.com/category/reports/

Chapter 2
Blockchain Use Possibilities: A Systematic Literature Review

Paulina Rutecka and Eduardo Parra-López

Abstract Blockchain and tourism have the potential to become a very advantageous combination, as this technology can provide security and transparency at certain key points. The blockchain can make the transfer and storage of this information easier and more secure, since the responsibility is shared by an entire network. The same goes for payments abroad, which increases the level of trust between all parties. This study aims to make a comparison between these two countries such as Spain and Poland with a significant growth of this technology in tourism. The results of the paper show that blockchain technology is still in an initial phase of its life and that there is a consensus regarding its potential; however, it is still very difficult to know how far it can go and how it will affect tourism.

2.1 Introduction

In the era of digitization, transactions are carried out over a long distance without the need for personal contact between the parties. In this case, trust in the exchange of information or payment [1] is essential. The development of the Internet [2–4], the Internet services availability [1, 4, 5], and also the increase in digital social competencies [6] translate into a gradually increasing number of available solutions [4], including also for tourism. It also promotes initiatives classified as smart tourism [1, 4, 7, 8]. At the same time, consumers' and entrepreneurs' scale of threats to security and privacy [5, 9] is growing.

Institutions like banks, insurance companies, hospitals, governments, and online intermediaries guarantee people that the promises made will be kept. However, especially in turbulent times, trust in various institutions changes. Economic crises, stock market collapses, epidemics, wars, or inflation can result in sudden changes in the

P. Rutecka (✉)
Department of Informatics, University of Economics in Katowice, Katowice, Poland
e-mail: paulina.rutecka@ue.katowice.pl

E. Parra-López
Department of Business Studies and Historic Economics, La Laguna, Islas Canarias, Spain
e-mail: eparra@ull.edu.es

J. L. Reis et al. (eds.), *Marketing and Smart Technologies*, Smart Innovation, Systems and Technologies 337, https://doi.org/10.1007/978-981-19-9099-1_2

situation, often forcing institutions to change the rules [10]. These changes may lead to a loss of trust in institutions [10, 11]. Due to the deteriorating trust in intermediaries and the costs of servicing such transactions [9, 12], a solution was sought [11] that could guarantee their security while omitting intermediaries. This solution, which is a form of an electronically confirmed civil contract concluded online [2], is called a "smart contract" [2, 8, 13–15]. In the tourism industry, an additional reason for looking for a solution to bypass intermediaries [12, 16] is the monopolization of the market by OTA [3, 17–19] and the freedom to use collected guest data by them [20]. As a consequence, OTAs may impose conditions on consumers or entrepreneurs that may be unfavorable to them. This results in a distrust of intermediaries [9, 12] and the need to reduce high brokerage fees [12, 15].

The main goal of this article is to analyze the possibility of using blockchain in the tourism industry context. Analysis was conducted based on content analysis of abstracts of scientific papers and literature research. We want to verify the practical application of blockchain technology in the business practice of tourism enterprises in Spain and Poland. To achieve the assumed goal, the following research questions were posed:

- RQ1: What are the possible applications of blockchain in the tourism industry?
- RQ2: Are there places in Spain and Poland where one can pay for services with cryptocurrencies?
- RQ3: Is there a relationship between the size of the: region's area, region's population or gross domestic product per capita, and the existence of places where cryptocurrency payments are possible?

The paper is organized as follows. Section 2.2 introduces blockchain issues, its possible use in tourism, and the limitations of implementing blockchain in tourism. Section 2.3 includes the methodology for data retrieval, while Sect. 2.4 presents the findings of the systematic literature review and data and the results of statistical analysis of empirical research. In Sect. 2.5, the authors highlight the research's contribution, discuss its limitations, draw conclusions about the results, and propose possible future research avenues.

2.2 Background

The first and most famous blockchain system is the bitcoin cryptocurrency [16, 21]. Therefore, blockchain technology is most often associated with cryptocurrencies [12]. Its appearance in 2008, at the time of the greatest economic crisis in the USA, gave this technology a strong start [11]. Blockchain, however, is a much broader concept than the cryptocurrencies themselves [15, 18]. This can be described as a decentralized, distributed database system [5, 7, 12, 15, 16, 21] with a high level of security [8, 14, 15, 19, 21], guaranteed by cryptographic algorithms [21] and alleviated the single point of failure problem [5, 18]. Blockchain is compared to an enormous book [12] of records of all operations organized in the form of "blocks"

[8, 13, 21]. Records can only be added to it, but they cannot be removed from it, nor can they be modified [7, 21, 22]. Each block is linked to the next block in the chain by the cryptographic hash of the previous block [23]. Each transaction is verified in the system by the consensus of the majority of its participants [19, 21]. Participants confirm the cryptographic compliance of the data [21] and agree on whether the transaction is trustworthy and can be added as another block in the chain [22] by solving complex cryptographic tasks [23]. The advantage of blockchain technology is that it integrates well with other modern technologies [14]. Blockchain is considered resistant to attacks [17] and less prone to hacking or corruption than centralized systems [5, 24]. As it is a decentralized system maintained by network participants, it belongs to no one [17]. The system also eliminates the necessity of a central institution responsible for transaction verification [5, 22]. The literature currently distinguishes three generations, also known as blockchain eras. These are blockchain 1.0, i.e., cryptocurrencies, blockchain 2.0, i.e., smart contracts, and blockchain 3.0, i.e., the possibility of user interaction with this technology via smartphones and browsers [5, 22, 25, 26]. Blockchain 4.0 operating in real time is also being considered [5].

2.2.1 *Obstacles to Blockchain Implementation in Tourism*

While blockchain technology could revolutionize many industries, including tourism, numerous obstacles to its successful implementation have been identified in the literature. An often stressed concern is the level of technical complexity [17, 25]. The current knowledge of entrepreneurs, including tourists, about blockchain is low [2, 24, 27]. During the qualitative interviews[17] with entrepreneurs, doubts arose that SMEs could keep up with the level of digitization when they can no longer keep up with digital marketing technologies [17]. This can deepen the inequalities between SMEs and large corporations. Corporations sometimes implement small blockchain projects for marketing purposes, not affecting the business model. This is to generate interest and reach a new target group [17]. SME owners who do not have the appropriate financial and human resources to be up to date with all technological novelties may also have more significant concerns about implementing blockchain in their businesses [17]. It may also result from the uncertain legal and regulatory status of this technology [22], although more and more countries are investing in blockchain and are gradually introducing regulations regulating it [24]. The indicated barrier is also scalability and speed of operation. Although bitcoin transactions can also be made on bank holidays, they can handle fewer transactions at one time than traditional payments [18]. As [28] notes, bitcoin can also be associated with criminal activity, as it was often referred to as a medium of payment used on the Dark Net.

The indicated barriers include tradition and attachment to a known form of carrying out processes, e.g., booking [17]. Restrictions may also occur in a market dominated by older adults who cannot use blockchain technology. The level of technical complexity aimed at greater transparency may, paradoxically, deepen the sense of information asymmetry [15] of consumer groups with weaker digital competencies. A challenge for blockchain is also the problem of ensuring the privacy of tourists [25]. This applies particularly to sensitive data [15], e.g., in medical tourism. Tourists may be concerned that their data, including sensitive data, is stored in a decentralized network and can be read by anyone who has it. It is also problematic to delete this data at the customer's request [15], because, according to the blockchain assumptions [25], the blocks are non-modifiable and impossible to delete.

One of the most frequently mentioned possibilities of using blockchain technology in tourism is direct peer-to-peer reservations of tourist services [17, 29], in particular accommodation. This allows you to remove the intermediaries [22] like OTA or banks, who have so far guaranteed the reliability of the concluded contract [29, 30]. However, there are also many doubts on this point. In particular, they relate to the compatibility of technology with older booking systems used in hotels [17]. In addition, it could interfere with using previously used marketing strategies and tools, including OTA, and the problem of accepting reservations directly (at the company's headquarters, by phone, or by e-mail) [17]. The systems would therefore require integration to prevent overbooking from different channels and from favoring customers from blockchain reservations when it occurs. This is related to the necessity to invest in system integration and, thus, again, the dominance of large enterprises: both OTA and hotel chains, which can afford to implement such technologies [17]. The nature of blockchain, as an ownerless system, carries a possible threat in which someone responsible for developing this technology for the needs of the industry will recognize the possibility of its monetization. This would mean that new intermediaries would replace the old intermediaries [17, 31], and the terms of cooperation with them are not possible to estimate at the moment.

The last objection to the use of blockchain is the high consumption of resources [18]. This applies to cloud spaces because data is impossible to delete. Their growth [32] will consume more and more resources necessary to store data blocks [17]. Increasing technology use entails a more significant energy consumption required for cryptographic verification of a greater number of operations [22, 25, 28, 32]. Following the success of bitcoin, many people started mining new blocks. At that time, there were visible increases in prices and insufficient availability of computer components used for cryptographic calculations necessary for the functioning of the technology.

2.3 Materials and Methods

A systematic literature review was conducted to answer the research questions posed. The search was made using the Web of Science and Scopus bibliographic database. The time range was limited to 2017–2022. Articles and Proceeding Papers in English were searched, for which topic contained the keyword "blockchain AND tourism." After restriction articles were rejected, 82 papers were obtained. Subsequently, texts for which full texts were not available were rejected thanks to the affiliation of the

University of Economics in Katowice. After applying exclusions, 54 publications were subjected to abstract analysis. Ultimately, up to 47 articles were used.

In the second part of the study, the use of cryptocurrencies was empirically verified. The website https://coinmap.org was used. This website contains a database and an interactive map of service points where it is possible to make cryptocurrency payments worldwide. It was decided to compare the results of Spain and Poland. These countries were selected because both are located in the European Union. Spain represents the highly developed countries of Western Europe, and Poland represents the less developed countries of Central and Eastern Europe. Both countries are large in area (third and eighth largest countries wholly located in Europe).

2.4 Findings

The research results were divided into two parts. The first is the result of a literature review, and its purpose is to answer RQ1. The second part consists of data obtained through an empirical study of coinmap.org, the results of statistical tests using this data, and statistical and demographic information about Spain and Poland.

2.4.1 Blockchain Application Possibilities in the Tourism Industry

Despite the many concerns, we have tried to outline above that blockchain technology offers unprecedented opportunities to solve urgent social and economic problems. It is essential to consider whether the chance of solving them is more important than the barriers. And also, consider removing the identified obstacles to successfully apply blockchain technology in tourism.

Payments With the Use of Cryptocurrencies

The primary function that can be fulfilled by BCT in tourism, as resulting from the history of BCT, is the possibility of paying with the use of cryptocurrencies [16, 21]. This function is indicated by many researchers in their works [1–3, 5, 12, 13, 18, 33–36]. This applies to the possibility of paying with cryptocurrencies for purchased services, such as, e.g., car rental or the purchase of travel insurance [19], making a hotel room reservation, or purchasing an air ticket [1, 3, 5, 9, 12, 13, 17, 19, 29, 31, 34, 35, 37]. Cryptocurrencies are a universal currency across borders. When making a payment, the tourist does not have to exchange currencies in different countries he travels to [11] and can make international transactions at no additional cost.

Elimination of the Bank's Intermediation in Transactions

Making transactions using cryptocurrencies [29, 30] is also not dependent on the bank's working days and hours, especially in the case of different time zones. Online transactions with BCT do not require support from a credit card service provider or payment gateway, reducing transaction costs [19]. Payment can be made in real time, and the availability of the user's funds is confirmed within seconds after placing the order. Using cryptocurrencies in payments also reduces the risk of corruption [13, 18, 19]. Unfortunately, the current cryptocurrency crisis at the time of writing, which has led to a dramatic decline in the value of bitcoin and other cryptocurrencies, may discourage people from accepting payments in this form. This is because it is associated with the storage of capital in the cryptocurrency market, where the value of their assets may drop by several dozen percentage points within a few days or the need to convert to traditional currency at an unfavorable rate.

Direct Booking of Tourist Services

Besides the possibility of paying for the reservations made, the BCT technology can also support the reservation system [1, 3, 5, 9, 12, 13, 17, 29, 31, 34, 35, 37]. Whether it concerns accommodation, car rental, or the purchase of airline tickets, it is possible to verify the availability of seats and book available seats quickly. This way, the risk of overbooking is minimized because each booking, i.e., a transaction in the chain, is permanently placed in it. However, to eliminate overbooking completely, there is a need to integrate BCT online booking systems with bookings accepted directly (in person, by phone, or by e-mail).

Elimination of OTA Brokerage and Commissions

BCT can be used to eliminate intermediaries and the commission they charge [11] because the guarantee of the availability of the booked service is confirmed in the blocks of the chain. BCT bases its trust on user recommendation systems. If the company offering the service did not exist or did not provide the service, this information would be quickly disclosed in the company's data. The trust function is thus decentralized, and no central intermediaries are needed [1–3, 5, 9, 17, 22, 31, 34, 36]. The possibility of connecting direct tourists and enterprises [19, 29, 30] and concluding their peer-to-peer transactions is increasing [16, 24]. However, the marketing function of the current intermediaries, the recognition of their brands, and their access to the database of potential tourists are invaluable [20]. New platforms that make reservations using BCT take over the marketing role of intermediaries, and there is a risk that they will also want to monetize their activities [17, 31]. Other platforms connect with the existing OTAs, such as TRAVALA with the booking service.

Verification of Tourists' Identity

BCT can assist with identity management and verification [5, 9, 18, 19, 31, 34]. The blocks can store information about the traveler, such as an identity card, driving license, or passport [9]. In this situation, the traveler is not required to carry these documents with him as there would be an exact and secure method of verifying his identity without a physical document. This could prevent the risk of the document being lost or stolen or using false documents. In addition, it is also possible to introduce, thanks to BCT, a solution based on biometric identification [9, 12], which can also successfully replace paper documents. Such verification could reduce the verification time and the queues at the checkpoints [12]. Importantly, the data is stored securely. In the case of a pandemic, as in the case of COVID-19, BCT can assist in verifying vaccination status [23, 38] and verify the certificate's authenticity. At the same time, the anonymity of the person holding the certificate is preserved [39] (e.g., by using biometrics), as well as the security of sensitive data [15].

Asset Tracking and Sharing
BCT can support guaranteeing the authenticity of products or services [17] and the security of asset flow. For example, the subsequent transactions saved in BCT may correspond to the next baggage locations in the air transport process [3, 13, 18, 19, 31, 34, 35]. The system based on BCT can check the availability of hotel rooms in real time and assign a digital room key to a hotel guest [9]. It may also be responsible for the availability of entire facilities, e.g., summer houses, on a similar principle [29, 30].

Control in Supply Chains
Thanks to BCT, it is possible to verify the authenticity of a product, e.g., food [1, 7–9, 11, 14, 28, 29, 31, 35, 40]. Thanks to the verification of a digital certificate that is not modifiable [17], the technology guarantees that the product is original and it is possible to check where it comes from. [7, 35]. Every stage of the production chain [11] and product life cycle can also be certified [14]. It is possible to track each step, manage credentials [17], and even monitor waste [14], especially hazardous waste. BCT allows for traceability, which may reduce the risk of food-borne diseases from other countries [11]. Thanks to BCT, it is possible to control the conditions in which food was stored or transported (e.g., temperature, time) and the compliance of parameters with the requirements [8]. It allows logistic flow control, including the environmental impact of transporting vehicles [14, 28, 41]. Closely following the process makes it possible to tailor better future services [29, 30].

Inventory and Warehouse Management
BCT enables inventory management, e.g., verification of the availability of goods in the warehouse [11]. It also allows for controlling orders from suppliers [17, 34]. It can help transfer inventory between points, e.g., restaurants or hotels, as excess is found at one point and missing at another (within one company or network). It can also help you manage your hotel inventory [29], e.g., clean hotel linen and the need to order services such as laundry.

Loyalty Programs
The support in the implementation of loyalty programs is one of the most frequently mentioned possibilities of using the BCT [1, 9, 11, 17, 19, 29–31, 35, 36, 42, 43]. At the same time, it guarantees that a specific consumer takes part in the program. The card for collecting points or the application cannot be transferred to another person. Consumers can collect points saved in an unmodifiable block, and the collected

and used points are recorded in the chain as subsequent transactions. Certain cryptocurrency units can be collected to be paid anywhere in a loyalty program. The second option is for companies or destinations to create their own cryptocurrency exchangeable for other products offered by a given company or network.

Promoting Sustainable Behavior

By using the loyalty program mechanism and rewards, it is possible to promote sustainable tourism [27]. It is possible to create a reward system for certain behaviors [14, 25, 27, 39, 41], such as saving water and energy in accommodation facilities, cleaning the area of rubbish [25, 41] choosing a bicycle instead of a car [25], or segregating waste that can be recycled [14]. The reward can be a specific unit of any popular cryptocurrency or your own currency created for the needs of Tourism Destinations [25, 41]. As pointed out by Varriale et al., whole destinations could be involved in global programs, which could thus improve their reputation or work together for a reward [14]. Entrepreneurs could also be rewarded for introducing sustainable solutions and engaging their guests. BCT can also be used for the protection of cultural heritage [32] and wild nature protection [44].

Creating Marketing Value

Promoting sustainable solutions in a destination is not the only marketing use of BCT. Thanks to technology, the local community could also access the data [25, 39]. Global action and improved reputation could benefit local communities [39] by increasing interest in the destination. Using data stored in chains, local producers and artisans could reach tourists and offer them their services and goods [25]. The use of BCT may allow for more accurate measurement of tourist traffic and tourists' interest in specific services [1, 45], as well as the exchange of knowledge between various tourist entities [1, 11, 45]. This, in turn, can be used to better serve guests based on preferences based on this data [19] and personalization of services [11, 45], including automated personalization [45]. Because BCT is very innovative, destinations and companies can attract new customers, including those belonging to a specific group of consumers (using BCT or interested in using it). The case of creating your own cryptocurrency can also attract the interest [25] of both consumers and the media.

Reliability of the Information

Because the data in the chain cannot be deleted or modified, BCT guarantees the possibility of verifying the authenticity of, e.g., products [40] or the product's compliance with the description based on opinions. It allows for traceability of other food products [39], along with its production process. Thanks to reliable information, BCT can reduce the problem of information asymmetry [15], which is essential in the tourism industry, where the purchased product cannot be touched and tested beforehand [17]. Raluca-Florentina et al. also propose using BCT as a central database where information about the company's services can be stored. Thanks to this solution, each data update would modify the information in real time in various places where it is displayed, such as a website and OTA content [34]. How such a solution works can be compared to product information management (PIM) solutions from e-commerce.

Credible Opinions

BCT allows you to create reliable and inviolable evaluation and review systems [1–3, 5, 9, 12, 18, 29, 30, 46, 47]. The key function of BCT, which guarantees the reliability of assessments, is that it cannot be modified or removed. Therefore, changing the ex-post review will not be possible [2]. The authors of the opinion can remain anonymous while verifying whether he has used the company's services for which he is writing the review [2]. So, it will not be possible to add several reviews for the same property (e.g., negative reviews), nor will it be possible to order positive ratings. The presence of the guest would have to be confirmed in the company's existing transactions.

Privacy of Tourists and Data Security

Due to complex cryptographic algorithms, BCT allows for high data privacy and anonymization. Travelers can simultaneously verify their identity, medical details, and financial capacity without revealing their personal information to the staff. This increases the security of sensitive data [15] and reduces privacy concerns [39]. At the same time, it guarantees the truthfulness of data and credentials. It also speeds up the processing of data. It improves the experience of tourists [39], who may feel, for example, embarrassed by the possibility of linking their name to the disease being treated (medical tourism).

2.4.2 *Payments with the Use of Cryptocurrencies in Spain and Poland in Tourism-Related Services*

As established from a literature review, payments are the most obvious use of cryptocurrencies. Among other possibilities, they are also the most widely used. To understand the actual level of cryptocurrency used to pay for services, data from https://coinmap.org/ was used. The website provides a map of the places that accept bitcoin divided into categories: Trezor Retailer, ATM, attraction, cafe, food, grocery, lodging, nightlife, shopping, sports, and transport. This study searched the map for places in the categories attraction, cafe, food, and lodging. The searches were made by regions of Spain and Poland. Then, a statistical analysis of the search results was performed.

Table 2.1 presents the search results for the number of places where bitcoin payment is accepted in individual regions of Spain and data on area, population, and GRDP per capita (EUR) in these regions. Analogous data for Poland is presented in Table 2.2. For the recalculation of GRDP from PLN to EUR, the exchange rate of 4.7 from 17/08/2022 was used.

In the analyzed categories in Spain, the largest number of places where it is possible to make payments with bitcoin belongs to the food (46% of the examined objects) and lodging (36% of the investigated facilities) categories. Most places are

Table 2.1 Data about places where cryptocurrencies can be used in Spain by region

Autonomous	Area (km^2)	Population (2020)	GRDP	Attraction	Food	Cafe	Lodging	Sum
Andalusia	87,268	8,464,411	19,107	3	7	4	7	21
Aragon	47,719	1,329,391	28,151	0	2	0	0	2
Asturias	10,604	1,018,784	22,789	0	2	0	2	4
Balearic Islands	4992	1,171,543	27,682	0	4	3	4	11
Basque Country	7234	2,220,504	33,223	0	1	1	1	3
Canary Islands	7447	2,175,952	20,892	1	4	1	14	20
Cantabria	5321	582,905	23,757	0	0	0	0	0
Castile and León	94,223	2,394,918	24,031	0	2	0	2	4
Castilla–La Mancha	79,463	2,045,221	20,363	0	0	0	0	0
Catalonia	32,114	7,780,479	30,426	2	14	2	5	23
Community of Madrid	8028	6,779,888	35,041	2	17	3	12	34
Extremadura	41,634	1,063,987	18,469	0	3	0	0	3
Galicia	29,574	2,701,819	23,183	0	2	1	1	4
La Rioja	5045	319,914	27,225	0	0	0	0	0
Navarre	10,391	661,197	31,389	0	0	0	0	0
Region of Murcia	11,313	1,511,251	21,269	0	1	0	0	1
Valencian Community	23,255	5,057,353	22,426	0	5	2	2	9

located in the Community of Madrid and further in the tourist regions of Spain: Catalonia, Andalusia, Canary Islands, and the Balearic Islands.

There are fewer places in Poland where it is possible to pay bitcoin for services. Most are in the Masovia region (where the country's capital is located) and Kuyavia-Pomerania (the coastal area). In Poland, it is possible to pay for services mainly in objects of the category: Cafe (41% of the examined) and lodging (37% of the examined) (Table 2.2).

The performed Shapiro–Wilk test showed statistical significance ($p < 0.05$). The data is not normally distributed. Therefore, non-parametric tests were performed. Correlation analysis was performed using the rho-Spearman method. The correlation between area, population, and GRDP of regions and places located in regions where it is possible to pay with cryptocurrencies was analyzed (Table 2.3).

The correlation analysis of the rho-Spearman pairs of the area, population, and GRDP parameters for the objects in the categories attraction, food, cafe, and lodging

Table 2.2 Data about places where cryptocurrencies can be used in Poland by region

Voivodeship	Area (km^2)	Population (2020)	GRDP*	Attraction	Food	Cafe	Lodging	Sum
Greater Poland	29,826	3,500,361	12,020	0	0	0	0	0
Kuyavia-Pomerania	17,972	2,069,273	8910	0	0	3	2	5
Lesser Poland	15,183	3,413,931	10,058	0	0	1	1	2
Lodzkie	18,219	2,448,713	10,240	0	1	0	0	1
Lower Silesia	19,947	2,898,525	12,176	0	1	1	0	2
Lublin	25,122	2,103,342	7598	0	0	0	0	0
Lubusz	13,988	1,010,177	9097	0	0	0	2	2
Masovia	35,558	5,428,031	17,686	0	0	3	2	5
Opole	9412	980,771	8740	0	1	0	1	2
Subcarpathia	17,846	2,125,901	7678	0	0	0	0	0
Podlaskie	20,187	1,176,576	7889	0	0	0	1	1
Pomerania	18,321	2,346,717	10,639	0	1	0	0	1
Silesia	12,333	4,508,078	11,416	0	0	0	0	0
Holy Cross	11,711	1,230,044	7866	0	0	1	0	1
Warmia-Masuria	24,173	1,420,514	7725	2	0	0	0	2
West Pomerania	22,897	1,693,219	9181	0	0	2	1	3

Table 2.3 Shapiro–Wilk test results

	Country	Statistics	df	Relevance
Attraction	ES	0.571	17	<0.001
	PL	0.273	16	<0.001
Food	ES	0.736	17	<0.001
	PL	0.546	16	<0.001
Cafe	ES	0.773	17	<0.001
	PL	0.682	16	<0.001
Lodging	ES	0.725	17	<0.001
	PL	0.732	16	<0.001

in Spain showed that for entities of all types, there is statistical significance in correlation with population. The results for pairs of features and area and GRDP are not statistically significant. In all cases, the correlation is strong and positive (Table 2.4).

Based on the analysis of the rho-Spearman correlation for Poland, it was found that the results for any of the pairs are not statistically significant (Table 2.5).

The results of the pairwise correlation analysis for all objects in Spain and Poland, between the area of the region and the facilities of any type show that there is no appropriate statistical significance. On the other hand, the results of the correlation of population–attraction, population–cafe, and GRDP–food pairs are statistically

Table 2.4 Rho-Spearman analysis for Spain

Spain		Attraction	Food	Cafe	Lodging	Sum
Area	Correlation coefficient	0.138	0.196	−0.105	−0.027	0.109
	Significance (two-sided)	0.596	0.45	0.688	0.919	0.678
	N	17	17	17	17	17
Population	Correlation coefficient	0.651**	0.716**	0.724**	0.673**	0.753**
	Significance (two-sided)	0.005	0.001	0.001	0.003	< 0.001
	N	17	17	17	17	17
GRDP	Correlation coefficient	0.018	−0.014	0.145	0.076	0.063
	Significance (two-sided)	0.945	0.958	0.578	0.771	0.81
	N	17	17	17	17	17

** Correlation significant at the level of 0.01 (two-sided)

Table 2.5 Rho-Spearman analysis for Poland

Poland		Attraction	Food	Cafe	Lodging	Sum
Area	Correlation coefficient	0.252	−0.188	0.117	−0.057	0.065
	Significance (two-sided)	0.346	0.486	0.666	0.833	0.81
	N	16	16	16	16	16
Population	Correlation coefficient	−0.196	0	0.209	−0.251	−0.168
	Significance (two-sided)	0.467	1	0.438	0.348	0.533
	N	16	16	16	16	16
GRDP	Correlation coefficient	−0.308	0.313	0.295	0.095	0.217
	Significance (two-sided)	0.246	0.238	0.267	0.726	0.42
	N	16	16	16	16	16

** Correlation significant at the level of 0.01 (two-sided)

Table 2.6 Rho-Spearman analysis for Spain and Poland

Spain and Poland		Attraction	Food	Cafe	Lodging	Sum
Area	Correlation coefficient	0.1	0.085	0.007	– 0.045	0.118
	Significance (two-sided)	0.581	0.637	0.968	0.803	0.512
	N	33	33	33	33	33
Population	Correlation coefficient	0.387*	0.34	0.523**	0.33	0.399*
	Significance (two-sided)	0.026	0.053	0.002	0.061	0.021
	N	33	33	33	33	33
GRDP	Correlation coefficient	0.166	0.569**	0.209	0.286	0.347*
	Significance (two-sided)	0.356	< 0.001	0.244	0.106	0.048
	N	33	33	33	33	33

*Correlation significant at the level of 0.05 (two-sided)
**Correlation significant at the level of 0.01 (two-sided)

significant. The correlation result is strongly positive for population–cafe and GRDP–food pairs. There is also a moderately strong relationship between the number of all places where you can pay with cryptocurrencies and population and GRDP at a significance level of $p < 0.05$ in the pooled analysis of both countries.

2.5 Discussion and Conclusion

BCT's wide range of possibilities for tourism requires practical implementations and network interconnections of various systems. As noted [37], a single-chain platform is not sufficient. To achieve consistency of information and the possibility of combining tourist services with the use of a customer ID, it is essential to be able to integrate individual services using BCT, as well as to use multi-chain architecture [37]. This can be guaranteed by stable and proven implementations of BCT-based systems. Unfortunately, some projects using BCT described in the literature, such as CoolCousin [37], are no longer continued. This may be evidenced, among other things, by the lack of understanding of how BCT technology could improve the condition of tourism, a small number of users, the lack of financial support, errors in the system functioning, or other detected. In the future, it would be worth analyzing the existing and closed BCT projects and conducting research that will explain the decision to complete the development of these projects.

The results of the statistical analysis of empirical data showed that in Spain, a country better developed economically and technologically than Poland, more enterprises operating in the accommodation, catering, and attractions industries allow customers to pay for services with cryptocurrencies. Payment for attractions is not popular in any of these countries, and the percentage of companies surveyed allowing payment for services is similar for lodging. In Spain and Poland, the largest number of such places is in the region where the capital of the country is located. The popularity of such places is greater in the tourist regions of both countries. In Spain, a strong positive correlation has been observed between the region's population and the number of places one can pay with bitcoin. No correlations were observed in Poland, which may be due to the small number of places offering such a possibility. The ability to pay with cryptocurrency is not a popular method in any of these countries.

The limitation of our study was that we only analyzed the results of two countries. It is also impossible to state whether coinmap.org has a complete and up-to-date database of service points where bitcoin payments can be made. In addition, the coin map service has a database of only those places where bitcoin payment is possible; it is unknown whether one can pay with other cryptocurrencies.

Future research will be conducted to verify other databases of enterprises and extend to a larger number of countries to increase the credibility of the verified assumptions. Further research should also look at the factors influencing companies' enabling cryptocurrency payments.

Apart from the theoretical contribution of identifying the possibilities of using blockchain in tourism, this study is also of practical importance. It has been shown that despite the great interest in blockchain technology, its implementation in the business practice of companies in the tourism sector in Spain and Poland is insignificant.

References

1. Filimonau, V., Naumova, E.: The blockchain technology and the scope of its application in hospitality operations. Int. J. Hosp. Manag. **87**, 102383 (2020). https://doi.org/10.1016/J.IJHM.2019.102383
2. Önder, I., Treiblmaier, H.: Blockchain and tourism: three research propositions. Ann. Tour. Res. **72**, 180–182 (2018). https://doi.org/10.1016/J.ANNALS.2018.03.005
3. Erceg, A., Sekuloska, J.D., Kelic, I.: Blockchain in the tourism industry—a review of the situation in Croatia and Macedonia. Informatics **7**, 5–7 (2020). https://doi.org/10.3390/INFORMATICS7010005
4. Wei C, Wang Q, Liu C (2020) Research on construction of a cloud platform for tourism information intelligent service based on blockchain technology. Wirel. Commun. Mob. Comput. 2020:. https://doi.org/10.1155/2020/8877625
5. Bodkhe, U., Tanwar, S., Parekh, K., et al.: Blockchain for industry 4.0: a comprehensive review. IEEE Access **8**, 79764–79800 (2020). https://doi.org/10.1109/ACCESS.2020.2988579
6. Cicha K, Rutecka P, Rizun M, Strzelecki A (2021) Digital and media literacies in the polish education system—pre-and post-covid-19 perspective. Educ. Sci. 11. https://doi.org/10.3390/EDUCSCI11090532
7. Baralla G, Ibba S, Marchesi M, et al (2019) A blockchain based system to ensure transparency and reliability in food supply chain. Lect. Notes. Comput. Sci. (including Subser Lect Notes Artif Intell Lect Notes Bioinformatics) 11339, LNCS:379–391. https://doi.org/10.1007/978-3-030-10549-5_30/FIGURES/2
8. Baralla, G., Pinna, A., Tonelli, R., et al.: Ensuring transparency and traceability of food local products: a blockchain application to a Smart Tourism Region. Concurr. Comput. Pract. Exp. **33**, e5857 (2021). https://doi.org/10.1002/CPE.5857
9. Calvaresi, D., Leis, M., Dubovitskaya, A., et al.: Trust in tourism via blockchain technology: results from a systematic review. Inf. Commun. Technol. Tour. **2019**, 304–317 (2019). https://doi.org/10.1007/978-3-030-05940-8_24
10. Yandle B (2010) Lost trust the real cause of the financial meltdown. Indep. Rev. 341–361
11. Willie, P.: Can all sectors of the hospitality and tourism industry be influenced by the innovation of Blockchain technology? Worldw. Hosp. Tour. Themes. **11**, 112–120 (2019). https://doi.org/10.1108/WHATT-11-2018-0077/FULL/PDF
12. Rashideh, W.: Blockchain technology framework: current and future perspectives for the tourism industry. Tour Manag **80**, 104125 (2020). https://doi.org/10.1016/J.TOURMAN.2020.104125
13. Valeri, M., Baggio, R.: A critical reflection on the adoption of blockchain in tourism. Inf Technol. Tour. **232**(23), 121–132 (2020). https://doi.org/10.1007/S40558-020-00183-1
14. Varriale, V., Cammarano, A., Michelino, F., Caputo, M.: The unknown potential of blockchain for sustainable supply chains. Sustain **12**, 9400 (2020). https://doi.org/10.3390/SU12229400
15. Parekh, J., Jaffer, A., Bhanushali, U., Shukla, S.: Disintermediation in medical tourism through blockchain technology: an analysis using value-focused thinking approach. Inf. Technol. Tour. **23**, 69–96 (2021). https://doi.org/10.1007/S40558-020-00180-4/FIGURES/5
16. Hawlitschek, F., Notheisen, B., Teubner, T.: The limits of trust-free systems: a literature review on blockchain technology and trust in the sharing economy. Electron. Commer. Res. Appl. **29**, 50–63 (2018). https://doi.org/10.1016/J.ELERAP.2018.03.005

17. Fragnière, E., Sahut, J.-M., Hikkerova, L., et al.: Blockchain technology in the tourism industry: new perspectives in Switzerland. J. Innov. Econ. Manag. **37**, 65–90 (2022). https://doi.org/10.1109/MCC.2017.3791019
18. Ahmad MS, Shah SM (2021) Moving beyond the crypto-currency success of blockchain: a systematic survey. Scalable Comput. Pract. Exp. 22:321–346. https://doi.org/10.12694/SCPE.V22I3.1853
19. Sharma, M., Sehrawat, R., Daim, T., Shaygan, A.: Technology assessment: enabling blockchain in hospitality and tourism sectors. Technol. Forecast. Soc. Change **169**, 120810 (2021). https://doi.org/10.1016/J.TECHFORE.2021.120810
20. Line, N.D., Dogru, T., El-Manstrly, D., et al.: Control, use and ownership of big data: a reciprocal view of customer big data value in the hospitality and tourism industry. Tour. Manag. **80**, 104106 (2020). https://doi.org/10.1016/J.TOURMAN.2020.104106
21. Wang, Q., Li, R., Zhan, L.: Blockchain technology in the energy sector: from basic research to real world applications. Comput Sci Rev **39**, 100362 (2021). https://doi.org/10.1016/J.COSREV.2021.100362
22. Tyan, I., Guevara-Plaza, A., Yagüe, M.I., et al.: The benefits of blockchain technology for medical tourism. Sustain **13**, 12448 (2021). https://doi.org/10.3390/SU132212448
23. Haque, A.B., Naqvi, B., Islam, A.K.M.N., Hyrynsalmi, S.: Towards a GDPR-compliant blockchain-based COVID vaccination passport. Appl. Sci. **11**, 6132 (2021). https://doi.org/10.3390/APP11136132
24. Aghaei, H., Naderibeni, N., Karimi, A.: Designing a tourism business model on block chain platform. Tour Manag. Perspect. **39**, 100845 (2021). https://doi.org/10.1016/J.TMP.2021.100845
25. Tyan, I., Yagüe, M.I., Guevara-Plaza, A.: Blockchain technology for smart tourism destinations. Sustain **12**, 9715 (2020). https://doi.org/10.3390/SU12229715
26. Boucher P, Nascimento S, Kritikos M (2017) How blockchain technology could change our lives In-depth analysis. Brussels
27. Özgit, H., Adalıer, A.: Can Blockchain technology help small islands achieve sustainable tourism? a perspective on North Cyprus. Worldw. Hosp. Tour. Themes **14**, 374–383 (2022). https://doi.org/10.1108/WHATT-03-2022-0037/FULL/XML
28. Karger, E., Jagals, M., Ahlemann, F.: Blockchain for smart mobility—literature review and future research agenda. Sustain **13**, 13268 (2021). https://doi.org/10.3390/SU132313268
29. Treiblmaier, H.: The token economy as a key driver for tourism: entering the next phase of blockchain research. Ann. Tour. Res. **91**, 103177 (2021). https://doi.org/10.1016/J.ANNALS.2021.103177
30. Kizildag, M., Dogru, T., Zhang, T.C., et al.: Blockchain: a paradigm shift in business practices. Int. J. Contemp. Hosp. Manag. **32**, 953–975 (2020). https://doi.org/10.1108/IJCHM-12-2018-0958/FULL/XML
31. Strebinger, A., Treiblmaier, H.: Profiling early adopters of blockchain-based hotel booking applications: demographic, psychographic, and service-related factors. Inf. Technol. Tour. **24**, 1–30 (2022). https://doi.org/10.1007/S40558-021-00219-0/TABLES/4
32. Trček, D.: Cultural heritage preservation by using blockchain technologies. Herit. Sci. **10**, 1–11 (2022). https://doi.org/10.1186/S40494-021-00643-9/FIGURES/4
33. Nuryyev, G., Wang, Y.P., Achyldurdyyeva, J., et al.: Blockchain technology adoption behavior and sustainability of the business in tourism and hospitality SMEs: an empirical study. Sustain **12**, 1256 (2020). https://doi.org/10.3390/SU12031256
34. Raluca-Florentina, T.: The utility of blockchain technology in the electronic commerce of tourism services: an exploratory study on Romanian consumers. Sustain **14**, 943 (2022). https://doi.org/10.3390/SU14020943
35. Viano C, Avanzo S, Cerutti M, et al (2022) Blockchain tools for socio-economic interactions in local communities. Policy Soc. https://doi.org/10.1093/POLSOC/PUAC007
36. Rashideh W, Alkhathami M, Obidallah WJ et al (2022) Investigation of the effect of blockchain-based cryptocurrencies on tourism industry. IJCSNS Int. J. Comput. Sci. Netw. Secur. 22 https://doi.org/10.22937/IJCSNS.2022.22.5.33

37. Zhang, L., Hang, L., Jin, W., Kim, D.: Interoperable multi-blockchain platform based on integrated REST APIs for reliable tourism management. Electron **10**, 2990 (2021). https://doi.org/10.3390/ELECTRONICS10232990
38. Mishra V (2022) Applications of blockchain for vaccine passport and challenges. J. Glob. Oper. Strateg. Sourc. ahead-of-print: https://doi.org/10.1108/JGOSS-07-2021-0054/FULL/XML
39. Luo, L., Zhou, J.: BlockTour: A blockchain-based smart tourism platform. Comput. Commun. **175**, 186–192 (2021). https://doi.org/10.1016/J.COMCOM.2021.05.011
40. Sharma M, Joshi S, Luthra S, Kumar A (2021) Managing disruptions and risks amidst COVID-19 outbreaks: role of blockchain technology in developing resilient food supply chains. Oper. Manag. Res., 1–14. https://doi.org/10.1007/S12063-021-00198-9/TABLES/11
41. Benedict S (2022) Shared mobility intelligence using permissioned blockchains for smart cities. New Gener. Comput., 1–19. https://doi.org/10.1007/S00354-021-00147-X/FIGURES/5
42. Pérez-Sánchez M de los Á, Tian Z, Barrientos-Báez A et al (2021) Blockchain technology for winning consumer loyalty: social norm analysis using structural equation modeling. Math 9:532. https://doi.org/10.3390/MATH9050532
43. Udegbe, S.E.: Impact of blockchain technology in enhancing customer loyalty programs in airline business. Int. J. Innov. Res. Adv. Stud. **4**, 257–263 (2017)
44. Dryga A, Tsiulin S, Valiavko M, et al (2019) Blockchain-based wildlife data-management framework for the WWF bison rewilding project. ACM Int. Conf. Proceeding Ser., 62–66. https://doi.org/10.1145/3358528.3358530
45. Kwok AOJ, Koh SGM (2018) Is blockchain technology a watershed for tourism development? 22:2447–2452. https://doi.org/10.1080/13683500.2018.1513460
46. Reyes-Menendez, A., Saura, J.R., Filipe, F.: The importance of behavioral data to identify online fake reviews for tourism businesses: a systematic review. Peer J. Comput. Sci. **2019**, e219 (2019). https://doi.org/10.7717/PEERJ-CS.219/SUPP-1
47. Zhou, L., Tan, C., Zhao, H.: Information disclosure decision for tourism O2O supply chain based on blockchain technology. Math **10**, 2119 (2022). https://doi.org/10.3390/MATH10122119

Part II
Business Intelligence Databases and Marketing

Chapter 3
Comparison of Semi-structured Data on MSSQL and PostgreSQL

Leandro Alves, Pedro Oliveira, Júlio Rocha, Cristina Wanzeller, Filipe Cardoso, Pedro Martins, and Maryam Abbasi

Abstract The present study intends to compare the performance of two Data Base Management Systems, specifically Microsoft SQL Server and PostgreSQL, focusing on data insertion, queries execution and indexation. To simulate how Microsoft SQL Server performs with key-value oriented datasets, we use a converted TPC-H lineitem table. The dataset is explored in two different ways, first using the key-value-like format and second in JSON format. The same dataset is applied to PostgreSQL DBMS to analyze performance and compare both database engines. After testing the load process on both databases, performance metrics (execution times) are obtained and compared. Experimental results show that, in general, inserts are approximately twice times faster in Microsoft SQL Server because they are injected as plain text without any type of verification, while in PostgreSQL, loaded data includes a validating process, which delays the loading process. Moreover, we did additional indexation tests, from which we concluded that in general, data loading performance degrades. Regarding query performance in PostgreSQL, we conclude that with indexation, queries become three or four percent faster and six times faster in Microsoft SQL Server.

L. Alves · P. Oliveira · J. Rocha
Polytechnic of Viseu, Viseu, Portugal

C. Wanzeller · P. Martins (✉)
CISeD—Research Centre in Digital Services, Polytechnic of Viseu, Viseu, Portugal
e-mail: pedromom@estgv.ipv.pt

C. Wanzeller
e-mail: cwanzeller@estgv.ipv.pt

M. Abbasi
CISUC—Centre for Informatics and Systems, University of Coimbra, Coimbra, Portugal
e-mail: maryam@dei.uc.pt

F. Cardoso
Polytechnic of Coimbra, Coimbra, Portugal
e-mail: filipe@isec.pt

J. L. Reis et al. (eds.), *Marketing and Smart Technologies*, Smart Innovation, Systems and Technologies 337, https://doi.org/10.1007/978-981-19-9099-1_3

3.1 Introduction

The purpose of this experiment is to test unrelated data in two relational databases, and for that, we use a dataset converted from the relational model TPC-h [1]. The reason for this is the need to have a dataset structure in key-value format, which will be tested in PostgreSQL and MSSQL. What we propose to test how relational databases handle semi-structured data.

This work summarizes a set of operations over a JSON dataset format on MSSQL versus PostgreSQL and key-value in MSSQL versus PostgreSQL. Indexes were applied to these three tables in the different databases. The workloads running on all databases, indexed and not indexed, are explained in more detail in the "experimental setup" section.

Results show that when using MSSQL, it is important to consider computed columns for the application of indexes, while PostgreSQL can handle semi-structured data without major changes/effort.

This paper is organized into five sections. The first section is this introduction, where the challenge or problem to be solved is addressed and what steps to follow; the second deals with the related work; the third is related to the experimental work, where the technologies and practices on these technologies are presented to solve the challenge/problem; the fourth goes through the presentation and analysis of the different results and, finally, the fifth section will be for the conclusions obtained about the present research.

3.2 Related Work

In research from authors [2], that analyze the performance in the same DBMSs, PostgreSQL and MSSQL, positioned in the cloud and they concluded that the most high-performance DBMS was MSSQL, in all the tests performed, but they never approach the theme of a dataset oriented to key-value, or JSON format. Another interesting work [3] also does load tests, records and analyzes the times with a dataset generated by TPCH, to tables on Oracle and PostgreSQL, and the results obtained differed depending on the method used for each load test. Some of the methods presented are used in our paper, like PostgreSQL "COPY" and "insert into" method.

Another research [4] demonstrates that performance on non-relational DMBS is higher, because MySQL includes a lot of complex queries which involves integrity constraints and joins, and NoSQL allows to eke more performance out of the system by eliminating a lot of integrity checks done by relational databases from the database tier.

In another research [5], the behavior of relational and non-relational database engines is also compared, when executing commands'insert,"update,"delete' and'select.' The comparison is made through the times analyzed when operating on 4 datasets of 100 records, 1000 records, 10,000 records and 100,000 records. According to the data presented, Redis obtains faster results when the volume of data is greater, specifically in'delete' and'insert' operations. For'update' and'select' operations, the results show that MariaDB can be faster than Redis. For data volumes of around 10,000 records, the results are very similar.

In another research [6], tests were carried out using'insert','update','delete' and'select' commands in relational DBMS, such as Oracle, MySQL and MSSQL, and non-relational DBMS, such as Mongo, Redis, Cassandra and GraphQl, and compared the times in obtaining results over a dataset of 10,000 and 100,000 records. To carry out the tests, a database of a railway with all the stations was used, but, to guarantee the quality of the tests, on a large volume of data, false records were added to the tables. The results show that non-relational database engines are more efficient, with MongoDB standing out as the fastest in all operations. In another research [7], MySQL was also compared with MongoDB. In all cases, with performance or load methods, NoSQL databases have better performance than relational databases and the reason for that is that MongoDB has a very powerful query engines and indexing features.

Not all operations are more efficient in non-relational databases. The research [8] compares the performance of relational and non-relational databases, running complex queries on a large dataset. Results show that'select' method is significantly faster in MongoDB; however, some math queries such as aggregate functions (sum, count, AVG) are better on Oracle RDBMS.

The proposed work differs from the others referenced is the fact that the performance tests are performed on computed columns in MSSQL. One of the key variables that need to be included in the equation is indexing. MSSQL lacks appropriate NoSQL data indexes and is not JSON friendly. We begin our quest for a method to make that happen. We discovered a mention of MSSQL computed columns that ought to serve as the key. We have discovered a method to use calculated columns and index those columns under this article's citation for SQL index [9]. Computed columns do not actually exist, as far as we know. On demand, the data is extracted from other derived columns. By including a non-clustered index in that column, the DBMS can avoid having to parse data again because the parsed-out information is stored in the index column table.

3.3 Experimental Setup

The main proposal of this experimental is to perform benchmark performance with NoSQL data on a SQL DBMS environment. MSSQL and PostgreSQL were used in this work, both working on Windows 10 Operating System. As for the hardware, the tests in the different DBMS were performed on a laptop computer, with an i7 processor, with 16 GB of RAM, and SSD, whose bandwidth flow, withstand by the interface, is up to 600 MB/s. Regarding the explored dataset, it was achieved through the benchmark TPC-H [10], in which eight tables are generated. However,

Fig. 3.1 Convert data table to JSON

```
INSERT INTO [_lineitem_json]
SELECT [value] FROM OPENJSON(
    SELECT TOP 1000000 *
    FROM [lineitem] FOR JSON AUTO
)
```

in our experimental setup, we only use one table (lineitem), the reason is related to the fact that SQL Server is a relational data model based, and semi-structured data models do not allow tables and relations, instead, make use of document model hierarchy. In contrast to relational databases, which store data as rows in a table, document databases store entities as documents or JSON documents.

After choosing the hardware and DBMS, the workload starts as described above:

- Generate JSON;
- Setup DBMS environment;
- Insert data;
- Query data;
- Update data.

3.3.1 Generate JSON

To convert data table to data JSON, we used "FOR JSON AUTO" command [11] that will produce an array of JSON based on a "SELECT" as shown on Fig. 3.1. In this script, first, we made a conversion using "FOR JSON AUTO." This instruction formats the output of the FOR JSON clause automatically, based on the structure of the SELECT statement, converting all columns into JSON properties.

3.3.2 Setup DBMS Environment

After generating JSON data, during the last Sect. 3.1, the next step is preparing the DBMS environment. First, we create a clean database on MSSQL and another on PostgreSQL. On MSSQL, add two tables, "lineitem no indx" and "lineitem." The reason for the two tables is to test workloads with and without indexes.

On PostgreSQL (hybrid DBMS that allows data-key-value and JSON), was added four tables. All tables have the same mission, test performance workloads with JSON and data-key-value, both, with and without indexes.

Indexing is one of the main variables that should be added to the equation. MSSQL is not JSON friendly and has no proper indexes for NoSQL data. With that in mind, we started a research to find a way to accomplish that. We found a reference to MSSQL computed columns that should be the key. Under this article [9], we have found a way to use computed columns and, index those columns. As we know, computed

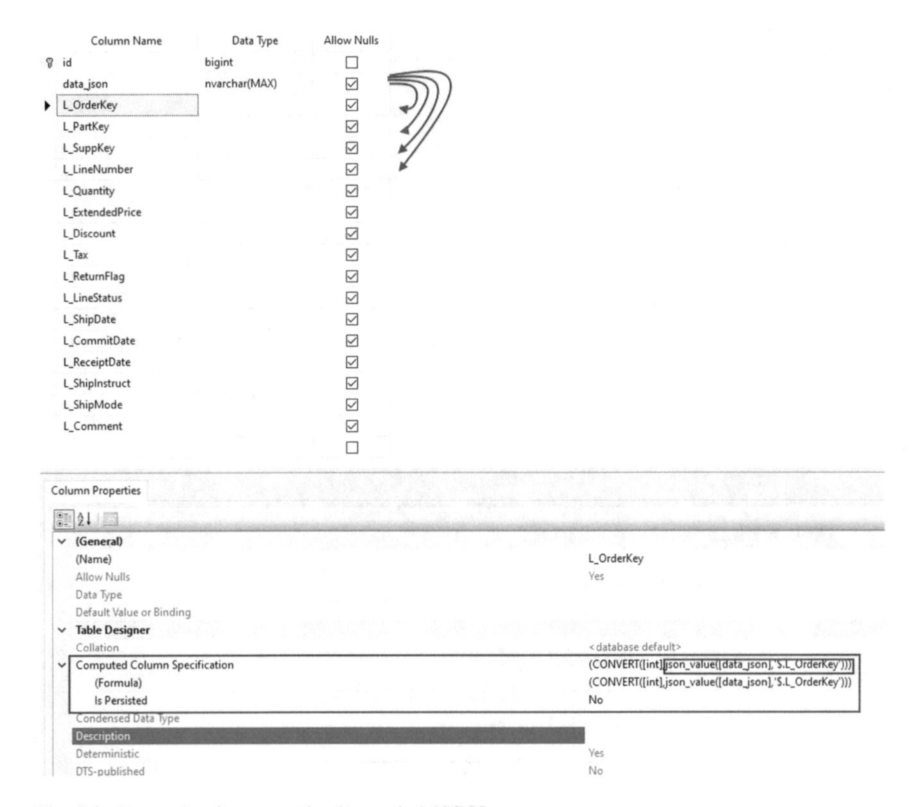

Fig. 3.2 Example of computed columns in MSSQL

columns do not exist physically. The data is parsed out on runtime from other derived columns. Adding a non-clustered index to that column, the parsed-out data is written to the index column table, and the DBMS does not have to parse data out again. We have created a computed column to each JSON property as shown on Fig. 3.2. The improvement is vast, as we show in the results and analyze Sect. 3.4. So, with this in mind, we start to add all JSON properties as computed columns and index the ones used on query where condition, as non-clustered indexes.

On PostgreSQL, we found the Generalized Inverted Indexes (GIN), specialized for semi-structured data, are quite useful when an index must map multiple values to a row, and good for array indexing values, as well as full-text search programs.

3.3.3 Insert Data

DBMS environment concluded that we perform the scripts to insert the same amount of JSON in each table, one million registries to be precise. Each insert should be executed three times. The lowest time registered should be considered. All scripts have been executed against the indexed and non-indexed tables. To collect times, turn on time collecting before the script "SET STATISTICS TIME ON" and turn it off after the insert "SET STATISTICS TIME OFF" as shown on Fig. 3.3.

Bulk insert is a faster way to insert data on MSSQL. The Fig. 3.4 is a script example of how it works. First, the table is defined on "BULK" instruction, and then the script point to a CSV file (JSON data generated in Sect. 3.1) where the data is located. Bulk only allows insertions over a file (Fig. 3.5).

For PostgreSQL, the process is identical. First, was made an insertion using method "INSERT INTO", as described in 1.4, and a second method, make use of massive insertion's instruction "COPY" on PostgreSQL, "BULK" equivalent on MSSQL as shown on Fig. 3.6.

```
SET STATISTICS TIME ON
    insert into [lineitem]
    select top 1000000 data_json from [_lineitem_json]
SET STATISTICS TIME OFF
```

Fig. 3.3 Insert using clause INSERT INTO

```
SET STATISTICS TIME ON
        BULK INSERT [ lineitem ]
      FROM 'C:\TPC-H\ json _postgress . csv '
      WITH (FORMAT = 'CSV'
                    ,FIRSTROW = 2
                    ,KEEPIDENTITY
                    ,KEEPNULLS
                   )
SET STATISTICS TIME OFF
```

Fig. 3.4 Insert using clause BULK

```
INSERT INTO lineitem_json (info)
SELECT info from lineitem_json_to_copy_source
```

Fig. 3.5 Insert using clause INSERT INTO on PostgreSQL

Fig. 3.6 Insert using clause COPY on PostgreSQL

```
COPY lineitem _json
FROM 'E:\ESTGV\ json _postgress . csv '
      DELIMITER ' , ' CSV HEADER;
```

```
SET STATISTICS TIME ON
        select * from [lineitem]
        where [l_shipmode] ='TRUCK'
              AND [L_LineNumber] = 1
              and L_lineStatus='O'
              and L_Quantity between 8 and 60
        order by L_ShipDate asc
SET STATISTICS TIME OFF
```

Fig. 3.7 Select on MSSQL

3.3.4 Query Data

This section describes the process adopted to test performance of both DBMS.

After run Sect. 3.3, all tables have been loaded with the exactly same data. The performed scripts, in the end, should return the exact same results, with same where conditions “L ShipMode = 'TRUCK' and L LineNumber = 1 and L LineStatus = 'O' and L Quantity between 8 and 60.” As shown on listing Figs. 3.7 and 3.8, this is a'sine qua non' condition, for a benchmark test.

```
select x.*
from lineitem_json,
        jsonb_to_record(info) as x(
                "L_OrderKey" int,
                "L_PartKey" int,
                "L_SuppKey" int,
                "L_LineNumber" int,
                "L_Quantity" int,
                "L_ExtendedPrice" decimal,
                "L_Discount" decimal,
                "L_Tax" decimal,
                "L_ReturnFlag" text,
                "L_LineStatus" text,
                "L_ShipDate" date,
                "L_CommitDate" date,
                "L_ReceiptDate" date,
                "L_ShipInstruct" text,
                "L_ShipMode" text,
                "L_Comment" text)
where (info->>'L_ShipMode' = 'TRUCK')
        and (info->>'L_LineNumber' = '1')
        and (info->>'L_LineStatus' = 'O')
        and (info->>'L_Quantity')::numeric between '8' and '60'
order by (info->>'L_Quantity')
```

Fig. 3.8 Select on PostgreSQL

Since we are using computed columns, as described in Sect. 3.2 and Fig. 3.2, no column's definition is needed, only "*" to show all. The where condition complies with criteria defined on beginning of Sect. 3.4, and times will be collected making use of "SET STATISTICS TIME ON" and "SET STATISTICS TIME.

OFF" instructions.

On this script, listing Fig. 3.8, we start to define all properties with correspondent data types, with where condition meeting the agreed criteria. To list data tabularly, we found function"jsonb to record", we pass all JSON record data, define data type for each JSON property, and a tabular view should be listed. This has been used for legible proposes.

3.3.5 Update Data

The last workload section updates have been performed with the same where conditions and same changes. Where condition, "L ShipMode = 'MAIL.'" Changes, update all "L ShipMode" equal to "MAIL" to "TRUCKKKKKK." In the end, all results have been verified.

On MSSQL listing Fig. 3.9, we make use of "JSON MODIFY" function, to navigate through JSON data on field "json data" and perform changes as listed on listing Fig. 3.9.

On PostgreSQL listing Fig. 3.10, same changes have been performed. First, we filtered all registries where property "L ShipMode" value equal to "MAIL", and change it to "TRUCKKKKKK."

```
SET STATISTICS TIME ON
UPDATE U
SET data_json =
    JSON_MODIFY(data_json , '$.L_ShipMode', 'TRUCKKKKKK')
FROM AEABD.dbo.[lineitem] as U
WHERE L_ShipMode = 'MAIL'
SET STATISTICS TIME OFF
```

Fig. 3.9 Update data on MSSQL

```
UPDATE lineitem_json
SET info = info || '{"L_ShipMode":"TRUCKKKKKK"}'
WHERE (info->>'L_ShipMode' = 'MAIL')
```

Fig. 3.10 Update data on PostgreSQL

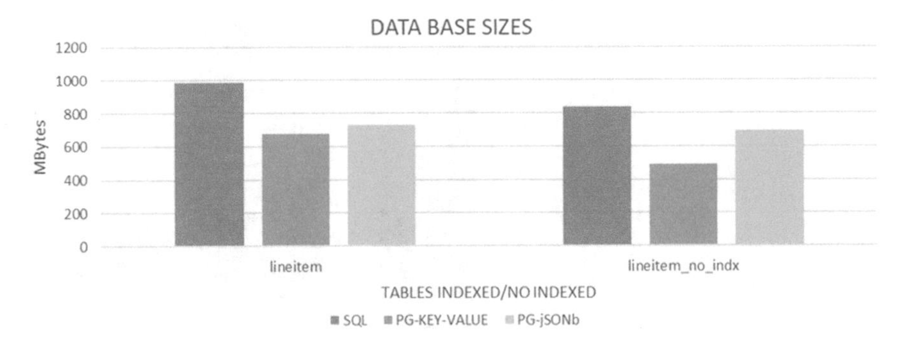

Fig. 3.11 Databases size

Table 3.1 Tables sizes

Table name	Rows	SQL	PG-KEY-VALUE	PG-JSONb
LineItem	1,000,000	988.21 MB	678 MB	726 MB
LineItem no indx	1,000,000	835.38 MB	493 MB	691 MB

3.4 Results and Analysis

3.4.1 Tables Sizes

On Table 3.1 and Fig. 3.11, it is clear that MSSQL, because is not optimized for unstructured data, spends more space to accommodate same data. The performed test only has been done in one table and the difference is huge.

Another evidence was on indexed tables. Because the indexes require space to organize and split table data, all indexed tables in all DBMSs are heavier than the ones without indexes. This means performance improvement will pay a high storage price.

3.4.2 Inserts

The data, loaded via "BULK" or "COPY", was slower than "INSERT INTO" or "SELECT INTO."

In the "INSERT INTO", the data was loaded from a select that already resides in memory due to DBMS optimizations. "SELECT INTO" is the fast one. When the script runs, the target table should not exist and will be created without any indexes. So, in the "INSERT INTO", the table already exists with the primary key; consequently, the primary key has a clustered index by default. The DBMS needs to manage it, creating a delay during the process, as revealed in Fig. 3.12.

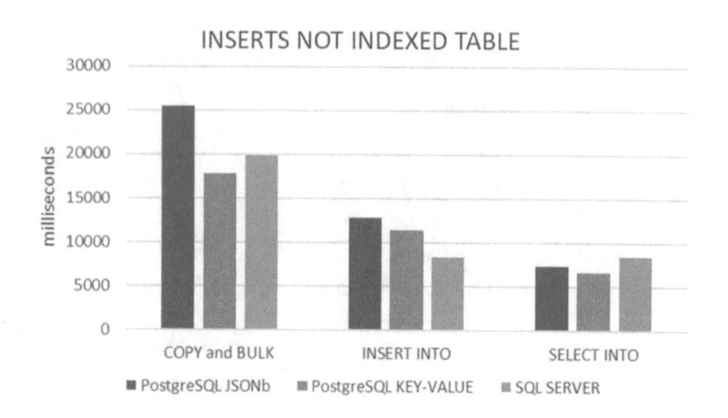

Fig. 3.12 Inserts not indexed tables

All indexed tables are slowest, as shown in Fig. 3.13, again, because of the need to manage all indexes. Therefore, adding indexes always be a trade-off, if any updates or inserts need to be executed on the database. We should not add unnecessary indexes to tables. The reason is related to the fact that more indexes mean more work, rebuilding and reorganize them during the insertions or updates and consequently more time-consuming.

On Fig. 3.14, we can figure out that PostgreSQL is much time-consuming during the inserts than MSSQL. MSSQL does not do any validation during the process, while PostgreSQL verifies each insertion to validate data. For that reason, has a poor performance.

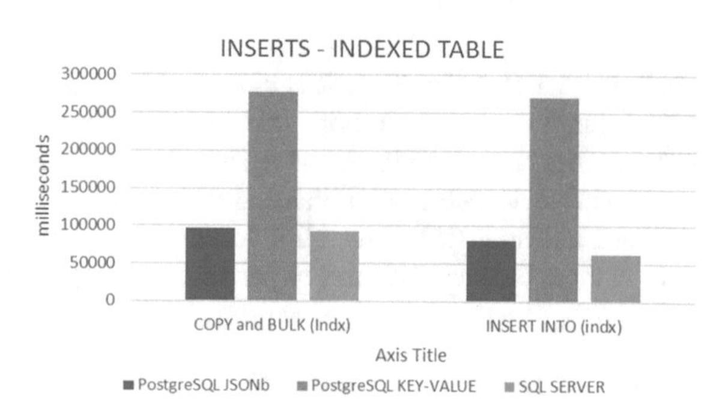

Fig. 3.13 Inserts in indexed tables

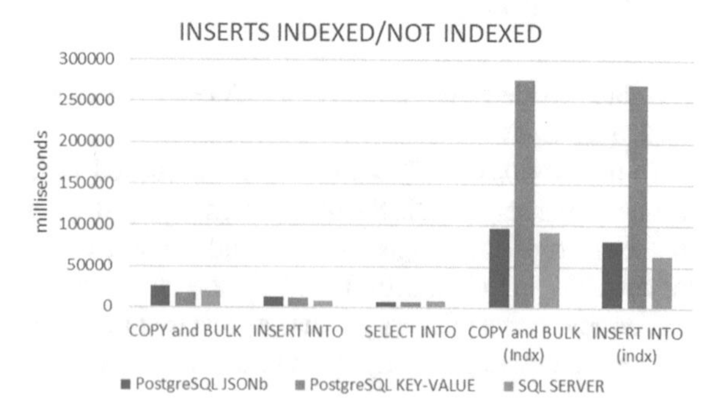

Fig. 3.14 Inserts in indexed and not indexed tables

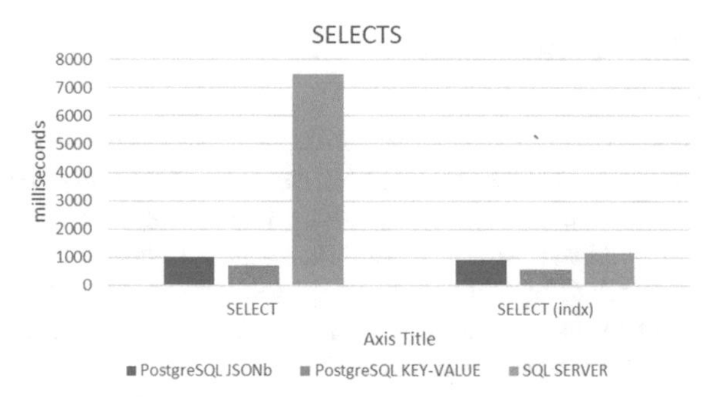

Fig. 3.15 Queries

3.4.3 Selects

All selects exposed a better performance on PostgreSQL than in MSSQL. PostgreSQL is optimized for unstructured data. That is a fact and is visible in Fig. 3.15. On MSSQL, the selects are prolonged on tables without indexes. On the other hand, MSSQL revealed, despite a worse performance, not too far from PostgreSQL, when indexed.

3.4.4 Updates

This section reveals the statistics against indexed and non-indexed tables during the update process. The problem mentioned above, regarding the management of the indexes, gets highlighted in Fig. 3.16. As we can see, all updates in both DBMS on indexed tables get the worst and worst performance every time a new index has been added. In conclusion, indexes are powerful, but we should moderate the way we add them to databases, since we can solve query performance issues, and get new ones on updates/inserts.

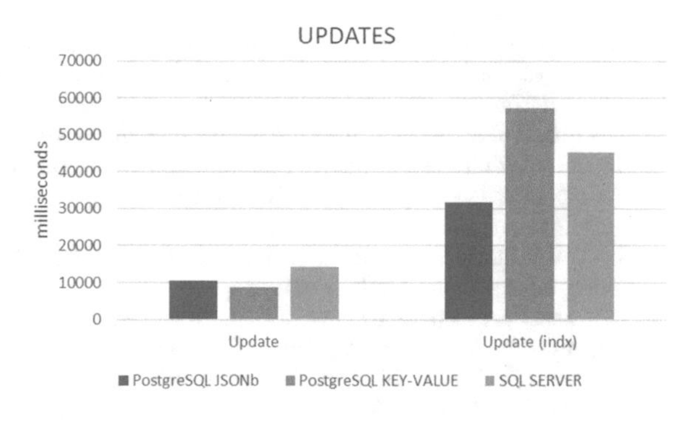

Fig. 3.16 Updates

3.5 Conclusions

MSSQL revealed a worse performance during the selects, but not so bad, when using indexes. This means, MSSQL made good improvements, in the last years, getting better results, making use of new features that we will explain above.

The insertions in MSSQL are generally faster, since the DBMS does not do any validation, but this could be a problem under production environments, since not well-formatted data could be inserted, triggering exceptions, and causing a huge pain on development teams.

Storage could be an issue too. MSSQL is not optimized for this kind of data, taking up much more space than PostgreSQL.

Generally speaking, PostgreSQL should be the right choice for accommodating semi-unstructured data, faster and better storage space managed.

Our major research, during this test case, was at MSSQL performance level, when we start to indexing computed columns. MSSQL, does not have a way to index NoSQL data as PostgreSQL. The way to accomplish that was indexing computed columns. Since all JSON data inserted on MSSQL is treated as text, we try to find a method to convert the data into tabular structure. Recurring to computed columns, as explained in Sect. 3.2, we have a chance to index does columns. At this level, the benefit was huge. The same query made an improvement of 82.4% faster than without indexes as shown on Fig. 3.15.

Acknowledgements "National Funds fund this work through the FCT—Foundation for Science and Technology, IP, within the scope of the project Ref UIDB/05583/2020. Furthermore, we would like to thank the Research Centre in Digital Services (CISeD), the Polytechnic of Viseu, for their support."

References

1. TPC-H. visited on 2022–06. https://www.tpc.org/tpch/
2. Vershinin, I.S., Mustafina, A.R.: Performance analysis of PostgreSQL, MySQL, microsoft SQL server systems based on TPC-H tests. In: *2021 International Russian Automation Conference (RusAutoCon)*, pp. 683–687 (2021). https://doi.org/10.1109/RusAutoCon52004.2021.9537400
3. Martins, P. et al.: A performance study on different data load methods in relational databases. In: *2019 14th Iberian Conference on Information Systems and Technologies (CISTI)*, pp. 1–7 (2019). https://doi.org/10.23919/CISTI8760615
4. Relational V S Key Value Stores Information Technology Essay. visited on 2022–06. 2018. https://www.ukessays.com/essays/informationtechnology/relational-v-s-key-value-stores-informationtechnology-essay.php
5. Wittawat Puangsaijai and Sutheera Puntheeranurak. "A comparative study of relational database and key-value database for big data applications". In: *2017 International Electrical Engineering Congress (iEECON)*. 2017, pp. 1–4. https://doi.org/10.1109/IEECON.2017.8075813
6. Cereˇsnˇ´ak, R., Kvet, M.: Comparison of query performance inˇ relational a non-relation databases. In: *Transportation Research Procedia* 40. TRANSCOM 2019 13th International Scientific Conference on Sustainable, Modern and Safe Transport, pp. 170–177

(2019). ISSN: 2352–1465. https://doi.org/10.1016/j.trpro.2019.07.027. https://www.sciencedirect.com/science/article/pii/S2352146519301887
7. Jose, B., Abraham, S.: Performance analysis of NoSQL and relational databases with MongoDB and MySQL. In: *Materials Today: Proceedings* 24. International Multi-conference on Computing, Communication, Electrical and Nanotechnology, I2CN-2K19, 25th and 26th April 2019, pp. 2036–2043 (2020). ISSN: 2214–7853. https://doi.org/10.1016/j.matpr.2020.03.634. https://www.sciencedirect.com/science/article/pii/S2214785320324159
8. Faraj, A., Rashid, B., Shareef, T.: Comparative study of relational and non-relations database performances using Oracle and MongoDB systems. Int. J. Comput. Eng. Technol. (IJCET) **5**(11), 11–22 (2014)
9. *One SQL Cheat Code For Amazingly Fast JSON Queries.* . https://bertwagner.com/posts/one-sql-cheat-code-for-amazinglyfast-json-queries/
10. Transaction Processing Performance Council: TPC-H benchmark specification. In: Published at http://www.tcp.org/hspec.html 21, pp. 592–603 (2008)
11. Microsoft. *Format JSON Output Automatically with AUTO Mode (SQL Server).* https://docs.microsoft.com/en-us/sql/relationaldatabases/json/format-json-output-automatically-with-automode-sql-server?view=sql-server-ver16

Chapter 4
Coolhunting Canvas: A Pedagogical Toolkit to Support Trendspotting and Sociocultural Innovation in Marketing

William Afonso Cantú and Nelson Pinheiro Gomes

Abstract Many approaches make use of specific technics to gather trend signals and analyse emerging sociocultural trends, but sometimes, it is difficult for students to explore or determine which is the most relevant information. The need for this research arose due to these issues within current pedagogical methodologies in trend research. There is a broad interest in the literature base for trend methodologies approaching several process paths that impact different areas, including Marketing. We conducted a narrative literature review to gather data on how to analyse trend signals. The results point to the benefit of presenting visual tools to support critical thinking, and for that reason, we present a visual framework named the Coolhunting Canvas. This framework proposal may lead to a better understanding of trend signals in an academic perspective and will help to improve and conduct future research regarding Trend Studies and coolhunting.

4.1 Introduction: Analysing Patterns in Culture

In recent years, Trend Studies has seen significant advances in terms of conceptual development as a scientific approach [1–5]. The study of trends is important to companies and brands, since it allows them to follow directions and make better decisions based on societal flow. It also helps to create bulletproof future product strategies, as Els Dragt mentions [4]. We clearly see this relation to marketing, since the research based on this methodological process allows for the creation of better strategies at this level [6].

The process of researching trends is not new. As a marketing approach, it appears around the 60 s in the USA to fulfil the gap of research that the crisis brought [7].

W. A. Cantú (✉) · N. P. Gomes
CEAUL/ULICES University of Lisbon Centre for English Studies, School of Arts and Humanities, University of Lisbon, Alameda da Universidade, 1600-214 Lisbon, Portugal
e-mail: williamcantu@campus.ul.pt

N. P. Gomes
e-mail: nelsonpinheiro@campus.ul.pt

J. L. Reis et al. (eds.), *Marketing and Smart Technologies*, Smart Innovation, Systems and Technologies 337, https://doi.org/10.1007/978-981-19-9099-1_4

Its importance is related to the contemporary analysis of society and the advantages that it brings when looking for innovation. The main advantage of this approach is that we can see how the world is, and how mindsets are representing the ideals of a period of time ("structure of feeling") through cultural patterns [8]. These advantages are particularly important from an applied perspective, because with that information, we have advantage in creating more relevant marketing strategies that impact communication, products, services, among others.

The continuous advances in Trend Studies as an approach are motivated by the latest advances in terms of methodological developments that explore trends in different perspectives, for example, analytical, resorting to content analysis and hermeneutical through semiotics [1, 2, 9–20]. In academia, the various processes to researching trends are designed to provide tools that enable students to act in the face of social changes and in the most diverse areas, especially in marketing. The wide range of methods creates a difficulty, as there are several methodological protocols. Trend research is still under debate with several studies suggesting if there is a specific methodology that should be followed. Examples of attempts to build new methodological paths can be found in the literature, and they serve as processes that systematize the knowledge in the field [1–4, 10, 14–20]. In this sense, it is important to constantly rethink approaches and methods, as well as systematize new literature and contributions to form new trend protocols, which is the aim of this paper.

There are several steps presented in the literature regarding trend research. Pioneering work on this subject has been carried out earlier by Els Dragt that presents a model divided into three main steps, namely: (a) scan (selection, categorization, and documentation of signals); (b) analyse (find clusters between the information documented and describe new trends); (c) apply (generate innovation using the knowledge from the data collected) [4]. We can also see Martin Raymond that emphasizes the collection of data and information in trend research process, proposing an applied "cultural triangulation" to trends. The model is divided into three phases: (a) interrogation (collecting data in large population mainly with quantitative data); (b) observation (collect specific and in-depth data in population using qualitative data to understand their lifestyles and way of life); (c) intuition (using personal feeling and insights to add viewpoints and information for the process of research) [17, pp. 120–124]. From another perspective, we can see the research process synthesized by Víctor Mártil throughout the Coolhunting Science Insights Model (CSI), divided into three steps: (a) coolhunting: get to know the background context of the research; (b) science: connection and analysis of the data to find correlation; and (c) insights: write about the trend and its implications in the business or organization [21, pp. 68–70].

As we can see, the research process is presented by the authors differently and their understanding may vary depending on the business, the problem, or the researcher itself. What is common to all is the existence of a collection of information for later analysis and that is where the meaning of coolhunting as a trend research methodology lies. Maioli, Presotto, and Palma mention that during coolhunting, it is important to observe the phenomena live through an ethnographic approach and participant observation. The authors also highlight desk research to collect data and its importance to understand issues already addressed by another research. We underline the authors' reference to coolhunting and the need for a recurrent practice of "looking for cool" given its importance in the context of trends studies and its fluidity in the sociocultural fabric [22, pp. 24–26].

However, we are faced with a research stage that is similar among the various authors: the identification of trend signals. This part of the research is important, as the collection of signals is one of the starting points for trend research. Based on that, this paper focuses on the construction of a toolkit that informs the process of analysing trend signals. It will contribute to a better analysis and interpretation of signals and their insights. For this, we used a narrative literature review [23] to structure the analytical guidelines in a Canvas format.

4.2 Mapping the Cool

Cool is a term used quite naturally in the field of trends. This is a concept inherent to coolhunting. It takes the form of signs that allow inquiring about changes that may be taking place in society, and it was mentioned by Gladwell in 1993 in the journal the New Yorker [24]. Looking for cool is, in essence, meeting changes and artefacts that can stand out in an environment full of meanings. Observing cool allows us to understand that some creations are more relevant than others. These cultural objects are more creative and visible to be analysed in the context of trends [25]. Several authors have already worked on this definition given its impact on the practice of coolhunting, which is constantly being improved. It is important to review some of them in this paper, since what is sought is to create a tool that contributes to its analysis. For Rohde, cool can be characterized as something attractive, inspiring and with growth potential [26, p. 15]. For Gloor and Cooper, cool should make the world a better place, it must be fresh, and must be part of a community bringing meaning to our lives [27, p. 7]. For the Laboratory of Trends and Culture Management at the University of Lisbon, cool can be defined as something relevant, viral, in the moment, irreverent, instigating, discontinuous [28]. The analysis and insights from the analysis of cool signals can generate leads that point to emerging consumer needs and wants and the creative basis that is hidden in each and can be extrapolated. This is particularly relevant when addressing change in consumer attitudes and developing both strategies and concepts in marketing. We can see patterns in the signals and act on them or use the signals directly as benchmarks for a marketing creative process.

It is important to underline that cool is subjective, in the sense that coolhunting is carried out by different individuals, with different life experiences and ways of looking at the world. However, as a rule, when a cool signal is identified, the vast majority of coolhunters agree with its cool nature. This leads us to think about practical issues, and how to contribute to this analysis, and the creation of an analytical structure seems to be a path that does not summarize the analysis in some parameters but contributes to a qualitative analytical roadmap. An overall summary of this topic shows us that mapping cool signals allow us to see innovation closer and find trends in an easier way, since it is the manifestation of the trend and represents the ideas that compose the trend itself [14].

4.3 Trend Research Tools: An Academic Perspective

The subjectivity of cool aligned with the academic practice of trends in marketing research focusing on sociocultural innovation led us to experience some difficulties in the analysis of cool signals. When performing coolhunting exercises, students can identify relevant signals for analysis, while some raise doubts as to their nature or relevance. In this sense, the main purpose of this work is to create a pedagogical tool to help the analysis of signals to, on the one hand, provide students with specific analytical skills and, on the other hand, compare signals and, in a broad way, verify those individual characteristics that each signal underlines, aiming at better decision-making in projects.

Through the conducted narrative literature review, we can see authors who present tools that can be used directly in trend research [4, 15, 17]. Meanwhile, others just present theoretical and conceptual models that were not translated into real tools but are equally useful in terms of conceptualization and guidance during the research [9, 10, 17, 29, 30].

It is important to draw attention to visual techniques that can help the development of trend analysis process. We see that in Rech and Felipe trends, project management occurs in a more fluid and interesting way when done through visual tools [1]. Although the process of trends management through this specific process may look complex, we must highlight the existence of practices and tools already in use such as TrendWatching's Consumer Trend Canva [31] or the DHL Trend Radar [32].

As aforementioned, we—trend professors—see difficulties during our trend classes when helping students to collect signals, as well as coordinating their analysis. It is the objective of this paper to create a tool to facilitate the process and to be tested in the future. In this way, this research presents a visual tool to analyse trend signals that can help in future development of marketing innovation based on the understanding of culture and society. This tool seeks to bring together several approaches that are considered by authors in the context of trends and contribute to thinking about the cultural role of drawings.

4.4 The Coolhunting Canvas

Given the emphasis on the search for specific signals, the collection of data in this academic context plays an important role, as it allows students to exercise the ability to find cultural manifestations that indicate new ways of seeing trends based on innovation. The steps in the proposed Canvas are described below and are the result of a narrative literature review that guided the draw of some methods to help the process during the analysis. It is also a collection of theories compiled by several academics for years and already synthesized by some authors [2, 14, 33, 34]. The Canvas (Fig. 4.1) is presented in the format of several parts that can be filled independently; however, for a better reading of the information and guiding the critical thinking of data, the following order is advised.

(a) **Signal name**. Indicate a suggestive name to the signal, as proposed by Gomes et al. [14]. It is particularly important to indicate a name for the signal so that it can be communicated later. The title of the signal must be representative of its characteristics.
(b) **Year**. The year of creation of the signal must be indicated. On one hand, we can identify when the signal was created, and on the other, we are able to perform a better chronological management of the information related to the signal.
(c) **Analysis of the signal context and its cool DNA**. At this stage, following Rohde's perspective, addressed by [14], the context of the signal must be analysed through its description. The real explanation of the object must be carried out through its characteristics (in the semiotics perspective, it is the denotative reading of the sign). A PESTEL analysis may help to structure a guideline to allow the understanding of the signal sociocultural context [35]. Regarding the cool ADN, one should consider the characteristics indicated by Carl Rohde [26] to describe the signal and its relevance. These are attractiveness, ability to inspire, and the growth potential of the idea behind the signal. This step is important because it is the first filter for trend researchers to think about the relevance of the signal for the analysis.
(d) **Coolest chart**. The Trends and Culture Management Lab at the University of Lisbon defined a set of characteristics to describe a cool signal. These characteristics, in an applied perspective, help to deepen the understanding of the cool factor of the signal. To be compared with each other and as a way of comparing signals and their own characteristics, we propose an analysis grid for the classification of these characteristics. Each of them should be rated from 1 (least present in the signal) to 6 (most present in the signal). It should be noted that, according to Trends Lab, a signal to be considered "cool" must have all the categories and, in this sense, the evaluation of the strength of each one serves for comparison, as all the cool signals will have the characteristics, with more or less impact [12]. Our academic practice also indicates that this type of analysis and the discussion generated around the attempt to quantify these characteristics is beneficial and works well as a comparative exercise.
(e) **References and observations**. This section of the Canvas is intended to be completed with some notes and bibliographical references about the observed signal [14]. It may be interesting to contextualize or frame a signal that has been observed during a walk, for example, and which will have many framing references. On the signals collected in the digital/internet environment, primary and secondary links must be kept that help to find more information about the signal.

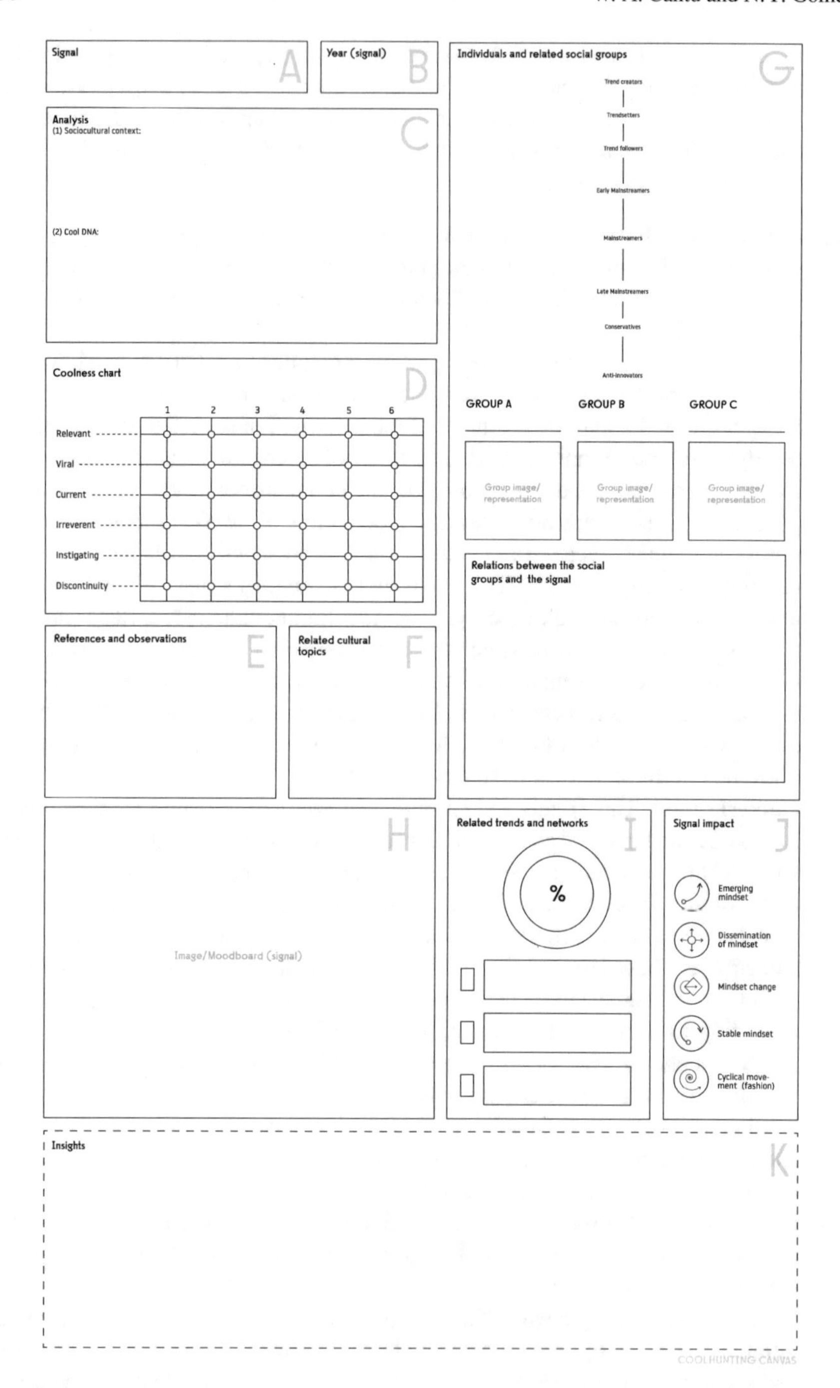

Fig. 4.1 Coolhunting Canvas. Developed by the authors

(f) **Related cultural topics**. At this stage of the signal analysis, Els Dragt already guides towards the correlation between similar signals, and Gomes et al. [2, p. 244] present a relationship between signals and cultural topics that can be understood as representative concepts of the current world. At this point, some themes and keywords that represent the main concepts of the sign should be thought of to correlate the signal with the cultural world.

(g) **Individuals and related social groups**. The relationship and analysis of dissemination groups are seen as advantageous step for understanding the movement of trends and its social behaviour, allowing for a better understanding of it as well as preparing for possible changes [4, 17, 30]. In this part of the Canvas, the analysis is based on the Diamond-Shaped Trend Model developed by Henrik Vejlgaard [30]. Here, among the social groups where trends are disseminated (trend creators, trendsetters, trend followers, early mainstreamers, mainstreamers, late mainstreamers, conservatives, anti-innovators) should be selected the one that best represents the positioning of the signal currently in social groups. One can think of which of these groups the actors related to the signal are. Next, you should think of two profiles that are related to the signal at any level (consumption, creation, communication, among others). Here, as referred by Gomes and Cantú [34, p. 473], "this specific exercise entails identifying if the narrative calls out to one or more tribes by (i) identifying elements that may address identities and shared passion/interests and (ii) finding the links between these elements". A name must be given to the identified group, and a visual representation of these groups must be made. Finally, some sentences must be constructed to summarize the relationship between the signal and the groups.

(h) **Mood board**. Citing William Higham [6], Gomes et al. [14] indicate that the signal must be represented through an image or video. In this stage of the Canvas, we propose the representation of the sign through a mood board, so that it can have its representation in a more expressive way [36].

(i) **Related trends and networks**. When identifying signals, it is important to understand whether they are already related to a previously identified trend. This can help in the initial phase of interpreting the signal. If there is a relationship between the signal and an already identified trend, it is possible to map eventual changes in the trend. If there is difficulty in relating the signal to a previously identified trend, we may be facing signs of a new emerging mindset [14]. At this stage, trends must be selected and an approximate value of the relationship between them and the signal should be considered. This is a perception of the analyst, and a focus group can be carried out to arrive at more approximate values. At this stage, it is also important to indicate the network/agency that identified the trend. In a pedagogical context, this phase of the work seeks to overcome some of the students' difficulties in defining which trends are behind certain identified signals.

(j) **Signal impact**. In this analytical phase, we will review the sociocultural impact of the signal. The trend agency Trend Hunter proposes six different "patterns" (movements) associated only with specific megatrends. They are "acceleration, reduction, convergence, cyclicality, redirection, divergence" [37]. It is important to refer that different trends may have similar or different behaviours that are always changing due to the nature of the cool signals which represent them. Trend Hunter associates the change with the trend. However, the behaviour of the change is directly connected with the signals of the trends and their meaning. A change of meaning in the signal may indicate a change in the trend, the representation of a cultural shift. Considering these contributions, we propose to analyse and think about the behavioural change represented nowadays by the signal (the indicator of the trend), selecting one of the five directions that, we believe, best frame the signal. They are emerging mindset (the sign is an indicator of a new way of thinking or acting in society); dissemination of mindset (the signal is an representation of social dissemination of a new way of thinking); mindset change (the sign indicates changes in the way of thinking about certain issues in society); stable mindset (the signal is a representation of concepts that are not changing nowadays); cyclical movement (as a fashion, the sign is a representation of previously defined meanings that are suffering a cultural peak again).

(k) **Insights**. Insights related to the signal should be drawn based on sociocultural analysis and the cool DNA of the signal. At this stage, we must pay attention to the fact that insight can be defined by the idea behind the signal and its ability to synthesize its meaning: its connotative message, as seen in [34]. Insights are important steps in signal mapping referenced by Dragt [4] and Gomes et al. [14]. Through the understanding of the messages that lay behind them, we are able to understand how trends are represented nowadays by material culture.

4.5 Conclusions

There has been a lot of work in Trend Studies trying to correlate different approaches in terms of research and different methodologies. Academia suffers from this ambiguous diversity. Students feel difficulties in understanding trend signals since the subjectivity of this type of data demands better guidance to confirm the signals and support the analysis.

We can observe from several trend methodologies that one of the phases which are common between them is the research/context/scan. This one is particularly important due to its initial role in the process of understanding trends. Because of that, we presented here a script to contribute to the analysis of cool signals. The visual tools are a good option to help to understand meanings and the contextualization of signals, and for that, we developed the structure in the format of a Canvas. The framework which guides this proposed model is composed by several steps that are drawn by authors from the area and intends to contribute for teaching trends, with focus on this specific part of the process (trendspotting). This can easily occur in applications like marketing strategies, due to the need of looking continuously for innovation in the area. This model has an obvious implication in terms of mapping signals of trends, which can take part as inspirations for the process of marketing creation.

This work suffers some limitations as the literature review process limits the range of publications that exists regarding this topic. On the other side, it gives us the opportunity to review and assess a broad field of study. Future research can confirm the real utility of this tool in the pedagogical scope in marketing and other related areas through its applicability in teaching. It will be also possible to approach the construction of new models using specific authors derived from a future systematic literature review.

References

1. Rech, S., Filippe, A.: Análise de tendências socioculturais and gestão visual de projetos: proposta de modelo conceitual para o Fashion LAB—coletivo/criativo Brasil. In: Raposo, D., et al. (eds.) Investigação e ensino em design e música. Castelo Branco, IPCB (2022)
2. Gomes, N., Cohen, S., Cantú, W., Lopes, C.: Roteiros e modelos para a identificação de tendências socioculturais e a sua aplicação estratégica em produtos e serviços. Moda Palavra. **14**(32), 228–272 (2021)
3. Powers, D.: On Trends: The Business of Forecasting the Future. University of Illinois, Illinois (2019)
4. Dragt, E.: How to Research Trends. BIS Publishers, Amsterdam (2017)
5. Cramer, T., van der Duin, P., Heselmans, C.: Trend analysis. In: van der Duin, P. (ed.) Foresight in Organizations: Methods and Tools. New York, Routledge (2016)
6. Higham, W.: The Next Big Thing. Kogan Page, London (2009)
7. Powers, D.: Thinking in trends: the rise of trend forecasting in the United States. J. Hist. Res. Mark. **1**(1), 2–20 (2018)
8. Williams, R.: The Long Revolution. Penguin Books, Harmondsworth (1961)
9. Gomes, N.: A análise cultural e o estudo de tendências na comunicação e gestão de marcas: Estudo de caso publicitário em contexto semiótico. Int. J. Mark. Commun. New Media. **7**, 56–79 (2020)
10. Gomes, N., Cantu, W.: Sociocultural trend reports as an intelligence tool of strategic cultural management. In: J. L. Reis et al. (eds.) Marketing and Smart Technologies, Smart Innovation, Systems and Technologies. Springer, Switzerland (2022)
11. Lobo, T., Cantú, W., Gomes, N.: The cork thread, a sustainable material: branding and marketing implications from a cultural perspective. In: Markopoulos, E. et al. (eds.) AHFE 2020: Advances in Creativity, Innovation, Entrepreneurship and Communication Design. Springer, Switzerland (2020)
12. Lobo, T., Cantú, W., Gomes, N.: A cultural mediation of meanings between consumer-goods, trends, and the culturally constituted world. In: Shin, C.S. et al. (eds.) AHFE 2021: Advances in Industrial Design. Springer, Switzerland (2021)
13. Silva, J.: Tendências Socioculturais: recorrências simbólicas do espírito do tempo no sistema publicitário. Tese de Doutoramento. Escola de Comunicação e Artes da Universidade de São Paulo, Brasil (2015)
14. Gomes, N., Cohen, S., Flores, A.: Estudos de Tendências: contributo para uma abordagem de análise e gestão da cultura. Moda Palavra **11**(22), 49–112 (2018)
15. Kjaer, L.: The Trend Management Toolkit. Palgrave MacMillan, London (2014)
16. Pedroni, M.: From fashion forecasting to coolhunting: prevision models in fashion and in cultural production. In: Berry, J. (ed.) Fashion Capital: Style, Economies, Sites and Cultures. Inter-disciplinary press (2019)

17. Raymond, M.: The Trend Forecaster Handbook. Laurence King Publishing, London (2010)
18. Mackinney-Valentin, M.: On the Nature of Trends. Doctoral Thesis. The Danish Design School, Copenhagen (2010)
19. Rech, S.: Estudos de Futuro and Moda: uma abordagem conceitual. Moda Palavra **6**(12), 93–110 (2013)
20. Rech, S.: Trends management: the qualitative approach as a methodology. In: Raposo, D. et al. (eds.) Perspective on Design. Series in Design and Innovation volume 1. Springer, Switzerland (2020)
21. Mártil, V.: Coolhunting: El arte y la ciência de decifrar tendências. Ediciones Urano, Barcelona (2009)
22. Maioli, F., Presotto, J., Palma, C.: Coolhunting: Métodos e Práticas. Vidráguas, Porto Alegre (2009)
23. Snyder, H.: Literature review as a research methodology: na overview and guidelines. J. Bus. Res. **104**, 333–339 (2019)
24. https://www.newyorker.com/magazine/1997/03/17/the-coolhunt
25. Cantú, W.: Tendências e MARCAS: Contributos Para a Análise Estratégica De Significados Em Produções Culturais. Tese de Doutoramento. Faculdade de Letras da Universidade de Lisboa, Portugal (2021)
26. Rohde, C.: SeriousTrendwatching. Fontys University of Applied Sciences and Science of the Time, Tilburg (2011)
27. Gloor, P., Cooper, S.: Coolhunting: Chasing Down the Next Big Thing. Amacon, New York (2007)
28. http://creativecultures.letras.ulisboa.pt/index.php/gtc-naturezadocool/
29. Kongsholm, L., Frederiksen, C.: Trend Sociology, v. 2.0. Pej Gruppen, Herning (2018)
30 Vejlgaard, H.: Anatomy of a Trend. McGraw-Hill, New York (2008)
31. https://www.trendwatching.com/toolbox/consumer-trend-canvas
32. https://www.dhl.com/global-en/home/insights-and-innovation/insights/logistics-trend-radar.html
33. Cohen, S.: Trend studies as an academic subject. Moda Palavra. **10**(20), 89–94 (2017)
34. Gomes, N., Cantu, W.: Cultural mediation between branding ans lifestyles: a case study based model for the articulation of cultural strategies and Urban tribes. In: Goonetilleke, R.S. et al. (eds.) AHFE 2021: Advances in Physical, Social & Occupational Ergonomics. Springer, Switzerland (2021)
35. https://pestleanalysis.com/
36. Rehn, A., Lindkvist, M.: Trendspotting, the basics. Booktango (2013)
37. https://www.trendhunter.com/

Chapter 5
Hotel Customer Segmentation Using the Integrated Entropy-CRITIC Method and the 2T-RFMB Model

Ziwei Shu, Ramón Alberto Carrasco González, Javier Portela García-Miguel, and Manuel Sánchez-Montañés

Abstract Customer segmentation helps the company better understand its target audience, which is vital to optimizing marketing strategies and maximizing the customer value for the company. This paper improves the original RFM model by including the potential loss to the hotel from a customer canceling their reservation in the indicator "Monetary" and adding a new indicator "Bonding" to indicate the degree of customer bonding with the hotel. The proposed model also includes the 2-tuple linguistic model to give hotel managers or decision-makers more easily understandable customer segmentation results. The aggregation of the four indicators (recency, frequency, monetary, and bonding) into a unique value is a Multi-Criteria Decision-Making (MCDM) problem. To generate the weights that can consider the relationship between various indicators and the level of data diversification contained in each indicator, the Entropy method and the CRiteria Importance Through Intercriteria Correlation (CRITIC) method have been integrated. Customer overall values are generated based on the 2T-RFMB model and the integrated Entropy-CRITIC method. Finally, various customer segments are obtained with K-means clustering. This proposal has been evaluated by a real dataset from a hotel in Lisbon. The results show that the proposed model can increase the linguistic interpretability of clustering results. It also demonstrates that the proposed model can provide hotel managers with more realistic customer values to assist them in allocating their Customer Relationship Management (CRM) resources efficiently.

Z. Shu (✉) · J. P. García-Miguel
Department of Statistics and Data Science, Faculty of Statistics, Complutense University of Madrid, Avenida Puerta de Hierro, S/N, 28040 Madrid, Spain
e-mail: ziweishu@ucm.es

R. A. Carrasco González
Department of Marketing, Faculty of Statistics, Complutense University of Madrid, Avenida Puerta de Hierro, S/N, 28040 Madrid, Spain

M. Sánchez-Montañés
Department of Computer Science, Universidad Autónoma de Madrid, 28049 Madrid, Spain

J. L. Reis et al. (eds.), *Marketing and Smart Technologies*, Smart Innovation, Systems and Technologies 337, https://doi.org/10.1007/978-981-19-9099-1_5

5.1 Introduction

Customers are crucial to any business; without them, it would be impossible to continue the company. Customer segmentation divides customers into similar groups, assisting business decision-makers in identifying the value of customers, targeting valuable customers, and developing suitable marketing strategies to retain them.

Customer behavior analysis is one of the most popular methods for customer segmentation. The RFM model is the most widely applied in customer behavior analysis [1–5]. It is also applied to estimate customer value [6]. However, some indicators from the initial RFM model should be modified, as customer behaviors differ by industry [7], and some industries do not entirely adhere to the RFM principle. In some cases, using only three criteria, the RFM is insufficient for creating successful marketing plans.

Therefore, this paper suggests an improved RFM model based on four indicators that directly affect the segmentation of hotel customers: recency, frequency, money, and bonding. It contributes a novel perspective to the original RFM model by taking into account the characteristics of the hotel industry. The proposed model includes the potential loss to the hotel from a customer canceling their reservation in the indicator "Monetary" and adds a new indicator "Bonding" to represent the degree of customer bonding with the hotel. This proposal also includes the 2-tuple linguistic model to give hotel managers more easily understandable customer segmentation results. The combination of two objective weighting methods to create a suitable weight distribution and calculate the 2T-RFMB overall value for each customer is another novelty of this paper. Finally, K-means clustering is employed to obtain diverse customer groups for a hotel in Lisbon. In this way, the proposed model not only can rank customers based on their 2T-RFMB value but also more effectively categorize them to determine which customer groups merit being allotted more Customer Relationship Management (CRM) resources.

The rest of this paper is organized as follows. The essential concepts on which the proposed model is built are introduced in Sect. 5.2. The application of the proposed model in a real hotel dataset is demonstrated in Sect. 5.3. Conclusions of this paper and future research are presented in Sect. 5.4.

5.2 Theoretical Framework

This section presents the essential concepts on which this proposal is based: the RFM model, the 2-tuple linguistic model, the 2T-RFMB model, the Entropy method, the CRITIC method, and the integrated Entropy-CRITIC method.

5.2.1 The RFM Model and Its Improvement

The RFM model, developed by Hughes, is a behavior-based model used to classify customers based on recency, frequency, and monetary [8]. Recency is the interval between the last purchase time and the analyzing time, frequency is the number of purchases during the established period, and monetary is the total amount a customer spent in a given period. The RFM model assists decision-makers in identifying valuable customers and developing relative marketing strategies. In a variety of industries, including banking [9, 10], tourism [11, 12], and retail [13–16], it has been successfully used and performed.

However, RFM values depend on the features of the products or services a business provides [17, 18]. Therefore, customer behaviors differ by industry, so it is necessary to modify or add some indicators from the initial RFM model.

Considering the characteristics of the hotel data studied in this paper, the indicator M (Monetary) in the RFM model has been modified to be the difference between the total amount a customer spent at the hotel and the potential loss of the hotel. Additionally, a new indicator B (Bonding) has been added to express the level of customer bonding with the hotel. Customer bonding refers to how a business establishes relationships with its customers. The stronger the bond, the more likely the customer will stay at the hotel again.

The definition of the proposed RFMB model is as follows.

Definition 1 Let $\text{Date}_q^{\text{first}}$ be the date that the qth customer checked in for the first time at the hotel, $\text{Date}_q^{\text{last}}$ be the date that the qth customer checked in recently at the hotel, and $\text{Date}_{\text{analysis}}$ be the date of analysis (i.e., the date of data extraction). The indicators R_q and B_q of the RFMB model can be calculated using Eqs. (5.1) and (5.2), respectively.

$$R_q = \text{Date}_{\text{analysis}} - \text{Date}_q^{\text{last}} \tag{5.1}$$

$$B_q = \text{Date}_q^{\text{last}} - \text{Date}_q^{\text{first}} \tag{5.2}$$

where R_q represents the number of days between the last check-in date of the qth customer and the analysis date; B_q represents the number of days the qth customer has established a relationship with the hotel during the analysis period. When B_q is equal to 0, it indicates that the customer has only been at the hotel once and has not yet developed a relationship with it. If B_q is not zero but also not very high, it means that the customer has just begun establishing a short-term relationship with the hotel. The larger B_q is, the better customer bonding with the hotel is.

Definition 2 Let $\text{Booking}_q^{\text{checked-in}}$ be the total number of times the qth customer checked in at the hotel, and $\text{Booking}_q^{\text{not-in}}$ be the total number of reservations that were made by the qth customer who either canceled (the hotel was informed of the cancelation) or did not show up (the hotel was not informed of the cancelation).

Assume that M_q^{room} represents the total amount the qth customer spent on lodging expenses, and M_q^{other} is the total amount the qth customer spent on other expenses (spa, beverage, food, etc.). The indicators F_q and M_q of the RFMB model can be calculated using Eqs. (5.3) and (5.4), respectively.

$$F_q = \text{Booking}_q^{\text{checked}-in} \tag{5.3}$$

$$M_q = M_q^{\text{room}} + M_q^{\text{other}} - \text{Booking}_q^{\text{not-in}} \times \frac{(M_q^{\text{room}} + M_q^{\text{other}})}{\text{Booking}_q^{\text{checked-in}}} \tag{5.4}$$

where F_q represents the number of times the qth customer stays at the hotel during the analysis period; M_q represents the net revenue that the hotel can earn from the qth customer, considering the loss of empty rooms caused by this customer. If $\text{Booking}_q^{\text{not-in}} = 0$, it means that the qth customer has never canceled a hotel reservation, so in this case, M_q is the total amount the customer spent on the hotel during the analysis period, same as its definition in the original RFM model. If not, the calculation of M_q needs to consider the potential loss, that is, multiply the number of customer "cancelation" by the average amount spent by the qth customer in the hotel during the analysis period.

5.2.2 *The 2-Tuple Linguistic Model*

Introduced by Herrera and Martinez, the 2-tuple linguistic model offers a convenient way to characterize linguistic assessments using linguistic variables and addresses the issue of information loss using symbolic translation [19]. This model is used in Multiple-Criteria Decision-Making (MCDM) problems to reduce the complexity of numerical calculations. The qualitative scales (i.e., linguistic terms) are closer to human thinking [20], which provides more precise and understandable results to decision-makers.

The 2-tuple linguistic model uses a 2-tuple value (s_i, α) to express the linguistic information, where $s_i \in S$ represents a linguistic term, and $\alpha \in [-0.5, 0.5)$ is a numeric value that represents the distance to the central value of s_i. The definition is as follows.

Definition 3 Let $S = \{s_0, \ldots, s_g\}$ be a linguistic term set, and $\beta \in [0, g]$ be a value that represents the result of an operation of symbolic aggregation. The function $\Delta : [0, g] \to \langle S \rangle = Sx[-0.5, 0.5)$ is used to convert β to 2-tuple value (s_i, α) as the Eq. (5.5):

$$\Delta(\beta) = (s_i, \alpha), \text{ with} \begin{cases} i = \text{round}(\beta) \\ \alpha = \beta - i, \alpha \in [-0.5, 0.5) \end{cases} \tag{5.5}$$

where round(·) is the rounding operation; s_i has the nearest index label to β; and α represents a numerical value of the symbolic translation.

The 2-tuple linguistic model can perform transformations between 2-tuple values and numerical values, since the function Δ is bijective. The inverse function of Δ is Δ^{-1}: $\langle S\rangle = Sx[-0.5, 0.5) \rightarrow [0, g]$; the 2-tuple value is converted into its equivalent numerical value as $\Delta^{-1}(s_i, \alpha) = i + \alpha = \beta$.

Furthermore, Herrera et al. proposed a method for comparing two 2-tuple linguistic values, the negation operator of a 2-tuple value, and aggregation operators for 2-tuple linguistic computing [21]. In this paper, the negation operator of a 2-tuple value should be discussed as it is always used to handle inverse indicators (e.g., the smaller the value of the recency, the better). The definition is as follows.

Definition 4 The negation operator of a 2-tuple value (s_i, α) is defined as Eq. (5.6):

$$\text{neg}((s_i, \alpha)) = \Delta(g - (\Delta^{-1}(s_i, \alpha))) = \Delta(g - \beta) \tag{5.6}$$

5.2.3 *The 2T-RFMB Model*

The original RFM model usually employs quantiles to determine customer value and make customer segmentation (i.e., the top 25% of customers have very high value, while the bottom 25% have very low value). However, these computing process results in information loss. Therefore, in this paper, the 2-tuple linguistic model has been included in the RFMB model to increase the accuracy and interpretability of the results. The definition is as follows.

Definition 5 Let U_{qj} be a set of values of the recency, frequency, monetary, and bonding of the qth customer, with $U_{q1} = R_q$, $U_{q2} = F_q$, $U_{q3} = M_q$, $U_{q4} = B_q$. For each element in U_{qj}, its 2-tuple value can be calculated using Eq. (5.7):

$$U_{\text{qj}}^{2T} = \Delta(U_{\text{qj}}) = \begin{cases} \Delta(\text{min\_max}(U_{\text{qj}})), \text{if } j \neq 1 \\ \text{neg}(\Delta(\text{min\_max}(U_{\text{qj}}))), \text{if } j = 1 \end{cases} \tag{5.7}$$

where min_max(·) represents the Min–Max normalization to scale data in the range [0,1]; $\Delta(\cdot)$ and neg(·) have been defined in Eqs. (5.5) and (5.6), respectively.

The 2T-RFMB overall value of the qth customer is calculated using Eq. (5.8):

$$V_q^{2T-\text{RFMB}} = \Delta(\sum_{j=1}^{n} w_j \cdot \Delta^{-1}(U_{\text{qj}}^{2T})) \tag{5.8}$$

where w_j represents the weight of each indicator of the 2T-RFMB model; $j = 1,2,\ldots,n$, n is the total number of indicators (in this paper, $n = 4$). Section 5.2.6

explains how to calculate the corresponding weights for each indicator (see Eq. (5.14)).

5.2.4 The Entropy Method

Introduced by Shannon, the Entropy method calculates the objective weights for each indicator by measuring the amount of valid information contained in known data [22]. The fundamental benefit of this method is that there is no human interference in the weighting process, making it more accurate to the distribution of the actual data [23, 24]. Furthermore, the Entropy method is very useful for analyzing contrasts between sets of data when there is no established preference among the numerous indicators [25].

The Entropy method divides the calculation of the objective weights into four steps as follows:

- Normalize the values in the dataset using the Min-Max normalization, since indicators measured at different scales do not contribute equally to analysis and result in inaccurate weight calculation.
- Compute the probability value of each entry under each indicator.
- Obtain the Entropy value and the degree of diversification for each indicator.
- Calculate the Entropy weights for each indicator.

The definition is as follows.

Definition 6 Let x_{qj} be the normalized value of the qth customer under the jth indicator, and p_{qj} be the corresponding probability value. The Entropy value E_j for the jth indicator is calculated using Eq. (5.9):

$$E_j = -K\sum_{q=1}^{m} p_{\mathrm{qj}} \ln p_{\mathrm{qj}} \tag{5.9}$$

where $K = \frac{1}{\ln m}$ is a constant that ensures $E_j \in [0, 1]$, m represents the total number of customers, $q = 1,2\ldots,m$; $p_{\mathrm{qj}} = \frac{x_{\mathrm{qj}}}{\sum_{q=1}^{m} x_{\mathrm{qj}}}$, $x_{\mathrm{qj}} \in [0, 1]$, $j = 1,2,\ldots,n$, n is the total number of indicators. The higher the value of E_j, the lower the information contained in the jth indicator, resulting in a smaller relative weight, and vice versa.

The degree of diversification D_j for the jth indicator is calculated using Eq. (5.10):

$$D_j = 1 - E_j \tag{5.10}$$

Definition 7 The objective weight of the jth indicator is calculated using Eq. (5.11):

$$w_j^{\mathrm{Entropy}} = \frac{D_j}{\sum_{j=1}^{n} D_j} \tag{5.11}$$

where D_j is complementary to Entropy value. The larger D_j is, the more weight is assigned to the *j*th indicator.

5.2.5 The CRiteria Importance Through Intercriteria Correlation (CRITIC) Method

Proposed by Diakoulaki et al., the CRITIC method calculates the objective weights for each indicator by considering the correlations between them [26]. Human interference is also not required in this method. The CRITIC method uses the standard deviations of each indicator to compute the contrast intensity and combines them with correlation analysis.

The CRITIC method divides the calculation of the objective weights into five steps as follows:

- Normalize the values of different indicators using the Min-Max normalization.
- Compute the contrast intensity of each indicator.
- Calculate the correlation coefficient between the indicators.
- Obtain the quantity of information on each indicator.
- Calculate the CRITIC weights for each indicator.

The definition is as follows.

Definition 8 Let S_j be the standard deviation of the *j*th indicator, and r_{jv} be the correlation coefficient between *j*th and *v*th indicators. The quantity of information contained in the *j*th indicator is calculated using Eq. (5.12):

$$C_j = S_j \sum_{v=1}^{n} \left(1 - \left|r_{\mathrm{jv}}\right|\right) \tag{5.12}$$

where $S_j = \sqrt{\frac{\left(\sum_{q=1}^{m} x_{\mathrm{qj}} - \overline{x_j}\right)^2}{m-1}}$, x_{qj} is the normalized value of the *q*th customer with respect to the *j*th indicator, $q = 1{,}2\ldots{,}m$, and m is the total number of customers; $\overline{x_j}$ is the mean of the *j*th indicator, $j = 1{,}2{,}\ldots{,}n$; $v = 1{,}2{,}\ldots{,}n$, n is the total number of indicators.

Definition 9 The objective weight of the *j*th indicator is calculated using Eq. (5.13):

$$w_j^{\mathrm{CRITIC}} = \frac{C_j}{\sum_{j=1}^{n} C_j} \tag{5.13}$$

where C_j represents the quantity of information contained in the *j*th indicator. The larger C_j is, the more weight is assigned to the *j*th indicator.

5.2.6 *The Integrated Entropy-CRITIC Method*

Although the weights obtained by the Entropy method are effective and informative, it ignores the horizontal influence between indicators, such as correlation. Therefore, to ensure that the weights acquired take into account both the data structure inside the indicator and their interdependency, it is necessary to combine the weight produced by the Entropy method with that computed by the CRITIC method.

The definition is as follows.

Definition 10 The integrated Entropy-CRITIC weight of the *j*th indicator is calculated using Eq. (5.14):

$$w_j = \frac{\left(w_j^{\text{Entropy}} \cdot w_j^{\text{CRITIC}}\right)^{\frac{1}{2}}}{\sum_{j=1}^{n} \left(w_j^{\text{Entropy}} \cdot w_j^{\text{CRITIC}}\right)^{\frac{1}{2}}} \tag{5.14}$$

where w_j^{Entropy} is the weight obtained by the Entropy method (see Eq. (5.11)), and w_j^{CRITIC} is the weight derived from the CRITIC method (see Eq. (5.13)). This equation is inspired by Jahan et al., who proposed a combinative weighting method to integrate the weights generated by various methods [27].

5.3 Application of the Proposed Model to Hotel Customer Segmentation

This section discusses how the proposed model was established and how it can be used to segment hotel customers. The five steps to building this model are shown in Fig. 5.1.

5.3.1 *Data Collection and Cleaning*

This paper uses the dataset published by [28] to make hotel customer segmentation. Extracted on December 31, 2018, this dataset includes the behavioral data of 83,590 customers at a hotel in Lisbon from 2015 to 2018.

Although this dataset contains 31 variables, considering the characteristic of the proposed model, only the following variables have been retained: CustomerID, LodgingRevenue, OtherRevenue, BookingsCheckedIn, BookingsCanceled, BookingsNoShowed, DaysSinceFirstStay, and DaysSinceLastStay. Table 5.1 gives the description of these variables.

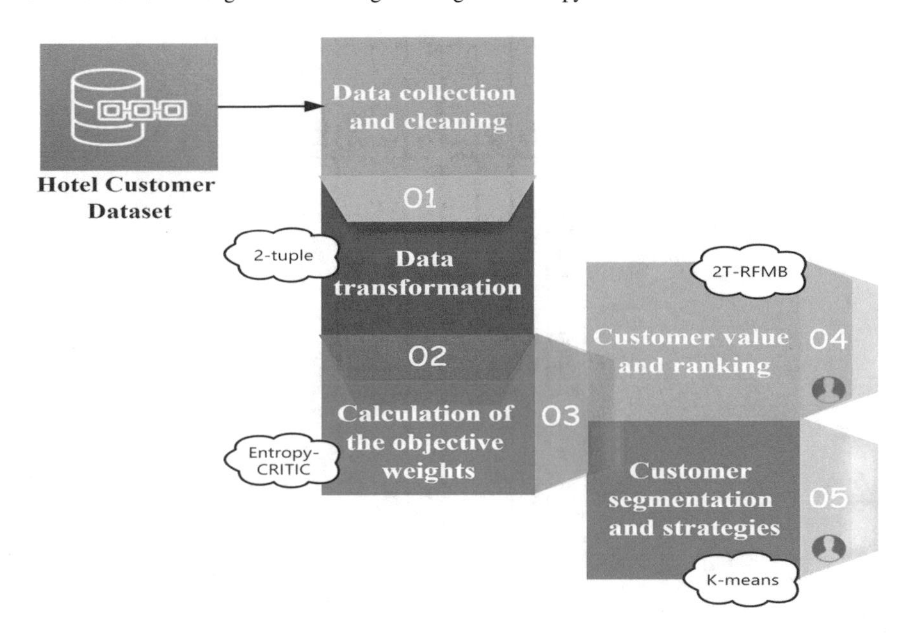

Fig. 5.1 Steps used to create the proposed model

Table 5.1 Variable description

Variable	Description
CustomerID	The customer is allocated a different number to identify him or her
LodgingRevenue	The total amount the customer spent on accommodation (in Euros), including costs for the room, crib, and related charges
OtherRevenue	The total amount the customer spent on other expenses (in Euros), including costs for food, drinks, spa visits, and others
BookingsCheckedIn	The total number of times the customer checked in at the hotel
BookingsCanceled	The total number of reservations a customer booked but canceled at the end (the hotel was informed of the cancelation)
BookingsNoShowed	The total number of reservations a customer booked but did not show up at the end (the hotel was not informed of the cancelation)
DaysSinceFirstStay	The number of days between the data extraction date and the first arrival date of the customer (of a checked-in booking)
DaysSinceLastStay	The number of days between the data extraction date and the latest arrival date of the customer (of a checked-in booking)

Besides, 19,920 customers in this dataset have only registered on the hotel website but have never stayed there. These people are potential customers of the hotel, who should be removed from the analysis as they have not done any real customer behavior yet. Therefore, only the 63,670 customers who stayed at the hotel are analyzed. An example of the dataset after data processing is shown in Table 5.2.

Table 5.2 Example of customer behavior data

CustomerID	3076	9865	54772	11426	54505	61562	33146	30506	31469	53974
LodgingRevenue	177.3	134.1	413	468.9	385	142.5	621	160.2	234	537
OtherRevenue	192	27.5	49	224	70.8	10	120	24	154.5	76
BookingsCheckedIn	1	1	5	1	2	1	1	1	1	7
BookingsCanceled	0	0	1	0	0	0	0	0	0	0
BookingsNoShowed	0	0	0	0	0	0	0	0	0	0
DaysSinceFirstStay	1022	905	242	883	245	187	529	565	551	249
DaysSinceLastStay	1022	905	118	883	177	187	529	565	551	167

5.3.2 Data Transformation

The parameters of the RFMB model are obtained using Eqs. (5.1)–(5.4). Table 5.3 shows some computation results for the customers' recency, frequency, monetary, and bonding.

In this paper, five terms—*Very Low (VL)*, *Low (L)*, *Average (A)*, *High (H)*, and *Very High (VH)*—form the linguistic term set *S* to demonstrate customer value. Let $S = \{s_0, \ldots, s_g\}$ with $g = 4$: $s_0 = VL$, $s_1 = L$, $s_2 = A$, $s_3 = H$, $s_4 =$ VH, as shown in Fig. 5.2. Thus, based on Eq. (5.7), the 2-tuple values of the customers' recency, frequency, monetary, and bonding are obtained. Table 5.4 illustrates how the customers' recency, frequency, monetary, and bonding are expressed in 2-tuple values.

The 2-tuple value distribution of recency, frequency, monetary, and bonding for the 63,670 customers is shown in Fig. 5.3, with the majority of their frequency and bonding belonging to the 2-tuple value with the linguistic term *VL*.

Table 5.3 Example of customer "RFMB" data

CustomerID	Recency	Frequency	Monetary	Bonding
3076	1022	1	369.3	0
9865	905	1	161.6	0
54772	118	5	369.6	124
11426	883	1	692.9	0
54505	177	2	455.8	68
61562	187	1	152.5	0
33146	529	1	741	0
30506	565	1	184.2	0
31469	551	1	388.5	0
53974	167	7	613	82

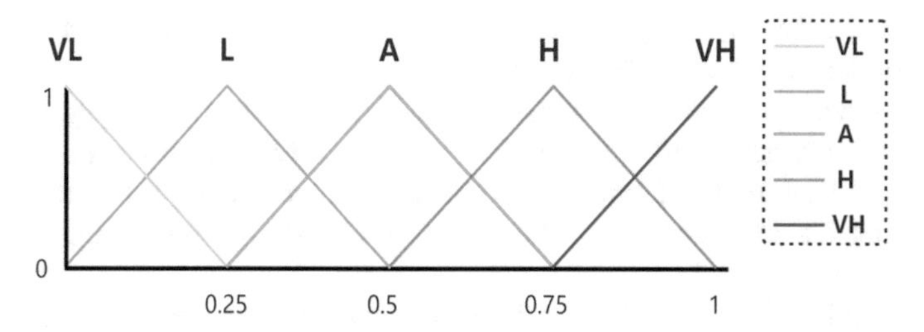

Fig. 5.2 Linguistic term set *S*

Table 5.4 Example of customer "RFMB" data expressed in 2-tuple values

CustomerID	Recency	Frequency	Monetary	Bonding
3076	(VL, + 0.2972)	VL	(A, -0.0476)	VL
9865	(L, −0.2788)	VL	(L, -0.4468)	VL
54772	(VH, −0.4276)	(VL, + 0.246)	(A, -0.0456)	(VL, + 0.4732)
11426	(L, −0.1992)	VL	(H, + 0.31)	VL
54505	(H, + 0.3588)	(VL, + 0.0616)	(A, + 0.4888)	(VL, + 0.2596)
61562	(H, + 0.3224)	VL	(L, -0.492)	VL
33146	(A, + 0.0832)	VL	(H, + 0.404)	VL
30506	(A, −0.0472)	VL	(L, -0.3036)	VL
31469	(A, + 0.0036)	VL	(A, + 0.0788)	VL
53974	(H, + 0.3948)	(VL, + 0.3692)	(H, + 0.1188)	(VL, + 0.3128)

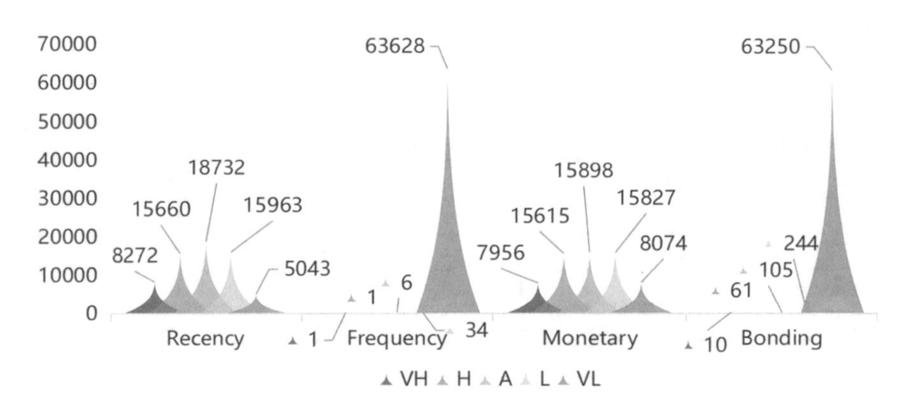

Fig. 5.3 Distribution of customer "RFMB" data expressed in 2-tuple values

5.3.3 Calculation of the Objective Weights for Each Indicator

The integrated Entropy-CRITIC method is applied to calculate the objective weights for each indicator in the 2T-RFMB model.

In the previous section, the 2-tuple values of the customers' recency, frequency, monetary, and bonding are obtained. Therefore, before calculating the weights of these four indicators, the 2-tuple values need to be converted into their numerical values using the function Δ^{-1}. Table 5.5 shows their objective weights obtained using Eq. (5.14).

As shown in Table 5.5, using the Entropy method or the CRITIC method alone would result in an uneven distribution of weights between indicators. The benefit of combining these two methods is that the weights obtained will be relatively more even, as it considers both the relationship between indicators (CRITIC method) and the degree of diversification of data contained in each indicator (Entropy method).

Table 5.5 Objective weight for each indicator of the 2T-RFMB model

Indicator	Weights obtained by the Entropy method (%)	Weights obtained by the CRITIC method (%)	Integrated Entropy-CRITIC weights (%)
Recency	1.62	45.4	20.28
Frequency	45.64	1.32	18.36
Monetary	2.02	48.18	23.33
Bonding	50.72	5.1	38.03

5.3.4 *Customer Value and Ranking*

This section shows how to calculate each customer's 2T-RFMB overall value and rank them. The values of the customers' recency, frequency, monetary, and bonding are aggregated using Eq. (5.8). Table 5.6 lists several customers' 2T-RFMB overall values and related rankings. The cluster IDs to which these customers belong are also included in Table 5.6. More details about the characteristics of each cluster can be found in Sect. 5.3.5 (see Tables 5.7 and 5.8).

The following example shows how to calculate the 2T-RFMB overall values for customer 54772:

$$\begin{aligned} V_{54772}^{2T-\text{RFMB}} &= \Delta(20.28\% \cdot \Delta^{-1}(VH, -0.4276) + 18.36\% \cdot \Delta^{-1}(VL, +0.246) \\ &\quad + 23.33\% \cdot \Delta^{-1}(A, -0.0456) \\ &\quad + 38.03\% \cdot \Delta^{-1}(VL, +0.4732)) \\ &= \Delta(20.28\% \cdot 3.5724 + 18.36\% \cdot 0.246 \\ &\quad + 23.33\% \cdot 1.9544 + 38.03\% \cdot 0.4732) \\ &= \Delta(1.406) = (L, +0, 406) \end{aligned}$$

Table 5.6 2T-RFMB overall value, ranking, and cluster ID for some customers

CustomerID	2T-RFMB overall value	Ranking	Cluster ID
3076	(L, −0.4842)	52,283	9
9865	(VL, + 0.2753)	60,512	3
54772	(L, + 0.406)	5980	6
11426	(L, −0.0654)	29,686	1
54505	(L, + 0.3718)	7084	4
61562	(L, −0.2077)	38,133	2
33146	(L, + 0.2166)	13,996	5
30506	(L, −0.4415)	50,284	7
31469	(L, −0.1087)	32,085	8
53974	(A, −0.3972)	1302	10

Table 5.7 2-tuple values for each cluster centroid

Cluster ID	Number of customers	Recency	Frequency	Monetary	Bonding	2T-RFMB overall value
1	6112	(L, −0.1638)	(VL, + 0.0016)	(H, + 0.2378)	(VL, + 0.0025)	(L, −0.074)
2	4951	(H, + 0.3257)	(VL, + 0.003)	(VL, + 0.4916)	(VL, + 0.0219)	(L, −0.202)
3	7330	(L, −0.277)	(VL, + 0.0005)	(L, −0.4096)	(VL, + 0.0018)	(VL, + 0.285)
4	5990	(H, + 0.3226)	(VL, + 0.0025)	(H, −0.4313)	(VL, + 0.0115)	(L, + 0.278)
5	6846	(A, + 0.0995)	(VL, + 0.0032)	(H, + 0.388)	(VL, + 0.0117)	(L, + 0.221)
6	5382	(H, + 0.3057)	(VL, + 0.0018)	(A, −0.4269)	(VL, + 0.0131)	(L, + 0.043)
7	7480	(A, −0.0138)	(VL, + 0.0008)	(L, −0.2442)	(VL, + 0.007)	(L, −0.418)
8	7160	(A, + 0.0093)	(VL, + 0.0009)	(A, + 0.1321)	(VL, + 0.0043)	(L, −0.093)
9	6539	(L, −0.2237)	(VL, + 0.0007)	(A, −0.1794)	(VL, + 0.0014)	(L, −0.417)
10	5880	(H, + 0.3805)	(VL, + 0.0133)	(VH, −0.4546)	(VL, + 0.0474)	(A, −0.467)
AoD*	63,670	(A, + 0.0912)	(VL, + 0.003)	(A, −0.0012)	(VL, + 0.012)	(L, −0.105)

*AoD shows the average of the 63,670 customers

5.3.5 *Customer Segmentation and Strategies*

Customer segmentation is done using K-means clustering. The Elbow Method indicates that the optimal number of clusters is 10. Table 5.7 shows the 2-tuple values for the cluster centroid of each customer group.

As can be seen in Table 5.7, the frequency and bonding of these ten customer groups are very low, indicating that this hotel does not maintain close contact with customers and does not engage in long-term management with them. In actuality, more than 97% of the customers in this Lisbon hotel dataset have only ever stayed there once (see Fig. 5.3). It means that this hotel does not have many repeat customers.

By analogy, the bonding of the customers is also very low. This hotel has not even established a long-term relationship with most of its customers. The probability that the customer will stay at the hotel again is very low. Therefore, to establish a long-term relationship, customers should be encouraged to stay at the hotel again, regardless of the cluster to which they belong. Strategies that could be done include,

Table 5.8 Description for each group of customers

Cluster ID	Description
1	It consists of customers who have not stayed in the hotel for a long time but used to spend much money in the hotel (just lower than the monetary value of clusters 5 and 10). The other two elements are virtually the same as the average of the 63,670 customers at this hotel. This type of customer has a high possibility of churn and needs to use some strategies to make them live in the hotel again
2	It consists of customers who recently stayed in the hotel but spent very little money during their stay. The other two elements are virtually the same as the average of the 63,670 customers at this hotel. The overall value of this group of customers is low
3	It consists of customers who have not stayed in the hotel for a long time and spent little money during their stay. The other two elements are virtually the same as the average of the 63,670 customers at this hotel. This group of customers has a lower overall value than the other nine groups. It can be temporarily set aside, so that limited CRM resources can be invested in more valuable customers
4	It consists of customers who recently stayed in the hotel and spent a lot of money there. The other two elements are virtually the same as the average of the 63,670 customers at this hotel. Although the overall value of this group of customers is not very high, the hotel should allocate some efforts to managing relationships with them to turn these new customers with relatively high spending power into hotel loyalty customers
5	It consists of customers who spent a lot of money in the hotel, although the other three elements are virtually the same as the average of the 63,670 customers at this hotel. The overall value of this group of customers is not high
6	It consists of customers who recently stayed in the hotel, although the other three elements are virtually the same as the average of the 63,670 customers at this hotel. The overall value of this group of customers is not high
7	It consists of customers who spent little money in the hotel, although the other three elements are virtually the same as the average of the 63,670 customers at this hotel. The overall value of this group of customers is low
8	It consists of customers who slightly spent above-average amounts of money at the hotel, although the other three elements are virtually the same as the average of the 63,670 customers at this hotel. The overall value of this group of customers is low
9	It consists of customers who have not stayed in the hotel for a long time, although the other three elements are virtually the same as the average of the 63,670 customers at this hotel. The overall value of this group of customers is low
10	It consists of customers who have recently been to the hotel and spent a large amount of money there. The other two elements are virtually the same as the average of the 63,670 customers at this hotel. This group of customers has a higher overall value than the other groups. Hence, it deserves to be invested in more CRM resources

for example, telling customers that they will receive a 20% discount on their second stay, giving them free flowers on their birthday, etc.

Table 5.8 shows the descriptions of the characteristics for each cluster, based on which hotel managers could take appropriate actions for each cluster to keep and strengthen the relationship with customers. For example, the values of recency, frequency, and bonding belonging to clusters 2 and 4 are very similar. However, the monetary value of customers in cluster 4 is much higher than that of customers in cluster 2. In this case, if the hotel's CRM resources are limited, it should prioritize providing more personalized service and in-depth communication to its most valuable customers to keep them coming back. That is, the limited resources should be invested in highly valued customers. In contrast, for those customers with low spending (e.g., cluster 2), relatively low-cost email mass coupons could be used to stimulate consumption.

5.4 Conclusions and Future Work

This paper improves the original RFM model by including the potential loss to the hotel if a customer cancels their reservation into the indicator "Monetary." In order to indicate the degree of customer bonding with the hotel, a new indicator "Bonding" is added to the RFM model. The proposed model also incorporates the 2-tuple linguistic model to give decision-makers more precise and understandable results. Furthermore, the Entropy and the CRITIC methods are integrated to obtain a more reasonable objective weight assignment to the four indicators of the 2T-RFMB model. Therefore, a novel method of segmenting and evaluating customer value has been developed utilizing the 2T-RFMB model, the integrated Entropy-CRITIC method, and K-means clustering.

The applicability of the proposed model has been assessed using a real dataset from a hotel in Lisbon. More than 63,000 customers are classified into ten groups based on their behavioral data during their stay at a hotel in Lisbon. The results show that the main benefit of the integrated Entropy-CRITIC method is that it calculates a more reasonable 2T-RFMB overall value for each customer by considering the relationship between different indicators and the level of data diversification contained in each indicator. Moreover, hotel managers can develop more efficient strategies to distribute their CRM resources by analyzing the 2-tuple value of each group of customers.

However, some drawbacks of this paper should be mentioned. First, in the 2T-RFMB model, customer demographic data like gender, age, and race are not considered. Second, the proposed model is unable to forecast the behavior of customers. Third, this paper only employs the objective weighting approaches to calculate the weight of each parameter in the 2T-RFMB model, without considering the decision-makers' opinions.

Therefore, some weaknesses of the proposed model should be addressed in later research. For instance, the proposed approach could be combined with machine learning forecasting algorithms to increase predictability. Additionally, the Entropy-CRITIC method could be combined with other methods to consider the subjective weights given by experts or decision-makers, such as the Analytic Hierarchy Process (AHP) method, Point Allocation method, Delphi method, etc. To gain a comprehensive picture of the customers, variables like age, occupation, family status, etc., could be incorporated into the customer segmentation.

Acknowledgements The author Ziwei Shu was supported by the Universidad Complutense de Madrid and Banco Santander (Project CT58/21-CT59/21). The co-author Ramón Alberto Carrasco González was funded by the Madrid Government (Comunidad de Madrid-Spain) under the Multi-annual Agreement with Universidad Complutense de Madrid in the line Excellence Programme for university teaching staff, in the context of the V PRICIT (Regional Programme of Research and Technological Innovation).

La autora Ziwei Shu fue financiada por la Universidad Complutense de Madrid y el Banco Santander (Proyecto CT58/21-CT59/21). El coautor Ramón Alberto Carrasco González fue financiado por la Comunidad de Madrid a través del Convenio Plurianual con la Universidad Complutense de Madrid en su línea Programa de Excelencia para el profesorado universitario, en el marco del V PRICIT (V Plan Regional de Investigación Científica e Innovación Tecnológica).

References

1. Alborzi, M., Khanbabaei, M.: Using data mining and neural networks techniques to propose a new hybrid customer behaviour analysis and credit scoring model in banking services based on a developed RFM analysis method. Int. J. Bus. Inf. Syst. **23**(1), 1–22 (2016)
2. Moghaddam, S.Q., Abdolvand, N., Harandi, S.R.: A RFMV model and customer segmentation based on variety of products. J. Inf. Syst. Telecommun. **3**(19), 155 (2017)
3. Haghighatnia, S., Abdolvand, N., Rajaee Harandi, S.: Evaluating discounts as a dimension of customer behavior analysis. J. Mark. Commun. **24**(4), 321–336 (2018)
4. Stormi, K., Lindholm, A., Laine, T., Korhonen, T.: RFM customer analysis for product-oriented services and service business development: an interventionist case study of two machinery manufacturers. J. Manag. Gov. **24**(3), 623–653 (2020)
5. Anitha, P., Patil, M.M.: RFM model for customer purchase behavior using K-means algorithm J. . King Saud Univ.—Comput. Inf. Sci. **34**(5), 1785–1792 (2022)
6. Heldt, R., Silveira, C.S., Luce, F.B.: Predicting customer value per product: from RFM to RFM/P. J. Bus. Res. **127**, 444–453 (2021)
7. Bacila, M.-F., Radulescu, A., Marar, I.L.: RFM based segmentation: an analysis of a telecom company's customers. Market. From Inf. Decis. **5**, 52–62 (2012)
8. Hughes, A.M.: Boosting response with RFM. Market. Tools **3**(3), 4–10 (1996)
9. Heidari, S., Radfar, R., Alborzi, M., Afshar Kazemi, M.A., Rajabzadeh Ghatari, A.: Clustering algorithm for electronic services customers: a case study of the banking industry. Int. J. Nonlinear Anal. Appl. **13**(2), 173–184 (2022)
10. Mosa, M., Agami, N., Elkhayat, G., Kholief, M.: A novel hybrid segmentation approach for decision support: a case study in banking. Comput. J., (2022)
11. Dursun, A., Caber, M.: Using data mining techniques for profiling profitable hotel customers: an application of RFM analysis. Touris. Manage. Perspect. **18**, 153–160 (2016)
12. Chen, A.H. L., Liang, Y.-C., Chang, W.-J., Siauw, H.-Y., Minanda, V.: RFM model and K-means clustering analysis of transit traveller profiles: a case study. J. Adv. Transp., (2022)
13. Doğan, O., AyçİN, E., Bulut, Z.A.: Customer segmentation by using RFM model and clustering methods: a case study in retail industry. Adminis. Sci., 8(1):1–19 (2018).
14. Yoseph, F., Heikkila, M.: Segmenting retail customers with an enhanced RFM and a hybrid regression/clustering method. In: IEEE International Conference on Machine Learning and Data Engineering 2018, iCMLDE, pp. 108–116. IEEE, Sydney (2018)
15. Acar, S., Köroğlu, F., Duyuler, B., Kaya, T., Özcan, T.: Customer Segmentation Using RFM Model and Clustering Methods in Online Retail Industry. In: Kahraman, C., Cebi, S., Cevik Onar, S., Oztaysi, B., Tolga, A.C., Sari, I.U. (eds.) Intelligent and Fuzzy Techniques for Emerging Conditions and Digital Transformation. Lecture Notes in Networks and Systems, pp. 69–77. Springer, Sydney (2022)

16. Tang, Y., Li, Y., Sun, G.: Research on E-Commerce Customer Churn Based on RFM Model and Naive Bayes Algorithm X. In: Sun, Zhang, X., Xia, Z., Bertino, E. (eds.) International Conference on Artificial Intelligence and Security 2022, LNCS, vol. 13338, pp. 371–381. Springer, Qinghai (2022)
17. Lumsden, S.-A., Beldona, S., Morrison, A.M.: Customer value in an all-inclusive travel vacation club: an application of the RFM framework. J. Hosp. Market. Manag. **16**(3), 270–285 (2008)
18. Yeh, I.-C., Yang, K.-J., Ting, T.-M.: Knowledge discovery on RFM model using Bernoulli sequence. Expert Syst. Appl. **36**(3), 5866–5871 (2009)
19. Herrera, F., Martínez, L.: A 2-tuple fuzzy linguistic representation model for computing with words. IEEE Trans. Fuzzy Syst. **8**(6), 746–752 (2000)
20. Zadeh, L.A.: The concept of a linguistic variable and its application to approximate reasoning—I. Inf. Sci. **8**(3), 199–249 (1975)
21. Herrera-Viedma, E., Herrera, F., Martínez, L., Herrera, J.C., López, A.G.: Incorporating filtering techniques in a fuzzy linguistic multi-agent model for information gathering on the web. Fuzzy Sets Syst., 148(1):61–83 (2004)
22. Shannon, C.E.: A Mathematical theory of communication. Bell Syst. Tech. J. **27**(3), 379–423 (1948)
23. Pomerol, J.-C., Barba-Romero, S.: Multicriterion Decision in Management: Principles and Practice. Kluwer Academic, Boston (2000)
24. Taheriyoun, M., Karamouz, M., Baghvand, A.: Development of an entropy- based fuzzy eutrophication index for reservoir water quality evaluation. Iranian J. Environ. Health Sci. Eng. **7**(1), 1–14 (2010)
25. Al-Aomar, R.: A combined ahp-entropy method for deriving subjective and objective criteria weights. Int. J. Ind. Eng.Theory Appl. Pract. **17**(1), 12–24 (2010)
26. Diakoulaki, D., Mavrotas, G., Papayannakis, L.: Determining objective weights in multiple criteria problems: the critic method. Comput. Oper. Res. **22**(7), 763–770 (1995)
27. Jahan, A., Mustapha, F., Sapuan, S.M., Ismail, M.Y., Bahraminasab, M.: A framework for weighting of criteria in ranking stage of material selection process. Int. J. Adv. Manuf. Technol. **58**(1), 411–420 (2012)
28. Antonio, N., de Almeida, A., Nunes, L.: A hotel's customers personal, behavioral, demographic, and geographic dataset from Lisbon, Portugal (2015–2018). Data Brief **33**, 106583 (2020)

Chapter 6
Intellectual Capital Versus Competitive Advantages: Together Which Underlines Some Relevant Literature?

Óscar Teixeira Ramada

Abstract The goal of this paper is to present the essential features of the literature review that relates the intellectual capital and the competitive advantages. On the other hand, it is still an idea, to know if there something that can be interpreted as being particularly noteworthy. In a synthetic way, this idea is that, only one, human capital, of the 3 components of intellectual capital, creates competitive advantages, playing an essential role, training and qualification, together with the innovation and entrepreneurial capacity of the same. If there is more qualification, better management of human capital, more and better competitive advantages will be obtained. The biggest gap in this research is related to the geographic contexts it covers (countries with a lower level of development) and the fact that it concerns with activity sectors of lesser relevance in terms of competitive advantages. Innovation needs further conceptual specification and its relationship with intellectual capital and competitive advantages.

6.1 Introduction

It has been generally recognized by several authors such as [10, 11] and [12], that one of the virtues of intellectual capital is to provide companies, in particular, with better performance and, above all, better competitive advantages. These can value the same companies, market shares and the position of nations in the international theater, with regard to international trade, both in products and services. If we add innovation to this, then this effect, which also stems from the intellectual capital, is even more pronounced.

Being of superior importance, the possession of competitive advantages, its improvement and increase, in any activity sector, of a country, it is necessary to know, what relationship exists, in the literature related to this topic, conjugated with the intellectual capital, among these pillars of knowledge, technological evolution, innovation, among others, namely.

Ó. T. Ramada (✉)
Instituto Superior de Ciências Educativas do Douro—ISCE Douro, Porto, Portugal
e-mail: oscarramada@gmail.com

J. L. Reis et al. (eds.), *Marketing and Smart Technologies*, Smart Innovation, Systems and Technologies 337, https://doi.org/10.1007/978-981-19-9099-1_6

It is observed that knowledge of this relationship, initially unknown, can often be seen as having associated other topics: see the case of [13–16] and [17]. In these cases, there is an association with the car production sector, with brands, with start-ups, in particular, which goes far beyond an approach limited only to the intellectual capital with the creation of competitive advantages.

This variety of topics, makes clarity and transparency less than would be desirable, the contribution to scientific knowledge, also less than would be expected in terms of its associated broadening, to its knowledge that allows application in the real world.

Furthermore, the 3 notions associated with the intellectual capital, definition-measurement-value and the notion of competitive advantages, are not always clarified in their semantic sense, so the scientific basis is not always the same: for some authors the definition, although having the same components, has different semantic meanings or has different components and also different semantic meanings, [18] and [19].

Thus, there is an obvious difficulty in applying the intellectual capital, combined with competitive advantages, so that it can be applied in practice and have a value of intellectual capital and competitive advantages as if it were a price of any product.

The *research question* that arises is: *what does the literature review that relate intellectual capital to competitive advantages say? Do you evidence any particular idea?*

About this pair, the essential idea is to know, at the heart of the brief literature review, what it underlines in particular, with the 2 concepts, simultaneously.

This research is divided into 3 Sections: in the first, an Introduction to the theme of intellectual capital and competitive advantages is carried out, framing these 2 topics with essential aspects associated with it; in the second, a Literature Review is carried out, in which 6 authors present their research on this pair, emphasizing the fact that this is not a mere collection of papers, but rather papers selected from the few existing ones center of the nucleus intended by the research; in the third, the Conclusions are presented, the essential points to be taken from the previous section are presented; also refer to the References, which served as the basis for the preparation of the paper.

6.2 Literature Review

Mubarik et al. [1] these authors address the relationship between the intellectual capital from the perspective of the relationship with competitive advantages. In fact, say the authors, nowadays, given the uncertainty and ambiguity of business activity, most companies strive to improve their degree of competitiveness. To achieve this, for these authors, they must develop capacities to be innovative, while exploring other capacities in such a way as to make companies ambidexterity. This research seeks to know the role that ambidextrous companies play in the relationship between these 2 aspects.

With regard to the methodology adopted, all medium-and-large-size companies in the Pakistani textile industry were considered. Companies with a workforce of between 100 and 250 workers are in medium size, while those with more than 250 workers are in large size. According to the sources consulted for this purpose by the authors, around 390 companies are included in this sample and are located in the cities of: Karachi, Lahore, Faisalabad and Gujranwala. A close-ended questionnaire was designed to collect data that allowed for constructs adopted by the previous literature consulted by the authors. The items had to do with the intellectual capital and its dimensions such as human capital, relational capital, structural capital, with competitive advantages and with ambidexterity in companies. A number of 580 questionnaires were distributed by 290 companies (2 questionnaires per company) selected at random. Regarding the respondents, directors, generic managers, human resources managers and managers with the tasks of planning activities or business tasks stand out.

Regarding the general conclusions, the authors emphasize that business ambidexterity is instrumental in channeling the effects of intellectual capital toward competitive advantages. The results also reveal that significant effects of human development capital and relational capital on competitive advantages were obtained. However, they observed a non-significant effect of the relational capital, per se. In general, the results were in line with previous literature. Even so, some results were divergent. It is a case of suggesting that the intellectual capital and business ambidexterity are important requirements to achieve competitive advantages. The human development capital, revealed itself as the biggest contributor, directly and indirectly. Which calls for companies to rethink their strategies in relation to intellectual capital.

As main implications, we highlight the fact that the factors of intellectual capital that lead to ambidexterity have been identified, as an instrumental capacity that allows achieving sustainable competitive advantages, which means that each component of intellectual capital needs to be systematically built within of business strategies.

Bakshi [2] this is an author who introduces research on the relationship between the intellectual capital and competitive advantages regarding the role of innovation and learning within organizations, in general, and companies, in particular.

In summary, research is based on the following 2 goals: to prove the level of intellectual capital in developing contexts and not in developed contexts, as is the case in most studies; on the other hand, knowing the impact of intellectual capital on competitive advantages being mediated by the role of innovation and moderated by the role of learning in organizations, which contributes to and strengthens the concept of intellectual capital in business literature.

With regard to methodology, the author constructed 4 constructs with items related to intellectual capital, innovation, organizational learning and competitive advantages, based on different studies, especially with regard to qualitative interviews, in which the respondents were the managers of 5 banks as well as 4 specialists. Items relating to intellectual capital cover human capital, relational capital and structural capital. Those of innovation, relating to radical innovation and incremental innovation. Self-development items were also included to measure bank workers' propensity for innovation. In order to measure organizational learning, a scale was used, having finished the items, with 64 on the intellectual capital, 10 on innovation, 16 on learning in organizations and 38 on competitive advantages. In terms of the sample used, it should be noted that it consisted of 144 branches of 21 public banks, 7 private banks operating in India, in the city of Jammu. From each branch, 3 executives (one manager and 2 senior workers), with more experience and knowledge, were contacted for the purpose of the study. A total of 5 questionnaires were distributed to respondents to answer, of which 339 questionnaires were answered, which corresponded to a response rate of 62.08%. As auxiliary analysis techniques, the author used Exploratory Factor Analysis (EFA) for data reduction, Confirmatory Factor Analysis (CFA), to confirm the measurement of the constructs.

In terms of more relevant conclusions, the author found that the intellectual capital affects competitive advantages and this relationship is moderate but significant. The results are consistent with other previous studies: This research established that human capital consisted of an important item such as worker creativity, staff commitment, training and education, experience, attitude and innovation of workers. The structural capital revealed to be composed of several items, such as the systems structure, information technologies, capabilities and culture, improvements and quality of service. It also concluded that at an organizational level of learning, it moderates the relationship between the intellectual capital and competitive advantages and that, seen individually, it improves human capital. The organizational culture proved to be innovative and strong, whereby the development of intellectual capital increases and is simultaneously influenced by learning in organizations and knowledge productivity. These are consistent practices across cities, regions and the culture in question, from all banks, can be considered similar.

Indiyati [3] in this research, its author analyzes the role of organizational culture, intellectual capital, competitive advantages in supporting government policies in education. Thus, the purpose is to examine and analyze the effects of organizational culture on the intellectual capital and its influence on competitive advantages in private universities in Indonesia, to support educational public policies.

With regard to the adopted methodology, it was the study of an explanatory survey, in which the type of research in question is the causal relationships. It focused on 157 private universities.

The sampling technique is that of sampling in probabilities in which the number of samples is decided by using the power of the power-analysis test, which resulted in 157 PTS in Region IV of Kopertis. The respondents were the heads of departments, or the deputy directors, or the deans or the students. Data collection was conducted using the technique of observation, interview and questionnaire. The analytical test used was the Structural Equation Model (SEM)-LISREL 8.3., via a second order approach.

In the main conclusions, the author concluded that organizational culture has a relevant influence on intellectual capital in private universities. This indicates that this improvement is strengthened if it is strengthened by the organizational culture via integration, direction, risk tolerance, individual initiative, control, identity, management support, reward system, communication patterns and tolerance of conflicts. The intellectual capital has a significant influence on the competitive advantages of private universities. This suggests that the production of innovation, innovation processes, innovation management and the quality of services provided improve and the cost is more efficient if the intellectual capital (human structure and customer) gives more value to consumers. Finally, the synergy between organizational culture and the intellectual capital can have a significant effect on improving the competitive advantages of private universities that are able to support government policies and have significant contributions to education in Indonesia. Relations with other universities and non-universities institutions are shown to be worthy of improving the reputation and motivation of its members.

Liu [4] this is an author who carried out research on the creation of competitive advantages and the perspectives of linking the learning of organizations, innovative behaviors and the intellectual capital, within the context of the hospitality industry in China. In the background, the author discusses the benefits of looking for market opportunities and maintaining competitive advantages in dynamic environments. The previous literature consulted to explain organizational performance and survival focused mainly on what organizations do but failed on how and what they should do.

From a methodological point of view, the hospitality of hotels in China provides a good example, insofar as this research requires open mindsets to absorb internal and external resources and even broaden their knowledge capabilities and increase innovation. Both this and the competitive advantages help to maintain the competitiveness that results from explorative and exploitative learning. Thus, hotel managers' perspectives were collected and the processes of how they use explorative and exploitative learning to influence workers' innovations and behavior and accumulate human capital in order to generate competitive advantages were examined. These traits proved to be critical for the hospitality industry in China. The research collects data from hotels with stars, suggesting that these provide insights to study the management of hotel human resources, and reflect the phenomenon of the development of hospitality sector in China. From a sampling point of view, data collection took place between July and August 2015 and involved 595 respondents (272 males and 323 females).

As main conclusions drawn by the author, we highlight the fact that this research of the literature on hospitality in the theorizing of learning in organizations, constitutes a fundamental attribute, in the transfer of innovative behaviors and in human capital, which makes the focus in hotel approaches can maintain competitive advantages. The research examines learning mechanisms as constituting a door that hotels can open to acquire information and knowledge that lead to positive results for the organization, in terms of their role as a buffer, social and organizational capital in the processes of performance creation. In the sample, research suggested that exploratory and exploitative learning can generate desirable results, such as the encouragement of creative and innovative behavior of workers, increasing the accumulation of human capital in organizations. The results also suggest that organizational learning can simultaneously increase opportunities to identify opportunities and competitive advantages through the behavior of innovation and human capital. On the other hand, the results also underline that the social capital of hotels and the capital of organizations play an important role in moderating these processes of creating complex competitive advantages. Social and organizational capital enhance the effects of innovative behaviors by external and internal linking resources in such a way that innovative behavior is related to increasing human capital only when there are high levels of social and organizational capital.

Dahash and Al-Dirawi [5] carry out research related to investment in the intellectual capital and the achievement of competitive advantages in the hotel sector in Iraq. The goal is to assess the role of intellectual capital in achieving competitive advantages, exploring the contributions of each component of intellectual capital that, in the end, improve business performance.

With regard to methodology and data analysis, the components of intellectual capital, human capital, relational capital and structural capital were considered and with the help of Partial Least Squares Method, the authors analyzed the contribution of each of these to the creation of advantages competitive. In terms of the sample, 4 and 5 stars hotels, were selected and data were collected through questionnaires based on past literature. In human capital (20 items), in relational capital (25) and in structural capital (16). Competitive advantages were obtained by developing a scale by Chahal and Bakshi [9]. The answers were rated on a Likert scale (5 points) with the hotels chosen in Iraq being the Basrah Hotel, Basrah International Hotel, Ishtar Sheratob Hotel Baghdad, Erbil International Hotel, Babylon Warwick Hotel Bagdhad, Al Mansour Melea Hotel, Royal Tulip Al Rasheed Hotel Bagdhad and the International Palestine Hotel Bagdhad. The data was based on the responses of top and mid-level managers as well as self-administrators. In total, 118 managers were chosen and 93 questionnaires were used.

In terms of conclusions, the authors emphasize that there was a lack of evidence in the hotel sector. Therefore, it was not obtained as a result that the intellectual capital had positive effects on competitive advantages. The human capital component revealed to have the greatest influence on competitive advantages and the structural capital the smallest. As Iraq is a developing country, the results obtained need further verification due to this. In different work contexts and socio-economic environments different results can be expected.

Jardon [6] is an author who has focused on obtaining and using the intellectual capital, by entrepreneurs, to obtain competitive advantages in regional small and medium-sized enterprises (SMEs), in the region of Galicia, in Spain.

From a methodological point of view, the author sought to isolate the effect of context dependence, reducing the geographic area of the sample. In fact, this area experienced a great development, especially from the 60's when the establishment of a multinational related to the manufacture of cars, increased the activities associated with fishing, food and transport of equipment. Thus, the author understood, as the best scientific approach, the comparison of theories of the development of comparative advantages. The economic structure of an area shows some characteristics, so the sample, in its design, should be conceived with this structure in mind. With different possibilities, the author chose the approach of business clusters, that is, the analysis of all activities associated with the same value chain of a product or service. Thus, the author categorized companies into 11 clusters in this geographic area. The target population included companies that had the particularity of having more than 9 and less than 250 workers.

A survey was carried out using a stratified random sample to obtain empirical data. The sample had the size of 400 companies, selected with a maximum error of 5%. In each cluster, the author selected 20 companies at random. The interviewers collected the questionnaires 2 weeks after they were sent to the companies. A total of 360 responses were valid, which corresponded to a response rate of 90%. The author found that the sample structure was consistent with the population study corroborated by the homogeneity test.

The core competencies were built using items on a likert scale from 1 to 5 (where 1 corresponds to not important as a competitive advantage and 5 as very important as a competitive advantage). SPSS Software (version 15) was used.

In the main conclusions, the author underlines that it is important to explain the processes through which entrepreneurs organize their intellectual capital within the scope of their core competencies in order to obtain benefits. It is important to explain these processes because they allow entrepreneurs to structure and organize the constituent elements to generate better performance. The effects of intellectual capital on performance did not occur directly but via core competencies generated by the entrepreneur. These, not grouped all the elements of intellectual capital in a single way, but rather, in combination with other resources and business capabilities.

The intellectual capital was used to build 2 core competencies: relationship management (RM), internal knowledge management (IKM) and innovativeness (INNO). It is the RM and IKM that improve INNO and, through it, the performance of companies. The effect of intellectual capital on INNO proved to be significant in this geographic area. The effect on RM was greater than the effect on IKM. Clusters proved to be important for the creation of regional innovation systems that are the basis of many innovation policies proposed in the same area. If these resources are scarce in companies, they limit their innovation potential. Financial and natural resources proved necessary for future investments in innovation. Thus, cooperation between them emerges as an adequate strategy to build networks for the development and improvement of products, processes and services.

The management of human and technological resources proved essential to promote INNO. This confirms that training, culture and technological processes are necessary to improve business innovation.

However, the effects of intellectual capital do not occur individually. It is necessary to combine with other elements of the companies, in order to specify the core competence. Business success is associated with know-how and the ability to combine, in a balanced way, elements, according to specific needs.

6.3 Conclusions

The literature, selected and relevant, from the few existing ones, involving these 2 interconnected concepts, states that the intellectual capital is composed of different components and each one also has different effects on the (creation) of competitive advantages of companies.

In addition to competitive advantages, innovation, translated essentially into the skills of employees, also makes a contribution. In particular, learning plays a more concrete role in moderating the relationships between the intellectual capitals versus competitive advantages.

Concretely, in the university context, the creation of competitive advantages through the intellectual capital involves the establishment of institutional relationships, especially by instilling motivation.

In the hotel sector in China, the role of learning is to be highlighted, in identifying and increasing opportunities, in the context of innovative behaviors and the component of intellectual capital that is human capital. This component moderates the processes that create advantages. It should be noted that in the case of the Iraqi hotel sector, the dominant note is the absence of empirical evidence that enables results in line with what is desired. Finally, it should be underlined, in another way, that the question of learning is present again, but from a collective point of view, that is, from the point of view of the management of human resources combined with technological ones. Thus, training, culture and technological processes prove to be crucial in business innovation. The effects of intellectual capital in obtaining competitive advantages are affected not directly, but via entrepreneurial action.

One of the limitations of this research is related to the fact that it focuses on countries that are not included in the most developed, except Spain. Thus, it would be very useful to cover countries such as Germany, France, United Kingdom, United States, among others. The notions of intellectual capital and innovation, perhaps, should be adjusted according to the activity sectors considered: more or less intensive in labor (or in capital).

With regard to the implications, they allude to the fact that they do not allow the extraction of great generalizations, since they are limited to countries that have poorly developed statistical information, which is the basis of results that are not very credible and very limited in their scientific scope.

As future avenues of research, it is worth mentioning, in addition to referring to countries, generally, developed, especially with regard to the European Union and the United States and, especially, to activity sectors where the aforementioned competitive advantages are more constituted as determining factors in the variation of market shares, in the leadership of the sectors, which involves considering sectors where each country has a stronger weight in the domestic market and in its relations with abroad (foreign trade).

Regarding the research question, the only evidence, which appears to be more notorious, in general terms and the idea to underline, is the fact that the intellectual capital creates competitive advantages not in its 3 components but rather via human capital and how much the more qualified it is, the greater the competitive advantages, with the contribution of innovation.

References

1. Mubarik, M., Naghavi, N., Mahmood, R.: Intellectual capital, competitive advantage and the ambidexterity liasion. Hum. Syst. Manag. **38**, 267–277 (2019)
2. Bakshi H (2015) Examining intellectual capital and competitive advantage relationship: role of innovation and organizational learning. Int. J. Bank Market., 1–35
3. Indiyati, D.: The role of organizational culture, intellectual capital and competitive advantage in supporting the government policies in education. Int J Econ Policy Emerg Econ **11**(1/2), 68–82 (2018)
4. Liu, C.: Creating competitive advantage: linking perspectives of organizational learning, innovation behavior and intellectual capital. Int J Hospital Manage **66**, 13–23 (2017)
5. Dahash, Q., Al-Dirawi, A.: Investment in intellectual capital and achievement of the competitive advantage in hotel sector. Manage. Sci. Let. **8**, 795–804 (2018)
6. Jardon, C.: The use of intellectual capital to obtain competitive advantages in regional small and medium enterprises. Knowl. Manage. Res. Pract. **13**, 486–496 (2015)
7. Todericiu, R., Stanit, A.: Intellectual capital—the key for sustainable competitive advantage for the SME's sector. Proc. Econ. Financ. **27**, 676–681 (2015)
8. Vatamanescu, E., Gorgos, E., Ghigiu, A., Patrut, M.: Bridging intellectual capital and SME's internationalization through the lens of sustainable competitive advantages: a systematic literature review. Sustainability **11**, 1–22 (2019)
9. Chahal, H., Bakshi, P.: Examining intellectual capital and competitive advantage relationship. Int. J. Bank Market. **33**(3), 376–399 (2015)
10. Gouveia L, Couto P (2017) A Importância Crescente do Capital Humano, Intelectual, Social e Territorial e a sua Associação ao Conhecimento. Atlântico Bus. J., 1, Número 0, outubro, 28–34
11. Chen, Y., Lin, M., Chang, C.: The positive effect of relationship learning and absorptive capacity on innovation performance and competitive advantage in industrial markets. Indus. Market. Manage. **38**(2), 152–158 (2009)
12. Pourmozafari, A., Heyrani, F., Moeinadin, M.: The examination of relationship between intellectual capital and financial performance according to the modulating role of competitive advantage. Int. J. Acad. Res. Account., Finance Manage. Sci. **4**(1), 188–200 (2014)
13. Subramaniam, M., Youndt, M.: The influence of intellectual capital on the types of innovative capabilities. Acad. Manage. J. **48**(3), 450–463 (2005)
14. Gupta, M., Bhasin, J.: The relationship between intellectual capital and brand equity. Manage. Labour Stud. **39**(3), 329–339 (2014)
15. Pang, L., Li, X.: Empirical analysis on Chinese self-brand automobile enterprises' competitive advantage. Int. J. Learn. Intell. Capital **7**(3/4), 374–393 (2010)
16. Matricano, D. (2019). Higher education and start-up's intentions: the role of intellectual capital in entrepreneuriship processes. Indus. Higher Educ., 1–9
17. Khan, A., Arafat, M., Raushan, M., Khan, M., Dwivedi, A., Khan, M.: Role of intellectual capital in augmenting the sart-up intentions of Indians—an analysis using GEM data. J. Glob. Entrepreneurship Res. **9**(25), 1–15 (2019)
18. Berzkalne, I., Zelgalve, E.: Intellectual capital and company value. Proc.—Soc Behav. Sci **110**, 887–896 (2014)
19. Gogan, L., Draghici, A.: A model to evaluate the intellectual capital. Proc. Technol. **9**, 867–875 (2013)

Chapter 7
Press Consumption in Chile During COVID-19: Digital Marketing Variables at Analysis

V. Crespo-Pereira, A. C. Vaca-Tapia, and R. X. Manciati-Alarcón

Abstract The media in general and the press in particular have to adapt to the exceptional situation caused by COVID-19. The pandemic forced to stop the paper version in favor of the digital version. The new ways of doing media, as well as the routines of readers cultivated in this context, lead us to ask many questions about the future of paper and the type of post-pandemic information consumption. This article analyzes the reading habits in the press at different times during confinement and post-confinement times and makes a comparison between native digital media and digital migrants in Chile. Following a descriptive approach, the article offers an average analysis of five variables out of 22 Chilean newspapers and a detailed analysis of the positioning of ten of them. The data reflect a preferential use of the smartphone device for news consumption, as well as a consolidated influence of Facebook and Twitter in the traffic to the digital press.

7.1 Introduction

The COVID-19 health crisis invites a deep reflection on the relevance and power of the traditional media in the consumption of information and its credibility in public opinion. Press problems, such as print runs, the economic situation, lack of credibility, or new consumption habits, are aspects that require a deeper comprehension and analysis in the context of post COVID-19 confinement. Portraying today's situation will allow us to understand future trends. These aspects are addressed below based on studies conducted in the context of the pandemic.

V. Crespo-Pereira (✉) · A. C. Vaca-Tapia
Universidade da Coruña, La Coruña, Spain
e-mail: veronica.crespo@udc.es

A. C. Vaca-Tapia
e-mail: anac.vaca@ute.edu.ec

A. C. Vaca-Tapia · R. X. Manciati-Alarcón
Universidad UTE, Quito, Ecuador
e-mail: roberto.manciati@ute.edu.ec

J. L. Reis et al. (eds.), *Marketing and Smart Technologies*, Smart Innovation, Systems and Technologies 337, https://doi.org/10.1007/978-981-19-9099-1_7

Digital platforms facilitate constant updating, which is highly beneficial for news coverage compared to the limited capacity of the printed press [1]. As a result, digital media consolidated itself as one of the most priority medium for information in pandemic times [2].

Press Websites and apps, which are moving forward new consumption habits, could have transformed and accelerated the announced decline of print newspaper versions [3] and the transition to new business models as well as technological, professional, and business convergence [4]. The economic situation of newspapers worldwide is worrying. Those newspapers that have not made progress toward digitalization before COVID-19 might be dramatically affected when it comes to neutralizing the loss of readers and print version profits.

Due to the pandemic, studies point out the renaissance of mass media consumption among citizens, in special in the digital press and television [8]. Mass media has repeatedly demonstrated to be relevant for their value to the public. Press has confirmed to be one of the main tools to promote informal education, and its educational and pedagogical capacity during 2020 health crisis are beyond doubt. Its value as a public service has been essential to educate and warn about a multiplicity of issues that can affect society as a whole: diseases, natural disasters… . Despite this, newsrooms are facing difficulties and impediments that affect the production of information and that are related to the explosion of the phenomenon of misinformation and fake news and sensationalism or the lack of journalists specialized in the area of health [2, 5, 6].

Journalistic rigor is an aspect continuously claimed by the profession. However, the emergence of new agents in the journalistic ecosystem modifies the existing status quo. The citizen journalism avoids the intermediation by information professionals and the press protocols. Despite the many benefits of this social movement, the lack of information of quality and rigorous health data on social networks was one of the challenges to be overcome during COVID-19 [7].

Today, the citizen becomes a gatekeeper characterized by the loss of trust in mass media. This disaffection is especially evident among young people [8]. The accessibility of news through social networks and private messaging is increasing [9] whereas competing with journalistic newsrooms. The citizen journalism movement in Latin American countries has been relevant during the health crisis for challenging traditional media bad habits. This social movement may become a suitable formula for countries with fragile democratic systems and political instability that affect journalistic action and censorship.

Despite the distrust of media, it is necessary to point out that, during COVID-19, social networks have served to generate and disseminate disinformation and to turn the thematic agenda for political purposes [10]. Fake news is an aspect of criticism toward social networks and citizen journalism [11]. In times of crisis, discernment in the detection of this type of news and disinformation is greater [11].

No doubt, COVID-19 has transformed media consumption. During confinement, the demand for information grows exponentially. News consumption is higher (it is not just reduced to the headline) and more frequent for at least 78% of citizens during confinement [12]. At the same time, subscription to digital media increased [2].

However, a greater journalistic consumption during confinement does not translate into an improvement in trust, on the contrary, periods of crisis point to a decline in this index [2].

Press does not seem to have capitalized on the demand for information in terms of trust and credibility. The editorial line of the media does not go unnoticed by citizens. A large part of society considers that the informative treatment is ideologically conditioned [2]. Although it is true that citizens have reservations about the accuracy of information in networks, this differs depending on the territories [9]. Moreover, the higher the educational level, the lower the credibility toward the media [11]. Certainly, newsroom routines have been modified due to the restrictions imposed by the pandemic. This imposed challenges have proved beneficial for journalism as a whole, since it increased the credibility and trust of the journalistic exercise in countries such as Chile [13].

7.2 Materials and Method

COVID-19 is the backbone of most of the scientific publications of the last two years. Academic research from all kinds of disciplines has treated the virus. In communication too. Studies researched about methodological approaches and validated scales to the role of the media and social networks in the pandemic [14, 15]; media consumption or misinformation in relation to social networks in the context of a pandemic [2, 8, 16].

In Latin America, journalism and COVID-19 are a topic lacking in the academia, and the existing research focuses on work practices, lack of media plurality, and professional ethics [13].

As mentioned above, confinement has driven alternative information consumption to print newspaper versions. Certainly, digital media consumption in the pandemic displaced the use of other mass media due to the capacity of the former to offer updated information. This research describes the patterns of press consumption in migrant and digital native newspapers during the COVID-19 crisis in Chile.

For this purpose, a descriptive and comparative research were carried out. Data were obtained through similar Web from a total of 22 newspapers, namely El Mercurio, La Tercera, La Cuarta, La Segunda, Publimetro, Diario Financiero, El Ovallino, El Andino, El Racangûino; Diario VI Región, El Centro, El Heraldo, La Discusión; Cónica Chillán, El Insular, El Divisadero, Hoy x Hoy, El Día, El Trabajo, El Proa, El Tipógrafo, and Diario Concepción. The data obtained will allow us to understand the evolution of the following variables: positioning of the newspaper, origin by channels, search traffic, social network traffic, and access device. The analysis period corresponds to December 2020 and January, February, March, April, and May 2021.

7.3 Data Analysis

COVID-19 changed, although for a short time, people's media consumption around the world. Informative demand and preferences shifted during the crisis, even among those individuals who usually are not interested in informing themselves [8]. Interest in the virus has prompted new routines among casual and regular readers. The pandemic emerged as an opportunity to launch new content strategies for connection and feedback between the media and society. Therefore, the pandemic forced newsrooms to rethink the objectives and strategies to connect directly to the needs of consumers via digital platforms.

Information management by media during COVID-19 was capitalized very effectively.

Given the right conditions, such as access to technological devices and Internet, digitization fosters total freedom of media consumption. Since the demand of global al local news increased exponentially during the health crisis, newspapers became the media of reference in Latin America. Media became the bridge between the media and society.

Trust, a key value for any media, also proved to be an essential element for attracting audiences in uncertain times. Some articles confirm these ideas: "*We almost doubled the rates of new digital subscribers and closed 2020 with 42 thousand digital subscribers*" *(La Tercera, Chile) (…) There was a record audience in digital traffic this year in Colombia, Chile, Brazil, Europe, and the United States. People turned to those big news media with their recognized brands that could generate trust, that which is called credibility* "*(El Tiempo, Colombia)*" [13].

As a consequence of the specialization of information on topics related to popularization and science and the verification of contents *(fact check),* some newspapers not only enriched their offer, but it was determined as a revulsive of great impact to maintain the audience's attention (see Table 7.1).

"*The pandemic forced us to concentrate and strengthen it a lot. Today, it is one of the most visited channels on the platform, which has to do with the demand for science (La Tercera, Chile) (…) The need to generate verification content had already happened to us with the social outbreak. I think this is also associated with one of the great aspects of the revaluation of journalism (La Tercera, Chile)*" [13].

Certainly, the crisis has driven new audience behaviors, among which media subscriptions and increased news consumption stand out. COVID-19 crisis triggered many media transformations that were to happen in the medium term. Some authors mention that: "*The health crisis included an increase in news consumption and production, so that the media had to adjust the operating and logistic model of journalists in relation to news coverage (…). It was a transformation we were going to make in a few years, but the pandemic forced us to anticipate it, mainly due to economic pressure. The income from subscriptions of the paper newspaper fell, it was reduced to 50% (La Tercera, Chile)*" [13].

Traffic to digital newspaper Websites comes from different sources. Organic research (traffic sent via organic (non-paid) results on search engines such as Google)

Table 7.1 Evolution of Chilean press traffic and engagement over time

Newspaper	Variables	2020	2021				
		December	January	February	March	April	May
El Mercurio	Category rank	3408	23,565	24,918	N/A	N/A	N/A
	Percentage increase in visits	−5.21%	134,59%	−17.24%	−12.00%	0.99%	−4.81%
	Monthly unique visitors	611,683	6461	5852	279,806	N/A	N/A
	Average duration of visits	0:05:52	0:33:34	0:45:08	0:06:14	0:06:33	0:06:22
	Pages/Visits	4.50	10.39	9.50	5.77	5.61	5.73
La Tercera	Category rank	439,00	129.341	111,925	399	N/A	N/A
	Percentage increase in visits	0.56%	−37.42%	60.69%	4.75%	−2.39%	4.86%
	Monthly unique visitors	9.787 M	5000	5.00 M	10.63 M	N/A	N/A
	Average duration of visits	0:03:28	0:03:44	0:00:16	0:03:31	0:03:44	0:03:41
	Pages/Visits	2.18	1.33	1.10	2.08	2.07	2.10
La Cuarta	Category rank	1371	1168	969	1138	N/A	N/A
	Percentage increase in visits	9.02%	21.51%	6.82%	−10.03%	−8.09%	−3.48%
	Monthly unique visitors	1.561 M	1.939 M	2.210 M	2.032 M	N/A	N/A
	Average duration of visits	0:01:55	0:01:55	0:02:06	0:02:02	0:02:04	0:01:54
	Pages/Visits	2.58	2.60	2.53	2.52	2.47	2.42
La Segunda	Category rank	7.919	7.870	7.793	N/A	N/A	N/A
	Percentage increase in visits	15.77%	14.15%	−15.14%	2.75%	−10.27%	−10.28%

(continued)

Table 7.1 (continued)

Newspaper	Variables	2020	2021				
		December	January	February	March	April	May
	Monthly unique visitors	178.619	189.109	163.219	86.771	N/A	N/A
	Average duration of visits	0:02:56	0:02:46	0:02:53	0:03:55	0:03:35	0:03:49
	Pages/Visits	2.80	2.52	2.63	3.45	3.31	3.30
Publimetro	Category rank	1.091	986	982	1.225	N/A	N/A
	Percentage increase in visits	−3.78%	15.93%	−14.37%	−10.76%	−14.28%	−3.77%
	Monthly unique visitors	2.583 M	2.912 M	2.491 M	2.190 M	N/A	N/A
	Average duration of visits	0:03:16	0:03:03	0:02:43	0:00:51	0:00:57	0:01:07
	Pages/Visits	2.84	2.82	2.80	2.48	2.47	2.55
Diario Financiero	Category rank	N/A	N/A	N/A	2.441	N/A	N/A
	Percentage increase in visits	116.76%	543.72%	−82.58	4.35%	−4.04%	5.96%
	Monthly unique visitors	< 5.000	7.525	5.000	1.640 M	N/A	N/A
	Average duration of visits	0:07:32	0:00:23	0:01:49	0:01:56	0:01:46	0:01:41
	Pages/Visits	1.52	1.05	1.13	1.58	1.51	1.60
El Ovallino	Category rank	25.746	27.863	24.835	27.908	N/A	N/A
	Percentage increase in visits	22.20%	−5.95%	12.22%	−9.25%	−21.85%	30.86%
	Monthly unique visitors	46.875	44.897	49.426	44.395	N/A	N/A

(continued)

Table 7.1 (continued)

Newspaper	Variables	2020	2021				
		December	January	February	March	April	May
	Average duration of visits	0:01:53	0:01:35	0:01:32	0:01:10	0:01:41	0:01:23
	Pages/Visits	1.89	1.96	1.82	1.76	1.97	1.78
El Andino	Category rank	48.481	54.609	64.124	42.147	N/A	N/A
	Percentage increase in visits	−35.17%	−19.30%	−37.43%	228.01%	N/A	N/A
	Monthly unique visitors	11.026	9.579	6.441	22.409	N/A	N/A
	Average duration of visits	0:01:50	0:02:17	0:01:41	0:01:51	N/A	N/A
	Pages/Visits	1.99	2.43	2.24	1.66	N/A	N/A
El Rancagûino	Category rank	18.224	16.851	15.212	17.120	N/A	N/A
	Percentage increase in visits	−14.56%	17.58%	0.81%	−8.74%	−8.19%	47.35%
	Monthly unique visitors	116.240	134.240	131.454	127.700	N/A	N/A
	Average duration of visits	0:00:55	0:01:13	0:01:02	0:01:11	0:00:39	0:01:16
	Pages/Visits	1.52	1.64	1.60	1.61	1.52	1.75
Diario VI Región	Category rank	33.098	53.913	62.030	77.202	N/A	N/A
	Percentage increase in visits	18.63%	−41.63%	−35.50%	−33.72	−19.59%	−5.78%
	Monthly unique visitors	15.343	7.848	5.709	5.00	N/A	N/A
	Average duration of visits	0:02:21	0:02:06	0:02:24	0:01:31	0:01:20	0:01:11
	Pages/Visits	5.00	2.37	2.13	2.02	2.44	1.84

Table 7.2 Evolution of Chilean press: marketing channels

Marketing channels	2020	2021				
	Dec (%)	Jan (%)	Feb (%)	March (%)	Apr (%)	May (%)
Direct	36.59	40.72	43.70	42.96	41.25	33.84
Email	1.30	1.59	1.43	1.17	1.01	1.22
Referrals	9.84	9.34	4.93	6.72	4.33	4.13
Social	5.98	4.95	6.67	5.59	5.08	4.40
Organic research	46.29	43.39	41.55	43.53	39.20	47.27
Paid research	0.00	0.00	0.00	0.04	0.02	0.03
Display adds	0.01	0.01	1.71	0.00	0.01	0.03

Table 7.3 Evolution of Chilean press: traffic search

Search traffic	2020	2021				
	Dec (%)	Jan (%)	Feb (%)	March (%)	Apr (%)	May (%)
Organic	100.00	95.45	95.45	99.89	0.00	99.89
Payment	0.00	0.00	0.00	0.11	0.00	0.11

is the most common formula to access Chilean digital press, followed by direct (traffic sent from users that directly entered a URL into a browser). Advertising (display adds) is the formula with the worst results when attracting people to the Website newspaper (Table 7.2).

The role of paid traffic is not common at all in press, whereas the organic traffic is the only viable option for consumers. Paid content has no positive repercussion for attracting audience in Chilean press (Table 7.3).

The appearance and democratization of instant messaging networks and platforms have generated profound changes in the traditional media-audience relationship and have probably affected media trust and usage. In fact, studies point out the consequences of social media such as the spread of false news, the fragmentation, or the excessive supply of information [8]. However, social networks have also opened the opportunity to digital newspapers to make the news visible and accessible. The social media platforms that attract most traffic in Chilean newspapers are Facebook and Twitter. Interestingly, although WhatsApp Web directs traffic to Chilean press in a minor volume, it will be very interesting to track their evolution in the years to come since the growing prominence of private instant messaging apps (Table 7.4).

It is worth mentioning people have shifted the ways they access the news daily, mobile devices gaining much significance (Table 7.5). Mobile phone device attracts most of the traffics to the digital editions nowadays.

Table 7.4 Evolution of Chilean press: social media traffic (traffic sent from social media sites to the analyzed domains)

Social media	2020	2021				
	Dec (%)	Jan (%)	Feb (%)	March (%)	Apr (%)	May (%)
DeviantArt.com	1.31	0.35	0.00	0.00	0.00	0.00
Facebook	49.41	47.98	45.02	51.31	52.95	53.55
Instagram	0.36	0.17	4.64	0.40	0.36	0.84
LinkedIn	0.52	0.00	0.36	1.45	1.76	1.15
Pinterest	0.07	0.01	0.10	0.00	0.07	0.00
Reddit	0.42	0.04	0.02	0.76	1.44	0.67
SoundCloud	0.00	0.00	0.00	0.00	0.00	0.00
Twitter	24.12	21.21	23.74	21.11	14.43	15.12
Whatsapp Web	2.24	4.04	4.42	2.05	2.01	2.68
YouTube	2.86	3.06	3.49	3.79	3.84	2.84
Otros	0.47	0.10	0.03	0.49	0.42	0.42

Table 7.5 Evolution of Chilean press: traffic per device

Access device	2020	2021				
	Dec (%)	Jan (%)	Feb (%)	March (%)	Apr (%)	May (%)
Desktop/computer	37.51	–	40.71	26.10	29.79	30.93
Mobile	62.49	–	59.29	55.72	61.12	59.98

7.4 Conclusions

COVID-19 health emergency led to a great increase in news consumption. The need of data and information to orient people has increased news demand in digital media and on Web pages. In a context characterized by citizen journalism and media distrust, news media proved themselves highly relevant for their high social value in uncertain times. Occasional audiences could have generated a positive precedent regarding the stimulation of local news access. For sure, the widespread increase in news consumption during the coronavirus accelerated the media digital transformation of media worldwide. In Chile too.

Chilean digital press has been monitored for months. Data offered in this paper allow researchers deeper analysis regarding Chilean native vs migrant newspapers.

Results obtained show that smartphone has become the main device to read digital newspapers; social medias such as Facebook and Twitter are the main platforms that lead audiences to the digital newspaper. Organic search is the only formula to gain readers since paid traffic has been demonstrated inefficient, at least at the moment. Newspapers brands need to focus on positioning themselves due to the relevance, the organic research and direct search (marketing channels) have on traffic. Being on the top of mind of Chilean, readers are a must for obtaining success.

References

1. Lázaro-Rodríguez, P., Herrera-Viedma, E.: Noticias sobre Covid-19 y 2019-nCoV en medios de comunicación de España: el papel de los medios digitales en tiempos de confinamiento. El profesional de la información **29**(3), e290302 (2020)
2. Masip P, Aran-Ramspott S, Ruiz-Caballero C, Suau J, Almenar E, Puertas-Graell D (2020) Consumo informativo y cobertura mediática durante el confinamiento por el Covid-19: sobreinformación, sesgo ideológico y sensacionalismo. El profesional de la información 29(3)
3. Meyer P (2004) The vanishing newspaper. Saving journalism in the information age. Columbia, Missouri: University of Missouri Press. ISBN: 978 0 8262 1877 3
4. Costa-Sánchez C, Rodríguez-Vázquez AI, López-García X (2015) Del periodismo transmedia al replicante. Cobertura informativa del contagio de ébola en España por Elpais.com. Profesional De La información 24(3):282–290. https://doi.org/10.3145/epi.2015.may.08
5. Toledo-Ibarra A (2020) The COVID-19 pandemic from the multidisciplinary visión of 28 University Professors of Nayarit, México. Revista Bio Ciencias 7:e976. https://doi.org/10.15741/revbio.07.e976
6. Andreu-Sánchez, C., Martín-Pascual, M.-Á.: Fake images of the SARS-CoV-2 coronavirus in the communication of information at the beginning of the first Covid-19 pandemic. El profesional de la información **29**(3), e290309 (2020). https://doi.org/10.3145/epi.2020.may.09
7. Jiménez-Sotomayor, M.R., Gomez-Moreno, C., Soto-Perez-de-Celis, E.: Coronavirus, ageism, and twitter: an evaluat ion of tweets about older adults and COVID-19. Am. Geriatrics Soc. **00**(00), 1–5 (2020)
8. Casero-Ripollés A (2020) Impact of covid-19 on the media system. Communicative and democratic consequences of news consumption during the outbreak. El Profesional de la Información 29(2):e290223. https://doi.org/10.3145/epi.2020.mar.23
9. Newman N, Fletcher R, Kalogeropoulos A, Nielsen RK (2020) Reuters Institute digital news report 2019. Oxford: Reuters Institute for the Study of Journalism (2020) http://www.digitalewsreport.org
10. Pérez-Dasilva JÁ, Meso-Ayerdi K, Mendiguren-Galdospín T (2020) Fake news y coronavirus: detección de los principales actores y tendencias a través del análisis de las conversaciones en Twitter. El profesional de la información 29(3):e290308. https://doi.org/10.3145/epi.2020.may.08
11. Casero-Ripollés A (2020) Impact of Covid-19 on the media system. Communicative and democratic consequences of news consumption during the outbreak. El Profesional de la Información 9(2):e290223. https://doi.org/10.3145/epi.2020.mar.23
12. Asociación de prensa de Madrid (2020) Aparecen algunos síntomas de saturación informativa en el consumo de medios durante el confinamiento. https://www.apmadrid.es/aparecen-algunos-sintomas-de-saturacion-informativa-en-el-consumo-de-medios-durante-el-confinamiento
13. Greene González MF, Cerda Diez MF, Ortiz Leiva G (2022) Prácticas periodísticas en tiempos de pandemia de coronavirus. Un estudio comparado entre Chile y Colombia. Revista de Comunicación 21(1), E-ISSN: 2227–1465 195
14. Dabbagh, A.: The role of Instagram in public health education in COVID-19 in Iran. J Clin Anesth **65**, 109887 (2020). https://doi.org/10.1016/j.jclinane.2020.109887

15. Chen, Q., Min, C., Zhang, W., Wang, G., Ma, X., Evans, R.: Unpacking the black box: how to promote citizen engagement through government social media during the Covid-19 crisis. Comput. Hum. Behav. **110**(2020), 106380 (2020)
16. Salaverría R, Buslón N, López-Pan F, León B, López-Goñi I, Erviti MC (2020) Desinformación en tiempos de pandemia: tipología de los bulos sobre la Covid-19. Profesional De La información 29(3). https://doi.org/10.3145/epi.2020.may.15

Chapter 8
Information Visualization (InfoVis) in the Decision Process

António Brandão

Abstract This work studies the importance of information visualization and visual analysis, in the decision process, in the context of marketing and its tools. Researchers, marketing strategies, brand positioning, financial analysis, and many other needs, have goals for a better understanding and analysis of data and information for understanding, growth, productivity, and innovation. Consumers and citizens analyze publicly and visually available data such as product specifications and price, via the web, websites, blogs, and online communities to choose the products they will be able to buy, when deciding to answer questions, voting through likes or other qualitative data, forms of voting and the search for related information. InfoVis and visual analytics technologies begin to provide an effective vision for the decisions, to deal with behavior analysis, general and specific patterns, data irregularities, lapses in information, and trends felt by users in society, in their intervention in social media, and in spaces physical and virtual.

8.1 Introduction

This work seeks to present the importance of information visualization and visual analysis, in the process of decision, in the context of marketing and its tools.

The different aspects of information visualization, at different levels, operational, tactical, and strategic, are revealed to multiple actors, in the most varied ecosystems, aggregating huge amounts of data, formulating numerous tables of indicators, synthesizing global indicators that allow performance comparisons, test and validate the application of strategies, policies and good practices.

The text focuses on the future challenges of information visualization (InfoVis) and visual analytics (VA), as a way of integrating visual, interactive, and automatic methods to support multifaceted analysis, in ecosystems and in dynamically changing networks.

A. Brandão (✉)
UAberta, Lisbon, Portugal
e-mail: ajmbrandao@gmail.com

J. L. Reis et al. (eds.), *Marketing and Smart Technologies*, Smart Innovation, Systems and Technologies 337, https://doi.org/10.1007/978-981-19-9099-1_8

The quality and relevance of InfoVis may allow a clear view of the different ecosystems that make up a business, its network, and relationships, which will imply a careful analysis of tools and frameworks that allow the proper dynamic exploration, understanding, and analysis of data through progressive, iterative, and multi-device visual exploration.

8.2 Objectives and Questions

The decision-making process with the use of information visualization techniques must be consistent and goes through defining the problem, the data to be represented, the dimensions necessary to represent the data, the data structures, the required interaction of the visualization, and the synthesis needed to determine which model will be the most effective for decision-making.

Decision-making is supported by exploring, discovering, and communicating visual information. Data analysis, the use of tools, techniques, and visualization methods provide applications for various areas such as business management, marketing, scientific research, financial analysis, market research, and many other applications. The questions start with the basic question: What is InfoVis and why do it? And, it goes through the following several others questions: Why have a human in the decision-making loop? Why have a computer in the loop? What resources are important? Why use an external representation? Why rely on vision? Why show data in detail? Why use interactivity? What is the design space of visualization languages? Why focus on tasks? How can we better be average? [1].

The central question in this work is to try to answer the question of how to simplify and optimize the visualization of information, to facilitate a decision-making cycle.

The interaction with the user, the decision variables, and the objectives to be achieved should define which techniques to use. Thus, it is necessary to determine the objective of action and the means to carry out that objective, and the degree to which the artifact provides representations that can be interpreted.

Effectiveness is the essential aspect, so you should define the objective correctly, the user or system tasks, apply the appropriate methods, and evaluate, promoting several iterations to achieve the defined objectives.

The exploratory tasks of treating raw data with its visualizations allow finding causality and recreating hypotheses for systematic decision-making. Resources should facilitate hypothesis finding and hypothesis testing. Techniques, with multiple views and coordinates, can improve the discovery process. These activities should be oriented toward obtaining empirical evidence to inform and facilitate decision-making processes.

8.3 Information Visualization (InfoVis)

The visualization of information reveals aspects combined in its definition of visualization as an image and knowledge as data synthesis. To detail the scope of this concept, we start by finding the definitions that were added over time and then characterizing the variants that this area can assume.

8.3.1 Definitions

The following are the various definitions, which can cover the different aspects of visualization and visualization of information, in which:

- It is the "transformation of the symbolic into geometry" [2].
- We are "finding the artificial memory that best supports our natural means of perception" [3].
- "Visualization is a method of computation. It transforms the symbolic into geometric research, allowing researchers to observe their simulations and calculations. Visualization offers a method for seeing the invisible. It enriches the process of scientific discovery and promotes deep,-- and in many fields, it is already revolutionizing the way scientists do science" —[3].
- "Is concerned with exploring data and information graphically to gain understanding and insight into the data" [4].
- "It means study, development, and use of graphic representation and support techniques that facilitate the visual communication of knowledge" —[5].
- This is "computationally intense visual thinking" [6].
- Information visualization (InfoVis) is the "communication of abstract data through the use of interactive visual interfaces" [7].
- "The use of computer-supported, interactive, visual representations of abstract data to amplify cognition" [8].
- "The interdisciplinary study of visual representation collects non-numerical information on a large scale, such as files and lines of code in software systems, libraries and bibliographic databases, networks of relationships on the Internet, and so on" [9].

From these perspectives, InfoVis and VA are research domain that aims to support users in exploring, understanding, and analyzing data through progressive and iterative visual exploration. The InfoVis produces interactive and visual representations of abstract data to enhance human understanding, allowing the observer to gain knowledge about the internal structure of the data and the causal relationships within it. It is the representation of information using spatial and graphical representations. But, abstract data can refer to heterogeneous data that do not have an inherent spatial structure and do not allow a direct mapping to any geometry.

The VA is the process of exploring, transforming, and representing data as images (or other sensory forms) to obtain information about phenomena and behavior. The VA and InfoVis use graphic techniques to communicate information, support reasoning, facilitate decision processing, and establish marketing strategies.

8.3.2 Visualization Goals

The objectives of information visualization are treating large sets of coherent data to provide information in a compact form; presenting information from multiple points of view; presenting information with various levels of detail, supporting visual comparisons, and telling stories about the data.

Visualization allows you to use the vision system to process tasks faster, and more naturally, and to improve memory using external representations that support understanding, lessening the load on working memory. The visual representation can be a more natural and efficient way of representing data or problem space, through visual languages or symbols.

8.4 Characterize the Variants of the Visualization Areas

The various variants or expressions used to designate different areas of visualization are information visualization, scientific visualization, data visualization, and we can add infographics and visual analysis. We detail the last three below, due to the possible specific use in certain aspects of marketing applications.

The visual analysis. Focusing on visual analytics as the science of analytical reasoning facilitated by interactive visual interfaces [9]. People use visual analysis tools and techniques to synthesize information and gain insights into massive data, dynamic, ambiguous, and often conflicting, to detect the expected and discover the unexpected, to provide timely, defensible, and understandable assessments, and to communicate assessments effectively for action [10].

Visual analysis is understood as the science of reasoning about visual information, using the "smart" of a computing machine with human intelligence, creativity, and visual representations.

Visual analysis, as a multidisciplinary field, includes several areas of focus, which include analytical reasoning techniques that allow users to obtain in-depth information that directly supports assessment, planning, decision-making, visual representations, and interaction techniques. They exploit the human eye and mind interaction to allow users to see, explore, and understand vast amounts of information. Simultaneously, by data representations and transformations that convert conflicting and dynamic data types into forms that support visualization and analysis and by techniques to support the production, presentation, and dissemination of analytical results to communicate information in the right context to a variety of target audiences [10].

Infographics. Information graphics or infographics are visual representations of information, data, or knowledge. Graphics are used where complex information needs to be explained quickly and clearly, such as in signs, maps, journalism, technical writing, and education and are used extensively as tools by computer scientists, mathematicians, and statisticians to facilitate the development process and communication of conceptual information [11].

Data visualization. Data visualization, as it is the study of the visual representation of data, translates into information that has been removed in some schematic way, including attributes or variables for the information units.

The basic idea of data visualization is to use computer-generated images as a means to obtain a greater understanding and apprehension of the information that is present in the data (geometry) and in their relationships (topology). According to [12], seven basic data types are considered: one-dimensional, two-dimensional, three-dimensional, multidimensional, temporal, hierarchical, network, and workspace.

Human Perceptions. Human perceptions go through the use of the eye for pattern recognition, in which users are able to scan, recognize, and remember images, through graphic elements that facilitate comparisons through size, shape, orientation, and texture, with animations that show changes over time and color that helps to make distinctions.

The importance of information visualization reflects the growing representation of data in digital format, the huge increase in data, and the number of types of data available, with more individualized information, making it difficult for a man to understand.

With the need for improved visualization, to improve the understanding of the data and the speed with which we can understand it, interactive visualizations under the user's control were added to customize their understanding. Information visualization forms support communication, and understanding that amplifies understanding, exploration, discovery, and decision-making by filtering or dynamic queries.

8.5 Visual Analysis in Dynamic Social Networks

Visualization and analysis of dynamic networks are applied in fields such as sociology or economics, which, due to the dynamic and multi-relational nature of this type of data, make it a challenge to understand the topological structure and its change over time. The visual analysis presented in [13], to analyze dynamic networks, integrates a dynamic layout whose balance is controlled by the user between stability and consistency, and used three temporal views in combinations of node-link diagrams, with the visualization of analysis metrics networks and with specific interaction techniques to track node trajectories and node connectivity over time. The integration of visual, interactive, and automatic methods allows supporting the multifaceted analysis of dynamically changing networks.

The approach followed for dynamic networks with the integration of interactive visualizations with analytical methods (algorithms and graph-theoretical metrics) was guided by basic perceptual principles. Also can be followed and explored through alternative representations, node-link diagrams, matrix-based representations, integration of visual and analytical methods, and moving from static metrics to dynamic algorithms. The application in social networks and in specific organizational networks can be extended to other concepts and ideas of dynamic networks in other contexts, such as businesses, in which relational and temporal aspects are important and graphic-graphic algorithms are applicable [13].

8.6 InfoVis Framework

The improvement of interactive visualization and the exploration of the growing amount of data, of collaborative and multiplatform systems, have as challenges for information visualization and visual analytics the data exploration scenarios: multi-user, multi-screen, and multi-device.

InfoVis application development frameworks and toolkits (D3.js, Chart.js, Grafana, Kibana, Google Charts, ChartBlocks, Datawrapper, Sigmajs, Infogram, Tableau, Polymaps) provide solutions for importing, storing data and a variety of visualization techniques, which manipulate or simulate aspects, practices, and "gestures" of touch that are required for mobile devices or cross-platform deployment. The example presented in [14], of the Tulip framework that comes with Tulip Graphics, has a complete OpenGL rendering engine adapted for the visualization of abstract data. This framework is efficient for creating research prototypes and developing applications for end users.

The combinable multi-user, multi-screen, and multi-device scenarios in the future require requirements, which include the "cross-platform" frameworks and standards that will be required for native deployment of InfoVis systems, and present four challenges, combined with interactive exploration systems of data and that according to [14] go through:

- Collaboration: To solve complex analysis tasks, the development of collaborative visual analysis systems, for users to share their knowledge, and cooperate during their work to find the best solutions for their problems or tasks.
- The different platforms: The available systems must create new development structures that include the possibility of cross-platform compilation and on different operating systems, allowing the user to choose the device he prefers to solve his problem.
- Synchronization: As InfoVis handles a large amount of data, the permanent transfer of analysis data will not be efficient and limited, overcoming through initial data synchronization with devices and followed by interaction synchronizations, where each interaction must have a timestamp, a user id, and an interaction type, also facilitating synchronization with devices with unstable network connections.

- History: With the synchronization described in the previous point, it will be possible for each device connected to the network group to have the same history in the backlog, making it easy to change devices, support undo, and do actions on each synchronized device.

8.6.1 *The Visualization Design*

Visualization design sits between what the designer does and the decisions the designer makes. The design process models are not linked to the models for visualization design decisions, and it is necessary to introduce the design activity framework, with a process model that is established with the nested, design decision, and visualization model. The four activities that overlap and characterize the design process go through understanding, ideating, doing, and implementing, in which each activity has a specific motivation to help place the visualization designer within the framework, with defined and tangible results.

The design activity framework presented in [15] is the result of reflective discussions of collaboration in a visualization redesign project, to ground it in a real design process. The two types of models focused on for visualization design are decision models and process models. The decision models capture what and why of design by characterizing the reasoning behind the decisions of what a designer does. The process models, on the other hand, capture the design mode, characterizing the actions a designer takes as a series of steps. Linking a process model to a decision model allows visualization designers to verify and validate the design decisions they make throughout each step of the design process.

The overall objective guides visualization designers through the design process, encouraging them to consider new design methods for generating, evaluating, and aiding communication.

Do visualization process models raise questions like where should I go in the next process? What is the best method for a given situation? When is the project known to be effective?

8.6.2 *InfoVis Evaluation*

According to [16], seven guiding scenarios for evaluating information visualization are defined, which are divided into four for understanding data analysis, which are: understanding work environments and practices, evaluating visual data analysis and reasoning, assessing communication through visualization, and evaluating collaborative data analysis.

Moreover, three scenarios for understanding visualizations, which are evaluating user performance, evaluating user experience, and evaluating visualization algorithms. These scenarios are described through their objectives, types of research questions, and distinguish different objects of study, mapping of specific scenarios and can be used to choose suitable research questions and objectives and for the most effective evaluation of a given information view. The work described in [16] analyzed works that included assessments that were coded according to 17 tags, condensed into the seven scenarios described above, and used as a practical context-based approach to explore assessment options, which serves as a starting point for broadening the range of assessment studies and open up new perspectives and ideas on information visualization assessment.

8.6.3 InfoVis Decision Support

The InfoVis evaluation has been referred to the user interface, to the perceptual function through the evaluation of the interface, of its cognitive support through the knowledge discovery process, but not to the evaluation of the effectiveness of InfoVis' decision support. The article [17] intended to characterize and categorize InfoVis evaluation theories, with perceptual, cognitive, and decision support. The theoretical characterization took into account that these supports are sequential, interconnected, and of dependent phases for an adequate conceptualization of InfoVis support evaluation, presenting the proposal of an evaluation framework for the effectiveness of InfoVis decision support and an evaluation process experimental.

The work analyzed the methodology for measuring the use and degree of integration of visual tools in integrated reports and examines the factors that affect visual disclosure, [18] applied to relevance for decision-making. It also analyzed the application of the four decision-making stages, in the visualization of knowledge: the uniform classification to obtain decision alternatives, group choice to determine the decision attribute, choosing the final hypothesis, reporting the cognitive learning process to facilitate the collaborative decision-making process, and its optimization [19].

The steps of the decision-making process can be summarized as follows: definition, identification, evaluation, weighting, implementation, and evaluation. InfoVis and VA can be used in different stages, with different tools, techniques, and methods.

8.6.4 Visual Comparison for Viewing Information

Data analysis involves comparing complex objects to greater amounts and complexity of data. Information visualization tools support these comparisons and allow the user to examine each object individually. The design of information views of complex objects must be studied independently of what those objects are. According to [20], the general taxonomy of visual designs for comparison groups designs into three basic categories, which can be combined: juxtaposition (showing different objects separately), superposition (overlaying objects in the same space), and explicit coding of relationships, where all designs are assembled from the building blocks of juxtaposition, overlay, and explicit coding, furthering the general understanding of comparative visualization and facilitating the development of more comparative visualization tools.

Comparison is not a single task, but a variety of tasks that a user may need to perform, given a series of related objects, where basic tasks can be enumerated, such as finding similarities, differences, and trends, detecting abnormal values, or determining the causality of the changes.

Scalability is a major challenge for the new comparison problems, as there are more items to compare, and more complexity in the items to compare. Other challenges will be the problems of greater diversity in the types of data and in the types of relationships that users can seek to understand in the data. Solutions for visual comparison are to break them down into basic elements and put them together with interaction and analysis [20].

8.6.5 InfoVis in Mobile Environments

The use of mobile devices, with the expansion and evolution of wireless communication networks, enhances access to a large amount of information available.

These devices are used in mobility scenarios in very different contexts of use, which may limit the tasks of viewing information on these devices, sometimes implying an increase in the users' cognitive load and making it difficult to perceive the data presented. The use of techniques that adapt the visualization to the context in which the user is to mitigate many of the existing problems will allow the reduction of the cognitive load necessary for a correct understanding of the information presented. In [21], an infrastructure was presented that intends to serve as a basis for the design of applications for adaptive visualization of information for mobile devices. This solution allows their reuse in different application domains and enables the use of a diverse set of contexts of use, with the use of a wider set of contexts and the calculation of more complex contexts, using the aggregation of contextual information from less complex contexts. Some aspects that could lead to future investigations were identified [21]:

- The reuse of these components in a modular and extensible way, which can be used, concurrently, by more than one application;
- The possibility of concurrent use raises issues that are particularly relevant for mobile applications for adaptive display of information, with the expansion of essential requirements, associating them with the privacy aspects of contextual information and its sharing with different applications;
- Adapting to the user's location and orientation, to exploit the automatic identification of the presence of points of interest in the device camera's field of view;
- The user's current position in the detection algorithm inside the building;

- Explore the integration of new contexts that allow capturing, in the greatest possible detail, the existing subtleties in the interaction of each user, in different situations.

8.6.6 Viewing Defined Sets and Data

In [22], the techniques are presented in seven categories according to the visual representations they use and the tasks they support, which are:

- Intuitive diagrams (based on Euler), these are diagrams that, when well adapted (training required), represent all the established standard relationships in a compact way, limited to a few sets due to disorder and design problems, and where desired properties are not always possible (e.g., convexity).
- Overlays, highlights elements, and define distributions according to other data resources (e.g., map locations), sometimes limited in several elements and sets, with unwanted layout artifacts (overlays, crossover, shapes, etc.).
- Node link diagrams visually highlight elements as individual objects, and show clusters of elements with similar set members, with limited scalability due to boundary crossings and no representation of established relationships in element set diagrams.
- Matrix-based techniques, which are reasonably scalable in both the number of elements and sets, do not suffer from boundary crossings or topological constraints, are limited in the established relationships they can represent, and the association patterns revealed are ordering sensitive.
- Based on aggregation, which is highly scalable in the number of elements, some techniques can show how attributes correlate with the defined association, do not highlight sets and elements as objects, and are limited in the established relationships they can represent.
- Scatter plots, which show groupings of sets according to mutual similarity, organized and scalable by showing sets only, do not represent standard set relationships, and points are often perceived as elements, not sets.

8.6.7 Advances and Challenges

Information visualization (InfoVis), the transformation of data into information and knowledge, through visual and interactive representations, for users, provides mental models of information. With big data, InfoVis was extended to several areas, through four main categories: empirical methodologies, user interactions, visualization frameworks, and applications.

The latest trends increasingly focus on methodologies and empirical applications deployed in the "real" world. The InfoVis techniques developed each have objectives, fundamental principles, recent trends, and an approach to the state of the art. Several of these techniques have been applied in various uses such as network visualization, text visualization, map visualization, and multivariate data visualization [23].

Researchers, brand strategies, financial analysts and human resource managers, and many others need a better understanding and analysis of data and information for growth, productivity, and innovation. Consumers and citizens analyze publicly available data, such as product specifications, via the web, on websites, blogs, and online communities, to choose the products they will be able to buy, when deciding by answering questions, voting through likes, or other qualitative forms of voting and the search for related information. InfoVis technologies begin to provide an effective vision to deal with the lapses of information felt by users, in their daily lives and their physical and virtual spaces.

8.7 Visual Analytics in the Future

Four main areas were identified in [24] for visual analysis: the science of analytical reasoning; visual representations and interaction techniques; data representations and transformations; and the presentation, production, and dissemination. These four areas provide the foundation for visual analysis as the science of analytical reasoning facilitated by interactive visual interfaces.

According to [24], the top 10 observations for visual analytics technologies and systems go through:

- The relationship between parts, where visual analysis systems show scale-independent visual representations of the entire information space to be analyzed, in addition to a detailed representation, by providing an interconnected context to the highest and lowest levels of understanding of the information and by involving multiple levels of abstraction of vision and interaction.
- Relationship discovery, where systems with interaction techniques allow the discovery of relationships between people, places, and times, through iterative queries or through a full multidimensional exploration. The discovery is accomplished through the exploration of high-dimensional spaces; temporary subsets; the identification of groups, clusters, and rare events; and the use of search techniques, including a Boolean keyword or phrase search and search for example. The combined exploratory and confirmatory interaction, where the exploratory interaction allows discovering relationships, developing and refining hypotheses, and confirming or refuting hypotheses, is the basis for a human cognitive model for analysis, which may include the beginning of predictive analyses.
- Multiple data types, where systems tend to be media type specific, with a focus on unstructured text, video, transactions, or issue-specific data. The interactions in these tools are designed specifically for the type of data or applications.
- Temporal views and interactions, where analytical systems have a degree of temporal dynamics, with representations of flow, timelines, and representations

of events and milestones, other systems are strongly geospatial in context and use maps and cartography as their organizing principles.

- Grouping and identification, in which the systems have analytical methods that allow the formation of individual items in groups and groups in macro groups with labeling and annotation.
- Multiple views, where systems have multiple views active in a simultaneous display, with actions in one (in what? “in one view”?) being represented within other views.
- Labeling, where systems develop extensive methods for labeling all information on screens, which conveys context and details that enable the analytical process, and multiple times labeling that is dynamic and that can provide users with control over items such as level of detail, size, and color.
- Reporting, critical to analytical assessment is the ability to capture analytical processes and results that can become part of an assessment report, presentation, web communication, or another form of communication.
- Interdisciplinary science, where these integrated systems and technologies are the products of highly interdisciplinary teams and often benefit from having direct and regular access to end users.

According to [24], the main challenges of visual analytics are:

- Device, network, or interaction independence, in that it does not depend on specific devices, network designs, or interaction schemes, and by allowing operation in the current or future multiplicity.
- Linking to data or information, where the usefulness of visual analytics capabilities is that they enable seamless use of various types, forms, and sources of data and information.
- Undefined or undetermined data, where the actual datasets or information in use varies at any given time and in their content, forms, and value will be unknown or uncertain, but the tools will have to allow assessments of its usefulness which are made in real time.
- Minimized transaction costs, where network bandwidth, computational processing power, and interaction and decision space are required for visual analytics capabilities, which must be minimized to allow immediate access and active use on multiple platforms.
- Trust, in which the origin and validity of data must be known, and the security of sources and the privacy of individuals must be guaranteed, even for dynamically established access and interaction.

According to [24], the top 10 applications in the future that can be leveraged through visual analytics are:

- Discourse on human information, in which the interaction that supports effective visual analysis in systems supported by “reasoning”, with interfaces that support interactions of mixed-initiative and multi-devices and in interactive platforms that are usable in various types of systems even on mobile devices.

- Collaborative analysis, where the new foundations of reasoning they support are not just evidential and confirmatory analyses, but also exploratory, hypothesis-driven, predictive, and proactive research.
- Holistic visual representations, where visual representations that tell a complete story with effective characterization must present multi-source, multi-type data, including structured and unstructured data from simulations, sensors, data structures, and broadcast data masses.
- Scale independence, where scale-tolerant mathematical and visual approaches to analysis allow reasoning across large and diverse information spaces to facilitate analysis and refinement of uncertainty.
- Information representations, where mathematically and semantically rich representations preserve data, the synthesis of information from all forms of data, including model and sensor data into interrelated knowledge structures, and the representation of human appreciation, with the inherent representations of techniques to maintain privacy and security.
- Information sharing, where effective decision-making toolsets support information sharing in secure, privacy-protected technologies, with dissemination and sharing between visual analytics components and people.
- Active information products, where the methods and science for capturing reusable analytical components in complete stories allow for effective communication of analytical results. The products must be active insofar as they must be able to support multiple levels of abstraction and allow users to get the logic within the product, add their reasoning and facts, and transform the results into a new product communication.
- "Lightweight" software architectures, where effective standards and support rapidly develop visual analytics applications and create analytics tools specific to new applications, domains, and data types, with sharing across visual analytics technologies and components.
- Utility assessment, where science, supporting framework, and data assess the utility of science, technology, and visual analytics systems, to provide basic methods for utility-based assessments that can be used to test applications for various audiences.
- Talent base support, where growing and sustainable talent enables research, design, and application development and operations support and training for new applications and visual analytics tools [24].
- These general challenges are globally applicable in several of the concepts of society, namely in the promotion of active participation and in the collaborative and informed aspects of the citizen.

The next visualization software, according to [25], goes through six categories: massive parallelization, emerging processor architectures, application architecture and data management, data models, rendering, and interaction, where each of these categories is advanced enough for the visualization software challenge. Visualization libraries are the foundation of most visualization applications such as the Visualization ToolKit (VTK), Advanced Visual Systems (AVS), and Open Data Explorer

(OpenDX), where visualization libraries provide three aspects: an execution model, a data model, and a collection of modules that input, process, or produce data.

8.8 Conclusion

Information visualization tools faced the challenge of supporting a wide variety of structured and unstructured data, from simple tables to complex graphs for unstructured data sampling. However, few visualization libraries are capable of addressing a larger subset of information visualization challenges by providing relatively simple data models along with flexible programming interfaces to shape the data structure to address a variety of problems.

The software challenges associated with these emerging requirements are to maintain a programming interface that allows algorithm programmers easily produce general or special purpose algorithms while hiding the complexity of the underlying data structures.

The process of decision-making requires the support of various tools and methods. The decision type has several types, namely based product focus, allocation of resources for marketing, and sales when market trends are identified. The decision type is based on customer focus based on allocation for marketing and sales when market trends are identified [26].

The impact of information visualization on decision-making can be seen as a debiasing technique, with cognitive adjustment, in a relationship between visual processing and decision-making processes, which use tools that combine symbolic and spatial, and narration, animation, and mental rehearsal for problem-solving [27].

Big Data is another challenge driving future changes in visualization software [25]. Data models can depend on changes in systems architectures, with core capabilities increasing dramatically, total system memory increasing, and the addition of heterogeneous processing units and multi-core devices.

The application of AI and machine learning also emerges as a challenge and potential to perform data transformation and assist in the generation and interpretation of visualization [28].

This work sought the themes presented to focus on the challenges that exist or may exist in the future and as future research paths for InfoVis and Virtual Analysis.

References

1. Munzner, T.: Visualization Analysis and Design. CRC press (2014)
2. Bertin, J.: Semiology of Graphics. University of Wisconsin press (1983)
3. McCormick, B.H.: Visualization in scientific computing. Comput. Graph. 21 (1987)
4. Brodlie, K.W., Carpenter, L.A., Earnshaw, R.A., Gallop, J.R., Hubbold, R.J., Mumford, A.M., Osland, C.D., Quarendon, P.: Scientific Visualization: Techniques and Applications. Springer Science and Business Media (2012)

5. Keller, P.R., Keller, M.M., Markel, S., Mallinckrodt, A.J., McKay, S.: Visual cues: practical data visualization. Comput. Phys. **8**, 297–298 (1994)
6. Rhyne, T.-M.: Scientific visualization in the next millennium. IEEE Comput. Graph. Appl. **20**, 20–21 (2000)
7. Keim, D.A., Mansmann, F., Schneidewind, J., Ziegler, H.: Challenges in visual data analysis. In: Tenth International Conference on Information Visualisation (IV'06). pp. 9–16. IEEE (2006)
8. Card, M.: Readings in information visualization: using vision to think. Morgan Kaufmann (1999).
9. Friendly, M., Denis, D.J.: Milestones in the history of thematic cartography, statistical graphics, and data visualization. URL Httpwww Datavis Camilestones. **32**, 13 (2001)
10. Thomas, J.J., Cook, K.A.: A visual analytics agenda. IEEE Comput. Graph. Appl. **26**, 10–13 (2006)
11. Carvalho, J., Aragão, I.: Infografia: conceito e prática. InfoDesign-Rev. Bras. Des. Informação. **9**, 160–177 (2012)
12. Carvalho, E.S., Marcos, A.F.: Visualização de informação. Centro de Computação Gráfica (2009)
13. Federico, P., Aigner, W., Miksch, S., Windhager, F., Zenk, L.: A visual analytics approach to dynamic social networks. In: Proceedings of the 11th International Conference on Knowledge Management and Knowledge Technologies. p. 47. ACM (2011)
14. Blumenstein, K., Wagner, M., Aigner, W.: Cross-Platform InfoVis Frameworks for Multiple Users, Screens and Devices: Requirements and Challenges. In: Workshop on Data Exploration for Interactive Surfaces DEXIS 2015. p. 7 (2015)
15. McKenna, S., Mazur, D., Agutter, J., Meyer, M.: Design activity framework for visualization design. IEEE Trans. Vis. Comput. Graph. **20**, 2191–2200 (2014). https://doi.org/10.1109/TVCG.2014.2346331
16. Lam, H., Bertini, E., Isenberg, P., Plaisant, C., Carpendale, S.: Empirical studies in information visualization: seven scenarios. IEEE Trans. Vis. Comput. Graph. **18**, 1520–1536 (2012). https://doi.org/10.1109/TVCG.2011.279
17. Akanmu, S.A., Jamaludin, Z.: Measuring InfoVis' decision support effectiveness: From theory to practice. In: Science and Information Conference (SAI), pp. 560–564. IEEE (2015)
18. Nicolò, G., Ricciardelli, A., Raimo, N., Vitolla, F.: Visual disclosure through integrated reporting. Manag. Decis. **60**, 976–994 (2021)
19. Zhao, N., Ying, F.J., Tookey, J.: Knowledge visualisation for construction procurement decision-making: a process innovation. Manag. Decis. (2021)
20. Gleicher, M., Albers, D., Walker, R., Jusufi, I., Hansen, C.D., Roberts, J.C.: Visual comparison for information visualization. Inf. Vis. **10**, 289–309 (2011)
21. Matos, P.M.C.P.P., et al.: Visualização de informação em ambientes móveis (2015)
22. Alsallakh, B., Micallef, L., Aigner, W., Hauser, H., Miksch, S., Rodgers, P.: Visualizing sets and set-typed data: State-of-the-art and future challenges. In: Eurographics conference on Visualization (EuroVis)–State of The Art Reports. pp. 1–21 (2014)
23. Liu, S., Cui, W., Wu, Y., Liu, M.: A survey on information visualization: recent advances and challenges. Vis. Comput. **30**, 1373–1393 (2014). https://doi.org/10.1007/s00371-013-0892-3
24. Thomas, J., Kielman, J.: Challenges for visual analytics. Inf. Vis. **8**, 309–314 (2009). https://doi.org/10.1057/ivs.2009.26
25. Childs, H., Geveci, B., Schroeder, W., Meredith, J., Moreland, K.: Research Challenges for Visualization Software (2013)
26. Odqvist, P.: Information Visualization of Assets under Management: A qualitative research study concerning decision support design for InfoVis dashboards in fund management. (2020)
27. Kellen, V.: Decision making and information visualization: research directions. Retrieved Novemb. **11**, 2005 (2005)
28. Ma, K.-L.: Intelligent Visualization Interfaces. In: 26th International Conference on Intelligent User Interfaces. pp. 2–3 (2021)

Part III
Customer Data Management and CRM

Chapter 9
CRM and Smart Technologies in the Hospitality

Rashed Isam Ashqar, Célia Ramos, Carlos Sousa, and Nelson Matos

Abstract For more than five decades, experiences have been a significant concept in both tourism production and research. Creating positive experiences has been described as the essence of the hospitality industry. Personalization allows customers to access customized information more efficiently at any time. From a company's perspective, personalization gives a way to build strong consumer relationships. The significance of using personalization in smart technology increased with the emergence of the COVID-19 pandemic. This paper aims to present the related literature regarding customer relationship management (CRM) and find the gaps in the literature. Also, we reviewed smart technologies and experience personalization in the hospitality industry context and present some recent techniques for using smart technologies and experience personalization. Most of the literature was limited to countries such as China, South Korea, and Malaysia or one case study such as a hotel in Switzerland. This research finds that the previous results might not be generalizable to all consumers or hotels around the world. Hence, it is important to fill the gap in the literature by examining cross-cultural studies or to expand the case study to multiple cases in order to establish the generalizability of the findings to a larger industry context.

R. I. Ashqar (✉) · C. Ramos · C. Sousa · N. Matos
ESGHT, University of Algarve, Faro, Portugal
e-mail: riashqar@ualg.pt

C. Ramos
e-mail: cmramos@ualg.pt

C. Sousa
e-mail: cmsousa@ualg.pt

N. Matos
e-mail: nmmatos@ualg.pt

C. Ramos · N. Matos
CinTurs, Faro, Portugal

C. Sousa
CiTUR, Lisbon, Portugal

J. L. Reis et al. (eds.), *Marketing and Smart Technologies*, Smart Innovation, Systems and Technologies 337, https://doi.org/10.1007/978-981-19-9099-1_9

9.1 Introduction

Recent techniques such as Internet-based technologies, social networking tools, and mobile technologies have allowed businesses and consumers to connect, interact, and create experiences on an unprecedented scale [34]. Therefore, the marketplace has experienced a shift toward consumers gaining increasing power and control [1]. Because the customers play an involved role in the production and consumption process [8], it has become important for businesses to use technology to engage consumers in a more personal way [41]. In this regard, [18] focuses on the potential of intelligent systems in tourism to meet tourists' personal and situational needs.

Previous research has considered the potential of technologies for more personalized experiences, such as the role of smartphones in travel and the mediation of the tourism experience [50, 51], the use of context-aware mobile applications in tourism [22], the use of high-tech for high-touch experiences [33], and the adoption of mobile tour guides for personalized routes and location-relevant information [44]. Businesses and consumers have become interconnected in the travel process through a range of hardware devices, software platforms, and applications, which led to more meaningful interrelations and convergence of people, technology, and more personalized tourism experiences [32].

The objective of this paper is to present the related literature regarding customer relationship management (CRM) and find the gaps in the literature. Also, we reviewed smart technologies and experience personalization in the hospitality industry context and present some recent techniques for using smart technologies and experience personalization. This paper is structured as follows. Section two presents the related literature regarding customer relationship management (CRM) and personalization. Section three shows the smart technologies and experience personalization and some recent techniques for using smart technologies and experience personalization. Section four concludes the paper.

9.2 Customer Relationship Management (CRM) and Personalization

Customer relationship management (CRM) has become a strategic imperative in tourism and hospitality, as it can address competition, create differentiation, and provide enhanced customer value [15, 46]. CRM is traditionally seen as a set of philosophies, strategies, systems, and technologies that help firms to manage their customers' transactions and relationships [17].

There are several benefits to collecting and integrating customer information for personalizing customer transactions, tourism and hospitality firms, which can exploit CRM for improving their customers' relationships and satisfaction [25, 54], identifying and retaining the most profitable customers and improving the profitability of less profitable customers [52, 54, 55], and increasing business performance such as customer lifetime value, customer satisfaction, and retention and business profits [14, 30].

CRM is the system that provides the empowerment of Hyper-Personalization at Hotels. There are lots of CRM systems, such as the generalists vs the CRM designed by and for Hoteliers, with obvious advantages to both sides of the "coin". There are lots of CRM Books, in and outside of the Hospitality Industry. CRM needs to interact, via EXtended Markup Language (XML) Webservice, with the PMS System, and the CRM needs, as well, to interact, via XML Webservice, with different systems such as Outlook, BI Portal, SMS/eMail Gateways, Social Media Systems, and so on. The more platforms connected via XML Interface Webservice, the more powerful the CRM system, and more closely is to the Hyper-Personalization in the Hotel or Hotel Chain. So, the CRM systems must be linked to (1) marketing staff systems, (2) commercial staff systems, (3) operations staff systems, and (4) guests' own mobile devices [20]. Table 9.1 presents the services offered to the guests when the Guest Mobile approach considered the Hyper-Personalization Services by [20].

Therefore, Hyper-Personalization, in this new normal in the Hotel industry, is a new way of modern Hotel Services, with real-time processes shared between the Guest and the Hoteliers, with a fundamental collaboration of a huge amount of XML Webservices interfaces. To achieve this new kind of Hotel Services, it's mandatory to implement XML Webservices interfaces, that provide the necessary "glue" that interconnects the inevitable growing set of hotel systems, hardware plus software solutions that increase the quality of Hotel Services and facilitate the processes arising from the hotel service, with guests, whether repeated or new ones [20].

9.3 Smart Technologies and Experience Personalization in the Context of the Hospitality Industry

There are several smart technologies that are used in Tourism such as the Internet of Things, the Internet of Everything, fifth-generation mobile network (5G), Radio Frequency Identification (RFID), mobile devices, wearable smartphones and devices, 3D printing, apps along with APIs, cryptocurrency and blockchain, sensor and beacon networks, pervasive computing, gamification as well as enhanced analytical capabilities supported by Artificial Intelligence (AI) and machine learning (ML) [9, 49]. All these technologies create the infostructure and the smart digital grid that support the seamless interoperability of all stakeholders [7].

In this section, we present one case from the literature regarding smart technologies. Neuhofer et al. [34] fill the gap between smart technologies and experience personalization to understand how smart mobile technologies can facilitate personalized experiences in the context of the hospitality industry. Therefore, [34] provided a practice example from Hotel Lugano Dante in Switzerland with its unique "Happy Guest Relationship Management system" (HGRM) platform. The HGRM system represents a comprehensive customer relationship management (CRM) database,

Table 9.1 Services offered to the guests with Hyper-Personalization

No.	Types	Services offered
1	Pre-check-in and validate their personal data	Starting from the very beginning of the Hyper-Personalization journey, upselling their Room typology, adding meal options, subscribing to SPA Services, and schedules for breakfast, lunch, or dinner. It is also possible, for the Guest, to acknowledge the GDPR Hotel Options. This kind of service also provides to a large cluster of guests, the confidence of knowing the booking is confirmed by the Hotel Services
2	Check-in	In this process, the Guest does not need to sign a significant set of papers, as he can validate, either on his mobile phone or on a tablet provided by the Hotel, the details of his reservation, digitally signing on the selected equipment. This digital functionality simplifies and improves data quality in the Check-In process, and the Check-In process in-self. There are a lot of Hoteliers that are adding another facility in the Check-In process: the Self-Check-In capability! Of course, there are much more hoteliers that consider that this hi-tech capability de-virtualizes the hotel service. If the Hoteliers consider applying this kind of solution, they need to implement XML interfaces between the Self-Check-In Kiosk, the PMS, the key-card machine (or smartphone mobile key-card solution), the ID-Card Reader machine, and the Payment Gateway
3	e-Mobility, room guest entry	In their Room, during the stay, with their smartphone, without the traditional Room key-card, minimizing the Guest's contact with hotel furniture. This kind of solution also provides the hotelier, with the capability of understanding, in real-time, if the Guest is, or is not, in their Room, improving in this way the speed and the quality of service of Housekeeping, Food and Beverage, Reception, Guest Service and Maintenance, if necessary. This smartphone mobile key-card solution, nowadays, is the new normal of hospitality, providing a modern and safe approach that the Guest tend to appreciate
4	Internet guest solutions	Via the captive portal, is an "old way" to manage the Internet Guest that allows the hoteliers to interact with the Guest, in a secure way, in cases of captive portals with a digital and secure certificate. This kind of solution allows the hoteliers to implement IoT solutions, to better understands where the Guest are, in the Hotel area, and to better communicate with the Guest, i.e., this kind of solution, allows the Hotel, for instance, to send push notifications to the selected Guest that are not at the Swimming Pool Bar, to send them promotion of Daiquiri, because the Terrace Bar has already an occupancy of 80%, and the Swimming Pool Bar has only 35% of occupancy

(continued)

Table 9.1 (continued)

No.	Types	Services offered
5	Hotel services portal	To provide to the guest the capability of buying hotel services, in complement to the staff interaction, and, of course, increase the Hyper-Personalization at the Hotel, because the Hotel knows, in real-time, the needs of each Guest, the times of service, the complaints and specific requests of each guest, whether for food and beverages or other hotel services, as well as the specific allergies and preferences of each guest! These solutions, can also if requested by the Hoteliers, managed satisfaction surveys, and the ability to manage the relationship with the social networks of guests who decided to identify themselves at the hotel, during or after their stay at the hotel

Source Gustavo et al. [20]

which functions as a meta-platform that combines several hotel operation systems. It merges the data received from the property management system (PMS) Fidelio, outlook, the guest's intranet site MyPage, and all operations platforms into one database. Therefore, the HGRM provides a centralized solution that unifies all internal and external information exchanges, transfers, and interactions among the hotel staff and between the hotel and its guests. Table 9.2 presents the HGRM system that covers processes of the entire customer journey consisting of pre-arrival, arrival, and post-departure stages at the hotel.

Also, [34] suggested three main technological requirements for personalized experiences, which include (1) information aggregation, (2) ubiquitous mobile connectedness, and (3) real-time synchronization of information. In addition, they suggested the goal of personalization is a twofold process of (a) customization of experiences and (b) one-to-one interactions that are facilitated by the support of mobile, dynamic, and smart technologies.

In addition, [45] suggested that the concept of Industry 5.0 is useful for the hospitality industry, as personalized service, an efficient supply chain, agility, a smart work environment, the use of big data for up-to-date information on customer preferences, highly customized services at a lower cost, and digital enhancement can influence customer satisfaction, loyalty, and perceived service quality. Therefore, every client is unique and needs personalized and customized service, and hotel employees have to provide these services [26]. The five design principles of Industry 5.0 are the following: interoperability, modularity, virtualization, real-time capabilities, and decentralization, which can be extended to the hospitality industry, leading to Hospitality 5.0 [40]. Moreover, Industry 5.0 focuses on the personalized demand of customers. With mass personalization, there is customer delight with higher value addition through Industry 5.0. There is also a shift from mass customization to mass personalization, especially to fulfill the requirements of an individual customer [35].

Moreover, [3] examined the types of systems used and hoteliers' main factors, drivers, and limitations to invest in information systems' maturity using 195 hotels in Portugal based on the Network Exploitation Capability (NEC) model by [39]. The

Table 9.2 HGRM system at the hotel

Time stage	Services offered
Pre-arrival stage	Guests receive an invitation upon confirmation to access their personalized guest website (MyPage). From this point onwards they are given a choice of whether or not they desire to personalize their stay. In case guests are willing to share personal information in exchange for experience personalization, they can independently manage their MyPage website to communicate with the hotel, virtually meet the team and engage with hotel employees, manage details of their stay, and select personal preferences. These include, for instance, the customization of room temperatures and beds, extra soft towels, organic bathroom sets, air cleaner, drinks and snacks in the mini-bar, special equipment for children, or the selection of the favorite newspaper
Arrival stage	A vast number of touchpoints are encountered in the different departments of the hotel, including the reception, housekeeping, restaurant, maintenance, bar, marketing, welcome, garage and parking. During these encounters, the hotel (and its individual employees) and the customer (the individual guest), interact for service experiences to be co-created (B2C). In adopting an employee-centric approach, each employee is empowered, equipped, and instructed to access and use the HGRM smart technology platform through dedicated mobile devices. In the service delivery process, the HGRM enables employees to retrieve guest names and profiles, service and communication history, room status, and personal preferences. By doing so, they can retrieve, modify and add up-to-date guest information obtained through one service encounter, which is instantly synchronized to all departments from one encounter to the next. In managing all service encounters on this integrated platform, employees are in full control to see what is happening and what action is required to turn a simple service routine into a personalized guest experience by proactively anticipating as well as dynamically responding to the emerging needs and preferences of the guest
Post-departure stage	Guests are sent a welcome-home message through their MyPage website, which includes a personalized thank-you note, a picture of the employee who has performed the check-out, a contact email address for concerns, and an invitation to leave a review on TripAdvisor. While in this stage no further personal information is collected, the principal purpose of this stage is to maintain the established relationship, reflect on the experience and keep the personal dialogue going on social media platforms

Source Neuhofer et al. [34]

NEC model has five levels: Basic, Operational, Integrated, Analytical, and Optimizing. António and Sá [3] found that hoteliers consider that their companies take more advantage of technology and information systems than they really do.

Another example is the WeChat app that permits one-to-one personalized and dedicated interaction between brands and users [11] and has been heralded as an effective tool for luxury CRM, with its functionality allowing brands to develop a range of marketing initiatives from immersive branding campaigns to boutique appointment systems and integrated WeChat e-commerce [12]. Users simply scan their WeChat QR code to connect with another individual (e.g., luxury sales personnel), without needing to provide any other contact details [43]. This permits highly personalized brand-consumer communications, which is paramount for high-end consumers.

In the following, we present some recent techniques for using smart technologies and experience personalization, namely Gamification, the Internet of Things (IoT), Chatbots, Augmented Reality (AR), and Virtual Reality (VR). Also, we offer smart technologies and experience personalization in the hospitality and customer knowledge framework.

9.3.1 Gamification

Deterding et al. [16] defined gamification as "The use of game design elements in a non-game context" (p. 9). Also, they include some mechanisms of gamification, namely: self-representation with avatars, three-dimensional environments, narrative context, feedback, reputation, rank, and levels, marketplace and economies, competition under rules that are explicit and enforced, teams, parallel communication systems, and time pressure.

There are some examples of online stores that have already used gamification such as the Starbucks rewards card is a prime example of the use of gamification to create consumer loyalty to the brand [47]. Also, the launch of Amazon Prime in the United Kingdom featured a promotion for one free delivery in exchange for signing up for a free one-month trial of their streaming services, in an attempt to expand their services portfolio and retain their current customers [2]. In addition, almost every bank in the world has an application for customers to manage their money, and 55% of European online banking users confirm that they had also used mobile banking services [23]. Loyalty or rewards programs were the most common previous use of gamification applied to consumer contexts [56], where consumers obtain points that they can redeem for products.

9.3.2 The Internet of Things (IoT)

IoT plays an important role in the conceptualization and implementation of smart environments by filling the gap between physical and mobile engagement [19]. Therefore, it provides interconnectivity among people, systems, and products [42] by integrating the digital world with physical infrastructure (cyber-physical systems). Ashton [5] indicated that IoT can enable computers to "observe, identify, and understand the world without the limitations of human-centered data". Also, IoT devices have been adopted rapidly by all domains of industry in the past decade to improve business operational efficiency and reduce costs [6].

An IoT device involves mainly three elements: processing and storage unit, appropriate sensors, and communication interface [38]. However, they are assumed to be resource-constrained in terms of computation capacity. An IoT ecosystem can be described as a network of things that comprises the following three major layers: (1) data collection (i.e., sensing), (2) network and technologies (i.e., transport and platforms); and (3) data processing and applications [31].

9.3.3 Chatbots

Chatbots have been used by destinations to integrate technology into customers' frontline experiences [36]. Therefore, chatbots are machine conversation systems developed to evoke humanlike interactions [28] and might be considered virtual service agents or "e-service agents" [48]. Using chatbots in the service experiences by tourism organizations is an important attribute that improves customer satisfaction [36, 48]. Moreover, chatbots are programmed with selected skills to help customers find restaurants, make hotel reservations, and purchase goods [28].

9.3.4 Augmented Reality (AR) and Virtual Reality (VR)

Although AR and VR are still new in the tourism field [13, 53], their popularity within the tourism industry, and the academic spheres, are rapidly on the rise [21]. AR indicates adding of digital information to the real environment [37] thus enabling consumers to view the real world in front of them, with the addition of a superimposed layer of information, including text and or images which enhances their experience. On the other side, VR is the use of digital technologies to create a simulated environment that customers can experience and explore through their various senses [37]. Therefore, both AR and VR involve interactive experiences, and VR goes beyond AR in creating a more immersive interaction [10].

Moreover, AR and VR can be considered radical changes in the tourism industry due to their ability to substitute actual tourism for virtual tourism [29]. From the perspective of destinations and tourism businesses, adopting VR provides opportunities for marketing, additional revenue generation, sustainability, and the preservation of heritage [49]. On the other side, from the perspective of tourists, AR and VR are useful as it improves tourism experiences through the simulation of a real physical environment interaction [29, 49, 53].

9.3.5 Smart Technologies, Experience Personalization in the Hospitality and Customer Knowledge Framework

Figure 9.1 presents the framework for smart technologies, experience personalization, and customer knowledge.

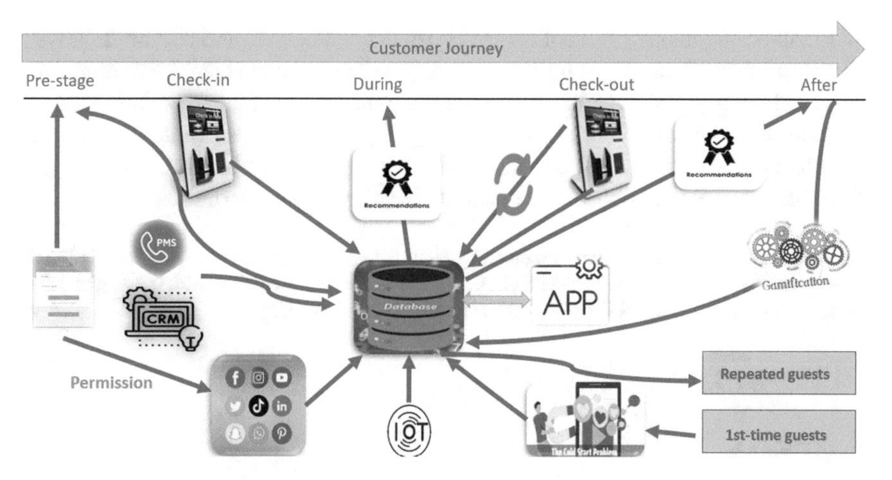

Fig. 9.1 The framework for smart technologies, experience personalization, and customer knowledge. *Source* Own elaboration

In Fig. 9.1, we showed the framework for smart technologies and experience personalization during the customer journey consisting of five stages (Pre-stage, Check-In, During, check-out, and After). Therefore, in the Pre-stage, the customers enrolled by an app on the phone or website and they permit to have access to their social media. This information and the data from the PMS and CRM transform into the Database. Also, the Check-In can transform the details of services used by the customers into the Database. During the journey, the Database can receive information from the recommendations and IoT. During the check-out, the system can refresh the information from the customers. After the trip, the system can provide recommendations by App for repeated guests. In addition, the system can use gamification to engage with consumers, employees, and partners to inspire collaborate, share, and interact. In the case of new guests, we suggest the system use the cold start problem to predict the needs of customers.

9.4 Conclusions

To conclude, previous studies show that personalization plays an essential role in hospitality. However, most of the literature was limited to countries such as China [27], South Korea [24], and Malaysia [4] or one case study such as a hotel in Switzerland [34]. Therefore, the previous results might not be generalizable to all consumers or hotels around the world. Hence, it is important to fill the gap in the literature by examining cross-cultural studies or to expand the case study to multiple cases in order to establish the generalizability of the findings to a larger industry context. We suggest that future research address the following: (1) capture the adoption, implementation, and impact of smart technological solutions in the coming years. (2)

Additional in-depth studies could focus on the emerged relational role of employee-consumer interactions to highlight the interdependence between employee empowerment, technology, and experience co-creation practices. (3) Should examine the use of mobile apps and their interplay with face-to-face interactions in hotels. (4) Should also continue to compare the effects of different types of personalization strategies. (5) Future studies may collect data from different groups of participants.

Acknowledgements This paper is financed by National Funds provided by FCT–Foundation for Science and Technology through project UIDB/04020/2020 and project Guest-IC I&DT nr. 047399 financed by CRESC ALGARVE2020, PORTUGAL2020 and FEDER.

References

1. Alt, R., Klein, S.: Twenty years of electronic markets research—looking backwards towards the future. Electron. Mark. **21**(1), 41–51 (2011)
2. Amazon (2018) Amazon prime. https://www.aboutamazon.co.uk/news/innovation/amazon-prime, last accessed 2022/06/27.
3. António, N., Sá, G.: Mapping information systems maturity: the case of the Portuguese hospitality industry. Tourism Manag. Stud. **17**(4), 7–21 (2021)
4. Ariffin, A.A.M., Maghzi, A.: A preliminary study on customer expectations of hotel hospitality: influences of personal and hotel factors. Int. J. Hosp. Manag. **31**(1), 191–198 (2012)
5. Ashton, K.: That 'internet of things' thing. RFID J. **22**(7), 97–114 (2009)
6. Breidbach, C., Choi, S., Ellway, B., Keating, B.W., Kormusheva, K., Kowalkowski, C., Lim, C., Maglio, P.: Operating without operations: how is technology changing the role of the firm? J. Serv. Manag. **29**(5), 809–833 (2018)
7. Buhalis, D.: Technology in tourism-from information communication technologies to eTourism and smart tourism towards ambient intelligence tourism: a perspective article. Touris. Rev. **71**(1), 267–272 (2019)
8. Buhalis, D., Law, R.: Progress in information technology and tourism management: 20 years on and 10 years after the Internet—the state of eTourism research. Tour. Manage. **29**(4), 609–623 (2008)
9. Buhalis, D., Sinarta, Y.: Real-time co-creation and nowness service: lessons from tourism and hospitality. J. Travel Tour. Mark. **36**(5), 563–582 (2019)
10. Chamboko-Mpotaringa, M., Tichaawa, T.M.: Tourism digital marketing tools and views on future trends: a systematic review of literature. African J. Hospital. Touris. Leisure **10**(2), 712–726 (2021)
11. Chen, P.-C.: The development path of WeChat: social, political and ethical challenges. In *The digitization of business in China* (pp. 127–152). Springer (2018)
12. China, C. (2018). *Luxury on WeChat : The Keys to Succeed in 2018—Retail in Asia.* https://retailinasia.com/in-trends/luxury-on-wechat-the-keys-to-succeed-in-2018/. Last accessed 11 Apr 2022
13. Cranmer, E.E., Tom Dieck, M.C., Fountoulaki, P.: Exploring the value of augmented reality for tourism. Tourism Manage. Perspect., **35**, 100672 (2020)
14. Daghfous, A., Barkhi, R.: The strategic management of information technology in UAE hotels: an exploratory study of TQM, SCM, and CRM implementations. Technovation **29**(9), 588–595 (2009)
15. Debnath, R., Datta, B., Mukhopadhyay, S.: Customer relationship management theory and research in the new millennium: directions for future research. J. Relationship Market. **15**(4), 299–325 (2016)

16. Deterding, S., Dixon, D., Khaled, R., Nacke, L.: From game design elements to gamefulness: defining "gamification". In: *Proceedings of the 15th International Academic MindTrek Conference: Envisioning Future Media Environments*, pp. 9–15 (2011)
17. Greenberg, P.: The impact of CRM 2.0 on customer insight. J. Bus. Indus. Market., **25**(6), 410–419 (2010)
18. Gretzel, U.: Intelligent systems in tourism: a social science perspective. Ann. Tour. Res. **38**(3), 757–779 (2011)
19. Gretzel, U., Ham, J., Koo, C.: Creating the city destination of the future: the case of smart Seoul. In *Managing Asian Destinations* (pp. 199–214). Springer (2018)
20. Gustavo, N., Mbunge, E., Belo, M., Fashoto, S.G., Pronto, J.M., Metfula, A.S., Carvalho, L.C., Akinnuwesi, B.A., Chiremba, T.R.: Emphasizing the digital shift of hospitality towards hyper-personalization: application of machine learning clustering algorithms to analyze travelers. In: *Optimizing Digital Solutions for Hyper-Personalization in Tourism and Hospitality* (pp. 1–19). IGI Global (2022)
21. He, Z., Wu, L., Li, X.R.: When art meets tech: the role of augmented reality in enhancing museum experiences and purchase intentions. Tour. Manage. **68**, 127–139 (2018)
22. Höpken, W., Fuchs, M., Zanker, M., Beer, T.: Context-based adaptation of mobile applications in tourism. Inf. Technol. Tourism **12**(2), 175–195 (2010)
23. Global Web Index (2017). *The online banking landscape in Europe.* https://insight.gwi.com/hubfs/Infographics/Online-Banking-in-Europe-Infographic-Q1-2017.pdf, last accessed 2022/06/27.
24. Kang, J.-W., Namkung, Y.: The role of personalization on continuance intention in food service mobile apps: a privacy calculus perspective. Int. J. Contemp. Hosp. Manag. **31**(2), 734–752 (2019)
25. Kasim, A., Minai, B.: Linking CRM strategy, customer performance measures and performance in the hotel industry. Int. J. Econ. Manage. **3**(2), 297–316 (2009)
26. Kuo, C.-M., Chen, L.-C., Lu, C.Y.: Factorial validation of hospitality service attitude. Int. J. Hosp. Manag. **31**(3), 944–951 (2012)
27. Lei, S.S.I., Chan, I.C.C., Tang, J., Ye, S.: Will tourists take mobile travel advice? examining the personalization-privacy paradox. J. Hosp. Tour. Manag. **50**, 288–297 (2022)
28. Leung, X.Y., Wen, H.: Chatbot usage in restaurant takeout orders: a comparison study of three ordering methods. J. Hosp. Tour. Manag. **45**, 377–386 (2020)
29. Li, S.C.H., Robinson, P., Oriade, A.: Destination marketing: the use of technology since the millennium. J. Destin. Mark. Manag. **6**(2), 95–102 (2017)
30. Lo, A.S., Stalcup, L.D., Lee, A.: Customer relationship management for hotels in Hong Kong. Int. J. Contemp. Hosp. Manag. **22**(2), 139–159 (2010)
31. Mercan, S., Cain, L., Akkaya, K., Cebe, M., Uluagac, S., Alonso, M., Cobanoglu, C.: Improving the service industry with hyper-connectivity: IoT in hospitality. Int. J. Contemp. Hosp. Manag. **33**(1), 243–262 (2020)
32. Neuhofer, B., Buhalis, D., Ladkin, A.: Conceptualising technology enhanced destination experiences. J. Destin. Mark. Manag. **1**(1–2), 36–46 (2012)
33. Neuhofer, B., Buhalis, D., Ladkin, A.: High tech for high touch experiences: a case study from the hospitality industry. In: *Information and Communication Technologies in Tourism 2013* (pp. 290–301). Springer (2013)
34. Neuhofer, B., Buhalis, D., Ladkin, A.: Smart technologies for personalized experiences: a case study in the hospitality domain. Electron. Mark. **25**(3), 243–254 (2015)
35. Özdemir, V., Hekim, N.: Birth of industry 5.0: making sense of big data with artificial intelligence, "the internet of things" and next-generation technology policy. Omics: J. Integrative Biol., **22**(1), 65–76 (2018)
36. Park, S.: Multifaceted trust in tourism service robots. Ann. Tour. Res. **81**, 102888 (2020)
37. Park, S., Stangl, B.: Augmented reality experiences and sensation seeking. Tour. Manage. **77**, 104023 (2020)
38. Patel, K.K., Patel, S.M.: Internet of things-IOT: definition, characteristics, architecture, enabling technologies, application & future challenges. Int. J. Eng. Sci. Comput. **6**(5), 6122–6131 (2016)

39. Piccoli, G., Carroll, B., Hall, L.: Network exploitation capability: Mapping the electronic maturity of hospitality enterprises, **11**(18), 4–16 (2011)
40. Pillai, S.G., Haldorai, K., Seo, W.S., Kim, W.G.: COVID-19 and hospitality 5.0: redefining hospitality operations. Int. J. Hospital. Manage., **94**, 102869 (2021)
41. Pine, B.J., Gilmore, J.H.: *The Experience Economy: Work is Theatre and Every Business a Stage*. Harvard Business Press (1999)
42. Porter, M.E., Heppelmann, J.E.: How smart, connected products are transforming competition. Harv. Bus. Rev. **92**(11), 64–88 (2014)
43. Sam, K.M., Chatwin, C.R.: Understanding Wechat users' motivations, attitudes and intention of reading promotional material. J. Inf. Technol. Manag. **30**(1), 25–37 (2019)
44. Schmidt-Rauch, S., Schwabe, G.: Designing for mobile value co-creation—the case of travel counselling. Electron. Mark. **24**(1), 5–17 (2014)
45. Shamim, S., Cang, S., Yu, H., Li, Y.: Examining the feasibilities of Industry 4.0 for the hospitality sector with the lens of management practice. Energies **10**(4), 499 (2017)
46. Sheth, J.: Revitalizing relationship marketing. J. Serv. Mark. **31**(1), 6–10 (2017)
47. Starbucks (2019) Starbucks card. https://www.starbucks.co.uk/starbucks-card. Last accessed 27 Jun 2022
48. Tussyadiah, I.: A review of research into automation in tourism: launching the annals of tourism research curated collection on artificial intelligence and robotics in tourism. Ann. Tour. Res. **81**, 102883 (2020)
49. Tussyadiah, I.P., Jung, T.H., tom Dieck, M..: Embodiment of wearable augmented reality technology in tourism experiences. J. Travel Res., **57**(5), 597–611 (2018)
50. Wang, D., Fesenmaier, D.R.: Transforming the travel experience: the use of smartphones for travel. In: *Information and Communication Technologies in Tourism 2013* (pp. 58–69). Springer (2013)
51. Wang, D., Park, S., Fesenmaier, D.R.: The role of smartphones in mediating the touristic experience. J. Travel Res. **51**(4), 371–387 (2012)
52. Wang, Y., Feng, H.: Customer relationship management capabilities: measurement, antecedents and consequences. Manag. Decis. **50**(1), 115–129 (2012)
53. Wei, W., Qi, R., Zhang, L.: Effects of virtual reality on theme park visitors' experience and behaviors: a presence perspective. Tour. Manage. **71**, 282–293 (2019)
54. Wongsansukcharoen, J., Trimetsoontorn, J., Fongsuwan, W.: Social CRM, RMO and business strategies affecting banking performance effectiveness in B2B context. J. Bus. Indus. Market. **30**(6), 742–760 (2015)
55. Wu, S.-I., Lu, C.-L.: The relationship between CRM, RM, and business performance: a study of the hotel industry in Taiwan. Int. J. Hosp. Manag. **31**(1), 276–285 (2012)
56. Zichermann, G., Linder, J.: Game-based Marketing: Inspire Customer Loyalty Through Rewards, Challenges, and Contests. John Wiley & Sons (2010)

Chapter 10
A Systematic Approach to Segmentation Analysis Using Machine Learning for Donation-Based Crowdfunding

Caroline Seow Ling Lim and Zhiguo Wang

Abstract Donation-based crowdfunding rose in popularity in the last five years. Yet the lack of customer provided information crowdfunding platforms challenged segmentation analysis for marketing decision-making. Our study solved a common problem that many donation-based crowdfunding platforms faced in segmentation analysis in the absence of donor descriptive attributes as well as unobserved customer heterogeneity. Using data from the world's only donation-based crowdfunding charity, we segment online donors' behavior with machine learning clustering algorithms by combining recency, frequency and monetary values (RFM) metric with donors' campaign preferences. We identify and compare the optimal number of donor clusters from different clustering algorithms. This segmentation analysis with machine learning, we termed 'RFMP' framework, offers a robust and novel approach to parsimoniously segment online behavior using observed and unobserved customer heterogeneity when extant segmentation strategies typically apply RFM metric in combination with demographic and socio-economic attributes to infer and predict customer behavior. The resulting donor clusters inform marketing decision-making in the design and implementation of crowdfunding campaigns for charitable causes. Our project contributes to the application of machine learning in marketing strategy.

10.1 Introduction

10.1.1 Background

Customer segmentation in marketing typically begins with collecting and analyzing customer socio-demographic, psychographic, behavioral and other descriptive information. The underlying assumption is that the differences across customer segments are observable and can be explained by demographic and socio-economic attributes

C. S. L. Lim (✉) · Z. Wang
Singapore University of Social Sciences, Singapore, Singapore
e-mail: carolinelimsl@suss.edu.sg

J. L. Reis et al. (eds.), *Marketing and Smart Technologies*, Smart Innovation, Systems and Technologies 337, https://doi.org/10.1007/978-981-19-9099-1_10

including personal characteristics like age, gender, ethnic groups, employment and marital status. In other words, customers who share common observable attributes are likely to behave similarly in response to marketing stimuli [1, 2]. The central aim is to employ the relevant strategy for each segment to improve customer retention, promote customer loyalty and enhance the customer lifetime value.

Crowdfunding platforms that conduct their transactions online generate tremendous digital footprint of their online users or contributors. The high volume of transactions makes for a rich reservoir of data for crowdfunding platforms to collect information about their users, segment them for insight generation and grow user lifetime value.

For pragmatic and practical reasons, most crowdfunding platforms do not collect sufficient demographic and socio-economic information about their platform users. The fact that these users are online and do not have physical interactions with the crowdfunding platform makes it resource-intensive to collect user data. Yet research using crowdfunding data mostly applied observed user characteristics like demographic and socio-economic attributes following traditional marketing segmentation analysis approach. These issues impact marketing decision-making particularly in user retention.

Besides observed customer heterogeneity, there are unobserved attributes like psychographic information including personality traits, values, beliefs, needs and motivations of their behavior. Information on unobserved customer heterogeneity is often not measured because they are costly to collect and thus inadvertently omitted.

Our goal in this project is to apply machine learning clustering algorithm to solve the problem with segmentation analysis in the absence of observed and unobserved customer attributes. We present a solution by using panel and user behavior data from a donation-based crowdfunding charity. By adopting clustering analysis algorithm that combined recency, frequency and monetary value (RFM) variables and individual preferences, we parsimoniously segment the platform donors into four distinct clusters.

We believe that this application of clustering tools, a form of batch and unsupervised learning algorithms, can enable firms to extract online user preferences from individual behavior on the internet platform. The characteristic profile of distinct user clusters improves marketing decision marketing such as in employing suitable engagement strategies to grow customer lifetime value on platforms.

In this paper, we begin with an illustration of the main challenge in segmentation analysis in the context of a donation-based crowdfunding platform. Next, we briefly review the literature related to our work. Following which, we present our RFMP framework, discuss various approaches to clustering analysis in machine learning and evaluate their limitations, respectively. We then apply our model in the context of the world's only donation-based crowdfunding charity. Finally, we conclude with a discussion of implications, managerial relevance and future direction of our research.

10.1.2 Donation-Based Crowdfunding Platforms

Based on data compiled by Statista, the global crowdfunding market valued at USD 13.64 billion in 2021 is projected to grow at a compound annual growth rate of 11.2% to a market size of USD 28.92 billion [3]. Through the minute contributions of many individuals online, project creators can finance their ideas in an open call through the internet without the need for financial intermediaries. There are four types of crowdfunding defined, namely equity-, reward-, lending- and donation-based [4].

Donation-based crowdfunding does not offer tangible returns or rewards to project donors or any expected return of equity or products. Instead, donation-based crowdfunding relies on the altruistic motivations of individuals. The projects on donation-based crowdfunding are created for philanthropic causes such as in humanitarian aid, poverty alleviation, children's education, shelter for the formerly incarcerated, health or medical emergencies and environmental conservation. The funds raised would contribute to these causes to benefit vulnerable and marginalized persons.

The explosion of COVID-19-related philanthropic, community and health-related online fundraising activities worldwide contributed to the jump of 160% in revenue on donation-based crowdfunding platforms between 2019 and 2020 [5, 6]. The project creators of donation-based crowdfunding can be individuals or organizations who are the direct beneficiaries of funds raised from the campaign or third parties motivated by philanthropic or civic causes. Donors who support these charitable crowdfunding campaigns mostly identified with the community; they are sensitive to community needs or responding to appeals from their social networks [7–11].

Donation-based crowdfunding democratized giving as almost everyone can participate in giving on the worldwide web; individuals can give in any amount they wish to in relation to the charitable causes through a variety of electronic payment modes.

10.1.3 Marketing Challenge in Donation-Based Crowdfunding

A significant challenge with donation-based crowdfunding platforms is the lack of robust and accurate information about the donors who supported and gave to the fundraising campaigns [12]. Unlike offline or physical charitable fundraising campaign, individuals who participated in online giving through a crowdfunding platform are physically unobservable and they need not disclose information about themselves.

To promote charitable giving, many crowdfunding platforms reduce the information required of donors and offer access to new donors without the need for account creation. Crowdfunding platforms also experience high donor attrition. A study that analyzed data from DonorsChoose.org reported that 74% of donors never return to the site after one donation [13]. With little information about their donors' profile, project creators rely on their social influence and social network to motivate prosocial giving.

To address this challenge, our research examined the transaction data of the world's only donation-based crowdfunding platform in Asia to profile donors from their online giving behavior and donation preferences. Like other donation-based crowdfunding platforms, there is no or minimal descriptive information about the online donors for customer segmentation to cultivate loyalty or increase retention.

Using clustering tools, we combined observed heterogeneity in donation behaviors with unobserved heterogeneous attributes in donor campaign preferences to identify distinct donor clusters that provide a systematic analysis for donor engagement. Extant literature usually adopted a single clustering algorithm or at most compare two clustering tools. These attempts disregarded many alternative clustering algorithms whose performance have not been explored. Our solution comprehensively compared the results of different clustering algorithms including K-means clustering, agglomerative, mean-shift, DBSCAN, OPTICS, GM and BIRCH, to determine an optimal number of donor clusters. We believe this is the first exploratory analysis in donor clustering for donation-based crowdfunding platforms which will provide a systematic guide to choosing the appropriate clustering algorithm in segmentation analysis.

10.2 Literature Review

10.2.1 Segmentation Analysis Using Socio-Demographic and RFM Metric

Recency, frequency and monetary value (RFM) metrics are widely used in marketing segmentation analysis to identify and target customers and donors based on historical transaction behavior. Recency measures the number of days, weeks or months since the last transaction took place. Frequency refers to how frequently a donor performed a transaction or presented a monetary gift to a non-profit organization. Monetary value totals the overall amount that the donor has contributed in the past.

Used in combination with customer descriptive variables like gender, age, education level and income level, the RFM metric can segment donors, model to identify prospective donors and predict (1) likelihood of donation; (2) value of donation [1, 14, 15]. The RFM metric has also been adopted to analyze donor lifetime value [16]. By computing donor lifetime value, marketers and fundraising managers of non-profit organizations gained insights into the donor segments for relationship management to inform donor engagement and stewardship strategy [17].

Observed donor attributes like demographic and socio-economic variables are often assumed to significantly predict new donors' propensity to give and donors' giving behavior. For example, extant literature identified men with higher educational level to be more generous in prosocial giving than women of similar educational background; elderly women are more generous as donors than women who are younger [18, 19]. Therefore, marketers or fundraising managers of non-profit organizations could be more efficient in meeting their fundraising goals by targeting donor segments of specific socio-demographic profile.

However, this underlying assumption that donors of similar observable profile behave similarly warrants further investigation. Specifically, customer segmentation based on socio-demographic variables has theoretical and practical limitations. We argue that demographic and socio-economic profiles only partially explain donor heterogeneity since there are unobserved donor heterogeneity in the data.

10.2.2 Theoretical and Practical Limitations

In segmentation analysis, donor demographic and socio-economic variables are combined with their transaction behaviors like RFM variables. This assumes that the sources of heterogeneity in donor behaviors are fully observed and explained by donor demographic and socio-economic attributes. This assumption can lead to sub-optimal marketing strategies in instances when we attribute the behavior of high donation frequency to higher income level, incremental donation value with a certain gender group.

Descriptive data like demographic and socio-economic information are observable heterogeneity; there are unobserved heterogeneity in customer data. By unobserved donor heterogeneity, we refer to systematic but not observed or unobserved, differences between donors and among donor segments that influence their donation behaviors.

Several factors are associated with unobserved heterogeneity. The donor heterogeneity is not observed because the marketer or analyst had not identified the effect of such an attribute in explaining donor behaviors; or not observed because such attributes are not measurable, hard to measure or too costly to collect and measure. For instance, moral identity and personality traits might explain an individual's propensity to donate in response to donation solicitations. Life experiences also shaped ones' propensity in prosocial giving such as one who donates regularly to causes in support of medical research in muscular dystrophy due to the loss of a loved one to the disease. Such unobserved differences among donors are hard to measure and quantify.

The second is a practical limitation. In practice, it is hard and costly to collect accurate demographic and socio-economic information from donors. Most online users would prefer not to disclose their personal information including data related to their demographic, socio-economic and psychographic profile. Hence, online platforms faced a real obstacle in collecting genuine insights into their user profile. Unlike face-to-face charitable fundraising events like a charity gala dinner, where the respective donor's identity is registered and known to the fundraising charity organization, online users need not disclose their personal information when giving to a charitable cause.

Online donors may prefer to maintain privacy and confidentiality. Alternatively, they may be concerned with data security and risks of data breach. In certain instances, online donors may wish to disclose their information for tax deduction purposes to obtain the donation receipts from crowdfunding platforms for file and submission to the relevant tax authorities. Irrespective, online donors have the option of being excluded from the database for customer relationship management since the provision of an individual's personal information is irrelevant to the fundraising campaign.

Should a donation-based crowdfunding platform or charitable organization make the data fields compulsory for donors to complete, some may furnish dummy data and it would be impossible to verify the data integrity. Thus, the analysis of online donors derived from unverified input data may no longer be reliable. There will be some online donors who prefer to give anonymously for various reasons and the donation platform often respect the donors' choice, where required request for donor information minimally satisfies regulatory and compliance purposes. There is legislation in certain jurisdictions to protect the privacy and safety of personal data online. In 2021, United Nations Conference on Trade and Development reported that 71% of 194 countries worldwide have implemented legislation to secure the protection of data and privacy. Thus, donation-based crowdfunding platforms and charity organizations operating in these countries or regions would need to implement robust data governance and cyber security systems and infrastructure when collecting critical personal information.

Given both theoretical and practical limitations, we propose an alternative model using RFM metric and preference; the 'preference' (P) pillar represents a donor's campaign preference represented by the campaign's target beneficiary.

10.3 Empirical Application and Methodology

10.3.1 Data and Model Specification

We obtained the data from the world's only donation-based crowdfunding charity for vulnerable and marginalized persons in Asia. This donation-based crowdfunding platform is founded in 2012 and based in Singapore.[1] As a registered charity, the platform does not charge donors a fee for giving to campaigns nor charge beneficiaries for funds disbursed. All donations raised from crowdfunding campaigns are transferred to the respective target beneficiaries. Project creators are not campaign

[1] The authors are grateful to the crowdfunding charity for providing the data used in this empirical application. Both authors received no financial support for the research, authorship and/or publication of this article.

beneficiaries but they are professionally trained social workers employed by of this crowdfunding charity, unlike other donation-based crowdfunding platforms.

The target beneficiaries of this crowdfunding charity include migrant workers, persons with life-limiting conditions, single mothers, families with sole breadwinner who had suffered incapacitation or sudden death. An example of a campaign for a migrant worker would be one who was hired and paid by the hour at the construction site but lost his sole source of income due to work injury; while awaiting repatriation to his home country, he needed funds to be disbursed to his family in urgent need.

Before the launch of a campaign, project creators conduct home visits and background verification of prospective beneficiaries to derive the fundraising goals. These beneficiaries are not eligible for other forms of state assistance and support or are in need of urgent short-term financial assistance. A family with young children may be in desperate need of financial assistance due to an untimely incapacitation of the family's sole breadwinner, while the surviving parent to seek employment.

With the information collected from home visit, checks and verification, the project creator prepares the campaign narrative, sets the campaign duration and launches the campaign on the charity's crowdfunding platform. Each campaign solicits donations from the wider public by social media posts or by email distribution to registered donors. Excerpts of each campaign are posted on the charity's social media channels and online users are channeled to the crowdfunding platform from these posts.

Individuals need not register with this donation-based crowdfunding charity to participate in their fundraising campaigns. Additionally, this crowdfunding charity does not collect demographic or socio-economic information from their donors. Recent attempts to survey existing donors for information about their socio-demographic profile did not return adequate response for statistical analysis.

We analyzed 41,083 recorded donations on the focal crowdfunding platform completed by 17,720 unique donors between January 1, 2020 and May 31, 2022. Donors included those who registered their email addresses as well as those who are not registered with the crowdfunding charity. We analyzed 466 campaigns equivalent to $2,224,122.32 in donation value. Table 10.1 gives a breakdown of these campaigns by beneficiary types or donor campaign preferences.

10.3.2 Model Development

In Fig. 10.1, we illustrate our proposed machine learning-based approach using RFM and P variables or 'RFMP' framework, in a flowchart that would support marketing decision-making.

The available input data included: (1) donor transaction data (e.g., donation date, donation amount, donation frequency) in RFM metrics, and (2) campaign preferences or beneficiary types (e.g., campaigns each donor had participated). The donor transaction data represented donor behavior patterns on the crowdfunding platform from which we derive the R, F and M variables while we drew insights into the

Table 10.1 Count of campaigns by beneficiary types or donor campaign preferences

Campaign theme	Count
Migrant workers	112
Single mothers	70
Economically disadvantaged	63
Caregivers	50
Formerly incarcerated	40
Persons with life-limiting/terminal illnesses	51
Sole breadwinner medically unfit for work/incapacitated	18
Children	16
Youth	11
Elderly	10
NGO	10
Miscellaneous including persons stranded overseas, homeless, freelance artistes	6
Foreign student	3
Overseas mission	2
Health care workers	2
Athletes	2

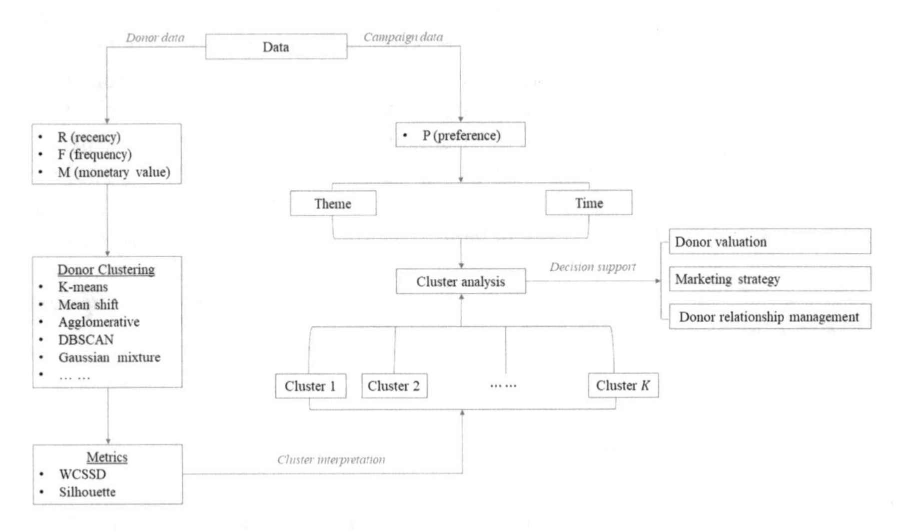

Fig. 10.1 RFMP Framework: A systematic analysis to segmentation analysis

donor's campaign preferences from the campaign categories and duration ('time'). The donor behavioral patterns are further analyzed and segmented into clusters using clustering algorithms.

We used both quantitative metrics and qualitative interpretations to evaluate the clustering performance. In the campaign details, each donor's campaign preferences contains two dimension, namely the campaign category (by beneficiary type like migrant worker, children, single mothers, etc.) and duration of campaigns. Analyses of donor's preferences were integrated with output from clustering algorithm. This complements the unobserved customer heterogeneity which are unobserved factors associated with donor behavior.

Following this model estimation, we develop a comprehensive understanding of the donor behavioral patterns (from RFM variables) as well as their respective preferences from the campaign categories (or target beneficiaries) the donor had supported. The resultant donor clusters would support marketing decision-making like donor valuation, marketing strategy and donor relationship management. We discuss the details about the proposed model in the following subsections.

10.3.3 Model Estimation Using RFMP Framework

Our estimated model clustered donors on four pillars in the RFMP framework:

- Recency: Time lapse (in days) since the last most recent donation made by the donor. It indicated individual donor's state of activity. The recency value (in days) is computed relative to the benchmark date June 26, 2022 when the data was initially ready for analysis. A long recency suggested a historical distant date from the benchmark date, while a short recency referred to a data nearer to the benchmark date.
- Frequency: Number of times a donor participated in any of the 466 campaigns launched by the crowdfunding charity. It indicated how frequently a donor participated in a crowdfunding campaign during the period of analysis. A higher frequency suggested a higher number of donation activity relative to a lower frequency figure.
- Monetary value: Total donation amount by a donor across all crowdfunding campaigns expressed in the domestic currency where the crowdfunding charity is based.
- Preference: Contains two dimensions. One dimension measured the campaign categories which is the beneficiary type whom the donors had supported. The second dimension measured the time axis of donation transactions made by the donors, i.e., date of the donation.

The RFMP framework reviewed online donation behaviors and preferences directly instead of relying on individual profile data akin to the analysis of customer preferences based on product purchased on e-commerce platforms.

We introduced pillar 'P' for preference which described the campaign for a beneficiary type. Extant literature found that empathic concerns positively impact one's donation decision [20, 21]. For instance, parents of young children would be more empathetic toward kids compared to adults without. This reflected a donor's motivation for prosocial giving, and explained the reasons for empathy [22]. The second dimension is the time axis. Studies have shown that donation behaviors can be related to seasonal effect; Christmas or year-end festivities tended to record higher donations [23]. Another example of the seasonal effect was the COVID-19 pandemic in 2020 [5, 6]. In some countries, charitable donations increased just before the end of tax paying season since donations could be used for tax rebates [24].

In the RFMP framework, the pillars, namely 'R' and 'F' are time-based. In contrast, a donor's demographic and socio-economic data can hardly capture and explain time-based behaviors. In fact, the proposed 'P' pillar which contained time preference may serve as a promising integration to the existing RFM framework.

'P' is more robust and reliable as it can be considered a collective reflection of a donor's behavior shaped by the donor's demographic profile and socio-economic status, personality traits, reasons for empathy, seasonal effect and possibly other unobserved factors associated with prosocial giving.

From a practical standpoint, data for the 'P' pillar is automatically recorded with each successful donation. This is unintentional data and easily available, which suggests that our proposed RFMP framework is applicable to all donation-based crowdfunding platforms without data restriction.

In summary, the proposed RFMP framework adopts actual transaction behavior data in return for a more robust and reliable results from segmentation analysis. With the adoption of unintentional data, our proposed model would benefit organizations that have yet access to their online users' profile. This RFMP framework could generate a wide range of applications given the surge in abundant data with digital transformation.

10.4 Model Performance and Result Analysis

10.4.1 Donor Segmentation

Machine learning-based clustering models are increasingly popular in the recent years as the model can learn to identify patterns on their own. This is known as an unsupervised learning algorithm as the classifiers are not scored or labeled. For instance, there are studies which explores the use of K-means clustering [25, 26], agglomerative clustering and Gaussian mixture clustering [27].

Our proposed clustering algorithm for donor segmentation aims to group donors with high degree of similarity by RFMP behavioral patterns into the same cluster, while distinguishing donors with low degree of similarity into distinct clusters. The goal is to understand donor behavior from hidden patterns across distinct clusters of donors, that would which facilitate the design and implementation of tailored marketing strategies and relationship management.

Suppose a data sample of n observations are given as $\boldsymbol{X} = X_1, \ldots, X_i, \ldots, X_n$, where each $\boldsymbol{X}_i = x_{i1}, \ldots, x_{\text{ij}}, \ldots, x_{\text{iq}}$, is a vector of dimension q. Denote the sets containing the observations in each cluster as $C_1, C_2, \ldots, C_K$, which need to satisfy the following two conditions: $(1) C_1 \cup C_2 \cup \cdots \cup C_K = X$; $(2) C_k \cap C_{k'} = \emptyset$.

The within-cluster sum of squared distance (WCSSD) is defined by

$$W(C_k) = \frac{1}{|C_k|} \sum_{X_i, X_{i'} \in C_k} \sum_{j=1}^{p} \left(x_{\text{ij}} - x_{i'j}\right)^2, \tag{10.1}$$

where $\frac{1}{|C_k|}$ represents the number of observations in cluster k. In words, Eq. (10.1) computes the sum of squared Euclidean distances between all pairwise observations in the kth cluster, divided by the size of the cluster. The key idea of donor clustering is that a good clustering is achieved when the total WCSSD across all clusters is minimized, i.e., the donor segmentation analysis is defined by solving the optimization problem.

$$\min_{C_1, \ldots, C_K} \left\{ \sum_{k=1}^{K} \frac{1}{|C_k|} \sum_{X_i, X_{i'} \in C_k} \sum_{j=1}^{p} \left(x_{\text{ij}} - x_{i'j}\right)^2 \right\}. \tag{10.2}$$

As can be seen from the mathematical formulation of this algorithm, it works generally well for convex and isotropic clusters and works poorly for elongated clusters. It also requires some data preprocessing like normalization, dimension reduction, to achieve better performance.

Evaluating the performance of a clustering algorithm is not as simple as counting the error rate in prediction. On the one hand, the knowledge of ground truth labels is almost never known in practice or requires manual assignment by analysts or human annotators [25] which usually rely on the customer observable attributes like demographic profile. On the other hand, an evaluation metric should not take the absolute values of the cluster labels into account. Instead, the clustering should define separations of the data similar to some ground truth set of classes or satisfying some assumption such that members belonging to the same class are more similar than members of different classes following some similarity metric. In view of this, three relevant metrics are identified to compare across clustering algorithms to determine the optimal clusters.

The silhouette coefficient is a measure of similarity within the same cluster (cohesion) and the dissimilarity across different clusters (separation). It is computed based on two scores, a and b. For any data point $X_i \in C_k$ (cluster k), the mean distance between X_i and all other data points within the same cluster is computed as

$$a_i = \frac{1}{|C_k| - 1} \sum_{X_j \in C_k, X_i \neq X_j} d\left(X_i, X_j\right), \tag{10.3}$$

where $d(X_i, X_j)$ is the distance between two data points.

The mean dissimilarity is defined as the smallest mean distance from a data point X_i to all points in other cluster $C_{k'}(C_k \neq C_{k'})$. That is

$$b_i = \min_{k \neq k'} \frac{1}{|C_{k'}|} \sum_{X_j \in C_{k'}} d(X_i, X_j). \quad (10.4)$$

The silhouette value of data point X_i is then computed as

$$s_i = \frac{b_i - a_i}{\max\{a_i, b_i\}}, \quad (10.5)$$

and the silhouette value of the entire data set is the mean value of s_i over all points.

10.4.2 Segmentation Analysis

Following the clustering results from segmentation analysis, the optimal number of clusters chosen is four, namely cluster 1, cluster 2, cluster 3 and cluster 4. Table 10.2 gives the key attributes of all four identified clusters and Fig. 10.2 visualizes the distribution of RFM values in each identified cluster in a bubble plot.

Figures 10.3 and 10.4 present two axes of donors' preferences, namely the campaign preferences and the time of donation during the period of analysis of each donor cluster. Figure 10.3 illustrates the distribution of preferred campaign categories by each donor cluster or the campaign categories that each cluster had supported by proportion. Figure 10.4 charts the distribution of donation activities over the months during the period of analysis.

On the crowdfunding platform, donors may choose to support campaigns for individual beneficiaries or groups of beneficiaries. For example, a campaign to crowdfund for a migrant worker would be classified as an individual beneficiary while one that crowdfunds for groups of migrant workers would be considered 'group'. Figure 10.5 is a diagrammatic representation of the number of 'individual' or 'group' campaigns supported by each donor cluster.

The next section analyzes each cluster's behaviors holistically.

Table 10.2 Attributes of different clusters

Cluster	Cluster size	R (in days)	F (per year)	M (in $)	Total donation ($)
1	674 (3.80%)	388.9	4.04	652.93	741,462.56
2	6203 (35.01%)	84.08	1.07	84.34	523,187.28
3	3908 (22.05%)	315.02	1.18	132.91	519,398.88
4	6935 (39.14%)	684.46	1.16	106.92	440,073.60

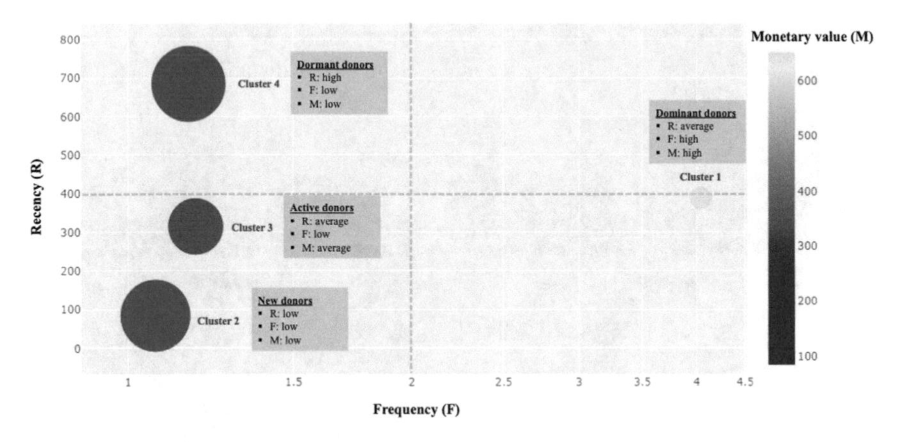

Fig. 10.2 Bubble plot of clusters. The optimal number of clusters chosen is 4, namely dominant, new, active and dormant donor clusters

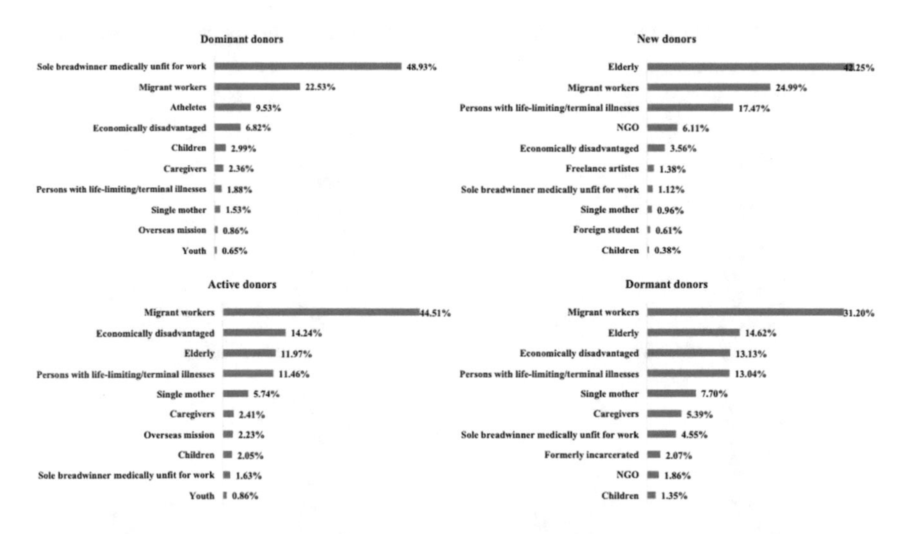

Fig. 10.3 Distribution of donor preferred campaign themes within each cluster

Cluster 1: Dominant donors. Cluster 1 is characterized by 'medium-to-long' in recency (R), high in donation frequency (F) and monetary value (M) following the RFM variables. We labeled this cluster of individuals—'dominant' donors. The recency value of cluster 1 averaged 388.9 days; this suggests that the last donation by a 'dominant' donor occurred more than a year ago from the benchmark date of June 26, 2022.

The donation frequency for cluster 1 averaged 4.04 or more than 4 donations annually; the highest among all four clusters. Each donor in the cluster donated $652.93 on average; representing 4–7 times higher than clusters 2–4. From Table 3, cluster 1 constituted the smallest proportion of donors (674 donors or 3.80%

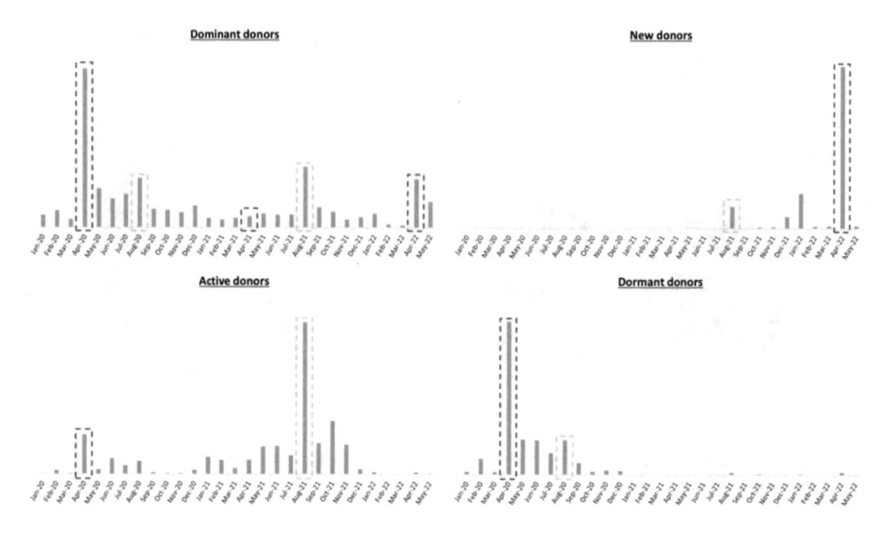

Fig. 10.4 Distribution of donation activities peaked in different months for each donor cluster

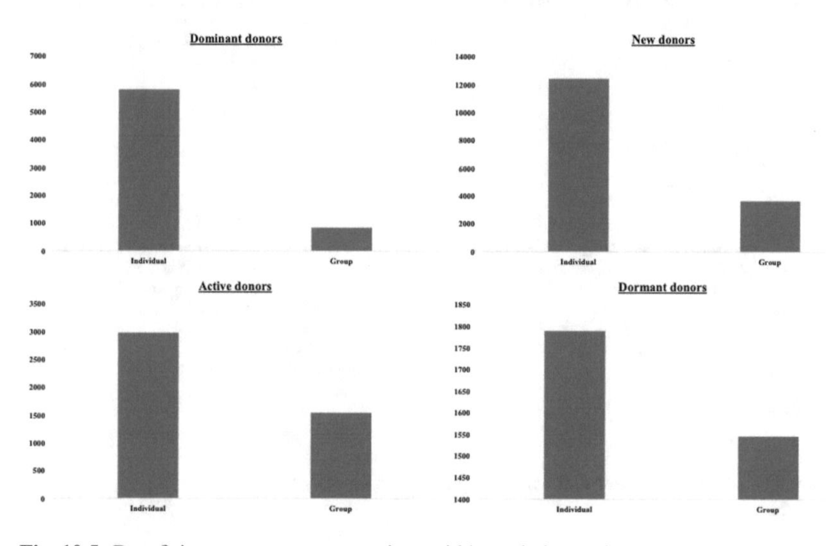

Fig. 10.5 Beneficiary types across campaigns within each donor cluster

of all donors). Although the smallest, this cluster contributed the largest amount of donations or one third of total funds raised on the crowdfunding platform in the period under analysis. Cluster 1 is therefore the most valuable donor segment and should be accorded top priority for engagement and donor stewardship by the crowdfunding charity.

Given the medium-to-long recency value of cluster 1, this crowdfunding charity is at risk of losing donors from this cluster. The donor stewardship strategy should aim at reducing the recency value. Hence, the campaign preferences of cluster 1 donors are helpful. These donors can be apprised of most recent crowdfunding activities especially those aligned to their preferences.

As evident from Fig. 10.3, cluster 1 donors preferred specific categories of crowdfunding campaigns, the top 3 preferred ones are: 'Sole breadwinner medically unfit for work' (48.93%), 'migrant workers' (22.53%) and 'athletes' (9.53%). Relative to other clusters, cluster 1 donors gave the least to groups of beneficiaries (Fig. 10.5).

We further analyzed the donation activities of cluster 1 and observed an even distribution of donation activities across all months as shown in Fig. 10.4. This explains the high frequency and 'medium-to-long' in recency behavior. From Fig. 10.4, the donation activities peaked in April 2020 and August 2021. The first peak corresponds to the commencement of lockdown measures that occurred during the pandemic which impacted vulnerable persons in the society. The second peak in August 2021 corresponds with campaigns that appealed to this donor cluster. More seasonality tests can be conducted with a longer period of data to draw conclusions from these observations.

Donors from cluster 1 gave throughout the year; an ongoing engagement would improve donor retention. Other objectives of engagement with cluster 1 donors may include updates of beneficiaries whom they had supported after campaign had closed.

Cluster 2: New donors. We label cluster 2 'new donors'. The average recency of this cluster is 84.08 days; this is the shortest relative to the other 3 clusters. This average recency value translates to a donation in the recent 2–3 months. The average frequency of donation in a year is 1.07 times. From the distribution of donation activities over the period of analysis, the activities of this cluster started only in August 2021.

This cluster of first-time or new donors to the crowdfunding charity has the lowest number of donations in a year. Besides donation frequency, the value averaged $84.34 was the smallest relative to the remaining three clusters or equivalent to one-eighth of average monetary value contributed by cluster 1. The total monetary value of donation from this cluster at $523,187.28 was under a quarter (or 23.52%) of total funds raised on the crowdfunding platform in the period under analysis. Following the RFM variables, the behavior of this 'new donor' cluster is low on all three RFM variables.

Irrespective, cluster 2 donors represented 39% of all donors. The crowdfunding charity was effective in attracting new donors to its platform. More data and further analysis are required to monitor and determine if these new donors were converted into recurring donors at a later stage. This is also the cluster at a high risk of attrition or remaining as only one-time or dormant donors. For the crowdfunding charity, the goal would be to convert these new donors to recurring ones like cluster 3.

From Fig. 10.3, almost three quarters or 74.22% of donors in cluster 2 had supported campaigns related to 'elderly' (42.25%), 'migrant workers' 24.99%, followed by 'persons with life-limiting/terminal illnesses' (17.47%).

Being new to the platform, there is not enough data about this cluster to infer any seasonal effect. Regardless the size of this cluster portends, the need for the crowdfunding charity to engage with new and first-time donors to drive recency and increase frequency of giving such as introducing impact stories of beneficiaries, launch of campaigns aligned with donor's preferences.

Cluster 3: Active donors. Cluster 3 referred to as 'active' donors, consisted of mostly individuals who have donated in the last 12 months. The average recency value of cluster 3 is 315.02 days, longer than the recency value of new donors. The mean frequency of cluster 3 at 1.18 suggests more than one donation in a year.

Donors in cluster 3 are recurring givers; the average monetary value of donation was $132.91, 58% higher than the new donors. Collectively, this cluster contributed $519,398.88 in donation value. By increasing the frequency and monetary value of donation, cluster 3 can be converted into 'dominant' donors like cluster 1. The goal of engagement for cluster 3 would be to keep them actively engaged with campaigns that they prefer. These 'active' donors have a preference for campaigns related to 'migrant workers' (44.51%), the 'economically disadvantaged' (14.24%) and the 'elderly'.

Over the period of analysis, we observed a significant peak in donation activities of cluster 3 donors in August 2021, but not in August 2020. This may coincide with campaigns launched in the period that are aligned with their preferences. For example, migrant workers were affected during the pandemic causing many to be medically unfit for work and suffered a loss of income. In response, this crowdfunding charity had intervened with crowdfunding campaigns to provide financial assistance to migrant workers who were impacted and hence the increased participation from cluster 3 donors.

Cluster 4: Dormant donors. The fourth cluster is the largest, representing 39.14% of total donors but the smallest in total donation value. The average recency value is the longest at 684.46 days—the most recent donation made by cluster 4 donors occurred more than one and a half years ago. The average frequency value is 1.16 times, or more than a donation within a year. Each donation averaged $106.92 or collectively $440,073.60 for the period under analysis. From RFM variables, long recency, low frequency and low monetary value, we labeled cluster 4 'dormant' donors.

There was minimal donation activity from cluster 4 donors since December 2020. Although dormant, cluster 4 donors can be converted into active ones by shortening the recency value and increasing the frequency of giving by engaging them with campaigns according to their preferences. As shown in Fig. 10.3, the donation preferences for cluster 4 donors were mostly for 'migrant workers' (31.20%), the 'elderly' (14.62%) and 'economically disadvantaged' (13.13%). Donors are motivated when they learned of how their donation had helped the beneficiaries. Such impact stories can be created for donors after the campaign has ended.

10.4.3 Campaign-Wise Analysis

We shortlisted the top eight campaign categories that represented more than 90% of campaigns analyzed and conducted a campaign-wise analysis by investigating into the donor clusters within each campaign category. These categories represented the target beneficiary of the respective campaign (refer Table 10.1).

Figure 10.6 presents four metrics from the donation activities in each campaign, namely campaign duration by days, proportion of actual funds raised against target campaign goal in percent, average amount of funds raised per day throughout the duration of each campaign and donor composition (in percentage) across each donor cluster.

In the first metric, the frequency distribution chart visualized the duration of each campaign in the same category from start to end in days. A campaign closes when the campaign goal is achieved, and this can occur before the scheduled campaign end date. Alternatively, a campaign would end when it no longer received any attention from prospective donors on the crowdfunding platform. A shorter duration suggested that a campaign attracted more donation and had achieved the campaign goal in a shorter time than those with longer duration. From the histograms, campaigns for 'migrant workers' lasted between 90 and 150 days (or 3–5 months). There were campaigns of under 90 days duration but few campaigns like those for 'children', lasted more than 180 days.

However, this metric on campaign duration can be biased as campaigns differ in scale by target donation value. A campaign of a low donation target should be more attainable within a short duration compared to another with a higher donation target, although there may be other drivers that impact the speed of campaign goal attainment.

Hence, the second and third metric are computed to mitigate the potential bias. The second metric is the fundraising percentage, computed as a ratio: $\frac{\text{actual amount of donation raised}}{\text{targeted amount of donation needed}} \times 100\%$. This measures the level of success of a campaign adjusting for its scale. The charity was successful in achieving most campaign goals since across all campaigns, and the fundraising percentage were mostly 100%.

The fundraising percentage exceeded 100% for a few reasons. By design, the project creators had to close their respective campaigns on the platform and might delay closing unintentionally after the campaign target was met. Another reason pertained to the monetary value of the last donation received for the campaign. The value of the last donation may be large causing the accumulative donation to exceed the campaign goal. The variation in the speed for attaining campaign goals was wide and a large portion of 'formerly incarcerated' campaigns failed to achieve the target campaign goal.

The third metric $\frac{\text{donation amount raised}}{\text{campaign duration}}$ computes the average donation collected per day on the crowdfunding charity platform. This standardized rate of donation is also a measure of fundraising speed on the crowdfunding platform. From the box plot, daily donation on this donation-based crowdfunding platform ranged from \$1 to about \$250. The small value is consistent with our understanding of donation-based crowdfunding platforms as crowdfunding democratized prosocial giving and donors give at discretion without the pressure to conform to social norms [11]. In the box plot of Fig. 10.6, the median value of campaign for the 'formerly incarcerated' at \$9.95 per day is the lowest compared to the rest of the campaign themes. The median

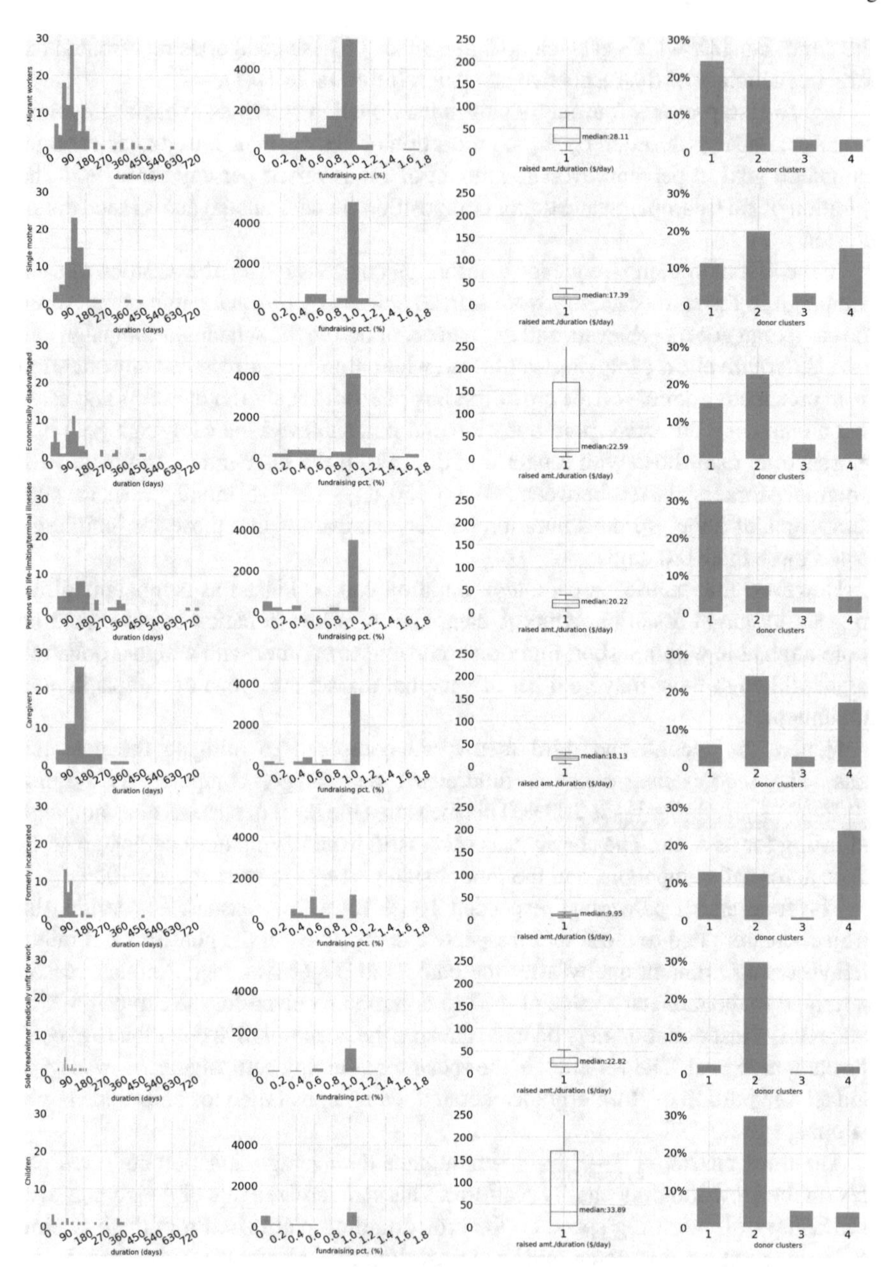

Fig. 10.6 Campaign-wise analysis of the top campaigns

donation value of campaigns for 'children' was the highest at $33.89 followed by 'migrant workers' at $28.11.

We then plotted the distribution of donor clusters by campaign preferences. Accordingly, 'dominant' donors dominated five of the 8 campaign categories, namely 'single mother', 'economically disadvantaged', 'caregivers', 'sole breadwinner medically unfit for work' and 'children'. The 'active' donors were the key donors for campaigns that benefitted 'migrant workers' and 'persons with life-limiting/terminal illnesses. The 'formerly incarcerated' drew most support from 'dormant' donors and 'new' donors represented the lowest proportion of donors across all campaign categories.

10.5 Conclusion and Future Research

We have developed a RFMP framework to parsimoniously segment customers into meaningful clusters to inform marketing decision-making for a crowdfunding charity, overcoming the problem with missing descriptive information about customers. Customer segmentation analysis typically uses customer provided data to inform customer relationship management strategy. In this project, we leveraged existing unintentional data on the donation-based crowdfunding platform to segment the online donors.

We further illustrated the benefits of clustering using the RFMP framework for managerial decision-making in developing marketing strategy, increasing donor lifetime value and donor relationship management. From four distinct clusters, we identified that the 'dominant' donors and the 'dormant' cluster, which is the largest in size among all 4 clusters, are at risk of attrition. This is consistent with extant literature where donor attrition rates on crowdfunding sites were significant [13]. The campaign-wise analysis performed offered insights into the campaign preferences of each donor cluster which the crowdfunding charity can use to promote retention and repeat donation.

The structure of the RFMP framework is scalable that allows for the use of real-time data feed from the crowdfunding platform to present managers with cluster performance on demand. By leveraging on donors' campaign preferences, the focal crowdfunding charity could launch campaigns align with the target donors' preferences to improve the efficiency and effectiveness of online fundraising.

With the rapid pace of digital transformation sweeping across industry sectors, firms can better leverage the data assets that are produced in abundance and unintentionally for customer insights. There is value to be created from the analysis of abundant available data generated online as customers grew accustomed to transacting in virtually.

While our findings using our RFMP framework for segmentation analysis highlights the application of machine learning to marketing, we acknowledge the limitations of our study. Firstly, our empirical application of the RFMP framework is confined to the focal donation-based crowdfunding charity. Our results are not yet generalizable to the worldwide population of online charitable donations although the outcomes contribute to the knowledge of applying machine learning to marketing.

Second, we described the application of our estimated RFMP model framework in donor clustering and infer the donor behavior from their historical campaign preferences. The estimated model, however, does not answer why different donor clusters have different preferences. For example, our results indicated that 'dominant' donors preferred campaigns for 'caregivers', while 'active' donors preferred campaigns that benefited 'migrant workers'. Our main goal of this project is to address the problem with the lack of customer information for marketing decision-making. Hence, the model we developed is descriptive and not yet predictive.

In this project, we also had not integrated data from other contact channels like email management and social media to investigate into each donor due to operational limitations in the focal firm. Nevertheless, future studies can integrate unstructured data across different channels for a more holistic understanding of donors by cluster.

Future research can additionally leverage our proposed RFMP framework to predict donor behavior toward the promotion of campaign categories based on their historical preferences. Our RFMP framework is computationally feasible to be packaged for application in different crowdfunding platforms. Moreover, a simulation model following the variables in RFMP can be developed that would quantify and predict the effect on overall campaign goal—target versus actual. Hence, enabling more individuals to contribute to campaign causes and empowering a more organizations in the community to effectively support needs of vulnerable and marginalized persons.

Our project contributes to the application of machine learning in solving marketing problems in a donation-based crowdfunding charity. The advent of computing power will fuel the applications of machine learning algorithms to overcome problems related to large-scale unstructured data generated from digital transformation today. Future application of deep learning models can examine the use of natural language processing like Bidirectional Encoder Representations from Transformers (BERT), to model campaign narratives that would multiply positive outcomes and improve crowdfunding efficiencies. Text, image, audio and video files that accompany the campaign narratives in crowdfunding platforms can be integrated to improve the interactions between project creators or beneficiaries and donors that would optimize crowdfunding outcomes.

It is our hope that this research can encourage more studies into the use of machine learning in marketing that can advance evidence-based strategies particularly in addressing the needs of the underserved and underprivileged.

References

1. Fader, P.S., Hardie, B.G.S., Lee, K.L.: RFM and CLV: using iso-value curves for customer base analysis. J. Mark. Res. **42**(4), 415–430 (2005)
2. Kumar, V.: A theory of customer valuation: concepts, metrics, strategy, and implementation. J. Mark. **82**(1), 1–19 (2018)
3. Statista https://www.statista.com/statistics/1078273/global-crowdfunding-market-size/. Last accessed 22 Aug 2022
4. Golder, P., Mitra, D.: Handbook of Research on New Product Development. Edward Elgar Publishing, US (2018)
5. McKitrick, M., Schuurman, N., Crooks, V.A., Snyder, J.: Spatial and temporal patterns in Canadian COVID-19 crowdfunding campaigns. PLoS ONE **16**(8), e0256204 (2021)
6. Cambridge Judge Business School https://www.jbs.cam.ac.uk/faculty-research/centres/alternative-finance/publications/the-2nd-global-alternative-finance-market-benchmarking-report/. Last accessed 18 Jun 2022
7. Sargeant, A.: Charitable giving: towards a model of donor behaviour. J. Mark. Manag. **15**(4), 215–238 (1999)
8. Dunn, E.W., Aknin, L.B., Norton, M.I.: Spending money on others promotes happiness. Science **319**(5870), 1687–1688 (2008)
9. Aaker, J.L., Akutsu, S.: Why do people give? the role of identity in giving. J. Consum. Psychol. **19**(3), 267–270 (2009)
10. Smith, S., Windmeijer, F., Wright, E.: Peer effects in charitable giving: evidence from the (running) field. Econ. J. **125**(585), 1053–1071 (2015)
11. Saxton, G.D., Wang, L.: The social network effect: the determinants of giving through social media. Nonprofit Volunt. Sect. Q. **43**(5), 850–868 (2014)
12. Burtch, G., Ghose, A., Wattal, S.: The hidden cost of accommodating crowdfunder privacy preferences: a randomized field experiment. Manage. Sci. **61**(5), 949–962 (2015)
13. Althoff, T., Leskovec, J.: Donor retention in online crowdfunding communities: a case study of donorschoose.org. In: Proceedings of the 24th International Conference on World Wide Web, pp. 34–44. Italy (2015)
14. Alet Vilaginés, J.: Predicting customer behavior with Activation Loyalty per Period. From RFM to RFMAP. ESIC Market Econ. Bus. J., **51**(3), 609–637 (2020)
15. Heldt, R., Silveira, C.S., Luce, F.B.: Predicting customer value per product: from RFM to RFM/P. J. Bus. Res. **127**, 444–453 (2021)
16. Durango-Cohen, E.J., Torres, R.L., Durango-Cohen, P.L.: Donor segmentation: when summary statistics don't tell the whole story. J. Interact. Mark. **27**(3), 172–184 (2013)
17. Bennett, R.: Predicting the lifetime durations of donors to charities. J. Nonprofit Public Sect. Mark. **15**(1–2), 45–67 (2006)
18. Bekkers, R., Wiepking, P.: A literature review of empirical studies of philanthropy: eight mechanisms that drive charitable giving. Nonprofit Volunt. Sect. Q. **40**(5), 924–973 (2011)
19. Einolf, C.J.: Gender differences in the correlates of volunteering and charitable giving. Nonprofit Volunt. Sect. Q. **40**(6), 1092–1112 (2011)
20. Lee, Y.K., Chang, C.T.: Who gives what to charity? characteristics affecting donation behavior. Soc. Behav. Personal. Int. J. **35**(9), 1173–1180 (2007)
21. Verhaert, G.A., Van den Poel, D.: Empathy as added value in predicting donation behavior. J. Bus. Res. **64**(12), 1288–1295 (2011)
22. Sisco, M.R., Weber, E.U.: Examining charitable giving in real-world online donations. Nat. Commun. **10**(1), 3968 (2019)
23. Ekström, M.: Seasonal altruism: how Christmas shapes unsolicited charitable giving. J. Econ. Behav. Organ. **153**, 177–193 (2018)
24. Rees-Jones, A., Taubinsky, D.: Tax Psychology and the Timing of Charitable-Giving Deadlines. Tax Policy and Charities Initiative (2016), https://repository.upenn.edu/fnce_papers/147. Last accessed 17 Mar 2022

25. Hosseini, M., Shabani, M.: New approach to customer segmentation based on changes in customer value. J. Market. Anal. **3**(3), 110–121 (2015)
26. Cheng, C.H., Chen, Y.S.: Classifying the segmentation of customer value via RFM model and RS theory. Expert Syst. Appl. **36**(3), 4176–4184 (2009)
27. Teuling, N.D., Pauws, S., van den Heuvel, E.: Clustering of longitudinal data: a tutorial on a variety of approaches. arXiv (2021), http://arxiv.org/abs/2111.05469. Last accessed 18 Sept 2022

Chapter 11
Models of Destination Loyalty at Heritage Sites: *Are We There Yet?*

Simona Mălăescu, Diana Foris, and Tiberiu Foris

Abstract Tourism heritage sites benefited in the last years from a consistent attention from scholars trying to explain when, why and how visitors form a destination loyalty. As a result, numerous models and variables in different roles were proposed. This paper aims at critically review the literature and argue for four directions of action. One argues for a need to explore in-depth the nature of the link between certain variables, and explaining the decision of using in new contexts less dimensions of the same construct tested in previous literature, before proposing a new comprehensive model. A second one pertains to building a volume of empirical data and arguments why previous models or variables does not apply to a certain context. Another one championships the idea of creating a taxonomy of categories inside the context of heritage sites more homogenous in nature, followed by the need to build models operating inside the different types of heritage tourism based on different tourists' satisfied needs, different resources, different motivations to visit and activities proposed at heritage sites. We also bring empirical evidence from visitors of 20 different heritage sites measuring place attachment, destination loyalty and other variables contextual in nature, using correlation and regression analysis to support the theoretical conclusions. Almost half of the variance registered in visitors' destination loyalty was explained by place dependence (as a functional dimension of place attachment), the level of national identification strength and perceived role of reenactment in enhancing patriotic feelings at heritage sites.

S. Mălăescu (✉)
Human Geography and Tourism Department, Babeș-Bolyai University, Cluj-Napoca, Romania
e-mail: simona.malaescu@ubbcluj.ro

D. Foris
Faculty of Food and Tourism, Transilvania University of Brasov, Brasov, Romania
e-mail: diana.foris@unitbv.ro

D. Foris · T. Foris
CiTUR—Centre for Tourism, Research, Development and Innovation, Peniche, Portugal
e-mail: tiberiu.foris@unitbv.ro

T. Foris
Faculty of Economic Sciences and Business Administration, Transilvania University of Brasov, Brasov, Romania

J. L. Reis et al. (eds.), *Marketing and Smart Technologies*, Smart Innovation, Systems and Technologies 337, https://doi.org/10.1007/978-981-19-9099-1_11

11.1 Introduction

Tourism and hospitality have been in recent years and still are in a process of profound transformation with increasing dynamics. In the context of increasing the diversity of services in the tourism industry and the requirements of tourists, continuous transformations are needed in the management of tourist destinations in order to increase their competitive advantage. Tourist Destination Loyalty (DL) received in the last years considerable attention from heritage tourism scholars [1, 2] after decades of research in consumers' loyalty in general, and on the "*sunny side of tourism*" (contexts like beach mass tourism and recreational tourism destinations). The range of research contexts for studying when and how tourists form a loyalty bond to a tourism destination is expending: national parks [3], fishing recreational destinations [4], religious tourism destinations [5, 6], adventure tourism [3], etc. Therefore, the number of variables related to DL is also growing. It is a natural process to adapt existent models or propose new models on how DL forms, with new antecedents, when the tourism resources in the destination or the type of tourism is different, even when the context is different but the resources and the tourism activities proposed in the destinations are similar. A salient number of studies in recent literature propose and test new models. However, the volume of empirical evidence testing and supporting core variables or arguing why certain antecedent variables of DL previously proposed by largely cited studies are not relevant for a certain context, and it is not growing in the same exponential proportion. Moreover, a recent meta-analysis examining the relationship between a largely acknowledged antecedent of DL on tourism destinations—place attachment (PA) concluded: "there is still insufficient evidence to prove the positive impact of local attachment on tourist loyalty" [2]. As meta-analysis usually do, the study took into consideration the statistical indexes (correlation and size effect) and based the conclusions on findings that contradict previous results such as PA has no positive effect on tourist loyalty [7] or the impact or role played by PA on DL, which also differed. Previous studies tested on consumers of beach tourism or skiing resorts, revealed that PA' components are antecedent to DL (mediated or not by tourist satisfaction) [8, 9]. Zou et al. [2] mentioned PA studied as a mediator in Strandberg et al. [10] work, respectively, as moderator [11] and the reverse role of DL: PA having DL as an antecedent (DL significantly and positively affects PA) in Plunkett et al. [12] study. Two decades ago, Chen and Gursoy [13] stressed the necessity that tourist loyalty should be determined according to the type of tourism products. Maybe so it should be when compare and interpret the results. Plunkett et al. [12] tested the previous and the new model on the consumers of a very different space in terms of *place dependence* [8, 14] or *functional attachment* [15] as the capacity of the place under scrutiny to provide conditions and features to support specific activities or goals [16]. They measured the park use frequency and proportion of use in relation to other settings in the case of visitors of eight urban parks in Manhattan Beach [12]. The cognitive component of both PA (*place identity*—PI) and DL (*cognitive loyalty*—CL) had conditions to result in different outcomes on the two different contexts due to different type of space' attributes and different consumer needs fulfilled. It might reveal the most natural argument that the impact of certain variable is relevant for a context but inconclusive for another. The needs that parks usually fulfil for their visitors, viewed by the perspective inspired by the transactional analysis, could be very different from the role played in self-identity forging (reflected by PI usually) by a heritage site for example. It is also different from the context of extreme sporting events on which Raggiotto and Scarpi [17] tested their model, or the context of beach tourism [8] where the authors placed in the model PA as an antecedent of DL and found a positive relation. From the same perspective, also, the context of cultural creative districts in the manufacturing hub (Redtory Art & Design Factory, China), where no positive impact of PA on consumer's loyalty was found [7], creates the premises for different outcomes in many aspects of PA and DL (as PI and CL) compared to the studies above, representing a particular case.

The number of dimensions assessed for PA and consumer loyalty manifested towards a space, in each study should also be considered when interpreting the conclusions of the studies. For the urban park, Plunkett et al. [12] were interested only in *the behavioural loyalty*. The loyalty towards a space usually is conceptualized as a bi- or three-dimensional concept (cognitive, behavioural and affective) being an attitude–as a psychological factor–attitude is conceptualized as a three-fold concept [18, 19]. Therefore, in the Plunkett et al. [12] study the impact of PA in findings should be interpreted as limited to the behavioural dimension of DL. To conclude, when testing and comparing models, it seems to be the right time to consider the research context in terms of type of (tourism) resources, the type of (tourism) activities proposed by the destination and tourism motivation to visit. By growing the corpus of research evidence in a certain tourism context, refine intra-context the variables and their role, we might obtain models that are more robust. Moreover, we argue for the necessity that the growth will be intra-contextual in literature and build on the previous literature bringing more empirical evidence and arguing for the (real) necessity of new models, because the limitless powers that statistics and big data mining bring nowadays to scholars could be luring and difficult to resist sometimes.

Rapid changes in the production and consumption of tourism services, increased competitiveness and the emergence of new tourist destinations, together with all the implications of information technology development, require a new approach to implementing modern management strategies and tools to ensure efficient management and sustainable development at the level of tourist destinations [20]. Although heritage tourism might seem as a homogenous context, in this category, the type of heritage, the emotions experienced by visitors at sites [21], and the personal relationship tourists have with the site, represent at least several factors determining different attitudes and different levels of DL for tourists. The heritage tourism literature mimics, in a smaller scale, the enthusiasm for proposing new models and the number of antecedents for DL is also growing. We argue in favour of refining this list of antecedents and mediators, more testing for the previous models instead of new ones, gather more empiric evidence and statistical support for the core variables of the model and shorten the list of variables related to DL. The methodology used in the recent heritage literature seems also dominated by model-building and testing analysis, especially structural equation modelling.

This paper aims to critically analyse the state of the art of heritage DL literature and bring an empirical argument for the need to explore more in-depth the nature of the link between certain variables in this context, such as PA and DL. It also argues for two aspects. On one hand, it stresses the need for reconsidering the evolving phase in which the research on heritage sites it really is, and the pertinence for multiplying the DL models on this particular context. On the other hand, it argues in favour of the imperative to build models operating inside the different types of heritage tourism, considering the satisfied needs, the resources, the motivation to visit and the activities proposed at heritage sites.

With this purpose in mind, the introduction of the paper will be followed by (1) the analysis of the recent literature on tourists' destination loyalty in terms of the antecedents, moderators and mediators evidenced by the studies and the methodologies used in order to reach their conclusions; (2) the analysis on the construct of DL in tourism literature; (3) the literature review on the construct of PA; (4) the description of the methodology of the study; (5) the results on the empirical evidence from the research field followed by conclusions and implications of the paper.

11.2 Recent Literature of Destination Loyalty on the Context of Heritage Tourism

11.2.1 A Bird-View of Destination Loyalty' Related Variables and Methodologies Used in Heritage Tourism

Heritage tourism literature was interested in factors increasing visitors returning intentions in a destination and studied DL mainly as a dependent variable. Therefore it was evidenced a list of variables connected to the genesis of tourists' DL or its antecedents (Table 11.1). Other studies evidenced also mediators and moderators of the antecedent variables on the DL. Tourist' satisfaction as a key antecedent of loyalty [8, 22] was confirmed also on the heritage context. Hernandez-Rojas et al. [21] studied perceived value, perceived heritage quality as moderators of visitor satisfaction and perceived cultural quality as moderator of visitors' satisfaction and DL.

The most popular variable connected with DL was visitor satisfaction [30, 40–42] tested in different roles, on different contexts and even found as having less impact [33]. The other antecedents considered and tested in more than one study were: perceived quality [30] as a mediator [23], perceived heritage quality as a moderator [21], perceived cultural quality as a moderator [21], perceived value [30] as a mediator [28], respectively, moderator [21], destination's brand [23, 28, 32], authenticity perception [34, 42], cultural identification with the destination [34, 39]. The majority of the studies took under consideration models with 3–5 variables or

Table 11.1 The antecedents and factors linked to destination loyalty in heritage tourism recent literature

Antecedents	Moderator/mediator variable/role	Context	Authors	Analysis method
Destination image Perceived health safety	Perceived quality–mediator (Med)	Lima (Peru)	[23]	Partial Least Square (PLS) path modelling
Covid		Cordoba (Spain)	[24]	PLS- Structural Equation Modelling (SEM)
Type of motivations		Mecca (Saudi Arabia), religious tourism	[5]	Factor and K-means, clusters analysis
Experiential loyalty		Religious destinations: Pakistan, Kedah, Malaysia	[6]	PLS-SEM
Visitor satisfaction	Perceived value Perceived heritage quality (moderator variable–Mod)	Visitors of The Citadel of the Catholic King (Cordoba Spain)	[21]	SEM using Warp-PLS 7.0
	Perceived cultural quality (Mod)			PLS-SEM
Satisfaction with traditional restaurants		Cordoba (Spain)	[25]	SEM
Tourists' emotions Image of the destination and the restaurants		Cordoba (Spain)	[26]	SEM
Tourists' heritage Brand experience	Cultural intelligence destination loyalty (Med)		[27]	SEM
Brand legitimacy	Brand authenticity perceived value (Med) Brand trust (Med)	Cultural heritage	[28]	
Experience of tourist Socio-cultural variables		Heritage destinations, Spain	[29]	SEM
Perceived quality and value of the visit	Visitor satisfaction	Cordoba (Spain)	[30]	SEM

(continued)

Table 11.1 (continued)

Antecedents	Moderator/mediator variable/role	Context	Authors	Analysis method
Customized authenticity		Heritage destinations, Old Town of Lijiang, (China)	[31]	SEM
Brand love and respect		Cultural heritage, night tourism ("Cultural Heritage Night", South Korea)	[32]	SEM
Satisfaction (less impact on)		Kali Besar Corridor (Jakarta, Indonesia)	[33]	Cross-tabulation analysis
Cultural identity, authenticity perception	Destination satisfaction	Intangible cultural heritage (Celadon Town, China)	[34]	SEM
Perceived quality and value	Satisfaction	Synagogue of Córdoba (Spain)	[35]	SEM
Customer-based loyalty (in customer-based brand equity)		Maimun Palace (Medan City)	[36]	Regression analysis
Mixed reality (interactivity, vividness)		Visitor attractions with XR technologies in Seoul (Korea)	[37]	PLS-SEM
Perception of the processions Evaluation of the elements of the destination		Intangible Cultural Heritage (Popayán (Colombia)	[38]	Mediation analysis (Smart PLS)
Tourist' engagement cultural identification with the destination		Intangible cultural heritage (Porto, Portugal)	[39]	
Tourist satisfaction		Tigrai, Ethiopia	[40]	
Self-congruity functional congruity	Tourist satisfaction	Shaoshan city (China)	[41]	
Constructive and existential authenticity	Tourist satisfaction	Hahoe village in South Korea	[42]	

under. Although we argue against treating all heritage sites as a homogenous context and in favour of the imperative to explore models on different heritage contexts, several variables should constitute, in our opinion, the salient core of the model on the heritage sites and should be the common denominator, and a few others could be varying depending on the context. This is not the case in the present in heritage literature, except maybe for one variable: tourist satisfaction. Although the empirical evidence came from different research contexts, ninety per cent of the studies tested their models on visitors of 1–3 objectives or, in just one heritage destination, so one can argue that we have limited evidence for extrapolation. Regardless of the fact that the number of variables in the model is reduced, and very few variables are common or tested on more than one destination, sixty three per cent of the studies are using Structural Equation Modelling simultaneously which it is pertinent to the objective to "model and estimate complex relationships among multiple dependent and independent variables" [43, p. 1], especially PLM which "focuses on explaining the variance in the model's dependent variables" [44] quoted in [43, p. 1]. The question is, in our opinion: *Are we there yet?* Do we have enough evidence from replicating the initial proposed models and the theoretical arguments to back-up the decision to exclude/include certain variables, or the consensus on a number of variables, which should represent the core-nucleus of the antecedents of DL in the model? The motive for which we did not propose and test a new model is that we don't think we are quite there yet. Apart from not having all the pieces of the puzzle, we also feel the need to have "the bridge studies" testing and explaining why the variables in the models that acquired a certain notoriety in other fields of tourism literature are not (all) relevant for heritage sites. For some variables the lack of relevance in not self-explanatory.

11.2.2 The Construct of Destination Loyalty in Tourism Literature

Previous studies were preoccupied with testing the dimensions of DL as much as to develop a model for its relevant interconnected variables. The construct of DL is usually conceptualized as a three-fold concept due to a sequential process in which consumers' firstly become loyal in a cognitive sense, furtherly in an affective sense, and later in a conative manner [45]. The first decade was however dominated by studies seeing DL as a unidimensional concept [8] and that was the behavioural intention to revisit [9, 22] or the word-of-mouth communication intention [46]. The three-fold conceptualization was used by numerous studies and the labels varied: *cognitive loyalty, affective loyalty, conative loyalty* [8], the cognitive and affective loyalty sometimes termed as *attitudinal loyalty* [8]. The three-fold conceptualization of DL acquires more empirical support considering that DL it is, after all, the attitude of tourists towards destination. An attitude as a construct in social and cognitive psychology is a three-fold conceptualization [18, 19]. The majority of individuals perceive, at the cognitive level, certain attributes/adjectives associated to the destination as an output of a comparative analysis of the respective destination against other competing destination (the cognitive component). All consumers manifest, in general, an affective component (positive or negative) of the attitude towards destination: for example, no matter how expensive, far or rocky we acknowledge a beach destination to be at the cognitive level, compared with other more affordable and accessible destinations, we could still feel great in that place, and we favour it against rational arguments. We could feel strongly about something, without even having the rational arguments. The attitude implies also the behavioural component (the return in that destination, the positive verbal appraisal, etc.). All these components are easy to discriminate and need to be assessed: at the cognitive level, a tourist can evaluate a destination as presenting the most favourable cost/benefit ratio, he/she felt great in that particular destination, and still feel nostalgic about it, but at the behavioural level he/she decide to go and see other destinations. In time the three components of DL received different labels and different conceptualizations: Oppermann [47] proposed a DL conceptualization consisting in: *attitude loyalty* (represented by intention to revisit), *behaviour loyalty* (which refers to consumers' repeat purchase behaviour) and *composite loyalty.* The cognitive component of loyalty is described as the destination attributes ordered on a ordinal scale after being compared and contrast with other attributes of similar destinations [8, 48]. The *affective loyalty* is seen as "the degree to which a customer 'likes' the destination and its services [8, 28] and *conative loyalty* referring to commitment and purchase intentions "used to determine if participants were conative loyal" [8, p. 278].

In general recent studies stress the importance of tourists' engagement [39] and the emotions experienced by tourists at heritage sites in relation with the genesis of destination loyalty [27, 29]. Hernandez-Rojas et al. [21] found that visitor's emotions are the most important factors in explaining the overall experience and tourists' loyalty.

Other studies moved the focus of the implication of reenactment performances during reenactment festivals at heritage sites in stirring visitors' emotions [49, 50] and in increasing their level of engagement with the destination [1]. Continuing to explore how the level of implication and emotions tourists experience at heritage sites reflects on the destination loyalty, Mălăescu [1] confirmed the relevance of PA' components as antecedents of destination loyalty also on the heritage sites context stressing the necessity of a closer examination of the intangible elements influencing destination loyalty formation at heritage sites. Moreover, the study brought evidence that tourists visiting different heritage sites (two subsamples: visitors at sites more linked with national historiography and other heritage sites), could register different level of motivation, implication, AL or perceived psychological impact and heritage sites are not as homogenous as we imagine. We have to treat these contexts consequently. Different heritage sites can steer different emotions in tourists and they can register different connotations for different people (in terms of social or ethnic groups). In the case of heritage sites strongly connected with national identity, variables like affective attachment (AA), place identity (PI), the strength of visitor's national identification and the destination's identification as a reenactment destinations seemed to explain near half of the variance of the visitors' AL for the destination. In the respective study, the visitors' level of AL, PI and independent motivation to visit registered was slightly higher.

Heritage sites closely linked with personal history and strongly implicated in the self-identity defining process (as PA literature revealed) could register different levels of correlation with certain components of DL as opposed to other cultural heritage destinations where the link and personal significance is not that profound.

The most recent studies, in the context of the Covid-19 pandemic, are preoccupied by the perceived health-risk and the destination image's impact on the DL [23], with perceived quality as a mediator. Other studies focussed on extending the context of DL research at recreational fishing destination [4] the variable emphasized as relevant to DL being the centrality of the angling activity. On the context of the parks Plunkett et al. [12] found self-development, strengthening interpersonal relationships with family and friends and the level of education being linked to positive image and intention to recommend. On the context of adventure tourism, Carvache-Franco et al. [3] emphasized learning, social, biosecurity, relaxation and competence mastery as variables impacting the intention to return, recommendation and saying positive things. Liu et al. [51] studied the variables involved in tourists' loyalty genesis in the case of minsu (homestays) accommodation (China) and found that spatial, cultural and social environment perception play the role of antecedent variables with emotional experience as mediator and personality traits as moderators of the impact of antecedents on consumer's loyalty.

11.2.3 Place Attachment Literature

When tourists visit a place they like, visitors form a long-term emotional bond with the place. Chen et al. [14] consider that this process is unfolding no matter how short the visit, and sometimes, it begins prior to their visit. The literature on PA registered a vibrant conceptual debate regarding the dimensions of PA (for a recent meta-analysis on PA see Dwyer [52]). PA was successively seen as a bi-dimensional concept, composed by the physical attachment and the social/interpersonal relationships [53, 54]. Later, the affective component gained recognition and PA became a three-fold concept [8, 55, 56]. Studies continued to restructure it in a four-dimensional and the more comprehensive model: six-dimensional construct [14]. The conceptualization gaining popularity seems to be the three-dimensional model [8, 14, 52] with place identity as a cognitive component, affective attachment (AA) and place dependence (PD) [8, 14] or functional attachment [15]. PI was seen as a component of self-identity including beliefs about one's relationship with the place [57]. From the three dimensions of PA, PI represents the component more vulnerable to modifications during visits at a heritage site [50], considering its role in self-identity definition, especially in helping an individual define his/her social identity [58]. In the particular case of heritage sites associated with crucial historical events for the definition/genesis of a social group or a nation, periodical trips in this destinations could contribute to the crystallization of self-identity and the preservation of self-congruity. Williams and Vaske [16, p. 831) saw the place as a "repository for emotions and relationships that give meaning and purpose to life". Wickham [58] emphasized the cyclic, interactional, emotional dynamic of the contribution of being at a particular place in the process of self-defining. In return, the identification with a place, and belonging to that particular place could reinforce positive appraisal of the destination [59].

11.3 Methodology

The main objective of the present research was firstly a theoretical review. The papers included in the review of heritage tourism literature where retrieved from Scopus bibliographic database using the key terms "destination" AND "loyalty" + "heritage" AND "tourism" in Title + abstract + keywords field. A hundred articles were automatically retrieved and after the primary abstract analysis, a number of 23 papers and documents where included in the analysis. We have retained in our preliminary database the independent variables, the antecedents, mediators and moderators when present, the dependent variables and selectively, when explicitly stated, the dimensions of the DL measured, the methodology used in data analysis, the authors and research context. For the integration of the heritage tourism DL review in the general state of the art on DL or PA, we based the reflections on the classical selection of relevant literature and our previous bibliographic analysis.

Secondly, the empirical data resulted from a preliminary analysis retrieved from a more complex, multi-phase research on tourists' experience at heritage sites where reenactment performances take place. For the present purpose, we used the data from a survey based on 258 intercept and online questionnaires on visitors at 20 tourism destination heritage sites in Romania presenting a historical and cultural heritage as a part of their tourism resources, from different historical periods (from Antiquity, and Middle Age to the creation of Romania as a modern state). The intercept survey was carried during reenactment festivals or celebratory events in general (such as the celebration of the National Day on the site Romania as a modern state is born) as the research technique most appropriate when capturing on-site tourists opinions and raw emotions during an event [60]. The sample of visitors was a convenience one based on the availability of the on-site subjects. In their case the availability for research purposes is scarce considering that during festivals, the schedule is almost packed with activities. The instrument included a close-ended section of items assessing the dimensions of PA and DL and variables related to the perception of the heritage sites, the events in the destination and personal characteristics that could alter or enhance visitors' experience at a site linked with their personal history or national historiography. For PA we used the three-fold conceptualization proposed by Chen et al. [14] and for DL the conceptualization proposed by Oliver [45] with the items adapted from the instrument that Yuksel et al. [8] used, adapted from Williams and Vaske [16] with modifications by Alexandris et al. [9] for PA, respectively, from Back [48] and Back and Parks [60] for DL.

11.4 Results and Discussion

The correlation analysis on the main components of PA, DL and other relevant variables for the research contexts (Table 11.2) showed rather medium correlations of PA with DL (0.64). When exploring the correlation indexes of the relevant variables in the case of visitors at heritage sites with different type of heritage, the results are prudent in encouraging further analysis. Several sites included in the survey, present a mix of cultural heritage with historic heritage, some possessing different period-layered historic heritage or, one uniquely connected with national historiography, others, during reenactment festivals on the antiquity period, even connected with nationalist manifestation [49]. The nature of heritage sites by default implies different meaning and different personal value and attachment for different individuals. That means that the cognitive and the affective component of PA and DL would differ. Poria et al., 2001 quoted by Timothy and Boyd [61, p. 7) define heritage tourism as a subgroup of tourism in which case the principal motivation for visiting a site is based on the heritage characteristics with respect for tourist's perception on his/hers own heritage. Smith [62, 63], sees it as a branch of cultural tourism preoccupied by the interpretation and representation of the past and includes as tourism resources castles, palaces, archaeological sites, entire cities, monuments and museums.

In the case of heritage tourism destinations, which present a range so various of tourism resources and tourism activities proposed in a destination, from reenactment performances to XR-based devices (meant to improve the experience at the site) the functional component of PA and the cognitive component of DL' genesis varies. As outputs of the contrast and compare process of the attributes in the respective destination (usually against alternatives and generating an ordinal scale of the possible evocated set of destinations to go) these components are definitely affected by the variance in attributes and functionalities when different heritage sites are treated as a homogenous categories.

Table 11.2 Correlation indexes of the relevant variables measured at the level of the entire sample

Variables	DL	PA	AL	AA	CgL	CnL	PI	Motv	RP
DL		0.64			0.86	0.56	0.43	0.41	
PA	0.64			0.85			0.56	0.43	
AL				0.51					
AA		0.85	0.51		0.55				0.4
CgL	0.86			0.55					
CnL	0.56								
PI	0.43	0.56		0.43			0.43	0.42	
Motv	0.41	0.43					0.42		
RP				0.4					

All correlations are significant at the level of 0.01 level (two tailed)

Previous studies showed that if visitor's data are categorized in more homogenous samples in terms of main type of heritage and tourist motivation, in the case of sites like the ones more strongly connected with national identity forging, the average level of PI, AL and independent motivation to visit registered were slightly higher than the average value on the sample of other types of heritage sites [1]. The emotional bond, but especially the identification with the place (measured in PI) when the place is deeply weaved as part of their identity, it is stronger, and visitors develop a stronger PI than in other cases due to the role that particular heritage site plays in helping an individual define his/her social identity [58]. In this particular case, where visits at heritage sites are a part of the self-identity forging process described in the literature [58, 64, 65] the transactional analysis of contrasting and comparing the attributes of a destination against a competitive one, will might not take place [50], conducting to different values of PD and CL registered by the visitors in these cases. The cognitive component of PA and DL is dominated by the unique significance the site holds for their social or cultural in-group, their history or emancipation in terms of social identity. However, this is not the case for all types of heritage sites. This is the main argument why, we argue in favour of creating more homogenous categories inside heritage tourism destinations (Table 11.3).

With no intention of revisiting the models on destination loyalty, but to explore what variables will be retain by the model in order to explain the variance of DL on this contexts, we run a regression analysis in exploratory purposes. The indexes revealed that almost half of the DL variance was explained by predictor variables relevant for the context of visitors at heritage sites (Table 11.4). The variable with the most explanatory power was place dependence, so the functionality of the destination in visitors' personal equation was the most important. With a diverse range of heritage, the sites visited by the tourists included in the sample, the common denominator of the variance in DL was the functionality of the visit in the destination. In the case

Table 11.3 Descriptive characteristics of the relevant variables measured at the level of the entire sample

Variables	Mean	Std. deviation
Destination loyalty	4.41	1.76
Place attachment	18.9	5.49
Affective loyalty	4.43	1.47
Conative loyalty	4.58	1.74
Cognitive loyalty	4.57	3.01
Independent motivation to visit	4.72	1.78
Exclusive reenactment motivation	3.63	2
Affective attachment	14.05	4.68
Place identity	4.74	1.74
The level of superposition of the destination with reenactment	9	2.68
The role of reenactment in the enhancement of national pride	4.81	1.86

Table 11.4 Destination loyalty' variability explained

Predictors in the model	Variable predicted	R	R^2	Adj R^2	Std. Est. Err	F	Sig.	Method
Place dependence, National identity strength, perceived role of reenactment in enhancing patriotic feelings	Destination loyalty	0.711	0.505	0.485	2.20296	25.165	0.000	Stepwise

of sites more connected with their personal identity the cognitive component could be more prominent, but in others had less relevance, the most relevant predictor of DL regardless of the type of heritage and its (personal) significance, prove to be the functional attachment or place dependence. The last two variables were represented by personal factors like the social identification (the level of national identity strength) and perceived role of reenactment in enhancing patriotic feelings at heritage sites. The strength of social identification has probably an impact on the cognitive component of DL, which again in the case of certain heritage sites register more unique personal significance but in others no. The level of national identity strength was adapted to the purpose of this study from [66] social identification strength. The results are consistent with previous studies that DL is dependent on the functionality met by different heritage sites, different sites fulfil different psychological, emotional or knowledge related needs of visitors.

11.5 Conclusions

In the research literature on tourism heritage destinations, the models and the list of variables connected with destination loyalty is continuously growing. In the meantime in destination loyalty research, some relationships with classic antecedents of DL like place attachment, are found inconclusive. The results of the paper are consistent with previous studies that satisfaction becomes a less significant step in loyalty formation as loyalty begins to set through other mechanisms which include the roles of personal determinism ("fortitude") and social bonding at the personal level and their synergistic effects [45, p. 1]. We argue that this is the case at least of heritage sites perceived as strongly connected with personal identity forging as a member of a cultural or ethnic group. In these particular cases, the place identity component and the cognitive loyalty will crystallize around unique personal significance instead of the ordinal top of best destinations as an outcome of the comparative analysis of the destination attributes that takes place in the case of beach tourism. The conclusions are consistent also with the need for more empirical evidence on the relationship between PA and DL [2], and the necessity to create more homogenous categories of heritage sites as a context for studying the relevant variables to destination loyalty [1]. The most predictive factor of destination loyalty on the different heritage sites the visitors considered in the sample was place dependence stressing the importance of the functionality of the destination in visitors personal equation. The place dependence was complimented by personal factors like the level of national identity strength and perceived role of reenactment in enhancing patriotic feelings at heritage sites consistent with previous results that the importance of tourists' engagement [39] and the emotions experienced by tourists at heritage sites are the most important factor when explaining the overall experience and tourists' loyalty [21]. The study stress the need to reconsider the evolving phase of the research the literature on heritage sites–respectively of the need for more in-depth exploring and testing of the relationships between variables—and the pertinence for multiplying the DL models on this particular context. On the other hand, it argues in favour of the need to build models operating inside the different types of heritage tourism, considering the satisfied needs, the resources, the motivation to visit and activities proposed at heritage sites. The potential added-value of the paper resides in launching a reflection theme regarding the necessity of a more strong empirical and theoretical argumentation in the case of future research proposing new DL models on similar contexts to prior proposed DL models and the existence in the future of a core variables representing the common denominator of all heritage context DL models besides other variables varying from context to context but more enlarged in nature research context within the heritage sites as a general framework.

References

1. Malaescu, S.: Visitors at heritage sites: from the motivation to visit to the genesis of destination affective loyalty. In: Katsoni, V., Șerban, A. (eds.) Transcending Borders in Tourism Through Innovation and Cultural Heritage. Springer Proceedings in Business and Economics 8th International Conference of the International Association of Cultural and Digital Tourism, IACUDIT, pp. 571–586. Springer (2022)
2. Zou, W., Wei, W., Ding, S., Xue, D.: The relationship between place attachment and tourist loyalty: a meta-analysis. Tourism Manag. Perspect. **43** (2022)
3. Carvache-Franco, M., Contreras-Moscol, D., Orden-Mejía, M., Carvache-Franco, W., Vera-Holguin, H., Carvache-Franco, O.: Motivations and loyalty of the demand for adventure tourism as sustainable travel. Sustain. **14**(14), Art. no. 8472 (2022)
4. van den Heuvel, L., Blicharska, M., Stensland, S., Rönnbäck, P.: Been there, done that? effects of centrality-to-lifestyle and experience use history on angling tourists' loyalty to a Swedish salmon fishery. J. Outdoor Recreat. Tour. **39** (2022)
5. Hassan, T., Carvache-Franco, M., Carvache-Franco, W., Carvache-Franco, O.: Segmentation of religious tourism by motivations: a study of the pilgrimage to the city of Mecca. Sustain. **14**, 7861 (2022)

6. Nisar, Q., Waqas, A., Ali, F., Hussain, K., Sohail, S.: What drives pilgrims' experiential supportive intentions and desires towards religious destinations? Tourism Manag. Perspect. **43**, 100997 (2022)
7. Luo, Q., Wang, J., Yun, W.: From lost space to third place: the visitor's perspective. Tour. Manag. **57**, 106–117 (2016)
8. Yuksel, A., Yuksel, F., Bilim, Y.: Destination attachment: effects on customer satisfaction and cognitive, affective and conative loyalty. Tour. Manag. **31**, 274–284 (2010)
9. Alexandris, K., Kouthouris, C., Meligdis, A.: Increasing customers' loyalty in a skiing resort. Inter. Contem. Hospit. Manag. **18**(5), 414–425 (2006)
10. Strandberg, S., Hultman, C., Strandberg, M., Styvén, M., Hultman, S.: Places in good graces: the role of emotional connections to a place on word-of-mouth. J. Bus. Res. **119**, 444–452 (2020)
11. Zhang, X., Chen, Z., Jin, H.: The effect of tourists' autobiographical memory on revisit intention: does nostalgia promote revisiting? Asia Pacific J. Tour. Res. **26**(2), 147–166 (2021)
12. Plunkett, D., Fulthorp, K.P., C.M.: Examining the relationship between place attachment and behavioral loyalty in an urban park setting. J. Outdoor Recreat. Tour. **25**, 36–44 (2019)
13. Chen, J.S., Gursoy, D.: An investigation of tourists' destination loyalty and preferences. Int. J. Contemp. Hosp. Manag. **13**(2), 79–85 (2001)
14. Chen, N., Dwyer, L., Firth, T.: Conceptualization and measurement of dimensionality of place attachment. Tour. Anal. **19**(3), 323–338 (2014)
15. Stokols, D., Shumaker, S.A.: People in places: a transactional view of settings. In: Harvey, J. (ed.) Cognition, Social Behavior and Environment. Erlbaum, Hillsdale, NJ (1981)
16. Williams, D.R., Vaske, J.J.: The measurement of place attachment: validity and generalizability of a psychometric approach. Forest Sci. **49**, 830–840 (2003)
17. Raggiotto, D.: Scarpi: this must be the place: a destination-loyalty model for extreme sporting events. Tour. Manag. **83**, 104–254 (2021)
18. Corsini, R.: Encyclopedia of Psychology, III, Wiley (1994)
19. Nelson, T.: Handbook of Prejudice, Stereotyping and Discrimination, 2nd edn. Psychology Press (2016)
20. Foris, D., Florescu, A., Foris, T., Barabas, S.: Improving the management of tourist destinations: a new approach to strategic management at the DMO level by integrating lean techniques. Sustainability **12**(23), 10201 (2020)
21. Hernandez-Rojas, R.D., Folgado-Fernandez, J.A., Palos-Sanchez, P.R.: Influence of the restaurant brand and gastronomy on tourist loyalty. a study in Córdoba (Spain). Int. J. Gastronomy Food Sci. **23** (2021)
22. Lee, J.: Examining the antecedents of loyalty in a forest setting: relationships among service quality, satisfaction, activity involvement, place attachment, and destination loyalty. Unpublished Dissertation, The Pennsylvania State University (2003)
23. Cambra-Fierro, J. Fuentes-Blasco, M., Huerta-Álvarez, R., Olavarría-Jaraba, A.: Destination recovery during COVID-19 in an emerging economy: Insights from Perú. European Res. Manag. Business Econ. **28**(3)
24. Huete-Alcocer, N., Huete-Alcocer, R., Hernandez-Rojas, R.: Does local cuisine influence the image of a World Heritage destination and subsequent loyalty to that destination? Int. J. Gast. Food Sci. **27**, 100470 (2022)
25. Hernández-Rojas, R.D., Huete-Alcocer, N. Hidalgo-Fernández, A.: Analysis of the impact of traditional gastronomy on loyalty to a world heritage destination, Int. J. Gastr. Food Sci. **30** (2022)
26. Hernandez Rojas, R.D. Folgado-Fernandez, J., Palos-Sanchez, P.: Influence of the restaurant brand and gastronomy on tourist loyalty. a study in Córdoba (Spain). Int. J. Gast. Food Sci. **23** (2021)
27. Rahman, M.S., Abdel Fattah, F.A.M., Hussain, B., Hossain, M.A.: An integrative model of consumer-based heritage destination brand equity. Tourism Rev. **76**(2), 358–373 (2021)
28. Chen, X., Lee, G.: How does brand legitimacy shapes brand authenticity and tourism destination loyalty: focus on cultural heritage tourism. Global Bus. Finance Rev. **3**(1), 53–67 (2021)

29. Sánchez-Sánchez , M.D., De-Pablos-Heredero C., & Montes-Botella, J.: A behaviour model for cultural tourism: loyalty to destination. Economic Research-Ekonomska Istraživanja (2020)
30. Del Rio, J.A., Hernandez-Rojas, R.D., Vergara-Romero, A., Millan, M.: Loyalty in heritage tourism: the case of Córdoba and its four world heritage sites. Int. J. Environ. Res. Public Health **17**(23), 1–201 (2020)
31. Shi, T., Jin, W., Li, M.: The relationship between tourists' perceptions of customized authenticity and loyalty to guesthouses in heritage destinations: an empirical study of the world heritage of Lijiang Old Town, China. Asia Pacific J. Tour. Res. **25**(11), 1137–11521 (2020)
32. Chen, N., Wang, Y., Li, J., Wei, Y., Yuan, Q.: Examining structural relationships among night tourism experience, lovemarks, brand satisfaction, and brand loyalty on "cultural heritage night" in South Korea. Tour. Recreation Res. (2020)
33. Nagari, B.K., Suryani, S., Pratiwi, W.D.: TOD tourism heritage district livability: user satisfaction in Kali Besar Corridor in Jakarta, Indonesia. Earth Environ. Sci. **532**(114) (2020)
34. Tian, D., Wang, Q., Law, R., Zhang, M.: Influence of cultural identity on tourists' authenticity perception, tourist satisfaction, and traveler loyalty. Sustain. **12**(16) (2020)
35. Del Rio, J.A., Hernandez-Rojas, R.D., Vergara-Romero, J.A., Hidalgo-Fernández, A.: The Loyalty of tourism in synagogues: the special case of the synagogue of Córdoba. Int. J. Environ. Res. Public Health **17**(12), 4212 (2020)
36. Siregar, O.M., Marpaung, N., & Abdillah, M.B.: Customer-based brand equity for a tourist destination (A study on Nusantara tourists at Maimun Palace, Medan City). Earth Environ. Sci. **452**(113) (2020)
37. Bae, S., Jung, T.H., Moorhouse, N., Suh, M., Kwon, O.: The influence of mixed reality on satisfaction and brand loyalty in cultural heritage attractions: a brand equity perspective. Sustain. **12**(71) (2020)
38. Santa, C.G., Lopez-Guzman, T., Pemberthy Gallo, L.S., Rodriguez-Gutierrez, P.: Tourist loyalty and intangible cultural heritage: the case of Popayán, Colombia. J. Cult. Heritage Manag. Sustain. Development **10**(2), 172–18819 (2020)
39. Vieira, E., Borges, A.P., Rodrigues, P., Lopes, J.: The role of intangible factors in the intention of repeating a tourist destination. Int. J. Tour. Policy **10**(4), 327–350 (2020)
40. Asmelash, A.G., Kumar, S.: Tourist satisfaction-loyalty Nexus in Tigrai, Ethiopia: implication for sustainable tourism development. Cogent Bus. Manag. **7**(11) (2020)
41. Zhou, M., Yan, L., Wang, F., Lin, M.: Self-congruity theory in red tourism: a study of Shaoshan City, China. J. China Tour. Res. **1**, 182020 (2020)
42. Park, E., Choi, B.-K., Lee, T.J.: The role and dimensions of authenticity in heritage tourism. Tour. Manag. **74**, 99–109 (2019)
43. Hair, J.F., Hult, G.T.M., Ringle, C.M., Sarstedt, M., Danks, N.P., Ray, S.: An introduction to structural equation modeling. In: Partial Least Squares Structural Equation Modeling (PLS-SEM) Using R. Classroom Companion: Business. Springer, Cham (2021)
44. Chin, W.W., Cheah, J.-H., Liu, Y., Ting, H., Lim, X.-J., Cham, T.H.: Demystifying the role of causal-predictive modeling using partial least squares structural equation modeling in information systems research. Ind. Manag. Data Syst. **120**(12), 2161–2209 (2020)
45. Oliver, R.L.: Whence consumer loyalty? J. Mark. **63**, 33–44 (1999)
46. Simpson, M. P., Siquaw, J.: Destination word-of-mouth: the role of traveller type, residents, and identity salience. J. Travel Res. (2008)
47. Oppermann, M.: Where psychology and geography interface in tourism research and theory. In: Woodside, A.G., Crouch, G.I., Mazanec, J.A., Oppermann, M., Sakai, M.Y. (eds.) Consumer Psychology of Tourism, Hospitality and Leisure. CABI Publishing, Cambridge, UK (2000)
48. Back, K.J.: The effects of image congruence on customers' brand loyalty in upper middle-class hotel industry. J. Hospitality Tourism Res. **29**(4), 448–467 (2005)
49. Popa, C.N.: The significant past and insignificant archaeologists. Who informs the public about their 'national' past? The case of Romania. Archaeol. Dialog. **23**(1):28–39 (2016)
50. Malaescu, S.: Place attachment genesis: the case of heritage sites and the role of reenactment performances. In: Katsoni, V., van Zyl, C., (eds.) Culture and Tourism in a Smart, Globalized, and Sustainable World, pp. 435–449. Springer Proceedings in Business and Economics (2021)

51. Liu S., Wang X., Wang L., Pang Z.: Influence of non-standard tourist accommodation's environmental stimuli on customer loyalty: the mediating effect of emotional experience and the moderating effect of personality traits. Tour. Manag. Perspect. **43**(19) (2022)
52. Dwyer, L., Chen, C., Lee, J.: The role of place attachment in tourism research. J. Travel Tour. Mark. **36**(5), 645–652 (2019)
53. Williams, D.R., Patterson, M.E., Roggenbuch, J.W., Watson, A.E.: Beyond the commodity metaphor: examining emotional and symbolic attachment to place. Leisure Sci. **14**, 29–46 (1992)
54. Brocato, E.D.: Place attachment: an investigation of environments and outcomes in service context. Doctoral Thesis, The University of Texas (2006)
55. Kyle, G., Mowen, A., Tarrant, M.: Linking place preferences with place meaning: an examination of the relationship between place motivation and place attachment. J. Environ. Psychol. **24**(4), 439–454 (2004)
56. Kyle, G., Graefe, A., Manning, R.: Testing the dimensionality of place attachment in recreational settings. Environ. Behav. **37**(2), 153–177 (2005)
57. Proshansky, H.M., Fabian, A.K., Kaminof, R.: Place identity: physical world socialization of the self. J. Environ. Psychol. **3**, 57–83 (1983)
58. Wickham, T.D.: Attachments to places and activities: The relationship of psychological constructs to customer satisfaction. Unpublished doctoral dissertation, The Pennsylvania State University, University Park, Pennsylvania (2000)
59. Veal, A.J.: Research Methods for Leisure and Tourism: A Practical Guide. Prentice Hall (2006)
60. Back, K.J., Parks, S.C.: A brand loyalty model involving cognitive, affective and conative brand loyalty and customer satisfaction. J. Hosp. Tour. Res. **27**(4), 419–435 (2003)
61. Timothy, D., Boyd, S.: Heritage Tourism, Pearson Education Ltd. (2003)
62. Smith, M.: Issues in Cultral Tourism Studies. Routledge, London and New York (2003)
63. Smith, M.: Issues in Cultral Tourism Studies, 2nd edn. Routledge, London and New York (2009)
64. Chen, N.C., Šegota T.: Conceptualization of place attachment, self-congruity, and their impacts on word-of-mouth behaviors. In: Travel and Tourism Research Association Conference (2016)
65. Park, H.J.: Heritage tourism Emotional Journeys into Nationhood. Ann. Tour. Res. **37**(1), 116–135 (2010)
66. Stephan, W.G., Stephan, C.W.: Cognition and affect in stereotyping: parallel interactive networks. In: Mackie, Hamilton (eds.), Affect, Cognition and Stereotyping: Interactive Processes in Group Perception, pp. 111–136. Academic Press, Orlando (1993)

Part IV
Gamification Technologies to Marketing

Chapter 12
Develop a Virtual Learning Environment (Eva) to Train Agents in Security and Private Surveillance

Nelson Salgado Reyes and Graciela Trujillo

Abstract Constitutes the planning of a software development project characterized mainly by the detailed identification of requirements, description of the activities of the process, distribution of efforts and tasks, scheduling of tasks, budget and analysis of risks. The project aims to cover the existing demand in Ecuador for Security Agents, the same ones that cannot currently be covered in the existing Training and Improvement Schools. The EVA Virtual Learning Environment is based on existing Learning Management Systems (LMS) that exploit web 2.0 tools, but with additional innovative and disruptive features such as biometric security and the use of 3D virtual reality environments for teaching and learning. e-learning learning. For total control of the EVA project from the Project Management, the PMBOK Guide was used; In terms of software development planning, Scrum, an agile framework, was used.

12.1 Introduction

The current world presents social, political, economic and cultural conditions that have allowed the development of the digital society, characterized by the use of Information and Communication Technologies (ICTs), the dominant form of communication, sharing of information and knowledge.

In the educational field worldwide, as well as in Ecuador, ICTs have produced great changes in the dynamics of traditional teaching and learning processes, through the design, creation and use of educational platforms, they have allowed the student to disappear the barriers of space and time. These educational platforms are based on the use of the Internet and involve the exchange of information between the Instructor and the students, in an asynchronous manner, where the student does not coincide in time or virtual space with the Instructor for the development of their activities,

N. Salgado Reyes (✉)
Pontificia Universidad Católica del Ecuador, 12 de Octubre, 1076 Quito, Ecuador
e-mail: nesalgado@puce.edu.ec

G. Trujillo
Instituto Superior Tecnológico Japón, Marietta de Veintimilla, 170120 Quito, Ecuador
e-mail: gtrujillo@itsjapon.edu.ec

J. L. Reis et al. (eds.), *Marketing and Smart Technologies*, Smart Innovation, Systems and Technologies 337, https://doi.org/10.1007/978-981-19-9099-1_12

or synchronously, where students attend classes live and match their classmates and the Instructor.

In the professional field, especially private security in Ecuador, there is still resistance to the use of technology and e-learning teaching and learning, maintaining face-to-face training in the Security training and improvement schools. However, the new vision of the National Association of Comprehensive Security and Investigation Companies ANESI seeks to streamline the process of registering, training and updating private security agents through the use of web software such as EVA virtual learning environments. This "Plan to develop a Virtual Learning Environment (EVA) to train agents in Security and Private Surveillance" is a proposal that aims to take advantage of Tics and web 2.0 tools to implement virtual teaching and learning processes, the same ones that have never been have implemented in this professional branch.

12.2 Methodology

The planning of the EVA project implemented the PMBOK Project Management Fundamentals Guide, which would allow, among its activities, mainly to identify requirements, establish communications between stakeholders, manage the scope, quality and risks of the project, as well as establish a schedule of work, budget and resources.

For the planning of the product development, the agile Scrum framework will be used, which will allow responding to unforeseen changes beforehand and being able to make much faster deliveries.

12.2.1 PMBOK

The Guide to the Fundamentals of Project Management PMBOK ("Project Management Body of Knowledge") [1], includes a set of good practices and standards for project planning.

With PMBOK, the entire life cycle of the project is managed with the achievement of process activities corresponding to Project Management, whose results define the end of one process and the start of another. As can be seen in Fig. 12.1, PMBOK concentrates the Project Management processes in 5 groups:

12.2.1.1 Start Process Group

They are a set of processes that officiate the beginning of the project, stating in a Project Constitution Act, relevant information such as the purpose and scope of the

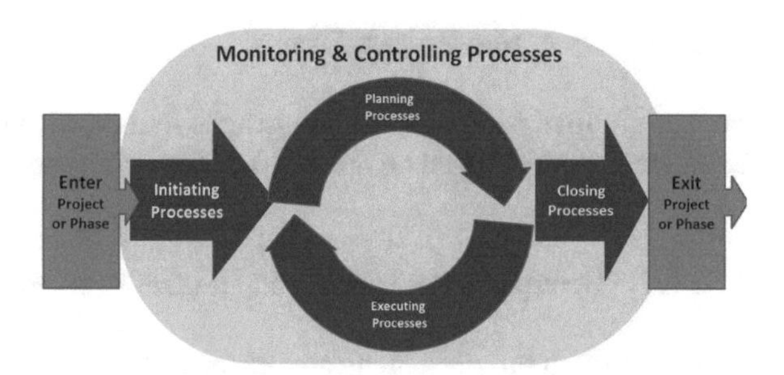

Fig. 12.1 PMBOK project management process groups

project; Identification of the Interested Parties; assumptions and restrictions; financial resources to commit and designation of the Project Director.

The information that appears in the start-up processes is of the utmost importance since any activity or decision that you want to take revolves around the business need that was established in the articles of incorporation.

12.2.1.2 Planning Process Group

It comprises a large set of processes that will define the scope of the project and detailed actions to be taken to achieve the objectives in terms of time, cost and quality.

The activities of the Project Management processes are:

- Project Scope Management: Plan Scope Management, Gather Requirements, Define Scope, Create WBS/WBS Work Breakdown Structure.
- Management of the project schedule: plan the management of the schedule, define activities, sequence the activities, estimate the duration of the activities and develop the schedule.
- Project cost management: plan cost management, estimate costs, determine budget.
- Project quality management: plan quality management.
- Project resource management: plan resource management, estimate activity resources.
- Project communications management, plan communications management.
- Project risk management: plan risk management, identify risks, perform quantitative risk analysis, plan risk response and perform qualitative risk analysis.
- Management of the Stakeholders of the project: plan the management of the interested parties.
- Project procurement management: Plan project procurement management.

12.2.1.3 Execution Process Group

For this group of processes, the Scrum agile framework will be implemented in order to obtain a product based on defined iterations with specific objectives and tasks.

12.2.1.4 Group of Monitoring and Control Processes

They are a set of processes dedicated exclusively to Monitoring project performance, taking action, reporting and disseminating; Control refers to comparing actual and planned performance, evaluating alternatives and recommending corrective actions.

The activities of the Monitoring and Control processes are:

- Project scope management: validate and control the scope.
- Project schedule management: controls the schedule.
- Project risk management: monitor risks.

12.2.1.5 Closing Process Group

They are processes that determine the formal closure of the project based on a verification of each process of each group.

12.2.2 Scrums

The Scrum Guide clearly and simply defines Scrum as a process framework that has been used to manage the development of complex products since the early 1990s, within which various processes and techniques can be used [1, 2]. Scrum is based on the Scrum Team and its roles, artifacts and events. We can see in Table 12.1 the roles, artifacts and events:

Table 12.1 Scrum framework

Scrum team	Artifact	Event
Product Owner	***Product Backlog**	*Sprint Planning
Scrum Master Development Team	***Sprint Backlog**	*Daily Scrum *Sprint review *Sprint retrospective

Table 12.2 Artifacts of Scrum

Artifact	Event
Product Backlog	It is the ordered and prioritized list of product requirements, it contains characteristics, functionalities, product requirements to be carried out. Some attributes of the Product Backlog are: description, order, estimate and value The Product Owner is responsible for managing the Product Backlog
Sprint Backlog	It is the list of elements of the Product Backlog selected for the Sprints, plus those product increments made by the Development Team

12.2.2.1 Scrum Team

The Scrum Team is made up of:

- The Product Owner is in charge of optimizing and maximizing the value of the product and the work of the Development Team. The Product Owner's decisions are respected and should be reflected in the content and prioritization of the Product Backlog.
- The Scrum Master has the main function that the entire Scrum Team understands and adopts the Scrum; its theory and rules. Maximizes the value created by the Product Owner and the Development Team.
- The Development Team or development team is a group of highly technical professionals responsible for creating a high-quality product.

12.2.2.2 Artifacts

They reference key information that maximizes work, provides transparency and opportunities for product development. A description of the artifacts is shown in Table 12.2.

12.2.2.3 Events

Events are blocks of time (time box) that allow the Scrum Team to hold work meetings. The Sprint represents a container of events and each event constitutes a temporary block (iteration), the result of the Sprint is an increment of finished product. Table 12.3 shows a description of the events:

12.2.2.4 Scrum Cycle

The Product Owner, Scrum Master and the Development Team are involved in the process. The characteristics of a product are reflected in the Product Backlog, whose order of priority is determined by the Product Owner, the Scrum cycle can be seen in Fig. 12.2

Table 12.3 Events in Scrum

Artifact	Event
Sprint planning	It is a meeting that allows planning the tasks that will be carried out in an iteration divided into 2 parts 1st part of the meeting with a maximum Timebox of 4 h • The Product Owner presents the prioritized Product Backlog • Name the goal of the iteration • Proposes the highest priority requirements to be developed in the iteration • The Scrum Team verifies the Product Backlog • Questions are asked to the Product Owner about concerns • The Scrum Team adds: satisfaction conditions and selects the highest priority objectives/requirements 2nd part of the meeting with a maximum Timebox of 4 h • The Scrum Team plans the iterations, prioritizing themselves to obtain the best possible result and with minimal effort • The Scrum Team defines the tasks that complete each requirement, the estimation of each task is made • The members of the Development Team assign themselves tasks according to their specialty
Daily scrum	They are Scrum Team meetings with a 15-min Timebox, whose purpose is to combine activities, transfer information and establish commitments to fulfill in a work plan for the next Daily Scrum The inevitable questions in this type of meeting are: What was done yesterday? What was done today? What will you do tomorrow? and what problems did you find?
Sprint review	They are informal meetings between the Scrum Team, Product Owner and Stakeholders with a Timebox of 4 h for Sprints of 1 month and shorter Timebox depending on shorter Sprints The meeting is for feedback on the Sprints and the Product Backlog
Sprint retrospective	They are formal meetings between the Scrum Master and the Development Team with a Timebox of 3 h for Sprints of 1 month and shorter Timebox depending on shorter Sprints The retrospective Sprint continually seeks to improve the performance and quality of the product, and improvements are defined for the next Sprint

The Product Owner subsequently transmits the Product Backlog to the Scrum Master and the Development Team in a Sprint Planning, where it is planned how to provide a solution to the product in the first phase, the result of which is the obtaining of a list of functionalities, Sprint Backlog; the product development process called Sprint continues in which the Scrum Master and the Development Team participate; During the Sprint, other events take place, such as the Daily Scrum, daily 15-min follow-up meetings. At the end of the Sprint, another event called Sprint Review occurs, which creates a moment for the Scrum Team to verify compliance with the goals and assigned development times; it ends with a last event called Sprint Retrospective that looks for improvements that can be applied to the next Sprint, the fulfillment of all the Sprints forms the final functional product.

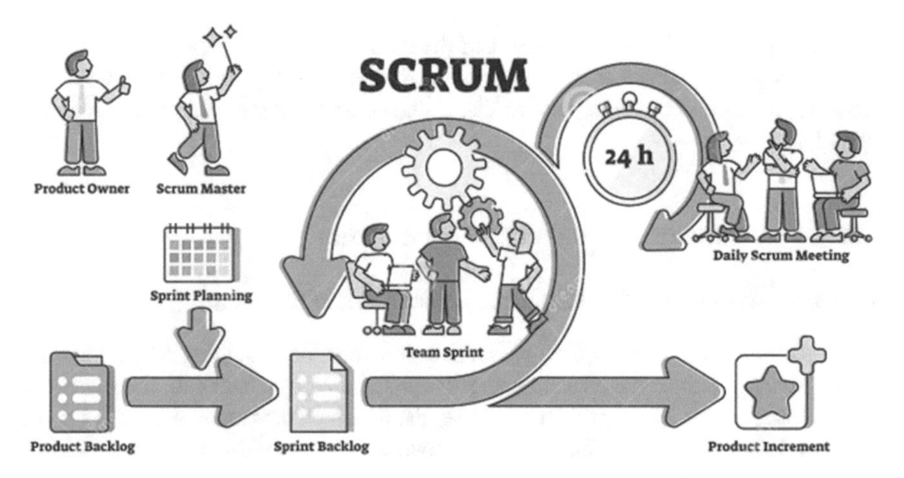

Fig. 12.2 Scrum cycle

12.3 Results

12.3.1 Start Phase

The National Association of Comprehensive Security and Investigation Companies—ANESI authorizes a project according to a business need and appoints Mr. Robert Cedeño as Project Director, who must comply with different processes that define the beginning of the project:

12.3.1.1 Project Constitution Act

The beginning of the project is endorsed or made official by the organization, the ANESI in a Project Constitution Act that generally contemplates:

- Description of the product to be obtained, purpose and scope of the project.
- Business need and cost–benefit analysis
- Internal agreements to ensure proper delivery of the project.

12.3.1.2 Purpose of the Project

The purpose of the project is to plan the development of a web platform that works in the cloud, provided with a set of computer tools that are easy to access and use for teaching e-learning to agents in Security and Private Surveillance.

The project must contemplate different pedagogical and technological components of an LMS learning management system, but with innovative implementations of biometric security and exploitation of 3D environments.

The implementation of the project in the cloud would imply the elimination of unnecessary geographical displacements on the part of the users, saving time and reducing travel and travel expenses; likewise, access to training in private security would increase.

12.3.1.3 Description of Roles and Responsibilities

The organizational structure for the project is shown in Fig. 12.3:

- Project manager: is the guide person, responsible for managing and evaluating the development of the project, achievements and compliance with the schedule; plans and monitors the project; manages the internal communication of the work team. Promote practices that reflect teamwork.
- Project Coordinator: is responsible for ensuring the scope and success of the project. Its functions are to coordinate all project activities, manage budget, ensure risks and quality of the project. Maintains communication with Stakeholders and Work Package Leaders.
- Stakeholders: The stakeholders of the project maintain communication with the Project Coordinator and the leaders of the WP work packages; The greatest contribution to the project is the promotion of participation, involvement and acceptance of work approaches with the purpose of achieving the success of the project.
- WP Work Package Leader: is the person responsible for the administration of each work package and all its tasks. Its functions are to organize and manage the tasks of the work package, coordinate delivery of work packages, ensure compliance with objectives and results of the work package. Delivery to the Project Coordinator information regarding the work package.
- Task Leader: is the person responsible for managing a task belonging to a WP work package. Its main functions are to coordinate delivery of tasks, maintain

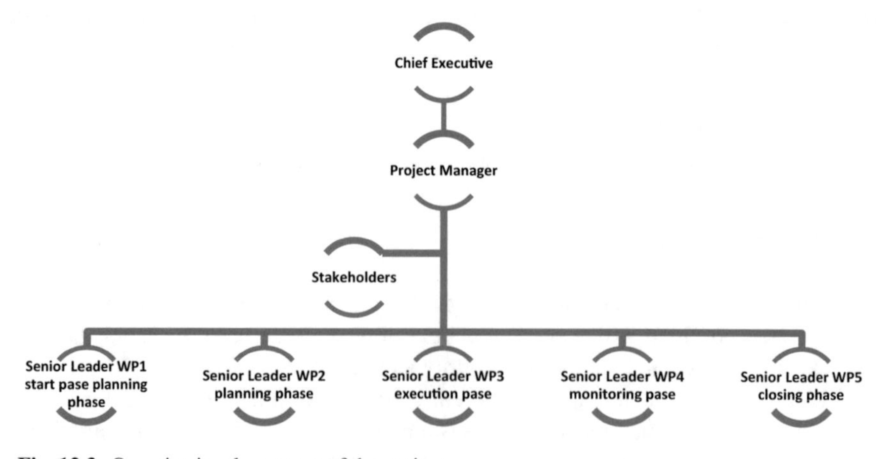

Fig. 12.3 Organizational structure of the project

communication with the WP Work Package Leaders by delivering information related to the task.

12.3.1.4 Assumptions and Restrictions

- The project aims to boost the education and training of security agents virtually, to provide a greater number of professionals in security and citizen surveillance available to private companies and civil society.
- The project implements new implementations to the existing LMS with the purpose of complying with the requirements of the Ministry of Government to the Training Centers of security and surveillance agents.
- There is a commitment by the Stakeholders to participate in the project in accordance with the provisions of the Work Plan.
- The start and end time of the project should not exceed 12 months.
- The project must consider functional and non-functional requirements.

12.3.1.5 Interested Parties

The Stakeholders or stakeholders of the project are the organizations with which ANESI has some type of interrelation, these organizations can be influenced by the implementation of the project, as shown in Fig. 12.4.

The level of interest in the project is analyzed once the stakeholders have been identified, for which it can be determined by using a Relevance Matrix [1, 3], see Fig. 12.5: Relevance Matrix.

Stakeholders with interest and influence in the project can also be identified according to the Influence-Impact Matrix, see Fig. 12.6.

12.3.2 Planning Phase

12.3.2.1 Scope Management Plan

Requirements Management

Through the active participation of the Stakeholders, it is possible to determine the needs in product requirements, the documentation of these conditions or characteristics of the product are known in the agile Scrum framework as User Stories. Based on the PMBOK Guide, Table 12.4 was built: Project Requirements, which includes the different functional and non-functional requirements of the project:

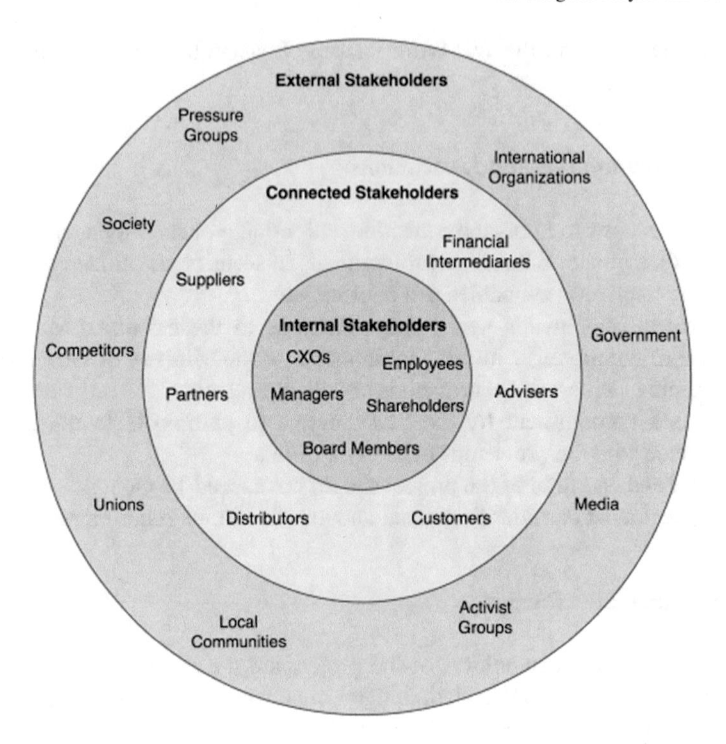

Fig. 12.4 The stakeholders

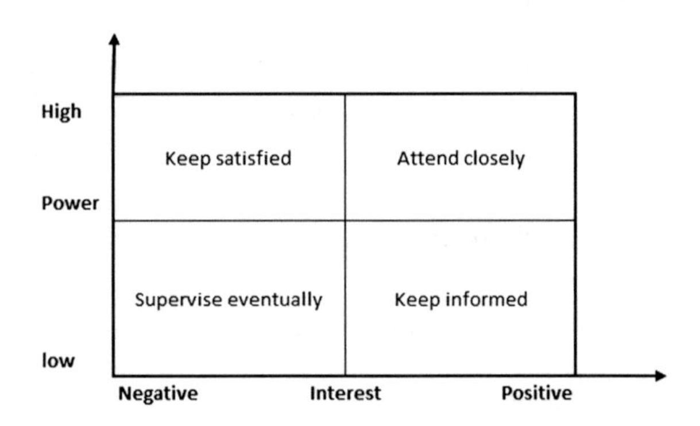

Fig. 12.5 Relevance matrix

Strategic Objectives of the Organization

Provide quality service: customers must receive a quality service through specialized and trained security and surveillance personnel, with a great sense of belonging.

Meet customer expectations: the organizational structure of ANESI [1] and the security companies with their personnel must demonstrate with actions the services they offer to customers to generate trust and fulfillment of expectations

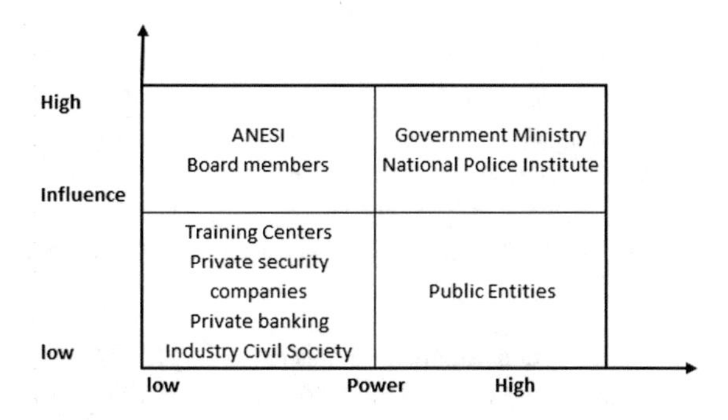

Fig. 12.6 Influence-impact matrix

for the service. Compliance with requirements: security companies must comply with current regulations in the preparation of their security personnel.

Maintain necessary infrastructure: ANESI must plan and invest in updating its technological infrastructure at the service of security companies.

Purpose of the Project

Systematization of services: the project will implement e-learning services available to users, replacing the face-to-face study paradigm currently in use.

Continuous service: the use of cloud services reduces ongoing costs of physical infrastructure, software update and guaranteed security for the operation of the web platform.

Additionally, it guarantees access to the system from any physical location and its continuous operation.

Implementation of services: e-learning services in this area of security and private surveillance are outstanding in the country.

Product Description

The EVA Virtual Learning Environment is designed to educate and train agents in Security and Private Surveillance. It is based on existing Learning Management Systems (LMS), but has additional features such as biometric security and the use of 3D virtual reality environments for teaching and e-learning.

The project is a web-type software that works with services in the cloud, taking advantage of the accessibility that it would allow from any geographical location, accessibility from any computing device, service availability 24 h a day, 365 days a year, availability of software updates, Hardware and Software scalability, data security.

For the design and subsequent development of the project, the use of free software has been planned to reduce costs.

Table 12.4 Project requirements

Functional requirement	Non-functional requirement
RF001-Authentication: user access to the study platform is through a username and password, through user biometrics	RNF001-Accessibility from any location: must have Cloud Computing services
RF002-User profile: information management of each user	RNF002-Accessibility from any computing device: accessible through desktop computers, laptops, tablets, smartphones
RF003-User administration: administration of user accounts, roles and permissions	RNF003-Service availability: with Cloud Computing services, service availability is 24 h a day, 365 days a year
RF004-Study cycle administration: the timing of the course to be taught is established	RNF004-Availability of updates: with Cloud Computing services, updates are instantaneous and indivisible for the client
RF005-Virtual Classroom Administration: virtual classroom management, user assignment according to role	RNF005-Hardware Scalability: Cloud Computing services, Infrastructure as a Service (IaaS) are scalable in terms of the characteristics of the servers, storage and data security
RF006-Administration of study subjects: management of study subjects, assignment of users according to role	RNF006-Software Scalability: Cloud Computing services, platform as a service (PaaS) maintain availability of components, services, configured APIs available at all stages of development and testing
RF007-Task management: it is the task management for the course	RNF007-Data security: maintain security in the data transport layer through TLS/SSL certificates
RF008-Activities administration: it is the activities management for the course	RNF008-Development with free software: base operating system, framework, server, database, web server and free code libraries will be used
RF009-Assessment administration: it is the evaluation management for the course	RNF009-Web design: the web design is responsive, friendly and easy to use
RF010-Administration of self-assessments: it is the management of self-assessments for the course	
RF011-Test administration: it is the test management for the course	
RF012-Exam administration: it is the administration of exams for the course	
RF013-Notes: control of notes of all tasks, activities, evaluations, self-assessments, tests and exams taken in the study cycle	
RF014-Resource Administration: includes the management for the administration of resources such as aids, questions, documents and outstanding works, digital libraries, digital repositories, wikis, blogs	

(continued)

Table 12.4 (continued)

Functional requirement	Non-functional requirement
RF015-Communication: includes asynchronous learning-oriented tools such as forum, email, calendar; and synchronous as chat, video conference	

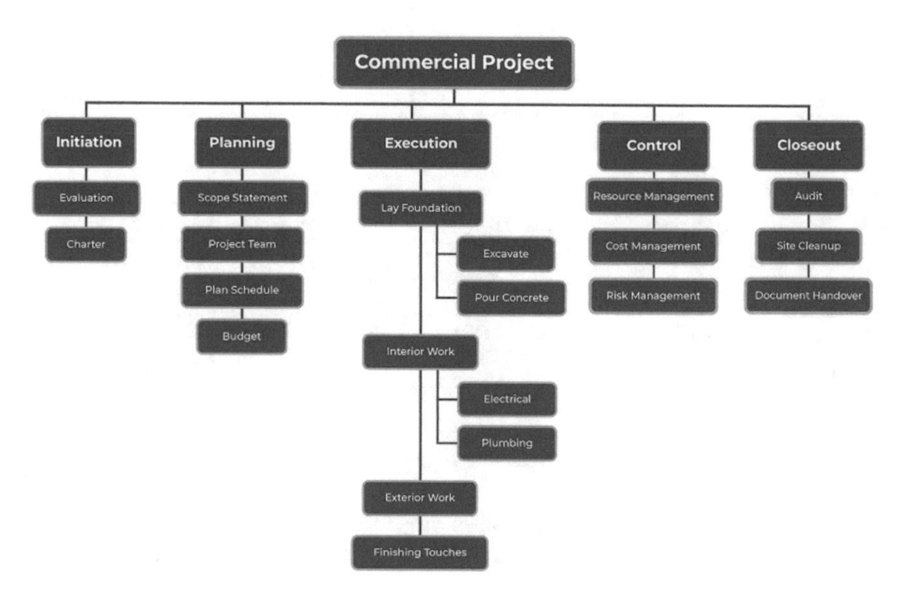

Fig. 12.7 Work Breakdown Structure (WBS)

The EVA Virtual Learning Environment will include an informative web portal, student registration and enrollment, as well as modules for Content Management, Administrative Management, Communication, Collaborative Work and Evaluation.

Work Breakdown Structure (WBS)

The work plan of this project is defined by a WBS—Work Breakdown Structure (EDT) divided into 5 Work Packages WP (Work Packages PT), the same ones that contain Task T, see Fig. 12.7:

WP1—Initiation Phase: the organization authorizes the planning of the project according to the business need. In a Constitution Act they define the purpose of the project, description of roles, success criteria and responsibilities, identification of interested parties and environmental factors.

WP2—Planning Phase: a Scope Management Plan is delivered containing the functional and non-functional requirements, strategic objectives of the organization and purpose of the project.

Table 12.5 Human resource

No.	Role	Quantity	Planned days	Cost days	Amount
1	Product management	1	119	$90.00	$10,710.00
2	Product Director	1	110	$80.00	$8800.00
3	WP Work Package Leader	5	100	$60.00	$30,000.00
4	WP Work Package Task Leader	4	100	$45.00	$18,000.00
5	Scrum Master	1	68	$60.00	$4080.00
6	Multidisciplinary professionals	3	68	$40.00	$860.00
	Total				$79,750.00

Table 12.6 Hardware resource

No.	Description	Quantity	Unit value	Amount
1	Laptops	12	$1200.00	$14,400.00
2	Server	1	$15,000.00	$15,000.00
3	Terminals	3	$1000.00	$3000.00
Total				$32,400.00

WP3—Execution Phase: the Scrum agile framework is used exclusively for software development planning.

WP4—Monitoring and Control Phase: project monitoring and control activities are displayed, the inputs are specified: scope control flow chart, schedule diagram and risk diagram.

WP5—Closure Phase: the formal closure of the project is established.

12.3.2.2 Budget Development

Budget Calculation

The budget estimate focuses on various costs of human resources, hardware resources, software resources and other expenses. The budget in Human Resources for the project is displayed in Table 12.5.

Regarding the Hardware Resource, the information is displayed in Table 12.6

12.3.2.3 Quality Management

PMBOK in its Quality Management addresses policies that seek to implement a Quality Management System in the context of the project, defining 3 clear processes as can be seen in Fig. 12.8.

Fig. 12.8 Quality management processes

However, the very nature of the Scrum agile framework, in the face of constant changes throughout the project life cycle, incorporates periodic controls in the Sprint Retrospective meetings with the purpose of discovering inconsistencies and quality problems early.

Plan Quality Management

The planning of the Quality Management of the project begins by identifying requirements by gathering information such as:

- Act of constitution of the project.
- Registration of interested parties.
- Documentation of requirements.
- Norms and evaluation standards.

Quality Assurance

The quality assurance of the project is carried out by defining and fulfilling the following activities shown in Table 12.7:

12.3.3 Monitoring and Control Phase

12.3.3.1 Scope Control

It is the process that monitors the status of the project and product scope. Scope Control is also used to manage changes, the same ones that are inevitable in any project, see Fig. 12.9 for the process and actions to be taken:

Table 12.7 Activities for quality assurance

No.	Activity	Description
1	Review of deliverables	• Review by a team member • Testing of deliverables by Development Team • Review to verify compliance with defined standards and checklist
2	Project fit review	• Review of critical products
3	Technical review	• Reviews in Scrum Team meetings
4	Critical documentation	Quality requirements on the use of the platform: a. Functionality b. Reliability c. Usability d. Efficiency e. Maintainability f. Portability

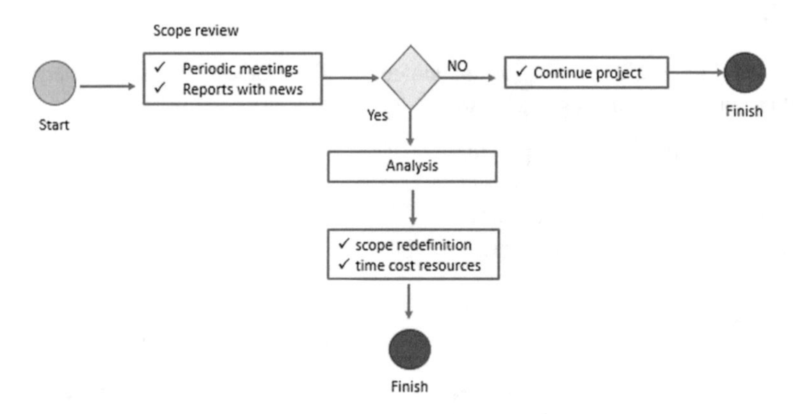

Fig. 12.9 Project scope control source: self-made

12.3.3.2 Schedule Control

It is the process that monitors all the activities planned in the work schedule in order to control and manage any changes, see Fig. 12.10 for the process and actions to be taken:

12.3.3.3 Risk Control

It is the process that allows risk response plans to be used, identified and monitored throughout the life cycle of the project. See Fig. 12.11 for the process and actions to be taken:

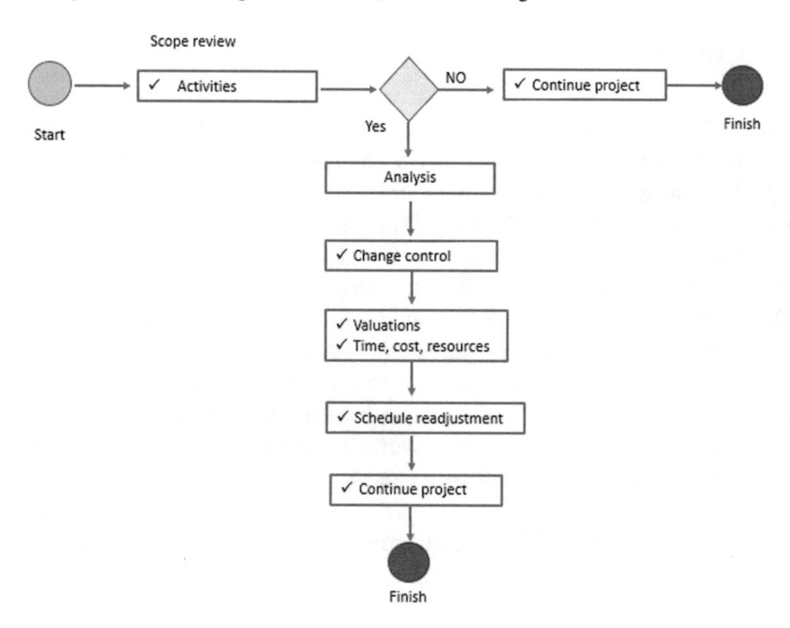

Fig. 12.10 Project schedule control

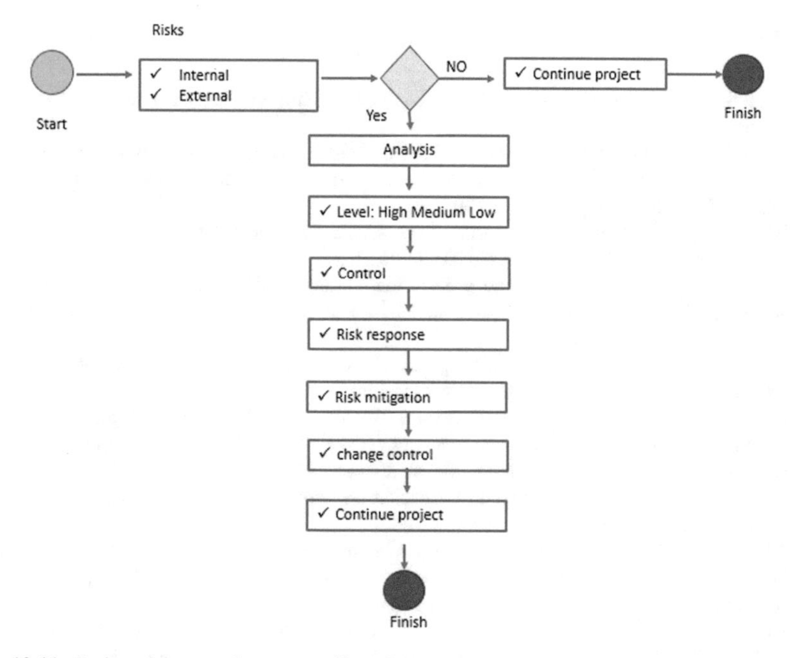

Fig. 12.11 Project risk control source: self-made

12.4 Conclusion

The ANESI with the fulfillment of the planning of the project in investment in technological, technical, logistical and pedagogical infrastructure will be able to become Education and training centers with the virtual study modality.

With the fulfillment of the planning of the present project, it will allow to establish a base work for build and implement a Virtual Learning Environment for the new Training Centers for Security and Private Surveillance agents.

The project meets the requirements and needs of users to be able to train virtually by integrating synchronous and asynchronous communication tools.

Implementing a Virtual Learning Environment in the country will promote the decentralization of the service, eliminating barriers of time and distance, economically benefiting users, reducing transportation costs and improving the flow of applicants for security and surveillance agents.

The project supplements with the use of 3D virtual reality immersion technology, practical modules of the curricular mesh that represents 10% of the required workload.

References

1. Argudín, M.L.: El conductismo (2016). Obtenido de http://hadoc.azc.uam.mx/enfoques/conductismo.htm
2. Schwaber, K., Sutherland, J.: La Guía de Srum. Obtenido de La Guía Definitiva de Scrum. Las Reglas del Juego, julio (2016). https://www.scrumguides.org/docs/scrumguide/v2016/2016-Scrum-Guide-Spanish.pdf#zoom=100
3. Newcomb, R.: Del cliente a las partes interesadas del proyecto. Obtenido de, 13 de mayo de 2017 https://doi.org/10.1080/01446190320000072137
4. Análisis Especializado de Riesgo y Seguridad Integral AERSIN CIA LTDA.: Galeria de Imagenes (2015). Obtenido de http://www.aersinsecurity.com/galeria-seguridad-privada-quito.html
5. ANESI.: Asociacion Nacional de Empresas de Seguridad Integral, 21 de octubre de 2005. Obtenido de http://anesi-ec.com/estatutosanesi.htm
6. Casallas, R., Yie, A.: Ingenieria de Software. Obtenido de Ciclo de Vida y Metodologias, 03 de diciembre 2017. https://web.archive.org/web/20131203234610/, http://sistemas.uniandes.edu.co/~isis2603/dokuwiki/lib/exe/fetch.php?media=principal:isis2603-m
7. CEFOSEG CÍA. LTDA.: Centro de formación en seguridad (2020). Obtenido de https://cefoseg.edu.ec/; https://cefoseg.edu.ec/
8. Aretio, G.: Bases, mediaciones y futuro de la educación a distancia en la sociedad digital. Obtenido de LMS. Plataformas Virtuales o Entornos Virtuales de Aprendizaje. Ventajas y funcionalidades (2018). https://aretio.hypotheses.org/3292
9. Hughey, D.: Comparación de análisis y diseño de sistemas tradicionales con metodologías ágiles (2018). Obtenido de http://www.umsl.edu/~hugheyd/is6840/waterfall.html
10. INEC.: Proyecciones poblacionales (2010). Obtenido de https://www.ecuadorencifras.gob.ec/proyecciones-poblacionales/
11. Ministerio de Gobierno.: Preparación de personal de seguridad privada(2020). Obtenido de https://www.ecuadorencifras.gob.ec/proyecciones-poblacionales/

12. Ministerio del Interior.: Ministerio del Interior. Programa de Capacitacion Para Los Guardias de Vigilancia y Segurid Privada, octubre (2015)
13. Jaramillo, M.M.: Plan de negocios para la creación de una empresa de capacitación y entrenamiento de guardias de seguridad en la ciudad de Quito (2019)
14. Perod, M.: La biometría, una apuesta por la seguridad (2018). Obtenido de https://www.muy interesante.es/tecnologia/articulo/la-biometria-una-apuesta-por-la-seguridad-751525347559
15. Pierre.: Pedagogía e internet: aprovechamiento de las nuevas tecnologías. Trillas, México (2018)
16. PMI.: Guía de los Fundamentos para Dirección de proyectos. Global Standard, Pensilvania (2020)
17. Pressman, R.: Ingeniería del Software (2017). Obtenido de Un enfoque práctico https://cotana. informatica.edu.bo/downloads/ld-Ingenieria.de.software.enfoque.practico.7ed.Pressman.PDF
18. Torres, M.R.: Desarrollo de Paginas Web Como Recurso Para Facilitar el Aprendizaje (2018). Obtenido de. https://Dialnet-DesarrolloDePaginasWebComoRecursoParaFacilitarElAp-271 9448.pdf
19. Sánchez, A.: Desarrollo de Entornos Virtuales para Web (2005). Obtenido de https://www.res earchgate.net/publication/280304213_Desarrollo_de_Entornos_Virtuales_para_Web
20. Siemens, G.: Connectivismo. Obtenido de A Learning Theory for the Digital Age, enero (2005). https://jotamac.typepad.com/jotamacs_weblog/files/Connectivism.pdf

Part V
Innovative Business Models and Applications for Smart Cities

Chapter 13
Internal Stakeholders' Readiness for Developing Smart Railway Services Through Crowd-Based Open Innovations

Nenad Stanisavljević, Danijela Stojanović, Aleksa Miletić, Petar Lukovac, and Zorica Bogdanović

Abstract The railway is large and complex traffic and business system. That is why solving specific problems in the business and functioning of railways through classic business processes is difficult, expensive, and slow. The development of the Internet of things has enabled the wider social community to identify problems and find solutions in many areas of life, including on the railways. Thanks to this, opportunities were created for the railway to approach the solution of specific problems through open innovation as a more flexible, faster, and cheaper business concept. While foreign railways apply open innovation to improve smart railway traffic in smart cities, this concept has hardly been applied to Serbian Railways until now. The authors of the paper researched railway companies in Serbia to analyze internal stakeholders' interest and readiness as possible participants in the process of open innovation in Serbian Railways to accept a new business concept and contribute to innovations in this area. The research showed that railway workers lack experience in applying open innovation and that there are unknowns in this area. They are also familiar with the possibilities, advantages, and importance of applying this business concept. With proper information to the employees, good conditions could be created to use open innovations on Serbian Railways successfully. The results of this research represent a reasonable basis for the application of open innovation on Serbian Railways.

N. Stanisavljević
Infrastructure of Serbian Railways, Belgrade, Serbia

D. Stojanović
Institute of Economic Sciences, Belgrade, Serbia

N. Stanisavljević · A. Miletić · P. Lukovac · Z. Bogdanović (✉)
Faculty of Organizational Sciences, University of Belgrade, Belgrade, Serbia
e-mail: zorica@elab.rs

J. L. Reis et al. (eds.), *Marketing and Smart Technologies*, Smart Innovation, Systems and Technologies 337, https://doi.org/10.1007/978-981-19-9099-1_13

13.1 Introduction

When introducing new technologies or business approaches into a company, it is beneficial to investigate the readiness of relevant stakeholders to accept the innovation [1, 2].

This paper investigates the readiness of internal stakeholders and possible participants in developing open innovation on Serbian Railways to show their interest in introducing and contributing to this new business concept.

The research aims to conduct a preliminary survey of the readiness of employees in railway companies in Serbia to introduce the business concept of "crowd-based" open innovation. By applying open innovations in railways, conditions are created for faster, easier, and better solving some specific issues in the functioning and business of this transport system [3, 4]. At the same time, open innovations in railways create opportunities for improving smart railway traffic, as one of the important services of a smart city [5, 6]. And finally, the application of "crowd-based" open innovations in railways provides the opportunity for citizens, the economy, and other external stakeholders to improve railway traffic in smart cities [7].

The authors described and defined internal stakeholders in railway companies, whose readiness depends on the acceptance and application of "crowd-based" open innovations on Serbian Railways and their place and role in developing this business concept. Then, the paper presents the model of application of open innovations in Serbian Railways, in which each of the internal stakeholders in the implementation of this project is defined.

In the paper, the authors presented research that included 54 employees of the "Infrastructure of Serbian Railways," and a railway company in the railway sector that is responsible for the safety and functioning of railway traffic has the most employees and implements the most significant infrastructure and investment projects in the railway sector, due to which is representative for the entire railway sector in Serbia. The research goal was to determine internal stakeholders' readiness for developing smart railway services through open investments. In the paper, the author was first presented with the methodology by which the research was carried out and then the research results. And finally, based on the results of the conducted study, the authors announced future projects and analyze that they will deal with in the field of application of "crowd-based" open innovations as a new business concept of Serbian Railways, in the function of developing smart cities and smart rail traffic in them.

13.2 Literature Review

Until two decades ago, innovative activities took place within companies and were realized with internal resources, knowledge, and capacities. For the first time in 2003, Chesbrough used the term "open innovation," defining it as the use of knowledge from the company and its environment, in order to increase internal innovation processes thanks to external knowledge, and to increase the market for plasma of existing internal innovations [8].

Thanks to the development of Internet intelligent devices, opportunities have been created for the wider social community in many areas of social life to define problems and offer solutions [9].

Thus, in the analysis of the readiness of telecommunication operators in Serbia for the application of crowd-based open innovations in the development of smart city services, internal and external stakeholders showed great interest precisely in smart traffic [10].

The railway sector has a significant impact on the European industry, so it is important that the railway follows the current innovative trends and adapts to the new digital revolution, in order to remain competitive. The RailActivation project investigated innovations in small and medium-sized enterprises for the needs of rail transport [3]. The development of the service sector and its improvement in meeting social needs, as well as the competitiveness of services, are directly related to innovations and the level of innovation [11].

When it comes to the railway sector, the authors investigated non-technological innovations, especially workplace innovations in 203 railway entities across the European Union, defining ways to implement them [12]. Some authors also propose the formation of an innovative transport-logistics cluster as an organizational basis for the collection and development of scientific ideas and knowledge for the needs of industrial railway transport companies [13].

Organizations must be able to attract external knowledge, which is extremely related to an open innovation approach. Through semi-structured interviews with eight railway traffic experts, it is confirmed that open innovation and knowledge management are complementary in various aspects, and that internal mechanisms at the railways can influence the integration of knowledge by using the online community [4].

Companies are reorienting from an exclusively internal type of organization of innovation activities to mainly external ones due to the mass digitization of all aspects of the activity and the reduction of the duration of the equipment and technology update cycle. That is why some authors researched the forms of innovation in selected world railway companies, with the aim of looking at the tools, methods and its most common forms [14]. The modernization of railway transport encourages the development of innovative activity in the wider social community, so the construction of the high-speed railway in China contributed to an increase in the number of registered patents in the cities connected to the high-speed railway [15].

The digitalization of the railway industry and its future challenges was among the main topics at the International Transport Technology Fair (Inno Trans) 2016. It is an opportunity for the future and competitiveness of the European railway industry, based on an intelligent railway system and artificial intelligence [16].

A true example of the connection between open innovation and artificial intelligence in the railway system is the "ATUVIS" project (autonomous robot for visual inspection of train undercarriage), which was implemented by professors from the

Faculty of Mechanical Engineering at the University of Niš, led by Professor Aleksandar Miltenović. The prototype was made at the request of private railway operators with the aim of controlling trains and was financed thanks to the European Union program [16].

And the application of open innovations contributes to the development of intelligent railway systems. Thanks to this, rail transport is becoming an important function of smart cities today.

13.3 Roles and Stakeholders in Crowd-Based Open Innovation in Railway Context

The beginning of open innovation on Serbian Railways dates back to the mid-eighties of the last century. Thus, in 1984, the Active Inventors of the Railway Transport Company "Belgrade" was founded, which in 1998 was transformed into the Organization of Inventors of Railwaymen of Serbia. The founders of this organization were inventors and authors of technical improvements to the railway.

Of course, then, the global development of the Internet and the Internet of things (hereinafter: IoT), and therefore also on the Serbian Railways, did not enable the application of the "crowd-based" business model of open innovation in today's classical sense.

Until two decades ago, their employees implemented innovative activities and development projects exclusively within the companies, under controlled conditions, and without competition.

At the end of the twentieth century, Serbian Railways lacked only Internet platforms to apply "crowd-based" open innovations. In designing, developing, and implementing innovations, the employees of Serbian Railways had complete independence, innovative activities were not part of their work duties and tasks, and experts from outside the railways, most often from business or educational institutions, could be involved in those development projects. With their innovations and technical improvements, they resolved issues of efficiency, safety, and working conditions on the railways. Their status was determined by normative acts of Serbian Railways, including monetary compensation. This resulted in close to a thousand technical solutions, of which about 60% were accepted, resulting in savings of several million euros in the previous three and a half decades [17].

However, the restructuring of "Serbian Railways" into four railway companies, which was implemented in 2015, was accompanied by a loss of business interest in innovative processes, so the three newly founded railway companies "Infrastructure of Serbian Railways," "Srbija Voz," and "Srbija Cargo" were not even until today defined the status of innovators and innovations in the railways. Unlike many other railways, which in the innovative field followed the global technological development of the Internet and IoT, this was not the case for Serbian Railways. All this resulted not only in the fact that the business system of open innovation did not take off but in the gradual shutdown of the innovative process on Serbian Railways. Nevertheless, past experiences in innovative work can certainly contribute positively to the future development of the "crowd-based" of open innovations in the railway sector in Serbia.

By adopting the "crowd-based" model of open innovation, Serbian Railways can simply and efficiently expand their opportunities for new ideas, research, innovation, and company development and increase savings and revenue.

13.3.1 A Model for Introducing the Concept of Open Innovation in Serbian Railways

For the concept of open innovation to be introduced in Serbian Railways, it is necessary to fulfill the following conditions:

1. *Formal–legal and normative*

(a) The Board of Directors of the railway company should adopt the rulebook on open innovation. This rulebook would fully regulate the area of open innovation in the railway company, among other things, the procedure for starting an initiative to solve specific issues through open innovation, the flow of innovative activities, then the deadlines for implementation, methods, and sources of financing, financial incentives for innovators, as well as the procedure for the practical application of innovations and copyright protection.

2. *Technical–technological*

(a) Construction of an Internet platform will enable the railway company to find sources of innovation and participants in the innovation process in the business environment and among citizens, not only among its employees.

The open innovation platform provider would have the task of providing the technical infrastructure and Internet platform. Internet services for the operation of the platform, setting up an internal organization and hiring human resources for marketing the platform, implementation of business processes, user support, service billing, as well as copyright and other rights protection participants in the work of the platform, which is a very complex set of tasks.

3. *Organizational and personnel*

(a) Establishment of the Department for Open Innovation.

The department for open innovation would be established based on the decision of the Board of Directors of the railway company on changes in organization and systematization. The department could be organized within the development or human resources sectors.

The department for open innovation would perform all administrative and professional tasks related to innovative activities and their application, which the decision would precisely define on changes to the organization and systematization of the railway company and the rulebook on open innovation.

(b) Appointment of the Expert Committee for Innovation.

The Innovation Commission would be formed by the Appointment Decision, which the General Manager makes of the railway company. The Appointment Decision defines the tasks, obligations, and responsibilities of the members of this Commission. The Commission would be composed of railway experts from various fields (traffic, electrical, mechanical, construction, legal, economic, information technology, etc.).

This Commission would have the task of defining issues and problems in the functioning and operation of railway systems, which should and could be solved through open innovation. Also, this Commission would resolve requests submitted in the open innovation system.

The Commission would submit all its decisions to the Board of Directors for final decision-making, and its work would answer to the general director of the railway company.

4. *Financial*

As its management bodies, the Board of Directors and the Assembly of the railway company would define the financial resources necessary for establishing and applying the principle of open innovation on an annual basis through the business program approved by the Government of the Republic of Serbia.

13.3.2 Stakeholders in Serbian Railways

For the "crowd-based" business concept of open innovation to be successfully implemented in Serbian Railways, it is necessary to define the stakeholders and their roles in this process.

Stakeholders in "crowd-based" open innovations on Serbian Railways can be internal and external.

Internal stakeholders

Internal stakeholders in Serbian Railways contribute to the launch of the business concept of open innovation by making appropriate decisions. They are looking for partners with ideas and knowledge for specific issues through the platform. With their work and actions, they contribute to the development of "crowd-based" open innovation.

The internal stakeholders of Serbian Railways are as follows:

(a) Management bodies of Serbian Railways,

- the general director, the Board of Directors, and the Assembly are management bodies that make appropriate decisions regarding the implementation of open innovations;

(b) Managerial level of the company,

- this management level should create conditions for the application of the concept of open innovation;

(c) Expert services of Serbian Railways,

- departments of traffic, construction, electrotechnical, investment, information, and other professional activities on railways, which should identify and define specific issues, problems, and areas that can be solved with the business concept of open innovation;

(d) Employees in the executive services of Serbian Railways,

- we are talking about employees who should point out technical–technological difficulties and problems when performing daily work and work tasks, the solution of which will improve efficiency, improve operations and functioning and increase savings and revenues of the railway company.

External stakeholders

External stakeholders represent individuals, groups, or organizations interested in transferring their innovative ideas and knowledge to Serbian Railways through the Internet platform, for money, or to achieve some other interest.

When it comes to Serbian Railways, external stakeholders can be:

(a) state authorities (Ministry, Government of the Republic of Serbia), which should accept the concept of open innovation as a principle for solving specific problems on the railways;

  (a) employees in the railway sector.

    - considering the complexity and multidisciplinarity of railway traffic, we are talking about railway workers who would be engaged outside of their regular business obligations and work tasks;

(b) Companies working for the needs of Serbian Railways–railway industry

- about two thousand domestic and foreign companies that work for the needs of Serbian Railways;

(c) Academic institutions (e.g., Faculty of Traffic, Mechanical Engineering, Civil Engineering or Electrical Engineering, Faculty of Organizational Sciences, Institutes, and the like);
(d) Public administration and local self-government;
(e) Entrepreneurs;
(f) Freelancers;
(g) citizens.

And while there is no implementation of an open innovation project without external stakeholders, the "crowd-based" open innovation business concept cannot even be launched and implemented without internal stakeholders.

13.4 Methodology

The railway is a significant and technically-technologically complex traffic and business system, which is why the definitive solution to specific problems in functioning is complicated, expensive, and slow. The improvement of infrastructure solutions and business processes in specific situations is usually limited and consequently difficult to apply globally on the entire territory of Serbian Railways.

Thanks to the IoT devices, the broader social community has been allowed to define problems and offer solutions in many areas of social life [18, 19] which also includes railway traffic [20–22]. This enabled railway companies to not only look for solutions to specific problems and situations in classic and traditional business processes.

Unlike classic solutions, the business concept of open innovation is much more flexible, effects can be achieved in a shorter period, and their implementation is easier to implement and costs less.

That is why, thanks to the business concept of open innovation, specific problems and specific situations in railway traffic can be solved much better, faster, and more manageable than classic business mechanisms. In this way, open innovations based on IoT, with their innovativeness and comprehensiveness, can significantly contribute to the functioning and business of the railway transport system [23, 24].

In foreign railways, open innovation has been represented for a couple of decades as an innovative and high-quality way of solving specific problems in functioning and is widely used to improve operations [4, 25–29].

Unlike foreign railways, the application of the concept of open innovation has not yet taken root on Serbian Railways.

The first research on the willingness to participate in open innovation projects on railways was conducted with students from the Faculty of Organizational Sciences of the University of Belgrade in cooperation with railway experts [23]. The research was conducted on solving characteristic situations on the railway in which traffic safety is threatened through open innovation.

Students have shown willingness to participate in this type of open innovation and similar projects in future. During this research, it was confirmed that prototype solutions based on open innovation and the Internet of things could contribute to increasing safety in specific segments of railway traffic [24].

However, for the business concept of open innovation to be introduced in Serbian Railways, it is necessary to research and analyze their readiness to adopt, organize, and implement this type of project.

The authors of this paper, bearing in mind the results of the previous research, surveyed the companies in the railway sector in Serbia dedicated to the analysis of their readiness to include an Internet platform for open innovation and the application of the "crowd-based" business model of open innovation in their services in future. The survey was conducted among employees of Serbian Railways in July–September 2022 and was in written form and anonymous.

The survey was dedicated to researching the readiness of possible participants in the process of open innovation in Serbian Railways to express their interest, accept this concept, and contribute to innovations in this area.

The survey covered 54 employees of the "Infrastructure of Serbian Railway," which is about 1% of the total number of employees in this railway company, which as of December 31, 2021 had 5,801 employees [30].

According to the structure, the employees who were surveyed belong to the managerial level of the company, which should create conditions for the application of the concept of open innovation on Serbian Railways, and professional services, which should identify issues and areas in which the quality and efficiency of work can be improved by applying t4 which is concept. Therefore, we are talking about internal stakeholders in Serbian Railways, whose influence is decisive in introducing and using the idea of open innovation in the business and practice of Serbian Railways.

Before the survey itself, the participants had the opportunity to familiarize themselves with the goal of this project and the definition, application, and importance of the concept of "crowd-based" open innovation through written information.

A total of 36 questions, grouped into twelve areas, were asked of the employees of the "Infrastructure of Serbian Railway." In the survey, railway workers answered questions about open innovation on the railways, which related to the improvement of innovation activity, the added value of open innovation, reputation, and social responsibility, then income and costs, effort and time spent, risk and loss of knowledge power, experiencing the value of open innovations, as well as resources for the application of open innovations and interest in participating in the development of specific services. And finally, the participants in the survey could present their initiatives and suggestions that the services needed by the railways could be realized through open innovation.

The participants in the survey could answer the questions with one of the five offered answers, which read: "completely agree," "agree," "not sure," "disagree," and "completely disagree."

In addition to demographic data, the surveyed railway workers had the opportunity to state whether they had the previous experience in open innovation, as well as what their desired role in the project was, with the answers offered: "participant—proposing ideas and solutions," "participant—solving tasks and developing prototypes," "organization and management of open innovation projects," and "manager—development of strategy and business models."

The survey and research aimed to examine railway companies' readiness in Serbia to apply the "crowd-based" open innovation business model.

13.5 Results

Out of 54 respondents, 15 were aged between 25 and 40, 24 were between 40 and 55, and 15 were over 50. There were 15 men and 39 women. Only six respondents had an education up to the secondary school level, and 48 had a high school education and above. Thirty-three respondents had earnings more elevated than 70 thousand dinars (Table 13.1).

Such demographic data are conditioned by the fact that the surveyed employees belong to the managerial level and professional services, which in the organization of the railway company significantly influences the application of the concept of open innovation.

Confirmation that in Serbian Railways, "crowd-based" open innovations have almost no application so far is the data from the research that out of 54 respondents, only 9 (16.7%) had the previous experience in this field. Even though they are experienced and well-positioned employees in the company's organization. It is precisely the lack of prior experience in open innovation on Serbian Railways that is most likely the reason that the most significant number of respondents—as many as 27 (50%), decided to be involved in future projects in this area only as a participant, who will propose ideas and solutions. Only nine respondents (16.7%) want to participate in open innovation projects as participants who will solve tasks and develop prototypes. In contrast, 15 of them (27.8%) opted for the functions of organizing and managing open innovation projects. And finally, only three respondents (5.5%)

Table 13.1 Demographic data

Years	25–40	40–55	<55
	15	24	15
Gender	M		F
	15		39
Experience	Yes		No
	12		48
Monthly earning	>70.000RSD		<70.000RSD
	21		33
Level of education	High school		Secondary school
	48		6

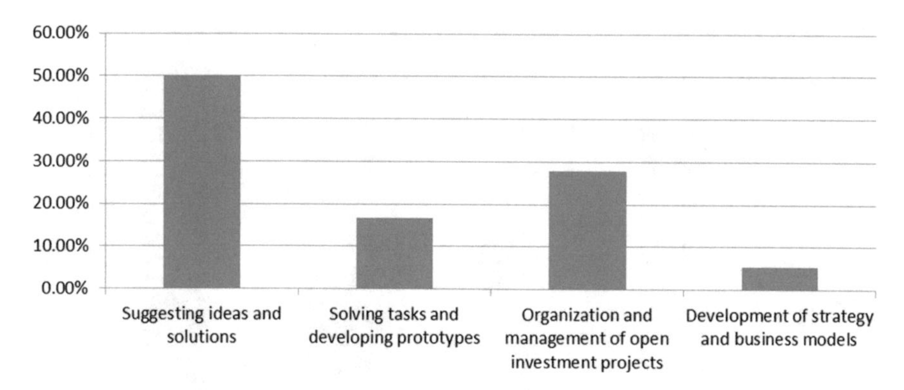

Fig. 13.1 Desired role in the project

would like to develop strategy and business models of open innovation on Serbian Railways as managers (Fig. 13.1).

Regarding the improvement of innovative activity on the railways, as many as 39 respondents (72.2%) declared in the survey that they fully agree or agree with the statements that open innovation improves the innovative activity and innovation potential of railway employees. In comparison, only 15 respondents (27.8%) stated that they were not sure of the accuracy of one or both of these statements. That the respondents know the impact of open innovation on innovation activity, and the innovation potential of railway employees is confirmed by the fact that none of them stated in the survey that they disagree with the above statements.

The respondents did not show such certainty regarding the additional values of open innovations. Namely, with claims that the implementation of the concept of open investments in the railway will increase the efficiency, quality, and volume of work on the railway, then the competitive advantages of the railway over other modes of transport, and finally increase the safety of railway traffic, with "I completely agree" or "I agree." Twenty-seven respondents (50%) answered. Not one of the respondents declared that they disagreed with the statements, which is why the remaining 27 respondents (50%) were unsure about any or all of the three statements. As many as 21 times, the respondents answered that they are not sure whether the application of the concept of open innovation increases the safety of railway traffic, and with the answer "I am not sure" 12 times, the respondents in the survey expressed their opinion on the claims that open innovations increase the efficiency, quality, and volume of work on the railways. That is, to improve the competitive advantage of railways over other modes of transport (Fig. 13.2).

In the survey, 36 respondents (66.7%) declared that they "completely agree" or "agree" with the statements regarding the impact of open innovation on the reputation of the railways. According to them, there is no doubt that participation in the development of innovative services has a positive effect on the importance of the railway, on existing railway services, and the improvement of relations with service users. Eighteen respondents for one or more such claims answered that they were "not

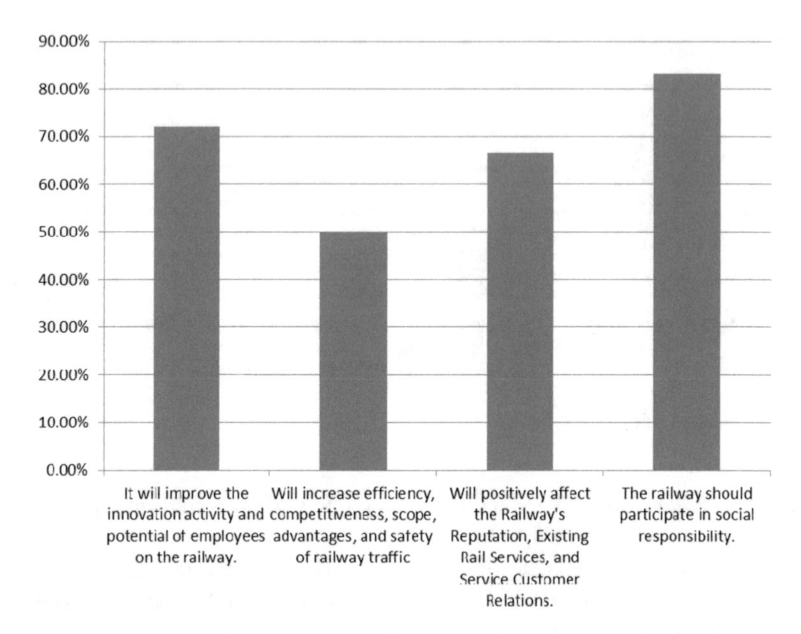

Fig. 13.2 Number of respondents who "completely agree" and "agree" with the following statements about open innovation in railways

sure" of their accuracy. Three respondents declared that they "completely disagree" or "disagree" with the statement that participation in developing innovative services contributes to improving relations with service users.

One of the statements in the survey where respondents had a minor dilemma refers to social responsibility. As many as 45 respondents (83.3%) fully agreed or agreed with the statement that railways should participate in the development of innovative services. Only nine respondents (16.7%) were unsure of such a statement. None of the interviewees disputed the accuracy of this statement and did not declare that they disagreed with it.

The most doubts among the respondents were caused by survey questions related to the income and costs of open innovation and the effort and time spent to implement the concept of open innovation on the railways. Thus, only 18 respondents (33.3%) in the survey declared that they "completely agree" or "agree" with the statements that the railways will generate income or savings from open innovations, as well as that the open innovation program will be able to finance through external grants. None of the respondents disagreed with any of these statements. However, as many as 36 respondents (66.7%) declared that they were "not sure" about the accuracy of one, two, or all three stated statements. As many as 27 respondents said that they were "not sure" about the accuracy of the statement that the railways will be able to finance the open innovation program through external grants, and 24 respondents were not sure about the accuracy of the information that the railways would achieve savings through the application of open innovations. In comparison, 18 respondents were unsure whether the railway would generate income from the open innovation program. This means that two-thirds of the respondents have significant uncertainties and dilemmas regarding revenues and savings from applying open innovation in railways.

On the other hand, when it comes to the costs required for the implementation of the open innovation program on the railways, the most significant number of respondents declared that a lot of investment will be needed to realize a business model based on open investments on the railways, but also that these investments will be cost-justified. The biggest dilemma for the respondents was whether it would be challenging to encourage interested parties within the railway to develop a business model based on open innovation. Thus, 33 respondents "completely agreed" or "agreed" that a lot of investment will be needed to implement a business model based on open innovation on the railways. Still, 42 respondents agreed that the investment in the open innovation program was cost-justified. 12 or 3 respondents did not agree with these two statements, and nine were unsure about the accuracy of these two statements. The respondents had the most dilemmas regarding the survey question of whether the income from the implementation of open innovations will justify the costs, so 15 respondents agreed that they would, that is, they would not, and 24 respondents were not sure of the accuracy of this statement.

In contrast to these dilemmas, the respondents in the survey were quite sure that a lot of effort will be necessary to implement open innovation, that it will be challenging to encourage railway stakeholders to implement and develop the business concept of open innovation, and that it will take a lot of time. Thus, 36, or two-thirds of respondents (66.7%), "completely agreed" or "agreed" with these three statements. Three respondents were not sure of the accuracy of all three of these statements, the same number of them were not sure whether it would take a lot of effort to realize the business concept of open innovations on the railway, and 12 answered in the survey that they were "not sure" in the statement that it would be challenging to encourage stakeholders within the railways to apply open innovation. Unlike the other respondents, three disagreed with the claims that it will be difficult to encourage those interested in the railway and that implementing this program will take a lot of time.

That the business concept of open innovation is new and unknown on Serbian Railways was also confirmed by the survey questions that concerned the possible risk and loss of power of knowledge due to the application of open innovation. Six respondents "completely agreed" or "agreed" with the statement that developing a business model based on open innovation will cause railways to lose specific knowledge, while 21 disagreed. While 12 respondents believed that the application of open innovations would threaten individual jobs on the railways, only three respondents disagreed with that statement. Unlike 15 who thought that open innovations are risky for the railways, 18 respondents had the opposite opinion. A total of 14 respondents answered 52 times to one of these statements that they were "not sure" of their accuracy.

Respondents had fewer dilemmas and unknowns regarding questions related to experiencing the value of open innovations. Thus, 30 respondents (55.6%) "completely agreed" or "agreed" with the statements that open innovations will bring value to the railway, as well as genuine value to the employee that it will be worth the effort invested and that it will pay off. Twelve respondents (22.2%) did not agree with these statements, while the remaining 12 respondents were "not sure" of the accuracy of the stated statements.

Respondents had even fewer dilemmas with the survey questions related to the resources needed to implement open innovations. A total of 36 respondents (66.66%) "completely agree" or "agree" with the survey statements that the railways have the necessary knowledge, skills, and resources for the application of open innovation, while nine respondents (16 0.67%). In contrast, the same numbers of respondents are "unsure" of the above statement.

In the survey, respondents were also offered the opportunity to state whether they are interested in developing specific services based on open innovations, solving particular problems and situations in the safety and functioning of railway traffic. Respondents were offered opportunities to develop the following services based on open innovation: traffic safety at road crossings, electric shock from the contact network above the railway, use of the railway for disabled people, protective fence along the railway, noise from railway traffic, theft of railway parts and equipment, accidents on the open railway, and the safety and quality of railway traffic functioning. In the survey, 42 respondents (77.8%) confirmed their interest in participating in the development of all these services, declaring that they "completely agree" and "agree" with the stated statements. Only six respondents said they are not interested in participating in developing a service, namely the one related to accidents on the open railway. A total of 12 respondents declared that they were "unsure" about their interest in participating in developing some of the services mentioned above (Table 13.2).

And finally, the respondents had the opportunity to propose services that the railways need, which could be realized through open innovation. The respondents mentioned three essential services for railway traffic safety that could be solved with open innovations. It is about detecting landslides on the railway, bursting rails, and deformations on the track, which could be done much more efficiently with the application of open innovations and IoT, with better and earlier detection than with current, traditional, and classic methods.

The research conducted among the employees of railway companies to analyze the readiness to accept open innovations showed that railway workers do not have much experience in applying this business concept and have a lot of dilemmas and unknowns. However, the analysis indicates that railway operators are familiar with the possibilities, advantages, and importance of applying the business concept of open innovation. With sufficient information related primarily to the impact of open innovation on the safety and functioning of the railway, the necessary costs and potential income from this business concept, as well as the risks to the railway and individual jobs, good conditions would be created for the successful application of the idea of open innovation on Serbian Railways.

Table 13.2 Respondent's interest in participating in the development of specific services

A proposal for the services that the railway needs	"Completely agree" or "agree"	"Not sure"	"Disagree" or "completely disagree"
Traffic safety at road crossings	54 (100%)	–	–
Electric shock from the contact network above the railway	54 (100%)	–	–
Use of the railway for persons with disabilities	48 (88.9%)	6 (11.1%)	–
Protective fence along the railway	48 (88.9%)	6 (11.1%)	–
Noise from railway traffic	48 (88.9%)	6 (11.1%)	–
Theft of railway parts and equipment	48 (88.9%)	6 (11.1%)	–
Accidents on the open track	42 (77.8%)	6 (11.1%)	6 (11.1%)
Functioning of railway traffic	48 (88.9%)	6 (11.1%)	–

The results of this research represent an excellent theoretical starting point for the practical application of the business concept of open innovation on Serbian Railways in the coming period.

13.6 Conclusion

This paper presents the research results into the readiness of the railway company "Infrastructure of Serbian Railways" and its employees to accept and apply the new business concept of "crowd-based" open innovation in solving specific difficulties and problems in the functioning and business of this complex technical–technological traffic system.

The research results showed that the surveyed railway workers had no experience applying open innovation and lacked confidence in using this business concept on Serbian Railways. However, the research showed that the employees of this railway company recognize the qualities, advantages, and opportunities of the concept of open innovation in solving problems on the railway and are ready not only to accept this business concept but also to participate in it.

For open innovation to be implemented as a concept for solving specific problems on the railways, it is necessary first to determine the readiness of critical internal and external stakeholders for this kind of business concept.

In subsequent research, the authors will analyze state authorities' readiness to accept open innovation to solve specific problems on the railways. This primarily refers to the Government of the Republic of Serbia and relevant Ministries, which should accept "crowd-based" open innovation as a new business concept for Serbian Railways. Apart from the fact that the state administration and the Serbian Railways would be presented to the public as modern, innovative, and progressive in this area, the application of open innovations would also create the conditions for allocating less money for the railway infrastructure in specific segments, and for specific problems to be solved more quickly and efficiently.

In future steps, it will be necessary to research and analyze other external stakeholders, primarily the railway industry and the academic community, to determine their readiness to participate in the business concept of open innovation of Serbian Railways.

After a critical analysis of the willingness of stakeholders to initiate and participate in the program concept of open innovations on Serbian Railways, the authors of this paper are preparing a complete pilot project and then evaluating the solutions obtained by the pilot project "crowd-based" open innovations.

Through the implementation of the research mentioned above and projects, the authors of this work gradually create conditions for applying the concept of open innovation on Serbian Railways, which can be considered the highest practical and theoretical goal of these activities.

In a practical sense, applying the concept of open innovation on railways can contribute to the development of specific products or economic activities in this transport system. In a scientific understanding, new theoretical research and assumptions can be presented through open innovations on Serbian Railways.

Funding This work was supported by the Ministry of Education, Science and Technological Development of the Republic of Serbia (institutional funding of FON and IEN).

References

1. KardanMoghaddam, H., Rajaei, A., Fatemi, S.: Acceptance of cloud computing in an airline company based on Roger's diffusion of innovation theory. Facta Univ. Ser. Electron. Energ. **34**(3), 461–482 (2021)
2. Karasev, A., Beloshitskiy, O., Shitov, A., Arkhipov, E., Tulupov, D.: Integral assessment of the level of innovative development of the railway industry companies. Open Transp. J. **16**(1) (2022)
3. Carranza, O., Petrini, G., Sanchez, G., De la Rua, B.: RailActivation project: the adoption of Workplace Innovation in the Rail Sector. Eur. J. Work. Innov. **7**(1), 29–52 (2022)
4. Babaei Ebrahimabadi, M., Radfar, R., Eshlaghy, A.T.: Knowledge management in railway industry: a conceptual model based on open innovation and online communities. Int. J. Railw. Res. **6**(1), 63–72 (2019)
5. Tokody, F., Flammini, F.: The intelligent railway system theory. Int. Transp. **69**(1), 38–40 (2017)
6. Alawad, H., Kaewunruen, S., An, M.: A deep learning approach towards railway safety risk assessment. IEEE Access **8**, 102811–102832 (2020)

7. Stanisavljević, N., Stojanović, D., Petrović, L.: Open innovation and crowdsourcing: challenges and opportunities for Serbian railways. E-Bus. Technol. Conference Proceedings **2**(1), 36–41 (2022)
8. Bogdanović, Z., Stojanović, M., Radenković, M., Labus, A., Despotović-Zrakić, M.: Mobile operator as the aggregator in a demand response model for smart residential communities. Lecture Notes on Data Engineering and Communications Technologies, 79, 58–67 (2021)
9. Chesbrough, H.: Open innovation: where we've been and where we're going research-technology management 55(4), 20–27 (2012)
10. Sarić, Ž., Obradović, V., Bogdanović, Z., Labus, A., Mitrović, S.: Crowd-based open innovation in telco operators: readiness assessment for smart city service development. Serbian J. vacije Management **17**(1), 179–196 (2022)
11. Saktaganova, G.S., Karipova, A.T., Legostayeva, L.V.: Innovative development of the railway service (2019)
12. de Loizaga, G.C.R., Gonzalez, B.S.: Innovation methodologies to activate inclusive growth in the organization. Adv. Decision Making, 79 (2022)
13. Shchepkina, N., Meshkova, N., Goigova, M., Maisigova, L., Tochieva, L.: Intellectual capital as a factor in ensuring the competitiveness of the railway transport enterprises. Transport. Res. Procedia **63**, 1444–1453 (2022)
14. Karasev, O.I., Zheleznov, M.M., Trostyansky, S.S., Shitova, Y.A.: Comprehensive analysis of forms of innovative activity of foreign railway companies. World Transp. Transport. **18**(2), 158–170 (2020)
15. Sun, D., Zeng, S., Ma, H., Shi, J.J.: How do high-speed railways spur innovation? IEEE Trans. Eng. Manag. https://doi.org/10.1109/TEM.2021.3091727
16. www.europa.rs/upoznajtezeleznickogrobota
17. Monografija "35 godina organizacije pronalazača železničara Srbije." Beograd (2020)
18. Petrović, L., Stojanović, D., Labus, A., Bogdanović, Z., Despotović-Zrakić, M.: Harnessing Edutainment in Higher Education: an Example of an IoT Based Game. In: The 12th International conference on virtual learning, Sibiu, Romania-Europe, pp 318–324 (2017)
19. Stojanović, D., Bogdanović, Z., Despotović-Zrakić, M.: IoT application model in secondary education. In: Reis, J.L., López, E.P., Moutinho, L., Santos, J.P.M.d. (eds) Marketing and Smart Technologies. Smart Innovation, Systems and Technologies, Springer, Singapore, pp. 329–338 (2022)
20. Fraga-Lamas, L., Fernández-Caramés, P., Castedo, T.M.: Towards the Internet of smart trains: A review on industrial IoT-connected railways. Sensors **17**(6), 1–44 (2017)
21. Righetti, F., Vallati, C., Anastasi, G., Masetti, G., di Giandomenico, F.: Failure management strategies for IoT-based railways systems. In: IEEE International Conference on Smart Computing (SMARTCOMP), pp. 386–391 (2020)
22. Zhong, B., Xiong, G., Zhong, K., Ai, Z.: Internet of things for high-speed railways. Intell. Converg. Networks, **2**(2), 115–132 (2021)
23. Stanisavljević, N., Stojanović, D., Bogdanović, Z.: Crowsourcing and IOT-based open innovations in increasing safety on railways. In: XVIII International Symposium Sustainable Business Management and Digital Transformation: Challenges and Opportunities in the Post-COVID Era, pp. 190–192 (2022)
24. Stanisavljević, N., Stojanović, D., Bogdanović, Z.: Fostering crowd-based open innovations in Serbian railways-preliminary readiness assessment. In: Mihić, M., Jednak, S., Savić, G. (eds) Sustainable Business Management and Digital Transformation: Challenges and Opportunities in the Post-COVID Era. Springer (2022)
25. Dodgson, A., Gann, M., MacAulay, D., Davies, S.: Innovation strategy in new transportation systems: The case of Crossrail. Transp. Res. Part A Policy Pract. **77**, 261–275 (2015)
26. Thurner, T., Gershman, M.: Catching the runaway train innovation management in Russian railways. J. Technol. Manag. Innov. **9**(3), 158–168 (2014)
27. Hanley, M., Li, D., Wu, J.: High-speed railways and collaborative innovation. Reg. Sci. Urban Econ. **93**, 103717 (2022)

28. Open innovation in railway: example of AlstomTM|ideXlab [Online]. Available: https://www.idexlab.com/open-innovation-railway/. Accessed: 19-Sep-2022
29. "RailActivation project" [Online]. Available: http://railactivation.eu/documents-2/. Accessed: 19-Sep-2022
30. Program poslovanja Akcionarskog društva za upravljanje javnom železničkom infrastrukturom "Infrastruktura železnice Srbije" za 2022. godinu (2022)

Chapter 14
A Model for Municipality Buildings Renting Auction on Algorand Blockchain

Miodrag Šljukić, Aleksandra Labus, Marijana Despotović-Zrakić, Tamara Naumović, and Zorica Bogdanović

Abstract Blockchain is one of the vital supportive technologies of smart city services. Thanks to its characteristics, blockchain impacts almost all smart city services. Algorand network started working in 2019. Its speed of transactions and low cost makes it the platform of choice for a smart city blockchain implementation. This paper proposes a model for blockchain auction for renting real estate owned by the municipality. This service is part of the overall smart city services and relies heavily on other smart city governance services like identity information management, land register, and others. Implementing such a system simplifies the process of land and real estate management in the smart city making it more transparent. It lowers the costs and increases the income municipality earns from renting the asset.

14.1 Introduction

Blockchain is a completely replicated distributed database that stores records of all network transactions [19]. It is the virtual chain of ordered blocks of data consisting of data about the transaction and the previous block [22]. Together these autonomous nodes make a P2P network with a mutually agreed protocol for intercommunication and validation [21]. Blockchain provides interoperability, decentralization, security, controllability, privacy, durability, and sustainability. It is considered that the technology itself is the root of trust, which enables the transfer of information between participants directly, without a mediator [4].

M. Šljukić · A. Labus · M. Despotović-Zrakić · T. Naumović · Z. Bogdanović (✉)
Faculty of Organizational Sciences, University of Belgrade, Belgrade, Serbia
e-mail: zorica@elab.rs

A. Labus
e-mail: aleksandra@elab.rs

M. Despotović-Zrakić
e-mail: maja@elab.rs

T. Naumović
e-mail: tamara@elab.rs

J. L. Reis et al. (eds.), *Marketing and Smart Technologies*, Smart Innovation, Systems and Technologies 337, https://doi.org/10.1007/978-981-19-9099-1_14

The ecosystem for the development of blockchain applications is defined by the choice of distributed database technology, the range of access to the distributed database, the blockchain platform, and the consensus algorithm. Support for application development includes an integrated development environment, software libraries, testing and debugging tools, programming languages for smart contracts creation, and applications and programs for data storage. The composition of every ecosystem depends a lot on the application's target area [12].

Blockchain is one of the critical technologies in the smart city. To comply with smart city requirements, blockchain technology must protect user privacy, enable fast processing of transactions, and comply with legal requirements. Citizens should have quickly learned about the use of technology [15]. The main areas of smart city services impacted by blockchain are healthcare, transportation, agriculture, employment, economy, education, energy and waste management, housing, and public privacy and safety [13].

This paper proposes a model for organizing a public competition for renting the municipality buildings as a part of lend and real estate smart city service. The context in which the auction is running is given, together with the detailed activity diagram presenting the way such a system is supposed to work. Characteristics of the Algorand blockchain and its underlying consensus protocol make it suitable as a platform for such an application.

14.2 Blockchain for the Smart City

Blockchain is a mechanism for achieving an agreement about the order of events between a set of entities. It solves the problem which originates from the ideas of distributed computer systems and cryptography [14]. The essential characteristic of a blockchain database is that most network nodes verify each transaction, which is immutable and guarantees transparency and security of transactions [4]. The nodes in the network manage the exchange of messages and make local decisions that keep the integrity of data in the network [19]. Nodes do not trust each other, and the tolerance level for dishonest nodes depends on the consensus algorithm [21]. Although there are a lot of distributed database technologies based on a consensus, blockchain is the only one that can provide a system resistant to censoring and does not require mutual trust or gaining permission for participation in a network [20].

A blockchain is a ledger in which the data can be added at the end, while previously added data cannot be modified [14]. Data is always added in the form of a block linked to the previous block, making the chain of the blocks. Adding a block to the chain is called a transaction, which can be "in progress" or "validated." The transaction validation process is called "mining," which is done by selecting the set of transactions, ordering it, and verifying if all transactions follow the defined consensus rules. Once this process successfully finishes, the transaction is considered valid and becomes immutable.

Table 14.1 Comparison of popular blockchain platforms

Platform	Consensus algorithm	Time to generate a block (in sec)	Number of transactions per sec
Bitcoin	PoW	~600	~4.6
Ethereum	PoW	~13	~7–15
Algorand	PPoS	~10	>1000

Source Medury, L., Ghosh, S.: Design and analysis of blockchain-based resale marketplace. Smart Inno. Syst. Tech. **251**(April), 481–490, p. 9 (2022). https://doi.org/10.1007/978-981-16-3945-6_47

The major characteristics of the network are decentralization, security, and scalability. Blockchain is considered decentralized if a single node cannot reach a consensus. The larger the number of participants in the consensus, the larger the degree of network decentralization. Blockchain security assumes that the attacker must control a large number of nodes participating in the consensus to change the correct functioning of the blockchain. Each consensus algorithm has its tolerance limit, which is usually about a third of the consensus participants. Scalability is the property of the blockchain to provide high throughput of transactions and future growth. Scalability is still a significant problem for blockchain networks. Scalability trilemma means that it is impossible to make the system scalable, decentralized, and secure at the same time, but concessions must be made in some aspects [21]. There are different ways networks fight for better performances, like limiting the number of new block creations, limiting the size of blocks, etc. [14]. The differences between different protocols and their variations and implementations on other platforms are significant. The comparison of the most popular platforms is shown in Table 14.1.

Depending on who can read and write the data in the blockchain, it can be public, private, or consortia. In public networks, everybody can access the data and start a transaction. There is no trust between the nodes. The throughput of this kind of blockchain is limited, costs are high, and privacy and anonymity are missing. Only authorized users are allowed to access the private blockchain. The level of trust between the nodes and the speed of transactions is high, but these networks are highly centralized. Consortia chains link multiple companies, grow easily, and are easy to manage, but the problem is the lack of unique industrial standards [28].

Similar to the previous classification, networks can be permissioned and permissionless. Only nodes with permission can participate in consensus in permissioned networks, while permissionless networks allow anybody to join the network. This causes different ways of enticing participants to participate in consensus: permissioned networks usually do not offer a reward since the users already have a clearly defined goal. In contrast, permissionless networks lack this goal and must be provided some kind of reward for the users to participate in consensus. This reward is usually the native cryptocurrency of the network [21]. Since permissionless networks lack central authority, reaching consensus in this type of network is not a trivial task [22].

The consensus is a mechanism that blockchain network uses to keep data integrity and consistency, prevents errors, and maintains record immutability. It is the key process in a blockchain because it provides chain security and reaches an agreement about its content [4]. The differences in consensus throughput, security, energy consumption, and scalability directly affect the chain's quality and usability for its intended purpose. The problem of consensus can be reduced to making a joint decision in the presence of a small group of traitors. Because of this assumption, a consensus has to be tolerant of errors [20]. There are many types of consensus algorithms, the most popular of which are proof of work (PoW) and proof of stake (PoS) protocols. Each of them has its variations [19]. PoW is one of the oldest and most widely accepted protocols on which Bitcoin is based. One of the main problems of this protocol is its speed and energy consumption. It is estimated that the total energy consumption per year caused by mining is more than that of countries like Austria or the Netherlands [8]. PoS is an alternative to PoW protocol which tries to overcome the weaknesses of PoW. Here, instead of solving the complex problem, participants are obliged to possess a certain amount of native cryptocurrency. Block rewards are smaller because the costs are smaller too. Opponents of this protocol point out the "nothing at stake" problem, which means that since the participant has nothing to lose, he might record any block, regardless of its validity. Scalability and the capability of processing a large number of transactions with insignificant growth of the costs, in particular, enable expansion of the usage of PoS protocol to the areas like asset tokenization. At the same time, it increases the efficiency of settling transactions [22].

One of the most important blockchain functionalities today is a smart contract. These are computer programs embedded in a blockchain that consists of a set of protocols and the conditions for their activation [4]. They can operate only over the data which resides inside the network [21]. Since they function inside the blockchain, they are trustworthy by definition, eliminating the need for a mediator and making the transactions cheaper [20]. Smart contracts create, verify, and take care of the execution of agreements when the conditions are fulfilled, which makes them applicable in finances, supply chains, intellectual rights, etc. [4].

Although blockchain technology can be considered revolutionary in many aspects, numerous problems prevent its more intensive application. These problems can be classified as technical and non-technical. The most significant technical problems are scalability and interoperability. Non-technical issues relate to the decision makers' lack of knowledge about technology and their reluctance to apply it until widespread, the lack of integration tools, and the lack of experts in this area. There is also a fear for data privacy and concern about the correctness of the data since, once added, they cannot be modified [12]. There are also a lot of legal problems, including anonymity of the participants, compliance with GDPR, and court jurisdiction [23].

14.3 Algorand Blockchain

Algorand is a new public blockchain platform established in 2019. Its native cryptocurrency is Algo. Algorand supports the creation of smart contracts and tokens that represent the property. Algorand standard assets (ASAs) are the concept that enables the simple creation of fungible (FT) and non-fungible (NFT) tokens and their management [16].

Consensus in Algorand is achieved using the pure proof of stake (PPos) protocol, whose author is also the founder of Algorand. It is based on Byzantine protocol but provides more equality than its predecessor [10]. The probability that a node in a network will be chosen to validate is proportional to its stake, i.e., to the amount of Algos, the user is willing to freeze for a certain period, giving them the right to validate. The goal of the freezing is to keep the currency on the platform while solving the "nothing at stake" issue. Two key concepts of Algorand's consensus are verifiable random function (VRF) and participation keys. VRF is a function that takes a user key as an input and a value, producing pseudorandom output that everybody can use to verify the result. It is used for the choice of a leader who proposes the block of transactions and the members of the voting committee. Accounts with higher stakes have a higher probability of being chosen. This ensures that the fraudster cannot get any benefit from fake accounts [1].

Algorand accounts are entities linked to specific data from the block. Account ID is the Algorand address created by the user's public key transformation. The account balance is initialized at the time of the first transfer of Algos to another account. The minimal balance is 100.000 microAlgos. It is possible to create an account using the TEAL program. Accounts can be online and offline accounts. Regarding the number of signatures, accounts can be single or multi-signature accounts requiring specific signatures from the list of all available signatures to initiate a valid transaction (*Overview*, n.d.-b).

Algorand smart contracts (ASC1) are small distributed applications that work at the first level and do different functions. There are two main categories of smart contracts: smart contracts (stateful contracts) and smart signatures (stateless contracts). Both kinds of contracts are written in TEAL programming language and executed inside Algorand's virtual machine (AVM). Since TEAL is an assembler-like programming language, the Python library was written to make programming easier. When uploaded to the network, smart contracts are called applications and can be activated by any node in Algorand's network. A special kind of transaction that activates them is called an application call. This is the way to handle the primary decentralized logic of DApps. Applications can modify the state of global and local variables and access the data in the blockchain, like account balance, asset parameter configuration, or the time of the last block validation. Applications are also used as a part of transaction execution logic, including application call transactions, which enables application chaining. Each application can have a linked account where Algos can be stored, ASA, or they can be linked to the account where Algos are kept. Smart signatures use logic to sign transactions in the situation of account rights delegation. The logic of a smart signature is activated at the transaction execution time; if it is not satisfied, the transaction will fail [11].

Stablecoins, loyalty points, system credits, game points, etc., are represented by ASA. Besides them, ASA supports NFTs, like collectibles, unique parts of supply chains, etc., and there is an option to restrict transactions with the asset, which enables the digitalization of securities and certificates. The type of asset depends on the parameters configurable at the time of the asset creation, while some parameters can be reconfigured later. Transactions with ASA include the creation, modification, acceptance, transfer, freezing, recalling, and canceling of the asset [2].

There are six types of transactions [26].

- payment, i.e., transfer of Algos between accounts
- key registration, i.e., account creation
- asset configuration, which includes creation, modification of parameters of removing the asset from the chain
- asset freezing, which gives or forbids the right to the user to transfer or accept the asset
- asset transfer, and
- application call.

Before transactions are sent to the network, they must be authorized by signing. Signatures can be single, multiple, or logical. Logical signatures are used for authorization of transactions linked to Algorand smart contracts in a situation where the spending is done from the account of the smart contract or delegated account [25]. Algorand transactions are valid at most during the 1.000 rounds. The minimal fee for transaction execution is 1000 microAlgos [26]. Transactions that are part of the transfer are executed in a group. If one of them fails, none is executed. This is the way to eliminate the mediator in a transaction. This way of transaction execution is called atomized transfer. It eliminates the need for more complex algorithms that exist in other blockchain networks. Atomized transactions are used in circular trades, group payments, decentralized exchanges, distributed payments, summary payments of fee when it is paid for multiple accounts from one account, etc. To implement atomized transfer, transactions are first grouped, and each of them is assigned a group ID. The individual transactions are separated again and sent to their creators for signing. In the final step, they are collected again and sent to the network as a group [3].

A new block is added in three steps executed by the randomly chosen committee: proposal, selection (also called soft vote), and validation (certify vote) of the block [9]. In the first step, a new leader is chosen, which will propose the block. Then, new committee is randomly chosen to reach the BA about the block proposed by the leader. A prescribed number of committee members digitally sign the block. In the last step, digital certificate and identity of the committee members who had signed the block, coupled with the hash, goes around the network so all the nodes can authenticate the new block. Compared to the PoW protocol, the Algorands protocol requires a minimal amount of energy, while the probability of forking is at the minimum [20]. It is estimated that Algorand is among the fastest permissionless networks with about 1200 transactions per second capacity. The block is created on average in less than 5 s [21].

Essential tools which provide nodes with the capability to perform their functions on the networks are [17].

- goal—a command-line tool that enables interaction with the node and execution of basic functions like account creation, searching the ledger, and transaction creation
- algod—the main process for blockchain management, which processes the messages between nodes, executes protocol, writes blocks to the disk, and enables access to the REST API server used for network access in applications
- kmd—background process responsible for interactions with private keys, transaction signatures, and communication with hardware wallets
- algokey—command-line tool for the generation, export, and import of keys
- carpenter— debugger which enables visualization and monitoring of protocol execution

Aside from these, there is a set of development tools that constitute the protocol and enable simple and efficient creation of applications for digital asset transfer, tokenization, issuing cryptocurrencies, transaction escrow, smart contract creations, access to historical data, metainformation addition, access rights delegation, access to the global variables or the combination of these functionalities.

The search is usually very slow in the blockchain. Algorand solves this problem with the tool called indexer. It is a Postgres database accessed through REST API [5].

14.4 Problem Statement and Solution Design

This paper focuses on the land and properties group of the smart city services, particularly the municipality buildings and their renting. To make use of the blockchain, a land registry needs to be put in the blockchain. It is useful to have all parcels of the land and all properties in the city in the register, not only the ones owned by the city itself.

Municipalities make income from renting the properties and land they own. The process of choosing a renter should be transparent and guided by clear rules. That is why blockchain is the technology of choice when it comes to organizing auctions.

Figure 14.1 shows the context in which the blockchain is applied to organize auctions. The main actors here are the municipality and bidder who communicate through the set of applications developed to enable citizens to lease the property.

Prerequisites of the auction process are as follows:

- The registry of all land and property intended for renting should be in the blockchain
- The bidder must be registered in the database and have an account in the blockchain
- The municipality must provide cryptocurrency in which the rent will be collected

The steps of the auction are presented in Fig. 14.2.

Auction starts when the municipality defines terms. The most important of them is the real estate that is the object of the auction, the time for registration, deposit

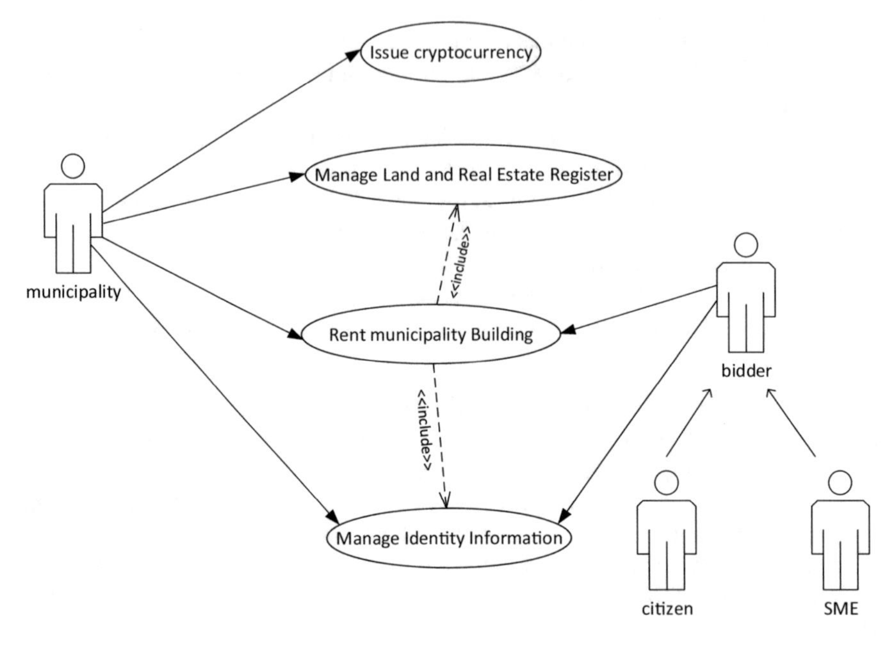

Fig. 14.1 Use case diagram for auction

amount, minimal price, starting and ending time for bidding, the types of participants who are allowed to participate in the auction, etc.

To register, bidders must have a blockchain account and deposit a prescribed amount of cryptocurrency. By getting bidder detail, the municipality will compare prescribed terms for the auction with statutory and historical data about the bidder and validate them. In that process, the municipality uses identity information retrieved from databases that are also on the blockchain. These databases contain personal information about owners, statutory information about MSP, and the history of their transactions with the municipality. If there are some special conditions for the auction, for example, the request for a bidder to have special licenses, that information will be obtained from the identity management database too. If the data satisfies the conditions for the auction, the participant is registered and can bid as soon as the auction starts.

At the scheduled time, the system should be ready to accept bids from registered users. Bids are put in the blockchain until the auction is closed. Upon the auction's closing, the best bid will be pronounced, and the smart contract will be signed between the municipality and the winner. The system will return the deposit to other participants. For the winner, the system will include the deposit in the bid amount and wait for the time scheduled for the payment. It can be periodical payment, or payment in advance at some time after the auction is closed.

The system will enable a municipality to monitor and audit the fulfillment of the contract. If the payment is not done following the contract, the contract will be

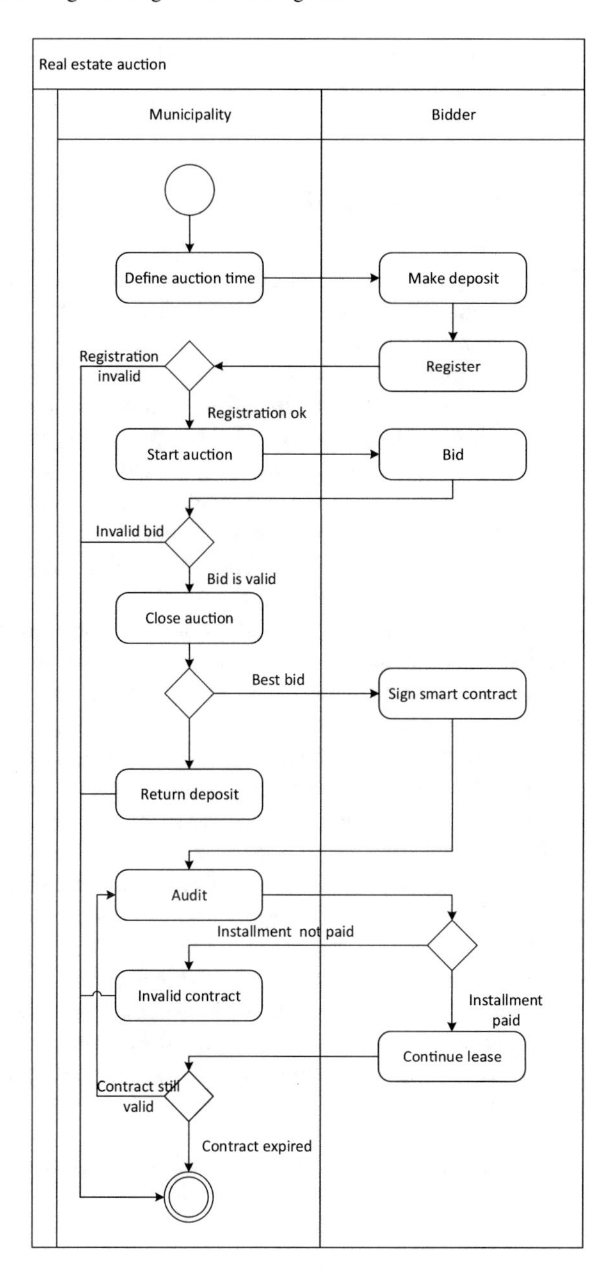

Fig. 14.2 Activity diagram for the auction of the real estate

canceled automatically by the system. This process will continue until the time of expiration of the contract.

14.5 Discussion and Conclusion

This paper presents a model of an auction for renting municipality buildings. The proposed model enables the municipality to better manage properties it owns, makes higher revenue, lowers the auction cost, and monitors fee collection. Using blockchain makes the process more transparent and simpler.

Similar solutions found in the literature can broadly be classified as general blockchain auction solutions and those put in the context of smart city services. Contrary to generalized solutions [18, 27, 24], the model presented here is put in the context of smart city services, particularly renting city property in relation to other smart city services. In the contrast to work related to smart city services [13, 15, 6], this research gives more details, while still trying to keep the proposed solution general enough to fit different auction techniques or business models. The proposed model is created having in mind Algorand platform and PoS protocol, while similar solutions found in the literature rely on PoW protocol [18].

The practical contribution of this research is the model of the auction process. By implementing a system like the one proposed in the paper municipality managers will improve the record and management of the municipality's properties and avoid conflict of interest in the process. It points out the necessity for transparency and automatization of the process, enabled by blockchain.

The research is based on an analysis of the auction process in a small city in Serbia. That poses the main limitation of this research in a view of possible generalization of the conclusion. Country regulation, city size, or different type of asset that is the subject of auction can cause different business models. Also, the pilot project requires an implementation to get reliable data.

Future research will be devoted to designing a more precise business model and developing a prototype implementation of the proposed system. Further research will also be focused on the integration of the proposed system with various smart city services [7].

Acknowledgements This work was supported by the Algorand Centres of Excellence programme managed by Algorand Foundation. Any opinions, findings, and conclusions or recommendations expressed in this material are those of the author(s) and do not necessarily reflect the views of Algorand Foundation.

References

1. Algorand consensus: (n.d.). https://developer.algorand.org/docs/get-details/algorand_consensus/
2. Algorand Standard Assets (ASAs): (n.d.). https://developer.algorand.org/docs/get-details/asa/
3. Atomic transfers: (n.d.). https://developer.algorand.org/docs/get-details/atomic_transfers/
4. Brotsis, S., Limniotis, K., Bendiab, G., Kolokotronis, N., Shiaeles, S.: On the suitability of blockchain platforms for IoT applications: architectures, security, privacy, and performance. Comp. Netw. **191**(January) (2021). https://doi.org/10.1016/j.comnet.2021.108005
5. Building Applications: (n.d.). https://developer.algorand.org/docs/archive/build-apps/apps/
6. Bustamante, P., Cai, M., Gomez, M., Harris, C., Krishnamurthy, P., Law, W., Madison, M.J., Murtazashvili, I., Murtazashvili, J.B., Mylovanov, T., Shapoval, N., Vee, A., Weiss, M.: Government by code? blockchain applications to public sector governance. Front. Block. **5**(June), 1–15 (2022). https://doi.org/10.3389/fbloc.2022.869665
7. Dankovic, D., Djordjevic, M.: A review of real time smart systems developed at University of Niš. Facta Universitatis. Series: Electron. Energet. **33**(4), 669–686 (2020)
8. de Vries, A.: Bitcoin's growing energy problem. Joule **2**(5), 801–805 (2018). https://doi.org/10.1016/j.joule.2018.04.016
9. Dimitri, N.: The economics of consensus in algorand. FinTech **1**(2), 164–179 (2022). https://doi.org/10.3390/fintech1020013
10. Ezekiel, A., Ahmad, A., Dakingari, M.: Introducing decentralized decision to our institutional platforms : Algorand and choice coin (2022)
11. Introduction: (n.d.). https://developer.algorand.org/docs/get-details/dapps/smart-contracts/
12. Jabbar, S., Lloyd, H., Hammoudeh, M., Adebisi, B., Raza, U.: Blockchain-enabled supply chain: analysis, challenges, and future directions. Multimedia Syst. **27**(4), 787–806 (2021). https://doi.org/10.1007/s00530-020-00687-0
13. Khanna, A., Sah, A., Bolshev, V., Jasinski, M., Vinogradov, A., Leonowicz, Z., Jasiński, M.: Blockchain: future of e-governance in smart cities. Sustainability (Switzerland) **13**(21), 1–21 (2021). https://doi.org/10.3390/su132111840
14. Kolb, J., Abdelbaky, M., Katz, R.H., Culler, D.E.: Core concepts, challenges, and future directions in blockchain: a centralized tutorial. ACM Comput. Surv. **53**(1), 1–40 (2020). https://doi.org/10.1145/3366370
15. Lukić, I., Miličević, K., Köhler, M., Vinko, D.: Possible blockchain solutions according to a smart city digitalization strategy. Appl. Sci. (Switzerland) **12**(11) (2022). https://doi.org/10.3390/app12115552
16. Mojtaba, S., Bamakan, H., Nezhadsistani, N., Bodaghi, O., Qu, Q.: A decentralized framework for patents and intellectual property as NFT in blockchain networks. Comp. Math. Theo. Comp. Sci., 11 (2021)
17. Node artifacts: (2022). https://developer.algorand.org/docs/run-a-node/reference/artifacts/
18. Omar, I.A., Hasan, H.R., Jayaraman, R., Salah, K., Omar, M.: Implementing decentralized auctions using blockchain smart contracts. Tech. Forecasting Soc. Change, **168**(May) (2020). https://doi.org/10.1016/j.techfore.2021.120786
19. Oyinloye, D.P., Teh, J.S., Jamil, N., Alawida, M.: Blockchain consensus: an overview of alternative protocols. Symmetry **13**(8), 1–35 (2021). https://doi.org/10.3390/sym13081363
20. Panarello, A., Tapas, N., Merlino, G., Longo, F., Puliafito, A.: Blockchain and iot integration: a systematic survey. Sensors (Switzerland) **18**(8) (2018). https://doi.org/10.3390/s18082575
21. Pennino, D., Pizzonia, M., Vitaletti, A., Zecchini, M.: Blockchain as IoT economy enabler: a review of architectural aspects. J. Sensor Actua. Netw. **11**(2) (2022). https://doi.org/10.3390/jsan11020020
22. Saleh, F.: Blockchain without waste: proof-of-stake. Rev. Finan. Stud. **34**(3), 1156–1190 (2021). https://doi.org/10.1093/rfs/hhaa075
23. Savin, A.: Blockchain, digital transformation and the law: what can we learn from the recent deals? SSRN Elect. J. (2019). https://doi.org/10.2139/ssrn.3198666

24. Shi, Z., de Laat, C., Grosso, P., Zhao, Z.: When blockchain meets auction models: a survey, some applications, and challenges, 1–33 (2021). http://arxiv.org/abs/2110.12534
25. Signature: (n.d.). https://developer.algorand.org/docs/get-details/transactions/signatures/
26. Structure: (n.d.). https://developer.algorand.org/docs/get-details/transactions/
27. Ullah, F., Sepasgozar, S.M.E., Wang, C.: A systematic review of smart real estate technology: drivers of, and barriers to, the use of digital disruptive technologies and online platforms. Sustainability (Switzerland) **10**(9) (2018). https://doi.org/10.3390/su10093142
28. Xu, J., Zhou, W., Zhang, S., Fu, J.: A review of the technology and application of deposit and traceability based on blockchain. J. High Speed Netw. **27**(4), 335–359 (2021). https://doi.org/10.3233/JHS-210671

Chapter 15
Rethinking Smart Mobility: A Systematic Literature Review of Its Effects on Sustainability

Pedro Rodrigues, Elizabeth Real, Isabel Barbosa, and Luís Durães

Abstract Smart mobility has been a subject of great interest among scholars, governments, and business. Although several studies have analyzed smart mobility strategies to develop sustainable cities, a gap has been identified about the effects of these initiatives. The purpose of this paper is to fulfill this gap by identifying the most frequent results on: (i) alternatives to private car; (ii) improvements of existing infrastructure; (iii) urban mobility policies; (iv) smart mobility environmental models. To achieve this purpose, a systematic literature review was performed using Web of science (WOS) database, and a total of 12 articles were reviewed. Results showed that reduction of traffic congestion, reduction of greenhouse gas emissions, and the improvement of life quality of citizens were the most recurrent effects on the promotion of sustainability. The present study has important implications for academics researching on the subject and for policymakers and business managers who can apply the identified outcomes in their context.

15.1 Introduction

Urban population is increasing globally, and its effects are becoming evident [1]. Cities around the world are facing some kind of environmental or society crises, such as climate changes, natural disaster, ecosystem destruction, and socioeconomic inequity [2]. As a consequence, pressure on local authorities and governments is pushing them to find innovative ways of supporting the city's economic development; at the same time, they preserve the environment and offer good living conditions to citizens. To make things worse, by 2050, it is expected that 88% of the world population will live in urban areas in high-income countries, aggravating the current urban development problems [3].

P. Rodrigues (✉) · E. Real · I. Barbosa · L. Durães
COMEGI, Universidade Lusíada do Norte, Porto, Portugal
e-mail: pedro.rodrigues@fam.ulusiada.pt

E. Real
e-mail: mereal@por.ulusiada.pt

J. L. Reis et al. (eds.), *Marketing and Smart Technologies*, Smart Innovation, Systems and Technologies 337, https://doi.org/10.1007/978-981-19-9099-1_15

However, recent technological advancements, more precisely, the information on communication technologies, are making the concept of smart cities as possible solution to the problems that arise in cities [4–6]. Nowadays, there are several smart cities initiatives worldwide, with substantial resources dedicated to them [7]. According to an online platform, investments in smart cities are expected to reach 189$ Billion in 2023 [8], which give us a strong confidence that smart cities interest will not slowing down in the foreseeable future.

Although the smart cities can be divided into different dimensions, namely economy, people, mobility, governance, environment, and living [9], we see smart mobility as one of the most prominents on creating sustainable cities since the transport sector is responsible for 80% of air pollution in developing countries [10]. Thus, smart mobility is focusing on the use of information and communication technologies infrastructures and sustainable transport systems to better serve the citizens and improve the quality of life.

Consequently, growing literature is being concerned about defining smart mobility, its area of action, and showing different types of smart mobility initiatives adopted by cities. However, the available literature do not pay explicit to attention to the association of smart mobility and sustainability. This way, our systematic literature review aims to investigate the potential contribution of smart mobility solutions and their impact on sustainability.

This paper is structured as follows: Sect. 15.2 presents the necessary background to understand our main topics; Sect. 15.3 describes the methodology of the study; Sect. 15.4 highlights the results from our systematic literature review; Sect. 15.5 discusses how smart mobility can impact the sustainable development of a city; the final section covers the main conclusions of the paper.

15.2 Background

15.2.1 Smart Mobility

Several interpretations have been proposed to the definition of smart city. Some scholars identify smart cities as the application of information and communication technology [11]. Others, see smart cities as an economic growth strategy, generating jobs and increasing business indicators of the city [12]. Most recently, there have been proposed interpretations of smart cities associated to sustainability [13, 14].

Meanwhile, one highly accepted interpretation is the one presented by ITU-T [15, p. 2], defining the concept smart city as: "an innovative city that uses information and communication technologies and other means to improve quality of life, efficiency of urban operation and services, and competitiveness, while ensuring that it meets the needs of present and future generations with respect to economic, social, and environmental aspects."

Being a dimension of smart city, smart mobility inherits some of its characteristics. This way, smart mobility is worried about the sustainability of the city and the enforcement of data and knowledge available to improve the efficiency and effectiveness of transport systems [16]. Moreover, smart mobility is a crucial topic of smart cities since it impacts in several dimensions and could improve the citizens' quality of life.

According to Lawrence et al. [17], smart mobility initiatives aim to reduce pollution, reduce traffic congestion, increase people safety, reduce noise pollution, improve transfer speed, and reduce transfer costs. In this context, cities like Amsterdam, Copenhagen, Seattle, Curitiba, and Songdo have been carrying out an important role when it comes to the development and implementation of initiatives [18].

15.2.2 Sustainability

Over the last 20 years, sustainability and sustainable development are becoming a popular topic not only for scholars but also for business, politicians, and governments. Due to the interdisciplinarity of domains contributing to the definition of sustainability, such as forest experts, political economists, natural scientists, ecologists, urban planners, and managers [19, 20], a common and unique definition could not be found [21].

One of the most popular and accepted definitions is the one present in the Brundtland Report where sustainability is the "development that meets the needs of the present without compromising the ability of the future generations to meet their own needs" [22]. It is also possible to recognize that environment, economy, and society are dimensions of sustainability well established in the literature [23, 24]. These dimensions are often presented in the form of three intersecting circles with sustainability being in the center of the intersections [19].

The environmental dimension describes aspects such as land use, waste management, and use of natural resources. The social dimension entails the satisfaction of basic human needs encompassing issues like working conditions, product safety, and human rights. The economic dimension focuses on solving the limitations of a sustainable society being concerned with aspects like profit, innovation, economic growth, and job creation [23, 25].

It should be noted that although contemporary sustainability discussion is centralized on UN's sustainable development goals, the three dimensions were explicit embedded in their formulation. The conception of these goals is grounded on economic development without compromising the prosperity of those with less resources and ensuring the conservation of the ecosystem [19].

Recent studies point out that smart cities have the potential to help cities achieving the UN's goals [26, 27]. However, most of them focus on the topic of environment, without considering social and economic aspects [28, 29]. This study aims to identify the main articles that provide solutions on the social, economic, and environmental dimensions.

15.3 Methodology

15.3.1 Research Strategy

A systematic literature review was conducted to find how smart mobility initiatives are helping cities achieving a better sustainable development.

This methodology is characterized which has being rigorous, reproducible, and auditable. It helps academics and practitioners developing their investigation and needs to increase their knowledge since it allows to screen the most related studies to answer a particular research question or topic; extract its main contributions; synthetize the data and its results; and report the findings with objectivity [30]. Thus, the systematic literature review helps to bring evidence on a specific phenomenon, gathering disperse sources of knowledge to form clusters [31]. Moreover, this methodology allows the researcher to reduce bias during the selection process, since a pre-defined criteria are followed [32].

According to Lim et al. [33], smart city literature has increased significantly in the last decade. Several factors contribute to this development, in particular, the need for solutions to overcome population growth and climate changes. Although it is possible to find systematic literature reviews about the topic of smart city and sustainability (examples of some), to the best of our knowledge, it was not possible to find any systematic literature review specifically concerned with the objective of this study. This study adopts the "preferred reporting items for systematic reviews and meta-analysis" (PRISMA) method and uses the Web of science database.

15.3.2 Screening Process

To identify the population of publications for review, we used the systematic review process present on Fig. 15.1. The filtering procedure that was used has four different and intercalated screening stages. Initial search was conducted using the Web of science database. In total, 172 documents were identified using the string query [("Smart Mobility") AND (Sustainab*)].

The first screening procedure that was used was based on three specific criteria. As a quality of measure, only articles written in English with full text availability

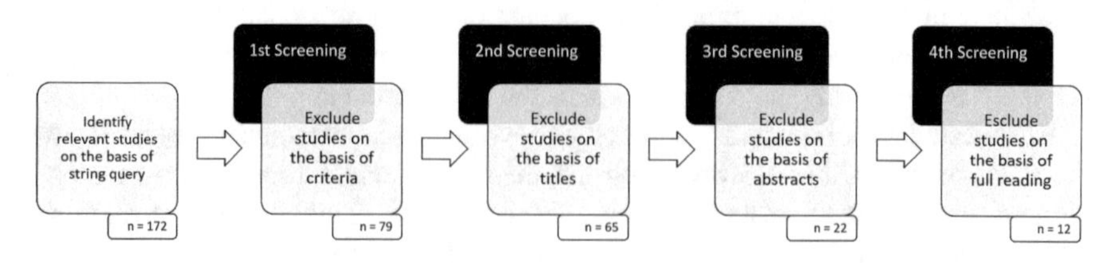

Fig. 15.1 Stages of the study selection process

and from academic journals were selected. Consequently, a substantial number of articles were excluded (93), and a total of 79 articles have moved forward.

From this moment, we copy the data if the first sample to an excel file, and we went through the titles of all studies in order to exclude the ones that were clearly not about smart cities, smart mobility, or sustainability. As a result, during the second screening procedure, 16 articles were excluded.

At the third screening procedure, abstracts were read, and the studies were excluded if their topic, or main topic, was not specifically concerned about smart mobility and sustainability. Thus, 41 articles were excluded, which left 22 articles for full reading. After reading these articles, we excluded 10 articles. Although the excluded articles talked about smart mobility, they did not present the results of the smart mobility initiatives, or the results clearly did not have nothing to do with sustainability. Thus, as a result of the last screening, 12 articles were selected for further analysis.

15.4 Results

15.4.1 General Observations

In order to ensure a standardized data extraction, a predesigned table (Table 15.1) was used during the analysis of the selected 12 articles. For each article, we noted the author, the year of publication, and the aim of the study, and in case of a case study, we also registered the region of the case study.

This classification brings to light that smart mobility and sustainability topics are relatively new, and it has increased over the last years. The oldest article of our analysis was published in 2014 and the newest in 2020. One quarter of the articles were published in 2018 ($n = 3$; 25%) and 2020 ($n = 3$; 25%), a bit over a one-seventh of them were in 2015 ($n = 2$; 16,7%) and the remaining in 2014 ($n = 1$; 8.3%), 2016 ($n = 1$; 8.3%) and 2017 ($n = 1$; 8.3%). These findings show a parallelism to other review works stating the growth of literature over the last years [49–52].

In terms of field of study, based on the initial categorization of Web of science, we can observe five major fields as follows: engineering, business and economics, computer science, transportation, and urban studies. On the other hand, when it comes to the analysis of the journals, we could not find any pattern because almost, all articles were published in different journals. The only exception was the transport policy and the journal of land use, mobility, and environment with 2 articles each one.

Almost, all of the articles have the case study as a principal research method. However, since no specific methods are excluded, we have 2 article that did not perform any case study. Among the 10 articles that conducted a single case study or a comparative case study, we identified a total of 15 regions. Following the categories of the World Bank, most of the regions in our sample are from the Europe ($n = 12$; 80%),

Table 15.1 Standardized table

No.	Literature	Journal	Aim	Region
1	[34]	Technologies	The main aim of the research is to show how blue–green mobility could play a role in creating modern urban areas that focus on sustainable development as a tool for making the world a better place for living	Songdo Copenhagen
2	[35]	Applied soft computing	This article proposes a mobility architecture to reduce gas emissions from road traffic in smart cities	Malaga Stockholm Berlin Paris
3	[36]	IMA journal of management mathematics	This article proposes a variant of vehicle routing to determine the best fleet of nonhomogeneous vehicles for delivering goods in urban areas, taking into proper account the government traffic restrictions	Genoa
4	[37]	Transport policy	This study investigates the relationship between the implementation of the smart city concept and the idea of sustainable transport, particularly with regard to the reduction of transport generated CO2 emissions	Warsaw
5	[38]	Chemosensors	The aim of this study was to test the SmartBus-NASUS IV approach in view of obtaining qualitative indications on the ambient air quality status along the smart ring track	Áquila
6	[39]	Journal of land use, mobility, and environment	The primary goal of the work was to make a review of policies, programs, and projects for sustainable urban mobility and of smart mobility solutions in Bari area. The second goal was to make an assessment on trends of urban mobility in order to evaluate its sustainability and smartness	Bari

(continued)

Table 15.1 (continued)

No.	Literature	Journal	Aim	Region
7	[40]	Transport policy	The authors explored whether the policy regulation can hold smart mobility providers accountable for their impacts on the urban environment, and if the accountability arrangements that are in place in each city can help local governments achieve their strategic goals for smart mobility	London Seattle
8	[41]	Sustainability	This paper provides a novel conceptual contribution that thoroughly discusses the scarcely studied nexus of AI, transportation, and the smart city assessing how this will affect urban futures	–
9	[42]	IEEE access	This paper proposes an optimal parking site selection scheme to alleviate CO2 emissions of the traffic flows for green urban road networks	Xiam
10	[43]	Scientific journal of silesian university of technology	The article is dedicated to the concept of the walkable city as an alternative form of urban mobility	Poland
11	[44]	Journal of school health	Investigate the role of WSBs in children's fitness related health, with a view to making recommendations for future studies	–
12	[45]	Journal of land use, mobility, and environment	The paper proposes a possible application of the inductive recharge technology to the public transport vehicles	Brescia

Asia ($n = 2$; 13.3%), and North America ($n = 1$; 6.7%) contexts. Not surprisingly, and in line with the rising literature, most of the case studies are focused on high-income or upper-middle-income regions. Furthermore, all the selected case studies are based on urban mobility solutions.

After carefully reviewing the selected 12 articles, they were clustered under four groups based on the main types of smart mobility initiatives. The reviewed literature was categorized into the followings: (1) alternatives to private car; (2) improvements of existing Infrastructure; (3) urban mobility policies; (4) smart mobility environmental models. It is important to note that although many of the articles could be related to other groups, only two of them were assigned to more than 1 group due to a high relation. The results of our research are presented in the following sections.

15.4.2 Alternatives to Private Car

One-third (33%) of the reviewed articles presented how different alternatives to private car could improve the sustainable development of the city. Cities worldwide are fighting to reduce climate changes by improving their mobility system. Predominantly, countries strive to reduce car ownership and to change the use of private car as principal mode of transport, since it is the principal cause of traffic congestion and, consequently, a great source of air, noise, and vibration pollution [41, 43]. Therefore, different alternatives to private car have been proposed and explored by different cities.

Moscholidou and Pangbourne [40] analyzed the potential impact of bikesharing, carsharing, and ridesharing in two different cities, namely London and Seattle. According to their findings, with a correct implementation of these initiatives, London could reduce the number of cars in streets, the greenhouse emissions, and the first-last mile travel time. Moreover, the use of this type of transports could increase the data shared across the city, the human health, and the efficiency of the urban space. On the other hand, the authors presented as potential contributions for Seattle, the increase of safety, dignity, race and social justice, and happiness. In addition, it was stated that bikesharing and carsharing could lead the Seattle's mobility system to a cleaner version.

Alternatively, the review of Nikitas et al. [41] has showed that with the advances of artificial intelligence over the past few years, new modes of transport have been proposed. In this way, the authors see that although in his infancy, personal aerial vehicles could be good solution to reduce traffic congestion in future, by making use of a free space in the air.

More audacious, many cities all over the world are trying to employ more walk habits in society as an alternative form of smart mobility. One good example of the implementation of this strategic plan is Poland. In the work of Turón et al. [43], we can see that, for instance, by having more streets where people and cyclist have priority over other means of transport, by using all-green traffic lights at the pedestrian crossings and improving the pavements in areas covered with cobblestones cities in Poland are experiencing better walk score index and, as a consequence, cultivating a sensation of safety on the streets, decreasing the environmental footprint, improving the attractiveness of public spaces, reducing spending on construction of road infrastructure, and improving the health of residents. Similarly, by reviewing the literature, Smith et al. [44] studied twelve walking school buses worldwide, involving a total of 9169 children. As a result, it was concluded that this way of traveling apart from saving car journeys could also improve children's activity levels and socialization.

15.4.3 Improvements of Existing Infrastructures

Among 12 articles, four of them have present smart mobility initiatives that could be used to improve the existing infrastructures in cities. Taking in mind that changing human mobility behavior can take many years, there have been suggested improvements of existing infrastructures in cities in an attempt to reduce their sustainable impact.

For example, Maternini et al. [45] after carefully review the solutions implemented in South Korea, Canada, and New York conducted a case study to analyze the possible application of inductive recharge technology in the city of Brescia. This case has shown that this type of system could improve the quality of life and reduce the air pollution since the working time of the electric buses will be higher.

Other prominent topic is the concept of connect autonomous vehicles. Despite the fact that this solution could not reduce the number of cars in streets, it is stated by Nikitas et al. [41] that using artificial intelligence to control our cars could eliminate the human error factor from driving leafing to traffic safety and security, reduce traffic congestion, saving time, reduce CO_2 and greenhouse gas emissions, decrease noise nuisance, and shrink energy consumption.

Another example of what is meant by improvements of existing infrastructure is the solutions for parking. Although we know the existence of more alternatives, our researcher just reveals the case developed by Shen et al. [42]. This may have happen because other parking models are mainly focused in travel distance or cost efficiency, instead of taking the CO_2 emission as the optimization goal. Using data from Xiam City, this case clearly demonstrates the effectiveness of the parking model when it comes to reduce CO_2 and traffic congestion relief.

It is well known that many diseases can be caused due to bad air quality, and recent literature has been focusing in quality monitoring [46].

This can be seen in the case of Villani et al. [38] where a case study has been performed in the city of Áquila. By equipping a bus with a low-cost sensor during 5 days, the authors were capable of identified the gas components experienced by buses users and the ones that most contributed for bad air quality status. With that data, city managers could take correction actions and citizens be aware of zones and times of the day with bad air quality. As a consequence, environmental and societal benefits can be attributed to this solution.

15.4.4 Urban Mobility Policies

Four of articles of our sample were related not only to one solution but to the overall city mobility strategies and policies. Thus, 6 cities were analyzed, and it can be stated that there is very important to have a clear and well-defined strategic plan to accomplish sustainable mobility goals.

The work of Mohammadian and Rezaie [47] explains how Songdo and Copenhagen are adapting their strategies to make the world a better place to live. When it comes to the mobility dimension, the U-city of Songdo is trying to use clean energy in their transportation infrastructure, sustainable and smart vehicles, and implement multi-modal transportation including walking and biking. On the other hand, in order to become a smart city, Copenhagen wants to be independent of fossil fuels, makes bike as the main transport of citizens, improves public transport systems, and uses intelligent transportations systems for fleet management. Overall, it is expected that both strategies could reduce environmental concerns such as air and noise pollution, water contamination, and climate change.

However, as stated in the study of Moscholidou and Pangbourne [40], cities regulations are important to achieve sustainable goals, but for better results, they should be directed to a specific type of mobility, clearly set out providers' responsibilities and what happens if they did not perform their obligations.

Niglio and Comitale [39] analyzed different actions taken at a regional, metropolitan, and municipal level that affected the mobility in Bari. More specifically, over the last two decades, different actions have been implemented. For example, the construction of park and rides lots, the development of an integration model for an overall parking system, the development of parking pricing zones, the introduction of restricted traffic zones in the city center, the increase of sharing services and extension of cycle paths network. According to the author, this actions might encourage behavioral change for users, and, as a consequence, the reduction of traffic congestion.

Based on three scenarios, the work of Zawieska and Pieriegud [37] calculated greenhouse gas emission levels in Warsaw from 2008 to 2050. The business as usual scenario assumed a continuation of the current state in Warsaw, and it was possible to conclude a future increasing (45%) on greenhouse gas emission. Alternatively, the optimistic scenario shares the same macroeconomic data with the previously scenario, however, it assumes faster implementation of innovative technologies. As a result, it could be expected a 7% of increase in emissions. On the other side, the innovative scenario might result in a 55% of emissions if the city adopts a radical and rapid technology advance.

15.4.5 Smart Mobility Environment Models

Slightly over 15% of our sample (2 articles) present different way of seeing and managing mobility in cities, to improve environmental sustainability. For instance, Stolfi and Alba [35] have developed and tested an algorithm, called Green Swarm, capable of reducing gas emissions, travel times, and fuel consumption. The proposed solution has used data from Malaga, Stockholm, Berlin, and Paris. In spite of the increasing in route length, the model achieved good results in more than 500 city scenarios.

In a completely different scenario, the work of Cerulli et al. [36] evaluated how distribution networks of a grocery company from Genoa could be improved within a smart city. To achieve this goal, the authors evaluated different distribution plans of e-commerce grocery orders, taking in consideration the air pollution cost and the marginal climate change cost of the included commercial vehicles. By doing so, it was possible to analyze which vehicle could less affect the environment.

15.5 Discussion and Conclusion

The systematic literature review conducted in this paper helps to further understand how smart mobility is helping cities achieving a better sustainable development. Current research reviewed the number of studies which focused on these issue. Smart mobility initiatives can have numerous applications, not only in turning transports more energy efficient, but also by providing news ways of transport, improving the quality of service, reducing travel times, and supporting behavioral changes.

The novelty of smart mobility initiative might indicate that we still need time to really observe and understand its results. Moreover, measures must be defined to help cities compare their initiatives' performance and results. To promote an accurate evaluation of these initiatives, it is important to select measures which indicators are reliable and trustworthy. For instance, considering the initiative of building a Cyclovia, we should measure the increasing number of users instead of evaluating the reduction of traffic congestion.

The findings of our study also advocate the importance of the strategies carried out by city governments. Before spending scarce results, it is necessary to access our priorities and what is most important and wanted in our city. Municipalities need to be informed about the city's problems and the type of person they want to reach in order to co-create solutions. More than ever governments, policymakers, public authorities, business, and citizens need to be involved in the improvement of the city.

Our findings further support the idea of Lim et al. [33]. Smart mobility is emerging more in high-income countries than in emerging countries. A possible explanation for this may be the high technology dependence of initiatives, which increase the cost of implementation. As stated by Yigitcanlar et al. [5, p. 360], "urban smartness is beyond technology smartness," and emerging countries must find smarter alternatives in order to overcome this economical barrier.

Another important pattern identified was that all the initiatives analyzed were related to urban mobility, corroborating the ideas of Porru et al. [48], who indicated that rural areas have fewer mobility options. Moreover, the effects of COVID-19 have shown that tourists are favoring less crowed and develop areas. This highlights the importance of improving mobility between urban and rural areas since it can increase the connectivity between them and, as a consequence, the economic growth of rural areas.

From the growing literature on smart mobility, it can be perceived that smart mobility can help cities achieving sustainability. Generally, the studies identified focused on alternatives to private car, improvements of existing infrastructure, urban mobility policies, or smart mobility environmental models. Overall, the principal results of smart mobility referenced by the literature were the reduction of traffic congestion and greenhouse gas emissions.

The most important limitation lies in the fact that only peer-reviews articles from the Web of science database were included in literature analysis and synthesis. It would be interesting to include literature from other databases, conference papers, monographies, edited volumes, and government reposts.

Further work needs to be done to establish better measures for smart mobility initiatives. Also, more research is needed in rural areas and emerging countries to better understand how this regions could benefit from smart mobility.

References

1. Yigitcanlar, T., Kamruzzaman, M.: Investigating the interplay between transport, land use and the environment: a review of the literature. Int. J. Environ. Sci. Tech. **11**(8), 2121–2132, 5 Nov (2014)
2. Caprotti, F.: Eco-urbanism and the eco-city, or, denying the right to the city? Antipode **46**(5), 1285–1303 (2014)
3. United Nations: World Urbanization Prospects: The 2018 Revision. New York (2019)
4. Bibri, S.E., Krogstie, J.: Smart sustainable cities of the future: An extensive interdisciplinary literature review. Sustainable Cities Soc. **31**, 183–212, 1 May (2017)
5. Yigitcanlar, T., Kamruzzaman, M., Foth, M., Sabatini-Marques, J., da Costa, E., Ioppolo, G.: Can cities become smart without being sustainable? a systematic review of the literature. Sustainable Cities Soc. **45**, 348–365, 1 Feb (2019)
6. Aldegheishem, A.: Success factors of smart cities: a systematic review of literature from 2000–2018. TeMA J. L. Use, Mobil. Environ. **12**(1), 53–64, Apr (2019)
7. Alexopoulos, C., Pereira, G.V., Charalabidis, Y., Madrid, L.: A taxonomy of smart cities initiatives. In: ACM International Conference Proceeding Series, 2019, vol. Part F1481, pp. 281–290
8. Statista.: Technology spending on smart city initiatives worldwide from 2018 to 2023 (2019) [Online]. Available: https://www.statista.com/statistics/884092/worldwide-spending-smart-city-initiatives/
9. Giffinger, R., Fertner, C., Kramar, H., Kalasek, R., Milanovic, N., Meijers, E.: Smart cities—ranking of European medium-sized cities (2007)

10. UNEP: Sustainable, resource efficient cities—making it happen (2012)
11. Capra, C.F.: The smart city and its citizens. Int. J. E-Planning Res. **5**(1), 20–38 (2016)
12. Hollands, R.G.: Will the real smart city please stand up? Intelligent, progressive or entrepreneurial? City **12**(3), 303–320 (2008)
13. Zhao, L., Tang, Z., Zou, X.: Mapping the knowledge domain of smart-city research: a bibliometric and scientometric analysis. Sustainability **11**(23), 6648 (2019)
14. Yigitcanlar, T., Kamruzzaman, M.: Smart cities and mobility: does the smartness of Australian cities lead to sustainable commuting patterns? J. Urban Technol. 26(2), SI, 21–46 (2019)
15. ITU-T: Overview of key performance indicators in smart sustainable cities (2016)
16. Benevolo, C., Dameri, R.P., D'auria, B.: Smart mobility in smart city action taxonomy, ICT intensity and public benefits (2016)
17. Frank, L., Kavage, S., Litman, T.: Promoting public health through smart growth building healthier communities through transportation and land use policies and practices. Vancouver (2016)
18. Cledou, G., Estevez, E., Barbosa, L.S.: A taxonomy for planning and designing smart mobility services. Gov. Inf. Q. 35(1), 61–76 (2018)
19. Purvis, B., Mao, Y., Robinson, D.: Three pillars of sustainability: in search of conceptual origins. Sustain. Sci. **14**(3), 681–695 (2019)
20. Trindade, E.P., Hinnig, M.P.F., da Costa, E.M., Marques, J.S., Bastos, R.C., Yigitcanlar, T.: Sustainable development of smart cities: a systematic review of the literature. J. Open Innov. Technol. Mark. Complex. **3**(3), 11 (2017)
21. Kidd, C.V.: The evolution of sustainability. J. Agric. Environ. Ethics **5**(1), 1–26 (1992)
22. WCED: Our common future—The Brundtland report. Report of the World Commission on Environment and Development, Oxford (1987)
23. Brown, B.J., Hanson, M.E., Liverman, D.M., Merideth, R.W.: Global sustainability: toward definition. Environ. Manage. **11**(6), 713–719 (1987)
24. Giddings, B., Hopwood, B., O'Brien, G.: Environment, economy and society: fitting them together into sustainable development. Sustain. Dev. **10**(4), 187–196 (2002)
25. OECD: Sustainable manufacturing toolkit - Seven steps to environmental excellence (2011)
26. Corbett, J., Mellouli, S.: Winning the SDG battle in cities: how an integrated information ecosystem can contribute to the achievement of the 2030 sustainable development goals. Inf. Syst. J. **27**(4), 427–461 (2017)
27. Ismagilova, E., Hughes, L., Dwivedi, Y.K., Raman, K.R.: Smart cities: advances in research—an information systems perspective. Int. J. Inf. Manage. **47**, 88–100 (2019)
28. Ismagiloiva, E., Hughes, L., Rana, N., Dwivedi, Y.: Role of smart cities in creating sustainable cities and communities: a systematic literature review. IFIP Adv. Info. Comm. Tech. **558**, 311–324 (2019)
29. Milakis, D., van Arem, B., van Wee, B.: Policy and society related implications of automated driving: a review of literature and directions for future research. J. Intell. Transp. Syst. **21**(4), 324–348 (2017)
30. Denyer, D., Tranfield, D.: Producing a systematic review. In: Buchanan, D.A., Bryman, A. (eds.) The Sage Handbook of Organizational Research Methods, pp. 671–689. Sage (2009)
31. Cooper, H.: The integrative research review: a systematic approach Sage Publications: Beverly Hills. Educ. Res. **15**(8), 17–18 (1986)
32. Fahimnia, B., Tang, C.S., Davarzani, H., Sarkis, J.: Quantitative models for managing supply chain risks: A review. European J. Operat. Res. **247**(1), 1–15, 16-Nov (2015)
33. Lim, Y., Edelenbos, J., and Gianoli, A.: "Identifying the results of smart city development: Findings from systematic literature review". Cities, **95**, 9–7 (2019). https://doi.org/10.1016/j.cities.2019.102397
34. Andrić, J.M., Mahamadu, A.M., Wang, J., Zou, P.X.W., Zhong, R.: The cost performance and causes of overruns in infrastructure development projects in Asia. J. Civ. Eng. Manag. **25**(3), 203–214 (2019)
35. Stolfi, D.H., Alba, E.: Green Swarm: Greener routes with bio-inspired techniques. Appl. Soft Comput. J. **71**, 952–963 (2018)

36. Cerulli, R., Dameri, R.P., Sciomachen, A.: Operations management in distribution networks within a smart city framework. IMA J. Manag. Math. **29**(2), 189–205 (2018)
37. Zawieska, J., Pieriegud, J.: Smart city as a tool for sustainable mobility and transport decarbonisation. Transp. Policy **63**, 39–50 (2018)
38. Villani, M.-G., Cignini, F., Ortenzi, F., Suriano, D., Prato, M.: The smart ring experience in l'Aquila (Italy): integrating smart mobility public services with air quality indexes. Chemosensors 4(4) (2016)
39. Niglio, R., Comitale, P.P.: Sustainbale urban mobility towards smart mobility the case study of Bari area, Italy. TEMA-J. Use Mobi. Environ. **8**(2), 219–234 (2015)
40. Moscholidou, I., Pangbourne, K.: A preliminary assessment of regulatory efforts to steer smart mobility in London and Seattle. Transp. Policy **98**, 170–177 (2020)
41. Nikitas, A., Michalakopoulou, K., Njoya, E.T., Karampatzakis, D.: Artificial intelligence, transport and the smart city: definitions and dimensions of a new mobility era. Sustain. **12**(7) (2020)
42. Shen, T., Hua, K., Liu, J.: Optimized public parking location modelling for green intelligent transportation system using genetic algorithms. IEEE Access **7**, 176870–176883 (2019)
43. Turon, K., Czech, P., Juzek, M.: The concept of a walkable city as an alternative form of urban mobility. Sci. J. SILESIAN Univ. Technol. Transp. **95**, 223–230 (2017)
44. Smith, L., Norgate, S.H., Cherrett, T., Davies, N., Winstanley, C., Harding, M.: Walking school buses as a form of active transportation for children—a review of the evidence. J. Sch. Health **85**(3), 197–210 (2015)
45. Maternini, G., Riccardi, S., Cadei, M.: Zero emission mobility systems in cities inductive recharge system planning in urban areas. TEMA-J. Use Mobil. Environ. No. SI, 659–669, Jun (2014)
46. Wahab, N.S.N., Seow, T.W., Radzuan, I.S.M., Mohamed, S.: A systematic literature review on the dimensions of smart cities. IOP Conf. Series: Earth Environ. Sci. **498**(1), 12087 (2020)
47. Mohammadian, H.D., Rezaie, F.: Blue-green smart mobility technologies as readiness for facing tomorrow's urban shock toward the world as a better place for living (Case Studies: Songdo and Copenhagen). Tech. **8**(3) (2020)
48. Porru, S., Misso, F.E., Pani, F.E., Repetto, C.: Smart mobility and public transport: opportunities and challenges in rural and urban areas. J. Traffic Transp. Eng. (English Ed.) **7**(1), 88–97, Feb (2020)
49. Yigitcanlar, T., Han, H., and Kamruzzaman, M.: "Approaches, advances, and applications in the sustainable development of smart cities: a commentary from the guest editors". Energies **12**, 4554, 29-Nov (2019). https://doi.org/10.3390/en12234554
50. Tomaszewska, E.J., Florea, A.: "Urban smart mobility in the scientific literature — bibliometric analysis". Engineering Management in Production and Services, **10**(2), 41–56 (2018). https://doi.org/10.2478/emj-2018-0010
51. Pérez, L.M., Oltra-Badenes, R., Oltra Gutiérrez, J.V., Gil-Gómez, H. A.: "Bibliometric Diagnosis and Analysis about Smart Cities". Sustainability, **12**, 6357 (2020). https://doi.org/10.3390/su12166357

Chapter 16
The Power of a Multisensory Experience—An Outlook on Consumer Satisfaction and Loyalty

Pedro Rodrigues, Elizabeth Real, and Isabel Barbosa

Abstract The design of a multisensory experience at a point of sale is in current times (and markets) essential for the success of a brand. This article aims to verify the influence of the sensory domain on the visual merchandising technique, understand the influence of the visual merchandising technique on consumer behavior, with a special emphasis on consumer satisfaction and loyalty, and finally, evaluate the effect of consumers' satisfaction on the loyalty felt by the consumer toward the brand. In order to achieve these objectives, a combined approach was implemented—quantitative (questionnaire; 252 respondents) and qualitative methodology (interviews; 10 interviews). As both of the methodologies implemented present identical results, it is possible to note that the sensory domain has a significant role in the visual merchandising technique, and the visual technique has a prevalent effect on consumer behavior, specifically on customer satisfaction and loyalty. As for the last objective, results show that customer satisfaction is a relevant antecedent for consumer loyalty.

16.1 Introduction

The continuous progress of markets on a global scale has encouraged organizations to view consumers differently, as they are faced with more demanding and educated consumers, which in turn "incumbent" organizations to provide multiple advantages, such as holistic engagement of the senses by providing multisensory experiences to them, to grant a unique and singular shopping experience to each consumer [1].

Nowadays, it is considered the functionality and quality of products as basic and elementary conjunctures of a given product, for this reason, organizations recognize that they must offer the consumer a memorable service, differentiated and that finally, is legitimized by them [2].

P. Rodrigues (✉) · E. Real · I. Barbosa
COMEGI, Universidade Lusíada do Norte, Porto, Portugal
e-mail: pedro.rodrigues@fam.ulusiada.pt

E. Real
e-mail: mereal@por.ulusiada.pt

J. L. Reis et al. (eds.), *Marketing and Smart Technologies*, Smart Innovation, Systems and Technologies 337, https://doi.org/10.1007/978-981-19-9099-1_16

Consequently, the topic of sensory has been gaining pertinence in management and marketing research since the scientific community has alluded that sensory can produce considerably positive consequences for organizations.

16.2 Literature Review

16.2.1 Sensory Marketing

The research allusive to sensory marketing was initiated in 1974 by Kotler. Throughout the years, numerous other authors [3–5] have defined the concept of sensory marketing by adding a more complex approach to the understanding of the subject.

The sensory aspect may be seen as a tool to persuade the emotional side of the consumer, thus allowing to conceive a bond between brand and client [6]. According to Lindsrom [7], sensory marketing is perceived as a tool to reach the consumers' subconscious, allowing the use of the senses in a more complex and rigorous way.

Lipovetsky [8] states that sensory marketing wants to enhance the sensitive, tactile, visual, sound, and olfactory attributes of products and the physical circumstance of purchase. Manzano et al. [9] define visual marketing as the strategies implemented by brands through the use of commercial or non-commercial stimuli, signs, and symbols to help them communicate and establish a relationship with their consumers. According to Blessa [10], vision is the most preponderant sense when it comes to the perception of reality; thus, it is a major player when it comes to the consumer's decision-making process, particularly in the process of purchasing a product or service.

As for the sense of smell, it allows the individual to feel, identify, and remember involuntarily smells and aromas. From a marketing perspective, a fragrance present at the point of sale allows the brand to reach the consumers' subconscious in a subtle and simple way [11]. In this context, smell also allows the brand to positively influence the consumers' actions inside the point of sale, reinforcing its positioning as the olfactory aspect is associated with a reduced cognitive effort in associating the stimuli of the store and stimulating involuntary (and subconscious) actions and thoughts in the point of sale [12].

The auditory sense does not require any cognitive effort [13], yet it remains quite useful from a marketing standpoint as it has a positive influence on the purchase circumstance [14]. Famous for being a stimulus that allows for memory recalling and creating associations [15], it is essential for audio branding, as the brand communicates through sounds (i.e., jingles, voices, ambient music) to the consumer, and, in turn, the consumer identifies and understands said communication [16].

On the other hand, the taste is a sensory marketing strategy based on the consumers' ability to understand the different tastes and flavors [17]; however, out of all the senses, it tends to be the one that is less investigated [18]. An exception to this is the gastronomic sector, especially when it comes to wine and coffee [9].

Finally, the sense of touch allows the individual to touch and make physical context with their surroundings, enabling the appreciation of products, particularly when it comes to their physical characteristics. Therefore, the sense of touch plays a very important role in the consumers' buying decision of a product [7]. According to Bardin [19], the sense of touch is considered in two main circumstances in intimate and commercial circumstances. The first one is when an individual touches another one in order to express emotions and feelings and connect with each other. The second one is when a consumer touches the product to check its physical condition, characteristics, and quality. Oliver [20] adds that touch plays a key role in consumer purchasing decisions, acting as a "bridge" to action whenever the customer is motivated or stimulated to make a purchase.

16.2.2 Customer Satisfaction

Satisfaction has various interpretations over the years. For [21], satisfaction can be defined as the refutation of the appreciation felt by the consumer between the previous expectations and the actual performance of the product that they are currently trying. Therefore, it is generally based on the principle of disconfirmation [22]. Thus, in other words, it is accepted that satisfaction happens through an intersection between past expectations and experiences, and the current performance perceived by the consumer [23].

Some authors [24, 25] have shown that visual merchandising positively influences consumers' behavior, particularly consumer satisfaction. The technique of visual merchandising is influential in consumer satisfaction [26], especially over a younger age group [27]. Cangusso et al. [28] points out that the visual merchandising technique is an antecedent in customer satisfaction, since the consumer perceives the point of sale in a more attractive way, being a preponderant technique for microenterprises, however, reiterate that in some cases, the commercial failure is due mainly to the lack of application of this technique by the brands.

In sum, it is plausible to conclude that the visual merchandising technique preponderantly influences consumer behavior and more specifically consumer satisfaction.

16.2.3 Customer Loyalty

One of the most accepted and cited definitions of loyalty is offered by Oliver [29], which defines loyalty as a form of deep commitment, from the consumer, to repeatedly purchase and use a product (or service) in future of the same brand, or set of brands, despite situational influences or marketing efforts from competitors were capable of changing their buying behaviors.

The concept of loyalty was later characterized according to two main criteria, those being the attitudinal and behavioral dimensions. Based on these advances, a more complex theory emerged, in which four dimensions are considered when analyzing loyalty, those being [29]: (i) cognitive; (ii) affective; (iii) behavioral; and (iv) action.

Visual merchandising can play a powerful role when it comes to consumer loyalty, as it conveys a profound message to the sensory receptions of the consumer, changing the perception about the shops' positioning and gradually increasing loyalty with the shop, and with the brand [30].

Based on [31], the visual merchandising technique is fundamental for the productive and profitable management of the point of sale, and therefore, it is necessary to develop an effective and impactful point of sale image. This point of sale image is perceived as fundamental as the visual merchandising increases the consumers' purchase intention and consequently, increases the loyalty to the point of sale and overall brand [32].

In the literature, through the focus on customer satisfaction, the need for brands to develop long-term relationships with the customers was found, as well as the instigation for the use of retention strategies, fostering the understanding of the concept of loyalty. For example [33] established a positive relationship between both concepts (customer satisfaction and customer loyalty). More [34] note their perception of consumer satisfaction as an important predictor of customer loyalty, in which, they ensure that there is a strong relationship between both concepts.

Although there is a large percentage of authors that consider the positive and strong relationship (linear relationship) between customer satisfaction and customer loyalty, there are also authors who analyze and show this relationship as a non-linear one. For example [35] established a model (hierarchy of consumer behavior) where the author observed an absence of a positive relationship between the concepts of customer satisfaction and customer loyalty. The authors reiterate the customer may have a high level of loyalty for the brand and its products, but a low level of satisfaction. To help clarify this perspective, we can point to [36] that state that customer satisfaction is more of an attitude and that there is no correlation between customer attitude and customer behavior.

16.3 Methodology

This research paper focuses on 3 main questions, those being: (i) Does the sensory domain positively influences the visual merchandising technique? (ii) Does the visual merchandising technique positively influences consumer behavior, particularly customer satisfaction and customer loyalty? And, (iii) does customer satisfaction contributes to customer loyalty?

In order to respond to these questions, a mixed methodology was applied. Regarding the quantitative methodology, a questionnaire was shared through customers at the fashion store, where 252 valid questionnaires were obtained. The questionnaire was composed of 7 sections characterized by validated constructs pertinent to the research objectives.

H1: Vision positively influences visual merchandising technique.

H2: Hearing positively influences visual merchandising technique.

H3: Touch positively influences visual merchandising technique.

H4: Smell has a positive influence on visual merchandising technique.

H5: Visual merchandising technique positively influences customer satisfaction.

H6: Visual merchandising technique positively influences customer loyalty.

H7: Customer satisfaction positively influences customer loyalty (Fig. 16.1).

As for the qualitative methodology, 10 structured interviews were performed with customers of a fashion store, based on a script that included 8 relevant questions. In order to draw out valid and comprehensive conclusions, a content analysis was performed subsequently, to collect and further analyze the relevant statements, according to the research objectives.

P1: Vision positively influences visual merchandising technique.

P2: Hearing positively influences visual merchandising technique.

P3: Touch positively influences visual merchandising technique.

P4: Smell has a positive influence on visual merchandising technique.

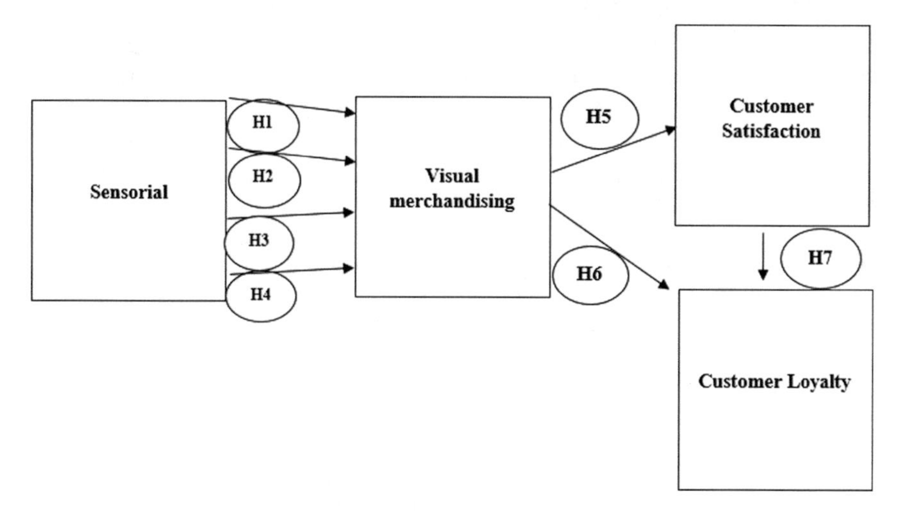

Fig. 16.1 Conceptual model

P5: Visual merchandising technique positively influences customer satisfaction.

P6: Visual merchandising technique positively influences customer loyalty.

P7: Customer satisfaction positively influences customer loyalty.

16.4 Results

See Tables 16.1 and 16.2.

16.5 Main Findings and Conclusions

This paper started off with 3 main research questions, those being: (i) Does the sensory domain positively influences the visual merchandising technique? (ii) Does the visual merchandising technique positively influences consumer behavior, particularly customer satisfaction and customer loyalty? And, (iii) does customer satisfaction contributes to customer loyalty?

In relation to the sensory influence on the visual merchandising technique of the fashion store, it was deduced that the visual sense is the most predominant sense in the visual merchandising technique, then the auditory sense and the olfactory sense presented identical results, being that the sense of touch was perceived by the customer as the least influential sense in the visual merchandising technique.

Table 16.1 Pearson correlation and consequent validation or rejection of hypotheses

Hypothesis	Interpretation	Pearson's correlation result	Result
H1	Vision positively influences visual merchandising technique	$r = 0.936$	Validated
H2	Hearing positively influences visual merchandising technique	$r = 0.758$	Validated
H3	Touch positively influences visual merchandising technique	$r = 0.748$	Validated
H4	Smell has a positive influence on visual merchandising technique	$r = 0.792$	Validated
H5	Visual merchandising technique positively influences customer satisfaction	$r = 0.689$	Validated
H6	Visual merchandising technique positively influences customer loyalty	$r = 0.623$	Validated
H7	Customer satisfaction positively influences customer loyalty	$r = 0.750$	Validated

Table 16.2 Interviewees' statements and consequent validation of propositions

Preposition	Interpretation	Statements by interviewees	Result
P1	Vision positively influences visual merchandising technique	**I1**: "During my visits to the fashion store in Oporto, I consider the visual sense as the most influential sense, since it is observed as the most influential sense for the consumer, as it is the sense that attracts consumers to the point of sale"	Validated
P2	Hearing positively influences visual merchandising technique	**I4**: "I consider that the fashion store's use of the auditory sense influences my shopping experience in this shop, it allows me to carry out my shopping in a serene way and allows me to concentrate in a better way on the products I visualize, unlike other clothing shops present in the same shopping center"	Validated
P3	Touch positively influences visual merchandising technique	**I2**: "…I think the tactile sense is what influences me the most, since, for me it is important to feel the products displayed, as well as the temperature of the point of sale, in which, in my opinion is important to enjoy a pleasant shopping experience, which is verified when I visit the same"	Validated
P4	Smell has a positive influence on visual merchandising technique	**I2**: "In the fashion store, I consider it essential to implement visual merchandising in order to persuade its customers, particularly their satisfaction, for me, it is an important condition nowadays in view of the high number of shops available"	Validated
P5	Visual merchandising technique positively influences customer satisfaction	**I2**: "In the fashion store, I consider it essential to implement visual merchandising in order to persuade its customers, particularly their satisfaction, for me, it is an important condition nowadays in view of the high number of shops available"	Validated
P6	Visual merchandising technique positively influences customer loyalty	**I4**: "I think that the visual merchandising implemented by the fashion store in its shop fosters customer loyalty in the medium and long term, given that as I mentioned, this technique stimulates customer satisfaction and ultimately customer loyalty"	Validated
P7	Customer satisfaction positively influences customer loyalty	**I1**: "In my opinion, the satisfaction provided by fashion store to its customers through the various elements of the point of sale allows it to transform ordinary customers into loyal customers to the shop, perhaps not in the short term, but in the medium and long term certainly"	Validated

Overall, it is stated that the sensory domain presents a positive influence on the visual merchandising technique.

As for the second research question, relating to the influence of the visual merchandising technique on consumer behavior, specifically in consumer satisfaction and consumer loyalty of customers (fashion store), it was concluded that the visual merchandising technique has a preponderant influence on consumer satisfaction.

The visual merchandising technique has a preponderant effect on the loyalty of the consumers of the fashion store, as each element presents a positive influence on loyalty, however, the layout and the lighting are the elements of the visual merchandising technique that have a greater influence on consumer loyalty. In general, the visual merchandising technique presents a positive influence on consumer behavior, regarding customer satisfaction and customer loyalty.

Relating to the third research question that regards verifying the influence of customer satisfaction on customer loyalty (fashion store), the results show that the customer satisfaction provided by the shopping experience is fundamental ensuring customer loyalty, which has a strong influence on customer loyalty.

Therefore, the visual merchandising technique is a relevant tool available to point of sale managers, since it has a preponderant effect on consumer behavior, specifically on consumer satisfaction and consumer loyalty.

16.5.1 Research Limitations

Upon the conclusion of this research, it is possible to point out some limitations. Based on these, future research should improve the study of this subject.

One of the limitations concerns the sample alluding to the quantitative methodology, in which, in the age range, a small presence of individuals aged 50 years old and over and individuals under 18 years was recorded, which may have some influence on the research results.

Another limitation recognized is the use of an online questionnaire as the data collection technique of choice for the quantitative methodology can further explain the few respondents from the older age range.

16.5.2 Recommendations for Future Research

The empirical research carried out, particularly the quantitative methodology, was distributed based on convenience criteria, and the sample obtained shows some discrepancy in relation to the younger age groups in relation to the older age group, in which a small presence of the over 50 years age group was recorded. Future research in this area should focus on a more homogeneous sample to reinforce the reliability of the results.

Funding This research work is financed through national funds from FCT - Fundação para a Ciência e Tecnologia - associated with the project "UIDB/04005/2020".

References

1. Pozo, V., Díaz, I., Frigerio, M.: Aplicación del modelo de Marketing Sensorial de Húlten, Broweus y van Dijk a una empresa chilena del retail. In: XVI Congresso Internacional de Contaduría Administración e Informática, Área de Investigación: Mercadotecnia, Cidad Universitaria, México (2011)
2. Peruzzo, M.: As três mentes do neuromarketing. Alta Books, Rio de Janeiro (2015)
3. Krishna, A., Schwarz, N.: Sensory marketing, embodiment, and grounded cognition: a review and introduction. J. Consum. Psychol. **24**(2), 159–168 (2014)
4. Krishna, A., Cian, L., Sokolova, T.: The power of sensory marketing in advertising. Curr. Opin. Psychol. **10**, 142–147 (2016)
5. Wörfel, P., Frentz, F., Tautu, C.: Marketing comes to its senses: a bibliometric review and integrated framework of sensory experience in marketing. Eur. J. Mark. **56**(3), 704–737 (2022)
6. Rathee, R., Rajain, M.: Sensory marketing—investigating the use of five senses. Int. J. Res. Fin. Market. **7**(5), 124–133 (2017)
7. Lindsrom, M.: A Lógica do Consumo. Nova Fronteira Participações S. A, Rio de Janeiro (2008)
8. Lipovetsky, G.: A Felicidade Paradoxal: Ensaio sobre a Sociedade do Hiperconsumo. Edições 70, Lisboa (2017)
9. Manzano, R., Gavilan, D., Avello, M., et al.: Marketing sensorial: comunicar con los sentidos en el punto de venta. Pearson Educación, Madrid (2012)
10. Blessa, R.: Merchandising no ponto-de-venda, 4th edn. Atlas, São Paulo (2011)
11. Hultén, B., Browdeus, N., Dijk, M.: Sensory marketing. Palgrave Macmillan, London (2009)
12. Herrmann, A., Zidansek, M., Sprott, D., et al.: The power of simplicity: Processing fluency and the effects of olfactory cues on retail sales. J. Retailing **89**(1), 30–43 (2013)
13. Elder, R., Aydinoglu, N., Barger, V., et al.: A sense of things to come – Future research directions in sensory marketing. In: Sensory Marketing–Research on the Sensuality of Products, pp. 361–376. Routledge, New York and London
14. Lund, C.: Selling through the senses: sensory appeals in the fashion retail environment. J. Design, Creative Process Fashion Indust. **7**(1), 9–30 (2015)
15. Lindstrom, M.: Brand sense, 1st edn. Gestão Plus, Lisboa (2013)
16. Hayzlett, J.: The language of audio branding. In: American Marketing Association, pp. 1–4 (2015)
17. Delwiche, J.: The impact of perceptual interactions on perceived flavor. Food Qual. Prefer. **15**, 137–146 (2004)
18. Kotler, P., Amrstrong, G.: Marketing: an introduction, 7th edn. Prentice Hall, Upper Saddle River (2005)
19. Bardin, L.: Análise de conteúdo. Edições 70, Lisboa (1987)
20. Petit, O., Velasco, C., Spence, C.: Digital sensory marketing: integrating new technologies into multisensory online experience. J. Interact. Mark. **45**, 42–61 (2019)
21. Oliver, R.: A cognitive model of the antecedents and consequences of satisfaction decisions. J. Mark. Res. **17**(4), 460–469 (1980)
22. Oliver, R.: Satisfaction: a behavioral perspective on the consumer, 2nd edn. Routledge, New York (2014)
23. Harris, L., Goode, M.: The four levels of loyalty and the pivotal role of trust: a study of online service dynamics. J. Retail. **80**(2), 139–158 (2004)
24. Thomas, A., Louise, R., Vipinkumar, V.: Impacto f visual merchandising, on impulse buying behavior of retail customers. Int. J. Res. Appl. Sci. Eng. Tech. **6**(2), 474–491 (2018)

25. Basu, R., Paul, J., Singh, K.: Visual merchandising and store atmospherics: An integrated review and future research directions. J. Bus. Res. **151**, 397–408 (2022)
26. Kumar, V.: Evolution of marketing as a discipline: what has happened and what to look out for. J. Mark. **79**(1), 1–9 (2015)
27. Kim, J.: A study on the effect that V.M.D (Visual Merchandising Design) in store has on purchasing products. Int. J. Smart Home **7**(4):217–223 (2013)
28. Cangusso, C., Garcia, G., Carrino, A.: O impacto das modificações do visual merchandising em relação a lealdade dos clientes em uma cantina escolar na cidade de Araraquara. Interface Tecnológica **17**(1), 232–243 (2020)
29. Oliver, R.: Satisfaction a behavioural perspective on the consumer. McGrawHill, New York (1997)
30. Marshall, N.: Commitment, loyalty and customer lifetime value: investigating the relationships among key determinants. J. Business Economics Res. (JBER) **8**(8), 67–84 (2010)
31. Sadachar, A., Konika, K.: The role of Sustainable Visual Merchandising Practices in Predicting Retail Store Loyalty. In: International Textile and Apparel Association Annual Conference Proceedings (2017).
32. Madhavi, S., Leelavati, T.: Impact of visual merchandising on consumer behaviour towards women apparel. Int. J. Manag. Res. Business Strat., 62–72 (2013)
33. Rahim, A., Ignatius, I., Adeoti, O.: Is Customer Satisfaction an Indicator of Customer Loyalty? Australian J. Business Manag. Res. **2**(7), 14–20 (2012)
34. Henning-Thurau, T., Gwinner, K., Gremler, D.: Understanding relationship marketing outcomes: an integration of relational benefits and relationship quality. J. Serv. Res. **4**(3), 230–247 (2002)
35. Heskett, J., Sasser, W., Schlesinger, L.: The Value Profit Chain: Treat Employees Like Customers and Customers Like Employees. The Free Press, New York (2003)
36. Cisneros, G., Moline, J.: Fidelizacion Efectiva: No Caiga En Los Errores mãs Frecuentes. Harvard Duesto Marketing y Ventas, 30–35 (1996)

Part VI
Mobile Marketing and Wearable Technologies

Chapter 17
Technology Acceptance: Does the Users Accept the Change of Operating System of Their Smartphone?

Ana Beatriz Palma and Bráulio Alturas

Abstract The main objective of this study was to understand how the users accepts the change of their operating system. Objectives were defined, like clarifying each of the operating systems, understand the preference of consumers' choice through a questionnaire and understand whether or not there is difficulty for users to change their operating system. Thus, a questionnaire was carried out to 204 participants aged between 16 and 72 years in the region of Lisbon. The main criteria were being over 12 years old and having a smartphone. Throughout the study, it was possible to understand that most people have a smartphone, that the satisfaction with the operating system they are currently using is quite high, that older consumers have more difficulty in changing operating system, and that product features are what most influences consumers to switch from an Android to an iOS or vice versa.

17.1 Introduction

Currently, smartphones have seen an increase in terms of demand, and one of the main differentiating features is their operating system. Smartphones can be considered as the most used means of communication worldwide both in less and more developed countries. This type of equipment requires an operating system that can support its services, such as voice calls, text messages, camera features. In early smartphones, operating systems were relatively simple since mobile phones also had more basic features. Today, the operating systems of new smartphones combine the functionalities of a personal computer with other functionalities, such as touch screen, Bluetooth, Wi-Fi, GPS, voice recognition, app stores, among others. Operating systems had to grow and adapt to new smartphone features [1].

A. B. Palma · B. Alturas (✉)
Instituto Universitário de Lisboa (ISCTE-IUL), ISTAR-ISCTE, Av. das Forças Armadas, 1649-026 Lisboa, Portugal
e-mail: braulio.alturas@iscte-iul.pt

A. B. Palma
e-mail: abppa@iscte-iul.pt

J. L. Reis et al. (eds.), *Marketing and Smart Technologies*, Smart Innovation, Systems and Technologies 337, https://doi.org/10.1007/978-981-19-9099-1_17

With this great demand for smartphones, several brands started to produce this type of equipment, such as Samsung, Apple, Xiaomi, Huawei, and many more. One of the best known is Apple. Apple Computer, Inc. was founded by Steve Jobs and Stephen Wozniak in 1976 in Jobs' garage. These two colleagues aimed to change the way people saw computers, so it was at this point that they decided to build their own. In 2007, at the MacWorld 2007 convention, the iPhone was unveiled by Steve Jobs [2]. Since then, the brand's sales have grown substantially, where in 2019 it had a gross revenue of 54.2 billion dollars, and where in the previous year of 2018, it had been the smartphone manufacturer with the highest recommendation rate [3].

The biggest competitor to iOS operating system is, without a doubt, Android. The Android operating system is present in around 1.6 billion smartphones [4]. Unlike iOS, this operating system, developed by Google, can be used on various types of smartphones, manufactured by multiple brands. Android Inc. was founded in 2003, by Andy Rubin, and purchased in 2005 by the American company Google. The main objective of this alliance was to build equipment according to their technologies and thus reduce costs, such as making Android an open-source scenario for developing software for mobile platforms [5].

The global smartphone market is expanding very quickly, and according to data from the GSMA [6] report, it is predicted that by 2025, 70% of the world's population will have a smartphone, predicting a number of 5.7 billion subscribers.

According to previous research, the most significant factor that impacts smartphone switching is price [7]. For example, Samsung's pricing strategy is to reduce prices in order to increase sales [8]. Another reason for switching smartphones is technology. Consumers are always looking for smartphones with the "State-of-the-art", and thus manufacturers of this type of equipment must always try to adapt to the latest technologies and mobile phone versions in order to attract the greatest number of customers [7].

The brand is also a very important feature when it comes to a mobile device. The characteristics of the brand can be what will lead to a future connection and satisfaction with the brand, or the other way around. Studies show that sometimes consumers are willing to pay a higher price for a product of a certain brand [9].

So, the present study intends to better analyze each of the operating systems discussed above (Android and iOS), as well as understanding through a questionnaire how the population feels satisfied with the operating system they use, if they feel satisfied with it, what features they tend to like most about their smartphone, as well as what is the main reason they believe it leads to a change from an Android smartphone to an iOS and vice versa.

17.2 Literature Review

17.2.1 Mobile Communications

The first mobile communications networks appeared in the late seventies and consisted of analog systems that granted only voice communications. In the early nineties, these networks were changed to 2nd generation digital networks, the so-called 2G.

This new network, represented by Global System for Mobile Communications (GSM) was enhanced by a European association that promised compatible voice services in many countries through a wide range of terminals. In addition to the voice service, it also completed a short message service (SMS) which was a great success. This new system has transformed the way people communicate and work [10].

With the purpose of being able to support the evolution of the Internet, effective access to multimedia services is granted with the appearance of the Universal Mobile Telecommunication System (UMTS), better known as the 3rd generation. With this new system (3G), which began to be installed in 2002, services such as web browsing, video and audio streaming, email and file transfers began to be used efficiently.

The fourth generation, nicknamed 4G, was created with the purpose of promoting a global mobile network, fully integrated and based on Internet Protocol (IP), in order to integrate voice, video and multimedia services for users. Access to this network does not necessarily have to be done by the mobile operator's network, it can be accessed, equally, by other wireless networks [11]. This generation, also recognized as Long Term Evolution (LTE), was designed, and made available in 2010 to meet the needs of users in high-speed video and services, with a less complex network architecture and authorization for transfer data at higher speeds and designed for multimedia communications [11].

The term "smartphone" began to be used in 1997 and symbolized a new era of mobile devices. Smartphones began to be considered universal portable computers that incorporated a telephone. One of the essential features of smartphones was their ability to run software programs, which later came to be known as "applications", allowing users to perform tasks that had not previously been foreseen when the phone concept was first manufactured [12].

According to Lee [13], a smartphone is defined as a mobile phone that offers advanced capabilities, often with features similar to a PC, and that is not limited to just making voice calls.

It is currently considered that many issues are resolved through a smartphone, so the need to, for example, go to the bank to consult a statement, make transfers, make purchases via virtual means, etc., is no longer necessary. Just like innumerous professional activities can be solved through apps, with the various functions that smartphones offer [14].

17.2.2 Mobile Operating Systems

An operating system is a program that manages a computer's hardware, as it provides bases for application programs and acts as an intermediary between the computer software and the computer hardware [15].

According to Tanenbaum and Woodhull [16], an operating system consists of a grouping of one or more programs that controls the computer resources, like processors, main memory, hard disks, printers, keyboard, mouse, monitor, network cards, and other input and output devices. Thus, operating systems are considered a piece of software, and deal with all the complexity of managing components, working with optimizations, abstracting from the user all the execution part, putting it in the background.

The main objectives of an operating system are run user programs and facilitate problem solving for them, make the computer system easy to use and efficiently use the computer system hardware. Operating systems offer services to both users and developers that make it possible to run a computer or other device without having to use low-level hardware controls as these are difficult to implement. These provide, relatively uniform, interfaces for accessing a wide range of devices that the computer interacts with, from input/output devices such as printers or digital cameras, to wired or wireless networking and components that ensure communication between computers [17].

Operating systems are largely responsible for the rise of mobile devices, as they are now being built to manage smartphone hardware resources, from managing simpler applications to managing more complex ones, such as calculating navigation routes, recognizing objects, using smart sensors, thus making it possible to perform activities that were previously impossible, making this part of our reality today [18].

These days there are a large number of OS for smartphones that try to be the best in the world, but the only ones that manage to reach the podium and cause the greatest impact are undoubtedly the iOS system and the Android system [19].

One of the operating systems for smart mobile devices is Android, which represents a technological alternative whose appearance generated a good impression on its group of users, being today a competitor facing other operating systems recently considered as leaders.

Android Inc. was founded by Andy Rubin in 2003, which after two years, in 2005, was bought by Google. Later, the team led by Rubin developed a mobile device platform powered by the Linux kernel, which was unveiled on November 5, 2007, by the Open Handset Alliance, a commercial alliance of several companies including Google, HTC, Intel, LG and 76 other companies [20].

The main objective of this alliance was to build equipment according to their technologies that could considerably reduce time and cost, as well as improve services and provide the best features to consumers [21]. Another objective of this alliance would be to turn Android into an open-source scenario for software development for mobile platforms [5].

According to data from the Statista website, from July 2020 [4] Android is the most used operating system with a number of users that rounds the 1.6 billion. The year of 2017 was the year that Android surpassed Windows, becoming the most popular operating system. Android in October 2020 continued with the leadership position as the leader of the mobile devices OS, controlling this market with a 72.92% share. Together, Google Android and Apple iOS have nearly 99% of the global market.

One of the main reasons for this OS to be so successful is the constant search for improvement in its numerous versions, with each one offering new and more advanced features, with faster Internet access. Another of Android's popularity is its strong collaboration with mobile device manufacturers [18].

The iOS abbreviation comes from the name "iPhone Operation System", developed by the well-known company Apple. This system was based on the MAC OS X operating system and designed to respond to the needs of mobile devices developed by this organization.

Apple can be considered one of the most successful companies in the world in recent years. According to the Forbes website, the company is the most valuable brand in the year of 2020, with an estimated value of 241.2 billion dollars and a revenue of 260.2 billion dollars [22].

Through the Net Promoter Score (NPS) model, studies indicated that for the smartphone market, Apple is the manufacturer with the highest NPS, with 60% in 2018, the same value that it had registered in the previous year [23].

The iOS operating system is restricted to hardware built by Apple. In this way, only Apple's own devices can successfully run the iOS operating system [24].

In January 2007, Steve Jobs introduced the iPhone during his speech at the Macworld Conference and Expo. Immediately that same year, there were soon sales of about 1.39 million units. The year with the highest number of sales between 2007 and 2018 was 2015 with 231.22 million units sold worldwide [25].

17.2.3 Technology Adoption

The development of the TAM model was due to an agreement between IBM Canada and the Massachusetts Institute of Technology, in the mid-1980s, in order to assess the market potential for the brand's new products and encourage an explanation of the determinants of computer use [26].

The TAM model was proposed by Davis [27], and intended to specifically explain the behavior of the use of IS (mainly the computer), focusing primarily on two fundamental dimensions:

1. The Perception of usefulness (PU), that is, the degree to which a person believes that the use of a certain system will improve their performance.
2. Perception of Ease of Use (PEU), relating to the degree to which a person believes that using a particular system will be effortless.

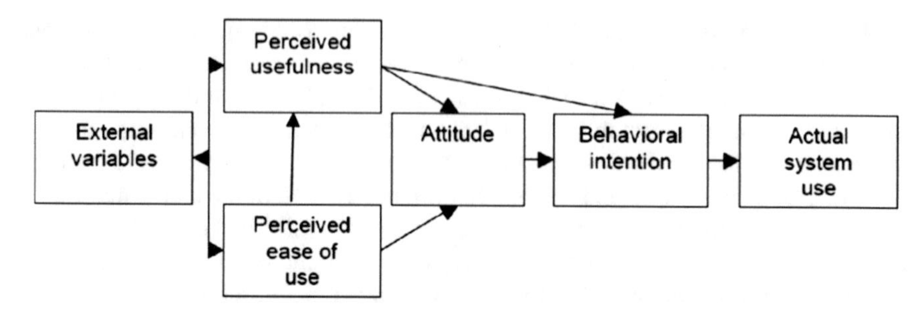

Fig. 17.1 Technology acceptance model [26]

According to TAM, users first consider the functions performed by computer systems (PU), and subsequently examine the ease or difficulty in using them (PEU). The behavioral intention of use (BI), is thus, defined by the person's attitude (A) in relation to the use of the system, as well as by the conviction that the IS will enhance performance [28]. This analysis can be represented in the diagram of Fig. 17.1 [26].

Several analyzes have concluded that this model gives a dominant role to behavioral intent, in the use of technology, compared to the perception of usefulness or the perception of ease of use. However, among these two variables, the one that best predicts the acceptance of use and technology is the perception of usefulness [29, 30].

Basically, the PEU is represented as the degree to which the person believes that the use of IS is effortless and the PU seeks to measure how much the person believes that the use of technology will contribute to an improvement in their performance. This perception on the part of the user appears to have a positive influence on the behavioral intention to use this technology [28].

Davis [27] presented the TAM model with the intention of focusing on why users accept or reject information technology and how to improve its acceptance, thus offering a support to predict and explain acceptance.

This model was created in order to understand the specific relationship between external variables of user acceptance and the actual use of the computer, thus seeking to understand the user's behavior through knowledge of its usefulness and perceived ease of use [27].

The TAM is useful for predicting, but also for characterizing, so that researchers and others can recognize why users do not accept a particular system or technology and, consequently, implement the appropriate corrections [26].

Lee et al. [31] led a literature search on the TAM model and generally found that the TAM presented cohesive results while preserving its security in explaining the acceptance of technology by users of information systems. This research was applied to different technologies, such as text processors, email, the Internet, banking systems, as well as in different situations (over time and in different cultures), with different control factors (gender, type and organizational structure) and different individuals (students and professionals), which makes us believe in its credibility [32]. The TAM model is still the most popular among researchers engaged in Acceptance and Use of Technology [33].

Nowadays, the smartphone is much more than a simple mobile phone that people use to make calls. Is a device used by different age groups that allows to be used as a work tool, for leisure, among other tasks [34].

Data from the 2018 "Marktest Telecommunications Barometer" study, showed that the smartphone penetration in Portugal continues to increase and is in the hands of 3 out of 4 mobile phone users. In the quarter of July 2018, this study counted about 6.9 million people who have a smartphone. This study also indicates that the penetration rate of smartphones is higher among males, residents of Greater Lisbon, the younger population and higher social classes [3].

According to data from the Statista website, it appears that the age group that most adopts smartphones in the United States of America, are the ages between 18 and 29 years old, and the one with the lowest values is the age group of individuals with over 65 years old [4].

One of the factors that may be connected with this difference in age groups may be social needs, which is one of the main factors for consumers' dependence on smartphones. These consist of an individual's social interaction needs. Represent the need for communication with friends, family, groups, clubs, churches and at work [35]. This need happens to smartphones because they have become much more versatile, allowing consumers to use them more in order to communicate and maintain relationships between individuals [36]. This statement can be complemented by the fact that some studies show that most people using smartphones are teenagers and young adults [34].

Another of the dependencies that can be mentioned, in addition to social needs, is the current dependence on the Internet and the modernization of mobile devices, where it was found that the smartphone together with the ease of mobile connection can also cause dependency [37].

17.3 Methodology

The methodology used was a quantitative and descriptive research. The purpose of the descriptive research is to describe or clarify characteristics of the group of participants in question and therefore a relationship is established between the questions to be asked and the object of study. When it comes to a descriptive research, as it is put into practice in this work, the researcher will conduct the interpretation and analysis of the study without interfering or manipulating the data [38].

This study was conducted with the goal of studying the user's acceptance on changing the operating system of their smartphone. In this study, is analyzed whether the operating system is a crucial factor or not in choosing a smartphone, as well as whether the users are satisfied with the OS they currently use, whether they recommend their OS to someone else, among other questions that were presented in a survey.

The survey was conducted for participants over the age of 16 who may or may not own a smartphone, and the smartphone could have as operating system an Android, iOS or other. Only participants who own a smartphone and one of the two OS under analysis (Android or iOS) were considered for the study.

The questionnaire was 204 respondents, but only 167 answers were considered valid. Of these 167 responses have ages ranging from 16 to 72 years old with various types of academic qualifications and net monthly incomes.

The questionnaires were sent via email or social media, depending on proximity to the respondents, and participants were asked to forward the questionnaire to other known contacts. The results were later analyzed using IBM SPSS 24 Statistics for statistical data processing.

The questionnaire was constructed so that in an initial phase (the first 6 questions) it would be possible to characterize the sample from a sociodemographic perspective, and in the remaining questions it would be possible to extract the participants' opinion on the theme.

The first 6 questions determine gender, age, education, net monthly income, whether they use a smartphone, and which smartphone they use. The remaining questions, based on TAM, were designed in such a way as to obtain conclusive answers with pre-defined answers on a Likert scale from 1 (Strongly disagree, Very dissatisfied, Not at all connected…) to 5 (Strongly agree, Very satisfied, Very connected).

17.4 Analysis and Discussion of Results

Once the data was collected, the results obtained in the empirical study were analyzed. It was found that the respondents were those who expressed interest in participating, which is therefore a convenience sample.

The present study had 167 participants ($N = 167$), of which 32.3% were male, 66.5% were female and 1.2% identified themselves as other.

Regarding the age of the participants, the majority is less than 25 years old with a percentage of 35.3%, followed by participants aged between 46 and 55 years (28.1%) and with a smaller percentage, participants older than 55 years with a percentage of only 6%.

As for academic qualifications, most of the respondents have a completed bachelor's degree (44.9%), and few have a completed doctorate (0.6%). Those who have completed only high school also occupy a large percentage of the chart (32.9%).

In terms of net monthly remuneration at the end of the month, the majority of the participants who answered this question, and there were 18 people who decided not to share this information, receive a remuneration in the range between 500€ and 1000€ at the end of the month (34.9%), and with a smaller percentage are the participants who have a remuneration above 2001€ at the end of the month (11.4%).

One of the crucial questions in this questionnaire was whether or not the participant uses a smartphone, where 160 participants said yes (96%) and only 7 respondents (4%) do not use this type of device.

To these 160 participants who answered "Yes" to the question "Do you use a smartphone?", they were asked the question what was the operating system that their smartphone operated, having as options the Android OS, the iOS OS and the third option was "Other", to which 63.1% replied that they had an Android, with a percentage of 33.8% the participants who have a iOS, and only 3.1% of the participants replied that they have another type of operating system, thus putting an end to the questionnaire for them.

For the participants who chose the options "Android" or "iOS", the question was asked whether they were satisfied with their operating system, to which 44.5% of the participants answered that they were satisfied and 38.1% very satisfied. Only 10.3% of respondents are very dissatisfied with the OS of their smartphone.

To better understand which users are the most satisfied with their operating system, a comparison of question 6 of the questionnaire "If yes, what is the operating system of your smartphone?" with question 7 "Are you satisfied with the operating system of your smartphone?" was performed. With this crossing of information, it was possible to draw that the percentage of participants satisfied with their operating system is higher in the Android OS (52.5%), but the percentage of participants very satisfied with their operating system is higher in respondents using the iOS operating system (53.7%). Regarding dissatisfaction, iOS users are most dissatisfied.

With the crossing of data, it was also possible to see that iOS users feel more emotionally attached to their OS than Android users, since for iOS users the answer with the highest percentage is "Connected" with 42.6% and then "Very Connected" with 29.6%, and for Android users the most given answer was the option "Neutral" with 34.7%.

On the question "Would you recommend other people to buy smartphones with the same operating system as yours?" most of the participants, with a percentage of 50.3% of the answers, answered that they agreed when recommending their smartphone to others. As for the answer "Strongly Disagree", no one choose this option. By cross-referencing this data with the users of the two types of operating systems under study, it was possible to extract that for Android users the most given answer was "Agree" (54.5%) and for iOS users it was "Totally agree" (44.4%).

To understand why users choose a certain smartphone, the question "Why did you choose your smartphone?" was asked, and it can be seen that price is a substantial factor when choosing a smartphone for Android users, who agreed with this option by 65.3%, unlike iOS users that only 11.1% agreed, with a higher percentage choosing the option "Disagree" with 33.3%. It was also retrieved in this question that the smartphone functions are one of the main reasons when choosing a smartphone for both users, that for iOS users' appearance is the key factor with the highest percentage of "Agree" (56.6%) and that for both users the operating system that their smartphone operates is a very important factor with Android users choosing the option "Agree" 61.9%, and iOS users choosing the option "Agree" and "Strongly Agree" more than 40% in both.

Since it was proven that the operating system is an important factor when choosing a smartphone, users were also asked whether this choice was made autonomously or under the influence of friends and family, to which the answer was mostly negative for both users. For Android users the most given answer was "Strongly disagree" with 29.7% and for iOS users the most answered was "Disagree" with 38.9%.

In order to understand what the most cherished features in the smartphones of the users are who took the survey, a question was asked for that purpose, where users could choose more than one option. Android users absolutely chose that for them the most cherished feature is "Image quality" with 100% of the users totally agreeing, while for iOS users, the most cherished feature is once again "Appearance" (68.5%). The least liked features for both users are "Color", "Touch" and "Sound quality".

Even if the brand is a more important factor when buying a smartphone, it was seen earlier that the operating system is also a very important factor, and with the question "Will I ever buy a smartphone with the same operating system as my current one again?", the answers were unanimous, and for both users, staying with the same operating system seems to be the right answer, since 52.5% of users using Android agreed to buy again a smartphone with the same OS, and 55.6% of iOS users totally agreed.

In another question, it was asked how often users think about changing operating systems. Since in the previous question, they said they would buy again a smartphone with the same OS as their current one, the willingness to change will be low, hence the most given answer to the question "How often do you consider changing operating systems?" was "Rarely", with 45.5% of answers given by Android users, and 46.3% with an answer given by iOS users.

Other question that was asked, was whether respondents believed that older consumers would find it more difficult to switch operating systems due to the new product's instructions being harder to re-learn, to which almost 54.2% of the participants said they agreed and 24.8% totally agreed. Only one person (0.7%) strongly disagreed with this matter.

When asked if "The greater the relationship with the brand of my smartphone, the less I intend to change", the participants mostly agreed with 59.2% of the answers. Relating this answer to the question "How often would you consider changing operating systems?" it can be seen that users of both operating systems feel a strong relationship with the brand of their smartphones, so they don't want to change them often.

At the same time, the answers to the question ""How often would you consider changing operating systems?" were not 100% "Strongly Disagree", because there are always some users who feel the need to change their smartphone or operating system with some frequency, so the question "Do you believe that the demand for variety has an impact on my intention to change operating systems" was asked, where the most given answer was "Neither Agree nor Disagree" with 36.6% of the answers given, but then comes right behind the option "Agree" with 30.1% of the votes.

Finally, the participants were asked what they believed to be the biggest influence on their intention to change from an iPhone to an Android smartphone (or vice versa), to which they agreed most on the product features (66.4%), then the price (55.9%) and lastly the brand (42.8%).

Then, proceeded the Pearson's linear correlation coefficient (Pearson's R) analysis between the new variables, taken from the principal component analysis (PCA). This analysis is done when is necessary to analyze the relationship between two variables. When exists a perfect correlation, that is -1 or 1, it means that it will know the value of one variable determining exactly the value of the other. If the value is 0, it means that there is no linear relationship between the variables [39].

It can be seen in Table 17.1 there are variables that are correlated and significant at the 0.01 level, and variables that are correlated and significant at the 0.05 level, and also others that have a very weak correlation to the point of not being significant.

When analyzing the "Price" column with the interaction of the "Satisfaction" variable, it can be seen that Pearson's correlation is negative. This means that if one variable increases, the other decreases, and vice versa. That said, the lower the price of the smartphone the higher the user satisfaction, and the higher the price of the smartphone the lower the user satisfaction. One can verify this type of correlation in the "Price"/"Features" variables because the user gives a lot of importance to the features of the cell phone, but always wants the lowest price possible. Also, in the "Price"/"OS". The correlation between "Price"/"Loyalty" is also negative, as is "Price"/"Influence". The correlation between the variable "Satisfaction"/"Influence" is also negative, which may indicate that it is not because third parties influence when choosing a smartphone or operating system that satisfaction is related.

The cell comparing the "Loyalty" and "Characteristics" variables has a Pearson correlation of 0.173. The "Influence" and "SO" variable has a correlation of 0.202. The correlation between the variable "Change" and the variables "SO", "Loyalty" and "Influence" has the following values of 0.020, 0.010 and 0.159, respectively. The variable "Price" has a correlation of 0.191 with the variable "Change". The variable "Satisfaction" with the variables "Characteristics", "SO", "Loyalty" and "Change" have the respective correlations 0.027, 0.158, 0.178 and 0.002. Thus, it can stated that the level of association between the previous continuous variables is low.

Looking at the column of the variable "Characteristics", it has a moderate correlation with the following variables "SO", "Influence" and "Change" with Pearson's correlations of 0.309, 0.367 and 0.352. Another correlation that is considered moderate, are the variables "Influence" and "Loyalty" with a correlation of 0.307.

The variable "Loyalty" and "OS" is the only one with a value greater than 0.5, thus, according to the author [40], it can be said that this is the only high correlation. Thus, there is correlation between all the variables present in the study (positively or negatively correlated).

Table 17.1 Pearson correlation for the variables under study

Correlations

		Features	OS	Loyalty	Influence	Change	Price	Satisfaction
Features	Pearson correlation	1						
	Sig. (2-tailed)							
	N	150						
OS	Pearson correlation	**0.309**[a]	1					
	Sig. (2-tailed)	0.000						
	N	150	150					
Loyalty	Pearson Correlation	**0.173**[b]	**0.563**[a]	1				
	Sig. (2-tailed)	0.036	0.000					
	N	148	148	152				
Influence	Pearson Correlation	**0.367**[a]	**0.202**[b]	**0.307**[a]	1			
	Sig. (2-tailed)	0.000	0.013	0.000				
	N	150	150	149	152			
Change	Pearson correlation	**0.352**[a]	**0.020**	**0.010**	**0.159**	1		
	Sig. (2-tailed)	0.000	0.814	0.899	0.053			
	N	148	148	152	149	152		
Price	Pearson correlation	**−0.106**	**−0.243**[a]	**−0.200**[b]	**−0.011**	**0.191**[b]	1	
	Sig. (2-tailed)	0.201	0.003	0.013	0.898	0.018		
	N	148	148	152	149	152	152	
Satisfaction	Pearson correlation	**0.027**	**0.158**	**0.178**[b]	**− 0.085**	**0.002**	**−0.170**[b]	1
	Sig. (2-tailed)	0.739	0.053	0.028	0.300	0.981	0.036	
	N	150	150	152	152	152	152	155

[a] The correlation is significant at the 0.01 level (2-tailed)
[b] The correlation is significant at the 0.05 level (2-tailed)

17.5 Conclusion

The present work had as a research question "In what way does the user accept the change of their operating system?". To try to answer this question, a study was conducted on the Android and iOS operating systems, where a questionnaire was conducted to understand which operating system the users preferred, if they were satisfied with them or, for example, if they intended to change their operating system in the future.

With the completion of the theoretical framework, it was possible to better understand both operating systems. The Android operating system is the most used operating system with several users that is around 1.6 billion, while the iOS operating system, according to the Forbes website, was the most valuable brand in the year 2020, with an estimated value of 241.2 billion dollars and revenues of 260.2 billion dollars.

A study conducted by Marktest in 2018 [3], indicated that 6.9 million people owned a smartphone, also indicating that the smartphone penetration rate was higher among males, the younger population, as well as people with a higher social class.

One of the factors for the Android operating system being the most purchased last year, is the fact that it has a strong collaboration with mobile device manufacturers, unlike iOS which operates only with its manufacturer Apple, that is, the user is more likely to like a non-iOS smartphone than an iOS one.

With the questionnaire applied to about 200 participants it was possible to draw only a few conclusions identical to those described above. The age group that took part in the study the least were participants over 55. So, one can agree with the fact that probably the younger population is more likely to own a smartphone because older consumers have had later contact with them.

Through the study, it was possible to extract that the TAM model presents at least three variables that lead a subject to accept or not the technology. According to this model, it will be the ease of use and the usefulness that will determine the intention to use, that is, if the user finds that a certain technology is not easy to use and is not useful, he will have no intention to use it.

A smartphone is already something so intrinsic to the daily life of the population that only 7 people from the sample of 167 (valid answers) answered that they did not use a smartphone in the questionnaire. Of the same 160 respondents who answered that they owned a smartphone, more than half answered that they owned a smartphone with the Android operating system. These data were already expected, because as previously mentioned, Android was the operating system with the largest number of users in 2020.

Through the application of the questionnaire, it was possible to learn that most users are satisfied with their operating system and have no intention of changing soon. This leads one to believe that the connection and trust that users have in their operating system, as well as the relationship they have with their smartphone brand, is becoming so great that they would rather stay comfortable with what they have than change.

Regarding what most influences the users when switching from an Android operating system to an iOS, or vice versa, the answer "Product Features" was the one that had the highest agreement where more than 60% of respondents said they agreed, while the answer "Brand", of the three possible answers, was the one that had a lower number of respondents to say they agreed.

Acknowledgements This work was undertaken at ISTAR-Information Sciences and Technologies and Architecture Research Center from Iscte-Instituto Universitário de Lisboa (University Institute of Lisbon), Portugal, and it was partially funded by the Portuguese Foundation for Science and Technology (Project "FCT UIDB/04466/2020").

References

1. Okediran, O.O., Arulogun, O.T., Ganiyu, R.A.: Mobile operating systems and application development platforms: a survey. J. Adv. Eng. Tech. **1**, 1–7 (2014). https://doi.org/10.15297/JAET.V1I4.04
2. Yuhesdi, A.: The influences of price, product features, brand image, and reference group towards switching intention from iphone to Android phone (A study on iPhone users in Indonesia). Jurnal Ilmiah Mahasiswa **8** (2020)
3. Marktest: Barómetro de Telecomunicações da Marktest [Marktest Telecommunications Barometer] (2018)
4. Statista: Population of internet users worldwide from 2012 to 2019, by operating system (2021)
5. Grønli, T.-M., Hansen, J., Ghinea, G., Younas, M.: Mobile application platform heterogeneity: Android vs Windows phone vs iOS vs Firefox OS. In: IEEE 28th International Conference on Advanced Information Networking and Applications (2014)
6. GSMA: The mobile economy 2021. London, UK (2021)
7. Ashfaq, H., Lodhi, S.: Factors leading to brand switching in cellular phones: a case of Pakistan. Int. J. Sci. Eng. Res. **6**, 1466–1483 (2015)
8. Sin, K.P., Yazdanifard, R.: The comparison between two main leaders of cell phone industries (Apple and Samsung) versus Blackberry and Nokia, in terms of pricing strategies and market demands. Center for Southern New Hampshire University (2013)
9. Aaker, D.A.: Managing brand equity: capitalizing on the value of a brand name. Free Press, New York, NY, USA (1991)
10. Ferreira, L.S., Correia, L.M.: Evolução e desafios das redes de comunicações móveis [Evolution and challenges of mobile communications networks]. Kriativ-tech (2018). https://doi.org/10.31112/kriativ-tech-2018-01-11
11. Xiuhua, Q., Chuanhui, C., Li, W.: A study of some key technologies of 4G system. In: 3rd IEEE Conference on Industrial Electronics and Applications, ICIEA 2008. pp. 2292–2295 (2008)
12. Campbell-Kelly, M., Garcia-Swartz, D., Lam, R., Yang, Y.: Economic and business perspectives on smartphones as multi-sided platforms. Telecomm. Policy **39**, 717–734 (2015). https://doi.org/10.1016/j.telpol.2014.11.001
13. Lee, S.Y.: Examining the factors that influence early adopters' smartphone adoption: the case of college students. Telematics Inform. **31**, 308–318 (2014). https://doi.org/10.1016/j.tele.2013.06.001
14. Pires, B.N.C., Monteiro, P.E.S., Ferreira, M.S.A., Duarte, M.B.A., Marçal, S.A.: A Influência da Marca no Processo de Decisão de Compra de Smartphones na Percepção de Estudantes de uma Instituição de Ensino Superior [The influence of the brand on the smartphone purchase decision process on the perception of students of a Higher Education]. Libertas: Revista de Ciênciais Sociais Aplicadas. **8**, 133–153 (2018)

15. Silberschatz, A., Galvin, P.B., Gagne, G.: Operating system concepts. Wiley (2004)
16. Tanenbaum, A., Woodhull, A.: Operating systems design and implementation. Pearson (2006)
17. Novac, O.C., Novac, M., Gordan, C., Berczes, T., Bujdoso, G.: Comparative study of Google Android, Apple iOS and Microsoft Windows Phone mobile operating systems. In: 14th International Conference on Engineering of Modern Electric Systems, EMES 2017, pp. 154–159 (2017)
18. Narmatha, M., Krishnakumar, S.V.: Study on Android operating system and its versions. Int. J. Scient. Eng. Appl. Sci. **2**, 439–444 (2016)
19. Lazareska, L., Jakimoski, K.: Analysis of the advantages and disadvantages of Android and iOS systems and converting applications from Android to iOS platform and vice versa. American J. Soft. Eng. Appl. **6**, 116 (2017). https://doi.org/10.11648/j.ajsea.20170605.11
20. Sheikh, A.A., Ganai, P.T., Malik, N.A., Dar, K.A.: Smartphone: Android vs IOS. The SIJ Trans. Comp. Sci. Eng. Its Appl. (CSEA) **01**, 31–38 (2013). https://doi.org/10.9756/sijcsea/v1i4/0104600401
21. Haris, M., Jadoon, B., Yousaf, M., Khan, F.H.: Evolution of android operating system: a review. Asia Pacific J. Contemp. Educ. Comm. Tech. **4**, 178–188 (2018). https://doi.org/10.25275/apjcectv4i1ict2
22. Swant, M.: Apple, Microsoft and other tech giants top forbes' 2020 most valuable brands list (2020)
23. Marktest: NPS—Net Promoter Score (2018)
24. Milani, A.: Programando para iPhone e iPad [Programming for iPhone and iPad]. Novatec (2014)
25. Statista: Unit sales of the Apple iPhone worldwide from 2007 to 2018 (2021)
26. Davis, F.D., Bagozzi, R.P., Warshaw, P.R.: User acceptance of computer technology: a comparison of two theoretical models. Manage. Sci. **35**, 982–1003 (1989). https://doi.org/10.2307/2632151
27. Davis, F.D.: Perceived usefulness, perceived East of use, and user acceptance of information technology. MIS Quart. **13**, 319–340 (1989). https://doi.org/10.2307/249008
28. Parreira, P., Proença, S., Sousa, L., Mónico, L.: Technology Acceptance Model (TAM): Modelos percursores e modelos evolutivos [Technology acceptance model (TAM): pathway models and evolutionary models]. In: Competências empreendedoras no Ensino Superior Politécnico: Motivos, influências, serviços de apoio e educação, pp. 143–166 (2018)
29. Legris, P., Ingham, J., Collerette, P.: Why do people use information technology? a critical review of the technology acceptance model. Inform. Manag. **40**, 191–204 (2003). https://doi.org/10.1016/S0378-7206(01)00143-4
30. Turner, M., Kitchenham, B., Brereton, P., Charters, S., Budgen, D.: Does the technology acceptance model predict actual use? a systematic literature review. Inf. Softw. Technol. **52**, 463–479 (2010). https://doi.org/10.1016/j.infsof.2009.11.005
31. Lee, Y., Kozar, K.A., Larsen, K.R.T.: The technology acceptance model: past, present, and future. Comm. Assoc. Inform. Syst. **12**, 752–780 (2003). https://doi.org/10.17705/1cais.01250
32. Silva, P.M., Dias, G.A.: Teorias sobre aceitação de tecnologia: por que os usuários aceitam ou rejeitam as tecnologias de informação? [Technology acceptance theories: why do users accept or reject information technologies?]. Brazilian J. Inform. Sci. **1**, 69–91 (2007). https://doi.org/10.36311/1981-1640.2007.v1n2.05.p69
33. Alturas, B.: Models of acceptance and use of technology research trends: literature review and exploratory bBibliometric study. In: Studies in Systems, Decision and Control, pp. 13–28 (2021)
34. Caracol, J.H.V., Alturas, B., Martins, A.: Uma sociedade regida pelo impacto do smartphone: Influência que a utilização do smartphone tem no quotidiano das pessoas [A society ruled by the impact of the smartphone: Influence that the use of the smartphone has in people's daily lives]. In: 14th Iberian Conference on Information Systems and Technologies, pp. 1–6. Coimbra, Portugal (2019)
35. Tikkanen, I.: Maslow's hierarchy and pupils' suggestions for developing school meals. Nutr. Food Sci. **39**, 534–543 (2009). https://doi.org/10.1108/00346650910992196

36. Lippincott, J.K.: A mobile future for academic libraries. Ref. Serv. Rev. **38**, 205–213 (2010). https://doi.org/10.1108/00907321011044981
37. Ortega, F.D.C., Corso, K.B., Moreira, M.G.: Dependência De Smartphone: Investigando a Realidade De Uma Prestadora De Serviço Do Sistema "S" [Smartphone dependence: investigating the reality of an "S" system service provider]. Revista Sociais e Humanas. **33**, 200–217 (2020). https://doi.org/10.5902/2317175837257
38. Santos, R., Alturas, B.: Factors influencing consumers to shop online for computer/telecommunications equipment. In: Reis, J.L., Peter, M.K., Cayolla, R., Bogdanovic, Z. (eds) Smart Innovation, Systems and Technologies, pp. 595–605 (2022)
39. Figueiredo Filho, D.B., Silva Júnior, J.A.: Desvendando os mistérios do coeficiente de correlação de Pearson (r) [Unraveling the mysteries of the Pearson correlation coefficient (r)]. Revista Política Hoje. **18**, 115–146 (2009)
40. Cohen, J.: Statistical Power Analysis for the Behavioral Sciences. Lawrence Erlbaum Associates (1988)

Chapter 18
Database Performance on Android Devices, A Comparative Analysis

Carolina Ferreira, Manuel Lopes, Luciano Correia, Cristina Wanzeller, Filipe Sá, Pedro Martins, and Maryam Abbasi

Abstract The number of mobile devices operating worldwide has reached almost 15 billion, and Android has become the most popular operating system. Consequently, massive amounts of user data need to be collected and processed, and mobile applications now have essential roles in knowledge management. Dealing with such an amount of data on a single mobile device running several applications simultaneously may be a challenging task, and database performance has a substantial impact on the global performance and, especially, on response times. While SQLite is extremely popular and used in the mobile industry since its appearance, new promising options are emerging like Room and greenDAO that, still relying on SQLite, were designed to overcome SQLite weaknesses like response time and lack of validation of SQL queries, or ObjectBox and Realm that abandon relational databases and adopt object-oriented and NoSQL approaches that do not require such deep knowledge of SQL. In this article, ObjectBox, Realm, SQLite with Room, and SQLite with greenDAO are reviewed, and a comparative analysis between them is presented alongside results

C. Ferreira · M. Lopes · L. Correia
Polytechnic of Viseu (ESTGV), Viseu, Portugal
e-mail: pv23783@alunos.estgv.ipv.pt

M. Lopes
e-mail: estgv17025@alunos.estgv.ipv.pt

L. Correia
e-mail: pv22382@alunos.estgv.ipv.pt

C. Wanzeller · P. Martins (✉)
CISeD—Research Centre in Digital Services, Polytechnic of Viseu, Viseu, Portugal
e-mail: pedromom@estgv.ipv.pt

C. Wanzeller
e-mail: cwanzeller@estgv.ipv.pt

M. Abbasi
CISUC—Centre for Informatics and Systems of the University of Coimbra, Coimbra, Portugal
e-mail: maryam@dei.uc.pt

F. Sá
Polytechnic of Coimbra (ISEC), Coimbra, Portugal
e-mail: filipe.sa@isec.pt

J. L. Reis et al. (eds.), *Marketing and Smart Technologies*, Smart Innovation, Systems and Technologies 337, https://doi.org/10.1007/978-981-19-9099-1_18

from tests performed on Android. Results revealed that Room and greenDAO have the lowest performance, while Realm has globally better performance regarding select queries when the number of records is exceptionally high, and ObjectBox provides the better performance regarding basic operations like create, update, load, and delete.

18.1 Introduction

Android is the most popular operating system in the world, with over 2.5 billion all around the globe [1]. Since Android mobile OS surpassed Windows OS as the most used Operating System in the world and continues to grow with hardware as powerful as any standard computing device, data computation from mobiles is also increasing [2] and, with that, comes the need to continuously increase performance.

Mobile databases are local storage operating on machines with limited computing and storage capabilities [3]. Most Android applications rely on database management systems, as this is the most common way of storing and managing data. Database performance plays a vital role in the success of applications [4] and has to cope with the workload to support desired user experience [5]. In this context, the industry offers a wide range of solutions [3].

Most mobile developers are probably familiar with SQLite, which has been around since 2000, and it is the most used relational database engine in the world [6]. However, new developers are now inclined to promising alternatives emerging more recently, like ObjectBox and Realm, since SQLite no longer can fulfill the needs that arise with technological evolution and user demands. SQLite is a robust and proven relational, but it is essential to keep up-to-date on current trends and needs [6]. Realm and ObjectBox are some of the biggest upcoming trends in recent years. They are databases designed from the ground up that, instead of relying on relational models, are based on object-store approaches [6].

Object-oriented database management systems (OODBMS), like Realm and ObjectBox, are designed to perform well as supporting tools when programming in object-oriented languages. That way, development becomes more accessible, faster, and safer since developers do not need to write heavy and complex queries, and one of the primary sources of errors [7].

Object relational mapping (ORM) has also gained popularity with the appearance of Room and greenDAO. It consists of a layer between the relational SQLite database and the object-oriented app to easily connect object code to a relational database [8, 9]. Room and greenDAO are examples of ORMs, both of which are frameworks that keep using SQLite underneath [4]. The main difference between them and SQLite for developers is that they no longer need to use basic SQL syntax for querying and manipulating data from the database [4].

This paper aims to give a complete analysis of the performance of the ObjectBox, Realm, SQLite, Room, and greenDAO databases concerning the execution time to perform basic data operations: create, load, update, delete, access, and also queries.

The ultimate goal of the analysis is to conclude whether these databases can achieve the demands of modern applications and which of them shows better performance.

This document is organized into six sections. Section 18.2 presents state of the art and an overview of related works with an emphasis on relevant conclusions. Section 18.3 presents individually each one of the database management systems (DBMS), focusing on the main technical features, behavior, strengths, and limitations. Section 18.4 explains the experimental setup, the used dataset, and the application used to perform testing and performance measurement. Section 18.5 presents the performance tests, variables, and an extended discussion about the results. Section 18.6 concludes the paper.

18.2 Related Work(s)

The development of complex applications is impossible without the use of powerful databases, and their application in Android development has its specifics caused by the features of mobile devices: fewer hardware resources, battery saving, and mobile application architecture [10].

Liu et al. [11] found that the most common performance issues are lagging by the graphical user interface (GUI). Users must wait for a task to complete before they can do anything else on the device. Energy leaks and memory exceeds when applications demand more hardware resources than the device has, namely, RAM and CPU [4]. These issues feature among the main reasons why users give low ratings for applications [4], and as shown by Guerrouj et al. [12], user app ratings are directly proportional to the success of applications. Thus, it is essential to build applications with reduced response times requiring little memory.

SQLite was first introduced in 2000 and quickly became one of the most popular DBMS, occupying the undisputed throne for several years until viable alternatives appeared. The first stable version of Realm for Android appeared in 2016 [13]. Its creators conceived Realm as a database that would avoid writing complicated standard code in SQL, enable work with data as objects, and reduce the response time of creating, reading, updating, and deleting (CRUD) operations [10].

SQLite remains the most frequently used DBMS in the development of Android applications [10]. Outside the context of relational databases, Realm and ObjectBox are two of the most popular DBMS for Android applications, and each of them has its place in the market of mobile applications [10].

ObjectBox, one of the newest DBMS, implements a NoSQL object-oriented DBMS build from scratch, like Realm. First presented in 2017, it was initially developed as a DBMS for mobile and IoT devices, and it presents super speed of operation and the convenience of integration into mobile applications [10].

Also, in 2017, Google introduced Room, that not being a database but a layer on top of SQLite, presented new customizations that SQLite lacked in its pure format and brought SQLite back into the spotlights [13]. Like Room, greenDAO is also an open-source Android ORM that makes developing for SQLite easier and faster since it prevents developers from dealing with low-level database requirements [9]. Despite having been released before Room, it was only in 2017 that it released a stable, compatible version with Android Studio [14].

Research to compare all the available databases for mobile development seems to be increasing since there are not many proven and published studies in this field. In order to compare Room, greenDAO, and SQLite, the author in [4] developed three versions of an application, one for each framework. Results suggest that greenDAO has better performance regarding CRUD operations, and the resulting application size was equivalent for both Room and greenDAO, but SQLite presented a 5% decrease. Regarding RAM and CPU usage, SQLite also presented better results. Regarding Realm and SQLite with Room and the added size to applications, Realm loses because it includes a separate database adding 4–5 MB, and Room only adds a small layer of a few KBytes. On the other hand, Realm is nearly ten times faster than Room when performing CRUD operations [13]. When it comes to security, Room and Realm are equally good since both of them allow 256-bit encryption of the database [13], meaning that 2^{256} different combinations are needed to break it, which is virtually impossible to be done even by the fastest computers [15].

New developers seem to choose Realm over SQLite since Realm is suitable for prototyping and easy to get started since it does not require a lot of configuration and SQL knowledge [13]. At the same time, the combination of SQLite and Room, despite being the most popular solution when working with structured data, requires knowledge of SQL basics and may become much more complex when manipulating objects, while Realm and ObjectBox, are based on an object-oriented approach to data organization, are significantly superior not only in speed but also in usability [10].

In [16], results of tests in simple CRUD operations considering different DBMS, including ObjectBox, Realm, Room and greenDAO, are presented and demonstrate that non-relational databases perform simple operations far better. ObjectBox seems to be the best when saving data, but Realm is incomparably faster when reading. For update operations, non-relational databases are also faster: ObjectBox outperforms Realm, followed by greenDAo and Room, which achieved similar results. When deleting data, greenDAo and Room achieved very similar results, with Realm slightly below them and ObjectBox being the worst.

18.3 Architectures

In this section, each of the DBMS in the study is presented individually with a focus on the main technical features, behavior, and already known strengths and limitations to enable a reasoned analysis and reach valid conclusions.

18.3.1 SQLite

SQLite is "an Open Source Database embedded into Android. SQLite supports standard relational database features like SQL syntax, transactions and prepared statements" [17]. Unlike other well-known DBMS, such as PostgreSQL and MySQL, SQLite does not require a separate database server and stores all information in one file, taking up little disk space. It was designed for devices with limited hardware capabilities [5] compared to the paradigm we face today, and the only supported data types are text, integer, and real, so other types must be converted before saving in the database.

Using a SQLite database in Android does not require a unique setup. After defining the SQL statements for creating and updating, the database is automatically managed and ready [17] for use. Although it is convenient to work with simple types, it requires additional time to implement relationships between objects [10], does not validate types, and access may be slow since it involves accessing the filesystem [17].

18.3.2 greenDAO

greenDAO is an open-source Android ORM that allows developers to use the database without worrying about converting objects into compatible formats for the relational database [9].

In its purest form, SQLite requires writing SQL and analyzing query results, which may be tedious and time-consuming tasks [18]. greenDAO was designed to free developers from those tasks by mapping Java objects to database tables. In addition to these, other advantages of this framework are its minimal memory consumption, the small library size (¡150 KB), and the decent performance [18].

18.3.3 Room

The Room persistence library is "an abstraction layer over SQLite to allow fluent database access while harnessing the full power of SQLite" [19]. Like greenDAO, it is an ORM for SQLite database in Android.

Room is now a better approach than SQLite since it allows working directly with objects by mapping and decreases the amount of low-level code. However, it still requires some basic SQL understanding and provides compile-time validation of SQL queries, contrary to SQLite. This validation means that an application will not be compiled if there is an SQL query error, preventing developers from encountering run time errors [13, 19, 20]. Another feature comparably to SQLite is that as the schema changes, there is no longer the need to update the affected SQL queries manually. Room solves that issue [19].

18.3.4 ObjectBox

ObjectBox is a superfast object-oriented database. It is built from scratch for storing data locally on mobile devices and is optimized for high efficiency using minimal CPU and RAM [21]. Like Realm, it implements a NoSQL approach in which the attributes of model classes and the relationships between them are directly written to the database [10]. This framework accelerates the development and release process and reduces costs. Another advantage of ObjectBox is that it can develop applications that work online and offline [22].

18.3.5 Realm

Realm is "a fast, effective, and user-friendly NoSQL database in which data is mapped to objects in a realm file, and therefore classes are used to define the schema" [13]. Despite being much more recent than SQLite, it is prevalent among new Android developers and works with Java, Kotlin, Swift, ObjectiveC, Xamarin, and React Native [23]. Although it presents significant advantages over SQLite like high speed, ease of use, a zero-copy of data policy, and excellent documentation for developers can easily find answers in case of doubt, Realm also has some limitations: it does not support auto-incrementing of id's, nested transactions, or inheritance [13], and it adds a far more extensive library (around 3–4 MB) than Room or greenDAO [23].

18.4 Experimental Setup

In order to generate fair and realistic evaluation results and ensure reliable and valid conclusions, test runs were performed on the same Android device, varying between ObjectBox, Realm, and SQLite sequentially, considering different numbers of records and the operations described in Fig. 18.1.

The application designed for tests was developed using Android Studio Chipmunk 2021.2.1 and installed on a Xiaomi Mi 9 device. The mobile device runs Android 11 and has an octa-core Snapdragon 855 CPU (2.84 MHz) with 6 GB of RAM. SQLite database engine came with the operating system, but two ORMs were installed: Room (2.4.2) and GreenDAO (3.3.0). ObjectBox (3.1.2) and Realm (10.9.0) had to be installed manually.

The initial application includes *EditText* fields to insert the number of tests to run and the number of records used in those tests. Besides that, four *CheckBoxes* allow the selection of the database system to be tested, and the available tests are displayed in a *ComboBox* to be easily seen and selected by users.

Finally, there is a button to test effectively, and the results are displayed in a list as soon as the tests finish. Each test was run ten times, and the final result is an average

Fig. 18.1 Test operations

1.1. Create
1.2. Update
1.3. Load
1.4. Access
1.5. Delete
2.1. Create - Indexed
2.2. Update - Indexed
2.3. Load - Indexed
2.4. Access - Indexed
2.5. Delete - Indexed
3. Query by String
4. Query by String - Indexed
5. Query by Integer
6. Query by Integer - Indexed
7. Query by Id

value considering all tests except the one with the most distant value from the rest to obtain a more realistic average value. Figure 18.2 displays the test application main screen.

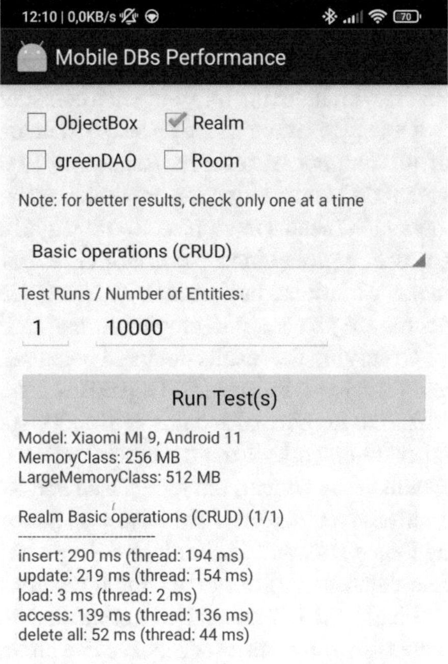

Fig. 18.2 Test application screen. The source code is available on: https://github.com/ManuelCLopes/mobile-db-performance

18.4.1 Dataset

The records generated along the tests followed a schema with one of the following attribute types: long, boolean, byte, short, int, float, double, string, and ByteArray. Since indexing makes columns faster to query by creating pointers to where data is stored within a database [24], to test on indexed data and evaluate the results, two indices were used: id and id plus all the other attributes.

18.5 Results and Analysis

Results strongly suggest that simple operations are far better when dealing with non-relational databases.

In the case of creating and updating data, ObjectBox presents the lowest time of performance, followed by Realm. Room and greenDAO are distinguishably slower. The results for these operations for indexed data are similar, but globally, with worst performances and are exceptionally sharp for Room and greenDAO.

Regarding the access, ObjectBox, Room, and greenDAO respond almost instantaneously while Realm does not. However, for load operations, Realm responds in instantaneous times, followed by ObjectBox, and with Room and greenDAO being much slower. The results on load with and without indexing are very similar, but access with indexing is faster when considering several records lower than 1,000,000.

ObjectBox remains the fastest for delete operations, followed by Realm, regardless of the number of records. Regarding Room and greenDAO, Room is faster than greenDAO considering lower numbers of records, but when increasing to 1,000,000, greenDAO seems to surpass Room slightly. For indexed data, the results are similar but with a global increase in response times. ObjectBox and Realm do not present major variations, but greenDAO and Room are negatively affected. The described results are presented in graphic format in Figs. 18.3, 18.4, and 18.5.

Observing the results obtained when performing queries in Figs. 18.6, 18.7, 18.8, and 18.9, it can be seen that ObjectBox has the better performance on indexed queries when the number of records is 10,000, 100,000, or 1,000,000, but loses for all the other frameworks when the number of records increases to 10,000,000. In that case, Realm is the winner, but greenDAO and Room present close results.

Moreover, Realm is the fastest on querying by string and by an integer, followed by ObjectBox and, distantly, by Room and greenDAO. When querying by ID, the four databases converge to similar results.

Finally, it is essential to notice that indexes will speed up queries. However, increasing the number of indexes in a table, the more extended operations like insert and update will take since these indexes will be automatically updated, which will cause slow performance in these operations [25].

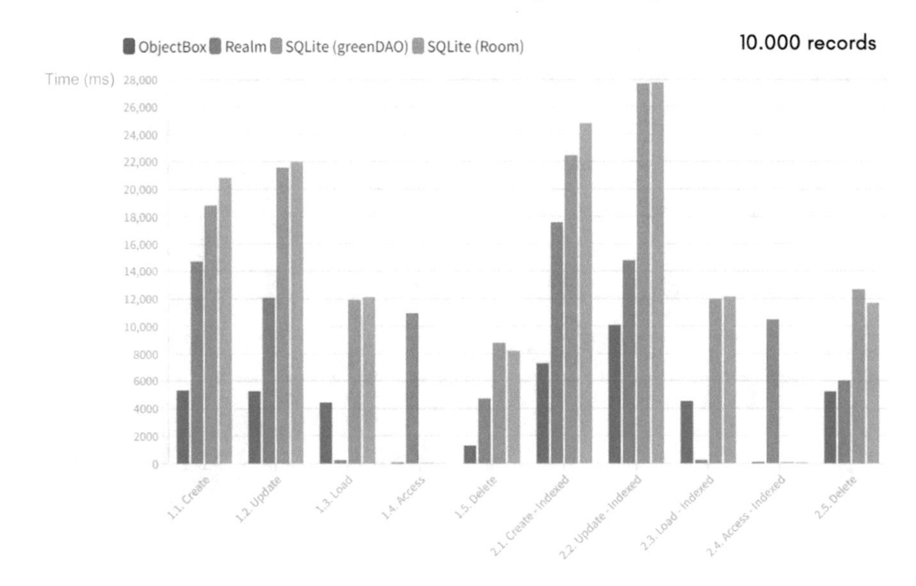

Fig. 18.3 Global results for 10,000 records

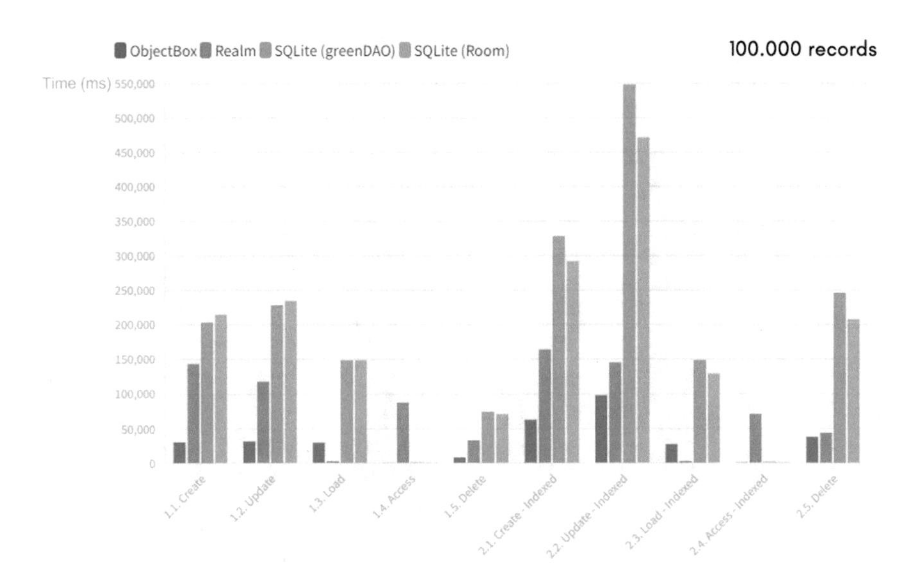

Fig. 18.4 Global results for 100,000 records

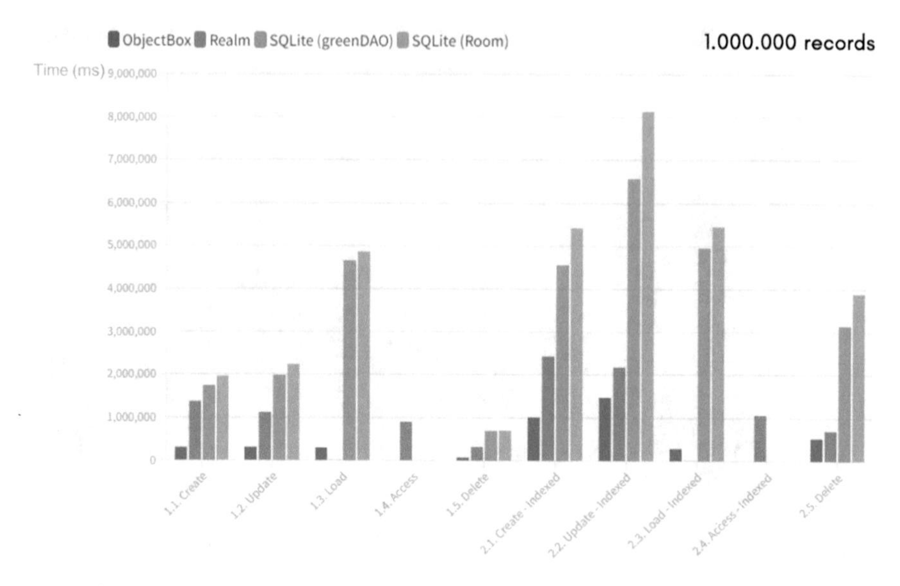

Fig. 18.5 Global results for 1,000,000 records

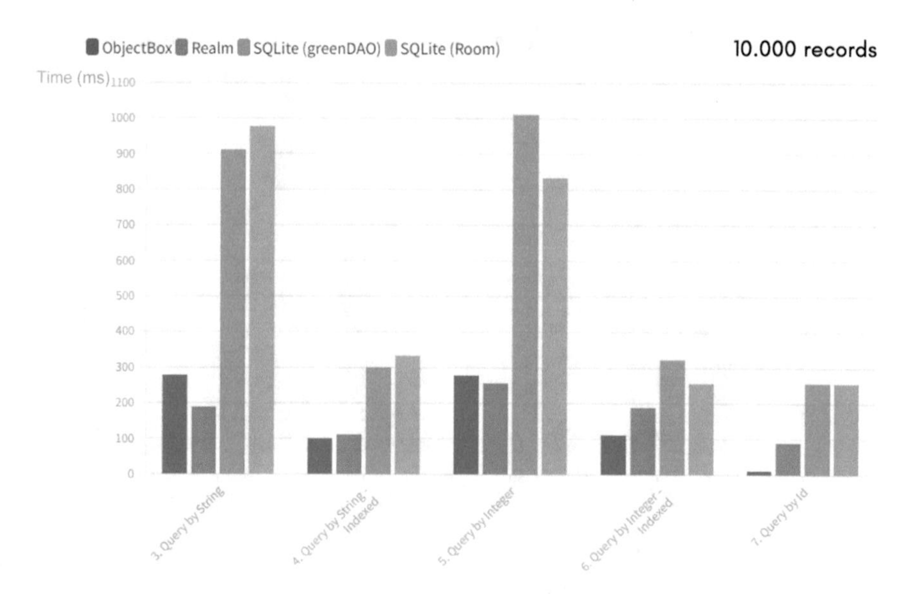

Fig. 18.6 Query results for 10,000 records

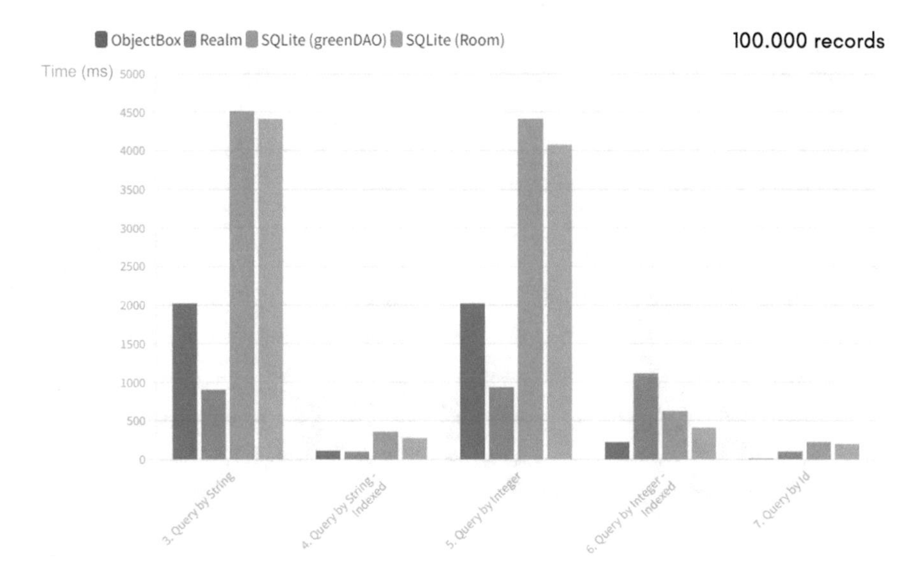

Fig. 18.7 Query results for 100,000 records

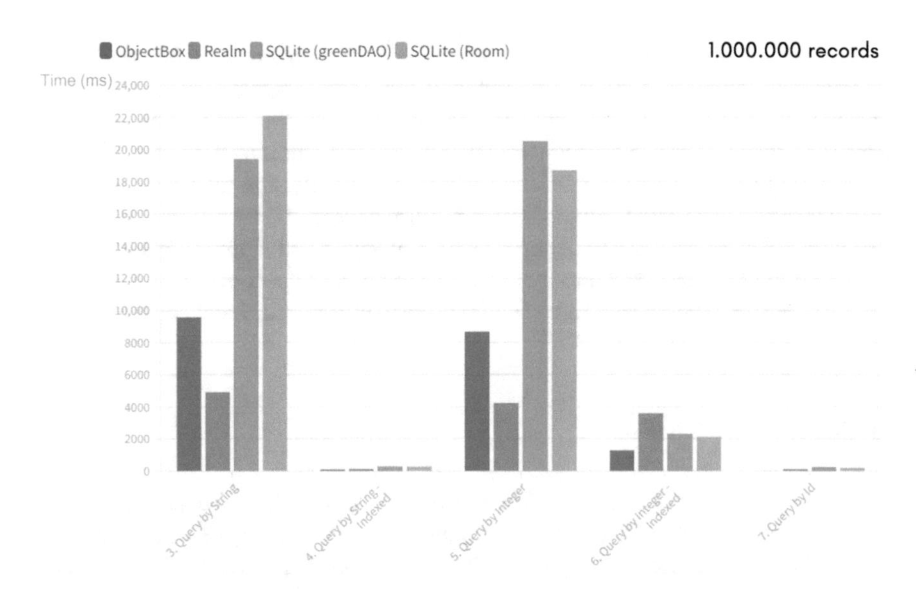

Fig. 18.8 Query results for 1,000,000 records

18.6 Conclusions and Future Work

Android developers can now choose from a wide range of database frameworks. However, new developers seem to be fleeing from relational to non-relational approaches like Realm and, more recently, ObjectBox.

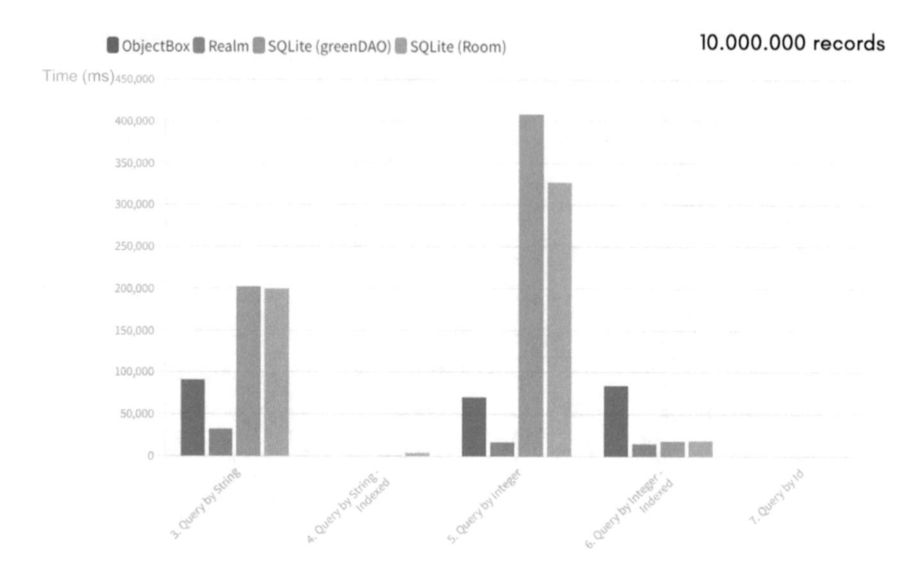

Fig. 18.9 Query results for 10,000,000 records

SQLite has clear limits, but it has the advantage of being a mature and well-understood technology, with many developers that fully understand it and use it well and wisely. Despite not being able to keep up with the growth in mobile device hardware, the appearance of Room and greenDAO brought it back to competition.

The results presented in this article suggest that, globally, non-relational databases like ObjectBox and Realm deliver the best performance regarding response times, but that does not mean that NoSQL approaches are always a better option. Many factors, including the intended structure for the project and the impact of the limitations of each framework when developing, should be carefully weighed and considered.

18.6.1 Future Work

The results presented in this article are limited to performance, considering only response times. Although a more profound analysis is presented in Sect. 18.2 concerning other metrics, in future work it would be interesting to include an analysis on memory consumption and CPU usage as well as replicate the present study on unstructured data considering Realm, ObjectBox, and other NoSQL options.

Acknowledgements "This work is funded by National Funds through the FCT-Foundation for Science and Technology, IP, within the scope of the project Ref UIDB/05583/2020. Furthermore, we would like to thank the Research Centre in Digital Services (CISeD), the Polytechnic of Viseu, for their support."

References

1. Business of Apps.: David Curry Android statistics (2022). https://www.businessofapps.com/data/android-statistics/
2. Parihar, T.A.: Five of the most popular databases for mobile apps (2017). https://blog.trigent.com/five-of-the-most-popular-databases-formobile-apps/
3. Kussainov, K., Kumalakov, B.: Mobile data store platforms: Test case based performance evaluation, 11 (2016)
4. Che LB.: Room vs greendao for android: a comparative analysis of performance of room and greenDao (2019)
5. Obradovic, N., Kelec, A., Dujlovic, I.: Performance analysis on Android Sqlite database. In: 2019 18th International Symposium Infotech-Jahorina (Infotech), pp 1–4 (2019)
6. Batista, M.G.R.: Realm is the best android database solution (2016). https://www.toptal.com/android/realm-best-android-database-solution.
7. Andersson, T.: Analysis and quantitative comparison of storage, management, and scalability of data in core data system in relation to realm (2018)
8. Techopedia.: Object-relational mapping(orm). https://www.techopedia.com/definition/24200/object-relational-mapping-orm
9. Sunday Akinsete. Integrating greendao into your android application (2017). https://www.codementor.io/@sundayakinsete/integrating-greendao-intoyour-android-application-yro5fzgtw
10. Redwerk.: Sqlite vs realm vs objectbox: complex data queries in android databases (2020). https://redwerk.com/blog/sqlite-vs-realm-vs-objectbox/
11. Liu, Y., Xu, C., Cheung, S.-C.: Characterizing and detecting performance bugs for smartphone applications. ICSE 2014, pp. 1013–1024, New York, NY, USA. Association for Computing Machinery (2014)
12. Guerrouj, L., Baysal, O.: Investigating the android apps' success: an empirical study. In: 2016 IEEE 24th International Conference on Program Comprehension (ICPC), pp. 1–4 (2016)
13. Solanki, M.: What to choose realm orsqlite with room? (2022). https://itnext.io/what-to-choose-realm-or-sqlite-with-room-e55c34b1675c
14. greenrobot. greendao changelog. https://greenrobot.org/greendao/changelog/
15. Techopedia.: 256-bit encryption (2022). https://www.techopedia.com/definition/29703/256bit-encryption
16. Lowski, M.K.: Android databases performance tests—crud (2019). https://proandroiddev.com/android-databases-performance-crud-a963dd7bb0eb
17. Vogel, L.: Android sqlite database and contentprovider-tutorial. Java, Eclipse, Android Web Program. Tutor. **8**, 5 (2010)
18. Abhishek, S.: Green dao: Android orm for your sqlite database (2020). https://abhiappmobiledeveloper.medium.com/green-dao-android-orm-foryour-sqlite-database-accee20d7065
19. AshishRawat: Using room database—Android jetpack (2019). https://medium.com/mindorks/using-room-database-android-jetpack-675a89a0e942
20. Kayere, P.: Introduction to room DB (2020). https://www.section.io/engineeringeducation/introduction-to-room-db/
21. Luo, R.: Objectbox, a future of mobile database (2019). https://medium.com/@rejaluo/objectbox-superfast-edge-database-for-mobileand-iot-60db3821c179
22. StackShare:. Objectbox. https://stackshare.io/objectbox
23. Alexander, G.: Realm vs sqlite: which database to choose in 2022 (2022). https://orangesoft.co/blog/realm-vs-sqlite
24. Barnhill, B.: Indexing (2021). https://dataschool.com/sql-optimization/howindexing-works/
25. Tutorialspoint.: Sql—indexes. https://www.tutorialspoint.com/sql/sqlindexes.htm

Chapter 19
QR Codes Research in Marketing: A Bibliometric and Content Analysis

Joaquim Pratas and Zaila Oliveira

Abstract The aim of this investigation is to analyze QR codes research main topics and evolution over time. A bibliometric study and content analysis were carried out on the documents published in Web of Science database. The findings show that research after the COVID-19 pandemic has been reinforced. The main topics are related to the willingness to use and buying intentions through the use QR codes in retail, communication, and collaborative supply chains; risks and privacy concerns with QR codes and loyalty programs; mobile payments with QR codes; and other industry-related specific topics. This study contributes to the area by providing scholars with the state of the art and trends in QR codes and can guide future research. From a managerial view, it gives insights on how to use QR codes.

19.1 Introduction

COVID-19 pandemic was a major force to enhance contactless economy. A 100% contactless experience is made possible through the use of native applications or a web page on connected devices [1] such as Quick response (QR) codes.

The contactless economy has been driven by both supply-side (with the rise of the digital technologies such as 5G, cloud platforms, artificial intelligence, and data analytics) and demand-side factors (with the need for convenience and heightened awareness for health and safety) [2].

With the environment that COVID-19 pandemic brought, namely the heightened awareness for health and safety, and the rise of 5G and cloud platforms, and the rise

J. Pratas (✉) · Z. Oliveira
Universidade da Maia—ISMAI, Maia, Portugal
e-mail: jmpratas@umaia.pt

Z. Oliveira
e-mail: d012069@umaia.pt

Research Unit UNICES, Universidade da Maia, Maia, Portugal

J. Pratas
Research Unit GOVCOPP, Universidade de Aveiro, Aveiro, Portugal

J. L. Reis et al. (eds.), *Marketing and Smart Technologies*, Smart Innovation, Systems and Technologies 337, https://doi.org/10.1007/978-981-19-9099-1_19

of new and more powerful mobile phones, smart watches, and other technologies, consumers have almost endless access to information in their pockets, anytime and anywhere. QR codes have experienced a renaissance due to the demand for a contactless economy, attracting a new wave of businesses, marketers, and customers who have adapted to a world where many aspects are becoming more digital.

With the increasing usage of mobile phones that can read QR codes, it is expected that QR codes tool will have a greater prominence as an informational, communicational, transactional, and even payments tool.

The mobile phones (or other mobile technologies) can have an application (app) that reads the QR code, but recent smartphones, such as Samsung Galaxy S20 can read or scan QR codes without having to download an app as the code reader is included in the camera software, enhancing simplicity and ease of use of QR codes.

Thus, the aim of this work is to analyze and assess the body of scientific literature that has been produced over time on the topic of QR codes. It is specifically meant to assess and synthesize the current state of the art of the topic and understand its main trends in research subtopics.

From a scholarly perspective, it is anticipated to contribute to the current status of QR codes study and develop a list of priorities for follow-up research. In terms of business, it should assist firms in adopting or improving this marketing and promotional tool.

19.2 Literature Review

19.2.1 QR Code Early Beginnings

A QR code is a type of bar code that is printed with tiny black-and-white squares that can be used to store data that can be scanned into a computer system. The black and white squares can represent any number between 0 and 9, any letter from A to Z, or a non-Latin character like a Japanese kanji [3].

QR codes can contain information such as text, Uniform Resource Locators (URL) links, automatic short message service (SMS) messages, or just about any other information that can be embedded in a two-dimensional barcode. This encoded data can be decoded by scanning the barcode with a mobile device that is equipped with a camera and QR reader software [4].

The Japanese company Denso Wave, a division of Denso, a division of the automaker Toyota Motor Corporation developed QR codes in 1994, to trace vehicle parts as they are being assembled [3].

The structure of the QR code has five areas [5]. The first area has been used to specify the QR code version. The information in the second zone is read first, and the second zone has degrees of mistake correction. The third region is used to store data. The finder, alignment, and timing pattern are the components of the fourth region. The detection model needed to read a QR code is the Finder pattern, which is present on three corners of the code. The alignment pattern is employed to fix a potential programming problem. For correcting the coordinates and the coding, the timing pattern is crucial. In order for the code to be read correctly, the fifth region must be empty [6].

The QR codes have several utilizations. They are still an emerging mobile technology that could have a high impact on the mobile marketing practices, including shopping, advertising, sales promotion, direct marketing, and customer relationship management [7].

For example, the URL of a website that has a coupon or other information about a product is regularly encoded into QR codes for use in advertising. They have also been noticed on tickets for athletic events and concerts [3].

Early in the new millennium, QR codes swiftly were goods standing in North America and Europe soon after the placement of QR codes by manufacturers and marketers on a variety of goods and services, including wine labels and bottles of shampoo to candy bars [8]. Fast-moving consumer goods (FMCG) rely heavily on inexpensive package-level technologies such as QR codes [9].

19.2.2 Future Projections for QR Codes After COVID-19

According to a research conducted in Portugal in 2015, the majority of respondents were eager to test out this new technology in the future, demonstrating that Portuguese consumers had a favorable attitude toward its use. In addition, many respondents state that they wanted to utilize the code in many situations since they find it to be easy to use and practical, even though due to budgetary reasons, many respondents lack a mobile device or an application (app) to read the code [7].

During a June 2021 survey, it was found that 45% of responding shoppers from the United States stated they had used a marketing-related QR code in the three months leading up to the survey. The share was highest among respondents aged between 18 and 29 years [10], showing that the acceptance of QR codes by younger generations is higher than in older generations.

In a survey released by [11], in February 2021, it was found that between March and December 2020, 8.74 million new mobile payment customers were enrolled. According to a comprehensive survey carried out by [12], 85% of all mobile payments were made via QR codes in 2020. A future without cash is predicted by the growing use of QR codes for payment in physical stores [13].

The number of smartphone users in the United States who used a QR code scanner on their mobile devices is expected to rise from 52.6 millions in 2019 to 99.5 millions in 2025, experiencing a constant growth [14].

Thus, it seems that QR codes are going to be an important tool in the future, due to its increased availability that can be used as communication, transaction, payment, and traceability among other possible functions.

19.3 Research Questions and Methodology

19.3.1 Research Questions

In order to achieve the outlined research objectives, the following research questions were formulated:

- Q1: Which scientific documents had the greatest impact in terms of their citations?
- Q2: How has the publication of scientific documents about QR codes evolved?
- Q3: What are the main research trends in the field of QR codes?

19.3.2 Methodology

The used methodology involved conducting a bibliometric analysis and a content analysis of documents about QR codes. In order to collect a representative sample of relevant knowledge, we retrieved data from Web of Science Core Collection (WoSCC) (all editions), in the categories of Business, Management, Economics and Communication, in August 28, 2022, searching on the options Title or Topic, the keywords: "QR code", "QR codes", "Quick response code" or "Quick response codes". The WoSCC is considered the most impactful and qualitative database with several source types such as journals, books, and conference proceedings. It also provides cleaner data, with less duplication than other databases (such as Scopus). No criterion has been established regarding the date of publication.

One hundred and twelve scientific documents were identified. We further read the title and abstract of all the documents, to guarantee that the sample was complete and accurate. We found that ten articles were not related to the topic we wanted to study. Thus, one hundred and two documents remained to be analyzed.

Using the Web of Science software, we retrieved all bibliometric information of these articles, such as the source or journal name, article title, authors, keywords, year and references. We further collected citation and co-citation data for each article. Then, manually, we normalized the authors' names and works and corrected for different editions of books. The data was analyzed with SciMAT, and networks were identified using the software VOS Viewer, to perform the analyses necessary to tackle the identified research questions.

19.4 Results

19.4.1 Most Cited Documents

To tackle first question "Which scientific documents had greatest impact in terms of their citations?" we have performed a citation analysis. Citation analysis quantifies the importance and influence of a published article within its field, as assessed by their citation frequency [15]. Jointly the 102 documents had 521 citations in WoS Core. This means that on average each published work had about 5 citations. Listing such a large list was considered unfeasible, and thus the list with 10 documents with more citations is listed in Table 19.1.

19.4.2 Evolution of Documents Publication About QR Codes

To address the second question "How has the publication of scientific documents about QR codes evolved?", we have done a bibliometric analysis that showed the publications by year. The results are shown in Table 19.2.

The findings show that the number of documents continuously increases. Between 2015 and 2019 the average per year was of ten documents, but after COVID-19 pandemic, the average number between 2020 and 2022 (to date—the data collection was made in August 28, 2022) was of eleven documents. This number will be higher considering the total documents published in 2022 that aren't considered in this analysis.

19.4.3 Main Research Trends in the Field of QR Codes

To answer the third question "What are the main research trends in the field of QR codes?", we have used a bibliographic coupling analysis. Bibliographic coupling between two documents occurs when there is an item used as reference by these two documents. Their bibliographic coupling strength is then the number of references they have in common [16]. A bibliographic coupling analysis was performed using VOS Viewer, in order to extant and enrich the knowledge about the intellectual ties between documents. The methodology used was based on a minimum number of citations per document of five, and fractional counting. Sixteen documents met these criteria assuring that these documents were connected to at least one other document (Fig. 19.1).

In this analysis were identified four different clusters. The documents in the first cluster are represented in red color and seem to analyze the attitude, interest, willing to use, and buying intentions through the use QR codes in physical, online or omnichannel retail, communication, and collaborative supply chains. In this cluster

Table 19.1 Top 10 most cited documents

Titles	Authors and data	Sources	Number of citations
User Behavior in QR Mobile Payment System: The QR Payment Acceptance Model	Liebana-Cabanillas et al. [27]	Technology Analysis & Strategic Management	68
Smart Shoppers? Using QR Codes and Green' Smartphone Apps to Mobilize Sustainable Consumption in the Retail Environment	Atkinson [17]	International Journal of Consumer Studies	50
Benchmarking the Use of QR Code in Mobile Promotion: Three Studies in Japan	Okazaki et al. [8]	Journal of Advertising Research	36
QR Code and Mobile Payment: The Disruptive Forces in Retail	Yan et al. [28]	Journal of Retailing and Consumer Services	34
Privacy Concerns in QR Code Mobile Promotion: The Role of Social Anxiety and Situational Involvement	Okazaki et al. [25]	International Journal of Electronic Commerce	28
Effects of the National Bioengineered Food Disclosure Standard: Willingness To Pay for Labels that Communicate the Presence or Absence of Genetic Modification	McFadden and Lusk (2018)	Applied Economic Perspectives and Policy	25
Applying Blockchain for Halal Food Traceability	Tan et al. [30]	International Journal of Logistics-Research And Applications	19
Performance Evaluation of Tracking and Tracing for Logistics Operations	Shamsuzzoha et al. [19]	International Journal of Shipping and Transport Logistics	18
Examining the Impact of QR Codes on Purchase Intention and Customer Satisfaction on the Basis of Perceived Flow	Hossain et al. [18]	International Journal of Engineering Business Management	15

(continued)

Table 19.1 (continued)

Titles	Authors and data	Sources	Number of citations
Innovators and Innovated: Newspapers and the Postdigital Future Beyond the Death of Print	O'Sullivan et al. (2017)	Information Society	15

Table 19.2 Number of documents per period

Year	2008–2009	2010–2014	2015–2019	2020–2022
Number	2	17	50	33 (to date)
Average number per year	1	3	10	11 (to date)

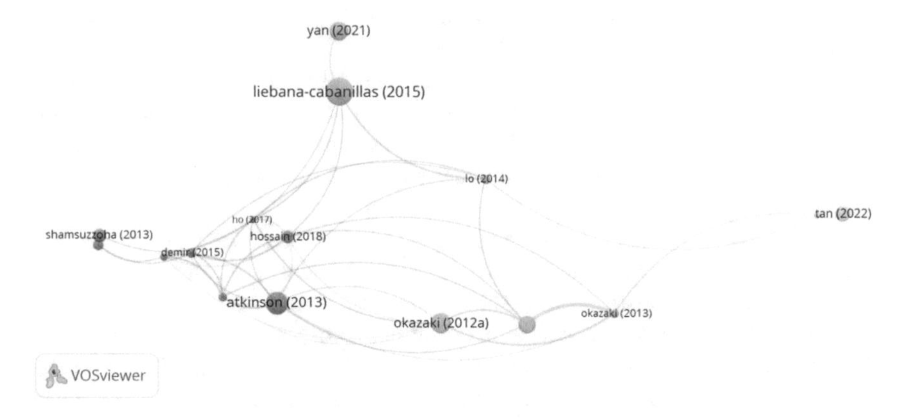

Fig. 19.1 Results of bibliographic coupling analysis. Research trends clusters

is also present the issue of tracking and tracing technologies associated with QR codes.

In the second in green color, the documents seem to analyze the risks and privacy concerns of cross-media advertising using QR codes, and impacts of QR codes in customers' loyalty.

In the third cluster, in blue, the document analyzes the topic of mobile payments using QR codes and characteristics of users that adopt this technology.

In the fourth cluster in yellow color, other topics are studied, namely the use of QR codes in specific industries, such as Halal food traceability and QR codes presence in front cover of newspapers. It seems that QR codes can change the business model and/or marketing strategies of these specific industries.

We have also performed a content analysis in order to analyze each document that was in each cluster. Two trained codes were used to classify each document and synthesize main issues in each document.

19.4.3.1 First Cluster: Attitude, Interest, Willingness to Use and Buying Intentions Through the Use QR Codes in Retail, Communication, and Collaborative Supply Chains

The first cluster contains eight documents.

Atkinson [17] investigated the causes of consumers' willingness to use QR codes in retail settings when making sustainable product purchases, using a representative sample of 401 American adults. The findings show that customer willingness to employ mobile phone-based QR code advertising is positively correlated with consumer trust in the government, buycotting, and market mavenism, but adversely correlated with consumer's trust in corporations.

Hossain et al. [18] studied the effect of QR codes on customer satisfaction and buying intention in the context of online shopping based on perceived flow, using a sample of 420 consumers who had purchased a product online using a QR code. The authors concluded that QR codes have a significant impact on customer satisfaction and buying intentions, confirming that QR codes affect perceived flow, which in turn influences online shoppers' contentment and, ultimately, purchase intention. Advertisers may communicate and impact consumer happiness and buying intentions by integrating QR codes into their ads. Consumers sharing information via QR codes will encourage other customers to interact with, and exchange information in the online community.

Shamsuzzoha et al. [19] studied how auto-ID technologies such as barcodes, QR codes, radio frequency identification (RFID) tags are widely used for monitoring the material flow in the logistics chain. These technologies help companies to trace and track products and manage supply chain operations across organizational boundaries. The authors developed a methodology to measure the performance of tracking and tracing technologies, using an experimental setup, helping manufacturing organizations to establish a plan before implementing tracking and tracing technologies and tools in business operations.

Albastroiu and Felea [20] analyzed the degree of usage, the willingness of the Romanian people to use codes in the buying process, and also their perception about the functionality and usefulness of QR codes in an omnichannel retail environment. The findings suggest that consumers know the applicability of QR codes, have used codes for accessing information about products and online purchases, and consider QR codes contribution to the improvement of the shopping experience.

Introini et al. [21] analyzed the Industry 4.0. technologies (using a temporal evolution of technologies applied in the food supply chain traceability, refer that technology environment is becoming increasingly complex due to the inclusion of new proposals (RFID, QR codes, NFC, Barcode, …)), that can be applied to traceability of different areas in the food sector, namely creating proposals for fruit, vegetables, meat, or fish. The authors also identified sectors that have not yet been approached by these new proposals, as well as technologies that are been applied in Industry 4.0 proposals but have not been used for traceability.

Good traceability systems help to minimize the production and distribution of unsafe or poor quality products. Therefore, traceability is applied as a tool to help ensure the safety and quality of food, as well as to achieve consumer confidence.

Demir et al. [22] investigated the current use and future intent of use of QR codes in mobile marketing among college students, in 2013. The findings indicate that while more than 80% of students recognized QR codes, only half of them had used QR codes before. Furthermore, while the interest in using QR codes is currently low, the likelihood of using them is slightly higher than the current interest.

Ho and Yang [23] used QR codes to explain the factors that affect users' usage intentions, using the theory of reasoned action (TRA) as a research framework. Using a mixed methodology with structural equation modeling analysis and qualitative analysis the authors determined that there were four reasons that individuals refer as impediments to scan QR codes, and six factors that imply limited usage, which explain customers' scanning willingness and are crucial to future practice.

Trivedi et al. [24] used commercial appeals with the moderating effects of product category participation to test the effects of QR codes in print advertising along five different stages of customer decision-making. The authors came to the conclusion that, when combined with an emotional appeal, QR codes can influence consumer decision-making in low-involvement product categories. When used in conjunction with an overall informational appeal, QR codes are beneficial for advertisements for high-involvement products. In a low-involvement product category, QR codes do not strengthen the persuading effects of informational appeals in commercials.

19.4.3.2 Second Cluster: Risks and Privacy Concerns of Using QR Codes, and Impacts of QR Codes in Customers' Loyalty

Okazaki et al. [8] studied the perceived risks connected with QR codes utilization in cross-media advertising, with three studies conducted in Japan. The authors found that consumers perceive risks differently depending on the context in which they scan QR codes and that consumers are most likely to use QR codes in print media to promote loyalty programs because they are convenient, cost-effective, and high quality.

Okazaki et al. [25] analyzed the effects of QR code mobile promotion in terms of information privacy concerns, defensive reactions (intention to protect, fake, or withhold), and loyalty, drawing on the utility maximization theory, and using a sample of 667 Japanese consumers. The findings show that social anxiety and situational involvement have significant interactions with both the goal to protect and the intention to fake personal information. Simultaneously, the authors found that QR code advertising is a helpful tool in building loyalty.

Okazaki et al. [26] analyzed the efficacy of QR code loyalty initiatives, using two studies. In study 1, key informant focus groups were used to evaluate the qualitative impressions of QR codes. In study 2, the authors investigate whether QR code loyalty programs can successfully reactivate dormant customers by encouraging them to make repeat purchases. Results from a scenario-based experiment indicate that delayed rewards—rather than instant rewards—induce more loyalty among inactive consumers when consumers are worried about privacy exposure. Likewise, a low-involvement service is more likely to foster loyalty than a high-involvement service. The timing of the reward and the degree of involvement also have a sizable interaction impact. However, neither main effects nor interaction effects are seen when consumers are unconcerned with privacy.

19.4.3.3 Third Cluster: Consumer Intentions to Use Mobile Payments Via QR Codes and Access Information

Using a behavioral model that extends the Technology Acceptance Model (TAM), in order to explain the intention to use mobile payment via QR codes, [27] examined users' intention to use QR code mobile payment systems. The findings indicate that the intention to utilize this technology is influenced by attitude, innovation, and subjective norms.

According to Yan et al. [28], mobile payment adoption still has a lot of space to grow, and QR codes and mobile payments are revolutionizing retail. According to the authors, perceived transaction speed and convenience have an impact on how helpful and user-friendly mobile devices are. The usefulness, simplicity, optimism, and individual inventiveness of mobile devices all have an impact on consumers' desire to utilize QR codes for mobile payments.

According to Lo [29], different personality qualities and attitudes toward innovations may affect a person's willingness to use QR codes to acquire information that would otherwise be challenging for them to find on their own. The authors indicate that laggards and late majority consumers have negative opinions toward innovation and are less likely to use QR code services after conducting an online poll with 689 participants. The degree to which attitudes toward innovation mediate the impact of personality factors on service acceptability varies among customer categories, according to a comparison model.

19.4.3.4 Forth Cluster: Other Topics

Tan et al. [30] analyzed traceability challenges for the food supply chain in Malaysia to comply with Halal requirements, proposing a novel traceability framework built on blockchain derived from real-life blockchain implementation in three distinct Halal supply chains from farm to fork, using mobility-related blockchain along with QR codes and other technologies to build a traceability system. A research made by [31] analyzed QR codes on the front pages of American newspapers. A qualitative research using interviews with newspaper executives has uncovered institutional isomorphism justifications for QR adoption as well as the opinion that the technology is not yet widely adopted by readers or the business.

19.5 Final Considerations

This investigation contributes to the study of QR codes, identifying the documents with the most impact, and highlighting the evolution and the growing interest in the recent years. Finally, research clusters were analyzed in order to establish future research guidelines.

The identified clusters contrast the QR codes generic utilization in retail, communication, and supply chain versus its utilization in specific industries such as Halal food supply chain or impacts in newspaper edition business. Another identified contrast is between the QR code utilization in mobile payments versus its utilization in advertising and customers loyalty programs (and its risks and privacy concerns).

Academic scholars can develop each of these research topics and join different research streams. Organizations can use this research to outline their strategies to increase QR codes utilization and improve business results. They can also have access to several real case studies that can be easily replicated or adapted to their strategies and actions.

However, there are several limitations in this study. We have only analyzed research documents published in sources covered in Web of Science database. Other research databases could be used in order to represent a broader research documents base. However, Web of Science database is the world's leading scientific citation search and analytical information platform. It is used as both a research tool supporting a broad array of scientific tasks across diverse knowledge domains and a dataset for large-scale data-intensive studies [32]. Also other bibliometric techniques could have been used.

References

1. Stegner, B.: How contactless tech and QR codes will bring us into the future (2021). Retrieved August, 28, 2022 from https://www.forbes.com/sites/forbestechcouncil/2021/08/17/how-contactless-tech-and-qr-codes-will-bring-us-into-the-future/?sh=210a10731eca
2. Mehrotra, M.: Perspectives contactless economy: are you prepared? (2020). Retrieved August, 28, 2022 from https://www2.deloitte.com/sg/en/pages/strategy/articles/contactless-economy.html?fbclid=IwAR2Yjv5amQVQcXdePCCp_bHT8qvno2X_tkO6alo7RiwMX0zKgEa0ihvUVzY
3. Gregersen, E.: QR code (2022). Retrieved August, 28, 2022 from https://www.britannica.com/technology/QR-Code
4. Rikala, J., Kankaanranta, M.: The use of quick response codes in the classroom. In: 11th Conference on Mobile and Contextual Learning. Helsinki, Finland, 148–155
5. Thomas, J., Goudar, R.: Multilevel authentication using QR code based watermarking with mobile OTP and Hadamard transformation. In: 2018 International Conference on Advances in Computing, Communications and Informatics (ICACCI), pp. 2421–2425. IEEE (2018)
6. Eroğlu, E., Özkoç, E.: A mobile QR code application for an article: QR-ticle. Computer Info. Sci. **13**(3) (2020). https://doi.org/10.5539/cis.v13n3p82
7. Santos, J.F.: QR code adoption and mobile marketing practices in Portugal: an empirical study. Int. J. Market. Comm. New Media **3**(5) (2015). https://doi.org/10.54663/2182-9306

8. Okazaki, S., Li, H., Hirose, M.: Benchmarking the use of QR code in mobile promotion: three studies in Japan. J. Advert. Res. **52**(1), 102–117 (2012). https://doi.org/10.2501/JAR-52-1-102-117
9. Delloite Insights: Capturing value from the smart packaging revolution (2018). Retrieved August, 29, 2022 from https://www2.deloitte.com/content/dam/insights/us/articles/4353_Smart-packaging/4353_Smart-packaging.pdf
10. Statista: Share of shoppers who used marketing-related QR codes in the United States as of June 2021, by age (2022). Retrieved August, 29, 2022, from https://www.statista.com/statistics/320655/qr-codes-usage/
11. CNNIC: The 47th China statistical report on internet development (2021). Retrieved August, 30, 2022 from http://www.cac.gov.cn/2021-02/03/c_1613923423079314.htm
12. China Union Pay: Mobile payment safety investigation report of China UnionPay in the year 2020 (2021). Retrieved August, 29, 2022 from https://www.mpaypass.com.cn/download/202102/01173414.html
13. Tu, M., Wu, L., Wan, H., Ding, Z., Guo, Z., Chen, J.: The adoption of QR code mobile payment technology during COVID-19: a social learning perspective. Front. Psychol. **12**, 798199 (2022). https://doi.org/10.3389/fpsyg.2021.798199
14. Statista: Number of smartphone users in the United States who used a QR code scanner on their mobile devices from 2019 to 2025, by age (2022). Retrieved August, 29, 2022, from https://www.statista.com/statistics/1297768/us-smartphone-users-qr-scanner/
15. Tahai, A., Meyer, M.: A revealed preference study of management journals' direct influences. Strateg. Manag. J. **20**(3), 279–296 (1999). https://doi.org/10.1002/(SICI)1097-0266(199903)20:3%3c279::AID-SMJ33%3e3.0.CO;2-2
16. Egghe, L., Rousseau, R.: Co-citation, bibliographic coupling and a characterization of lattice citation networks. Scientomet. **55**, 349–361 (2002). https://doi.org/10.1023/A:1020458612014
17. Atkinson, L.: Smart shoppers? using QR codes and green' smartphone apps to mobilize sustainable consumption in the retail environment. Int. J. Cons. Stud. **37**(4) (2013). https://doi.org/10.1111/ijcs.12025
18. Hossain, M., Zhou, X., Rahman, M.: Examining the impact of QR codes on purchase intention and customer satisfaction on the basis of perceived flow. Int. J. Eng. Bus. Manag. **10** (2018). https://doi.org/10.1177/1847979018812323
19. Shamsuzzoha, A., Ehrs, M., Addo-Tenkorang, R., Nguyen, D., Helo, P.: Performance evaluation of tracking and tracing for logistics operations. Int. J. Shipping Transp. Logistics **5**(1) (2013). https://doi.org/10.1504/IJSTL.2013.050587
20. Albastroiu, I., Felea, M.: Enhancing the shopping experience through QR codes: The perspective of the Romanian users. Amfiteatru Econ. **17**(39) (2015)
21. Introini, S., Boza, A., Alemany, M.: Traceability in the food supply chain: review of the literature from a technological perspective. Direccion Y Organ. **64** (2018)
22. Demir, S., Kaynak, R., Demir, K.: Usage level and future intent of use of quick response (QR) codes for mobile marketing among college students in Turkey. In: Proceedings of the 3rd International Conference on Leadership, Technology and Innovation Management, 181 (2015). https://doi.org/10.1016/j.sbspro.2015.04.903
23. Ho, C., Yang, J.: Factors affecting users' mobile technology usage intentions: an example of QR code scanning for mobile commerce. Int. J. Mobile Comm. **15**(2) (2017). https://doi.org/10.1504/IJMC.2017.10001842
24. Trivedi, R., Teichert, T., Hardeck, D.: Effectiveness of pull-based print advertising with QR codes Role of consumer involvement and advertisement appeal. European J. Market. **54**(1) (2020). https://doi.org/10.1108/EJM-06-2018-0383
25. Okazaki, S., Navarro-Bailon, M., Molina-Castillo, F.: Privacy concerns in quick response code mobile promotion: the role of social anxiety and situational involvement. Int. J. Elect. Comm. **16**(4) (2012). https://doi.org/10.2753/JEC1086-4415160404
26. Okazaki, S., Navarro, A., Campo, S.: Cross-media integration of QR code: a preliminary exploration. J. Elect. Comm. Res. **14**(2) (2013)

27. Liebana-Cabanillas, F., de Luna, I., Montoro-Rios, F.: User behaviour in QR mobile payment system: the QR Payment Acceptance Model. Techn. Anal. Strategic Manag. **27**(9) (2015). https://doi.org/10.1080/09537325.2015.1047757
28. Yan, L., Tan, G., Loh, X., Hew, J., Ooi, K.: QR code and mobile payment: The disruptive forces in retail. J. Retailing Cons. Serv. **58** (2021). https://doi.org/10.1016/j.jretconser.2020.102300
29. Lo, H.: Quick Response Codes around us: personality traits, attitudes toward innovation, and acceptance. J. Elect. Comm. Res. **15**(1) (2014)
30. Tan, A., Gligor, D., Ngah, A.: Applying blockchain for halal food traceability. Int. J. Logis. Res. Appl. **25**(6) (2022). https://doi.org/10.1080/13675567.2020.1825653
31. Roberts, C., Saint, K.: A slow response to quick response: diffusion of QR technology on US newspaper front pages. Journal. Mass Comm. Quart. **92**(1) (2015). https://doi.org/10.1177/1077699014554036
32. Li, K., Rollins, J., Yan, E.: Web of Science use in published research and review papers 1997–2007: a selective, dynamic, cross-domain, content-based analysis. Scientometrics **115**(1), 1–20 (2018). https://doi.org/10.1007/s11192-017-2622-5

Part VII
Omnichannel and Marketing Communication

Chapter 20
How Brand Marketing Communications Affect Brand Authenticity for Fast-Moving Consumer Goods

Novalia Mediarki and Yeshika Alversia

Abstract The research examines how consumer perceptions of brand marketing communications influence brand authenticity. During the COVID-19 outbreak, most industries experienced the negative consequences of the pandemic. However, the FMCG market in Indonesia continues to grow and is not negatively affected by the pandemic. One example is the increase in milk purchases which can be seen from the panic buying phenomenon of a milk brand in 2021. People aggressively buy milk products because they believe that the product can cure COVID-19. Hence, the authors chose a specific milk brand as the object of research. The authors investigated the direct effects of brand marketing communication on brand authenticity as well as the mediating effects of clarity of positioning. In addition, the authors analyze by controlling other marketing mix variables. This research uses a quantitative approach with a total of 326 respondents. The data is analyzed using Partial Least Squares-Structural Equation Modeling (PLS-SEM). The results of this study imply that there is a direct positive influence from brand marketing communications and an indirect influence through the clarity of positioning that can affect brand authenticity. Then, from all marketing mix variables, only brand satisfaction, price deals, and price images show the effect on brand authenticity.

20.1 Background

Authenticity is defined as the feeling or character of a brand that is associated with a particular sense of historical heritage [13]. Further, authenticity has evolved as a consumer-desired attribute, expressing a sense of a brand that is distinctive, authentic, or original [14] and helping marketers to be able to differentiate their businesses from other brand competitors. In business, many products and brands strive for competitive distinction despite lacking inherent authenticity as defined in influential books. Those brands are produced massively and are consistently bought or consumed by people, which was classified as fast-moving consumer goods (FMCG) [20].

N. Mediarki · Y. Alversia (✉)
Universitas Indonesia, Jakarta 16464, Indonesia
e-mail: yeshika@ui.ac.id

J. L. Reis et al. (eds.), *Marketing and Smart Technologies*, Smart Innovation, Systems and Technologies 337, https://doi.org/10.1007/978-981-19-9099-1_20

In the Indonesian context, the consumption of fast-moving consumer goods (FMCG) in modern retail continues to slowly grow by 2% throughout 2020, largely as a result of the COVID-19 pandemic [11]. The findings of a study by Kantar Indonesia in 2021 [57] about consumer behavior in Indonesia during the pandemic suggest that the industries of food and beverage, home care, and body care have experienced significant growth. This is supported by the results of a study conducted by Badan Pusat Statistik in 2021, which demonstrate that persons who resided in both urban and rural areas made the most purchases of ready-made food and drink items [5].

In more detail, out of all food and beverage products, milk and dairy products exhibited the biggest increase in consumption. Meaning that the need for fresh milk in Indonesia continues to increase during the COVID-19 pandemic [28]. This is evidenced by the emergence of one of the phenomena associated with rising milk consumption, namely, the panic buying phenomena for one of the dairy products in Indonesia, Bear Brand, in the middle of 2021 [47]. Bear Brand was sought after by the general public due to the posting of one of Facebook users claiming that Bear Brand has benefits in curing the coronavirus, which later sparked word-of-mouth among social media users and led many to assume the allegations were accurate [55].

Historically, Bear Brand is one of the products produced by Nestle company, as it was first distributed in Indonesia in the 1930s [55]. The brand has known for its uniqueness in terms of its product and branding strategy. First, although being marketed as a 'bear's milk', Bear Brand is actually made from sterilized cow's milk, and later on the company brands their product as a sterile milk [9]. Second, Bear Brand has long been claimed as a medicinal milk that can cures disease. In the past, Bear Brand has been trusted for generations by the community as a product that can restore stamina and neutralize toxins in the body [4]. Furthermore, despite the brand name implying a bear or a cow, Bear Brand primarily depicted the image of a dragon for most of their commercials [19] and consistently promoted their own jargon known as "Feel the Purity" [23]. Driven by its lengthy history in the market, Bear Brand is able to be perceived by consumers as an authentic goods, which leads to people having perception that Bear Brand is a health milk, despite the facts that all milk products are equally beneficial for health. With this argument, the authors decided to employed Bear Brand as the research object.

Brand authenticity itself has become a critical component of brand success because it contributes to the development of a distinct brand identity [37]. Nowadays, authenticity has emerged as one of the key factors that might influence consumers' preferences for a brand, with consumers favoring brands that seem more authentic. This definition leads to the rationale behind why this study was conducted. People nowadays live in the age of distrust, where they no longer fully trust the advertising of a brand. Therefore, brand authenticity matters to retain brands and create a sense of connection with the customers [1]. Additionally, customers' started to have growing scepticism toward the company's marketing and advertising effort and have grown justifiably wary of advertisements as a result of years of pop-up ads, outdated marketing stunts, and outright objectionable content. According to a research by the American Association of Advertising Agencies only 4% of customers think that marketers and advertisements uphold integrity [51]. Another problem is that the FMCG industries are fiercely competitive with numerous companies competing against each other and producing similar or identical products [39]. In order to maintain brands and build a sense of connection with customers, brand authenticity is therefore necessary [1]. It also becomes crucial for businesses to be able to compete with the brand competitors by offering brand differentiation and authentic perception.

It is presumed that consumer views of brand authenticity may be influenced by how consumers evaluate the brand marketing communications. There are two methods in which authenticity can be affected by brand marketing communication, namely the direct effect and indirect effect via the mediating role of clarity of brand's positioning [20]. Therefore, this study will attempt to discover whether the marketing communication effort of Bear Brand can directly increase brand authenticity or whether authenticity may be accomplished through the indirect means.

As the study's chosen object, Bear Brand has implemented many marketing communications strategies, including sponsorship, social media, corporate responsibility, and sales promotion. These strategies will serve as the foundation for an analysis of brand marketing communication.

Additionally, studying brand authenticity matters because recent research finds that brand authenticity positively influences brand trust and loyalty [44]. When consumers are uncertain or anxious about a buying decision, they seek trustworthy brands. Meaning that trust developed as a result of considering possible risks [18]. Authenticity has been discovered as a way to ease consumer confusion [14] and to be the ideal cure for firms that are looking to rebuild consumer trust [21]. Authentic brands are dedicated to keeping their promises, and customers are more likely to trust brands they believe will be committed [14]. Lastly, understanding the views of brand authenticity also contributes to developing a better strategy for segmenting business markets [21, 38].

The authors made two major contributions as a result of this investigation. First, the impact of consumer evaluation of brand marketing communication on brand positioning clarity and perceived brand authenticity in FMCGs has received little attention, and this study was among the first to study about the brand authenticity for FMCG brands, particularly for Bear Brand as the study's object. Second, the authors proceed to add additional variable to assess brand marketing communication, which is sales promotion. Next, the authors designed and tested the direct and indirect effects of consumer perceptions of brand marketing communications on brand authenticity with the use of mediation effects of clarity of positioning. Additionally, the authors proceed to examined these effects by controlling for the effects of other elements of the marketing mix [20].

20.2 Theoretical Framework and Research Hypothesis

20.2.1 *Brand Authenticity*

The term *authenticity* is examined as brand authority, consistency, innovation, provenance, and heritage [17]. It is also used to describe a product or other object that is an authentic, genuine article, and not an imitation [16]. Another definition of authenticity was also studied by prior research that developed brand authenticity in the context of a brand extension, which upholds the brand's standards or style, brand heritage, brand essence, and minimizing exploitation [53]. Additionally, brand authenticity can be defined as brand's evaluation based on the heritage, expectations, and views of customers. The level to which customers believe a brand to be faithful and true toward itself and its consumers, and to support consumers being genuine to themselves [37]. Subjective definitions of authenticity include the consumer's perception of authenticity as a part of a brand's image as well as the tendency for consumers to judge companies based on the behavioral norms that are expected of other people, namely a breach of a norm and brand promise, that might result in negative attitudes and judgments [2, 20].

20.2.2 *The Direct Influence of Brand Marketing Communication on Brand Authenticity*

Brand marketing communications represent a vital element of a brand's marketing strategy [33] and it is defined as all of the communication efforts that will enable an organization to explain what it has to offer, what it can provide, and what its existence relates to and its products and services [12]. There are five brand marketing communication used in this study, which are advertising, social media, sponsorship, corporate social responsibility [20], and additional variable of sales promotion.

In understanding the direct influence of brand marketing communication and its ways to affect the customers' perception of brand authenticity, it can be comprehended by paying attention to the concept of cultural meaning transfer process. The culturally relevant meaning that was attributed to each of the consumer items was being transmitted to specific brands through the use of marketing communication [35]. Later on, the culturally constructed meaning is then consumed by consumers when they use those brands [40]. As a result, a brand gets associated with particular acceptable cultural values or brand meaning [38], which customers ingest through using their preferred product.

It is important to note that brand marketing communications are not only a physical representation of a brand's marketing efforts [54], but it also conveys the voice of the brand [29]. With that in mind, consumers' positive evaluations of the brand marketing communication might have ramifications for consumer perceptions of authenticity. If consumers favorably perceive the entirety of the marketing communication activities, this may alter emotional judgments among customers, which, in turn, may affect the perception of brand authenticity. Therefore, motivated by the preceding arguments from previous studies, the authors seeks to demonstrate if Bear Brand's efforts at brand marketing communication can indeed contribute to brand authenticity with developing the following hypothesis:

H1. Brand marketing communication positively influences the formation of brand authenticity.

20.2.3 The Indirect Influence of Brand Marketing Communications on Brand Authenticity Through the Clarity of Positioning

Furthermore, this study assesses brand authenticity through the mediating role of clarity of positioning. A previous study by Dwivedi and McDonald [20] argues that competitive companies within a category use the message of marketing communication tactics to portray an authentic identity, with a focus on the clarity of the consumer-perceived positioning. The variable of clarity of positioning is chosen as a mediator because it was argued that several brands are able to stand-out from their competitors driven by the clarity of positioning strategy that is being associated in the customers' minds. The clarity of positioning here emphasized brand differentiation, driven by the ability of clarity of positioning to enhance the value and message of a brand [52]. Further, clarity of positioning is linked to how strong brand associations are in consumer's memory. Thus, a strong brand connection between customers and a brand causes the brand to be remembered easily, contributing to the establishment of brand authenticity in the minds of consumers [30].

A prior study also conceptualized that there is an indirect influence between the consumer judgment of brand marketing communication towards brand authenticity that is caused by the mediating role of clarity of positioning [20]. The indirect effect was based on the theory of associative network memory. This was derived from a study showing that consumer understanding of a brand that is represented in their memory is fundamentally illustrated as a web or network that stored all the information nodes [30]. Further, these nodes are connected by associations that are linked to brand attributes and brand advantages, and yet others may represent consumers' brand experiences [31]. When these nodes are triggered, they have the potential to activate adjacent nodes, and therefore, they might influence the recall of brand information from memory [20]. Therefore, it is hypothesized that clarity of positioning bridged the brand marketing communication and brand authenticity and should be considered for this study:

H2. Clarity of positioning positively mediating the formation of brand authenticity.

20.2.4 *Control Influences of Place, Product, and Price of Marketing Mix*

This study employs control influences that were initially not a part of the research hypothesis and were not relevant to the study's objectives. Yet, it is only used as conceptual information in order to acknowledge that there could be the factor other influences. The control influences used in this study are other elements of the marketing mix, which comprises of Product, Place, and Price.

Since the entire study is devoted to examining the impact of brand marketing communication on brand authenticity, the authors have concentrated the study solely on the promotion aspect of the marketing mix. However, marketing communication is not the only marketing mix strategy that can be developed by the company to enhance authenticity, and it is only one among four marketing mix elements. Therefore, in addition to assessing the effect of brand marketing communication on brand authenticity, the authors decided to also analyze the other related marketing mix as well and employ them as the control influences. These control influences consist of other related 4Ps marketing mixes, which are Place operationalized with distribution intensity and store image, Product which is measured with brand satisfaction, Price is measured with price image and price deals, as well as brand popularity.

20.3 Methodology

20.3.1 *Sample and Data Collection*

The respondent of the study comprises of Indonesian citizen aged 19 to 45 who are likely aware of Bear Brand's marketing communication which has secured the title as the number two top brand of packaged milk in Indonesia according to by Top Brand Award in 2021 [5]. The choosing of respondents on that range of age is due to the fact that people aged 19–45 are the targeted consumers of Bear Brand [43], who initially aim to target their products to young and working adults [15].

Further, the study was carried out using a quantitative model and was tested with PLS-SEM method. The authors developed questionnaires in order to obtain information regarding the brand marketing communication in the perception of consumers as one of the drivers to understand the brand authenticity of Bear Brand's brand. In obtaining complete and specific information, the questionnaire given is a structured questionnaire. The questionnaire in this study is in the form of a self-administrative questionnaire that can be filled out by respondents online. Respondents will be given a link to the questionnaire page in the form of a Google Form that can be filled in directly by the respondent. The authors employed an online survey to increase response rate in terms of reach and eliminate any potential concerns that could have arisen from sources of errors in the survey such as coverage error, sampling error, and nonresponse error [3, 24]. An online survey link was distributed via email and social media sites such as Line, WhatsApp, Facebook, Telegram, and Instagram to collect online responses. The authors also encouraged respondents to distribute the survey link on their own connections to increase the number of potential respondents that cannot be reached by the authors.

This study consists of three stages, including the wording test, pretest, and main test. At the wording test stage, the authors ensure that the use of the words and sentences in the research questionnaire has been in accordance with the intent and meaning and can be understood by the respondents. The wording test was conducted with 6 (six) respondents. In the pretest stage, 33 respondents were selected to fill out the research questionnaire, and the results were analyzed to determine whether the research questionnaire could be regarded as valid and reliable. By April 2022, 313 valid respondents were gathered. Lastly, the authors proceed to the main test analysis. The results of the main test data were processed using SPSS 26 and Smart-PLS 3.0.

20.3.2 Measurement

The measurement indicators for advertising are adapted from Yoo et al. [58]; Villarejo-Ramos and Sánchez-Franco [56] the measurement indicators for sales promotion from Pappu and Cornwell [42] and Simmons and Becker-Olsen [52], social media was operationalized using Schivinski and Dabrowski [50], sponsorship measured using Pappu and Cornwell [42] Simmons and Becker-Olsen [52], and corporate social responsibility is adapted from Menon and Kahn [36]. Next, in order to measure the clarity of positioning adapted from Pappu and Cornwell [42], and brand authenticity was measured using Schallehn et al. [49]. As for the control influences, the brand satisfaction is adapted from Aurier and N'Goala [7]; Homburg et al. [27], store image, price image, and price deals, and the distribution intensity scales are measured with items from Yoo et al. [58], and brand popularity from Dwivedi and McDonald [20].

The questionnaire item in this research used structured questions with close-ended types of questions. The survey is using a 7-point Likert scale with an interval data scale for all the measurement questions except the screening questions, the confirmation question, and the question within the respondents' profile section that asked about the demographical data of the respondents. Respondents were asked to answer all scale items related to Nestle's Bear Brand. Respondents who choose 1 mean they "Strongly Disagree" and people who score 7 mean they "Strongly Agree".

20.4 Results

20.4.1 Measurement Model

Considering the hierarchical order construct of brand marketing communication, the authors evaluated the outcomes of higher-order structures in PLS-SEM by utilizing the repeated indicator approach and the two-stage approach in the calculation of the measurement model [48].

An internal consistency test was conducted to determine the reliability of the measurement model. Based on the results of internal consistency testing, the result shows that Cronbach's alpha value of each variable has met the requirements by which have values above 0.7. Likewise, the composite reliability value for each variable that meets the requirements has a value above 0.7.

Furthermore, by assessing the convergent validity results, it is possible to verify the accuracy of the analysis of the measurement model. Convergent validity can be analyzed from the outer loadings of the indicators and the average variance extracted (AVE). First, the author proceeds to analyze the repeated indicator approach. The calculation of convergent validity and discriminant validity in this approach shows that all the variables contained in this study have met the minimum requirements for the AVE value. In detail, the variable of corporate social responsibility appears to achieve the highest AVE value, which accounts for (0.918), followed by social media with an AVE score of (0.910). Furthermore, clarity of positioning has an AVE value of (0.902), sponsorship (0.887), and brand authenticity (0.875). The smallest number of AVE occurs to happen to sales promotion variable by (0.854) and advertising with (0.822).

In the two-stage approach, all higher-order components are now considered as indicators for brand marketing communication. All the outer loading values for brand marketing communication, clarity of positioning, and brand authenticity have an outer loading value above 0.70. This finding indicates that variables have good convergent validity and demonstrates that it measures what it intends to measure. Brand authenticity has the highest AVE of (0.875), clarity of positioning has a value of (0.795), and brand marketing communication with (0.742).

Based on the calculation of internal consistency and convergent validity for repeated Based on the calculation of internal consistency and convergent validity for repeated and two-stage approaches, it seems that the authors had no problems ensuring the reliability and validity of each indicator. Therefore, for the following subsections, the authors continued with the calculation of the repeated approach only. This is driven by the arguments that the repeated approach might result in an accurate depiction of the second-order construct's noticeable impact on the higher-order component because here the authors must also establish the links between the preceding construct and the lower-order components, rather than looking at the direct connection between the preceding construct and the higher-order construct [10].

Lastly, the authors tested the discriminant validity to determine whether the research model represents phenomena that are not represented by other variables.

The results of cross-loading value show that each indicator has the greatest value than any values of the cross loadings within each indicator and implies that it has the most significant correlation with its own constructs and not with other constructs.

20.4.2 Structural Model

Collinearity testing is done by using a structural model (inner model), and the results of data processing seen in the collinearity test are the results of the inner VIF (Inner Variance Inflation Factors) value. The requirement to be said to have a good VIF value is the inner VIF value < 5.0 [26]. From the results of the study, it was found that all relationships between variables in the study had good VIF values because they had values below 5.0.

The next step in the analysis of the structural model is to evaluate its feasibility of the structural model. This is done by testing the results of the coefficient of determination or R^2. From the analysis, it is described that all values of R^2 are ranging between 0 and 1, with the highest value of accuracy coming from brand marketing communication by (1.000), followed by brand authenticity with a value of (0.720), and the smallest R^2 occurs to the clarity positioning with the number of (0.586). This shows a good result because the values of R^2 are typically ranging from 0 to 1, by which those who have a higher number or value are believed to have a high level of accuracy.

The significance was evaluated by conducting a bootstrapping technique on Smart-PLS using 5000 subsamples [45]. Furthermore, the bootstrapping calculation was done by using a one-tailed test type with a significant level of 0.05. The one-tailed type is chosen for this study because it is more suitable to assess a specific and directional hypothesis [59], especially when the authors want to evaluate the direct and indirect impact of brand marketing communication toward brand authenticity. For the significance, the authors focused on looking at the value of T-Statistics of the independent variable towards the dependent variable. For the relationship to be considered positive significant, the T-Value should be ≥ 1.645, meanwhile, if the T-Value is ≥ -1.645, it will be considered as having a negative significance or negative influence.

Figure 20.1 summarizes the result of the hypothesis testing of all variables and control influences by paying attention to the value of original sample coefficient. First, the direct relationship between brand marketing communication to brand authenticity shows a positive effect, and thus, hypothesis H1 is supported. The significance of brand marketing communication to brand authenticity is proven by the T-Statistic value that is accounted for (4.177), which is larger than the required value of above 1.645.

Among the five types of brand marketing communication evaluated by the authors, corporate social responsibility (CSR) yield the highest original sample coefficient by accounting for (0.262) meaning that it is the most variable that represents the perception of brand marketing communication. Madhavaram [33] believes that consumers

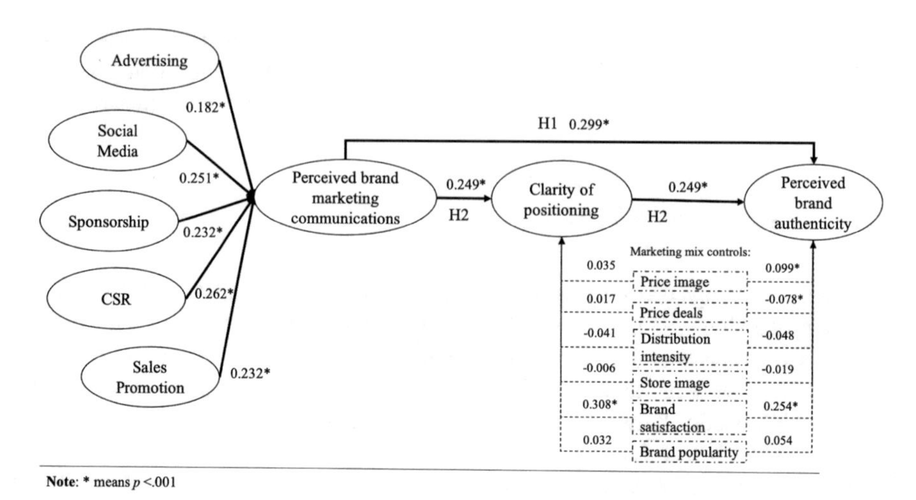

Fig. 20.1 Research results

form an overall perception of a company's brand marketing communication effort, and this innovative approach is built on the synergistic utilization of brand marketing communications to present a consistent identity across several communication channels. With that in mind, this research finds that consumers can generate holistic impressions of brand marketing communications [20]. Therefore, consumers can develop their own evaluations of brand authenticity based on the marketing communication activities done by Bear Brand, such as advertising, social media, corporate social responsibility, sales promotion, and sponsorship.

Subsequently, the authors tested the indirect path coefficient through a bootstrapping method to study the mediating effect of clarity of positioning on the formation of a brand. The result indicates that the relationship between brand marketing communication toward brand authenticity with the mediating role shows a T-Value of (5.088), which is also significant because the scores are larger than the required value of above 1645, meaning that hypothesis H2 is supported. Hence, it can be understood that the strategic positioning of a brand, especially Bear Brand, might be able to encourage the perception of unique or distinct characteristics of a brand in consumers' minds. Simmons and Becker-Olsen [52] argued that several brands can stand out from their competitors driven by the clarity of positioning strategy that is associated with the customers' mindset. The clarity of positioning here emphasized brand differentiation, driven by the ability of clarity of positioning to enhance the value and message of a brand.

If we do a comparison between the result of direct and indirect methods of brand marketing communication toward brand authenticity, the relationship between brand marketing communication to brand authenticity with the mediating role of clarity of positioning yielded the original sample score of (0.249). Meanwhile, in the calculation of the direct path coefficient, the result of the direct effect from brand marketing communication to brand authenticity shows the original sample accounted for (0.299), which is bigger than the value of indirect effect. Therefore, it can be concluded that the direct effect of brand marketing communication is stronger to enhance authenticity. Brand marketing communication variable is considered a good variable where it still can be self-sustaining even without the mediation variable and have the power to affect authenticity on its own. Yet, the role of clarity in positioning a mediating variable is still considered important to form brand authenticity. Hence, we can imply that clarity of positioning is considered as having a partial mediation [8].

Furthermore, the calculation of Smart-PLS in assessing the other types of the marketing mix (Price, Place, and Product) shows that not all variables can influence the clarity of a brand's positioning. Among all the other types of marketing mix variables used in this study, only brand satisfaction is shown to have a significant impact on the clarity of positioning driven by the T-Value of (4.017) which is above 1.645, and the original sample value of (0.308).

On the other hand, the assessment of other types of marketing mixes toward the formation of brand authenticity shows that there are three variables of the marketing mix, namely, brand satisfaction, price deals, and price image that have an influence on building the brand authenticity of a brand. Among the three variables, some positively affected authenticity, while one negatively affected authenticity. Brand satisfaction and price image have a positive influence on authenticity. This conclusion was obtained by looking at the T-Value of brand satisfaction and price image which have a score of (4.419) and (2.537) respectively, and exceed the minimum required value of above 1645. Both of the variables also have a positive number of the original sample. Meanwhile, for the case of price deals, even though it has a significant *T*-value of 1.754, the value of the original sample is negative by (−0.078). This means that price deals have a significant influence on authenticity, but the effect is negative.

20.5 Discussion

20.5.1 Implications

Among the five types of brand marketing communication evaluated by the authors, corporate social responsibility (CSR) yields the highest original sample by accounting for (0.262) meaning that it is the most variable that represents the perception of brand marketing communication to authenticity. Hence, CSR activities are brand marketing communication that marketers can emphasize more to build brand authenticity. There are several indicators that the company should emphasize for its CSR programs. First, Bear Brand to be should focus on creating social programs that are highly involved in community activities. This can be done through aid in fundraising and volunteering for charity. Also, marketers should emphasize creating a CSR program that shows the company is very concerned with environmental issues.

It can be done by utilizing recycled materials for the brand's packaging [46] and started to offering an environment & sustainable education to customers [41].

Next, although the direct effect of brand marketing communication on brand authenticity is proven to have a stronger effect compared with the indirect effect, the mediating role of clarity of positioning is still important to consider by the marketer. By putting the focus on the clarity of positioning, the marketing strategy of Bear Brand should emphasize giving or conveying a clear image in all of the company actions to customers' minds. Bear Brand should be able to have a true understanding of who their prospective customers are so that the company can adjust the brand messaging accordingly. Additionally, the company should investigate how visual components such as the brand logo, company fonts, and packaging can effectively draw more attention to your value proposition [34].

In terms of other types of marketing mix (Product, Price, and Place), there are three control influences that are important to the clarity of positioning and brand authenticity, which are brand satisfaction, price image, and price deals. First, marketers should emphasize how the brand can make their customers satisfied with the products and how the brand can meet the expectations of its buyers. To increase satisfaction, marketers can try to understand customer needs by establishing effective customer service. Further, brand satisfaction can also be done by conducting surveys to learn about what are potential customers' needs and desires. This will assist the company in understanding how to meet customers' expectations and adjust to new trends [6].

Furthermore, there are several ways that companies can do to control how people perceive their pricing strategy. Companies should monitor the competitor's pricing as frequently as possible to ensure that the current market position corresponds to the brand and intended price image. Second, if businesses pay close attention to what constitutes a price image and keep committed to the personality of the brand, companies will be well on their way to developing the impression they desire [22].

Lastly, this research shows that price deals are significant in affecting brand authenticity, but it affects authenticity in a negative way. Presumably driven by the fact that people perceive Bear Brand as a premium product driven by the high prices, any promotion activities may then be perceived as conflicting with its premium image. Hence, if Bear Brand still wants to consider offering price deals as their strategy, they can reduce the intensity of their promotions and promote only on a certain occasion.

20.6 Limitation and Future Research

In terms of the object selection, the authors chose a low-involvement product which is Bear Brand's packaged milk. Consequently, it is argued that high-involvement products such as branded products or clothing lines may react differently to the aspects of brand marketing communication [20]. Therefore, future research might focus on assessing consumer perception of the companies within these categories. Second, this research didn't look into the impact of message content on the brand marketing strategy. This research suggests that the message of marketing strategy was desegregated, which focused on customer evaluations of marketing communications. Hence, future studies could include message elements in the research model, which are the knowledge that the marketer wishes to convey to the consumers. Next, future research might focus on the distribution of respondents. The questionnaire might be more evenly distributed in each city in Indonesia and with a number of samples that are more representative of the population. Lastly, future research might as well examine the brand experience elements of brand marketing communication, such as the volume or frequency of being exposed or being exposed to the marketing communications [20]. This should be added because consumers' perceptions of authenticity may be influenced by these factors.

20.7 Conclusion

Bear Brand's marketing communications indeed have a direct effect on brand authenticity, with corporate social responsibility emerging as the most significant action in reflecting brand marketing communication to affect authenticity compared to other marketing communication activities. Second, the clarity of positioning can mediate the relationship between brand marketing communications and brand authenticity, even though it is classified as partial mediation. Lastly, for the control variables, the results show that brand satisfaction and price image have a positive effect, while price deals have a negative effect on brand authenticity. In conclusion, brand marketing communication can be considered crucial to overcoming challenges that are now arising in the market or business rivalry. Through leveraging brand marketing communication, companies can start to differentiate themselves or increase loyalty by developing a strategy centered on authenticity.

References

1. Adiwaluyo, E.: Pentingnya Menjadi Authentic Brand di Era Millenial, December 6 (2018). www.marketeers.com. Retrieved October 9, 2022, from https://marketeers.com/pentingnya-menjadi-authentic-brand-di-era-millenial
2. Aggarwal, P.: The effects of brand relationship norms on consumer attitudes and behavior. J. Consumer Res. **31**(1), 87–101 (2004). https://doi.org/10.1086/383426
3. Alsnih, R.: Characteristics of web based surveys and applications in travel research. Travel Surv. Method., 569–592 (2006). https://doi.org/10.1108/9780080464015-032
4. Amarasthi, N.: Ketahui 5 Fakta Susu Bear Brand yang Mendadak Diburu Saat COVID-19 Melonjak. VOI—Waktunya Merevolusi Pemberitaan (2021). https://voi.id/lifestyle/64315/ketahui-5-fakta-susu-i-bear-brand-i-yang-mendadak-diburu-saat-covid-19-melonjak
5. Annur, C.M.: Merek Produk Susu Terfavorit di Indonesia, Siapa Juaranya? Katadata (2021). Retrieved October 10, 2022, from https://databoks.katadata.co.id/datapublish/2021/07/05/daftar-produk-susu-terfavorit-di-indonesia-siapa-juaranya
6. Arakelyan, L.: 10 ways to meet and satisfy your customer's needs. Customerly (2021). https://www.customerly.io/blog/marketing/2019/11/15/meet-and-satisfy-customers-needs/

7. Aurier, P., N'Goala, G.: The differing and mediating roles of trust and relationship commitment in service relationship maintenance and development. J. Acad. Mark. Sci. **38**(3), 303–325 (2010)
8. Baron, R.M., Kenny, D.A.: The moderator–mediator variable distinction in social psychological research: conceptual, strategic, and statistical considerations. J. Pers. Soc. Psychol. **51**(6), 1173–1182 (1986). https://doi.org/10.1037/0022-3514.51.6.1173
9. BEAR BRAND.: Nestlé (n.d.). Retrieved October 10, 2022, from https://www.nestle.co.id/produk/minumansiapminum/bearbrand
10. Becker, J.M., Klein, K., Wetzels, M.: Hierarchical latent variable models in PLS-SEM: guidelines for using reflective-formative type models. Long Range Plan. **45**(5–6), 359–394 (2012). https://doi.org/10.1016/j.lrp.2012.10.001
11. Bisnis Indonesia.: Konsumsi Membaik Penjualan FMCG Mulai Naik (2021). https://bisnisindonesia.id/article/konsumsi-membaik-penjualan-fmcg-mulai-naik
12. Bozkurt, İ.: Bütünleşik Pazarlama İletişimi (2000). MediaCat, Ankara
13. Brown, S., Kozinets, R.V., Sherry, J.F.: Teaching old brands new tricks: retro branding and the revival of brand meaning. J. Mark. **67**(3), 19–33 (2003). https://doi.org/10.1509/jmkg.67.3.19.18657
14. Bruhn, M., Schoenmüller, V., Schäfer, D., Heinrich, D.: Brand authenticity: towards a deeper understanding of its conceptualization and measurement. Adv. Consum. Res. **40**, 567–576 (2012)
15. Campaign Portofolio.: Case study: Getting milk into the Filipino adult diet (2016). https://www.campaignasia.com/agencyportfolio/CaseStudyCampaign/428157,case-study-getting-milk-into-the-filipino-adult-diet.aspx#.Ye1yVfVBzvU
16. Chhabra, D., Kim, E.G.: Brand authenticity of heritage festivals. Ann. Tour. Res. **68**, 55–57 (2018). https://doi.org/10.1016/j.annals.2017.11.007
17. Choi, H., Ko, E., Kim, E.Y., Mattila, P.: The role of fashion brand authenticity in product management: a holistic marketing approach. J. Prod. Innov. Manag. **32**(2), 233–242 (2015)
18. Delgado-Ballester, E., Luis Munuera-Alemán, J.: Brand trust in the context of consumer loyalty. European J. Market. **35**(11/12), 1238–1258 (2001). https://doi.org/10.1108/eum0000000006475
19. Dinisari, M.C.: Asal usul Bear Brand: Susu Sapi Berlogo Beruang, Iklan Bergambar Naga. Bisnis.com (2021). https://lifestyle.bisnis.com/read/20210703/106/1413332/asal-usul-bear-brand-susu-sapi-berlogo-beruang-iklan-bergambar-naga
20. Dwivedi, A., McDonald, R.: Building brand authenticity in fast- moving consumer goods via consumer perceptions of brand marketing communications. European J. Market. (2018). EJM-11-2016-0665. https://doi.org/10.1108/EJM-11-2016-0665
21. Eggers, F., O'Dwyer, M., Kraus, S., Vallaster, C., Güldenberg, S.: The impact of brand authenticity on brand trust and SME growth: a CEO perspective. J. World Bus. **48**(3), 340–348 (2013). https://doi.org/10.1016/j.jwb.2012.07.018
22. Ellsworth, M.: What is price image? quick guide for brands and retailers. Wiser Retail Strategies (2021). https://blog.wiser.com/what-is-price-image-quick-guide-for-brands-and-retailers/
23. Fauziah, L.: Mengapa Susu "Beruang" Bear Brand Logonya Beruang Iklannya Naga? Kabar Lumajang (2021). https://kabarlumajang.pikiran-rakyat.com/iptek/pr-422162899/mengapa-susu-beruang-bear-brand-logonya-beruang-iklannya-naga?page=3
24. Groves, R.M., Lyberg, L.: Total survey error: past, present, and future. Public Opin. Q. **74**(5), 849–879 (2010). https://doi.org/10.1093/poq/nfq065
25. Hair, J.F., Hult, G.T.M., Ringle, C.M., Sarstedt, M.: A primer on partial least squares structural equation modeling (PLS-SEM), 2nd edn. Thousand Oaks, Sage (2017)
26. Hair, J.F., Sarstedt, M., Ringle, C.M., Mena, J.A.: An assessment of the use of partial least squares structural equation modelling in marketing research. J. Acad. Mark. Sci. **40**(3), 414–433 (2012)
27. Homburg, C., Koschate, N., Hoyer, W.D.: Do satisfied customers really pay more? a study of the relationship between customer satisfaction and willingness to pay. J. Mark. **69**(2), 84–96 (2005)

28. Jannah, K.M.: Konsumsi Susu Naik selama Pandemi, Peluang Buat Peternak Lokal (2021). https://economy.okezone.com/. https://economy.okezone.com/read/2021/11/18/455/2503872/konsumsi-susu-naik-selama-pandemi-peluang-buat-peternak-lokal?page=1
29. Keller, K.L.: Building strong brands in a modern marketing communications environment. J. Mark. Commun. **15**(2–3), 139–155 (2009). https://doi.org/10.1080/13527260902757530
30. Keller, K.L.: Conceptualizing, measuring, and managing customer-based brand equity. J. Mark. **57**, 1–22 (1993). https://doi.org/10.2307/1252054
31. Krishnan, H.: Characteristics of memory associations: a consumer-based brand equity perspective. Int. J. Res. Mark. **13**(4), 389–405 (1996). https://doi.org/10.1016/s0167-8116(96)00021-3
32. LNCS Homepage: http://www.springer.com/lncs, last accessed 2016/11/21
33. Madhavaram, S., Badrinarayanan, V., McDonald, R.E.: Integrated marketing communication (IMC) and brand identity as critical components of brand equity strategy: a conceptual framework and research propositions. J. Advert. **34**(4), 69–80 (2005). https://doi.org/10.1080/00913367.2005.10639213
34. MasterClass: Brand image: how to create a positive brand image. MasterClass (2022). https://www.masterclass.com/articles/brand-image-explained#how-to-create-a-positive-brand-image
35. McCracken, G.: Culture and consumption: a theoretical account of the structure and movement of the cultural meaning of consumer goods. J. Cons. Res. **13**(1), 71 (1986). https://doi.org/10.1086/209048
36. Menon, S., Kahn, B.E.: Corporate sponsorships of philanthropic activities: when do they impact perception of sponsor brand? J. Consum. Psychol. **13**(3), 316–327 (2003)
37. Morhart, F., Malär, L., Guèvremont, A., Girardin, F., Grohmann, B.: Brand authenticity: an integrative framework and measurement scale. J. Cons. Psychol. **25**(2), 200–218 (2015). https://doi.org/10.1016/j.jcps.2014.11.006
38. Napoli, J., Dickinson, S.J., Beverland, M.B., Farrelly, F.: Measuring consumer-based brand authenticity. J. Bus. Res. **67**(6), 1090–1098 (2014)
39. Nuraini: Strategi Pemasaran Digital Industri FMCG yang Tak Boleh Dilewatkan di 2022. Tada (2022). Retrieved May 14, 2022, from https://blog.usetada.com/id/strategi-pemasaran-digital-industri-fmcg
40. O'Guinn, T., Allen, C., Semenik, R.J., Scheinbaum, A.C.: Advertising and integrated brand promotion. Cengage Learning, Stanford, CT (2014)
41. Okutoyi, P.: The role of consumer education in achieving sustainable consumption behavior. Africa Sustain. Matt, 13 May (2019). https://africasustainabilitymatters.com/the-role-of-consumer-education-in-achieving-sustainable-consumption-behavior
42. Pappu, R., Cornwell, T.B.: Corporate sponsorship as an image platform: understanding the roles of relationship fit and sponsor–sponsee similarity. J. Acad. Mark. Sci. **42**(5), 490–510 (2014)
43. PDF Coffee: Analisis segmenting, targeting, positioning Pada Produk Minuman Susu Steril "Bear Brand" (n.d.). Retrieved February 24, 2022, from https://pdfcoffee.com/analisis-stp-bear-brand-pdf-free.html
44. Portal, S., Abratt, R., Bendixen, M.: The role of brand authenticity in developing brand trust. J. Strateg. Mark. **27**(8), 714–729 (2019). https://doi.org/10.1080/0965254X.2018.1466828
45. Preacher, K.J., Hayes, A.F.: Contemporary approaches to assessing mediation in communication research. In: Hayes, A.F., Slater, M.D., Snyder, L.B. (eds.) The Sage Sourcebook of Advanced Data Analysis Methods for Communication Research, pp 13–54. Sage. https://doi.org/10.4135/9781452272054.n2
46. Queensland Government.: The benefits of an environmentally friendly business. Business Queensland, 18 Jan (2021). www.business.qld.gov.au/running-business/environment/environment-business/benefits
47. Rabbi, C.P.A.: Susu Beruang Jadi Rebutan, Nestle Indonesia Optimalkan Produksi. Katadata (2021). https://katadata.co.id/safrezifitra/berita/60e2cec99259f/susu-beruang-jadi-rebutan-nestle-indonesia-optimalkan-produksi

48. Sarstedt, M., Hair, J.F., Cheah, J.H., Becker, J.M., Ringle, C.M.: How to specify, estimate, and validate higher-order constructs in PLS-SEM. Australas. Mark. J. **27**(3), 197–211 (2019). https://doi.org/10.1016/j.ausmj.2019.05.003
49. Schallehn, M., Burmann, C., Riley, N.: Brand authenticity: model development and empirical testing. J. Product Brand Manag. **23**(3), 192–199 (2014)
50. Schivinski, B., Dabrowski, D.: The effect of social media communication on consumer perceptions of brands. J. Mark. Commun. **22**(2), 189–214 (2016)
51. Shane, D.: 96 percent of consumers don't trust ads. Here's how to sell your product without coming off Sleazy. Inc.Com (2020). https://www.inc.com/dakota-shane/96-percent-of-consumers-dont-trust-ads-heres-how-to-sell-your-product-without-coming-off-sleazy.html
52. Simmons, C.J., Becker-Olsen, K.L.: Achieving marketing objectives through social sponsorships. J. Mark. **70**(4), 154–169 (2006). https://doi.org/10.1509/jmkg.70.4.154
53. Spiggle, S., Nguyen, H.T., Caravella, M.: More than fit: brand extension authenticity. J. Mark. Res. **49**(6), 967–983 (2012)
54. Stern, B.B.: A revised communication model for advertising: multiple dimensions of the source, the message, and the recipient. J. Advert. **23**(2), 5–15 (1994). https://doi.org/10.1080/00913367.1994.10673438
55. Utami, F.A.: Peminat Susu Bear Brand Masih Tinggi, Ternyata Ini Sejarah "Susu Beruang" Sejak 115 Tahun Lalu!Warta Ekonomi (2021). https://www.wartaekonomi.co.id/read349703/peminat-susu-bear-brand-masih-tinggi-ternyata-ini-sejarah-susu-beruang-sejak-115-tahun-lalu#:%7E:text=Pada%20tahun%201930%2Dan%2C%20susu,di%20Thailand%20pada%20tahun%201980
56. Villarejo-Ramos, A.F., Sánchez-Franco, M.J.: The impact of marketing communication and price promotion on brand equity. J. Brand Manag. **12**(6), 431–444 (2005)
57. Wulandari, D.: Selain Belanja Kebutuhan Rumah Tangga Meningkat 5%, Konten FMCG di Platform Digital pun Bertumbuh. MIX (2021). https://mix.co.id/marcomm/news-trend/selain-belanja-kebutuhan-rumah-tangga-meningkat-5-konten-fmcg-di-platform-digital-pun-bertumbuh/
58. Yoo, B., Donthu, N., Lee, S.: An examination of selected marketing mix elements and brand equity. J. Acad. Mark. Sci. **28**(2), 195–211 (2000)
59. Zar, J.H.: Biostatistical analysis, 4th edn. Prentice Hall, Upper Saddle River (1999)

Chapter 21
Digital Innovation Hubs: SMEs' Facilitators for Digital Innovation Projects, Marketing Communication Strategies and Business Internationalization

Amalia Georgescu, Mihaela Brîndușa Tudose, and Silvia Avasilcăi

Abstract In the context of the SME's digital transition in the European Union (EU), scientific literature and practice started to give importance to the Digital Innovation Hubs (DIHs), considered to be a true help in testing before investing in digital technologies. The present study assesses an overview of DIHs and the services they provide, in order to identify possible gaps in their service portfolio. Adapting these services to the needs of EU' economy could enable SMEs to improve their digital development and innovation capabilities by accessing better and faster the DIHs' services. We analyzed data from the European Commission's Smart Specialization Platform (S3P) by applying benchmark and multivariate clustering analysis. We focused on understanding what services do DIHs offer mainly for SMEs. We noticed that those related to marketing are not well represented, but have a great potential to be developed in the current digital transformation period. This study is practical, mapping the actual status of DIHs, their expertise and how they are nationally impacting SMEs in various industries and also is adding research to the scientific literature gaps on this subject. By leveraging the results of the study, decision makers can better understand the benefits offered by DIHs services and what countries have untapped opportunities for business and digital ecosystem growth. Research results provide valuable contributions toward the DIHs role in improving SMEs performance in the EU, but also shows where there is space to develop additional DIHs in specific sectors with specific services, especially in Central and Eastern Europe.

A. Georgescu (✉) · M. B. Tudose · S. Avasilcăi
Gheorghe Asachi" Technical University of Iaşi, Bd. D. Mangeron 29, Corp TEX 1, Iaşi, Romania
e-mail: amalia.georgescu@student.tuiasi.ro

M. B. Tudose
e-mail: mihaela-brindusa.tudose@academic.tuiasi.ro

S. Avasilcăi
e-mail: silvia.avasilcai@academic.tuiasi.ro

J. L. Reis et al. (eds.), *Marketing and Smart Technologies*, Smart Innovation, Systems and Technologies 337, https://doi.org/10.1007/978-981-19-9099-1_21

21.1 Introduction

Associative and non-associative business structures interested in fulfilling objectives of common interest represent the most appropriate way to wisely use the resources, the market expertise and to achieve results with multiple impact dimensions in various economic and academic fields [1].

Nowadays, innovation hubs are important actors which have been assigned multiple roles: facilitating and valorizing the results of innovation, supporting the business environment, developing the national economies, facilitating digital transformation transition processes, etc. Within them, Digital Innovation Hubs (DIHs) are distinguished as outputs of the European Union (EU) Digital Europe Program. Their main objective is the development of key areas, such as artificial intelligence, high-performance computing, cybersecurity, advanced digital skills and digitalization of public administration, interoperability and introducing digital software into economic process optimization of small and medium-sized enterprises (SMEs) [2].

Depending on the coverage area of the services they offer, DIHs are the result of a regional, national or European policy initiative. From a legal and organizational point of view, DIHs are associative structures that have the role of facilitating access of businesses or other entities interested in new digital technologies. From the point of view of their main goals, DIHs are non-profit entities, with a legal personality, created by an organization or a group of organizations which make their services available to SMEs and companies with mid-sized capitalization. DIHs offer specialized services, oriented toward digital transformation, transfer of expertise and know-how, development of key areas but also of advanced digital skills [2].

A DIH can offer one or more services (from a predefined set of 16 types of services according to EU standards) and cover one or more sectors (from a predefined set of 36 sectors). Moreover, the services offered by DIHs are managed on 9 technological readiness levels (TRL). Depending on the capabilities of the founders and partners, the DIHs offer specialized services adapted to the needs of the stakeholders [3]. For example, if a founder/partner has market intelligence expertise, this service can be provided upon request to all interested DIH clients (Fig. 21.1).

Depending on the types of activity and the competences of the DIHs' founders or partners, the services offered can cover one or more sectors of activity. For example, a DIH created by organizations whose interests are in the textile industry (with a TRL of 9, in this case, the DIH/industry has a model/technology that is validated and ready for commercialization) will be able to offer dedicated services for SMEs in this sector of activity.

The strength of a DIH is given by the number and diversity of capabilities of its partners and founders. The more they cover a greater number of activity sectors (and, implicitly, the more they have skills for providing more types of services suitable for more levels of technological readiness), the better the DIH will cover the needs of businesses from a region/country/group of countries [1–3].

The role of DIHs in society and economy has been recently the subject of several scientific studies, according to the scientometric analysis in Table 21.1. Among the

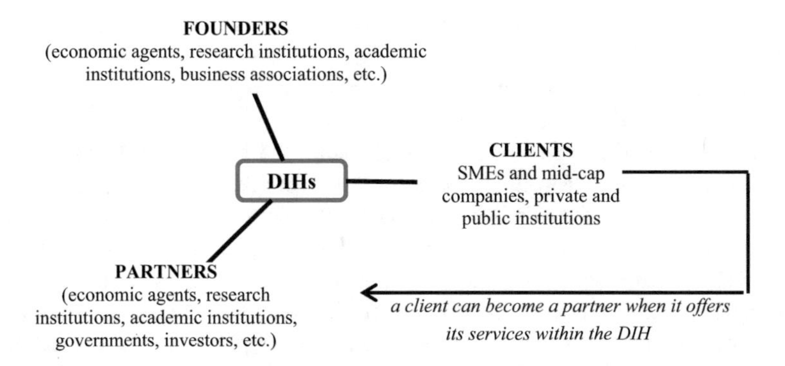

Fig. 21.1 DIHs actors in the business ecosystems

23 selected articles, identified as the most recent, the work by [4] is the most cited paper according to a literature review on the Web of Science. In the research, the authors identified the most important variables of a competition platform (price and quality) based on the notion that "hub" platforms support the creation and capture of value in the overall business ecosystems [4].

Most importantly, the paper from [5] evaluates the extent to which DIHs can be considered knowledge brokers, contributing to the digital transformation of SMEs through specific practices of open innovation. Processing primary data, the authors highlight the fact that DIHs have roles that go beyond the competences of knowledge brokers, being considered true incubators for training skills and capabilities of specialists serving SMEs enrolled in digital transformation [5]. Other relevant studies are those focused on case studies which aim to: (a) evaluate the geography of an emerging creative digital cluster (combining the analysis of spatial data with the analysis of the company's behavior) [6]; (b) to provide evidence regarding the functionality and usefulness of on-line communities, considering innovation incubators as a foundation for civic platforms [7]; and (c) to evaluate the extent to which government funding (for the initial support of the hubs) support results in terms of the economic environment and to as many local or regional actors as possible [8].

DIHs are not subject of research only in recent studies. Some DIHs exist for more than half a century and literature exists, but the new element is represented by the new responsibility assigned to them through the Digital Europe Program, by supporting the digital transformation process, especially for European SMEs. This

Table 21.1 Scientometric analysis

Identified articles in web of science	No
Number of publications (all fields)	129
Number of publications focused on the fields of economy, management and business entrepreneurship	37
Number of publications published between 2018 and July 2022	23

research project at the EU level tries to understand if and why an uneven geographical distribution of DIHs is assumed to exist. Moreover, they seem to have different sizes and different economic impacts, they have various service portfolios and might not support SMEs in all sectors, in a balanced way.

As an identified benchmark research problem, the present study has two objectives. First, it aims to map the DIHs in the EU from the following points of view: density (assessed by the number of DIHs per member state), age (assessed by the number of years of operation), sectors served, services portfolio and level of technological maturity. Secondly, the study aims to evaluate the extent to which DIHs (through specific services, such as commercial infrastructure, ecosystem building and networking, market intelligence, voice of the customer, product consortia) contribute to the integration of SMEs in the EU's market development, by facilitating access to digital innovation projects, business internationalization and the optimization of marketing communication strategies.

Through its structure and content, the present study facilitates the understanding (scientific and practical) of the important role of DIHs in today's society. The results of the study might be useful both for DIHs, which are looking for opportunities to develop their service portfolio and for SMEs, which are looking for opportunities for appropriate development in an environment driven by digital transformation initiatives.

To achieve this objective, three benchmarks are taken into account: (1) the maturity level of DIHs (assessed by the number of years of activity—assuming that DIHs with more years of operation can offer more alternatives for developing businesses); (2) the services provided by DIHs (with a focus on services that facilitate the incorporation of digital market strategies into overall business strategies); and (3) the sectors of activity in which they operate (to identify the sectors that need most support).

21.2 Literature Review on DIHs' Role in Business Development

DIHs are entities that provide support for the digital transition of SMEs [9] and for the digital development of the European economy [10]. From an organizational point of view, DIHs are associative structures [11], with the role of knowledge brokers [5] and with the objective of increasing the competitiveness of its clients and partners [12, 13]. Through the services they offer, DIHs respond to specific needs identified at the regional/national or global level and adopt business models oriented toward creating value for relevant stakeholders [11].

Studies regarding the role of DIHs address issues related to facilitating production processes in the digital era [14], offering new products and services [15], ensuring the development of rural areas [16] and developing hub networks to support SMEs in Europe [17], among others.

The current literature as identified for instance analyzes DIHs from the perspective of entities that ensure the sharing of knowledge and the transfer of technical skills that facilitate production processes (including creative arts) [14]. The authors believe that the cultural side of creative processes provides support for social innovation, thus contributing to the satisfaction increase of both market and social needs. On the other side, [15] are concerned with the development of new products and services to the real estate market. Therefore, innovation centers must adopt a holistic approach, creating opportunities to collaborate and innovate in the entire ecosystem represented by the industry, from product or service ideation to the overall client experience. In terms of innovation and digital transition, [16] focus on identifying the most important opportunities and challenges associated with DIHs in Europe, especially in rural areas. They consider information technology and communications (IT&C) as the most appropriate tool that can contribute to the improvement of the rural business environment (also facilitating digital transition).

The analysis of a network of twelve DIHs and research organizations, focusing on cross-border cooperation, was performed by [17]. These DIHs generate benefits for participants (such as funding/co-funding opportunities, providing access to knowledge and equipment, information on foreign markets, transferability assessment, etc.). The authors, concerned with creating competitive advantages for SMEs, showed that SMEs with a lower level of digitization are more willing to work in a collaborative system and hence, with DIHs. In addition, previous research has shown [18] that collaborative work within associative and non-associative business structures are an opportunity for SMEs development in the digital age. DIHs, as associative structures, share knowledge, gain insights and collaborate for new business development opportunities and can therefore provide access to new markets.

21.3 Context, Data and Methods

To build a better understanding of DIHs in the economic system of the EU, the first research step was the construction of a database. For this task, the Smart Specialization Platform (S3P) provided a rich dataset [2, 3]. The purpose built database is represented by the information related to the 625 DIHs registered in the 27 states of the EU for which the following data is available: the category they represent (fully operational, in preparation or potential DIHs from H2020); the geographical coverage it provides (global, international, European, national or regional); the funds accessed for financing and the source of these funds (European, national, regional, private financing or from members); the average annual number of clients and turnover volume; the market they serve, specifying the activity sectors and the technological readiness level (TRL)—36 sectors and 9 levels of technological maturity are defined in the data; and the services offered according to the defined object of activity (a DIH offers one or more services from 16 predefined categories). The database created by the authors refers to the DIHs registered in the platform on the August 1, 2022.

Examples of DIHs in the EU include

1. The Pannonia DIH (Croatia) was registered in 2021 as is listed in the fully operational category. This DIH has an annual turnover of less than EUR 0.25 million and offers services, at a regional level, for more than 50 clients. The funding sources used by this DIH are represented by funds attracted through projects, to which is also added the financial support from the European, national and regional level. This DIH offers 6 of the 16 possible types of services (awareness creation; collaborative researches; ecosystem building, scouting, brokerage, networking; education and skills development; mentoring; other) and serves 5 of the 36 sectors (agriculture and food; education; manufacture of electrical and optical equipment; manufacture of machinery and equipment; public administration). In terms of technological maturity, this DIH covers TRL1–TRL5 (out of 9 levels); this means that its partners can provide technologies validated either by investigation (TRL4) or in a relevant environment (TRL5). Therefore, this DIH will not be able to provide services for an SME that plans, for instance, to prototype a technology (meaning, one of the services related to levels TRL6–TRL9).
2. The Trakia DIH (Romania), also in the category of newly established DIHs, has a turnover level and number of clients comparable to the Croatian DIH, but differs because it provides greater sector coverage (covering 16 of the 36 sectors) and offers all possible services (16 services). In terms of technology maturity level, this DIH provides complete and validated systems/models at the end of the potential development scale (from TRL1 to TRL8).
3. The Jožef Stefan Institute (Slovenia) is an established DIHs, incorporate before 1960. This DIH, classified as fully operational, has an annual turnover of more than EUR 5 million. It serves 28 of the 36 sectors, offers 15 of the 16 possible services and covers, through its services, all nine levels of technological maturity.

The second step in the research process was to provide macro-level metrics from the DIHs database, focusing on the total number of DIHs per country (Fig. 21.2) and their age (Fig. 21.3), the degree of sector coverage (Fig. 21.4), the structure of services (Fig. 21.5) and the degree of technological maturity (Fig. 21.6).

According to Fig. 21.2, Spain is the country with the highest number of DIHs. Italy, Germany, France and the Netherlands represent the group of countries that have a number of DIHs between 46 and 73. Regarding the average age of DIHs (Fig. 21.3), Greece, Luxembourg, Belgium, Slovenia, France and Poland are the countries where DIHs have the longest experience (greater than 15 years). In the list of countries with the youngest DIHs are the Czech Republic and Bulgaria (for which DIHs have an average age of 4.6 and 6.3 years).

The sectors' structure for which DIHs provide services is shown in Fig. 21.4. From the perspective of the existing DIHs, the best covered sectors with services offered by more than 200 DIHs) are manufacture of machinery and equipment (S19 covered by 291 DIHs); education (S8—280 DIHs); transport and logistics (S35—274 DIHs); life sciences and health care (S12—248 DIHs); manufacture of electrical and optical equipment (S16—248 DIHs); agriculture and food (S2—232 DIHs); other manufacturing (S29—216 DIHs); energy and utilities (S9—209 DIHs).

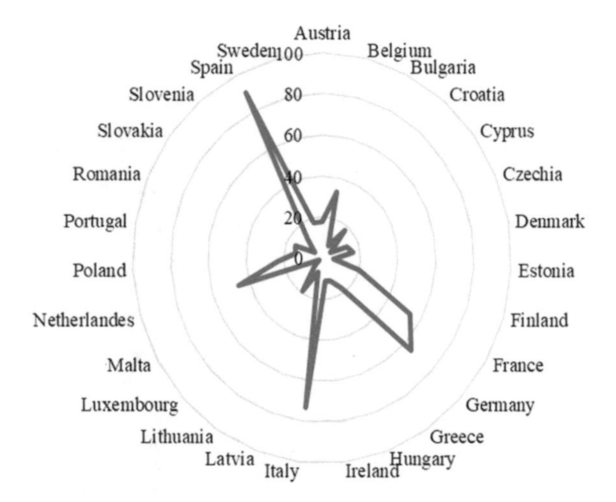

Fig. 21.2 Number of DIHs per country

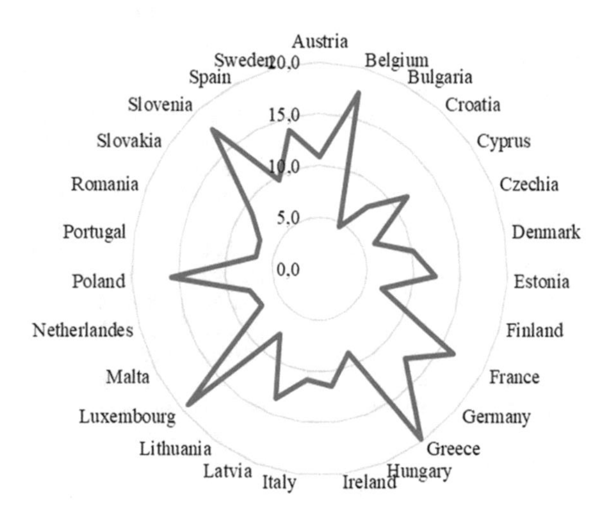

Fig. 21.3 DIHs' average years of existence

On the opposite side, the sectors least served by the services offered by DIHs are defense and security (S7—5 DIHs); aeronautics and space (S1—6 DIHs); professional, scientific and technical activities (S30—12 DIHs); telecommunications, information and communication (S33—13 DIHs); mobility (incl. automotive) (S28—14 DIHs); culture and creative industries (S6—15 DIHs); consumer goods/products (S5—18 DIHs); environment (S10—22 DIHs); mining and quarrying (S27—50 DIHs).

Figure 21.5 indicates that some services are offered by almost all DIHs, while others are offered by less than a fifth of them. The services provided by most DIHs are ecosystem building, scouting, brokerage, networking (EB offered by 486 DIHs);

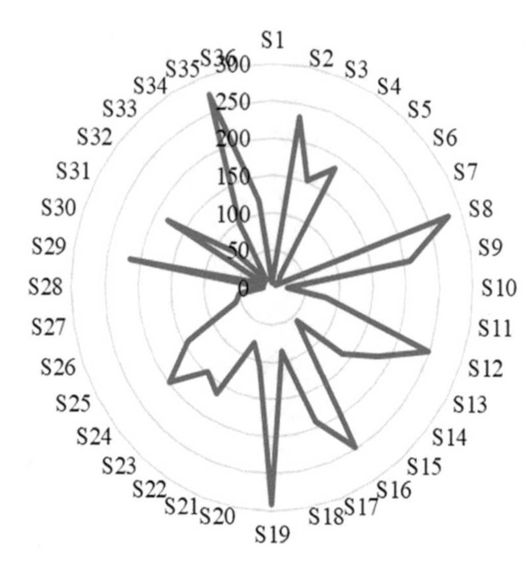

Fig. 21.4 Sector coverage. *Legend* S1—aeronautics and space; S2—agriculture and food; S3-community, social and personal service activities; S4—construction; S5—consumer goods/products; S6—culture and creative industries; S7—defense and security; S8—education; S9—energy and utilities; S10—environment; S11—financial services; S12—life sciences and health care; S13—manufacture of basic metals and fabricated metal products; S14—manufacture of chemicals, chemical products and man-made fiber's; S15—manufacture of coke, refined petroleum products and nuclear fuel; S16—manufacture of electrical and optical equipment; S17—manufacture of food products, beverages and tobacco; S18—manufacture of leather and leather products; S19—manufacture of machinery and equipment; S20—manufacture of other non-metallic mineral products; S21—manufacture of pulp, paper and paper products; publishing and printing; S22—manufacture of rubber and plastic products; S23—manufacture of textiles and textile products; S24—manufacture of transport equipment; S25—manufacture of wood and wood products; S26—maritime and fishery; S27—mining and quarrying; S28—mobility (incl. automotive); S29—other manufacturing; S30—professional, scientific and technical activities; S31—public administration; S32—real estate, renting and business activities; S33—telecommunications, information and communication; S34-tourism (incl. restaurants and hospitality); S35—transport and logistics; S36—eholesale and retail

collaborative researches (CR—468 DIHs); education and skills development (ES—464 DIHs); awareness creation (AC—433 DIHs); concept validation and prototyping (CV—418 DIHs).

On the opposite side, the services that are found slightly in the portfolio of DIHs are market intelligence (MI—215 DIHs); commercial infrastructure (CI—141 DIHs); pre-competitive series production (PP—141 DIHs); voice of the customer, product consortia (VC—122 DIHs); other (O—112 DIHs).

Regarding the technological readiness level (Fig. 21.6), the analysis of EU DIHs indicates an orientation toward the intermediate levels: TRL5–TRL7. Of the 625 DIHs, 440 have expertise on the TRL6 level; 420 on TRL7 level and 410 on TRL5 level. The least covered are the levels TRL1 (212 DIHs), TRL9 (224 DIHs) and TRL2 (275 DIHs).

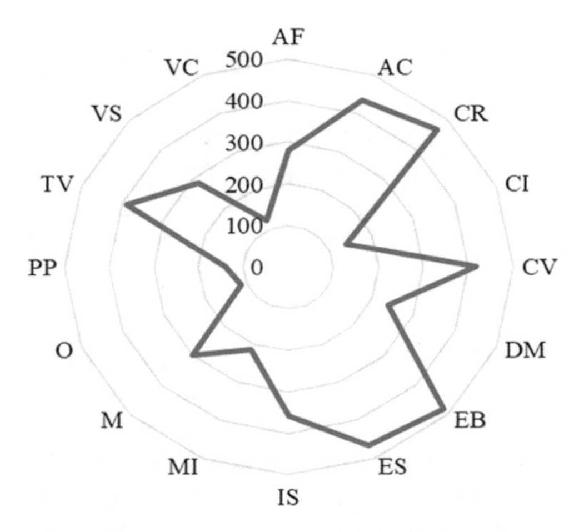

Fig. 21.5 Services offered by DIHs. *Legend* AF—access to funding and investor readiness services; AC—awareness creation; CR—collaborative researches; CI—commercial infrastructure; CV—concept validation and prototyping; DM—digital maturity assessment; EB—ecosystem building, scouting, brokerage, networking; ES—education and skills development; IS—incubator/accelerator support; MI—market intelligence; M—mentoring; O—other; PP—pre-competitive series production; TV—testing and validation; VS—visioning and strategy development for businesses; VC—voice of the customer, product consortia

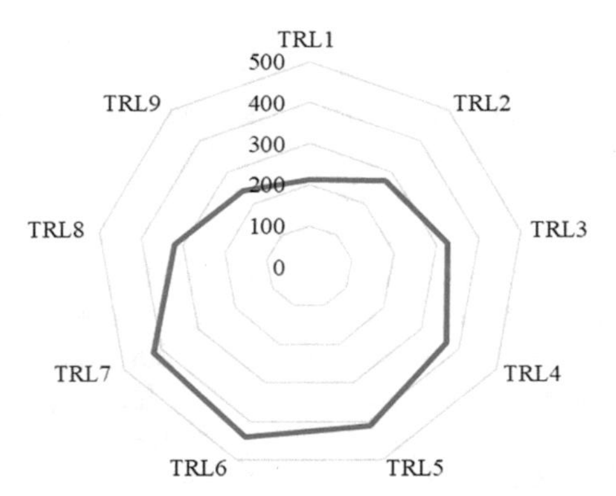

Fig. 21.6 DIHs' TRL status. *Legend* TRL1—basic principles observed and reported; TRL2—technology concept and/or application formulated; TRL3—analytical and experimental critical function and/or characteristic proof of concept; TRL4—component and/or breadboard validation in laboratory environment; TRL5—component and/or breadboard validation in relevant environment; TRL6—system/subsystem model or prototype demonstration in a relevant environment; TRL7—system prototype demonstration in an operational environment; TRL8—actual system completed and qualified through test and demonstration; TRL9—actual system proven through successful mission operations

Table 21.2 New emerged database—regarding four services offered by DIHs

Database no	Selection criteria	Number of DIHs
1	Only DIHs that have the CI services in their portfolio	141
2	Only DIHs that have the MI services in their portfolio	215
3	Only DIHs that have the VC services in their portfolio	122
4	Only DIHs that have the EB services in their portfolio	486

This preliminary analysis (the context evaluation and database general evaluation) highlighted the preliminary research problem of this article: the existing DIHs do not ensure all SMEs (regardless the activity and the sectors in which they operate) fair access to digital innovation projects, business internationalization and expertise for marketing communication strategies in the European ecosystem.

In order to answer the research problem identified, attention was directed to the four services that have the role of supporting SMEs in the process of business internationalization and implementation of marketing communication strategies: commercial infrastructure (CI); market intelligence (MI); voice of the customer, product consortia (VB); ecosystem building, scouting, brokerage, networking (EB). A reorganization of the existing database was carried out, and four separate databases were created (Table 21.2).

To carry out the analyses, there was used a statistical method known as multivariate clustering analysis. This analysis applies a principle that aims to reduce the sets of large databases to a summary in the form of illustrated typologies, results based on common features. The particularities of the resulting typologies can be interpreted based on the boxplots obtained from the analysis.

A boxplot is based on presenting a variable using six values: minimum, maximum, median, mean, first quartile and third quartile. It should be noted that the average of the multivariate clusters is not represented by the general average of the variable used, the averages in this case being represented by the average of the group. In the present study, DIHs were organized into 4 groups with different typologies.

Using this method, three global analyses were carried out, at the level of all DIHs registered at the EU level. These analyses were doubled by specific analyses at the level of the services selected and relevant for this research.

The first analysis presents aspects of the diversity of the sectors DIHs activity, highlighting where DIHs operate in the same sectors of activity and can form a specific cluster. The second analysis, with a more exploratory role, tries to identify a series of common particularities based on the variables inserted in the analysis. All variables used for the second analysis were standardized. According to the collected data, a mapping of the DIHs was carried out by the authors as a benchmark of the four selected services, on one hand, and four exploratory variables, on the other hand, the number of DIHs, the age of the DIHs, the sectors on which they cover the TRL.

The ultimate goal of these analyzes was to identify useful information such as countries (in the EU) where the selected services are offered by more or less DIHs;

the activity sectors covered by the DIHs providing the selected services; the TRL associated with the 4 selected services.

The intended practical utility for this research can be translated as follows:

- From SMEs' perspective, they will have an actual status of the EU DIHs and an orientation toward the DIHs that best meet their needs (from the perspective of the sectors or the efficiency of omnichannel and marketing communication services offered by DIHs); by viewing the maps based on the diversity criteria, SMEs can choose (from the multitude of possible choices) the entities that best suit their needs;
- From DIHs' perspective, they will know the potential of other DIHs and can adopt strategies to diversify/restrict services in order to better adapt to the market, innovation projects and new digital marketing strategies for entering or serving new markets.

The global database, corresponding to the 625 DIHs registered in the EU, was used to create the maps. Excel, XLSTAT, ArcGIS Pro and Philcarto programs were used for the database creation, analysis and graphic representation.

21.4 Results and Discussions

21.4.1 Global Perspective on European Union DIHs

At the European Union level, there are 625 Digital Innovation Hubs, most of which are located in Spain (90), Italy (73) and Germany (65). The three states together own 36.5% of the digital innovation hubs present at European level.

At the opposite pole, the states with the fewest digital innovation hubs are: Malta (2), Slovakia (5), Luxembourg (5) and Cyprus (5). Together, these four states have 17 DIHs, less than 3% of the total number of DIHs.

Regarding the years of activity, it can be observed that the oldest DIHs in Europe are those in Greece, with an average of approximately 20 years, 9 years above the average of 11 years registered at the level of the European Union. In general, considering the situation at EU level, DIHs represent a fairly recent field, most entities being developed in the last 10 years (10% of DIHs were registered before the year 2000; 34% were established in the period 2000–2015 and 44% in the period 2016–2021; 12% did not specify the year of establishment).

A global perspective of DIHs in the EU is presented in Figs. 21.7 and 21.8.

Following the analysis to identify some common features regarding the diversity of sectors where DIHs operate, four typologies/classes were obtained (Fig. 21.8). In the first typology (class I), there were included the countries in which DIHs have the most balanced distribution at the level of sectors. This selection indicated two countries: France and Germany (which have, on average, 10 DIHs per sector). For the second typology, there were selected the DIHs that have the largest number of

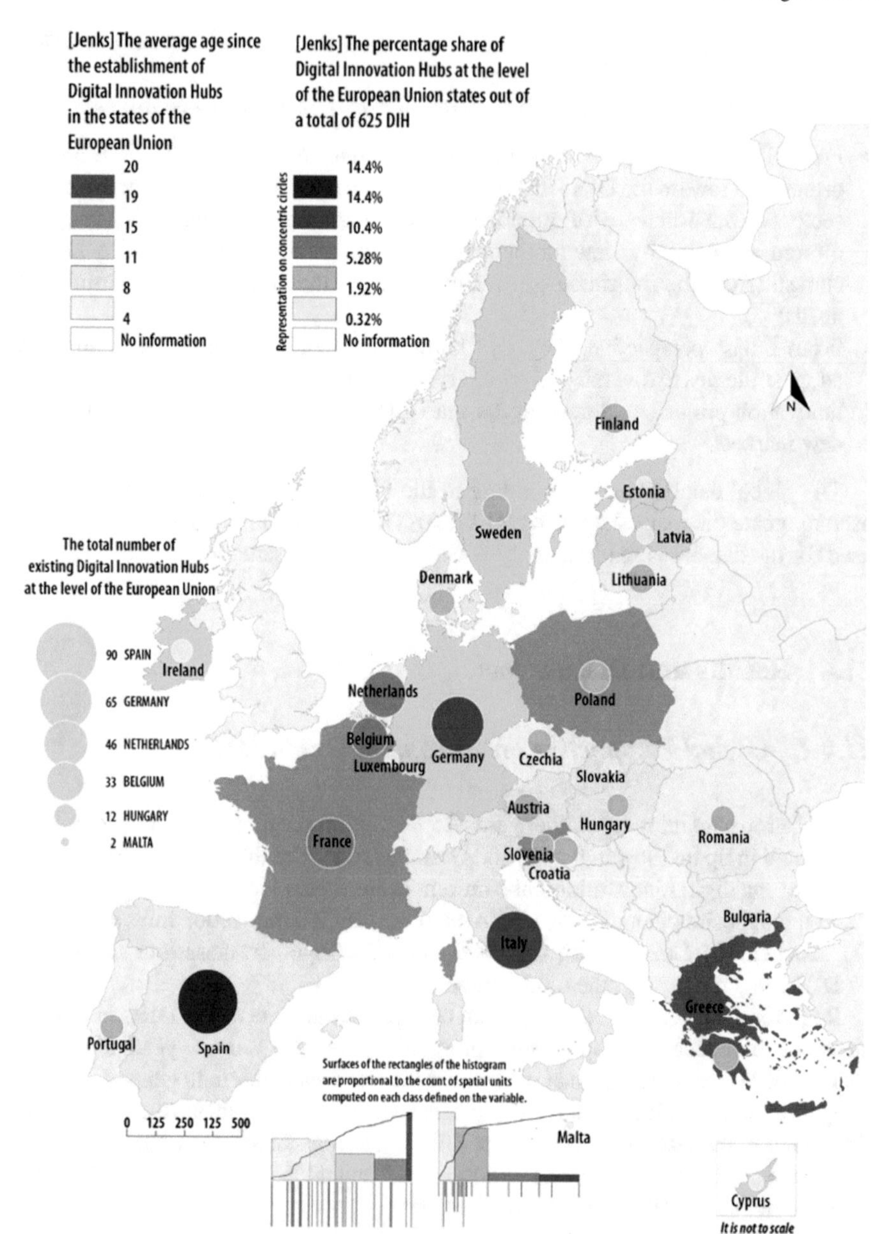

Fig. 21.7 DIHs average age and percentage share in UE

hubs divided by sector. This typology included Spain (where the S17 sector is served by the largest number of DIHs—48; no DIH operates on the S7 sector) and Italy (where in S12 operate 39 DIHs and in S35 operate 38 DIHs; sectors S5, S6 and S7 are each served by one DIH). At the level of these two states, the large number of DIHs and the diversity of distribution by sector, place them as outliers in the analysis carried out.

(a) DIHs types according to the 4 typologies identified by the authors

Fig. 21.8 DIHs mapping according to the sectors covered in EU

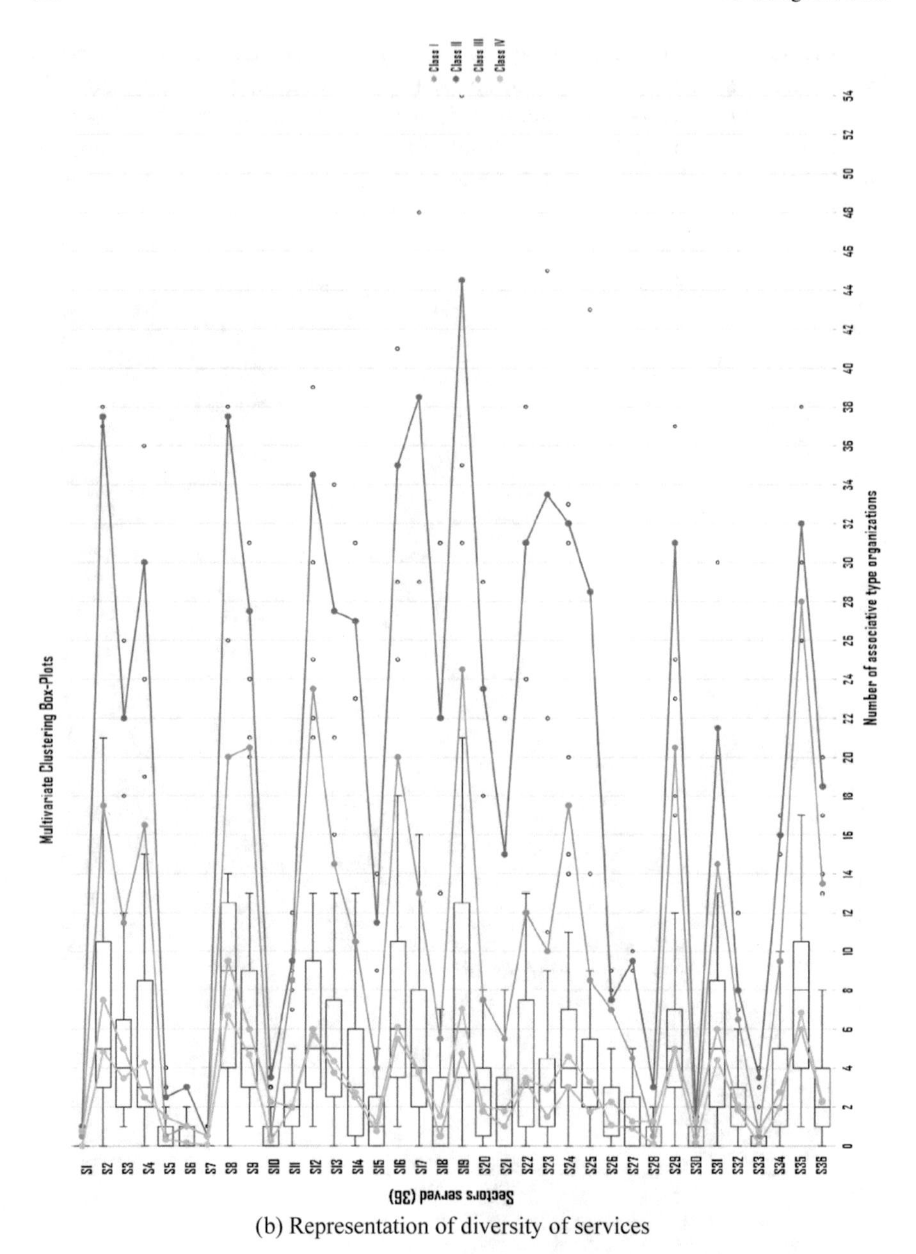

(b) Representation of diversity of services

Fig. 21.8 (continued)

Several attempts to identify new typologies resulted into new grouping of states (not included in previous typologies) as follows: class III included the states whose average number of DIHs per sector is predominantly below the median value at the sector level: Poland, Slovenia, Greece and Cyprus; class IV included the states whose average number of DIHs per sector is predominantly above the median value at the sector level; most of the states answered this criterion. The last two classes represent the most common typologies within the European Union.

The representation in section (b) of Fig. 21.8 provides a detailed picture of the four typologies identified. Class II (represented by the red axis) stands out for the fact that it exceeds the other typologies, grouping the states with the largest number of DIHs operating at the level of a sector (Spain and Italy). Diversity in these two states is so high that they have an outlier character in the analysis, which is easily linked to a strong concentration of digital innovation hubs in the two states. Class I (represented by the blue axis) is positioned between Class II and Class III, grouping states with a more balanced distribution of DIHs by sector. Classes III and IV (represented by the green and yellow axes) include countries with a smaller number of IHLs, but where the sectoral distribution of DIHs is more balanced. Class III includes the states where the best covered sectors of the national economies are served by a maximum of 10 DIHs. Class IV includes the states where the best covered sectors of the national economies are served by a maximum of 7 DIHs. This last class also includes states where some sectors remain not covered by DIHs.

The last analysis carried out, based on the multivariate cluster method, has the role of presenting the four types of services specifically targeted by this research, with some exploratory variables also added: the average age of DIHs, the total number of DIHs and a composite index showing the level of maturity of the services offered by all DIHs present at the European level (Fig. 21.9).

The four realized typologies have the following particularities:

Class I is represented by the typology with most of the variable values located below the median. At the same time, this resulting typology includes the least developed states in terms of DIHs that offer these 4 analyzed services: CI, EB, MI and VC. Most states in this typology are located in Eastern Europe.

The second class is a typology of positive outlier type, according to all the variables used in the analysis. Spain stands out here, the country with the most DIHs that offer the most varied range of these four services in the analyzed sectors. Although Spain has the largest number of DIHs, their average age is 9.6 years (below the EU average). Moreover, the median age of DIHs in Spain is 6 (only 40% of DIHs are more than 6 years old).

The third classification included the states with high values regarding the variables used, many of these values being outliers. This class also includes states with a long tradition in digital innovation (Germany, France and Italy—each having 7 DIHs older than 30 years). Most of the states in this class are located in Western Europe.

The last class included the states with values above the median of the boxplots in the analysis. This class includes the states where most DIHs have the four services in their portfolio: CI, EB, MI and VC. The average age of these DIHs is higher than the EU median (10.8 years). Through the services offered, by these DIHs, most of the sectors of the national economies are covered (this class having values above the average and the median at the EU level). The services offered by these DIHs mostly cover the nine levels of technological maturity. In this class were included several states from Eastern Europe, some Central European states, Scandinavia, Greece and the Netherlands.

Analyzing the four services, that can help SMEs to innovate faster in the digital area, to better communicate in their marketing strategies and internationalize their businesses, we have noticed some differentiators explained in the following section.

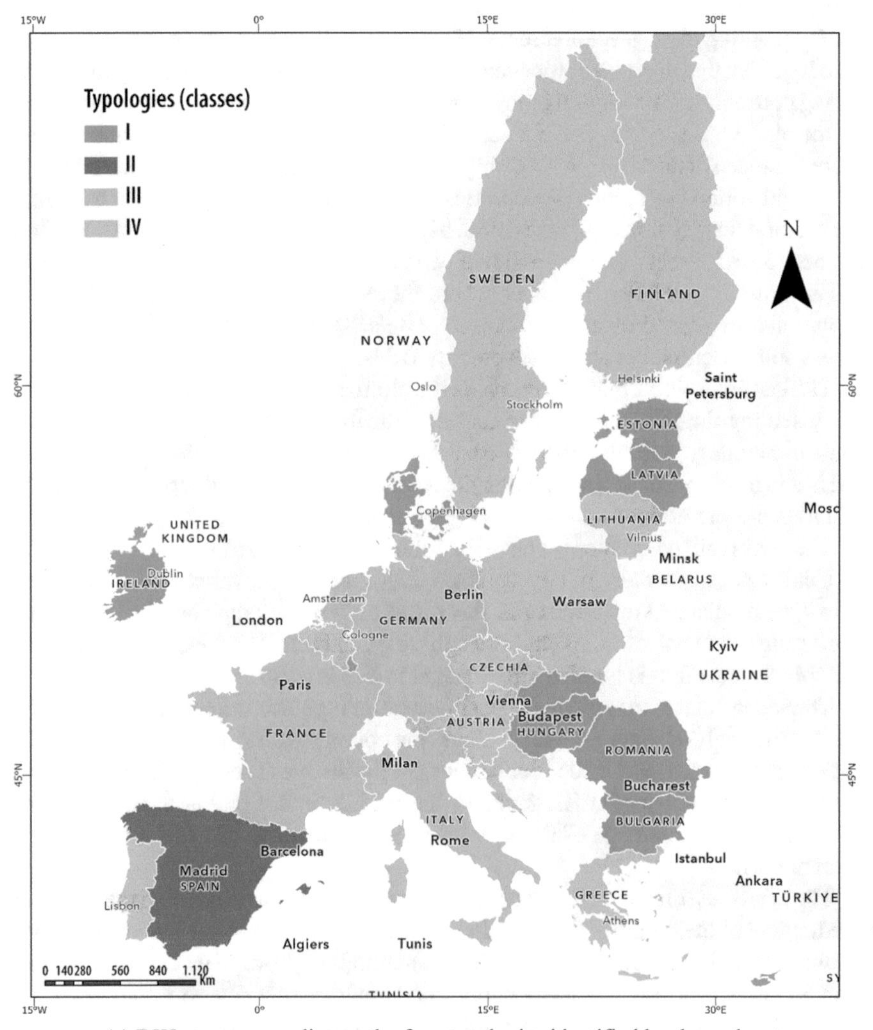

(a) DIHs types according to the four typologies identified by the authors

Fig. 21.9 DIHs mapping according to the 4 services (CI, EB, MI and VC) covered in EU

(b) Representation of specificity of the analyzed services

Fig. 21.9 (continued)

21.4.2 *Commercial Infrastructure—DIHs Support Services for SMEs*

The components of a commercial infrastructure are transport networks (air, land and sea), communications and power generation, systems logistic for institutions responsible with research and education, ensuring the functionality of markets and legal standardization [19].

The performance of a commercial infrastructure also depends on the "natural" features specific to each country. The following are considered: the availability and accessibility of natural resources—which facilitate exploitation, transport and commercialization; climatic characteristics; countries that have favorable weather for both commercial transport and trade are favored; geoeconomic particularities; some countries have access to water transport, others are integrated into road transport networks, etc.; the efficiency of the production factor markets (including the labor market)—these markets can present geographic concentration, degree of openness/freedom and specialization; cultural and social norms—which promote positive attitudes toward work, education, trade and legal institutions [19].

Also, trade infrastructure (along with other specific determinants) has a very important role in innovation-oriented countries with a high level of competitiveness. Therefore, DIHs that offer services associated with commercial infrastructure can be

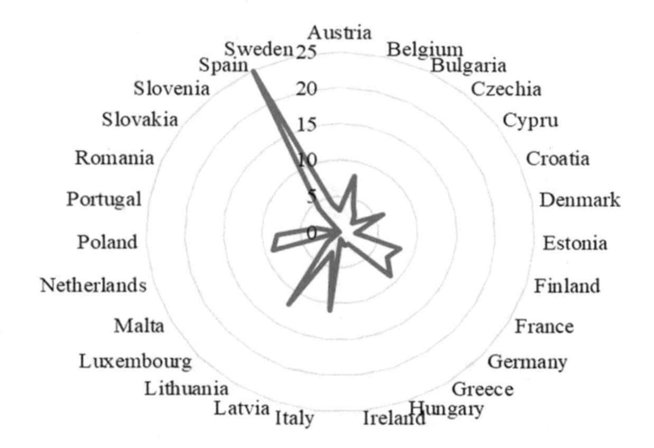

Fig. 21.10 DIHs offering services on commercial infrastructure (CI)

of real interest to SMEs looking for opportunities to place their products and services on new markets [20].

According to our analyses (Fig. 21.10), only 22.56% of the total number of DIHs have this service in their portfolio. Spain is the country with the biggest number of DIHs offering specialized services and know-how in the commercial infrastructure area (25 Spanish DIHs). Then, there are Latvia (12 DIHs) and Italy (11 DIHs). The countries that do not offer this service are Slovakia, Romania and Malta.

21.4.3 Market Intelligence—DIHs Support Services for SMEs

Market intelligence is a very important benchmark for planning and implementing market-oriented strategies. Considering market intelligence as a foundation of marketing, the studies highlighted the most representative practices: distribution (reports, presentations, emails and newsletters), resource centralization (centralized intelligence database), consultative selling (customized reports and presentations for each user group), empathic learning (ethnographic stories, videos and personas), experiential learning (consumer immersion, individual market contact) [21].

Providing information on changes in the business environment, the market intelligence represents the premise of implementing the most appropriate on-line and off-line marketing programs and penetrating new markets [22]. The efficiency of market intelligence services depends on several factors: individual, environmental, organizational and extra-organizational. Furthermore, it has been shown that market intelligence strategies depend on the organization's size, strategic approach and organizational resources.

In the case of SMEs—as small medium-sized entities that do not always have sufficient financial and human resources with specific expertise and skills—DIHs

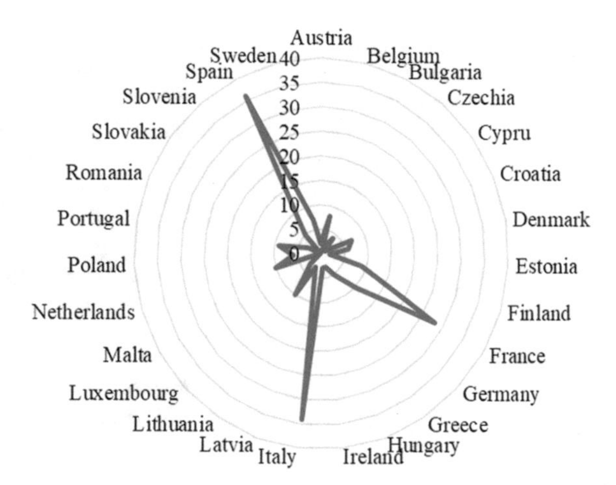

Fig. 21.11 DIHs offering services on market intelligence (MI)

have an important role in providing specific market intelligence services, through appropriate market studies, consulting and specialized support in this field.

According to Fig. 21.11, only 34.40% of the 625 DIHs at EU level have the competence to provide these services. Spain (33 DIHs), Italy (33 DIHs) and France (28 DIHs) have the most DIHs that can offer marketing intelligence services to SMEs.

21.4.4 Voice of the Customer, Product Consortia—DIHs Support Services for SMEs

Voice of the customer (VC) is a market research term for the process of collecting, analyzing and implementing customer feedback data according to its needs, wants, expectations and preferences [23]. VC becomes a product-development technique based on the customers' insights and the innovation and production capabilities of the organization [24].

DIHs can provide this type of services that help SMEs to understand and capitalize the voice of customers and improve the market information regarding its products, thus facilitating the adaptation according to the needs and requirements of the market. This helps also to improve the customer relationships and developing new products and services. The analysis of the vocal behaviors (proactive and/or prohibitive) of customers has great impact over the enterprise's marketing approach and innovation or optimization of products or services in any industry [25].

Only 19.52% of the DIHs registered at the EU level offer services that allow SMEs to capitalize on the "voice of the customer" or to create products consortia to better respond to customer needs (Fig. 21.12). Spain (22 DIHs), France (13 DIHs) and Netherlands (12 DIHs) are the countries that are more offering these services.

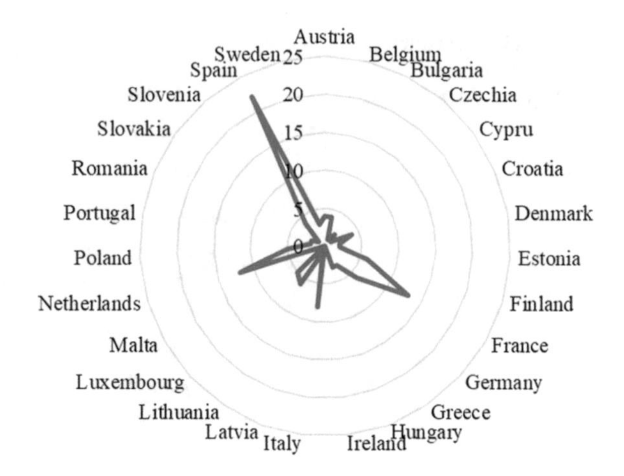

Fig. 21.12 DIHs offering services on voice of the customer, product consortia (VC)

21.4.5 Ecosystem Building, Scouting, Brokerage, Networking—DIHs Support Services for SMEs

Creating and developing digital business ecosystems is another responsibility assigned to DIHs. Suuronen et al. [26] conducted a literature review and emphasized the importance of integrating digital ecosystems into business ecosystems. The authors raise an alarm signal about the efficiency of business ecosystems in the digital age (which lose the necessary capabilities for production development). To be effective, digital business ecosystems must integrate (in addition to the two ecosystems—business and digital) digital platforms through which to facilitate access, interaction, leadership and value creation.

Digital business ecosystems provide a series of benefits, such as new business opportunities, the participation of all members in value co-creation, the promotion of innovation, the creation of competitive advantages, the joint exploitation of resources (including knowledge), risk reduction, increased cost management efficiency, better satisfaction of customer needs [26].

Ecosystem building is also analyzed for the utility of a digital service ecosystem. An ecosystem model can be focused on five aspects: connection, content, computation, context and commerce. But, while desired by ecosystem members, platform-oriented business models are not yet within everyone's reach [27].

The analysis at the EU level reveals that the vast majority of DIHs offer specialized services for building ecosystems, for research, intermediation and networking (Fig. 21.13). Of the 625 registered DIHs, 77.76% offer such services. Spain (75 DIHs), Italy (61 DIHs), France (47 DIHs) and Greece (45 DIHs) are the countries more specialized in offering these types of services.

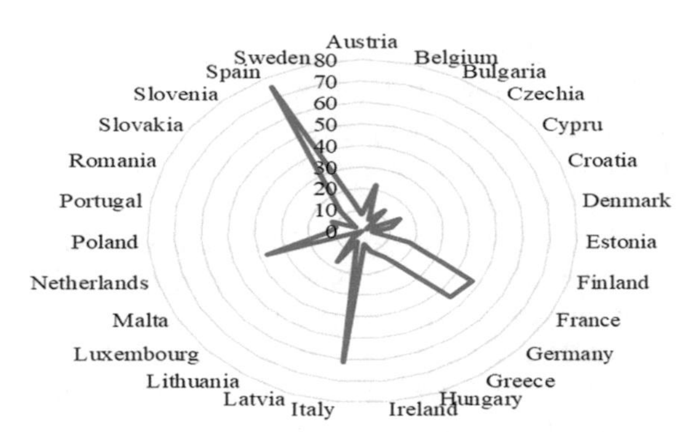

Fig. 21.13 DIHs offering services on ecosystem building, scouting, brokerage, networking (EB)

21.5 Conclusions and Future Research

DIHs' main objective is the development of key areas, such as artificial intelligence, high-performance computing, cyber security, advanced digital skills and digitalization of public administration, interoperability and introducing digital software into economic process optimization of SMEs and other entities. Due to these digital areas, SMEs can innovate more, incorporate new technologies to improve their products and services, develop new markets and increase their annual revenue.

The present research revealed that countries like Spain (90), Italy (73), Germany (65), France (54) and the Netherlands (46) have the majority of the DIHs in European Union (52.5%), offering the great majority of the services and give access to technology according to their level of technological maturity.

Central and Eastern Europe countries, at a relatively beginner stage in the DIHs actively helping SMEs to develop easier, testing before invest and incorporate new technology in the digital transformation process, could learn from these Western Europe DIHs and see how they can adapt locally and regionally examples of good practice. Also, there is big potential for the new DIHs (in preparation or proposed for H2020) in Eastern Europe to access new regions and new partners or clients in order to be immediately operational and able to offer as many services in many sectors as possible. Eastern European SMEs also have the opportunity to learn to work together in different projects with other entities for their personal or collective good of others, and DIHs are the answer to help then faster transition the digital transformation process, necessary especially during and after the pandemic period.

The services provided by most DIHs are EB, CR, ES, AC and CV. On the opposite side, the services that were found rarely in the portfolio of DIHs are MI, CI, PP and VC. These services are offered mainly by DIHs from Spain, France, Italy, Germany and The Netherland. These countries are known to have great experience and knowledge in these marketing services (off-line and on-line marketing strategies integrated with the help of different digital platforms), by helping companies to develop long-term strategies, innovate more and developing their businesses at national and international level.

Our analysis highlighted once more the preliminary research problem of this article: the existing DIHs do not ensure all SMEs (regardless the activity and the sectors in which they operate) fair access to digital innovation projects, business internationalization and expertise for marketing communication strategies in the European ecosystem.

This means a great opportunity for actual and future DIHs to expand their services in this area of digital marketing services, internationalization, market intelligence and offer more expertise in the area of the voice of the consumer and product consortia. Co-working in this area could bring new market opportunities (consultancy services, studies, testing, financing), new cooperations and help SMEs develop faster on-line and off-line worldwide.

From our comparison, only the ecosystem building, scouting, brokerage, networking services was offered by almost 78% of the analyzed DIHs (486 of the 625 DIHs in the EU). This is most likely to be accessed by SMEs because integrating digital ecosystems into business ecosystems, it is a trend and a necessity also for a company to remain relevant on the market nowadays. These ecosystems are also easier to be built because they involve co-creation, joint innovation and there are multiple actors that can bring value, not only for one SME, but for the use of many more. In the same time, the potential of the other services that are not now in the DIHs' focus could be easily integrated in the digital marketing strategies, by testing and developing new platforms in communicating better with potential clients all over the world (e-commerce platforms, social media strategies, AI, virtual reality, 3D presentation platforms, live interactions, chat boxes etc.). Also, these could be easily integrated with the main focuses of DIHs' key areas of interest.

From a literature point of view, our research fills the gaps in researching this subject, showing the importance of DIHs in developing the digital ecosystem and also becoming a new business model, focused to support business development and encourage innovation of new products, services or technologies.

The future research has into consideration the quantitative and qualitative analyses in understanding better why DIHs (from Central and Eastern Europe) do not focus on these four services related to marketing strategies and business internationalization. Also, we can interview SMEs from these regions to understand if these services are of importance for their business development or if they want to develop them internally to enhance their strategic advantage, form a commercial, digital marketing strategic point of view.

References

1. Georgescu, A., Peter, M.K., Avasilcăi, S.: Associative and non-associative business structures: a literature review for the identification of business development opportunities for SME in the digital age. In: Reis, J.L., Peter, M.K., Cayolla, R., Bogdanović, Z. (eds.) Marketing and Smart Technologies Smart Innovation, Systems and Technologies, vol. 280, pp. 337–348. Springer, Singapore (2022)

2. European Commission, Digital Europe Programme, https://digital-strategy.ec.europa.eu/en/activities/digital-programme, last accessed 2022/08/15
3. European Commission, Smart Specialisation Platform (S3P), https://s3platform.jrc.ec.europa.eu/digital-innovation-hubs-tool, last accessed 2022/08/15
4. Rietveld, J., Schilling, M.A.: Platform competition: a systematic and interdisciplinary review of the literature. J. Manag. **47**(6), 1528–1563 (2021)
5. Crupi, A., Del Sarto, N., Di Minin, A., Gregori, G.L., Lepore, D., Marinelli, L., Spigarelli, F.: The digital transformation of SMEs—a new knowledge broker called the digital innovation hub. J. Knowl. Manag. **24**(6), 1263–1288 (2020)
6. Evans, G.L.: Emergence of a digital cluster in east London: birth of a new hybrid firm. Compet. Rev. **29**(3), 253–266 (2019)
7. Maciuliene, M., Skarzauskiene, A.: Building the capacities of civic tech communities through digital data analytics. J. Innov. Knowl. **5**(4), 244–250 (2020)
8. Campagnolo, G.M., Nguyen, H.T., Williams, R.: The temporal dynamics of technology promises in government and industry partnerships for digital innovation: the case of the Copyright Hub. Technology Analysis & Strategic Management **31**(8), 972–985 (2019)
9. Asplund, F., Macedo, H. D., Sassanelli, C.: Problematizing the service portfolio of digital innovation hubs. In: 22nd Working Conference on Virtual Enterprises (PRO-VE 2021), pp. 433–440. France (2021)
10. Maurer, F.: Business intelligence and innovation: a digital innovation hub as intermediate for service interaction and system innovation for small and medium-sized enterprises. In: Working Conference on Virtual Enterprises, Smart and Sustainable Collaborative Networks 4.0, vol. 629, pp. 449–459 (2021)
11. Dalmarco, G., Teles, V., Uguen, O., Barroso, A.C.: Digital innovation hubs: one business model fits all? In: 22nd IFIP WG 5.5 Working Conference on Virtual Enterprises (PRO-VE 2021), pp. 441–448. France (2021)
12. Queiroz, J., Leitão, P., Pontes, J., Chaves, A., Parra-Domínguez, J., Perez-Pons, M.: A quality innovation strategy for an inter-regional digital innovation hub. Adv. Distrib. Comput. Artif. Intel. J. **9**(4), 31–45 (2020)
13. Georgescu, A., Avasilcăi S., Peter, M.K.: Digital innovation hubs—the present future of collaborative research, business and marketing development opportunities. In: Rocha, Á., Reis, J.L., Peter, M.K., Cayolla, R., Loureiro, S., Bogdanović, Z. (eds.), Marketing and Smart Technologies. Smart Innovation, Systems and Technologies, vol. 205, pp. 363–374. Springer (2021)
14. Di Roma, A., Minenna, V., Scarcelli, A.: Fab labs. New hubs for socialization and innovation. Design J **20**(1), S3152–S3161 (2017)
15. Richner, P., Heer, P., Largo, R., Marchesi, E., Zimmermann, M.: NEST—una plataforma para acelerar la innovación en edificios. Inf. Constr. **69**(548), e222 (2017)
16. Gernego, I., Dyba, M., Onikiienko, S.: Challenges and opportunities for digital innovative hubs development in Europe. Manage Theory Stud Rural Bus Infrastruct Devel **43**(2), 298–306 (2021)
17. Volpe, M., Veledar, O., Chartier, I., Dor, I., Silva, F.R., Trilar, J., Kiraly, C., Gaffuri, G., Hafner-Zimmermann, S.: Experimentation of cross-border digital innovation hubs (DIHs) cooperation and impact on SME services. In: 22nd IFIP WG 5.5 Working Conference on Virtual Enterprises, PRO-VE 2021, 10 p. France (2021)
18. Georgescu, A., Peter M.K., Avasilcăi, S.: Associative and non-associative business structures: a literature review for the identification of business development opportunities for SME in the digital age, In: Reis, J.L., Peter, M.K., Cayolla, R., Bogdanović, Z. (eds.), Marketing and Smart Technologies. Smart Innovation, Systems and Technologies, vol. 280, pp. 337–348. Springer, Singapore (2022)
19. Yen, M.-F., Wu, R., Miranda, M.J.: A general equilibrium model of bilateral trade with strategic public investment in commercial infrastructure. J. Int. Trade Econ. Devel. 1–20 (2019)
20. Martínez-Fierro, S., Biedma-Ferrer, J.M., Ruiz-Navarro, J.: Entrepreneurship and strategies for economic development. Small Bus Econ **47**, 835–851 (2016)

21. Gebhardt, G.F., Farrelly, F., Conduit, J.: Market intelligence dissemination practices. J Market 1–19 (2019)
22. Rahchamani, A., Ashtiani, B.R., Vahedi, M.A.: The impact of marketing intelligence and business intelligence on acquiring competitive advantages. Revista Gestao Tecnologia J. Manage. Technol. **19**(5), 52–70 (2019)
23. Amsler, S: Voice of the customer (VOC), TechTarget, https://www.techtarget.com/searchcustomerexperience/definition/voice-of-the-customer-VOC, last accessed 2022/08/15
24. Griffin, A., Hauser, J.: The voice of the customer. Mark. Sci. **12**(1), 1–27 (1993)
25. Chen, G., Li, S.: Effect of employee-customer interaction quality on customers' prohibitive voice behaviors: mediating roles, of customer trust and identification. Front. Psychol. J. **12**, 773354 (2021)
26. Suuronen, S., Ukko, J., Eskola, R., Semken, R.S., Rantanen, H.: A systematic literature review for digital business ecosystems in the manufacturing industry: prerequisites, challenges, and benefits. CIRP J. Manuf. Sci. Technol. **37**, 414–426 (2022)
27. Xu, Y., Ahokangas, P., Turunen, M., Mäntymäki, M., Heikkilä, J.: Platform-based business models: insights from an emerging AI-enabled smart building ecosystem. Electronics **8**(10), 1150 (2019)

Chapter 22
Omnichannel Marketing in Ambato's SMEs

Juan Carlos Suárez Pérez

Abstract Globalization and the pandemic showed the shortcomings that exist in the management of small and medium-sized enterprises (SMEs) in Ambato, especially in marketing issues, because they do not allocate the necessary resources to design and implement strategies that allow them to satisfy the demands of consumers who, thanks to the Internet, access large information bases, and that also provides the option to buy products and services from a wide range of companies, whose mission is to differentiate themselves from competitors through the offer of unique experiences. In this sense, the objective is to determine the importance of the integration of communication and distribution channels in improving the customer experience. The methodology used is mixed, because it allows to describe and explain phenomena based on qualitative and quantitative data, which allow to understand reality in a comprehensive way. The results obtained underpin the value of omnichannel and digitalization of businesses, and the creation of company profiles to determine how they can take advantage of this type of strategy and improve consumer satisfaction levels. Finally, it was possible to know that at present, there is a hyper-connectivity of consumers, and therefore, marketing trends focus on strategies in which traditional and digital communication and distribution channels coexist and thus improve the company-customer relationship.

22.1 Introduction

Technological advances and the pandemic highlighted the weaknesses and empiricism with which small and medium-sized enterprises (SMEs) at the local level manage their administrative processes. This was exacerbated during the pandemic, as many companies could not counteract the effects of the pandemic, since they did not have solid management processes, therefore, they lost positioning in the market, sales

J. C. S. Pérez (✉)
Carrera de Administración de Empresas, Facultad de Ciencias Administrativas Y Económicas, Universidad Indoamérica, Ambato, Ecuador
e-mail: juansuarez@uti.edu.ec

J. L. Reis et al. (eds.), *Marketing and Smart Technologies*, Smart Innovation, Systems and Technologies 337, https://doi.org/10.1007/978-981-19-9099-1_22

were reduced, and there were several business closures, including some emblematic ones that were characterized by their trajectory and prestige.

SMEs are especially considered because of the contribution they make to the national economy; according to the National Institute of Statistics and Censuses (NISC), these represent 99.5% of the business network, and therefore, they are vital to generate wealth and jobs. Tungurahua, according to NISC, is the sixth province of Ecuador with the highest number of SMEs with a total of 39.608, of which 3.538 are microenterprises, 952 are small and 118 are medium-sized. Of these, 86.8% are concentrated in Ambato [1]. Local companies are mostly engaged in agro-industry, with sectors such as short-cycle fruit production, metal bodywork for buses, and manufacturing, especially clothing and everything related to leather products [2].

However, they face a severe crisis, due to, factors such as high competition with lower prices, tax burdens, political and economic instability, high financing rates [1], and other factors that worsened during the pandemic, causing the closure of several businesses. At the national level, before the pandemic there were 882,200 companies, but during 2022 they were reduced to 842,400, that is, there was an average variation of −4.5%. Small companies were the most affected with a variation of −18.5%, followed by medium-sized type A and B, with −13.6%, and finally, micro-companies that presented a variation of −3.2% [3].

The crisis facing SMEs is not new; administrative and management empiricism has been in force since past years, because they do not invest what is necessary in training and innovation, and therefore, there are gaps in the management of the areas and processes of companies. Other problems are the lack of liquidity and the difficulty in accessing sources of financing; they face technological backwardness and lack of innovation, which causes the possibilities of expansion to be minimal, therefore, their capacity for growth, being sustainable and developing is very complex according to the current dynamism of the markets worldwide [4–6].

Regarding marketing, the problems are more evident, only 9.6% of companies see marketing as a strategic area. 32% of entrepreneurs make decisions based on adequate marketing planning, 21% conduct market research and, for the most part, SMEs [7] are focused on production (marketing 1.0) and do not adapt to new marketing trends, which seek to integrate as many channels as possible to give a better experience to consumers, facilitating access to information and products and services without restrictions of space and time [8].

Channel integration is known as omnichannel, which refers to the evolution of multichannel (presence in several channels that work separately), that is, it creates an environment in which channels work together to allow consumers to access digital media to investigate and experience a purchase process easily from any place and without time restrictions. This strategy includes the optimization of the available channels of the companies [9] and thus provide customers with an exclusive experience that will be reflected in higher levels of conversion or sales [10].

Omnichannel is a challenge for SMEs, since consumers are more demanding and request extraordinary experiences during the customer journey. To get involved in this strategy, it is essential to understand how consumers behave in each channel, and in this way, they take advantage of their own resources and tools, the efficiency and effectiveness of the channel is achieved. In that sense, in the context of COVID-19, the fear of contagion considerably changed the behavior of consumers, who do not only demand quality products and services [11], but also seek experiences that make them feel safe during the purchase process [12].

It should be mentioned that, worldwide, consumers are channel and brand agnostics, that is, they do not maintain their preference for only one, which hinders the processes of customer satisfaction and loyalty. Therefore, strategies must be designed focused on the new profile of the online consumer, which is characterized by prioritizing their personal security needs, interact less with people, are more aware of value propositions, and acquire products that offer quality, value, and tranquility. Strategies should be customer-centric, which reduces consumer distrust during potential lockdowns, isolations, and post-COVID-19 [13, 14].

According to the firm PriceWaterhouseCoopers, Internet purchases in Ecuador have increased by 15 times since the pandemic, so air services, non-personal goods such as clothing, cosmetics, electronics, and household products are the most demanded [15]. The pandemic showed that Ecuadorians have an interest in online purchases, if companies offer guarantees of return and/or confidentiality of data, more information about products and services, and attention or assistance during the purchase.

Other figures demonstrate the importance of channel integration, stating that 1 in 3 shoppers chose the online channel for the first time to generate their purchases and reported an 800% growth in website visits and 44% in orders compared to 2019. In the business case [16], the companies that have benefited from the increase in digital consumers are businesses to consumer (B2C) highlighting the food sectors, followed by home, then personal goods, games, and entertainment, that is, e-commerce and everything related is an opportunity for all types of business [17].

This means that the digitization of business is a reality that should not be overlooked, and that companies must adapt to this opportunity, otherwise, they could disappear from the market. To benefit from this, the offer must be professionalized, that is, trained so that a better experience can be provided to users; understand the demand, identify and differentiate the products that users prefer to buy online and which in person; strengthen public–private relations, which benefit electronic commerce; encourage governments to improve the scope of banking, to facilitate access to and trust to digital transactions; and access to technologies.

In this sense, the research aims to answer the following questions: What is the current situation of local SMEs with respect to marketing? How are local companies profiled to integrate communication and distribution channels in the new digital ecosystems? and What is the importance of omni-channeling in the competitiveness of companies?

22.2 Methodology

The research allowed to investigate and specify the most relevant characteristics about the phenomenon under study; it seeks to identify and demonstrate the importance of omnichannel as a current marketing trend in Ambateño SMEs. The approach is mixed: qualitative because it allows to deepen, co-contextualize and investigate the natural, holistic and flexible information on current marketing trends, especially those focused on SMEs, since as an engine of the local economy, they want cutting-edge strategies that satisfy the needs of consumers who are more demanding and request better experiences in the purchasing processes [18].

The qualitative, interpretative, and bibliography approach aimed to establish the process of selection, access, and registration of the documentary sample. A bibliographical review involves the analysis and explanation of all concepts. For this, it was used Boolean search is a way to improve search results with components, such as omnichannel marketing, SMEs, Ecuador, and digital marketing. The articles were synthesized and analyzed using bibliographic and analytical matrices designed in Excel, which made it possible to relate the articles and determine the advantages and disadvantages of integrating the communication and distribution channels, can you see in Table 22.1.

In summary, omnichannel represents an opportunity for companies, however, it is not an easy task to execute this type of strategy due to the empiricism with which most processes in SMEs are managed, in that sense, companies are required to train their staff, or in turn hire professionals who have the skills to integrate the channels available to the company and add those necessary for the execution of this type of strategies.

With regard to the quantitative approach, a descriptive analysis was carried out for the collection, grouping, presentation, and analysis of data in a clear and simple way, on the current situation of SMEs [29].

For the descriptive analysis, 1076 small and medium-sized companies were considered as population, however, 18.50% and 13.60% were discarded, respectively, due to the closure of companies in the context of the pandemic, in that sense, 883 SMEs were worked with; in addition, only companies that are registered in the database of the Chamber of Commerce of Ambato were considered, and a total of 355 companies were obtained. Subsequently, a sampling was performed with a confidence level of 95% and an error of 5%, obtaining a sample of 185 SMEs. The samples were distributed in service companies (68%), commerce (24%), and manufacturing (8%); the years of operation have an average of 12 years, and together, they create 2469 jobs, with an average of 13 employees per company.

In addition, factor analyses were carried out on the organizational aspects of the planning of marketing and the importance of investing in marketing strategies; the objective was to discover clusters of variables that are correlated and thus identify company profiles according to the importance they give to each of the items of the data collection instrument.

Table 22.1 Advantages and disadvantages of omnichannel according to different authors

Advantages	Disadvantages
Integrate multiple channels to keep consumers informed	If the channels are not synchronized properly, it can lead to confusion [19]
It allows interaction with customers through various channels, adapts the offer and affects the consumer experience	It is incompatible with cross-selling, that is, this strategy that usually gives good results is lost [20]
Contributes to the achievement of leads	It requires the implementation of several channels, which represents a high investment, and not all SMEs have the resources [21]
Creates a global communication experience between consumers and brands	There are limitations due to the necessary resources [22]
Companies connect and empathize with customers (empower customers)	The obsolescence of the channels is immediate due to technological advances [23]
It allows you to manage data and know consumer behavior	Companies often confuse the multichannel idea, they have several channels, but are not integrated [24]
There is an incalculable potential of customers who are hyper-connected, and therefore, it is crucial to use the channels where they are	It represents costs that not all companies can cover [25]
It allows the effective coexistence of on and off-line channels, to improve the experience of consumers	It requires the expertise of marketers in omnichannel issues, which in the local environment is uncommon [26]
It helps predict and project sales more accurately and reliably	Demand for prior knowledge of data analysis and other techniques that are little known [27]
It allows direct interaction between companies and consumers	It requires a full-time person to serve the people who contact the company through each of the available channels [28]

Factor analysis was performed in the SPSS statistical software; it began with the correlation matrix, to determine the relevance of the analysis and was integrated by the Bartlett sphericity test that was 0.000, determining that the variables are correlated; the Kaiser–Meyer–Olkin index (KMO) obtained was 0.939 and 0.758, values that are between 0.5 and 1; therefore, the sample is adequate, and the analysis is appropriate. The rotation of the initial factors that facilitates the interpretation of the results, for this the Varimax rotation method is used which allowed to group the variables and obtain two profiles of SMEs [30].

The data collection instrument applied was an adaptation of the marketing audit tool proposed by McDonald and Leppard [31], which consists of several aspects that measure the effectiveness of marketing management in companies [32], however, 14 items referring to key factors of the organization were taken into consideration related to the planning of the marketing area, because others are not adaptable to the reality of SMEs. Aspects related to traditional marketing versus the digital part, and the importance of investing in digital assets, were increased to the collection instrument.

The survey items were structured according to the McDonald and Leppard instrument, and refer to: (a) interest in planning; (b) planning for decision making; (c) allocate sufficient resources; (d) information and data; (e) set aside enough time; (f) detect opportunities in the market; (g) members contribute to planning; (h) sales staff supports planning; (i) priority aspect; (j) preparing to learn; k) frequent research is used; (l) periodic evaluation; (m) control mechanism; and (n) meeting objectives. The scale ranged from 1 to 5, with 1 being "if you strongly disagree with the statement" and 5 being "if you strongly agree with the statement".

On the other hand, the items added are divided into two parts: first, the importance of the marketing mix variables in the planning of content for the marketing strategy; and second, the importance of investing in the following variables: (a) new marketing channels; (b) banners, fliers, etc.; (c) paid media or channels; (d) digital assets; (e) online advertising; (f) off and online strategies; (g) e-mail marketing; and (h) marketing departments. The scale ranges from 1 to 5, with 1 being "very important" and 5 being "not important at all".

Finally, questions were asked about the type of strategy used by the SMEs, whether they have a marketing department or an external manager, whether the campaigns are carried out based on strategic planning, what digital assets they currently have, among other questions that provide information about the current situation of the companies.

22.3 Results

According to the diagnosis made to SMEs in the city of Ambato, the following results are obtained, so that the items referring to the importance of the variables of the marketing mix, it was possible to know that the determining variable is the price (76.8%), followed by communication (63.8%), the product (56.2%), and finally, distribution (49.2%). It is important to mention that those who responded in each "very important" item were taken into consideration. In addition, the average was 4.5, which means that entrepreneurs are aware of the importance of variables.

Additionally, SMEs were enquired about the type of marketing strategies they used, and it was found that 42% use traditional and digital strategies, the same percentage apply digital strategies, 10% have opted for traditional strategies, and 6% have not applied any type of strategy. The variation of type of strategies is given by the lack of knowledge that exists about the benefits of each of them; also, many entrepreneurs focus on the digital part and have neglected other traditional alternatives that according to the target audience can be of great benefit, for example, the PoP material.

Likewise, companies are investigated if they have an internal marketing team, or if they are looking for external advisors, it was found that in 56% of SMEs they have one internal person in charge of marketing strategies of the company, while 44% hire external advice, however, at the city level there are no companies that really provide marketing services—most of them are graphic design agencies, which are only responsible for the visual part and do not focus on the strategic part, therefore, the results are not usually as expected and this is what has caused entrepreneurs not to trust 100% in the implementation of marketing strategies.

In relation to the digital assets that companies have, it could be evidenced that at the local level, there is a predominance of social networks: 168 companies have a presence in networks, 88 companies have web pages, 68 have databases, 21 have some type of application, and 7 companies have assets related to the subject of artificial intelligence. Obviously, social media marketing is the most used by local SMEs, and Facebook is the most used network, followed by Instagram and WhatsApp Business, and with less presence YouTube and TikTok.

The average investment in marketing strategies in general is $703, but as mentioned in the previous paragraph, it is mostly done solely on social media. It was evidenced that companies use informative content and have not managed to create communities in their networks, because most of them show on average less than 1000 followers in their accounts, therefore, it can be understood that they do not take advantage of the tools of each of the networks, and this makes it difficult to implement omnichannel marketing.

With respect to the factor analysis on the organizational aspects within the marketing planning in SMEs, a value of 0.939 higher than the level of significance 0.5 was determined in the KMO and Bartlett test, that is, there is a correlation between variables and the analysis could be carried out, for which a "rotated component matrix" was elaborated that allowed to identify two profiles of SMEs. In Table 22.2, the first profile shows the correlation between the variables: meeting objectives, control mechanism, frequent research is used, set aside enough time, allocate sufficient resources, members contribute to planning, information and data, sales staff supports planning, and priority aspect; while, second profile shows a correlation between the variables, set aside enough time, allocate sufficient resources, members contribute to planning, preparing to learn, interest in planning, planning for decision making, detect opportunities in the market, information and data, sales staff supports planning, and priority aspect (Fig. 22.1).

The analysis allowed to correlate the 14 variables of the applied audit tool; in this case, Table 22.2 shows that profile 1 is made up of SMEs that strongly agree with periodic evaluation, meeting objectives, control mechanism, frequent research is used, etc. On the other hand, Profile 2 is made up of SMEs that strongly agree with: preparing to learn, interest in planning, planning for decision making, followed by other variables with less impact. In both profiles, there is a positive correlation between the variables, and they differ in certain cases, for example, in Profile 1, the determining variable is periodic evaluation, while in Profile 2, the most important variable is interest in planning. In summary, factor analysis allows us to profile the companies based on the importance they give to each of the variables.

With respect to the average in both profiles was 3.16 and 3.69, respectively, it is the reflection of the lack of knowledge or empiricism with which marketing strategies are worked, since in some components such as the issue of resources, contribution of the members of the company, among others, a considerable percentage answered that do not know if they agree or disagree. However, they recognize the importance

Table 22.2 Rotated component matrix to (organizational aspects)

Variables studied	Profile 1	Profile 2
Periodic evaluation	0.913	
Meeting objectives	0.872	
Control mechanism	0.868	
Frequent research is used	0.851	
Set aside enough time	0.699	0.524
Allocate sufficient resources	0.618	0.610
Members contribute to planning	0.539	0.465
Preparing to learn		0.873
Interest in planning		0.870
Planning for decision making		0.797
Detect opportunities in the market		0.771
Information and data	0.602	0.641
Sales staff supports planning	0.424	0.632
Priority aspect	0.515	0.621

Extraction method: principal component analysis
Rotation method: Varimax with Kaiser normalization
a. The rotation has converged in 3 iterations

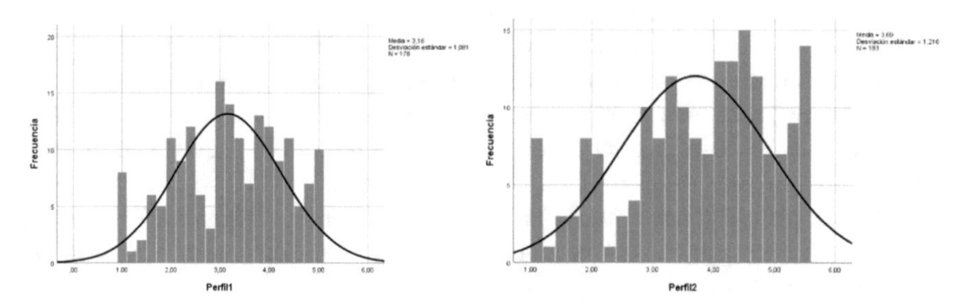

Fig. 22.1 Histogram profile 1 and 2 of SMEs

of all organizational aspects within marketing planning, but as has been insisted, it is not applied technically, therefore, it is complex to talk about marketing in local SMEs.

Additionally, six items were added to the collection instrument, related to the importance of investing in traditional and digital strategies, these refer to online advertising, on and off-line strategies, digital assets (web pages, databases, applications, etc.), payment and communication channels, and advertising material such as fliers or banners. It was important to add these items, as they are the most used at the local level, and the objective is to know the current situation, to have a base that allows companies to integrate their current channels and assets, with new alternatives that are at the forefront and allow companies to be competitive.

In this sense, a factor analysis was carried out on these components, and the results obtained in the KMO and Bartlett test were 0.758 higher than the significance level 0.5, therefore, there is a correlation between variables; in the first profile, the correlation exists between the variables online advertising, on and off-line strategies, digital assets, new marketing channels; while in the second profile, the correlation is between the variables new marketing channels, channels or means of payment, banners, flier, among others. And the analysis could be continued, the "rotated component matrix" was applied and two SME profiles were formed (Fig. 22.2) (Table 22.3).

Based on the factor analysis, it was possible to identify two profiles of companies according to the level of importance they give to the use of marketing strategies, the average was 4.31 and 3.61, respectively, this allows to identify that in profile one there is security to affirm the importance of each of the components, and the profile was constituted by entrepreneurs who consider online advertising decisive, online and off-line strategies, digital assets, and new marketing channels. On the other hand, profile two, there is evidence of a certain degree of doubts in the selection of the components, in this case, the profile is constituted by those entrepreneurs who consider the new marketing channels, the means of payment and the banners.

Both profiles show an acceptable percentage of the acceptance of the importance of marketing for companies today, however, sometimes unawareness and lack of resources could be the cause why no type of planning or strategies were applied. It

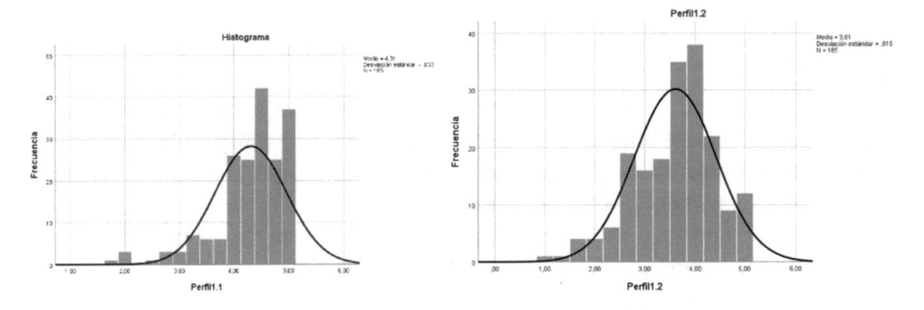

Fig. 22.2 Histogram profile 1.1 and 1.2 of SMEs

Table 22.3 Rotated component matrix (importance of investing in digital strategies)

Variables studied	Profile 1	Profile 2
On-line advertising	0.826	
On and off-line strategies	0.812	
Digital assets	0.733	
New marketing channels	0.547	0.404
Channels or means of payment		0.867
Banners, flier, among others		0.734

Extraction method: principal component analysis
Rotation method: varimax with Kaiser normalization
a. The rotation has converged in 3 iterations

was evidenced that certain strategies have been executed that are mostly focused on social networks, which should only be part of the strategy, since the integration of the largest number of channels in an appropriate way allows to improve the experience of consumers and, therefore, makes the company stand out from its competitors.

22.4 Discussion

The findings obtained in this study show that the current situation of local SMEs with respect to marketing presents several drawbacks, especially due to the empiricism with which most of these types of companies are managed.

In other research, this is ratified, as they mention that 80% of SMEs do not reach five years and 90% do not exceed ten years, the main cause, from the point of view of entrepreneurs are high financing costs, excessive government control, and high tax rates. On the other hand, several analysts state that the reason is the lack of administrative management capacity [33].

Regarding the marketing area, in a previous study carried out on 127 SMEs, it was determined that less than 50% of companies have a real interest in carrying out marketing planning, and most of them are still focused on marketing 1.0 trends, that is, they focus all their attention and resources on the product, therefore, the other variables of the mix marketing have not been efficiently integrated, which is reflected in the loss of competitiveness and other problems that were exacerbated during the pandemic [8].

It is important to note that the application of omnichannel marketing strategies represents an opportunity for companies to generate unique experiences for consumers and profitable relationships for business; it also allows a direct relationship with customers [28]; it helps to make predictions of sale or demand more accurately [27]; and due to hyper-connectivity of consumers, it increases the ease of viralization of information and brand positioning.

However, there are still difficulties in creating synergy between the channels available to companies, therefore, it is required that managers or managers of the marketing area are trained and incorporate in their businesses various theories that transcend international and cross-cultural perspectives, in order to explore and implement new trends related to omnichannel environments [34], without neglecting the traditional part, since it is necessary that distribution and communication channels coexist and be integrated to improve consumer experiences [26].

It is important that companies take into account that the application of this type of strategy has had great results in companies such as Starbucks, Walgreens, Bank of America, among others, however, these are multinationals that have sufficient economic resources to be able to invest in omnichannel marketing strategies but the local reality is different because Ambateño SMEs may not have available the necessary resources to implement them, therefore, a more exhaustive analysis of the cost–benefit ratio must be made, to determine if it is practical or not to invest in this type of marketing [25].

The implementation of omnichannel marketing strategies must be carried out progressively, that is, SMEs can use the available channels to carry out pilot tests, and at the same time increase new channels or strategies; in addition, they can analyze the cost–benefit ratio of the results obtained for future planning. Some alternatives to start working with these strategies are the buyer persona, CRM, and market research, and this allows you to have information about the preferences and behaviors of consumers, to define the appropriate channels and, based on this, manage the use that will be given to each one.

After the analysis, strategies can be carried out, for example, online communication and sales channels in which the consumer can know the availability of a product or book an appointment to acquire a service, these channels can also receive traffic from the company's social networks or through QR codes that are scanned from the point of sale or a means of communication channel such as the newspaper.

22.5 Conclusions

The main one is that local SMEs manage their processes empirically, therefore, the results they obtain are negative, and this was evidenced in the closure of several businesses during the pandemic, which did not have the strategic capacity to deal with this situation and therefore had to close their businesses.

It was determined that omnichannel is essential for all companies, as they provide a variety of opportunities that improve the quality of the user experience, thus, for example, it allows predicting consumer behaviors, and projecting demands accurately, in addition, it allows companies to have direct contact with consumers, which helps to obtain information about their behaviors and preferences and consequently, identify the ideal channels for planning strategic marketing.

Additionally, it allows companies to adapt to the new scenario that is a product of the pandemic and that requires the presence and use of digital media, for all marketing processes, however, it must be considered that investments are required both in knowledge, that is, training, as well as in digital assets and other tools that improve the effectiveness of strategies.

As mentioned, the traditional part should not be neglected, it is still essential that traditional marketing strategies are applied, as these serve as support for the digital part and vice versa. The coexistence of off and online strategies means that the results of planning are as expected.

Finally, two profiles of SMEs were formed, the difference between these is the importance they give to each of the variables related to the organizational aspects of marketing planning, as well as the categorization of the variables related to digital marketing; however, in both cases, it is evident that there is an interest in designing marketing strategies, which is why entrepreneurs must train and innovate their businesses.

References

1. Valle, T., Sánchez, A., Vayas, T., Mayorga, F., Freire, C.: Empresas y establecimientos en Tungurahua. Observatorio Económico y Social de Tungurahua (2021)
2. Paredes, Á., Gallardo, W.: Administración estratégica en las pymes de Tungurahua y su impacto en la reactivación post pandemia. Visionario Digital **6**(2), 6–22 (2022)
3. Instituto Nacional de Estadística y Censos: Directorio de Empresas y Establecimientos 2020, INEC, Quito (2021)
4. Lovato, S., López, M., Acosta, M.: Incidencia de las Herramientas Administrativas y el Marketing en el Desarrollo Microempresarial de la Provincia de Santa Elena de Ecuador. Revista Espacios **39**(24), 21 (2018)
5. Rodríguez-Mendoza, R., Aviles-Sotomayor, V.: Las PYMES en Ecuador. Un análisis necesario. Digital Publisher **5**(5), 191–200 (2020)
6. Solis, L., Robalino, R.: El papel de las PYMES en las sociedades y su problemática empresarial. INNOVA Res J **4**(3), 85–93 (2019)
7. Oller, M., Játiva, E.: La competitividad en las pequeñas y medianas empresas (PyME). In: Tendencias y retos del Marketing en Ecuador 2015, pp. 22–42. Macasar Ediciones (2016)
8. Suárez, J.C., Pérez, O.: Tendencias y perspectivas del marketing en las pymes. Contabilidad y Negocios **16**(32), 129–142 (2021)
9. Mosquera, C.O., Juaneda, E.: Understanding the customer experience in the age of omni-channel shopping. Icono 14 **15**(2), 166–185 (2017)
10. Moncayo, M.: Omnicanalidad. Revista: Caribeña de Ciencias Sociales **4** (2018)
11. Mulqueen, T.:Is "Omnichannel" actually possible? Debunking omnichannel marketing myths. Forbes (2018)
12. González, J.: Customer journey, elemento clave para afrontar el future. Tecnohotel 30–31 (2020)
13. Vergara, M.: El comportamiento del consumidor post covid-19: oportunidad o desafío para los emprendedores. Rev. Col. Ciencia **3**(2), 102–112 (2020)
14. Ortega, M.: Efectos del Covid-19 en el comportamiento del consumidor: Caso Ecuador. RETOS. Revista de Ciencias de la Administración y Economía **10**(20) (2020)
15. PriceWaterhouseCoopers, "PriceWaterhouseCoopers" (2020)
16. Ekos: En 2021, el comercio electrónico mantendrá un crecimiento sostenido en Ecuador (2021)
17. Cámara de Comercio Electrónico, "Situación de las empresas durante el COVID-19 Ecuador" (2020)
18. Hernández Sampieri, R., Fernández, C., Baptista, M.: Metodología de la Investigación, McGraw-Hill / Interamericana Editores, S.A. de C.V. (2014)
19. Santos, V., Mendoza, J.: Relación del marketing omnicanal y las comunicaciones integradas en una empresa peruana. Economía y Negocios **3**(2), 10–22 (2021)
20. Alonso, J., Suárez, A., Trespalacios, J.: El papel del vendedor en un entorno omnicana. In: Omnichannel marketing: las nuevas reglas de la distribución y el consumo en un mundo omnicanal, pp. 35–48. Cátedra Fundación Ramón Areces (2019)
21. Caycho, C., Mel, J.: Marketing Omnicanal y su impacto en el Customer Journey de la empresa Representaciones FBA SAC-Lima 2021, Universidad César Vallejo (2021)
22. da Cruz, J.: A Comunicação Integrada de Marketing na era do Marketing Omnicanal. Casos do setor automóvel português, Universidade de Lisboa (2019)
23. Martínez, E.: Marketing en el sector de la moda, ante un consumidor omnicanal, experiencial y concienciado medioambientalmente, Universidad del País Vasco (2021)
24. Estella, C.: Estrategia de Zara, Colegio Universitario de Estudios Financieros (2019)
25. Haitao, T., Ghose, A., Halaburda, H., Raghuram, I., Pauwels, K., Sriram, S., Tucker, C., Venkataraman, S.: Informational challenges in omnichannel marketing: remedies and future research. Am. Market. Assoc. **85**(1), 103–120 (2021)
26. Lorenzo-Romero, C., Encarnación, M., Martínez, A., Mondejár, J.: Omnichannel in the fashion industry: a qualitative analysis from a supply-side perspective. Heliyon **6**, 1–10 (2020)

27. Zimmermann, R., Weitzl, W., Auing, A.: Identifying sales-influencing touchpoints along the omnichannel customer journey. ScienceDirect **196**, 52–60 (2021)
28. Shankar, V., Kushwaha, T.: Omnichannel marketing: are cross-channel effects symmetric? Int. J. Res. Mark. **28**(2), 290–310 (2021)
29. Rendón-Macías, M., Villasís-Keever, M., Miranda-Novales, M.: Estadística descriptiva. Revista Alergia México **63**(4) (2016)
30. Montoya, O.: Aplicación del análisis factorial a la investigación de mercados. caso de estudio. Scientia et Technica **13**(35), 281–286 (2007)
31. McDonald, M., Leppard, J.: La auditoría de marketing : cómo pasar de la teoría a la práctica del marketing, Díaz de Santos (1994)
32. Cancino, Y., Torres, J., Bautista, Y., Palacios, J.: Determinación de variables de evaluación para la auditoria de marketing asociadas al modelo de trazabilidad de objetivos. Revista Venezola de Gerencia **26**(96), 1413–1434 (2021)
33. Baque-Cantos, M., Cedeño-Chenche, B., Chele-Chele, J., Gaona-Obando, V.: Fracaso de las pymes: Factores desencadenantes, Ecuador 2020. FIPCAEC **5**(5), 3–25 (2020)
34. Menser, E., Peltier, J., Barger, V.: Omni-channel marketing, integrated marketing communications and consumer engagement. A research agenda. J. Res. Interact. Mark. **11**(2), 185–197 (2017)

Chapter 23
Relationship Marketing, The Way to Customer Satisfaction and Loyalty

Adriano Costa and Joaquim Antunes

Abstract In a highly competitive and at the same time very volatile market, it is essential for brands not only to be able to win more customers, but to keep current and profitable ones. Thus, it is imperative to develop positive relationships with customers and that they are long-term. This investigation aims to identify the factors that precede satisfaction and how it influences brand loyalty in the Portuguese market for personal hygiene products. The methodology used was based on a literature review on relationship marketing, brands, satisfaction and loyalty. Subsequently, an online survey distributed through social networks was applied. There were 608 valid responses. The results show that there are three factors that precede satisfaction and that influence it in a positive way. It was also found that there is a very strong and positive relationship between satisfaction and loyalty. Finally, the main conclusions of the work, limitations of the study and guidelines for future research are presented.

23.1 Introduction

Bearing in mind that today's markets are highly competitive, but at the same time very volatile, and that the speed of information transmission has increased significantly, it is very important for brands not only to win new customers, but also to have the ability to keep current and, especially those that are highly profitable, over long periods of time. That is why it is essential to understand and identify which factors may be the basis of high satisfaction and, indirectly, may contribute to brand loyalty. It is these relationships established between organizations and their customers that are fundamental to their success. Hence, the emergence of relationship marketing oriented toward maintaining and creating long-term relationships. Transactional relationships are boosted with new, more efficient communication channels and the new

A. Costa (✉)
CITUR/UDI/IPG –Guarda, Guarda, Portugal
e-mail: a.costa@ipg.pt

J. Antunes
CISeD/CITUR/IPV- Viseu, Viseu, Portugal
e-mail: jantunes@estv.ipv.pt

J. L. Reis et al. (eds.), *Marketing and Smart Technologies*, Smart Innovation, Systems and Technologies 337, https://doi.org/10.1007/978-981-19-9099-1_23

approach inherited from industrial and service marketing, reorients the company's way of relating to customers. In this way, it is sought to ensure lasting and faithful relationships [1]. The relationship, in addition to allowing the sale, also allows the knowledge of the consumption experiences of each customer, which leads to the knowledge of consumer buying trends [2].

The notoriety influences the purchase decision process [3, 4]. A strong reputation allows the brand to enjoy the following advantages: (i) it inhibits the evocation of other competing brands, (ii) it fosters a relationship of familiarity, which increases the probability of the brand being considered, in the absence of the usual brand or in the lack of sufficient motivation to make the purchase, (iii) it demonstrates to the consumer its commitment to remain in the market, strengthening its image and indicating its quality due to the high number of buyers, (iv) it provides the company with arguments to motivate and force the distribution itself, (v) and constitutes an indicator of the marketing effort, when interpreted as a result of its actions [4]. Thus, brands with a high notoriety can be the key to success for high customer satisfaction and, consequently, customer loyalty.

In this sense, the objective of this investigation is to identify the factors that precede satisfaction and how it influences brand loyalty in the Portuguese market of personal hygiene products. This market is increasingly competitive, with new products appearing regularly on the market, with more and more aggressive marketing actions.

From the point of view of relationship marketing and business practice, this research provides clues to identify the antecedents of satisfaction in mass consumption products, which can be controlled by organizations, allowing the prioritization of investments for the most relevant sources of value.

23.2 Literature Review

The literature review of the present investigation is fundamentally based on the marketing concept, namely relationship marketing, on the value of brands and their contribution to customer satisfaction and, consequently, customer loyalty.

23.2.1 Marketing

Marketing is a concept that is constantly evolving in order to respond to the different changes that arise over time, so there is a need to define several variants that complement and strengthen the strategy defined by organizations, adapting to different situations, needs, channels, businesses and consumers [5].

Thus, it is common to talk about traditional marketing, digital marketing, green marketing, relationship marketing, among others. As a result of globalization and the technological development that took place from the 1990s onwards, the ways in which organizations work with their customers have changed profoundly. Access to any information and increasingly interactive communication have affected and changed the way consumers deal with organizations [6]. Hence the appearance of the concept of relationship marketing. The first reference to relationship marketing took place in 1983 by Leonard Berry [7–9] in which his work contained a chapter called. For Berry [10] companies, when defining their strategies should focus greater efforts on developing relationships with current customers, paying attention not only before the moment of purchase, but continuously after the purchase, as it can make all the difference in sustainability of the business and the continuity of the company, while enabling the creation of trust.

23.2.2 Brand

A brand is a name and/or a symbol whose purpose is to identify a seller's product or service in order to differentiate it from rivals [11]. Therefore, consumers can recognize a product or service through a brand [12]. Through brands, consumers can distinguish between products from the same supplier. A brand is sometimes an intangible asset whose value is much higher than other physical assets, such as land, buildings and also production machinery. Sometimes it is the brands that arouse consumer confidence. In this way, brand loyal consumers are willing to pay more for the brand, because they feel they are getting something, or the brand gives them a value that is unique and that they cannot obtain in other brands. Thus, the brand can be used as a marketing strategy that adds value to goods or services companies [12].

Brand reputation represents people's opinion about a brand or organization. As such, it is essential that companies convey a positive image of trust and security to consumers with regard to the quality of their goods and services and the security of payment in the deals carried out [13].

As such, the brand must create a pleasant and satisfying experience in the eyes of the consumer, equating the performance of its goods or services with the performance expected by consumers, since this is what allows for satisfied customers. Hence, the value of brands can be evaluated by the degree of loyalty consumers which have toward a particular brand [14].

23.2.3 Satisfaction

Satisfaction has been seen as an antecedent of loyalty [7, 15–17]. Thus, a company has the potential to gain long-term relationships, it must give priority to customer satisfaction, since satisfied customers tend to buy more products or services from the same supplier [18].

In this way, satisfaction can be seen as the result of comparing their perceptions of the benefits with the expectations they would have received [19]. If the perceptions exceed expectations, the customers are satisfied and attribute quality to the service and/or product. If perceptions are lower than expectations, there will be customer dissatisfaction [20, 21].

According to Lestari and Likumahua [12], customer satisfaction is defined as the level of happiness or regret obtained by a person when comparing the expectations of the purchased product with the original product. Therefore, customer satisfaction is one of the fundamental aspects in the management of any business, as it compares the assessment of consumer expectations of a particular product or brand with the actual quality of the product or brand [22]. Thus, satisfaction is a fundamental variable for attracting and maintaining customers.

23.2.4 Loyalty

Customer loyalty has been studied in many investigations and industries. In this way, creating loyalty to a product or brand is to make our customers perceive that it is the best alternative or the only acceptable alternative [23]. The meaning and interpretation of the concept of consumer loyalty differs in the literature, since a multiplicity of approaches to define brand loyalty prevails [5]. Authors such as Frank [24] and McConnell [25] considered the precursors of the concept, defined and determined customer loyalty as the repetitive purchase process. This view of the concept of loyalty and its measurement was subject to multiple criticisms [18, 26, 27]. For these authors, loyalty should not only be evaluated by repeat purchases, since this repetition can be triggered by the influence of other factors, such as the absence of available alternatives/options, very high cost of switching and the existence of a leader of low price, which can make this change very difficult. These criticisms show that consumer loyalty must incorporate a favorable attitude of the customer toward the brand, together with a repetitive purchase behavior.

So, the definition of loyalty must encompass a primary behavior (frequency and amount of effective purchase), as well as other behaviors such as customer referral and positive word of mouth [28–30].

Thus, the concept of loyalty should be considered as a multidimensional concept that includes aspects related to behavior (repeated purchase) as well as elements related to attitude (customer referral and positive word of mouth). Therefore, the use of a composite measure increases the predictive power of the construct, as each cross-variable validates the nature of a true loyalty relationship.

23.3 Methodology

The methodology followed for the present investigation was based on the literature review related to relationship marketing, brands, customer satisfaction and loyalty.

The objective of the study was to analyze which factors influence satisfaction and, consequently, loyalty in mass consumption products, more specifically in personal hygiene products, in the Portuguese market.

For the construction of the scale of the relationship marketing construct, the investigations of Antunes and Rita [7] were considered. In order to measure overall satisfaction with the shampoo brands used, items from the literature review by Antunes and Rita [7], Costa [28] and Lestari and Likumahua [12] were used. To analyze the level of loyalty to shampoo brands, items included in the literature review by Antunes and Rita [7], Costa [28] and Zhou et al. [30].

The survey was applied through social networks, disseminated essentially through Facebook, WhatsApp and emails. We collected 608 valid surveys for analysis. However, this online dissemination generated some limitations due to the sociodemographic profile of the interviewers being essentially younger people, with more possibility of using digital tools, which does not happen with older people.

Pre-tests were carried out on a small sample and researchers related to marketing, in order to better refine the questionnaire. To process the data, Statistical Package for the Social Sciences (SPSS) version 27.0 was used.

23.4 Results Analysis

Next, we will proceed to a socioeconomic characterization of the sample based on sex, age, educational qualifications, income and employment status.

From the analysis of the Table 23.1, we can conclude that the sample is mostly female, aged between 18 and 24, with higher education, students and income between €1000 and €1500.

In order to know the notoriety of the existing shampoo brands in Portugal, an analysis of the frequencies of the different brands was carried out. Of the more than sixty brands identified, we can say that the four most notoriety brands are Pantene, Garnier, L'Oreál and Tresemmé, as shown in the Table 23.2.

A Pearson chi-square test was performed between brand awareness and age and gender demographic variables. It can be seen that there are significant relationships between spontaneous top of mind notoriety and age (X2 = 504.817, Sig. 0.000), noting that the Garnier brand is better known among young people aged 18–24, the L´Oreál brand also is well known among 18–24 year olds, but also considerably among 25–34 year olds. Regarding the Pantene brand, it is known by different age groups. With regard to gender, there are also significant differences in relation to the three most cited brands (X2 = 188.666 and sig. 0.000), with greater notoriety on the part of females in relation to the three most cited brands.

Then, the importance attributed to each item related to the purchase of shampoos, which precede satisfaction, was analyzed. Thus, Table 23.3 presents the mean and standard deviation of the 17 items considered. From the analysis of the table, it can be concluded that the item "This shampoo brand offers quality products" is the most valued by respondents with a value of 4.22 on a five-point Likert scale. In the opposite

Table 23.1 Socioeconomic characterization of the sample

Variable		N	%
Gender	Male	81	13
	Female	*527*	*87*
	Total	608	100
Age	18–24 years	342	56
	25–34 years	68	11
	35–44 years	95	16
	45–54 years	70	11
	55–64 years	28	5
	More de 65 years	5	1
	Total	608	100
Literary qualifications	Elementary education	11	2
	High school	262	43
	Higher education	*335*	*55*
	Total	608	100
Income	Até 665€	54	9
	665–750€	69	12
	751–1000€	110	19
	1001–1500€	*148*	*26*
	1501–2000€	91	16
	Mais de 2000€	100	18
	Total	572	100
Employment status	Employed on behalf of another	205	34
	Self-employed	29	5
	Entrepreneur	14	2
	Unemployed	37	6
	Student	*299*	*50*
	Domestic	6	1
	Reformed	12	2
	Total	602	100

Table 23.2 Notoriety of shampoo brands

	First answer		Second answer		Third answer		Global total	
Brand	N	%	N	%	N	%	N	%
Pantene	176	28.9	122	20	76	12.5	374	20.5
Garnier	119	19.5	108	17.7	74	12.2	301	16.5
L'Oreál	49	8.0	72	11.8	57	9.4	178	9.7
Tresemmé	33	5.4	38	6.2	59	9.7	130	7.1

direction, that is, the least valued item is "It is a family tradition to consume that brand of shampoo", with a value of 2.04.

Next, a factor analysis of the principal components of this set of items was carried out in order to find combinations of variables (factors) that explain the correlations between all pairs of variables. For the application and validation of this technique, it is necessary to evaluate the correlations between the variables in order to know if it is legitimate to carry out a factor analysis.

The Kaiser–Meyer–Olkin value (KMO = 0.957), which presents the value of the suitability measure, considers the analysis to be very good. Bartlett's test, which tests the null hypothesis that the correlation matrix is an identity matrix, has the value 9184.287 and an associated probability of 0.000, which rejects the null hypothesis that the correlation matrix is an identity matrix (Table 23.4). These tests indicate that the 17 variables are suitable for carrying out a factor analysis.

Through the Kaiser criterion (eigenvalue greater than 0.8) four factors were found that explain 77.505% of the total variance in the set of 17 variables analyzed. For a better interpretation of the factors, a rotation of the axes was carried out using the varimax method. The proportion of variance explained by the components remains constant, it is just distributed differently so that the differences between the combinations of variables are maximized: increasing those that contribute the most to the formation of the factor and decreasing the weights of those that contribute the least.

Table 23.3 Descriptive statistics of the items related to the purchase of shampoos

Items	Average	Std. deviation
5.1 Whenever possible I buy that shampoo brand	4.01	1.140
5.2 That brand of shampoo offers quality products	4.22	0.924
5.3 I am proud that others know that I am a client of that shampoo brand	3.18	1.358
5.4 That shampoo brand keeps its promises	3.79	1.08
5.5 That shampoo brand is suitable for my needs	4.07	0.932
5.6 That brand clearly presents the information on its products	3.89	1.005
5.7 It is family tradition to consume that shampoo brand	2.04	1.218
5.8 I really trust that shampoo brand	3.70	1.003
5.9 I believe that shampoo brand defends my interests	3.51	1.097
5.10 That brand assumes a positive contribution to society	3.17	1.090
5.11 That brand cares about its customers	3.45	0.938
5.12 That brand is concerned with producing products based on technological innovations	3.48	1.063
5.13 I have high reliability in products of that brand	3.77	1.016
5.14 The price charged is adjusted to the value of the product	3.51	1.075
5.15 The quality of the product justifies the amount paid	3.65	1.016
5.16 The brand is concerned with conveying a credible image	3.87	0.987
5.17 That brand is stable and well implemented in the market	4.02	0.987

Table 23.4 KMO and Bartlett test

Kaiser–Meyer–Olkin measure of sampling adequacy		957
Bartlett's sphericity test	Approx. Qui-quadrado	9184,287
	Gl	136
	Sig	000

The internal consistency of each factor was then analyzed using Cronbach's Alpha (α). The values of α vary from 0 to 1 and the closer to 1, the greater reliability between the indicators. In this case, the first three factors have a strong internal consistency (greater than or equal to 0.90) and the fourth factor, which consists of a unique item, does not need to be evaluated for consistency.

Table 23.5 shows the weights of the variables in each factor. Thus, factor 1, which we call "Trust/Compromise", is composed of 7 variables and is the factor that most contributes to the explanation of the total variance (25.8%). Factor 2 is composed of 5 items which are related to "Image", contributes 25.6% of the total variance, factor 3, called "Quality/Price Relation", explains 19.8% of the total variance and the factor 4, which consists of a single item and which gives the name to it, which is Family Tradition, explains 6.36% of the total variance. Only weights greater than 0.5 were considered for easier interpretation and so that each variable had only weights in a single factor.

Regarding the satisfaction construct, it shows a high level of satisfaction of the respondents to their brand consumed. As can be seen in the Table 23.6, the averages vary between 4.08 and 3.90 (1 strongly disagree to 5 strongly agree). On the other hand, the medians of all items present high values.

Furthermore, we can say that the results obtained through exploratory factor analysis point to the unidimensionality of the satisfaction scale, based on KMO, Bartlett's test and item weight (Table 23.7).

This position is further reinforced by the analysis of the aforementioned scale, since Cronbach's Alpha for the 4 items has a value of 0.965.

With regard to the Loyalty construct, the data show that there is a high level of loyalty among respondents, as the average values of the items considered for their assessment vary between 3.81 and 4.00 (1 strongly disagree to 5 strongly agree), in addition to that the median in all items is also 4 (Table 23.8).

Then, an exploratory factor analysis was carried out to confirm the unidimensionality of the loyalty scale, considering the 4 items used. From this analysis, it can be concluded that the data point to this unidimensionality and that it is reinforced by the high Cronbach's Alpha obtained for the set of items, which is 0.92 (Table 23.9).

Next, we sought to find out which factors influenced satisfaction and in what way. For this, we will use the linear regression model and we will consider the satisfaction factor as a dependent variable and the factors that influence satisfaction as independent variables. The coefficient of determination (R2) has a value of 0.572, which represents a good measure of fit of the model to the data. Also the analysis of variance to the model through the Snedecor F test presents satisfactory

Table 23.5 Matrix of components after varimax rotation

Items	Component			
	1	2	3	4
Factor 1—trust/compromise				
5.1 Whenever possible I buy that shampoo brand	0.822			
5.2 That brand of shampoo offers quality products	0.793			
5.4 That shampoo brand keeps its promises	0.690			
5.5 That shampoo brand is suitable for my needs	0.760			
5.6 That brand clearly presents the information on its products	0.584			
5.8 I really trust that shampoo brand	0.568			
5.17 That brand is stable and well implemented in the market	0.539			
Factor 2—image				
5.3 I am proud that others know that I am a client of that shampoo brand		0.644		
5.9 I believe that shampoo brand defends my interests		0.723		
5.10 That brand assumes a positive contribution to society		0.836		
5.11 That brand cares about its customers		0.764		
5.12 That brand is concerned with producing products based on technological innovations		0.623		
Factor 3—quality/price relation				
5.13 I have high reliability in products of that brand			0.513	
5.14 The price charged is adjusted to the value of the product			0.858	
5.15 The quality of the product justifies the amount paid			0.817	

(continued)

Table 23.5 (continued)

Items	Component			
	1	2	3	4
5.16 The brand is concerned with conveying a credible image			0.583	
Factor 4—family tradition				
5.7 It is family tradition to consume that shampoo brand				0.975
Eigenvalue	10.386	1.086	0.874	0.831
Variance explained (%)	25.8	25.6	19.8	6.3
Alpha de Cronbach	0.939	0.919	0.900	–

Rotation method: varimax with Kaiser Normalization. Converged rotation in 6 iterations

Table 23.6 Measurements of univariate statistics of satisfaction

Items	Average	Std. deviation	Median
I am satisfied with my decision to use shampoo from that brand	4.08	0.914	4
My choice of using shampoo from that brand was the right one	3.90	1.003	4
If I had to choose a shampoo again, I would make the same choice	3.89	1.024	4
I'm sure I made the right choice to use that shampoo brand	3.92	0.994	4

Table 23.7 Result of the factor analysis of the satisfaction construct

Satisfaction	KMO	T. Bartlett	No factors	Variance explained	Weight	Commonality
I am satisfied with my decision to use shampoo from that brand	0.877	0.000	1	90.658	0.939	0.882
My choice of using shampoo from that brand was the right one					0.959	0.921
If I had to choose a shampoo again, I would make the same choice					0.949	0.900
I'm sure I made the right choice to use that shampoo brand					0.961	0.924

Table 23.8 Measurements of univariate loyalty statistics

Items	Average	Std. deviation	Median
I will say positive things about the products of that brand	3.94	0.60	4
I will recommend that brand's shampoo to anyone who seeks my advice	3.97	0.975	4
I consider that brand as my first choice in terms of shampoo	3.81	1.103	4
I intend to continue using shampoo from that brand	4.00	0.976	4

Table 23.9 Result of the factor analysis of the loyalty construct

Loyalty	KMO	T. Bartlett	No factors	Variance explained	Weight	Commonality
I will say positive things about the products of that brand	0.797	0.000	1	81.223	0.898	0.807
I will recommend that brand's shampoo to anyone who seeks my advice					0.918	0.843
I consider that brand as my first choice in terms of shampoo					0.883	0.780
I intend to continue using shampoo from that brand					0.905	0.819

values (F = 201.335 with an associated significance level of less than 0.001). The following table presents the linear regression coefficients. The coefficients of factors 1, 2 and 3 present a positive value of 0.513, 0.446 and 0332, respectively. These values are statistically significant, since the t test has significance of 0.000. Factor 4 is not considered relevant, and the significance level is greater than 0.05, so it is not statistically significant (Table 23.10).

Considering that from the literature review satisfaction influences loyalty, we will also use the linear regression model in which we consider the loyalty factor as the dependent variable and the satisfaction factor as the independent variable. The coefficient of determination (R2) has a value of 0.842, which represents an excellent measure of data adjustments. Also the analysis of variance to the model, through Snedecor's F test, presents satisfactory values (F = 3234.202 with an associated significance level of <0.001). Table 23.11 presents the linear regression coefficient, which is positive and high (Beta = 0.918). This value is statistically significant (t = 56.870 and sig = 0.000).

Table 23.10 Dependent variable satisfaction—linear regression coefficients

	Non-standardized coefficients		Standardized coefficients		
	B	Erro	Beta	t	Sig
(Constant)	−7.19E−17	0.027		0.000	1.00
Factor 1_trust/compromise	0.513	0.027	0.513	19.248	0.000
Factor 2_image	0.446	0.027	0.446	16.721	0.000
Factor 3_quality/price relation	0.332	0.027	0.332	12.457	0.000
Factor 4_family tradition	−0.006	0.027	−0.006	−0.242	0.809

a. Dependent Variable: Satisfaction

Table 23.11 Dependent variable loyalty—linear regression coefficients

	Non-standardized coefficients		standardized coefficients		
	B	Error	Beta	t	Sig
(Constant)	3.61E−16	0.016		0.000	1.00
Satisfaction	0.918	0.016	0.918	56.870	0.000

a. Dependent Variable: Loyalty

23.5 Conclusion

After analyzing the data, we can conclude that the brands with the most notoriety are Pantene, Garnier and L'Oréal. There were significant differences in the notoriety of these brands between the variables gender and age. Organizations must take these results into account, adjusting their marketing actions in order to improve awareness among different target audiences.

Four factors that precede satisfaction were also identified, namely factor 1—Trust/Compromise, factor 2—Image, factor 3—Quality/Price Relation and factor 4—Family Tradition. However, only the first 3 factors influence satisfaction in a positive way, with factor 1—Trust/Commitment—with a standardized coefficient of 0.513, followed by factor 2—Image—having the greatest weight, followed by factor 2—Image—and whose weight is of 0.446 and finally factor 3—Quality/Price Relation—with a weight of 0.332 and which is in line with other studies carried out.

Finally, it was found that there is a strong positive relationship between the satisfaction and loyalty of shampoo consumers (massive consumer products) and that this is in line with other research works carried out.

Thus, brands must focus on offering products that satisfy their customers that have the quality they desire that fulfill the promises made that convey their information in a clear way so that customers can trust the brand.

In addition, they must develop innovative actions in the creation of new products, as well as actions that take sustainable development into account.

Finally, they should not neglect the variable price, offering products at a fair price, that is, that the quality is adequate with the price they practice.

This study presents a limitation that the sample may be biased, given the way in which the data were obtained. Thus, it is suggested in the future studies to continue the work study, replicating the questionnaire to a larger and more diversified sample in terms of socioeconomic profile.

Acknowledgements This work is supported by National Funds through FCT—Fundação para a Ciência e a Tecnologia, I.P., under the project Refª UIDB/05583/2020 and UIDB/04470/2020. We would also like to thank the Centro de Investigación em Serviços Digitais (CISeD) and the Politécnico de Viseu for their support.

Thanks are also due to the Center for Research, Development and Innovation in Tourism (CiTUR), UDI—Research Unit for the Development of the Interior and the Polytechnic Institute of Guarda.

References

1. Antunes, J., Rita, P.: O Marketing Relacional como novo paradigma: Uma análise conceptual. Revista Portuguesa e Brasileira de Gestão **7**(2), 36–46 (2008)
2. Brambilla, F.R., Pereira, L.V., Pereira, P.B.: Marketing de Relacionamento: Definições e Aplicações. INGEPRO—Inovação, Gestão e Produção **02**(12), 1–9 (2010)
3. Keller, K.L.: Conceptualizing, measuring and managing customer-based brand equity. J. Mark. **57**, 1–22 (1993)
4. Serra, E.M., Gonzalez, J.A.V.: A Marca: Avaliação e Gestão Estratégica, Lisboa: Editorial Verbo (1998)
5. Martins, A.: Os determinantes para a fidelização de clientes nas empresas. Dissertação de mestrado em Marketing e Negócios Internacionais, Instituto Superior de Contabilidade e Administração de Coimbra, Coimbra (2022)
6. Martins, M.J.: Marketing Relacional e Qualidade do Serviço na Satisfação do Cliente. Dissertação mestrado em Marketing, Instituto Superior Contabilidade e Administração de Aveiro, Aveiro (2013)
7. Antunes, J., e Rita, P.: O marketing relacional e a fidelização de clientes: estudo aplicado ao termalismo português. Economia Global e Gestão, Agosto 109–132 (2007)
8. Brito, C.M.: Marketing Relacional—Das origens às atuais escolas de pensamento. Revista Portuguesa de Marketing **26** (2011)
9. Fonseca, D.H.: Ferramentas de Comunicação Relacional e o CRM na Promoção da Satisfação e Fidelização de Clientes. Publicidade e Relações Públicas, Universidade da Beira Interior, Covilhã, Dissertação de Mestrado em Comunicação Estratégica (2020)
10. Berry, L.: Relationship marketing. In: Berry, L. Shostack, G., Upah, G. (eds.), Emerging Perspectives in Services Marketing, American Marketing (1983)
11. Stanton, J., Etzel, J., Walker, J.: Fundamentos de Marketing. 14º edição, MacGraw-Hill Interamericana, México (2007)
12. Lestari, R, Likumahua, C.E.J.: The Influence of Customer Relations, Pricing Strategy, and Branding Identity on Customer Satisfaction and Its Impact on Sales. Budapest International Research and Critics Institute-Journal (BIRCI-Journal) Volume 5, N. 3, August 2022, Page: 25311–25319 (2022).
13. Mota, R.P.P.A.: Fatores antecedentes do uso de um consumidor num mercado digital. Tese de mestrado em Gestão; Instituto Universitário de Lisboa-ISCTE, Lisboa (2020)

14. Aaker, D.: Commentary: do brands compete or coexist? By Sheth and Koschmann. From brand to subcategory competition. Eur. J. Market. **53**(1), 25–27 (2019)
15. Anbori, A., Ghani, S.N., Yadav, H., Daher, A.M., Su, T.: Patient satisfaction and loyalty to the private hospitals in Sana'a Yemen. Int. J. Qual. Health Care **22**(4), 310–315 (2010)
16. Sahut, J.M., Moez, K., Mutte, J.L.: Satisfaction et fidélisation aux services d´internet Banking, quelle influence sur la fidélité à la banque? Revue Management & Avenir **47**(7), 260–280 (2011)
17. Sharma, V.: Patient satisfaction and brand loyalty in health-care organizations in India. J. Asia Bus. Stud. **11**(1), 73–87 (2017)
18. Marques, A.: Marketing relacional – como transformar a fidelização de clientes numa vantagem competitiva. (2ª. ed.). Edições Sílabo, Lisboa (2014)
19. Oliver, R.L.: Satisfaction—a behavioral perspective on the consumer, 2nd ed. eBook, Routledge (2014)
20. Chiusoli, C., Kolodi, E., Ladislau, F., Egler, J., & Morais, J.: Avaliação da satisfação sob a ótica do consumidor. Revista Estudos e Pesquisas em Administração **4**(1), 106–123 (2020)
21. Esteban, I.: Marketing de los Servicios. 3º edição, ESIC Editorial, Madrid (2000)
22. Rahayu L., Siti F. M.: Pengaruh Kualitas Produk, Kualitas Layanan, dan Persepsi Harga terhadap Kepuasan Pelanggan melalui Brand Trust Minuman KOI The Cabang Mall Plaza Indonesia. Syntax Literate: Jurnal Ilmiah Indonesia p–ISSN: 2541–0849 eISSN: 2548–1398 Vol. 7, No. 3 (2022).
23. Rivas, J.A., Nogales, A.F., Arrizabalaga, I.G., Salinas, E.M., Liano, L.G.R, Maya, S.R., Moro, M.L.S.: Comportamiento del Consumidor". ESIC Editorial, Madrid (1999)
24. Frank, R.E.: Correlates of buying behavior for grocery products. J. Mark. **31**(4), 48–53 (1967)
25. McConnell, J.D.: The development of brand loyalty: an experimental study. J. Mark. Res. **5**(1), 13–19 (1968)
26. Kotler, P., Kartajaya, H., Setiawan, I.: Marketing 4.0: Mudança do Tradicional para o Digital. Conjuntura Actual Editora, Coimbra (2019)
27. Moliner, M.A.: Loyalty, perceived value, and relationship quality healthcare services. J. Serv. Manag. **20**(1), 76–97 (2009)
28. Costa, A.: Perfil e Motivações do Enoturista do Pólo de Turismo do Douro. Universidade de Aveiro, Aveiro, Tese de doutoramento (2014)
29. Kozak, M., Rimmington, M.: Tourist Satisfaction with Mallorca, Spain, as na off-season holiday destination. J Travel Res **38**, 260–269 (2000)
30. Zhou, W.J., Wan, Q.Q., Liu, C.Y., Feng, X.L., Shang, S.M.: Determinants of patient loyalty to healthcare providers: an integrative review. Int. J. Qual. Health Care **29**(4), 442–449 (2017)

Part VIII
Social Media and Networking

Chapter 24
Behaviour of the Adolescents and Their Parents in Relation to the Micro-Influencers in Instagram

Diana Soares and José Luís Reis

Abstract Influence marketing on social networks is a strategy that has emerged in recent years, especially in Instagram. Communication through digital channels is more effective than traditional advertising, influencing individuals' purchasing decisions and thus representing a great opportunity for companies. The micro-influencer is defined as an individual with a great impact through the word-of-mouth method in digital media. Micro-influencers are not considered traditional celebrities, but individuals who are exclusively dedicated to categories of true knowledge, passion, and authenticity, in addition to being seen as reliable sources when recommending purchasing situations, usually tend to generate more impact than the average expert by the power of dissemination of information and impact more consumers. This research aimed to explore the perspectives of adolescents as receptors of commercial messages. It also aims to understand and investigate the position of adolescents' parents, whose role can be decisive in this scenario. A quantitative study was carried out through an online survey of adolescents aged between 10 and 19 and a qualitative study through individual interviews with the parents of adolescents. The results obtained confirm the influence of digital micro-influencers on teenagers' choices, who often follow their suggestions and consumption patterns, as well as their lifestyle. This proximity is because micro-influencers appear several times to use the product they promote, giving more credibility. The results show weaknesses in parents' control of adolescents' social networks these days because they believe their children can recognize these types of advertising.

D. Soares · J. L. Reis (✉)
IPAM - Portuguese Institute of Marketing Administration, R. Manuel Pinto de Azevedo 748, Porto, Portugal
e-mail: jlreisg@gmail.com

J. L. Reis
University of Maia - ISMAI, Avenida Carlos de Oliveira Campos - Maia, Maia, Portugal

LIACC - University of Porto, FEUP, Rua Dr. Roberto Frias, Porto, Portugal

J. L. Reis et al. (eds.), *Marketing and Smart Technologies*, Smart Innovation, Systems and Technologies 337, https://doi.org/10.1007/978-981-19-9099-1_24

24.1 Introduction

Social networks have been increasingly used by consumers as a means of connecting them, but also with companies and brands. Influence marketing on social networks is a strategy that has emerged in recent years, especially in Instagram. Regarding the products promoted by micro-influencers in social networks, when compared to digital influencers or celebrities, usually show greater credibility, as they have a closer attitude towards adolescents and greater transparency in the message they want to convey, on the other hand, they usually promote products that are part of their daily lives [1].

This study analysed the behaviour of the young Portuguese population and their parents, with the aim of understanding how they are influenced by the Instagram social network and by the micro-influencers associated with it, and to what extent they are persuaded to buy a certain product, what kind of behaviour they adopt towards this promotion and how their parents deal with this problem.

This work, in the first part, presents information about consumer behaviour, influence marketing, and influencing agencies, including information about influencers and micro-influencers. In the second part, the research methodologies are presented, including de hypotheses, the conceptual model, and the data collection method. This is followed by an analysis of the results of the questionnaires and interviews and the validation of the hypotheses. Finally, the conclusions, limitations, and future work are presented.

24.2 Consumer Behaviour and Influence Marketing

Consumers use social networks on an ongoing basis to get information about the different products they need to buy. Consumers could influence other consumers by constantly conducting evaluations on social networks, making this marketing strategy an integral part of any company [2]. Online presence is crucial in the purchasing decision process. Today it is common to access information about products before accessing the brand's website, looking for information on digital platforms and reviews of these products [1].

Customer sovereignty is challenging; however, marketing strategies can affect their motivation and behaviour towards a certain product or service, realizing whether it is designed to meet their needs and expectations. The same happens in the purchasing decision process, where the consumer chooses according to the financial resources and information about a certain product or service [3].

The concept of influence marketing focuses, essentially, on strategies that act to persuade their consumers, through the sharing of content by digital influencers, who have a great capacity to add value to the products [4].

Traditionally, influence marketing focused essentially on numbers. It was assumed that a larger number of followers had more impact and was more attractive to brands.

Nowadays, we realize that strategies that incorporate influencers with smaller but more committed audiences leave a mark with more loyal followers; they are the "micro-influencers" [5].

24.2.1 Micro-influencers

Digital influencers define themselves as content generators and opinion leaders with high numbers of followers on social networks, able to influence consumer choice by using their publications to advertise a brand or product and thus persuade the purchasing decision of those who follow the publication. Its main objective is to capture the attention of followers and other users of social networks, creating a strong brand of its own that results in a symbiosis between them and the brands, since they are rewarded with unique products or experiences and the brands reach a higher level of recognition [6].

There is no specific number for a person to be considered a digital influence. They can count on tens of millions of followers, or they can count on only a few hundred. There are several classifications for influencers according to the number of followers they have, and they can be classified as follows [7]:

- Mega-influencer, over one million.
- Macro-influencer, from 500,000 to one million.
- Intermediary, from 100,000 to 500,000.
- Micro-influencer, from 10,000 to 100,000.
- Nano-influencer, up to 10,000.

Among digital influencers, digital micro-influencers stand out, translating into ordinary people producing content in which they transmit their ideals and consumption patterns to a specialized target audience. They have a smaller number of followers compared to celebrities, which translates into a greater proximity to their target audience [8].

24.2.2 Influencing Agencies

In influence marketing, the creation of content is sovereign, promoting certain brands with the objective of obtaining followers, and brand recognition. With the constant expansion of social networks, identifying relevant influencers can become a difficulty, so different forms of evaluation stand out to help companies distinguish the most prominent influencers, which will allow the organization to collect more value. In this way, the variables that allow us to measure quantitatively and qualitatively the preponderance of the influencer are views per month, popularity of inbound and outbound links, the frequency of posts, industry score, social participation, engagement rate, and related theme [9].

Table 24.1 Agencies of influencers on social networks

Blog Agency	Smart Influencer	Klear	Samyroad
Leading Portuguese agency, has Carolina Patrocínio as its agent and counts with the partnership of Caras Magazine	It has an "influence academy" where you can have Masterclasses on how to "Create, manage and monetize a successful YouTube channel"	Big data influencer Marketing software platform. Helps brands build, scale, and measure influencing programmes	Marketing platform for influence brands based in Madrid. More 25,000 creators and influencers from all over the world

Influencing agencies in social networks play a key role in supporting companies that define influencing marketing investment strategies. There are more and more agencies of influencers in social networks in Table 24.1 which are presented some agencies of this type.

24.3 Instagram Social Network

Instagram is a photo and video sharing application accessed mostly through a smartphone and is considered one of the most recognized social networks in the world, with about 1 billion users, an estimated 80% of whom are teenagers [10].

Instagram was made public on 6 October 2010 by Mike Krieger and Kevin Systrom in order to simplify the Burbn application, an application for iPhones created by the same engineers in order to allow, through geolocation, the individual to locate future visits. Faced with the difficulty of using the Burbn application, the two engineers concentrated all the functionalities in only one resource, thus appearing the "insta", which means instant camera, in turn "gram" derives from telegram. This social network can be used in several functionalities such as marketing, social, branding, the most common being the exclusive use for entertainment, possessing a set of features used for different purposes: publication of photos and videos, Instagram web, Instagram direct, news feed, activity pages, hashtags, and Instagram stories [11].

The hashtags were introduced into the Instagram environment, characterized by the symbol "#" before each word and functionally represent the way to relate publications to each other, grouping them dynamically under the same theme. As a social component, this functionality encouraged the production of categorized content with the aim of reaching the largest number of followers and stimulated hashtags research in order to know different views about the world from other users [12].

24.4 Adolescence and Social Networks

Adolescence is a transitional phase between childhood and adulthood. According to the World Health Organization, adolescence begins at the age of 10 and ends at 19, characterized by a period of numerous physical, psychological, and social changes [13]. The adolescent public is made up of consumers more vulnerable than the adult public in consumer relations, sometimes resulting in the use, by brands, of the still incomplete discernment of adolescents [14].

Adolescents spend too much time on social networks, corresponding to the phase of their life in which they develop physically and morally and vindicate their personality [15].

Surfing the Internet, especially through social networks, has become one of the activities with a primordial place in the lives of adolescents. Most become involved from an early age with mobile phones and social networks during early adolescence, a time in their lives when they become more concerned about their image and the friendships they begin to create, and their focus begins to be their image and what they look like abroad [16].

24.4.1 Parental Intervention

We can identify two types of intervention by parents regarding the exposure of adolescents to information transmitted by micro-influencers: active intervention and restrictive intervention. In the active intervention, parents help their children to understand the content and intent of the advertisements, while in the restrictive intervention, parents limit the amount of advertising their children see [17].

Parents' behaviour is understood as the set of behaviours, attitudes, and emotional climate in the relationship between parents and children (body expression, tone of voice, mood). In addition, they involve parental educational practices most used in such interactions [18].

It is essential that parents, as educators, focus exhaustively on the opportunities that arise, but also on the risks inherent in the adolescent's intensive use of social networks. It is essential for parents to play an active role in this sense, since young people are not able to ascertain the credibility of certain information [19].

Parents perceive an increased and extremely important role in monitoring and tracking their children's online behaviour, with a focus on promoting safer Internet use. It is essential that educators adapt to the resulting technologies [20].

24.4.2 Research Methodology

In this section, the methods used in this research are identified and described, which is divided into three parts; in the first part, the objectives of the research, in which the problem and the general and specific objectives to be achieved will be evaluated, in the second part, the research hypotheses and the conceptual model are presented, and in the third part, the method of data collection used, in which the methodology that will respond to the problem will be selected and finally, the techniques of data collection appropriate to meet the objectives.

24.4.3 Objectives

The main objective of this investigation is related to the current perception and behavioural analysis of parents and adolescents in the face of the presence of digital micro-influencers in the Instagram social network, in the face of a disproportionate technological evolution. Regarding the specific objectives of the research, which stems from the general objective, it focuses on five distinct objectives in order to answer the question set out in the general objective:

- Understand what kind of consumer behaviour is influenced by digital micro-influencers;
- Understand what kind of gender is most active in the search for recommendations of digital micro-influencers;
- Understand the relationship between the number of digital micro-influencers followed and the influence on their activity in the social network Instagram's.
- Identify teenagers' and parents' habits regarding presence on social networks;
- Understand the position of parents regarding the control of their children's activity on social networks.

24.4.4 Research Hypotheses and Conceptual Model

The formulation of hypotheses is intended to respond to the problem raised by the research, consisting of a supposed and provisional response and a conjecture of the relationship between variables. In this way, the hypotheses function as hypotheses put forward as provisional solutions to the problem in question and can later be confirmed with the follow-up of the study [21].

The following research hypotheses have been formulated, and following the objectives proposed for the elaboration of the research.

H1: Teenagers use the Instagram social network.

H2: Teenagers are fans of digital micro-influencers.

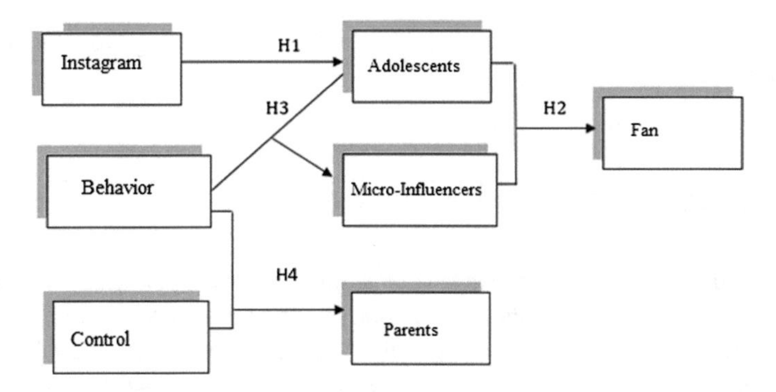

Fig. 24.1 Conceptual research model

H3: Teen behaviour is influenced by digital micro-influencers.

H4: Parents control the activity of adolescent children on the social network Instagram.

According to the literature review that served as the basis for the research, the conceptual model that supports the research was constructed in accordance with the objectives and hypotheses defined—see Fig. 24.1.

According to Fig. 24.1, it is intended to check whether the use of Instagram has an impact on adolescent behaviour. On the other hand, it is considered that being a fan of micro-influencers can have an impact on adolescents. It is important to understand whether adolescents' behaviour is directly influenced by the choices of micro-influencers, and the concept of behaviour associated with micro-influencers emerges.

As a result of adolescent activity on social networks, the concept of control, associated with parents, emerges, and it is necessary to verify whether this control is carried out. The hypotheses defined were validated through the analysis of the survey results.

24.4.5 Data Collection Method

Research methods are the path to scientific knowledge and are defined by the set of procedures that serve as tools for achieving the purpose of research [22].

The development of tools for interview construction and data collection by survey was followed by the adaptation of qualitative and quantitative methodology, as the aim was to approach the subject under study and establish generalizations on the subject.

The methodology adopted for the research consisted of a mixed approach, being quantitative at an initial stage, translated into the carrying out of surveys of adolescents, with a sample of 232 respondents and qualitative in the second stage, including interviews with three mothers and three fathers of adolescents aged between 10 and 19 years.

24.5 Analysis of the Survey Results

Within the framework of the research, the validation of the hypotheses with adolescents was carried out through the construction of a semi-structured survey, composed of closed and very simple questions, given the age group of the target public and the subsequent statistical treatment and analysis of the answers, allowing conclusions to be drawn, according to the hypotheses listed above.

The answers to the survey were given via the Internet, from a link generated in Google Forms, made available to young people in the first two weeks of May 2020. Social networks were used to increase response rates to the questionnaire.

The survey was answered by 232 adolescents. Of these, 32 are aged between 10 and 14 years. The remaining 200 are aged between 15 and 19 years.

Regarding the frequency of use of the social network, 96.5% of the adolescent respondents say they have access to Instagram daily, highlighting their increasing contact with the digital environment.

Relatively to the number of micro-influencers they follow, most young people (43.17%) follow between one and five micro-influencers, followed by followers of twelve or more (29.07%), revealing that young people are increasingly in contact with influence marketing.

Regarding the control of adolescent activity in social networks, according to the data presented in Table 24.2, most parents do not control the activity, however, the highest incidence of parents controlling their children's activity in Instagram occurs in the 16–20 age group.

Regarding the reasons that lead to the purchase of products, the identification of the adolescent with the product (60.09%) is the most impacting, followed by the price (42.60%) and the opinion of the influencer about the product (31.39%), reflecting the strong preponderance of the opinion of the micro-influencers in the mind of the adolescent consumer. It is followed by brand awareness (29.69%) and finally confidence in the opinion of the influencer (28.70%). As can be seen the last three reasons have percentages very close to 30%.

According to the results presented in Table 24.3, most adolescents are not interested in the activity of micro-influencers in Instagram, not considering themselves

Table 24.2 Frequency of parental control of children's activity on social networks

		5–10 years	11–15 years	16–19 years	Total
Monitoring of Instagram activity by parents	Yes	1	5	21	27
	No	4	26	170	200
Total		5	31	191	227

Table 24.3 Interest of adolescents in micro-influencers

		Female	Male	Total
Interest in micro-influencers	Yes	44	11	55
	No	114	58	172
Total		158	69	227

Table 24.4 Search for micro-influencers recommendations

		Female	Male	Total
Search for micro-influencers recommendations that follows	Yes	115	26	141
	No	43	43	86
Total		158	69	227

fans. Only 44 girls consider themselves as fans of the work of micro-influencers, for males only 11 adolescents—see Table 24.3.

The study aimed to understand which gender (male/female) is the most active in the search for digital recommendations from the digital micro-influencers that follow. It can be concluded, as shown in Table 24.4, that most female adolescents look for recommendations (115), only 43 do not. Regarding male gender, the number of adolescents not seeking recommendations (43) overlaps with those seeking (26), resulting in girls' greater concern with the image and follow-up of current products.

Concerning the knowledge of products through digital micro-influencers, the results show a minimum percentage of difference, 55.1% of adolescents did not know products through micro-influencers and 44.9% did.

Pearson's correlation coefficient is used to analyse the intensity and in turn the linear relationship between two variables, in which we can observe their resistance and direction [23].

In the results obtained, it was found that the resistance is moderate negative, that is, as the average of the micro-influencers followed by the respondents' increases, the influence on their activity in Instagram decreases. Thus, the analysis of Table 24.5 shows that the greater the number of micro-influencers followed, the less influence they have on Instagram users' activity, probably due to the dispersion of attention, which not only focuses on one, but on several, leading to less engagement being created.

In the construction of the survey by questionnaire, the respondents being 9–19 years old, attention was paid to the way it was composed and to its simplicity. In order to understand if evaluation of the reasons why respondents follow micro-influencers, the internal consistency of the questions and answers was measured through the Alpha de Chronbach reliability degree, and the value of 0.916 was obtained, which means that it is quite positive and precise.

Table 24.5 Pearson correlations

		Average of micro-influencers followed	Influence of the presence of digital micro-influencers on Instagram activity
Average of digital micro-influencers followed	Pearson correlation	1.0	−0.3*
	Sig. (2 extremities)		0.0
	N	227.0	227.0
Influence of the presence of digital micro-influencers on the social network Instagram	Pearson correlation	−0.3*	1.0
	Sig. (2 extremities)	0.0	
	N	227.0	227.0

*The correlation is significant at level 0.01 (2 extremities)

24.6 Analysis of Interviews

To analyse what is intended, the researcher must have the capacity to build a real relationship with the participant, which creates trust between them, thus encouraging them to give honest answers. The interviewee feels free to say what they think without succumbing to any potential pressure or distraction from other participants [24]. In view of the vulnerability of adolescents in certain issues, interviews were conducted with parents (female and male) to learn about their opinion regarding their children's activity on social networks and the phenomenon of digital micro-influencers.

Parent interviews were chosen for the Zoom application, given the current situation of COVID-19, as the most viable alternative to ensure the safety of the interviewees and the interviewer.

The choice of participants comprised a weighted selection, whose main objective was to choose parents with different academic degrees and backgrounds, to obtain different perceptions. It was also intended to select parents with different genders (female and male), personalities, and points of view, so that the final answers were different, resulting in interesting research.

The interviews were conducted with a pre-defined script and included six participants, four mothers and two fathers and lasted about ten minutes each and were transcribed before their individual analysis.

Regarding presence on social networks, everyone recognized having Instagram, except for mother 1, the others confirmed that their activity is less than that of the students. Regarding the activity of their children on the social network, only two mothers affirm that their pupil does not consider, giving preference to TikTok, another social network often mentioned by the parents.

As for the recognition of the phenomenon of digital micro-influencers and, consequently, advertising on social networks, everyone claims to be familiar with this issue, including the mother who has no active account on the social network. Most parents claim to follow at least one micro-influencer and follow their work, they also claim to have already come across the use of digital micro-influencers as a marketing strategy. Mothers say they essentially follow fashion, beauty, cosmetics, and decoration pages.

Regarding parents' views on engagement, created by micro-influencers with teenagers, opinions were different and there were different points of view. "With my children I have never seen attitudes of this kind, but as a mother I would not allow it. Although I don't directly control the activity in Instagram, I try to understand how they use it and for what they use it, without being too controlling" (mother 1). "Kids are deluded by the apparent easy life presented by micro-influencers. They feel that their life is to appear and enjoy luxury services and they want to achieve these products and services effortlessly, translating into something negative" (mother 2).

As regarded, parental control of children's social network activity and more appropriate parental intervention, opinions again take on different proportions and there is a gap in the opinions of different family members.

In the case of mother 3, she sometimes controls the parental intervention, stating that "as everything in life has positive and negative points. I find it funny to see my daughters' taking ideas from micro-influencers about the way they wear certain clothes, however, we have to understand if it fits the age, the posture we want our children to have in society or not. Everything has its positive and negative sides". For his part, father 2 says he controls, although not exhaustively, recognizing that his children "must be accompanied exhaustively by their parents, be called to attention in the face of today's reality, understand what they are following and why". Mother 1 does not declare her children's activity, but she declares that "the parents must make them aware; they must act in such a way as to dissuade them if they think they should not buy certain products. They should talk a lot with them and make them realize that products are often an illusion and futility".

Parents also gave their views on the safety of Instagram for adolescents, generating different opinions among them. "I don't think it's very safe. My child still doesn't have Instagram because I don't think it's safe, he still doesn't have a strong enough mentality to make decisions and understand what's real and what's not, and who his real friends are and sometimes they can lead to extreme cases" (father 2). "I do. I feel that Instagram is not safe for teenagers. In adolescence, and I speak from experience, kids don't see evil in anything at all, I as a mother do. There should be a way in terms of technology to control certain access. I feel that Instagram is a useful tool, but it can be extremely dangerous" (mother 3). "I do. From my point of view Instagram is considered safe for teenagers, it is up to them to have some responsibility to use it in the best way and to be aware of the dangers of the internet" (mother 4).

24.7 Validation of the Hypotheses

This research was carried out using quantitative and qualitative data to validate or reject the hypotheses defined, in accordance with the research objectives, and the following evaluations were obtained.

H1—Adolescents use the Instagram social network.

The question related to hypothesis 1 (H1) was intended to prove the relationship between adolescents and their presence in the Instagram social network. H1 can be considered validated, as 97.4% of the adolescents surveyed are active in the Instagram social network.

H2—Teenagers are fans of digital micro-influencers.

According to the results presented, this hypothesis was rejected, as most young people do not consider themselves to be fans of digital micro-influencers (172 adolescents).

H3—Adolescent behaviour is influenced by digital micro-influencer.

Based on the results obtained, H3 is validated, as more than 30% of adolescents follow the recommendations of micro digital influencers.

H4—Parents control the activities of their adolescent children in the Instagram social network.

According to the results presented in Table 24.2, as well as the answers obtained in the interviews, most parents do not control their children's activity in the Instagram social network. Thus, hypothesis H4 is rejected.

24.8 Conclusions, Limitations, and Future Work

The results obtained by this research allow us to conclude that adolescents recognize the promotions that are presented by micro-influencers, besides realizing that currently, the number of followers is considered attractive as a marketing approach in social networks. In addition, parents have revealed that their children are still unaware of the proportions that this phenomenon of marketing through micro-influencers is taking and are not aware of the possibility of these partnerships having monetary rewards, which could be reflected in less credibility on the part of the micro-influencers.

The results of the research conclude that adolescents do not fully understand how they are influenced, nor are they aware that partnerships made by brands and micro-influencers usually have an economic return. On the other hand, the revealing hashtags of partnerships go unnoticed by most teenagers.

Regarding parental intervention, the interviews conducted as well as the data obtained reveal that parents have some difficulty in controlling their children's activity on social networks, since they have very busy daily lives dedicated to their professional activities and because they believe that their children are familiar with the pros and cons.

The limitations of the work are that the sample was obtained for convenience and is not significant. Thus, as research is directed at adolescents, it would be important to conduct face-to-face surveys, with a briefing on the topic, which has not been possible to do given the current COVID-19 pandemic. However, in future research and work, a focus group with parents and adolescents should be carried out. On the other hand, given the growing importance of the subject of this research for companies and brands, the survey should be extended to a significant sample so that the results obtained can have an adequate degree of confidence.

References

1. Voxburner: Why influencer marketing is the way to go for addressing youth generations. https://www.voxburner.com/blog-source/2017/3/6/why-influencer-marketing-is-the-way-to-go-for-addressing-youth-generations, last accessed 2020/07/12
2. Kataria, M.V.: Tween consumers: a study of the impact of social media on tween's buying decisions. Int. J. Market. Finan. Manage. **5**(1), 01–11 (2017)
3. Blackwell, R.D., Engel, J.F., Miniard, P.W.: Comportamento do Consumidor. Thomson Learning, São Paulo (2005)
4. Oliveira, S.: Marketing de Influência O fenómeno dos blogs de moda e beleza em Portugal, https://ubibliorum.ubi.pt/bitstream/10400.6/6363/1/5122_10005.pdf, last accessed 2020/07/14
5. Rutherford, G.: Porque é que os Micro Influenciadores são mais envolventes? https://www.apan.pt/os-micro-influenciadores-sao-envolventes/, last accessed 2020/06/06
6. Silva S., de Brito P.Q.: The characteristics of digital influencers and their ethically questionable attitudes. In: Rocha Á., Reis J., Peter M., Bogdanović Z. (eds.), Marketing and Smart Technologies. Smart Innovation, Systems and Technologies, vol. 167. Springer, Singapore (2020)
7. Politi, C.: Influenciador digital: o que é e como classificá-lo? https://www.influency.me/blog/influenciador-digital/, last accessed 2020/07/10
8. Santos, G.P.: Identificação de Microinfluenciadores no Instagram, https://www.cin.ufpe.br/~tg/2018-1/ghps-tg.pdf, last accessed 2020/07/10
9. Booth, N., Matic, J.A.: Mapping and leveraging influencers in social media to shape corporate brand perceptions. Corporate Commun Int J **16**(3), 184–191 (2011)
10. Steinmetz, K.: Instagram's challenge. Time Magazine, 46–51 (2019)
11. Guarda T., Lopes I., Victor J.A., Vázquez E.G.: User behaviour: the case of Instagram. In: Rocha, Á., Reis, J., Peter, M., Bogdanović, Z. (eds.), Marketing and Smart Technologies. Smart Innovation, Systems and Technologies, vol. 167. Springer, Singapore (2020)
12. Bessa, A.R.: Influenciadores em redes sociais digitais: uma análise aplicada ao Instagram. Escola de Comunicações e Artes, Universidade de São Paulo, São Paulo, Dissertação de Mestrado (2018)
13. Eisenstein, E.: Adolescência: definições, conceitos e critérios, https://cdn.publisher.gn1.link/adolescenciaesaude.com/pdf/v2n2a02.pdf, last accessed 2020/07/29
14. Affornalli, M.M.: A Publicidade e a Proteção do Consumidor Infanto-Juvenil: Breve denúncia da Violação de Garantias. Emancipação **6**(1) (2006)
15. Tartari, E.: Benefits and risks of children and adolescents using social. Eur. Sci. J. ESJ **11**(13) (2015)

16. Common Sense Media: Children, Teens, Media, and Body Image - A Common Sense Media Research Brief, pp. 96–98 (2015)
17. Calvert, S.L.: Children as consumers: advertising and marketing. Future Child. Spring **18**(1), 205–234 (2008)
18. Reppold, C., Pacheco, J., Hutz, C.: Comportamento agressivo e práticas disciplinares parentais. In: Hutz, C. (Org.). Violência e risco na infância e adolescência: Pesquisa e Intervenção, pp. 9–42 (2005)
19. Valcke, M., De Wever, B., Van Keer, H., Schellens, T.: Long-term study of safe Internet use of young children. Comput. Educ. **57**(1), 1292–1305 (2011)
20. Prensky, M.: Digital natives, digital immigrants part 1. On Horizon **9**(5), 1–6 (2011)
21. Reis, F.: Como elaborar uma dissertação de mestrado. Pactor, Lisboa (2010)
22. Coutinho, C.P.: Metodologia de Investigação em Ciências Sociais e Humanas, 2ª Ed. Edições Almedina, Lisboa (2013)
23. Pereira, AM.: Guia prático de utilização do SPSS: análise de dados para ciências sociais e psicologia, 7ª Ed. Edições Sílabo (2006)
24. Best Practices: Market Research with Children & Young People. https://info.angelfishfieldwork.com/market-research-fieldwork-blog/best-practices-qual-mr-children-young-people, last accessed 2020/07/06

Chapter 25
Consumer Profile and Behavior in Specific Marketing Contexts: A Study on Luxury Brands

Rosa Barbosa, Bruno Sousa, and Alexandra Malheiro

Abstract The luxury sector has aroused more and more interest, this fact is not only due to the economic importance, and it has in some countries but also due to its growth trend in emerging markets. Through the bibliographic analysis it is verified that the assets have a high degree of intangible, psychological and social values. Increasing the competitiveness of markets and internationalization requires that brands are increasingly oriented toward the market, with the analysis of consumer behavior being the most effective method to generate more value for the customer and in this way, attract and retain them.

25.1 Introduction

Luxury has changed over time. Although the construct is difficult to define given its subjective and personal nature, [1] argue that the term is shaped by high price, high quality and the effect of the brand, showing that luxury is related to the exclusive, the rare and limited, which consequently has a high cost.

More recent studies show that currently, luxury does not value so much the ostentation of goods, but rather the search for experience and individual pleasure, as mentioned by [2], "what we see today is attraction for the luxury of the senses, of the pleasure and sensitivity felt in the intimacy of each individual and not the outward luxury of display and opulence, which aims simply to demonstrate status". Nowadays, there is a growing desire for luxury products, where consumers are increasingly willing to pay for it. This fact shows that luxury products go beyond traditional consumers, no longer being inaccessible and exclusive to be part of the dreams of individuals in general, causing the luxury market to register a great growth in recent

R. Barbosa · B. Sousa (✉) · A. Malheiro
Polytechnic Institute of Cávado and Ave (IPCA), Barcelos, Portugal
e-mail: bsousa@ipca.pt

A. Malheiro
e-mail: amalheiro@ipca.pt

CiTUR – Centre for Tourism Research, Development and Innovation, Coimbra, Portugal

J. L. Reis et al. (eds.), *Marketing and Smart Technologies*, Smart Innovation, Systems and Technologies 337, https://doi.org/10.1007/978-981-19-9099-1_25

years. The luxury market is very important and has shown some immunity to all adverse economic conditions, which has aroused some curiosity. The continuous growth of this market has placed great challenges and changes in the management of brands that operate in this sector. In this sense, it is crucial to know the desires, motivations and external variables that influence the consumer's intention to purchase luxury brands.

In this context, an exploratory approach based on literature review will be carried out. This review includes the analysis and interpretation of scientific articles and books, in order to carry out a state of the art and address the most recent studies on luxury, luxury products, luxury brands and the luxury brand consumer. This literature review aims to know and evaluate the most relevant aspects that influence the intention to purchase luxury goods, so that brands develop more effective strategies in order to attract and retain customers.

25.2 Luxury and Marketing

The concept of luxury is attractive and elegant [3] and difficult to define. According to [4], this fact is due to the subjective and personal nature of the construct, which depends on its social, economic and cultural evolution. Each author ends up having their own notion of luxury, however, [1] argue that the term is molded to high price, high quality and the effect of the brand, showing that luxury is related to the exclusive, the rare and limited which, consequently, has a high cost. For [5], luxury is about pleasure, refinement, perfection and rarity, as well as appreciation, but not necessarily about price. Luxury still has the values of tradition, history, craftsmanship and excellence in details, as its greatest assets. Paraíso [6] argues that luxury has a fascinating magnetism that attracts attention, curiosities and desires of belonging, in addition to, it attracts for its visible and absolutely irreproachable quality, but above all for its intangibility, mystery, class, innovation, history, tradition, exclusivity, sophistication and pleasure for the senses that makes them accessible by few and desired by many.

Tynan et al. [7] argue that luxury can be defined as non-essential products and services, with high prices and high quality, where exclusivity and prestige offer a high symbolic level to consumers. However, [8] argue that the physical characteristics of products are no longer associated with luxury, with the brand and lifestyle being considered important characteristics related to luxury. Products with premium prices, limited offers and high demand were once considered attributes of luxury goods where only a few people, those who had money, enjoyed the privilege of being able to purchase luxury products, that is, they were bought by individuals who they had a comfortable life financially. The definition of luxury is no longer based only on the wealth of consumers, since, currently, any individual can belong to the world of luxury, where he increasingly desires luxury and is willing to pay for it [9]. This growing desire for luxury products that goes beyond traditional customers gives rise to a new concept, the "democratization of luxury". This means that luxury is no longer inaccessible and exclusive, to become part of the dreams of individuals in general.

Luxury is no longer what it used to be, it has evolved over time. Less recent studies show that luxury is strongly linked to social stratification [10], a notion of utility, as well as decisions related to the distribution of wealth, as defended by [11]. However, in more recent studies, there is a shift from a luxury more focused on emotions and less on appearance, stating that what we see today is attraction to the luxury of the senses, pleasure, sensitivity felt in the intimacy by each individual and not the exterior luxury, of exhibition and opulence, which aims simply to demonstrate status. This means that luxury currently values more the character of the experience, being focused on the individual pleasure of feeling different and for the admiration of others, devaluing its elitization that aimed solely and exclusively at ostentation. Luxury does not represent a need or a requested expense but a personal, authentic, emotional and experimental nature. Luxury is mainly based on the emotions, it provides than on the good that is purchased, thus demonstrating that luxury is behavioral and showing that consumers buy it for emotional reasons, always in a logic of individual choice but above all in a way of expression.

The value of the luxury object is the result of social perception and not, properly, of the characteristics of the good: luxury is essentially a social invention, a creation of man. It is, fundamentally, a set of meanings attributed to certain objects and activities.

With regard to the accessibility of luxury, there are several authors who argue that it is hierarchical. This means that there is not just one luxury category, but different levels for different consumers. Thus, making the inaccessible accessible is one of the main and most complex challenges facing the luxury market, without neglecting the prestige and magic of a particular product [12].

For a consumer to remain in the highest category of luxury, that is, inaccessible luxury is a complicated task. On the one hand, consumers at this level have an increasingly higher purchasing power, in the same way that the number of intermediate and affordable luxury consumers is larger, however, it is not usual for them to drill into the inaccessible luxury category. Thus, [13] argues that luxury products can be ranked according to three levels, based on the degree of accessibility.

25.3 Luxury Products and Services

For centuries and in a generalized way around the world, consumers have been satisfied with the purchase of luxury products. However, nowadays, this phenomenon has been gaining intensity, as consumers are increasingly purchasing luxury products, seeking that they provide them with greater satisfaction and that they are more and more distinct, that is, more luxurious. Demonstrating that, currently, one of the main goals of luxury products is to impress others and position the owner in a specific social class to which he wants to belong [14].

Luxury finds its peak when the product has limited production or when something seems extraordinary to us. This limited purchase of the product gives the consumer the privilege of having something unique and unusual [15]. However, luxury is not, by definition, something unique, but something that arouses an increased desire to acquire it, that is, a product becomes more and more luxurious as the desire to have it increases. According to [16], a luxury product is a good that the individual wants, but does not need it and which, in addition to comfort, provides pleasure. In everyday life, the term luxury is used to refer to products, non-essential services that contribute to a luxurious life, behind the bare minimum. Luxury can be defined as everything that is consumable or not, which goes beyond our everyday reality and which has a strong symbolic content of personal pleasure or social admiration and which has as key elements, a strong element of human involvement, very limited supply and recognition of value by others [17].

Kapferer [10] defines luxury products as:

> (…) well-crafted pieces, hedonistic and aesthetic objects, with prices exceeding their functional utility, sold in exclusive stores that provide personalized service and a unique experience to the consumer, usually from a brand with history, heritage, the set provides a rare feeling of exclusivity. (p. 43)

Luxury products are characterized by being governed by certain characteristics. For [18], a luxury product has excellent quality and durability, good taste and refinement status, high price, scarcity and uniqueness, strong aesthetic appeal, history and tradition. Quality, strong identity, limited production, low availability, rarity, differentiation, selective and limited distribution, the symbolic sphere of dream and magic, appeal to the senses and high price are attributes considered fundamental for a product to be considered luxurious. A luxury product must be something personalized, technically perfect, aesthetically beautiful. Adding the fact that this is rare, out of the ordinary completes the necessary attributes to be considered a luxury object, and its rarity and scarcity is the justification for the price difference.

25.4 Luxury Brands

In recent years, the consumption of luxury brand products has represented considerable sales all over the world, in several product categories, namely clothing, accessories, high quality watches and jewelry [19]. This globalization of luxury is slowly causing luxury products and services to become very similar and lose value. In this sense, luxury brands must permanently invest in employees who convey friendliness and charm, in order to provide pleasure and happiness to their customers and, in this way, differentiate themselves from other brands luxurious.

However, the perception of luxury brands does not have a concrete definition in different market segments, as luxury is a subjective concept, which depends on the perception of value by each consumer [4]. However, it is possible to define luxury brands as the top of the prestige brands that essentially reflect psychological, social, hedonic values, products with excellent quality [20] that are based on tradition, connection with the designer and creativity [21]. Thus, there is a positive relationship between the brand's perceived value and the desire to consume it.

Luxury brands, unlike "normal" brands, evoke feelings and emotions in the people who buy them. In addition to presenting a limited production [22] and, consequently, a scarce supply, they manage to generate a certain recognition among others. According to this thought, luxury brands are seen as unique and differentiated by their unique characteristics that cannot be found in non-luxury brands. However, [23] scarcity is the characteristic that defines a luxury good and for this reason they are attractive to consumers who have a great need for exclusivity. High prices, excellent quality, aesthetics, rarity and symbolic meaning are crucial attributes in luxury products. However, they are not enough for consumers to consider luxury, as they need a marketing strategy that enhances these characteristics to customers.

The psychological benefit, which may be superior to the functional one, is the main distinguishing factor between luxury and non-luxury brands. The brand is a guarantee of luxury, that is, the consumer seeks a certain brand because it is associated with luxury. It is found that brands are increasingly consumed according to what they represent, whether in lifestyle, concept or even emotions and not so much related to the product [24]. Consumers choose a certain luxury brand according to their personality and values affinity with it, instead of a functional comparative process. Luxury brands aim to create strong associations in consumers' minds about a high level of benefits and social status. Luxury brands are tied to images created in the minds of consumers associated with high price, quality, aesthetics, rarity, extraordinary and symbolic value.

25.5 Luxury Consumer

Luxury consumption has always been present at all times, as it works as a way of marking the difference between the favored social classes and the rest of the population. Consumption is rational and objective choices, always aiming at an "ideal" purchase—one in which maximum utility and lowest monetary expenditure are combined. However, despite the fact that the act of consumption has been technically explained for a long period, behind it there are motivations, desires, culture and other influencing factors. The consumer has personality and, for this reason, each person identifies and understands the world in different ways, reacting automatically to the context through the senses. This means that each consumer is a particular case, who has their own motivations and desires, which lead them to consume luxury products that provide them with greater satisfaction [24].

Consumers with great purchasing power always have very high expectations, appreciate the unexpected and how much they are surprised, so they rarely question the price. Thus, constant creativity is one of the key factors for any luxury product or service to be truly so. Luxury consumers are those who aim to build a positive social identity by satisfying their need to belong to a social group, which is the basic reason why they purchase luxury products.

In the contemporary luxury market, it is possible to observe essentially two types of luxury consumers. There are consumers who seek to reproduce purchasing behaviors from social groups they would like to belong to, which they call aspirational consumers. These consumers are characterized by having conspicuous consumption, that is, they tend to acquire luxury brands with great emphasis on the visibility of the brand logo, in order to demonstrate their wealth to society. Traditional consumers are from traditionally wealthy families who want to maintain status. In the contemporary luxury products market, it can be seen that there are consumers who are attracted by the highly visible markers of the brand in order to display its wealth [25] and it is possible to verify the existence of an increase in demand by the sophistication and subtlety of the luxury brand, demonstrating a shift in consumer attitudes away from social class and the quest for status. Furthermore, consumers are currently more informed, more demanding and, at the same time, more sensitive to price and less sensitive to the brand [24].

25.6 Final Considerations and Next Steps

The literature review on the concept of luxury revealed the lack of a consensual definition of the construct, since it has a subjective and personal nature, which depends on its social, economic and cultural evolution, making each author has his own notion of luxury. However, luxury is molded to high price, high quality and the effect of the brand, showing that luxury is related to the exclusive, rare and limited, which, consequently, has a high cost. This financial barrier makes luxury have a social function, that is, it helps in social representation. Luxury has changed over time, with a growing appreciation of emotions, that is, consumers value more the character of experience and pleasure than the luxury product provides them, devaluing its ostentation.

Luxury brands essentially reflect psychological, social, hedonic values, products with excellent quality that are based on tradition, connection with the designer and creativity. Thus, there is a positive relationship between the brand's perceived value and the desire to consume it. However, the globalization of luxury is slowly causing luxury products and services to become very similar and lose value. In this sense, luxury brands must permanently invest in employees who convey friendliness and charm, in order to provide pleasure and happiness to their customers and, in this way, differentiate themselves from other brands luxurious. Thus, the exploratory approach carried out allowed us to conclude that the future of luxury brands involves the ability to consume luxury goods to provide pleasure and evoke emotions to consumers, in addition to serving as a social marker, i.e., positioning the owner in a specific social class. This requires that luxury brands define their target audience, so that they create strategies that meet the needs and desires of their consumers.

Acknowledgements This work is financed by national funds through FCT—Foundation for Science and Technology, IP, within the scope of the reference project UIDB/04470/2020.

References

1. Lipovetsky, G. & Roux, E. (2012). *O luxo eterno: da idade do sagrado ao tempo das marcas.* Lisboa, Portugal: Edições 70.
2. Vilhena, M. & Passos, J. (2018). *Religião e consumo: relações e discernimentos.* São Paulo, Brasil: Paulinas.
3. Kapferer, J., Bastien, V.: The specificity of luxury management: turning marketing upside down. J. Brand Manag. **16**(5–6), 311–322 (2009)
4. Phau, I., Prendergast, G.: Consuming luxury brands: The relevance of the 'Rarity Principle.' J. Brand Manag. **8**, 122–138 (2000)
5. Roux, E. & Floch, J. (1996). Gérer l'ingérable: La contradiction interne de toute maison de luxe. *Décisions Marketing*, 9 (Setembro – Dezembro): 15 - 23.
6. Paraíso, A. (2013, outubro 16). *Quo vadis, Luxus?.* Disponível em: http://macppg.blogspot.com/2013/
7. Tynan, C., McKechnie, S., Chhuon, C.: Co-creating value for luxury brands. J. Bus. Res. **63**(11), 1156–1163 (2010)
8. Shao, W., Grace, D., Ross, M.: Consumer motivation and luxury consumption: testing moderating effects. J. Retail. Consum. Serv. **46**, 33–44 (2019)
9. Danziger, P. (2005). *Let them eat the cake: marketing luxury to the masses: as well as the classes.* Londres, Inglaterra: Kaplan Business.
10. Kapferer, J. (2010). Luxury after the crisis: pro logo or no logo. *The European Business Review,* (Setembro - Outubro): 42–46.
11. Kapferer, J. & Bastien, V. (2012). The luxury strategy: break the rules of marketing to build luxury brands.
12. Galhanone, R. (2005). *O mercado do luxo: aspetos de marketing.* São Paulo
13. Allérès, D. (2006). *Luxo… Estratégias, marketing.* Rio de Janeiro, Brasil: Editora FGV
14. Husic, M., Cicic, M.: Luxury consumption factors. J. Fash. Mark. Manag. **13**(2), 231–245 (2009)
15. García, S. (2003). *El universo del lujo: una vision global y estrategica de los amantes del lujo.* Madrid, Espanha: Mc Graw – Hill.
16. Carnevali, F.: Luxury for the masses: Jewellery and jewellers in London and Birmingham in the 19th century. Entrep. Hist. **46**(1), 56–70 (2007)
17. Cornell, A. (2002, abril 27). Cult of luxury: the new opiate of the masses. *Australian Financial Review.*
18. Dubois, B., Laurent, G. & Czellar, S. (2001). Consumer rapport to luxury: Analyzing complex and ambivalent attitudes, *Consumer Research Working Article*, nº. 736, HEC, Paris.
19. Bian, Q., Forsythe, S.: Purchase intention for luxury brands: A cross cultural comparison. J. Bus. Res. **65**(10), 1443–1451 (2012)
20. Prendergast, G., Wong, C.: Parental influence on the purchase of luxury brands of infant apparel: An exploratory study in Hong Kong. J. Consum. Mark. **20**(2), 157–169 (2003)
21. Nueno, J., Quelch, J.: The mass marketing of luxury. Bus. Horiz. **41**(6), 61–68 (1998)
22. Vigneron, F., Johnson, L.: Measuring perceptions of brand luxury. Journal of Brand anagement **11**, 484–506 (2004)
23. Kim, E., Drumwright, M.: Engaging consumers and building relationships in social media: How social relatedness influences intrinsic vs. extrinsic consumer motivation. Comput. Hum. Behav. **63**, 970–979 (2016)

24. Lourenço, V., Catarino, A., Fonseca, M. & Sousa, B. B. (2020). *Consumer-brand relationship and use of the website in virtual communication in the luxury furniture industry*. De Rodrigues, P. and Borges, A. P. (Eds.), Building Consumer-Brand Relationship in Luxury Brand Management (pp.). IGI Global.
25. Chan, W., To, C., Chu, W.: Materialistic consumers who seek unique products*: How does their need for status and their affective response facilitate the repurchase intention of luxury goods?*. J. Retail. Consum. Serv. **27**, 1–10 (2015)

Chapter 26
The Impact of YouTube and TikTok Influencers in the Customer Journey: An Empirical Comparative Study Among Generation Z Users

Paulo Duarte Silveira, Fábio Sandes, and Duarte Xara-Brasil

Abstract The present paper aims to analyze the impact of YouTubers and TikTokers influencers in the customer journey phases, among Generation Z users. To do so, a quantitative deductive empirical study was carried out. The respective data collection was made via an online questionnaire survey, obtaining a valid sample of 529 participants. The results show that both type of influencers might influence the customer journey, but mainly in the first stages of the process. It was also found that YouTubers tend to have a higher influence in each of those stages than TikTokers. Although the topic of social media influencers has already been studied in several research papers, no study was found addressing separately the stages of the customer journey decision-making process, neither comparing influencers of both platforms in that context.

26.1 Introduction

Social media users have been increasing across the world. In 2021, there were 2.26 billion users, and it is expected that in 2027, the global number of users will reach 5.85 billion [1]. The use of social media platforms is especially relevant among the younger generation: in 2021, in the US, 70% of the citizens stated that they use social network sites (SNS), and 84% of adults aged between 18 and 29 say they use, mainly YouTube, Instagram, Facebook, Snapchat, and TikTok [2].

P. D. Silveira (✉)
Instituto Politécnico de Setúbal, ESCE, Setúbal, Portugal
e-mail: paulo.silveira@esce.ips.pt

CEFAGE, Universidade de Évora, Évora, Portugal

F. Sandes
Universidade Lusófona, CICANT Research Centre, Lisbon, Portugal
e-mail: fsandes@gmail.com

D. Xara-Brasil
Instituto Politécnico de Setúbal, ESCE, CICE, Setúbal, Portugal
e-mail: duarte.brasil@esce.ips.pt

J. L. Reis et al. (eds.), *Marketing and Smart Technologies*, Smart Innovation, Systems and Technologies 337, https://doi.org/10.1007/978-981-19-9099-1_26

Aligned with the growing importance of Internet and specifically around SNS, digital influencers expanded exponentially in numbers and credibility, and firms diverted billions of dollars in their marketing budgets [3]. An increasing number of companies are allocating higher budgets on digital marketing and digital influencers due to their relevance and persuasive power [4, 5]. These digital influencers represent an independent third-party endorser who shape audiences' attitudes, creating and sharing content, endorsing brands and products through different social media platforms [4]. Digital influencers may have different typologies, from nano-influencers to macro-influencers and celebrities' endorsers [5], different backgrounds, and different purposes. Therefore, they can be effective across many different market segments by presumably being authentic and relevant in their content creation and in the interaction with their followers, attracting attention for products and services and often compel their followers to action [6]. In fact, they are being used along with traditional celebrity endorsers in digital marketing activities because they appear to be more authentic, like regular consumers [7]. So, digital marketing strategy and digital influencers may have an important impact on the different stages of the customer journey, due to their credibility, expertise, intimacy, and authenticity with their followers. Given that Generation Z consumers grew up in an "always on" technological mode, and research has already shown that this brings dramatics shifts in youth behaviors, attitudes, and lifestyles [8], the influencers' impact might be particularly relevant within that generation.

Therefore, based on the available literature, this study intends to understand how individuals belonging to Generation Z perceive the level of influence of YouTubers (YTrs) and TikTokers (TTrs) in their consuming decisions, not restricting the approach to decisions related to buy/not buy, but rather understanding the impact in each stage of the customer journey process. Consequently, this study addresses the following two research questions, focusing on Generation Z users:

- RQ1: Does YTrs and TTrs influence the phases of the customer journey process in the light of users' perception?
- RQ2: Are there perceived differences between YTrs and TTrs in their ability to influence each phase of the customer journey?

To address those research aims, this paper is structured as follows. After this introduction, the theoretical background is presented, addressing the characterization of Generation Z and its use of SNS. Then, the literature related to digital influencers and their impact in the customer journey is revised. This is followed by the description of the methodology adopted in the empirical study, describing the survey, variables used, and sample obtained. Next, the statistical results are presented. Finally, in the last section, the conclusions report the main findings, and the implications for managers and researchers are drawn, as well as the study limitations.

26.2 Theoretical Background

26.2.1 Generation Z and the Use of Social Network Sites

According to Pew Research [9], the year of birth, and therefore age, is an important predictor of differences in attitudes and behaviors among individuals, being the reason why cohort analysis is important, leading to grouping individuals in generations. Some generations are already widely accepted and mentioned in literature, like Baby Boomers, Generation X, Millennials, or Generation Z (also known as "Centennials" or "GenZ"). There is not a unique definition of who belongs to GenZ, but according to Pew Research Center (2019), most part of the literature considers that GenZ is composed of individuals who were born between the 1996 and 2010 [10]. More important than the birth year is what has shaped each generation and its main characteristics. GenZ is the first generation that has never known a world without the Internet, a technology that has become a natural part of their lives, being true digital natives [10, 11]. Therefore, one defining characteristic of GenZ is that they are the first generation that grew up using SNS, like YouTube, Instagram, Snapchat, Twitter, and TikTok.

GenZ is the cohort with the biggest population in the consumer world nowaday [11–13], and anecdotal evidence suggests that these consumers may behave differently than consumers from previous generations. It is expected that these consumers would "do everything" online, but marketing practitioners seem to figure out that, even though these consumers have been connected online throughout their whole lives, their shopping behavior demands a tactile experience [14, 15], where they appreciate the presence of physical stores. This demand, however, is not due to lack of trust of what they see online, but it is more related to a thrive for a different consumption experience in retail settings. Even though GenZ consumers are familiar with most of SNS channels, some are more popular among these consumers, namely Instagram, YouTube, and TikTok [16]. Furthermore, YouTubers influencers are especially popular among youngers, namely, pre-teens, teens and young adults [17, 18]. The same is true for TikTok [19–22]. In fact, according to the latest data available from Datareportal [23], TikTok users aged 18–24 account for the largest share of its advertising audience.

However, the reason why people use TikTok seems to be different from the reasons related in using YouTube. According to Montag et al. [22], the use of TikTok by individuals might be explained by gratification theory, social impact theory, and self-determination theory. Besides, TikTok has been used for several purposes, including for marketing products [24].

All those facts lead to the proposition that YouTube and TikTok might pose different challenges for marketers, especially in what relates to savvy Internet users like GenZ. Therefore, a comparative study of both platforms was designed in the present study.

26.2.2 Customer Journey and Touchpoints

The concept of consumer journey was created by Court et al. [25] in the practitioner market, and it has been discussed and used in the market by marketing professionals ever since [26]. Lemon and Verhoef [27] adapted this concept to the academic literature by connecting the concept of the journey with the concept of customer experience, mentioning that the journey is influenced by previous journeys, and it will influence future journeys. The authors, then, proposed the concept of customer journey, an academic version of the consumer journey proposed by Court et al. [25] that incorporates the notion of past and future experiences a customer experience throughout their journeys. There is not a unique "state of the art" definition for customer journey in the literature, as the discussion related to its elements and terminologies are extensive throughout different perspectives, but from the systematic literature review performed by Folstad and Kvale [28] it is clear that the customer journey incorporates previous customer experiences, and it is a consumption journey that can be separated into three main stages. The (i) prepurchase, where individuals recognize/identify needs, look for options, acquire information, analyze and compare options. The (ii) the purchase stage, in which the individuals analyze and choose an option to buy, not buy, or postpone. The (iii) post-purchase stage, where they use, discard, and comment their experiences on using the product or service. It is important to notice that the customer journey framework considers the buying process not as a fragmented flow process, but as a dynamic and iterative process in which the individuals have multiple touchpoints to interact with brands and products, and this journey is connected to past and previous customer experiences [27].

So, throughout this journey, the customers go through several touchpoints—points of formal and informal interaction between the customer and brands— where they access and exchange information with brands and other consumers [27–29]. In these interactions, customers access information, view, touch and try products or services, buy, pay, return, complain, connect with other customers, socialize with their friends, learn about a brand, among several other activities [30]. These touchpoints may be online, offline, and are either controlled by the brand (i.e., their online and offline stores), or are out of their control (i.e., SNS channels, such as TikTok, Twitter, and Instagram) [31].

Resuming, the concept of customer journey has been vastly discussed in the marketing literature, as it entails researchers to focus on the experience customers have in each stage, and in the connections between previous and future experiences challenging the traditional view of the buying process. So, it is highlighted that the journey is a continuous, dynamic, and iterative process, where consumers interact with brands through a "myriad touchpoints in multiple channels and media, and customer experiences are more social in nature" [27].

26.2.3 *Digital Influencers in the Customer Journey*

Recent previous studies (e.g. [32, 33]) have explored the relevance of digital signs and SNS in the customer journey with different focus, like in specific markets where the presence of electronic word-of-mouth and user generated content is essential, such as tourism (e.g. [34, 35–37]) or clothing (e.g. [6, 38, 39]). Among all these studies found an agreement related to the fact that digital influencers have an important influence in consumers' decisions.

Kozinets et al. [40] studied how the creation of desire occurs in the use of SNS. They performed an in-depth analysis of how content is created and disseminated in SNS and extended the theory of desire by conceptualizing desire creation and critically questioning the role of SNS in the creation of a technologically enhanced desire in consumers. Specifically, they analyzed the desire for food, with the analysis of the phenomena of #foodporn on Instagram. This complex research is an example of how consumers are influenced by what they post and see in SNS and how it affects their customer journeys. Complementary, Jiménez-Castillo and Sánchez-Fernández [41] discussed how digital influencers influence consumers' purchase intention, engagement and expectation. These are examples of how the literature is approaching SNS and digital influencers are present in different stages of the consumer journey.

26.3 Methods

To address the research problem, a quantitative empirical study was carried out with primary data, inquiring a sample of YouTube and TikTok Generation Z users, via an online questionnaire built specifically for this purpose using Google Forms. The questionnaire was disseminated in 2021, via Facebook, Instagram, WhatsApp, and LinkedIn, between October and November.

To be included in the sample, each respondent must have had already seen product/brand reviews in YouTube and TikTok. Users that had seen product reviews in only one of those social media platforms were excluded from analysis. Following that procedure, a valid sample of 529 individuals was the final basis for analysis. The data was analyzed with IBM SPSS Statistics 25 and the mean age of respondents was 19.49 (standard deviation 2.288), 70% female and 30% male. Besides the filtering and demographical questions, the questionnaire contained questions directed to obtain information to provide statistical testing basis to answer the hypothesis established. The variables used to do so are summarized in Table 26.1. All those variables were answered by respondents in a four-option frequency scale ("1-never", "2-rarely", "3-sometimes", "4-very often").

Table 26.1 Main variables in the questionnaire

Variable	Platform	Cust. journey stage	Operationalization
YT-Awr	YouTube	Pre-purchase (Need recognition)	Know about new products or brands due to YouTubers
TT-Awr	TikTok	Pre-purchase (Need recognition)	Know about new products or brands due to TrikTokers
YT-Lrn	YouTube	Pre-purchase (Need recognition)	Learn more about products in YouTube without being that the initial searching purpose
TT-Lrn	TikTok	Pre-purchase (Need recognition)	Learn more about products in TikTok without being that the initial searching purpose
YT-SrchBfr	YouTube	Pre-purchase (Information search)	Search for YouTubers product reviews videos before a purchase
TT-SrchBfr	TikTok	Pre-purchase (Information search)	Search for TikTokers product reviews videos before a purchase
YT-SrchAft	YouTube	Pre-purchase (Alternatives evaluation)	Search for more informations about products after watching YouTubers
TT-SrchAft	TikTok	Pre-purchase (Alternatives evaluation)	Search for more informations about products after watching TikTokers
YT-Purch	YouTube	Purchase	Purchase after watching YouTubers reviews
TT-Purch	TikTok	Purchase	Purchase after watching TikTokers reviews
YT-Cup	YouTube	Purchase	Used YouTubers cupons
TT-Cup	TikTok	Purchase	Used TikTokers cupons
YT-Rgrt	YouTube	Post-purchase	Regret having followed advices of YouTubers product reviews
TT-Rgrt	TikTok	Post-purchase	Regret having followed advices of TikTokers product reviews

Source survey output

26.4 Results

A comparative analysis of YTrs and TTrs was made using parametric tests. In this study, even if several variables do not present normal distributions (as seen in the following Table 26.2 with K-S tests $p < 0.05$), according to Marôco [42] the parametric testing techniques were possible to be done because the sample was not small and the variables' distributions were not extremely skewed or flat (i.e., symmetry $sk < 3$ and flatness values $|ku| < 10$), as presented in Table 26.2.

Table 26.2 Descriptive statistics of the main variables in the study

Variable	N	Mean	St. Dev.	*Sk*	*ku*	K-S test
YT-Awr	529	3.25	0.740	−0.604	−0.382	0.000
TT-Awr	529	2.86	0.945	−0.353	−0.850	0.000
YT-Lrn	529	2.98	0.287	−0.527	−0.492	0.000
TT-Lrn	529	2.54	1.026	−0.062	−1.124	0.000
YT-SrchBfr	529	2.83	1.033	−0.418	−1.000	0.000
TT-SrchBfr	529	1.89	0.995	0.773	−0.593	0.000
YT-SrchAft	529	2.79	0.899	−0.331	−0.650	0.000
TT-SrchAft	529	2.11	0.995	0.453	−0.889	0.000
YT-Purch	529	2.50	1.015	0.042	−1.101	0.000
TT-Purch	529	1.83	0.948	0.814	−0.468	0.000
YT-Cup	529	2.19	1.160	0.339	−1.394	0.000
TT-Cup	529	1.62	0.976	1.329	0.389	0.000
YT-Rgrt	529	1.70	0.885	0.989	0.038	0.000
TT-Rgrt	529	1.60	0.845	1.232	0.511	0.000

Source Survey output

Considering the middle and neutral point of the response options (threshold 2.5), not all the variables present a higher mean than that. In Table 26.3, that same statistical significance tests are shown.

As mentioned, Table 26.3 presents the individual statistical testing of the perceived impact of each variable studied, by social media platform. It is not totally clear neither straightforward to draw conclusions and generalize from that, but it seems that both YTrs and TTrs have a higher influence in the earlier stages of the customer journey. It is also noticeable that YTrs tend to have a higher impact in each of those stages, when compared to TTrs.

The already mentioned statistical parametric testing made to compare the perceived influence of YTrs and TTrs in each variable of the customer journey is presented in Table 26.4. It is possible to observe significant differences in each variable, leading to the insight that YTrs and TTrs in fact do not have the same impact in the customer decisions. Moreover, the perceived influence of YTrs is significantly higher than the influence of TTrs in all variables and stages of the customer journey.

26.5 Conclusions, Limitations, and Implications

This study aimed to contribute to the discussion of the impact of digital influencers in marketing, more specifically to address and compare the influence of YouTubers and TikTokers along the customer journey. The study was focused on GenZ

Table 26.3 Perceived impact of YTrs and TTrs in the customer journey

Variable [a]	t	df	Sig. 2-tailed	Mean difference to threshold [b]
YT-Awr	31.176	784	0.000*	0.813
TT-Awr	7.794	786	0.000*	0.267
YT-Lrn	14.955	684	0.000*	0.504
TT-Lrn	−0.019	682	0.000*	−0.001
YT-SrchBfr	8.456	785	0.985	0.304
TT-SrchBfr	−16.374	646	0.000*	−0.642
YT-SrchAft	9.809	666	0.000*	0.341
TT-SrchAft	−12.317	783	0.000*	−0.439
YT-Purch	−1.042	769	0.298	−0.038
TT-Purch	−23.839	769	0.000*	−0.784
YT-Cup	−7.717	750	0.000*	−0.324
TT-Cup	−28.108	770	0.000*	−0.955
YT-Rgrt	−29.044	750	0.000*	−0.850
TT-Rgrt	−33.006	756	0.000*	−0.973

Source Survey output
[a]Variables measured from 1 to 4 (1-never, 2-rarely, 3-sometimes, 4-very often)
[b]Difference to middle point of the response options (2.5)
*Significance 2-sided test 0.05

consumers, aiming to understand how these consumers are influenced by YouTubers and Tiktokers on their consumer journeys.

Regarding the first research question (Does YTrs and TTrs influence the phases of the customer journey process in the light of users' perception?), the results indicate that both YouTubers and TikTokers influence various stages of the customer journey, but that fact is much more visible in the prepurchase stage of the journey. Concerning the second research question (Are there perceived differences between YTrs and TTrs in their ability to influence each phase of the customer journey?), the results also show that such power is not equal between those two social networks, since YouTubers tend to have a higher perceived influence. This is consistent with the fact that videos from YouTube are longer than in TikTok, and therefore, contains more information that would help consumers in acquiring more information about a given product in the prepurchase stage.

In what relates to implications for managers, the results suggest that both YouTube and TikTok are relevant tools to consumers, especially in the prepurchase stages of their journeys, and if brand managers critically analyze the digital influencers in these SNS they may find a way to connect, interact and promote their brands to GenZ consumers in this stage. YouTubers can be used in marketing materials with more dense information, and TikTokers may be used in short videos with one or two essential information about a brand or product, that may be useful in the pre-purchasing stage of consumers' journeys.

Table 26.4 Equality of means T-test between YTrs and TTrs influence in the customer journey

How frequently[a]…	T	df	Sig. (2-tailed)	Difference means	Std. error difference
Know about new products or brands due to YT versus TT	14.805	784	0.000*	0.549	0.037
Learn more about products without being that the initial searching purpose YT versus TT	11.525	682	0.000*	0.505	0.044
Search for product reviews videos before a purchase YT versus TT	20.169	645	0.000*	0.997	0.049
Search for more information about products after watching YT versus TT	16.957	666	0.000*	0.732	0.043
Purchased after watching YT versus TT	18.699	747	0.000*	0.737	0.39
Used cupons YT versus TT	16.707	768	0.000*	0.633	0.038
Regret followed advice of product reviews YT versus TT	5.323	743	0.000*	0.122	0.023

Source Survey output
[a]Variables measured from 1 to 4 (1-never, 2-rarely, 3-sometimes, 4-very often)
*Asymptotic significance 2-sided test 0.05

Regarding the limitations and implications for research, the study was conducted with a large, but non probabilistic sample, which poses limitations to generalizations. Also, it is important to stress that respondents recalled their previous experience when answering about previous impact on their customer journeys, and therefore, memory of past events may not be totally accurate. Therefore, future studies could address these matters by capturing respondents' actual responses and behaviours, without the limitation of memory. Furthermore, the SNS context is rapidly and constantly changing, so it is important to study how the influence of SNS and digital influencers change when the context changes. More specifically, we suggest that future studies may offer tools that help monitoring the influence of specific SNS and digital influencers in the customer journey stages and touchpoints.

Acknowledgements The authors acknowledge financial support from Instituto Politécnico de Setúbal (IPS) and National Funds of the FCT–Portuguese Foundation for Science and Technology within the project "UIDB/04007/2020". Authors also gratefully acknowledge the help of IPS Marketing Research course students in gathering the primary data.

References

1. Statista page https://www.statista.com/statistics/278414/number-of-worldwide-social-network-users/. Last accessed 2022/08/1 (2022)
2. Auxier, B., Anderson, M.: Social media use in 2021 page https://www.pewresearch.org/internet/2021/04/07/social-media-use-in-2021/. Last accessed 2022/07/11 (2022)
3. Leggett, B.R.: Social media influencers: an examination of influence throughout the customer journey (Doctoral dissertation, University of South Alabama) (2022)
4. Freberg, K., Graham, K., McGaughey, K., Freberg, L.A.: Who are the social media influencers? A study of public perceptions of personality. Publ. Relations Rev. **37**(1), 90–92 (2011)
5. Zeng, J.: 5 marketing strategies—and missed opportunities—during COVID-19 American Marketing Association page https://www.ama.org/marketing-news/influencer-marketing-grows-up-in-2020, last accessed 2022/07/15 (2019)
6. Abidin, C.: Aren't these just young, rich women doing vain things online? Influencer selfies as subversive frivolity. Social Media+ Society **2**(2), 1–17 (2016)
7. Barker, S.: What's the difference between celebrities and influencers—and which does your brand need? Page https://smallbiztrends.com/2018/02/influencers-vs-celebrities.html. Last accessed 2022/08/1 (2021)
8. Dimock, M.: Defining generations: where millennials end and generation Z begins. https://www.pewresearch.org/fact-tank/2019/01/17/where-millennials-end-and-generation-z-begins/. Last accessed 2022/07/15 (2019)
9. Pew Research Center page The Whys and Hows of Generations Research: https://www.pewresearch.org/politics/2015/09/03/the-whys-and-hows-of-generations-research/. Last accessed 2022/07/15 (2015)
10. Cervi, L.: Tik Tok and generation Z. Theatre Dance Perform. Train. **12**(2), 198–204 (2021)
11. Francis, T., Hoefel, F: 'True Gen': generation Z and its implications for companies. Mckinsey & Company (2018)
12. Djafarova, E., Bowes, T.: 'Instagram made Me buy it': generation Z impulse purchases in fashion industry. J. Retail. Consum. Serv. **59**, 102345 (2021)
13. Dolot, A.: The characteristics of Generation Z. E-mentor **2**(74), 44–50 (2018)
14. Angus, A., Westbrook, G.: Top ten consumer trends 2022, Euromonitor webpage, https://go.euromonitor.com/white-paper-EC-2022-Top-10-Global-Consumer-Trends.html. Last accessed 2022/09/08, released on 2022/01/18 (2022)
15. CRITEO: Relatório Sobre a Geração Z, Criteo webpage, https://www.criteo.com/br/wp-content/uploads/sites/5/2018/08/18-GenZ-Report-BR.pdf. Last accessed 2022/09/08, released on 2018/08/18 (2018)
16. Kemp, S.: Digital 2022: global overview report, WeAreSocial webpage, https://datareportal.com/reports/digital-2022-global-overview-report. Last accessed 2022/09/08, released on 2022/01/26 (2022)
17. Aran-Ramspott, S., Fedele, M., Tarragó, A.: YouTubers' social functions and their influence on pre-adolescence. Comunicar. Media Educ. Res. J. **26**(2), 71–80 (2018)
18. Pereira, S., Moura, P., Fillol, J.: The youtubers phenomenon: what makes Youtube stars so popular for young people? Fonseca J. Commun. **17**(2), 107–123 (2018)
19. Bossen, C.B., Kottasz, R.: Uses and gratifications sought by pre-adolescent and adolescent TikTok consumers. Consumers **21**(4), 1747–3616 (2020)

20. Haenlein, M., Anadol, E., Farnsworth, T., Hugo, H., Hunichen, J., Welte, D.: Navigating the new era of influencer marketing: how to be successful on Instagram, TikTok, & Co. California Manage. Rev. **63**(1), 5–25 (2020)
21. Kennedy, M.: 'If the rise of the TikTok dance and e-girl aesthetic has taught us anything, it's that teenage girls rule the internet right now': TikTok celebrity, girls and the Coronavirus crisis. Eur. J. Cult. Stud. **23**(6), 1069–1076 (2020)
22. Montag, C., Yang, H., Elhai, J.D.: On the psychology of TikTok use: a first glimpse from empirical findings. Front. Public Health **9**, 62 (2021)
23. Datareportal page https://datareportal.com/essential-tiktok-stats. Last accessed 2022/07/11 (2022)
24. Anderson, K.E.: Getting acquainted with social networks and apps: it is time to talk about TikTok. Library Hi Tech News **37**(4), 7–12 (2020)
25. Court, D., Elzinga, D., Mulder, S., Vetvik, O.J: The consumer decision journey. McKinsey Quart. **3**(3), 96–107 (2009)
26. Richardson, A.: Innovation X: Solutions for the New Breed of Complex Problems Facing Business. Wiley (2010)
27. Lemon, K.N., Verhoef, P.C.: Understanding customer experience throughout the customer journey. J. Mark. **80**(6), 69–96 (2016)
28. Følstad, A., Kvale, K.: Customer journeys: a systematic literature review. J. Service Theory Pract. (2018)
29. De Keyser, A., Verleye, K., Lemon, K.N., Keiningham, T.L., Klaus, P.: Moving the customer experience field forward: introducing the touchpoints, context, qualities (TCQ) nomenclature. J. Serv. Res. **23**(4), 433–455 (2020)
30. Roggeveen, A.L., Grewal, D., Schweiger, E.B.: The DAST framework for retail atmospherics: the impact of in-and out-of-store retail journey touchpoints on the customer experience. J. Retail. **96**(1), 128–137 (2022)
31. Kranzbühler, A.M., Kleijnen, M.H., Verlegh, P.W.: Outsourcing the pain, keeping the pleasure: effects of outsourced touchpoints in the customer journey. J. Acad. Mark. Sci. **47**(2), 308–327 (2019)
32. Demmers, J., Weltevreden, J.W., van Dolen, W.M.: Consumer engagement with brand posts on social media in consecutive stages of the customer journey. Int. J. Electron. Commer. **24**(1), 53–77 (2020)
33. Schweidel, D., Bart, Y., Inman, J., Stephen, A., Libai, B., Andrews, M., Rosario, A., Chae, I., Chen, Z., Kupor, D., Longoni, C., Thomaz, F.: How consumer digital signals are reshaping the customer journey. J. Acad. Market. Sci. 1–20 (2022)
34. Guerreiro, C., Viegas, M., Guerreiro, M.: Social networks and digital influencers: their role in customer decision journey in tourism. J. Spatial Org. Dyn. **7**(3), 240–260 (2019)
35. Hamilton, R., Ferraro, R., Haws, K.L., Mukhopadhyay, A.: Traveling with companions: the social customer journey. J. Mark. **85**(1), 68–92 (2021)
36. Pop, R.A., Săplăcan, Z., Dabija, D.C., Alt, M.A.: The impact of social media influencers on travel decisions: the role of trust in consumer decision journey. Curr. Issue Tour. **25**(5), 823–843 (2022)
37. Shen, S., Sotiriadis, M., Zhang, Y.: The influence of smart technologies on customer journey in tourist attractions within the smart tourism management framework. Sustainability **12**(10), 4157 (2020)
38. Gomes, M.A., Marques, S., Dias, Á.: The impact of digital influencers' characteristics on purchase intention of fashion products. J. Global Fashion Market. 1–18 (2022)
39. Sudha, M., Sheena, K.: Impact of influencers in consumer decision process: the fashion industry. SCMS J. Indian Manage. **14**(3), 14–30 (2017)
40. Kozinets, R., Patterson, A., Ashman, R.: Networks of desire: how technology increases our passion to consume. J. Consumer Res. **43**(5), 659–682 (2017)

41. Jiménez-Castillo, D., Sánchez-Fernández, R.: The role of digital influencers in brand recommendation: examining their impact on engagement, expected value and purchase intention. Int. J. Inf. Manage. **49**, 366–376 (2019)
42. Marôco, J.: Análise Estatística Com o SPSS Statistics. Report Number. Pero Pinheiro, Portugal (2011)

Chapter 27
Relationship Marketing on Higher Education Institutions (HEI) in Times of Pandemic (Covid-19)

María Paula Espinosa-Vélez, Mayra Ortega-Vivanco, and Daysi Garcia-Tinisaray

Abstract The current research aims to analyze the relational marketing application in higher education institutions. Quantitative descriptive research was done, composed of two variables and ten indicators that explain these effects. The results describe the level of relation and student service contentment through digital marketing tools and communication channels for students enrolled in face-to-face and online classes. It has been determined that the digital tools with a more significant relation level to online students in academic matters are Instagram and YouTube. In the communication field, the Facebook tool and web portals show the highest percentage as significant, on the other hand, the most used communication channels to establish a relationship with the university during pandemic times were: Call Center, WhatsApp, virtual rooms, and every online service.

27.1 Introduction

In difficult times (Covid-19), relationship marketing (through social media) plays an important role in higher education students' satisfaction [1].

The economic, political, and social crisis triggered in the world due to the Covid-19 pandemic has generated important impacts on educational organizations, the same ones that had to restructure the administrative planning and implement emergent strategies that allow them to continue their operations. It is estimated that the pandemic affected the global economy by $90 trillion; this is considered the worst crisis in the last hundred years [2] with Latin America being the most affected region a [3].

M. P. Espinosa-Vélez (✉) · M. Ortega-Vivanco · D. Garcia-Tinisaray
Técnica Particular de Loja University, San Cayetano, Ecuador
e-mail: mpespinosa@utpl.edu.ec

M. Ortega-Vivanco
e-mail: mjortega@utpl.edu.ec

D. Garcia-Tinisaray
e-mail: dkgarcia@utpl.edu.ec

J. L. Reis et al. (eds.), *Marketing and Smart Technologies*, Smart Innovation, Systems and Technologies 337, https://doi.org/10.1007/978-981-19-9099-1_27

In the higher education case, a sector that has been facing changes at a rate without precedent in a highly competitive university environment, the pandemic's presence carried universities, on one side, to adapt teaching and learning strategies; and on the other hand, to develop strategies to grant sustainability and be able to differentiate from the wide and global high educational offer strengthened by the Covid-19.

Considering the current context, the Higher Educational Institutes (HEI) are motivated to find new modus operandi of their activity in different action fields: academic, research, and administrative. In the administrative field, the management model plays a significant relevance and accentuates the need of differentiating from competence, before which the importance of counting on professional management structures and entrepreneurial and innovative organizations gets ratified [4]. Among the new skills and competencies that HEI must strengthen, the resources and users' management stand out, in which relationship marketing gets strong, allowing related universities from that perspective a higher emphasis on satisfying the student's expectations and needs. On the other hand, in this competitive and globalized context, digital marketing is a powerful tool that can enhance relationship marketing through knowledge and understanding of the environment, generating value on different interactions [5].

This research work is integrated into the following sections, on section number two the literary revision is done and shows some empirical evidence of related investigations with relationships and digital marketing on organizations. In section number three the applied methodology is shown and the results, and finally the work's conclusions are presented.

27.2 Theoretical Framework

In order to accomplish the research's purpose and solve the questioning: how the relationship marketing can be applied to Higher Educational Institutes? and What digital tools are used by college students to communicate and tale academic processes in times of crisis? The current theoretical framework was done.

27.2.1 Relationship Marketing

Through the years, marketing had a constant evolution, going from being focused mainly on the product, to focusing on the customer, even more in the new market's dynamic, where it can be seen how companies wide their business spotlight toward more human affairs, where the profits must be more balanced with the corporate responsibility [6].

The relationship marketing perspective is based on the notion that besides the product value or the exchanged services in a transaction, there is a value created by the relationship that is established between the client and a service provider, as [7]. Kotler and Keller [8] argues that relationship marketing has an objective to establish satisfactory mutual relations and long-term relations with the participants or interested targets from the organization (consumers, providers, distributors, and other marketing associates) with the purpose of maintaining and increasing the business, while [9] defines relationship marketing as a group of activities which pursue to establish, maintain, develop, and when needed, also conclude client relations and other interested parties, so the goals of the involved people can be accomplished. On the other hand, [10] complements that this relationship is also analyzed as a central element of the communication between the organization and its stakeholders, while [11] defines relationship marketing as a targeted strategy to establish and sustain long-term relations between the company and its clients, turning itself into an oriented practice toward a competitive sustainable advantage.

In these scenarios, the relevance of relationship marketing lies in its incidence on the satisfaction of its stakeholders, a special way in clients, and the reputation of the organization [12].

27.2.2 HEI Relationship Marketing

As the most innovative international commerce theories point out, price is no longer the only determining element in the product or service choice, giving place to other characteristics, such as quality, and diversification. In the HEI case, the election from a university is given by several factors, such as academic reputation, college organization, quality of materials used in class and learning methods applied, student-university interaction, experience development, service satisfaction, etc. [12, 13]. On the other hand, the way that HIE tends to develop or interact with their clients—students and relatives, alumni and donors, teachers and members of the staff—could be a determining factor due to the benefits that a loyal student can mean in the long term [13].

Diverse variables identify as keys to the relationship process in the service field. de Juan-Jordán et al. [14] highlight: trust, value, commitment, or loyalty, and, in the case of universities, they point out: satisfaction, loyalty, the interaction between student–teacher quality in the service encounter, identification of the graduate with the university, and loyalty. The relationship concept in HEI, from a relation communication model, considers also the relation and communication perspectives. The relationship field stands out: goals and shared knowledge and mutual respect, while in the communication field: frequent, precise, appropriate, and oriented to problem fixings [1].

From the communication perspective, relationship marketing considers two fields: relation bonds and communication bonds, before which it is necessary to understand the dynamic of coordination between relationships and communication dynamics in organizations, in order to obtain better results in organizational management [5, 15].

27.2.3 Relationship Marketing Strategies

Digital Marketing

As [16] quotes, digital marketing uses technology as a marketing channel, enhancing communication and information exchange with its clients, potential customers, providers, investors, competence, and other relevant factors in the company's life. Its strength lies in the capacity of providing a marketing model oriented for the client, given the bidirectional interaction that facilitates, in contrast with the traditional media. Likewise, according to [17], since 2011, the HEI has integrated the use of digital social media as a way to build relationships, securing the fidelity and participation of the students, and turning them into active agents in collaborative environments, participative, and interactive. If business actions oriented to the digitalization of the processes triggered by Covid-19 are added to this scenario, the university has to permanently reinvent itself in order to answer proactively to the challenges of a changing environment that implies adopting the reality of a digital world [18].

Digital marketing allows organizations to (a) identify the most discussed topics by clients, (b) identify the key areas for improvement based on the higher quantity of negative comments, and (c) determine the relation between important concepts and loyalty improvement, adding value in content or services for the client [19]. Diverse tools stand out for their capacity of building relations, such as blogs, mail, web portals, mobile apps, and social media like Facebook, Instagram, LinkedIn, Twitter, and more.

On the other hand, the convergence of media and content that it's been produced on the web opens big possibilities for universities to count on new communication channels and new marketing strategies, for internal communication and external communication toward society in its group [20]. Among these channels, traditional channels can be identified (call center), digital (WhatsApp, SMS, chatbot, and virtual rooms), and online services (intranet).

Innovation

Due to data and information contribution, the relation with the client contributes to the organization, turning into an important scenario for innovation of their products or services, placing the client/student as the center of their operation [12], ratifying Philip Kotler's visión on relationship marketing, mainly focusing on building customer loyalty, but also basing the co-creation of value and through these relations, the client contributes to the organization [17]. Some scenarios might be evaluated in the college innovation context, as innovation of relationship marketing is: (a) product development, (b) virtual environments, and (c) collaborative work, focused scenarios on optimizing and improving the student's experience and within the quality of the service [7].

There are other relevant strategies based on digital transformation; one of the main tendencies is big data, which can be used to enhance client loyalty [21] facilitating loyalty strategies innovation, thanks to the client's data analytics through its diverse interactions and exchange with the organization. Another featured strategy is Customer Relationship Management (CRM) which is defined as an integrated group of practices that provides a consolidated and systematic vision to the client, with a better perspective of all linked areas in its attention to assure a better service level [22].

27.3 Methodology

27.3.1 Data and Simple

Geographically, the study was carried out in Ecuador, and the analysis unit were the students from the university of the two study modalities: Face-to-face; and Open and Online (MAD) of the Private Technical University of Loja, an educational institution with a presence in the four regions: Coast, Sierra, Oriente and Insular, in which it is present through ninety university centers. The sample was calculated with a trust level of 96% and an error range of 6%, getting a total of 615 validated polls.

The database was built with obtained information through an online poll, 10 indicators were elaborated (questions). Each value was tested considering the Likert scale, extremely significant (5), very significant (4), moderately significant (3), hardly significant (2), and insignificant (1), which constitutes one of the most used scale instruments in social science.

The quiz structure is integrated by general information and by two variables (See Table 27.1) (i) relationship level; and (ii) satisfaction of the student attention (communication channels) that will allow determining the relationship grade through the use of digital tools (Facebook, Instagram, Twitter, e-mail, YouTube, web portal, mobile app, linked, blog, and intranet) and traditional communication channels, digital and online services.

27.3.2 Analysis Methods and Results

This research has a quantitative focus, in order to analyze the relationship between marketing and student satisfaction cross-sectional descriptive statistics were used to specify and measure the behavior of students from both study modalities, through just one measure of the variables in a unique moment of time.

Table 27.2 starts analyzing the relationship level (satisfaction and loyalty) thanks to the use of digital tools to establish a relationship with the university in academic procedures for the communicational environment. For the procedure field, students from MAD tended to use, in a larger proportion, Instagram, Twitter, and YouTube.

Table 27.1 Variables, sub-variables, and indicators

Variables		Indicators	Sub-variables
Relationship level (satisfaction and loyalty)	01	Digital marketing tools are applied	Academic administrative procedures and communication
	02	Digital tools and relationship level: academic procedures and communication	
Student service satisfaction	01	Communication channels and univer-sity relationship	
	02	Message meaning (content of value)	Relationship features (attention channels)
	03	The way HEI transmits the message is attractive and answers the expectations (quality)	
	04	Message is concrete and appropriate (quality)	
	05	Level of interaction of the message—informative (promotes the conversation and answers the concern of the stakeholders)	
	06	The level of interaction of the message —academic administrative service (promotes the conversation and answer concerns)	
	07	Messages come with an environmental commitment or sustainability	
	08	Communication is personalized	

In the communication field, the students from MAD use, in higher proportion, Instagram, Twitter, and YouTube, and students from face-to-face modality Facebook, Instagram, Twitter, and YouTube.

The collected data also allows us to determine in Table 27.3 (students from online modality) and in Table 27.4 (students from face-to-face modality) the relationship features for academic procedures and university-student communication. This analysis allows us to know the behavior according to the features of the relationship with every digital tool. The level of relationship is set by 1 for insignificant (NS) and 2 for significant (s).

Table 27.3 presents the relationship percentages with respect to communication. The results show that in most of the functionalities the e-mail tool is considered insignificant and the Facebook tool and web portal are the ones with a larger percentage as significant, this behavior changes for the custom communication

Table 27.2 Use of digital marketing tools (%)

Tools	Academic procedures				Communication			
	MAD		Face-to-face		MAD		Face-to-face	
	Yes	No	Yes	No	Yes	No	Yes	No
Facebook	39	61	1	99	38	62	99	1
Instagram	99	1	2	98	97	3	98	2
Twitter	90	10	9	91	85	15	91	9
E-mail	83	17	86	14	76	24	14	86
YouTube	90	10	5	95	86	14	95	5
Web portal	37	63	94	6	38	62	6	94
Mobile apps	49	51	65	35	85	15	35	65
Linked	14	86	26	74	71	29	75	25
Blog	8	92	12	88	74	26	88	12
Intranet	75	25	73	27	78	22	27	73

feature, given that the students point out that Twitter, e-mail and web portal are the most significant tools, which reflects the orientation, every time higher, towards onmiquality and service customizing, quoting [23] are characteristic elements of the relationship strategies, which evolve over the basis of a major knowledge for students.

Table 27.4 contains the results of the face-to-face modality students. It is determined for the case of the academic procedures, the relationship level that is generated as insignificant in a larger percentage belongs to the digital tool Linked, while there's a positive relationship with mobile app and web portal. As long as communication, the relationship level is insignificant in a larger percentage with digital tools such as Blog and Linked, and the tool that's considered by the students with the largest use percentage is Instagram.

The second analyzed variable focuses on the satisfaction of the student's attention, it is important to mention that during pandemics, the university, subject of study, focused its attention on traditional channels, digital, and online services, In Table 27.5, the proportion of use according to these and the study modalities are shown.

Within the traditional channels, the most used one by the students of both modalities was the call center. On what concerns digital attention, students from online modalities mostly used communication channels WhatsApp, while most of the students from face-to-face modalities chose virtual counters. Finally, online services had the use of more than 90% of both modalities. This behavior could obey the fact that students from the online modality were already using WhatsApp as an attention channel since before the pandemic, while students from the face-to-face modality can easily spot access to virtual counters, which work in a similar way as physical counters, which they're using to [23].

Table 27.6 analyzes the level of relationship of the features with every communication channel, considering a traditional attention channel (call center) and digital

Table 27.3 Digital tools and MAD relationship-student features (%)

Tools	Significant (content of value)		Attractive and answers to expectations (quality)		Concrete and appropriate (calidad)		Interaction level (promotes conversation and answers to concerns)		Comes with environmental commitment or sustainability		Custom communication	
	NS	S	NS	S	NS	S	NS	S	NS	S	NS	S
Academic procedures												
Facebook	7.1	18.8	6.6	19.9	8.2	19.9	6.4	19.1	7.7	19.1	9.8	16.4
Instagram	10.3	19.7	8.2	20.3	8.7	20.4	10.3	19.8	11.4	19.2	11.1	19.9
Twitter	11.6	19.5	25.1	19.1	13.5	19.6	14.3	18.2	15.1	18.9	14.3	19.6
E-mail	1.6	13.6	1.6	14.2	1.6	12.9	2.4	19.4	4.5	20.6	1.6	15.1
YouTube	3.7	20.6	4.5	20.7	5.3	20.4	7.4	20.8	6.6	20.8	10.1	19.7
Web portal	1.6	14.9	2.4	16.3	1.6	15.4	1.6	17.1	0.8	18.1	0.8	18.2
Mobile apps	2.4	15.5	1.6	18.4	2.4	17.8	2.4	18	3.2	17.6	4	15.1
Linked	5.8	20.7	5.8	15.5	2.1	19.6	11.9	17.6	1.1	18.6	1.9	18
Blog	2.4	16.2	1.6	19.3	2.4	19	2.4	18.9	4.8	17.3	5.3	15.3
Intranet	3.7	11.2	6.1	19	9	19.9	1.6	17.5	3.7	20.6	4.5	20.7
Communication												
Facebook	5.3	20.4	7.4	20.8	6.6	20.8	10.1	19.7	7.7	19.1	9.8	16.4
Instagram	7.1	18.8	6.6	19.9	8.2	19.9	6.4	19.1	7.7	19.1	9.8	16.4
Twitter	10.3	19.7	8.2	20.3	8.7	20.4	10.3	19.8	11.4	19.2	11.1	19.9
E-mail	11.6	19.5	13.5	19.1	13.5	19.6	14.3	18.2	15.1	18.9	14.3	19.6
YouTube	1.6	13.6	1.6	14.2	1.6	12.9	3.2	20.2	5	20.8	1.6	17.3
Web portal	3.7	20.6	4.5	21.1	5.3	20.4	7.4	21.1	6.6	20.8	10.1	19.7
Mobile apps	4	15.3	2.4	16.3	1.9	13.4	1.6	17.2	0.8	18.1	1.1	18.5

(continued)

Table 27.3 (continued)

Tools	Significant (content of value)		Attractive and answers to expectations (quality)		Concrete and appropriate (calidad)		Interaction level (promotes conversation and answers to concerns)		Comes with environmental commitment or sustainability		Custom communication	
	NS	S	NS	S	NS	S	NS	S	NS	S	NS	S
Linked	2.4	16.1	2.4	18.7	2.4	18.3	2.9	18.4	3.2	18.2	4	15.6
Blog	1.9	13.4	2.4	19.4	5.3	21.3	1.9	15.9	3.7	20.6	0.8	18.4
Intranet	2.4	15.6	1.6	18.4	2.7	18.1	2.4	18.2	3.2	17.5	4.5	16

*Significant (S), Insignificant (NS)

Table 27.4 Digital tools and relationship attributes-face-to-face (%)

Tools	Significant (content of value)		Attractive and answers to expectations (quality)		Concrete and appropriate (quality)		Interaction level (promotes conversation and answers to concerns)		Comes with environmental commitment or sustainability		Custom communication	
	NS	S	NS	S	NS	S	NS	S	NS	S	NS	S
Academic procedures												
Facebook	32.1	14.6	25.4	13.2	28.3	14.8	25	11.6	23.2	14.1	16	14.7
Instagram	27.4	14.3	26.2	12	28.7	13.8	28.3	11.5	25.3	13.6	28.7	13.2
Twitter	39.2	11.8	36.3	10.1	38.8	10.9	39.2	9.7	40.5	10.3	40.1	11
E-mail	1.3	16.9	0.9	16.8	0.8	20.1	1.3	18	1.3	22.4	1.3	18.8
YouTube	33.3	12.8	34.6	12.8	42.2	11.9	35.4	12.2	33.8	12	35.4	11.7
Web portal	5.9	19.1	4.6	21.4	4.6	21.1	4.2	22.2	5.9	20.9	5.9	20.6
Mobile apps	3.4	21.2	2.5	21	3	21.9	2.1	22	3.4	21.7	4.2	21.7
Linked	59.9	6.7	64.1	4.8	62	6.3	63.3	4.8	61.2	5.7	65	5.3
Blog	57.4	8	51.3	6.9	54.9	8.5	50.6	7.9	54.4	7.2	56.5	7.7
Intranet	5.9	20.7	4.2	20.6	5.1	22	4.2	21.4	5.1	21.7	5.5	20.2
Communication												
Facebook	1.7	17.5	1.3	19	1.3	20.7	1.3	21.2	1.3	21.7	2.1	19.9
Instagram	3.8	23.1	2.5	23.3	5.1	22.6	3.8	22.5	3.8	22.8	5.5	22.4
Twitter	21.5	17.9	20.3	17.2	23.6	17.5	21.5	18	24.5	17.8	22.8	17.5
E-mail	4.2	21.4	1.3	21.9	2.5	21.5	1.7	22.2	2.1	23.1	3	21
YouTube	7.6	21	5.5	19.4	7.6	20.5	8	20.3	7.6	19.3	8.9	20.4
Portal Web	5.5	20	3.8	21.3	5.1	20.9	4.2	21.7	4.6	21.6	5.5	21.3
App movil	5.5	20.7	4.6	22.8	5.9	21.5	5.1	22.1	6.3	21.9	6.3	21.8

(continued)

Table 27.4 (continued)

Tools	Significant (content of value)		Attractive and answers to expectations (quality)		Concrete and appropriate (quality)		Interaction level (promotes conversation and answers to concerns)		Comes with environmental commitment or sustainability		Custom communication	
	NS	S	NS	S	NS	S	NS	S	NS	S	NS	S
Linked	54.4	8.4	49.8	8.7	52.3	9	51.7	8.2	49.8	8.5	49.4	8.4
Blog	57	8.2	60.3	7.4	57.4	8.3	58.2	7.6	61.2	7.2	55.3	8.6
Intranet	6.3	20.6	6.4	21.8	5.9	21.5	5.1	22.5	5.5	22.4	5.9	21.6

*Significant (S), Insignificant (NS)

Table 27.5 Use of communication channels (%)

Type	Channels	Online		Face-to-face	
		Yes	No	Yes	No
Traditional attention channels	Call center	79	21	93	7
	Physical counter	58	42	91	9
	Both	68	32	88	12
Digital attention channels	Whatsapp	81	19	46	54
	SMS	46	54	11	89
	Chatboot	31	69	69	31
	Virtual counter	75	25	97	3
Online services	Registration and enrollment	95	5	99	1
	File information consultation	91	9	91	9
	Academic administrative procedures	96	4	98	2

attention channels (WhatsApp and Chatbot). It is observed that in the MAD students all features of the relationship are more significant with the Chatbot, while with face-to-face students WhatsApp is more significant.

27.4 Conclusions

Higher education institutions should concern about service quality by giving students satisfaction to their needs. If the work is not done consistently in these processes, the education success may be affected in long term.

The most used digital tools to establish a link with the university in the academic procedures and communication fields in students from MAD were, in a larger proportion, Instagram, Twitter, and YouTube, while in face-to-face students were e-mail and web portal, Facebook, Instagram, Twitter, and YouTube.

Digital tools that have a significant relationship level to online students in matters of academic procedures are Instagram and YouTube. In the communication field, Facebook and Portal Web are the larger percentage as significant tools.

The most used communication channels used to establish a link with the university during pandemic times were: the call center, WhatsApp, virtual rooms, and every online service.

Digital tools that have a significant relationship level for face-to-face modality students in academic procedures are Mobile-app and Web Portal. In the communication field, Instagram is the tool that represents a larger percentage as significant.

Table 27.6 Communication channels and relationship features

Channels	Significant (content of value)		Attractive and answers to expectations (quality)		Concrete and appropriate (quality)		Message informative (promotes conversation and responds to stakeholder concerns)		Message administrative academic service (promotes conversation and responds to your concerns)		Comes with environmental commitment or sustainability		Custom communication	
	NS	S	NS	S	NS	S	NS	S	NS	S	NS	S	NS	S
Online														
Centro de contactor	3	12	3	12	4	14	3	14	3	14	3	17	3	14
Whatsapp	4	15	4	16	4	16	4	16	5	16	6	16	5	15
Chatboot	9	18	9	19	9	19	10	19	10	19	11	19	11	19
Face-to-face														
Centro de contactor	13	17	15	17	15	16	17	14	13	17	13	14	17	17
Whatsapp	16	19	15	17	16	18	16	19	18	18	16	18	18	18
Chatboot	50	11	49	11	47	11	46	11	47	11	49	12	44	11

*Significant (S), Insignificant (NS)

The relationship percentage regarding the communication and based on the obtained results, in most of the features the e-mail tool is considered insignificant, and the Facebook tool and Web portal represent the higher percentage as significant.

It is worth mentioning that one of the main limitations was collecting information in times of pandemic, so it was necessary to look for other approach mechanisms, especially with distance learning students.

From the research results, several questions are opened for future investigations; for example, What are the most used features of relationship (attention channels) in Ecuadorian Higher Education Institutions? What innovative strategies the university center is using in order to satisfy students' needs? What are the advantages that relationship marketing shows as a strategy to generate differential value compared to the competition? and, what are the effects of relationship marketing on college students?

References

1. Jain, V., Mogaji, E., Sharma, H., Babbili, A.S.: A multi-stakeholder perspective of relationship marketing in higher education institutions. J. Mark. High. Educ., 1–19 (2022). https://doi.org/10.1080/08841241.2022.2034201
2. Sarıışık, M., Usta, S.: Global effect of COVID-19.In: COVID-19 and the hospitality and tourism industry: a research companion, pp. 41–59 (2021). https://doi.org/10.4337/9781800376243.00008
3. Backes, D.A.P., Arias, M.I., Storopoli, J.E., Ramos, H.R.: Los efectos de la pandemia de Covid-19 en las organizaciones: una mirada al future. Revista Ibero-Americana de Estratégia **19**(4), 1–10 (2020). [Online]. Available: https://periodicos.uninove.br/riae/article/view/18987
4. Alves, H., Mainardes, E.W., Raposo, M.: A relationship approach to higher education institution stakeholder management. Tert. Educ. Manag. **16**(3), 159–181 (2010). https://doi.org/10.1080/13583883.2010.497314
5. Lacayo-Mendoza, A., de Pablos-Heredero, C.: Cómo gestionar las relaciones y comunicaciones de manera eficiente a través de las redes sociales digitales en instituciones de educación superior: Una propuesta desde el modelo de coordinación relacional. DYNA (Colombia) **83**(195), 138–146 (2016). https://doi.org/10.15446/dyna.v83n195.49296
6. Gómez-Bayona, L., Arrubla-Zapata, J.P., Aristizábal Valencia, J., Restrepo-Rojas, M.J.: Análisis de las estrategias de marketing relacional en instituciones de educación superior de Colombia y España. Retos **10**(20), 343–359 (2020). https://doi.org/10.17163/ret.n20.2020.09
7. Arosa-Carrera, C.R., Chica-Mesa, J.C.: Innovation in the paradigm of relationship marketing. Estudios Gerenciales **36**(154). Universidad Icesi, pp. 114–122, Mar. 01, 2020. https://doi.org/10.18046/j.estger.2020.154.3494
8. Kotler, P., Keller, K.L.: Marketing Management, 12th ed. New Jersey (2006)
9. Grönroos, C.: The relationship marketing process: communication, interaction, dialogue, value. J. Bus. Ind. Market. **19**(2), 99–113 (2004). https://doi.org/10.1108/08858620410523981
10. Lock, I.: Explicating communicative organization-stakeholder relationships in the digital age: a systematic review and research agenda. Public Relat. Rev. **45**(4), 101829 (2019). https://doi.org/10.1016/j.pubrev.2019.101829
11. Schlesinger, M.W., Cervera Taulet, A., Iniesta Bonillo, M.Á., Sánchez Fernández, R.: Un enfoque de marketing de relaciones a la educación como un servicio: aplicación a la Universidad de Valencia. Innovar **24**(53), 113–125 (2014). https://doi.org/10.15446/innovar.v24n53.43919
12. Espinosa-Vélez, M.P., Cárdenas-Carrillo, R.: Relational marketing applied to higher education institutions in the field of digital society: approach from a strategic perspective.

In: 6th Iberian conference on information systems and technologies (CISTI), pp. 1–7 (2021) [Online]. Available: https://ieeexplore.ieee.org/stamp/stamp.jsp?tp=&arnumber=9476438&isnumber=9476219
13. Prima Zani, A.Y.: The increasing of students' satisfaction and loyalty by the use of customer relationship management (CRM) (a case study in medical education institutions, nursing study program in Jakarta, Bogor, Tangerang, Depok and Bekasi). IOSR J. Bus. Manage. **7**(4), 28–34 (2013). https://doi.org/10.9790/487x-0742834
14. de Juan-Jordán, H., Guijarro-García, M., Hernandez Gadea, J.: Feature analysis of the 'customer relationship management' systems for higher education institutions. Multidiscipl. J. Educ. Social Technol. Sci. **5**(1), 30 (2018). https://doi.org/10.4995/muse.2018.9232
15. Gallego Sánchez, M.C., De-Pablos-Heredero, C., Medina-Merodio, J.A., Robina-Ramírez, R., Fernandez-Sanz, L.: 20Relationships among relational coordination dimensions: impact on the quality of education online with a structural equations model. Technol. Forecast. Soc. Change **166**, no. January (2021). https://doi.org/10.1016/j.techfore.2021.120608
16. Chan-Tien, L., Dan-Sheng, W., Shun-Fa, H.: The conceptual framework for applying digital community marketing and marketing practices into educational relationship marketing model of private technical high school in Taiwan. In: ACM international conference proceeding series, pp. 14–19 (2019). https://doi.org/10.1145/3341042.3341050
17. Varona Aramburu, D., Pérez Escolar, M., Sánchez Muñoz, G.: Teoría del framing y protoperiodismo. Estudio de los atributos asociados a la figura de Magallanes en los diarios de Pigafetta y Francisco Albo. Latina, Revista de Comunicación **74**, 734–747 (2019). https://doi.org/10.4185/RLCS
18. Espinosa-Vélez, M.P., Armijos-Buitrón, V.A.: La transformación digital y su incidencia en el e-commerce en Ecuador. In: CICIC 2022—Decima Segunda Conferencia Iberoamericana de Complejidad, Informatica y Cibernetica en el contexto de the 13th international multi-conference on complexity, informatics, and cybernetics, IMCIC 2022—Memorias, no. Cicic, pp. 169–174 (2021). https://doi.org/10.54808/CICIC2022.01.169
19. Zhan, Y., Han, R., Tse, M., Ali, M.H., Hu, J.: A social media analytic framework for improving operations and service management: a study of the retail pharmacy industry. Technol. Forecast Soc. Change **163**, no. April 2020, 120504 (2021). https://doi.org/10.1016/j.techfore.2020.120504
20. Universidad, Papel De Las Tic En El Nuevo Entorno Socioeconómico (2020)
21. Fernández-Rovira, C., Álvarez Valdés, J., Molleví, G., Nicolas-Sans, R.: The digital transformation of business. Towards the datafication of the relationship with customers. Technol. Forecast Soc. Change **162**, no. June 2020, 120339 (2021). https://doi.org/10.1016/j.techfore.2020.120339
22. Seeman, E.D., O'Hara, M.: Customer relationship management in higher education: using information systems to improve the student-school relationship. Campus-Wide Inform. Syst. **23**(1), 24–34 (2006). https://doi.org/10.1108/10650740610639714
23. Espinosa-Vélez, M.P., Armijos-Buitrón, V.-A., Mora, M.E.E.: Digital transformation in HEIs and its impact on the user experience/student service processes. In: 17th Iberian Conference on Information Systems and Technologies (CISTI), pp. 1–6 (2022). https://doi.org/10.23919/CISTI54924.2022.9820207

Chapter 28
Analysis of #YoDecidoCuando (I Decide When) Campaign on TikTok, as Educational Communication for Teenage Pregnancy Prevention

Kimberlie Fernández-Tomanguillo, Melina Mezarina-Castilla, and Eduardo Yalán-Dongo

Abstract The purpose of this research is to analyze how # YoDecidoCuando campaign on TikTok is perceived as an educational communication for teenage pregnancy prevention in Peru. A qualitative case study methodology was applied. Using the homogeneous sampling technique, 15 semi-structured interviews were carried out with young women aged 16–19 years, who had a smartphone and followed gynecologists accounts on TikTok. As a result, we observed that this type of campaign on TikTok encourages people to become aware of health issues, because it generates interest and motivation to learn more about sex education. In this sense, the findings serve as a starting point for applying educational communication strategies on TikTok by institutions and organizations looking for new strategies to target a young audience.

28.1 Introduction

Since the beginning of the pandemic in 2020, the correct use of communication tools about health issues was promoted, where mainly the WHO carried out multiple actions to fight against the misinformation on social networks [5]. Thus, during their break hours, WHO professionals that dealt with COVID-19 used TikTok to post educational videos that were focused on raising citizen's awareness about health care [15]. Likewise, TikTok provided reliable information in an exclusive tab for COVID-19, which included videos from seven accounts from official institutions, becoming a window of opportunity to educate about health issues during this pandemic context [4].

A digital campaign that used the same platform was #edutok, which had an educational approach and was characterized for encouraging the attendance to workshops

K. Fernández-Tomanguillo (✉) · M. Mezarina-Castilla · E. Yalán-Dongo
Princeton University, Princeton, NJ 08544, USA
e-mail: pierina.fernandez10@gmail.com

Springer Heidelberg, Tiergartenstr. 17, 69121 Heidelberg, Germany

J. L. Reis et al. (eds.), *Marketing and Smart Technologies*, Smart Innovation, Systems and Technologies 337, https://doi.org/10.1007/978-981-19-9099-1_28

in six states of India during six months, from October 2019 to March 2021 [24]. Its aim was to provide users with live broadcasting tutorials to encourage education and learning from home about issues regarding ways to cook, sing, learn mathematics or study English, exploring new fields of knowledge, and training. This campaign obtained more than 127 trillion views during its exhibition [17].

This context allows us to get close to possible educational communication uses and strategies on this social network. As Machuca pointed out [15], TikTok is a platform where interesting knowledge is shared, where users can learn through their favorite TikTokers in an enjoyable way that allows to educate the adolescent population about different issues.

#YoDecidoCuando campaign, at public health level, was among the most highlighted TikTok campaigns that went viral in Peru, and it was included as part of this research. This campaign was carried out between September 26 and 28, 2020 and was launched by the Mayor's Office of Medellin, Colombia. Yina Rose, who is a popular Colombian singer, internationally recognized by young teenagers, starred in this campaign. The #YoDecidoCuando campaign was focused on teenagers and young people, in order to make them aware of their sexual and reproductive rights. Likewise, it invited them to use long-term contraception methods which would allow a more conscious planning of their life project. The campaign challenge on TikTok social network was to invite users to personalize and dramatize the song that identified the campaign. The four videos with more "likes" would be awarded [14]. The video reached more than 240 thousand likes, as Yina Rose's account shows, and it allowed spreading the hashtag and song, getting the message out quickly. It is important to remark that the campaign on TikTok got the reaction of users of several countries, such as Chile, Argentina, and Peru, among others. Although the challenge could not be developed in Peru, the video commentaries showed big interest for viewing, supporting and disseminating the campaign in the said region.

Based on this context and on the referred research, this work is focused on the health education campaigns on social networks, specifically centered in teenage pregnancy prevention campaigns spread throughout Peru. The interest in the Peruvian case is because in the last years it has been confirmed that there was no improvement in the fall of the teenage pregnancy rate, despite the continuous campaigns, mainly boosted by Peru through traditional and digital media.

In this context, "La decisión es mía" (The decision is mine) Campaign (April 2022) was highlighted. It was elaborated for the purpose of empowering teenagers, teachers, and parents of Maynas and Yurimaguas, in order to have a proactive attitude toward cases of teenage pregnancy. The programs carried out in educational institutions were spread on social networks; however, there were only 2828 Facebook followers and 243 Instagram followers with about ten interactions per post [7].

Likewise, the contest for the teenage pregnancy prevention week (October 2021) was carried out by the Ica Assistance Network of EsSalud, on TikTok, but without a profile. The campaign's aim pretended to spread and sensitize about the teenagers' health rights in favor of the teenagers' pregnancy reduction in the country. The winning video was posted on the official Facebook account of EsSalud Ica, but only obtained 346 likes and 45 congratulation messages for the female student who did the video [9].

Facing this panorama, in Peru, it is observed that health campaigns using social networks have not shown a significant acceptance in their application. The latter is evidenced by the low acceptance of the campaigns that were focused on the reduction of teenage pregnancy percentages in Peru. In that sense, considering that #YoDecidoCuando campaign on TikTok went viral very quickly and had a high acceptance, and taking into account this social network tendencies to spread contents virally, due to videos' shortness and motivation to view them frequently [3], identifying how this campaign was perceived by the female teenagers aged 16–19 years as an educational communication for the teenage pregnancy prevention was set out as an objective.

28.2 State of the Art

Social network videos are considered the new ways of communication for teenagers, where TikTok has recently become one of the most striking apps for this social group, used daily by 90% of its users, as stated by Suárez and García [23]. According to Ballesteros [3], short videos can be created, edited, and posted on TikTok, which allows its users to actively interact, and beyond the upload of content, it promotes challenges where people can participate doing their own version. In accordance with Ahlse et al. [2], entertainment is the main motivation for social network users, being necessary to delve into the different types of it and then choose the best one that may adapt to a campaign.

In recent years, because of technological progress, social life, for many people, revolves around social networks, modifying their way of socialization and boosting the emergence of online health communities, which usually pick up and confront the people's experience, as mentioned by Piedra [19]. In that sense, Xiaolin et al. [26] add that TikTok, being a community where all can create contents, areas related to health information can communicate healthy lifestyle and share knowledge, to the extent that users share and interact with posted information.

In that regard, Becerra and Taype [4] state that a diversity of people and institutions look for information to post on TikTok, showing the challenge faced by health institutions while combining users' expectations on TikTok (humoristic replicable short videos) with educational contents for a wider reach. In this sense, Kong et al. [13] add that TikTok has drawn attention of healthcare providers, mainly boosted by COVID-19 pandemic, where the content posted on app was widely viewed and shared, observing that the higher quality contents were those posted by non-profit organizations.

28.2.1 Teenage Pregnancy

Teenage pregnancy is pregnancy in a female aged 10–19 years, independent of the gynecologic age [25]. There are diverse factors at individual, family, and social levels related to this context, which are mainly determined by age, social context, belief system, culture, traditions, and socioeconomic status, among others. Therefore, the study and understanding of this issue involve an integral biopsychosocial study that considers, simultaneously, the following elements: teenage, sexuality, and reproduction [10].

Social networks can be used as a tool for exchange and collaboration with different sectors, which will allow generating an ecosystem of participation reached by the created posts, considering that people interact and share posts of their liking or interest, popularizing some contents [16]. This is because social networks show viability to spread content, specifically for young people's education, promoting an adequate practice of sexuality [21]. Thus, [22] add that, in the current context, social networks are key to spreading information to pregnant women as well as in the preconception period, because they use these in order to be informed.

28.3 Research Method

An empirical method was proposed for this research in order to review the perception of the female young followers of educational communication regarding teenage pregnancy prevention, using #YoDecidoCuando campaign on TikTok social network and, in this way, to study in more detail the perceptions, motivations, and interactions that this campaign generated on interviewees at educational settings on this social network.

Regarding the participants, the sample was made up by women between 16 and 19 years old because, in accordance with INEI's figures between 2019 and 2020, this age group showed the highest percentage of pregnancies. Moreover, the target interviewees were also selected by the type of socioeconomic status they belonged to, considering that TikTok is an app mainly used in Smartphone technology and that 84% of urban population had Smartphones, from which 97% belonged to the higher socioeconomic status [12]. Interviewed residents of Metropolitan Lima from the districts of Jesús María, Lince, Pueblo Libre, Magdalena, and San Miguel were considered, which according to the APEIM [1] coincide with these social strata. In addition, the interviewees had to have knowledge of the #YoDecidoCuando campaign and had to follow gynecologists' accounts on TikTok at least for 6 months.

#YoDecidoCuando campaign was carried out between September 26 and 28, 2020 by the Mayor's Office of Medellin (Colombia). Its aim was to get closer to young people in order to make them aware about their sexual and reproductive rights and invite them to use long-term contraception methods, which would allow a more conscious planning of their life project. The challenge on TikTok invited users to personalize and dramatize the song that identified the campaign. The four videos with more "likes" would be the winners (Mayor's Office of Medellin, 2020). This campaign was requested by several Latin-American countries, because Yina Rose, a singer well-received by young people starred in it. More than 900 comments supporting the campaign were viewed and registered between September 26 and 28, which not only were Colombian but also from other countries. Evidence of the latter was mainly shown by the emoji of the country's flag that accompanied their comments or an explicit mention of their home country. In Peru, more than 20 supportive comments were registered, considering both emoji flags and the mention of the home country.

The structured interview methodology was used to collect the data, in order to generate comparisons among the answers. The tool consisted of a guide of thirty-four questions that resulted in an interview of about 45 min, which was recorded by audio on Google Meet platform (Table 28.1).

28.4 Results and Discussion

28.4.1 TikTok and TikTokers

For all the interviewees, TikTok is an entertainment app, easy to use, but, in addition, it is also an educational option, because they find contents related to their careers, interests and hobbies. As noted by Ballesteros [3], this platform allows interaction in an active way and promotes challenges that invite users to share contents.

Regarding the motivation behind the following of different TikTokers, the interviewees agreed that TikTokers dealt with issues of their interest and spread contents empathetically. As mentioned by Diaz [8], TikTokers are people that offer contents according to their point of view, knowing that they have earned the trust of a community.

28.4.2 Perception of TikTok in the Areas of Education and Health

About the knowledge of educational and health activities carried out on TikTok, all the interviewees said that they know or have watched contents related to these issues and comment that this type of campaigns are very useful, dynamic, and easy to understand. This is related to the statement of Xiaolin et al. [26], who affirmed that being TikTok a community where all create contents, the health area can communicate healthy lifestyles and share knowledge to the extent that users share and interact with the posted information.

Table 28.1 Topics and guiding questions for in-depth interviews

Topics	Semi-structured questions
TikTok roles	1. What do you think about the social network TikTok and the use of it? 2. What motivations do you consider that make TikTok different from other social networks? 3. What activities do you do more often on TikTok? 4. What do you think about the comments left on the videos of the accounts that you follow? 5. What do you consider to be the added value to a video on TikTok to be shared?
Perception of TikTokers	6. Which TikTokers do you follow and why? 7. What motivates you to interact and share the TikToker's posts?
Perception of TikTok in the educational and health field	8. What's your opinion on the educational and health campaigns made on TikTok? 9. What campaigns of this type do you remember? 10. Which are the positive and negative aspects that you will highlight from them? 11. What motivations/reactions do you think that these campaigns generate? 12. What do you stand on the TikToker that promote educational campaigns and health ones?
Educational campaigns for the prevention of adolescent pregnancy through the TikTok social network	13. Which campaigns do you know about teen pregnancy prevention on TikTok? 14. Which TikTokers (gynecologist) that made teen pregnancy prevention campaigns do you follow and why? 15. What's your opinion on the educational campaigns about teen pregnancy prevention on TikTok? 16. What do you consider that generates a teen pregnancy prevention campaign on TikTok? 17. What advantages and disadvantages do you consider that exist for this type of campaigns on TikTok?
Case evaluation: #YoDecidoCuando campaign	18. What's your opinion on the proposal shared for the campaign #YoDecidoCuando? 19. Which people do you consider the message about the campaign #YoDecidoCuando it is aimed? 20. Which do you think are the main motivations/reactions that will make sharing this campaign? 21. Why will you participate on the challenges proposed for the campaign #YoDecidoCuando? 22. What do you think about the TikToker (singer) that starred in this campaign? 23. Why will you promote the teen pregnancy prevention through this campaign?

Regarding TikTokers that promote these educational and health campaigns, a positive perception is observed from the interviewees. They are considered people concerned with issues of social interest, because they use this media app in a proper way, they inform and generate a positive influence. Piedra [19] coincides with the above mentioned, he establishes that, in recent years, technological progress has caused the social life of many people to revolve around social networks, modifying the ways they socialize and boosting the emergence of virtual healthcare communities, in this case, the TikTokers' accounts.

Regarding motivations generated by the use of TikTok in educational and health campaigns, the interviewees agreed that there are several motivations. Among the most mentioned were the desire to be constantly informed about this issue; to share information with whoever needs it and to debate points of view with other platform users. Restrepo et al. [20] consider that these aspects stimulate changes in individual and societal harmful behaviors, achieving, through the creation of communities of people with common interests, to favor emotional support, the exchange of information, experiences and advice related to self-help.

28.4.3 TikTok in Educational Campaigns for Teenage Pregnancy Prevention

In relation to the knowledge that interviewees have about campaigns for teenage pregnancy prevention on TikTok, most of them mentioned that they viewed or searched contents related to the issue on this social network. They mainly mentioned: contraceptive methods, apps to control the menstrual cycle and to measure the fertility, also, information about health centers where they could obtain contraceptives. Related to this Ojeda and Montero [18] stated that there are diverse individual, domestic, and social factors, which are associated with teenage pregnancy problem, and understanding these involves an integral biopsychosocial study that considers simultaneously adolescence, sexuality and reproduction.

With respect to the following of TikTokers that carry out prevention campaigns, most of the interviewees mentioned that they follow gynecologists' accounts and consider them professional TikTokers, passionate about their job, because they provide information in a very creative and ludic way. They also mentioned that on TikTok they have the opportunity to ask in the comments and obtain answers from the specialists, agreeing with Herrera and Campero [11] approach, who state that to prevent adolescent pregnancy it is necessary to speak openly about sexuality and contraception with teenagers.

For the interviewees, this type of educational campaigns for teenage pregnancy prevention can be carried out on TikTok, due to its wide scope and because it is focused on teenagers who are the ones who need this information. The above mentioned coincides with Chamba et al. [6], who note that the health sector should adjust to the new social media, in this case to TikTok, developing strategies which boost the relationship and interaction with the community.

28.4.4 #YoDecidoCuando Case

Regarding the #YoDecidoCuando campaign, all the interviewees mentioned that it is a very innovative educational campaign, because it deals with this subject in a creative way, without losing the objective of informing on these issues. Some interviewees commented that they have never seen a structured campaign such as this one and, considering that users are invited to participate, they would like this campaign to be replicated throughout Peru.

When they were asked about the campaign's target group, a significant part of the interviewees answered that the target group would be teenagers (male and female) because they need this information. However, another part of the interviewees thought that it targets the general public, because the message involves a responsibility that should be everyone's role. Likewise, some interviewees mentioned that this type of educational campaign strengthens the women's voice, who still perceive a macho society. Such as Herrera and Campero [11] point, awareness raising campaigns should be outlined, where the emphasis is on healthy lifestyles and reproductive health.

About the motivations to share, #YoDecidoCuando campaign with TikTok' contacts and with other social networks, the interviewees answered that it stimulates the desire that more people are informed, think and become aware about this subject. Moreover, they added that they had perceived that through this content the necessity to look for more information was encouraged. As noted by Becerra and Taype [4], there are diverse people and institutions that provide information on TikTok, therefore, it is shown that health institutions on TikTok face a big challenge.

28.5 Conclusions

To answer the research question, how #YoDecidoCuando campaign on TikTok was perceived by the teenagers between 16 and 19 years old as an educational communication for teenage pregnancy prevention? It can be observed that the campaign on this social network had a positive perception, considering that information was presented in a creative and fast way, which generated, as indicated in the interviews, a greater impact in memory and the motivation to apply those recommendations.

The motivations generated by this campaign were mainly related to the encouragement oriented to staying informed. This is because the dynamic generated by the TikTok platform: the debate occurred in the comment section of the platform, the sharing of the video on different social networks, the identification of themselves with a social problem from several countries and the possibility to make it visible through its emoji flag and spread it by joining the challenge, so other people could be aware of the campaign.

Referred to the tone in the communication of this campaign, the empathic use of the language was highlighted. The campaign achieved the attention of young people to the prevention of teenage pregnancy, an issue that usually deserves to get more attention and understanding. Another achievement was that followers did not feel that the information came as an order, but just as a recommendation, which allowed reflecting deeply about this issue.

Finally, about the objective of the campaign, which is teenage pregnancy prevention, it was remarked that it is an issue that is necessary to be spoken about openly with teenagers. It is important to provide them with information in a creative and ludic way, so that they understand and become aware of the cares that they should have and, in this sense, this platform allows users to generate a solid, direct, and persuasive communication.

The main limitation was in the recruitment of the sample, as since it was a very personal issue, not many adolescents agreed to be interviewed.

For future research, it would be recommended to evaluate the new proposals for adolescent pregnancy prevention campaigns through social networks at the national level, in order to identify the perception of them and thus improve the development of communication strategies.

References

1. APEIM: "Niveles socioeconómicos 2021" 16 (2021). http://apeim.com.pe/wp-content/uploads/2021/10/niveles-socioecono%CC%81micos-apeim-v2-2021.pdf
2. Ahlse, J., Nilsson, F., Sandström, N.: It's time to TikTok: exploring generation Z's motivations to participate in #Challenges. Digitala Vetenskapliga Arkivet 1–77 (2020). http://www.diva-portal.org/smash/record.jsf?pid=diva2%3A1434091&dswid=1742
3. Ballesteros, C.: La propagación digital del coronavirus: Midiendo el engagement del entretenimiento en la red social emergente TikTok. Revista Española De Comunicación En Salud suplento **1**, 171–185 (2020). https://e-revistas.uc3m.es/index.php/RECS/article/view/5459
4. Becerra, N., Taype, A.: TikTok: ¿una nueva herramienta educativa para combatir la COVID-19? Acta médica peruana **37**(2), 249–251 (2020). https://dialnet.unirioja.es/servlet/articulo?codigo=7778151
5. Cender, P.V., Meza, M., Moquillaza, V.: COVID-19: Una pandemia en la era de la salud digital. Tesis de maestría. Universidad Cayetano Heredia (2020). https://preprints.scielo.org/index.php/scielo/preprint/view/164/195
6. Chamba, C., Altamirano, V., Yaguache, J.: Gestión de la comunicación 2.0 en las campañas de salud pública en la Comunidad Andina. Revista de comunicación volumen **20**(1), 49–65 (2021). https://dialnet.unirioja.es/servlet/articulo?codigo=7937111
7. Coordinadora de entidades extranjeras de cooperación internacional (coeeci). Lanzamiento de la campaña comunicacional: La decisión es mía (2022). https://coeeci.org.pe/lanzamiento-de-la-campana-comunicacional-la-decision-es-mia/
8. Díaz, L.: Soy marca: quiero trabajar con influencers, influencer marketing. Ámbitos: Revista internacional de comunicación **39**, 3–16 (2017). https://dialnet.unirioja.es/servlet/articulo?codigo=6337585

9. EsSalud: Premian a escolar por utilizar las redes sociales para prevenir el embarazo adolescente (2021). http://noticias.essalud.gob.pe/?inno-noticia=essalud-ica-utilizando-redes-sociales-previenen-embarazo-adolescente
10. Favier, M., Samón, M., Ruiz, Y., Franco, A.: Factores de riesgos y consecuencias del embarazo en la adolescencia. Revista Información Científica **97**(1), 205–214 (2019). http://www.revinfcientifica.sld.cu/index.php/ric/article/view/1805/3517
11. Herrera, C., Campero, L.: Decir a medias: límites percibidos por los adultos para involucrarse en la prevención del embarazo adolescente en México. Nueva antropología **31**(88), 134–154 (2018). https://dialnet.unirioja.es/servlet/articulo?codigo=7126333
12. IPSOS Perú. "El smartphone consolida su avance". Infografía Perfiles Ipsos (2019). https://www.ipsos.com/sites/default/files/ct/publication/documents/2019-11/elcomercio_2019-11-18_04_2.pdf
13. Kong, W., Song, S., Zhao, Y., Zhu, Q., Sha, L.: TikTok as a health information source: assessment of the quality of information in diabetes-related videos. J. Med. Internet Res. **23**(9), 1–8 (2021). https://www.ncbi.nlm.nih.gov/pmc/articles/PMC8444042/
14. Loaiza, K.: El Despacho de la Gestora Social de Medellín presenta campaña de prevención del embarazo adolescente. Medellín joven (2020). https://www.medellin.gov.co/irj/portal/medellin?NavigationTarget=navurl://bdbedc9695755c5c5f2e7c2dc449c5c8
15. Machuca, G.: Tik Tok: la red social usada para relajarse por médicos y enfermeras que batallan contra el COVID-19. Diario El Comercio. 12 de Julio del 2020. 18 de octubre del 2020. https://elcomercio.pe/somos/historias/tik-tok-la-red-social-usada-para-relajarse-por-medicos-y-enfermeras-que-batallan-contra-el-covid-19-noticia/
16. Mercado, F., Huerta, V., Urias, J.: Redes sociales virtuales y Salud. Una experiencia participativa sobre la enfermedad renal crónica. Revista Interface volumen **23**(1), 1–11 (2019). https://www.scielo.br/j/icse/a/rQwzQjBjNBht7z4Fsp8FYjz/?lang=es
17. Naranjo, M.: TikTok lanza #EduTok con tutoriales en vivo para descubrir y aprender desde casa. Noticias Redes sociales El Grupo Informático, 24 de Marzo del 2020, 20 de Octubre del 2020. https://www.elgrupoinformatico.com/noticias/edutok-tiktok-tutoriales-vivo-t77070.html
18. Ojeda, A., Montero, L.: Adolescencia, sexualidad y reproducción: tres dimensiones fundamentales para la comprensión del fenómeno del embarazo adolescente. Universidad de Cartagena **19**(2), 36–53 (2019). https://repositorio.unicartagena.edu.co/handle/11227/14271
19. Piedra, J.: Redes sociales en tiempos del COVID-19: el caso de la actividad física. Sociología del deporte **1**(1), 41–43 (2020). https://dialnet.unirioja.es/servlet/articulo?codigo=7505057
20. Restrepo, A.M., Muñoz, Y., Duque, M.A.: Análisis de los elementos de mercadeo social implícitos en campañas de prevención de embarazo en adolescentes. Revista Facultad Nacional de Salud Pública **36**(6), 18–27 (2018). https://dialnet.unirioja.es/servlet/articulo?codigo=6709937
21. Sandoval, V., Garcia, L., Ramirez, D.: El uso de herramientas de comunicación digital como estrategia de educación sexual para jóvenes universitarios. Revista Reflexiones Pedagógicas **25**, 2–12 (2021). https://pure.urosario.edu.co/es/publications/el-uso-de-herramientas-de-comunicaci%C3%B3n-digital-como-estrategia-de
22. Skouteris, H., Savaglio, M.: The use of social media for preconception information and pregnancy planning among young women. J. Clin. Med. **10**(9), 1–11 (2021). https://www.ncbi.nlm.nih.gov/pmc/articles/PMC8123806/
23. Suárez, R., García, A.: Centennials en TikTok: tipología de vídeos. Análisis y comparativa España-Gran Bretaña por género, edad y nacionalidad. Revista Latina de Comunicación Social **79**, 1–22 (2021). https://dialnet.unirioja.es/servlet/articulo?codigo=7954969
24. TikTok: TikTok Democratizes eLearning With the Launch of the #EduTok Program. Cision Pr Newswire (2019). https://www.prnewswire.com/in/news-releases/tiktok-democratizes-elearning-with-the-launch-of-the-edutok-program-864546835.html

25. WHO. El embarazo en la adolescencia (2020). https://www.who.int/es/news-room/fact-sheets/detail/adolescent-pregnancy
26. Xu, X., Chen, J., Zhang, W., Zhu, C., Evans, R.: How health communication via Tik Tok makes a difference: a content analysis of Tik Tok accounts run by Chinese provincial health committee. Int. J. Environ. Res. Publ. Health **17**(1), 192, 1–13 (2019). https://pubmed.ncbi.nlm.nih.gov/31892122/

Chapter 29
Social Media Followers: The Role of Value Congruence and the Social Media Manager

Concepción Varela-Neira, Zaira Camoiras-Rodríguez, and Teresa García Garazo

Abstract Despite organizational interest in leveraging participation in social media, the impact of the social media manager in its success has been neglected in scholarly research. Through a moderated mediation process, this research examines the impact of value congruence of the social media follower with the brand social media presence on his/her willingness to pay premium price, considering the mediating role of the social media manager's perceived authenticity and the moderating role of perceived task competence. This study employs data from 327 social media brand followers. The results support the model and show that social media followers' value congruence influences perceptions of social media manager's authenticity, and eventually the followers' willingness to pay a premium price. Finally, social media managers' perceived task competence enhances the relationship between the value congruence and perceived authenticity and between perceived authenticity and willingness to pay a premium price. These findings highlight the need for social media managers to learn about the values of their target audience before engaging with them and, once they have done so, to align the social media messages and content to these values. This value congruence will result in superior judgments and behavioral intentions.

29.1 Introduction

Hootsuite: The Global State of Digital 2022 [1] shows that 58% of the global population in 2021 is an active user of social media, which implies an increase of more than 10% in the number of users this last year, while the time dedicated to social media

C. Varela-Neira · T. García Garazo
Universidad de Santiago de Compostela, Santiago de Compostela 15782, España
e-mail: Conchi.varela@usc.es

T. García Garazo
e-mail: mariateresa.garcia@usc.es

Z. Camoiras-Rodríguez (✉)
Universidad de Santiago de Compostela, Lugo 27002, España
e-mail: Zaira.camoiras@usc.es

J. L. Reis et al. (eds.), *Marketing and Smart Technologies*, Smart Innovation, Systems and Technologies 337, https://doi.org/10.1007/978-981-19-9099-1_29

each day is also rising. This numbers are forcing organizations to progressively integrate social media marketing into their business strategies to stay competitive [2]. However, despite the potential returns of an active presence on social media, key questions stay unanswered [3].

Since most of prior research on social media has investigated how to adapt the organization's marketing mix to this context [4], other relevant topics, such as the role and impact of social media managers, have been neglected. This is surprising, as positions linked to digital marketing are currently the most demanded in marketing and expected to increase, with 61% of organizations specifying their interest in recruiting these positions [5]. Furthermore, social media managers have role specifies that require an individualized study. Social media mangers are involved in defining and executing a firm's brand communication strategy on social media [e.g., 6], for which they must rely on their knowledge of the particularities of social media platforms and from social media monitoring and analytics [e.g., 7]. Moreover, [8] claim that social media managers must articulate employee policies and policing, enlist employees, collaborate internally, try technology, detect and deal with crises, build and preserve reputation, assess social media campaigns success, and implement programs for developing customer relationships. Nevertheless, even though social media managers would traditionally be included in the back-office employees' category [9], they may be considered front-office employees, as they have direct contact with customers in social media by posting material and replying to followers' comments and queries. Social media managers have an important boundary-spanning role, as they must build community, engage internal and external audiences, develop customer relationships, and build and preserve brand reputation through their interactions in social media [7]. Still, social media managers have larger role in the organizational setting that differentiates them from conventional customer-contact front-office employees. Given that brand attitude and loyalty depend on all distinct areas of a customer's interaction with the brand [10], this investigation contributes to fill this gap by studying the social media managers' impact on the behavioral intentions of the social media followers.

As "being authentic" is a fundamental characteristic of social media content, the perceived authenticity of an organization's social voice is a key measure of a social media managers' success [6]. In other words, since social media managers personify the organization's social media voice, they need to transmit authenticity to their social media audiences to get them to commit. Hence, this study examines the antecedents and consequences of the followers' perceptions of authenticity of the social media manager. Especially, this study analyses the impact of value congruence of the social media follower with the brand social media presence on their perception of the social media manager's authenticity and the effect of this perception on their willingness to pay premium price. In doing so, we consider the mediating effect of the social media managers' authenticity as well as the moderating effect of their task competence. The moderated mediation model proposed in this study makes several contributions to knowledge.

First, this work differs from previous research on social media in that it examines the impact of social media managers, specifically the customers' perception of the social media manager's authenticity and task competence, on social media followers' behavioral intentions, whereas prior studies have centered on the effect of social media content or strategies on brand perceptions [e.g., 11, 12] and on customer behavior and/or firm performance [e.g., 13–15].

Second, this research adds to the literature as it focuses on customers' perceptions of authenticity, which have received little attention [16]. This is surprising since observers' responses are influenced by the authenticity inferred from the one observed [17, 18]. Still, whether its outcomes may be positive or negative is yet unknown, and consequently, there have been calls for research to study purchase behaviors on models of authenticity in a frontline context [16].

Moreover, to our knowledge, this is the first study to consider the moderating effect of task competence perception on the relationship between employee authenticity and willingness to pay a premium price for the brand. Since the relevance of authenticity may differ across consumers [19, 20], there is a need to investigate moderating effects in the relationship between authenticity and its outcomes. Lately, social cognition has been drawing attention as a crucial factor explaining consumers' responses [21, 22]. However, the competence dimension of social cognition in technology-based service interactions has only just started to be noticed [23–25].

Finally, as far as the authors know, the antecedents of customers' perceptions of social media manager authenticity have not been examined. This research centers on value congruence of the social media follower with the brand social media presence, since "advertising is probably the most visible, recognizable, and memorable element of organizational communication" [26, p. 5], and it innately sends messages about the brand and organization values that impact the consumer's perceptions [27]. However, prior investigations on value congruence have mainly focused on employee-organization relations [28–31], whereas only a limited number of studies have centered on consumer-brand relations and the organization's external communication [32, 33].

29.2 Research Background and Hypotheses

29.2.1 The Social Media Manager's Perceived Authenticity

Perceived authenticity differs in its operation when it refers to brands [e.g., 19, 20, 34, 35] compared to individuals [e.g., 36–40]. The current study focuses on customer perceptions of the social media manager's authenticity, where authenticity refers to when people behave according to their own words and internal values [41–44]. Consistent with this, the social media manager's perceived authenticity is defined as the degree to which a follower perceives that the social media manager is genuine and is acting without pretense as his/her own person [45, 46].

Although previous marketing literature has mainly studied brand authenticity [e.g., 19, 47, 48], the perceived authenticity of boundary-spanning employees is a key aspect to be investigated [49], as it has a relevant role in crafting the customer experience and shaping brand perceptions [50, 51].

Since these employees must behave professionally, which may mean, in authentically, due to the profit motivations [e.g., 51, 52], when customers perceive that they are being authentic, portraying their "true" self in the interaction, they do not come across as being in a "selling" mode [53]. Thus, when followers perceive that the social media manager is authentic with them, they feel that he/she has a sincere interest in them as individuals and believe that he/she is sincerely willing to help, which would result in greater levels of followers' confidence toward the service [16, 54]. In addition to its impact on the followers' confidence, perceptions of authenticity would also trigger feelings of social benefit, where followers believe that the social media manager, and by extension the brand, desires a more personal and meaningful relationship, where their value is recognized beyond the dollar return [54, 55]. These beliefs would lead to customer loyalty [56]. Thus,

H1: The perception of the social media manager's authenticity is positively related to the willingness of the brand's follower on social media to pay a premium price.

29.2.2 The Follower's Value Congruence

Rokeach [57] described values as "enduring beliefs that a specific mode of conduct is personally or socially preferable to an opposite or converse mode of conduct or end state of existence" (p. 5). Values are moral compasses that show individuals what they should do or not do, guiding their attitudes and decisions in various contexts [58–60]. As a result, values and value congruence are central topics in organizational research [61–63].

Ad value congruence refers to the similarity between personal values and values highlighted in an ad [64]. Given that an ad or message is full of implicit or inferred values, it conveys information (even at a distance) about the moral positions of the person responsible for this message [65]. Thus, the interaction with social media managers through messages and social information processing helps followers gain an understanding of the manager and by extension the brand's values, and they will compare these values with their self-relevant values.

When followers hold similar values to those portrayed in social media by the social media manager, this value congruence will enable a better understanding and smoother communication [66, 67] due to comparable cognitive schemas [66, 68]. As such, ambiguity and the likelihood of interpersonal frictions can be avoided [69, 70], and high-quality interpersonal relationships can be gradually developed [71]. Furthermore, research on self-affirmation [72] and self-justification [73] suggests that when a person makes a positive assessment of the group with which he/she identifies, a positive view of the self is reinforced. Consequently, similarity in values would give rise to positive sentiments and liking, promoting positive perceptions of the other party [74], including attributions of competence and benevolence [75], whereas dissimilarity could lead to negative emotions [76]. Therefore, we expect value congruence to lead to a more positive perception of the social media manager, in particular, an enhanced perception of his/her authenticity. Moreover, customers are likely to use brands that show a higher congruence with their personal values [77, 78]. This is in line with previous research on value congruence that links this construct to several positive outcomes, such as satisfaction, affective commitment, trust, and high-quality exchange relationships [32, 71]. Thus,

H2: The social follower's value congruence is positively related to (a) the perception of the social media manager's authenticity and (b) willingness to pay premium price.

29.2.3 The Social Media Manager's Task Competence

Knowledge of how customers perceive front-office employee performance is crucial [25], since, according to social cognitive theory, the perceived capabilities of parties in an interdependent relationship matter [79, 80]. Similarly, marketing literature has argued that when customers believe employees to be able and competent to solve problems, they are more satisfied [81]. However, when it comes to judging others, in particular when customers evaluate employees, they usually do not analyze their idiosyncratic merits and flaws but make general judgments regarding their ability to identify and meet their needs and their capacity to understand them [82].

Consequently, a type of competence acquires special relevance in employee-customer relationships: task competence, which refers to the employee's knowledge, ability, and concern to fulfill a task based on expertise about the customers' needs, and it entangles the delivery of a correct core service [83–86]. While definitions differ, competence judgments involve confidence, effectiveness, intelligence, capability, skillfulness, and competitiveness [e.g., 87–90]. Despite previous research mainly having studied customers evaluations of competence in traditional service encounters [91, 92], it has been suggested that competence can also be assessed in the online service context [25]. Due to the multifaceted nature of social media management responses, different sets of cues in the service encounter and communication could prompt followers' perceptions of task competence [93, 94]. Consequently, this study considers task competence from an observer's perspective, in other words, it focuses on the task competence attributed to the social media manager due to the cues embedded in his/her social media responses, and examines its moderating role on the previously argued relationships.

According to the resource-based view, personnel-based resources comprise technical know-how and other skills and assets, such as employee training and loyalty [95, 96]. Task competence and employee authenticity can therefore be considered personnel-based resources. At the same time, intangible resources include aspects such as reputation and brand image [97, 98]. Customer relationships are also viewed as an important firm resource [99], since "the potential exists for any organization to develop intimate relations with customers to the point that they may be relatively rare and difficult for rivals to replicate" [100, p. 779]. The same can be said about consumers perceiving value congruence with the organization.

According to the economic theory of complementarities, resource combinations have a super-additive value where complementary resources are mutually supportive and their joint adoption results in greater returns than the sum of their individual returns [101]. The complementarity concept can be applied to the case of the customer's value congruence and the employee's perceived task competence and authenticity. During the social media experience process, the task competence of the social media manager reinforces the value congruence of the social media message in aiding the organization to generate a favorable attitude in the customer and, at the same time, bolsters the impact of the social media manager authenticity on the behavioral intentions of the customer. For example, higher levels of the social media manager task competence involve a more skillful handling of the internal tasks required for a successful social media management as well as a better provision of product or service-specific information to support followers' interactions. In addition, task competence plays an important role in several stages of problem resolution, one of the main chores of social media managers. These improvements in the social media manager's performance will color and help strengthen the impact of a congruence of values between the social media message and the follower on the follower's attitudes. Moreover, in line with [54] findings that customers' perceptions of employee authenticity interact with customer judgments to affect customer loyalty, we argue that when followers judge social media managers as more competent and authentic, their willingness to pay premium price for the brand is accentuated. Thus,

H3: The perception of the social media manager's task competence moderates the effect of the follower's value congruence on the perception of the social media manager's authenticity.

H4: The perception of the social media manager's task competence moderates the effect of the perception of the social media manager's authenticity on the willingness to pay a premium price of the brand's follower on social media.

29.3 Methodology

29.3.1 Sample

To test the research hypotheses, this study involved 327 social media brand followers. Regarding the social media brand page's profile, 26.9% were following the brand on Facebook, 61.8% on Instagram, 7% on Twitter, 2.8% on LinkedIn, 1% on YouTube, and 0.3% on Google. Regarding the social media followers' characteristics, 54% were women, with an average age of 37.35 years (SD = 10.33), an average seniority as follower of the firm on social media of 4.96 years (SD = 4.35), and an average interaction frequency of around once per month.

29.3.2 Measures

The constructs considered in this study were measured using Likert scales adapted from previous investigations. Willingness to pay premium price was measured using a scale adopted from [12]. The measure for social media manager authenticity was adapted from [16]. Value congruence was measured based on [64], and for task competence, we relied on [102]. Two control variables were included in the model: the follower's age (the natural logarithm of age) and gender (0 = male, 1 = female). The results from confirmatory factor analysis, descriptive statistics, and correlations are provided in Tables 29.1 and 29.2.

Regarding reliability, all constructs manifest a composite reliability (CR) and average variance extracted (AVE) greater than the recommended threshold values of 0.7 and 0.5, respectively [103]. Moreover, Cronbach alphas exceed the 70 cut-off. Convergent validity is supported as all standardized coefficients are significant and greater than 0.5. Discriminant validity is also supported among the follower-related constructs as correlations among all variables show confidence intervals that do not include the unit value, and their squared value does not exceed the corresponding AVE. This provides evidence of the validity and reliability of the measurement instrument [104].

Table 29.1 Measurement scales and properties

Constructs
Willingness to pay premium price (α = 0.88; CR = 0.89; AVE = 0.72). Adapted from: [12]
The price of this brand would have to increase quite a bit before I would switch to another brand I am willing to pay a higher price for this brand than for other brands I am willing to pay a lot more for this brand than for other brands
Value congruence (α = 0.90; CR = 0.90; AVE = 0.75). Adapted from: [64]
I'm proud of the values expressed in this Brand social media community The values featured in this Brand social media community are similar to my own values The organizational values implied by the contents of this Brand social media community are the values and beliefs we should be highlighting
Social media manager authenticity (α = 0.86; CR = 0.90; AVE = 0.82). Adapted from: [16]
The social media manager seems phony or fake (r) The social media manager says things that make it seem true thoughts are not being expressed (r)
Task competence (α = 0.91; CR = 0.91; AVE = 0.67). Adapted from: [102]
The social media manager is very capable The social media manager is efficient The social media manager is organized The social media manager performed as expected The social media manager meets followers needs

Note CR = Composite reliability; AVE = Average variance extracted; α = Cronbach alpha coefficient

Table 29.2 Descriptive statistics and correlations

	Obs	Mean	S.D.	Min	Max	1	2	3
1. Task competence	327	6.04	1.05	1	7			
2. Willingness to pay premium price	327	3.69	1.72	1	7	0.271**		
3. Value congruence	327	4.85	1.29	1	7	0.546**	0.616**	
4. Social media manager authenticity	327	4.50	2.04	5	7	−0.042	0.356**	0.244**

**$p < 0.01$; *$p < 0.05$

29.4 Results

Our research model implies a moderated mediation process [105, 106]. To simplify the model, we replaced the constructs by the average score of the indicators. Moreover, we mean centered the variables involved in the interaction terms. Finally, to assess the proposed moderated mediation process, we carried out path analysis using Stata 14.0.

The results reported in Table 29.3 reveal that the relationship between social media manager authenticity and willingness to pay a premium price is positive and significant ($b = 0.15$, $p < 0.01$). The results also show that value congruence has a positive and significant direct relationship with social media manager authenticity ($b = 0.58$, $p < 0.01$) and willingness to pay a premium price ($b = 0.76$, $p < 0.01$). The findings also reveal that the interaction between social media manager authenticity and task competence has a positive and significant relationship with willingness to pay a premium price ($b = 0.09$, $p < 0.05$). These results indicate that social media manager authenticity has a stronger positive relationship with willingness to pay a premium price for higher levels of task competence. Finally, the results reveal that task competence also strengthens the positive relationship between value congruence and social media manager authenticity ($b = 0.16$, $p < 0.01$). We plotted these interactions in Figs. 29.1 and 29.2 following the process proposed by [107].

However, to establish a moderated mediation process, the strength of the mediation via social media manager authenticity must vary across different levels of the moderator. Consequently, to further assess the moderated mediation, we calculated the significance of the indirect relationship between value congruence and willingness to pay premium price via social media manager authenticity for social media managers with low/moderate/high task competence (see Table 29.4). The results reveal that value congruence is positively and indirectly related to willingness to pay premium price via social media manager authenticity when the social media manager's task competence is high or moderate but not when it is low. Moreover, the results also reveal that the positive indirect relationship between value congruence and willingness to pay premium price via social media manager authenticity is greater for customers who perceive greater task competence from the social media

Table 29.3 Path analysis results

	Willingness to pay premium price		
Variables	*Model* 1	*Model* 2	*Model* 3
Ln_age	−0.906**	−0.781**	−0.533*
Gender	0.028	0.002	−0.110
Manager authenticity (MA)			0.150**
Value congruence (VC)		0.860**	0.766**
Task competence (TC)		−0.094	−0.059
VC × TC			0.016
MA × TC			0.092*
	Social media manager authenticity		
Ln_age	−1.576**	−1.324**	−1.275**
Gender	0.468*	0.489*	0.458*
Value congruence (VC)		560**	0.580**
Task competence (TC)		−0.424**	−0.112
VC × TC			0.158**
Log likelihood	−1604.224	−2454.236	−3695.198
Coefficient of determination	0.089	0.475	0.475
CFI	0.415	0.924	0.971
SRMR	0.101	0.037	0.022

**$p < 0.01$; *$p < 0.05$

manager. Therefore, this positive indirect relationship is stronger the greater the task competence.

29.5 Discussion

Nowadays, to stay competitive, organizations have to include social media marketing into their business strategy [2]. However, there are still key questions that remain unanswered regarding what organizations must do to successfully engage in social media [3]. This investigation contributed to the literature in social media marketing

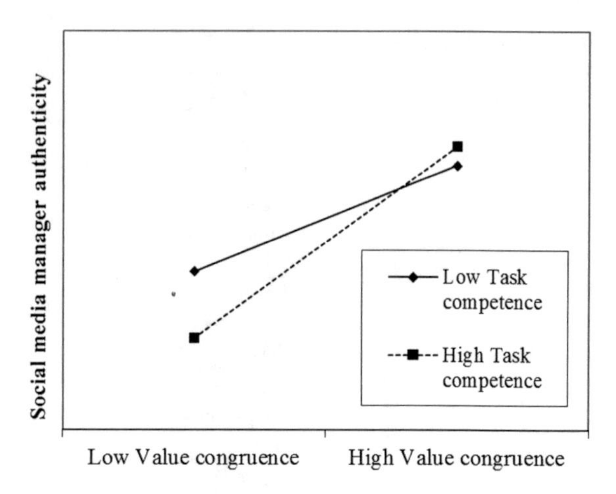

Fig. 29.1 Interaction between value congruence and task competence

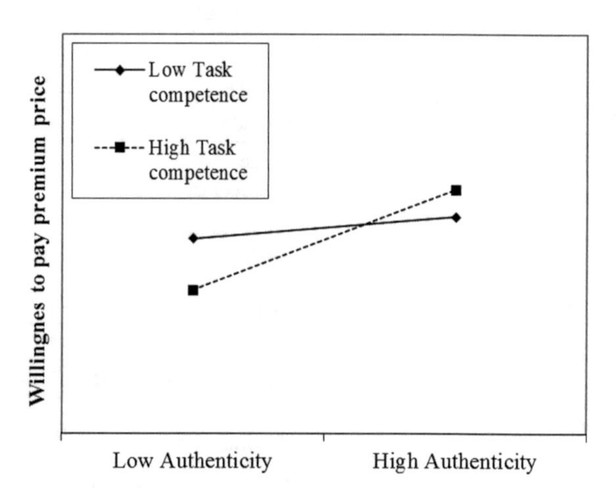

Fig. 29.2 Interaction between social media manager authenticity and task competence

Table 29.4 Indirect effects on willingness to pay a premium price via social media manager's authenticity

Effect	Task competence	Coef	SE	LLCI	ULCI
Value congruence	Low (−1 SD)	0.022	0.034	−0.037	0.100
	Moderate (Mean)	0.087^{**}	0.033	0.035	0.170
	High (+1 SD)	0.184^{**}	0.069	0.075	0.342

$^{**}p < 0.01$; $^{*}p < 0.05$

by examining the impact of social media managers on the willingness of social media followers to pay a premium price. To do so, it analyzed the mediating role of the social media manager perceived authenticity in the relation between the social media follower congruence in values with the social media brand page and his/her willingness to pay a premium price, and the enhancing role of the social media manager perceived task competence in the proposed relationships. The results obtained offer wide support to our model and make several contributions to the literature.

First, the findings showed that the fit between the values of the social media follower and those presented in the social media brand page gives rise to positive judgments and behavioral intentions on the social media follower. Therefore, we add to the literature on value congruence by supporting the relevance of this construct in consumer-organization relationships in a social media environment, and to the literature on social media by suggesting that, to achieve positive outcomes through these media, it is key to post messages and content with matching values to those of the target audience.

Second, the results revealed the importance of the social media manager role. The social media manager's skills and abilities, conveying to brand page followers authenticity and competence, have a significant impact on their willingness to pay a premium price. As far as we are aware, this is one of the first investigations to analyze the influence of organizational determinants on social media followers. However, our findings imply that scholarly research should fix this gap and devote attention to the characteristics of the social media managers, as they seem to drive important and positive outcomes. Moreover, the results also support the need to understand the social media followers' beliefs regarding the social media manager, in particular, his/her task competence and authenticity. Surprisingly, there are a lack of investigations on judgments of other's competence as well as on perceptions of employee's authenticity in a technology driven environment, such as social media.

29.5.1 Managerial Implications

Consumers want to engage with authentic voices in social media. When this happens, it drives consumers to be willing to pay a premium price for the brand with which they interact in social media. This investigation helps managers understand how to prompt perceptions of authenticity in a social media context as well as how to enhance the effect of these perceptions. The results indicate that value congruence between the consumer and the brand fuels perceptions of social media manager's authenticity and the willingness to pay a premium price. Accordingly, social media managers must carefully analyze the conversations taking place on each platform that the organization intends to use to understand its target audience and their values. In other words, they must get to know the target audience and their values before engaging with them. Afterwards, messages in social media must be designed to imply values that properly align with those of the target audience, and the content to be communicated must also take into consideration the target audience values. Moreover, the positive effects of value congruence and authenticity are enhanced when consumers perceive that the social media manager is competent and efficient in the task. This suggests that supervisors should hire social media managers who have deep understanding of the social media environment, with experience and/or training in the field.

29.5.2 *Limitations and Future Research*

Finally, we need to present both the limitations of the research as well as the opportunities for future research. First, this is a cross-sectional study, as a result, causality cannot be established. Moreover, all measures in this study were reported by the social media follower, which may result in common method variance. On the other hand, there are several possibilities regarding future research. First, to analyze the impact of another key competence: the social competence of social media managers. Second, to compare the model between different services, social media platforms, and social media motivations. Lastly, to study which cues in a social media context result in beliefs of task competence.

References

1. Hootsuite: The Global State of Digital 2022. Retrieved on June 19th, 2022, from https://www.hootsuite.com/resources/digital-trends (2022)
2. Dwivedi, Y.K., Ismagilova, E., Hughes, D.L., Carlson, J., Filieri, R., Jacobson, J., Jain, V., Karjaluoto, H., Kefi, H., Krishen, A.S., Kumar, V., Rahman, M.M., Raman, R., Rauschnabel, P.A., Rowley, J., Salo, J., Tran, G.A., Wang, Y.: Setting the future of digital and social media marketing research: perspective and research propositions. Int. J. Inf. Manage. **59**, 102168 (2021)
3. Borah, A., Banerjee, S., Lin, Y.T., Jain, A., Eisingerich, A.B.: Improvised marketing interventions in social media. J. Mark. **84**(2), 69–91 (2020)
4. Marchand, A., Hennig-Thurau, T., Flemming, J.: Social media resources and capabilities as strategic determinants of social media performance. Int. J. Res. Mark. **38**(3), 549–571 (2021)
5. McKinley Marketing Partners: 2019 Marketing hiring trends: an in-depth report on factors shaping demand for marketing and creative talent. Retrieved on June 24th 2022 from https://cdn2.hubspot.net/hubfs/4517345/2019%20Marketing%20Hiring%20Trends%20Report.pdf (2019)
6. Bossio, D., McCosker, A., Milne, E., Golding, D., Albarrán-Torres, C.: Social media managers as intermediaries: negotiating the personal and professional in organisational communication. Commun. Res. Pract. **6**(2), 95–110 (2020)
7. Bossio, D., Sacco, V.: From 'selfies' to breaking Tweets: how journalists negotiate personal and professional identity on social media. J. Pract. **11**(5), 527–543 (2017)
8. Neill, M.S., Moody, M.: Who is responsible for what? Examining strategic roles in social media management. Publ. Relations Rev. **41**, 109–118 (2015)
9. Lariviere, B.: Personal communication. In: International service research scholars conference, Karlstad University, Karlstad Sweden, Fall, 2014 (2014)
10. Brexendorf, T.O., Mühlmeier, S., Tomczak, T., Eisend, M.: The impact of sales encounters on brand loyalty. J. Bus. Res. **63**(11), 1148–1155 (2010)
11. Dabbous, A., Barakat, K.A.: Bridging the online offline gap: assessing the impact of brands' social network content quality on brand awareness and purchase intention. J. Retail. Consum. Serv. **53**, 101966 (2020)

12. Godey, B., Manthiou, A., Pederzoli, D., Rokka, J., Aiello, G., Donvito, R., Singh, R.: Social media marketing efforts of luxury brands: influence on brand equity and consumer behavior. J. Bus. Res. **69**, 5833–5841 (2016)
13. Li, Y., Xie, Y.: Is a picture worth a thousand words? An empirical study of image content and social media engagement. J. Mark. Res. **57**(1), 1–19 (2019)
14. Ordenes, F.V., Grewal, D., Ludwig, S., Ruyter, K.D., Mahr, D., Wetzels, M.: Cutting through content clutter: how speech and image acts drive consumer sharing of social media brand messages. J. Consumer Res. **45**(5), 988–1012 (2019)
15. Tellis, G.J., MacInnis, D.J., Tirunillai, S., Zhang, Y.: What drives virality (sharing) of online digital content? The critical role of information, emotion, and brand prominence. J. Mark. **83**(4), 1–20 (2019)
16. Yoo, J., Arnold, T.: Frontline employee authenticity and its influence upon adaptive selling outcomes: perspectives from customers. Eur. J. Mark. **53**(11), 2397–2418 (2019)
17. Campbell, M.C., Kirmani, A.: Consumers' use of persuasion knowledge: the effects of accessibility and cognitive capacity on perceptions of an influence agent. J. Consumer Res. **27**(1), 69–83 (2000)
18. Yoon, Y., Giirhan-Canli, Z., Schwarz, N.: Beliefs, attitude, intention and behavior: an introduction to theory and research. In: The Effect of Corporate Social Responsibility (CSR) Activities on Companies With Bad Reputations (2006)
19. Morhart, F., Malär, L., Guèvremont, A., Girardin, F., Grohmann, B.: Brand authenticity: an integrative framework and measurement scale. J. Consum. Psychol. **25**(2), 200–218 (2015)
20. Moulard, J.G., Raggio, R.D., Folse, J.A.G.: Brand authenticity: testing the antecedents and outcomes of brand management's passion for its products. Psychol. Mark. **33**(6), 421–436 (2016)
21. Grandey, A.A.: When "the show must go on": surface acting and deep acting as determinants of emotional exhaustion and peer-rated service delivery. Acad. Manag. J. **46**(1), 86–96 (2003)
22. Scott, M.L., Mende, M., Bolton, L.E.: Judging the book by its cover? How consumers decode conspicuous consumption cues in buyer–seller relationships. J. Mark. Res. **50**(3), 334–347 (2013)
23. Li, X., Chan, K.W., Kim, S.: Service with emoticons: how customers interpret employee use of emoticons in online service encounters. J. Consumer Res. **45**(5), 973–987 (2019)
24. Wu, J., Chen, J., Dou, W.: The internet of things and interaction style: the effect of smart interaction on brand attachment. J. Mark. Manag. **33**(1–2), 61–75 (2017)
25. Huang, R., Ha, S.: The effects of warmth-oriented and competence-oriented service recovery messages on observers on online platforms. J. Bus. Res. **121**, 616–627 (2020)
26. Ewing, M.T., Pitt, L.F., De Bussy, N.M., Berthon, P.: Employment branding in the knowledge economy. Int. J. Advert. **21**(1), 3–22 (2002)
27. You, L., Hon, L.C.: Testing the effects of reputation, value congruence and brand identity on word-of-mouth intentions. J. Commun. Manag. **25**(2), 160–181 (2021)
28. Christiansen, N., Villanova, P., Mikulay, S.: Political influence compatibility: fitting the person to the climate. J. Org. Behav. Int. J. Ind. Occupational Org. Psychol. Behav. **18**(6), 709–730 (1997)
29. Enz, C.A.: The role of value congruity in intraorganizational power. Admin. Sci. Quart. 284–304 (1988)
30. Jehn, K.A., Mannix, E.A.: The dynamic nature of conflict: a longitudinal study of intragroup conflict and group performance. Acad. Manag. J. **44**(2), 238–251 (2001)
31. Lau, D.C., Liu, J., Fu, P.P.: Feeling trusted by business leaders in China: antecedents and the mediating role of value congruence. Asia Pacific J. Manage. **24**(3), 321–340 (2007)
32. Zhang, J., Bloemer, J.M.: The impact of value congruence on consumer-service brand relationships. J. Serv. Res. **11**(2), 161–178 (2008)
33. Lee, S.A., Jeong, M.: Enhancing online brand experiences: an application of congruity theory. Int. J. Hosp. Manag. **40**, 49–58 (2014)
34. Grayson, K., Martinec, R.: Consumer perceptions of iconicity and indexicality and their influence on assessments of authentic market offerings. J. Consumer Res. **31**(2), 296–312 (2004)

35. Napoli, J., Dickinson, S.J., Beverland, M.B., Farrelly, F.: Measuring consumer-based brand authenticity. J. Bus. Res. **67**(6), 1090–1098 (2014)
36. Arnould, E.J., Price, L.L.: River magic: extraordinary experience and the extended service encounter. J. Consumer Res. **20**(1), 24–45 (1993)
37. Beverland, M.B., Farrelly, F.J.: The quest for authenticity in consumption: consumers' purposive choice of authentic cues to shape experienced outcomes. J. Consumer Res. **36**(5), 838–856 (2010)
38. Deci, E.L., Ryan, R.M.: A motivational approach to self: integration in personality. Persp. Motivation **38**, 237–288 (1991)
39. Kernis, M.H., Goldman, B.M.: A multicomponent conceptualization of authenticity: theory and research. Adv. Exp. Soc. Psychol. **38**, 283–357 (2006)
40. Wood, A.M., Linley, P.A., Maltby, J., Baliousis, M., Joseph, S.: The authentic personality: a theoretical and empirical conceptualization and the development of the authenticity scale. J. Couns. Psychol. **55**(3), 385–399 (2008)
41. Driscoll, C., McKee, M.: Restorying a culture of ethical and spiritual values: a role for leader storytelling. J. Bus. Ethics **73**(2), 205–217 (2007)
42. Erickson, R.J.: The importance of authenticity for self and society. Symb. Interact. **18**(2), 121–144 (1995)
43. Jackson, K.T.: Towards authenticity: a Sartrean perspective on business ethics. J. Bus. Ethics **58**(4), 307–325 (2005)
44. Liedtka, J.: Strategy making and the search for authenticity. J. Bus. Ethics **80**(2), 237–248 (2008)
45. Price, L.L., Arnould, E.J., Deibler, S.L.: Consumers' emotional responses to service encounters: the influence of the service provider. Int. J. Serv. Ind. Manag. **6**(3), 34–63 (1995)
46. Schaefer, A.D., Pettijohn, C.E.: The relevance of authenticity in personal selling: is genuineness an asset or liability? J. Market. Theory Practice **14**(1), 25–35 (2006)
47. Fritz, K., Schoenmueller, V., Bruhn, M.: Authenticity in branding–exploring antecedents and consequences of brand authenticity. Eur. J. Mark. **51**(2), 324–348 (2017)
48. Guèvremont, A., Grohmann, B.: The brand authenticity effect: situational and individual-level moderators. Eur. J. Mark. **50**(3/4), 602–620 (2016)
49. Yagil, D., Medler-Liraz, H.: Moments of truth: examining transient authenticity and identity in service encounters. Acad. Manag. J. **56**(2), 473–497 (2013)
50. Keller, K.L.: Conceptualizing, measuring, and managing customer-based brand equity. J. Mark. **57**(1), 1–22 (1993)
51. Sirianni, N.J., Bitner, M.J., Brown, S.W., Mandel, N.: Branded service encounters: strategically aligning employee behavior with the brand positioning. J. Mark. **77**(6), 108–123 (2013)
52. Gammoh, B.S., Mallin, M.L., Pullins, E.B.: The impact of salesperson-brand personality congruence on salesperson brand identification, motivation and performance outcomes. J. Prod. Brand Manage. **23**(7), 543–553 (2014)
53. Horn, B.A.: Six reasons why authenticity results in more sales. Retrieved on June 19th, 2022, from www.huffingtonpost.com/brian-horn/6-reasons-why-authenticit_b_6070622.html (2014)
54. Groth, M., Hennig-Thurau, T., Walsh, G.: Customer reactions to emotional labor: the roles of employee acting strategies and customer detection accuracy. Acad. Manag. J. **52**(5), 958–974 (2009)
55. Jeong, M., Lee, S.A.: Do customers care about types of hotel service recovery efforts? An example of consumer-generated review sites. J. Hosp. Tour. Technol. **17**(1), 5–18 (2017)
56. Dagger, T.S., David, M.E., Ng, S.: Do relationship benefits and maintenance drive commitment and loyalty? J Services Market. (2011)
57. Rokeach, M.: The nature of human values. Free Press, New York, NY (1973)
58. Bardi, A., Schwartz, S.H.: Values and behavior: strength and structure of relations. Pers. Soc. Psychol. Bull. **29**(10), 1207–1220 (2003)

59. Fritzsche, D., Oz, E.: Personal values' influence on the ethical dimension of decision making. J. Bus. Ethics **75**(4), 335–343 (2007)
60. Van Quaquebeke, N., Graf, M.M., Kerschreiter, R., Schuh, S.C., van Dick, R.: Ideal values and counter-ideal values as two distinct forces: exploring a gap in organizational value research. Int. J. Manag. Rev. **16**(2), 211–225 (2014)
61. Edwards, J.R.: 4 person–environment fit in organizations: an assessment of theoretical progress. Acad. Manag. Ann. **2**(1), 167–230 (2008)
62. Joyner, B.E., Payne, D.: Evolution and implementation: a study of values, business ethics and corporate social responsibility. J. Bus. Ethics **41**(4), 297–311 (2002)
63. Schneider, B.: Fits about fit. Appl. Psychol. **50**(1), 141–152 (2001)
64. Celsi, M.W., Gilly, M.C.: Employees as internal audience: how advertising affects employees' customer focus. J. Acad. Mark. Sci. **38**(4), 520–529 (2010)
65. Pillai, R., Williams, E.A., Lowe, K.B., Jung, D.I.: Personality, transformational leadership, trust, and the 2000 US presidential vote. Leadersh. Q. **14**(2), 161–192 (2003)
66. Meglino, B.M., Ravlin, E.C.: Individual values in organizations: concepts, controversies, and research. J. Manag. **24**(3), 351–389 (1998)
67. Suazo, M., Turnley, W.H., Mai-Dalton, R.R.: Antecedents of psychological contract breach: the role of similarity and leader–member exchange. Academy of Management, Best Paper Proceedings (2005)
68. Erdogan, B., Kraimer, M.L., Liden, R.C.: Work value congruence and intrinsic career success: the compensatory roles of leader-member exchange and perceived organizational support. Pers. Psychol. **57**(2), 305–332 (2004)
69. Jehn, K.A., Chadwick, C., Thatcher, S.M.: To agree or not to agree: the effects of value congruence, individual demographic dissimilarity, and conflict on workgroup outcomes. Int. J. Conflict Manage. (1997)
70. Meglino, B.M., Ravlin, E.C., Adkins, C.L.: A work values approach to corporate culture: a field test of the value congruence process and its relationship to individual outcomes. J. Appl. Psychol. **74**(3), 424–433 (1989)
71. Ashkanasy, N.M., O'connor, C.: Value congruence in leader-member exchange. J. Soc. Psychol. **137**(5), 647–662 (1997)
72. Steele, C.M.: The psychology of self-affirmation: Sustaining the integrity of the self. In: Advances in Experimental Social Psychology, vol. 21, pp. 261–302. Academic Press (1988)
73. Staw, B.M.: Rationality and justification in organizational life. Res. Org. Behav. **2**, 45–80 (1980)
74. Pulakos, E.D., Wexley, K.N.: The relationship among perceptual similarity, sex, and performance ratings in manager-subordinate dyads. Acad. Manag. J. **26**(1), 129–139 (1983)
75. Turban, D.B., Jones, A.P.: Supervisor-subordinate similarity: types, effects, and mechanisms. J. Appl. Psychol. **73**(2), 228–234 (1988)
76. Byrne, D.: The attraction paradigm. Academic Press, New York (1971)
77. Elbedweihy, A.M., Jayawardhena, C., Elsharnouby, M.H., Elsharnouby, T.H.: Customer relationship building: the role of brand attractiveness and consumer–brand identification. J. Bus. Res. **69**(8), 2901–2910 (2016)
78. Hu, Y., Ma, Z., Kim, H.J.: Examining effects of internal branding on hospitality student interns' brand-supportive behaviors: the role of value congruence. J. Hosp. Tour. Educ. **30**(3), 144–153 (2018)
79. Bandura, A.: Self-efficacy: the exercise of control. Freeman, New York (1997)
80. Bandura, A.: Social cognitive theory: an agentic perspective. Annu. Rev. Psychol. **52**(1), 1–26 (2001)
81. Bitner, M.J.: Evaluating service encounters: the effects of physical surroundings and employee responses. J. Mark. **54**(2), 69–82 (1990)
82. Cuddy, A.J., Glick, P., Beninger, A.: The dynamics of warmth and competence judgments, and their outcomes in organizations. Res. Org. Behav. **31**, 73–98 (2011)
83. Rod, M., Ashill, N.J., Gibbs, T.: Customer perceptions of frontline employee service delivery: a study of Russian bank customer satisfaction and behavioural intentions. J. Retail. Consum. Serv. **30**, 212–221 (2016)

84. Williams, K.C., Spiro, R.L.: Communication style in the salesperson-customer dyad. J. Mark. Res. **22**(4), 434–442 (1985)
85. Van Dolen, W., Lemmink, J., De Ruyter, K., De Jong, A.: Customer-sales employee encounters: a dyadic perspective. J. Retail. **78**(4), 265–279 (2002)
86. Homburg, C., Stock, R.M.: Exploring the conditions under which salesperson work satisfaction can lead to customer satisfaction. Psychol. Mark. **22**(5), 393–420 (2005)
87. Aaker, J.L.: Dimensions of brand personality. J. Mark. Res. **34**(3), 347–356 (1997)
88. Grandey, A.A., Fisk, G.M., Mattila, A.S., Jansen, K.J., Sideman, L.A.: Is "service with a smile" enough? Authenticity of positive displays during service encounters. Organ. Behav. Hum. Decis. Process. **96**(1), 38–55 (2005)
89. Judd, C.M., James-Hawkins, L., Yzerbyt, V., Kashima, Y.: Fundamental dimensions of social judgment: understanding the relations between judgments of competence and warmth. J. Pers. Soc. Psychol. **89**(6), 899–913 (2005)
90. Yzerbyt, V., Provost, V., Corneille, O.: Not competent but warm... really? Compensatory stereotypes in the French-speaking world. Group Process. Intergroup Relat. **8**(3), 291–308 (2005)
91. Lemmink, J., Mattsson, J.: Warmth during non-productive retail encounters: the hidden side of productivity. Int. J. Res. Mark. **15**(5), 505–517 (1998)
92. Wang, Z., Mao, H., Li, Y.J., Liu, F.: Smile big or not? Effects of smile intensity on perceptions of warmth and competence. J. Consumer Res. **43**(5), 787–805 (2017)
93. Kirmani, A., Hamilton, R.W., Thompson, D.V., Lantzy, S.: Doing well versus doing good: the differential effect of underdog positioning on moral and competent service providers. J. Mark. **81**(1), 103–117 (2017)
94. Wojciszke, B., Bazinska, R., Jaworski, M.: On the dominance of moral categories in impression formation. Pers. Soc. Psychol. Bull. **24**(12), 1251–1263 (1998)
95. Grant, R.M.: The resource-based theory of competitive advantage: implications for strategy formulation. Calif. Manage. Rev. **33**(3), 114–135 (1991)
96. Huang, S.M., Ou, C.S., Chen, C.M., Lin, B.: An empirical study of relationship between IT investment and firm performance: a resource-based perspective. Eur. J. Oper. Res. **173**(3), 984–999 (2006)
97. Boyd, B.K., Bergh, D.D., Ketchen, D.J., Jr.: Reconsidering the reputation—performance relationship: a resource-based view. J. Manag. **36**(3), 588–609 (2010)
98. Roberts, P.W., Dowling, G.R.: Corporate reputation and sustained superior financial performance. Strateg. Manag. J. **23**(12), 1077–1093 (2002)
99. Bendapudi, N., Berry, L.L.: Customers' motivations for maintaining relationships with service providers. J. Retail. **73**(1), 15–37 (1997)
100. Srivastava, R.K., Fahey, L., Christensen, H.K.: The resource-based view and marketing: the role of market-based assets in gaining competitive advantage. J. Manag. **27**(6), 777–802 (2001)
101. Tanriverdi, H., Venkatraman, N.: Knowledge relatedness and the performance of multibusiness firms. Strateg. Manag. J. **26**(2), 97–119 (2005)
102. Di Mascio, R.: The service models of frontline employees. J. Mark. **74**(4), 63–80 (2010)
103. Bagozzi, R.P., Yi, Y.: On the evaluation of structural equation models. J. Acad. Mark. Sci. **16**(1), 74–94 (1998)
104. Bollen, K.A.: Structural equations with latent variables. Wiley, Inc., New York (1989)
105. Bauer, D.J., Preacher, K.J., Gil, K.M.: Conceptualizing and testing random indirect effects and moderated mediation in multilevel models: new procedures and recommendations. Psychol. Methods **11**(2), 142 (2006)
106. Edwards, J.R., Lambert, L.S.: Methods for integrating moderation and mediation: a general analytical framework using moderated path analysis. Psychol. Methods **12**(1), 1 (2007)
107. Dawson, J.F.: Moderation in management research: what, why, when, and how. J. Bus. Psychol. **29**(1), 1–19 (2014)

Chapter 30
Hybrid Entrepreneurship: A Systematic Review

Maria I. B. Ribeiro, Isabel M. Lopes, José A. M. Victor, and António J. G. Fernandes

Abstract This study intends to be a further contribution to the growing academic research on hybrid entrepreneurship. Therefore, the main objective of this study was to investigate the current status of hybrid entrepreneurship. In order to achieve this objective, a computer-assisted bibliographic search was performed on 28 July 2022. This search included all publications in the scientific areas of Business, Management and Accounting and documents of the "Article" type, available in the Scopus and Web of Science (WoS) databases. The search was based on the words "hybrid entrepreneurship". The articles that include groups of individuals who combine business activities with salaried work were considered. Subsequently, for each article, it was collected information about authorship and publication date, country where the study took place, type of study, sample size, methods, objectives of the study and findings. Twenty articles were selected. The publications aimed to explore the reasons that lead individuals to be hybrid entrepreneurs, the influence of passion around this career, the uncertainty associated with salaried work, the conflict of roles between salaried work and hybrid entrepreneurship, the factors that promote or hamper the

M. I. B. Ribeiro
Centro de Investigação de Montanha (CIMO, Instituto Politécnico de Bragança, Campus Santa Apolónia, 5300-253 Bragança, Portugal
e-mail: xilote@ipb.pt

Laboratório Associado Para a Sustentabilidade e Tecnologia em Regiões de Montanha (SusTEC), Instituto Politécnico de Bragança, Campus de Santa Apolónia, 5300-253 Bragança, Portugal

I. M. Lopes (✉) · J. A. M. Victor (✉)
Escola Superior de Tecnologia e Gestão, Instituto Politécnico de BragançaUnidade de Pesquisa Aplicada em Gestão – IPB, Campus de Santa ApolóniaBragança, Portugal, Centro ALGORITMI da Universidade Do Minho, Braga, Portugal
e-mail: isalopes@ipb.pt

J. A. M. Victor
e-mail: javemor@ismai.pt

A. J. G. Fernandes
Instituto Politécnico da Maia, Maia, Portugal: Universidade da Maia, Maia, Portugal
e-mail: toze@ipb.pt

J. L. Reis et al. (eds.), *Marketing and Smart Technologies*, Smart Innovation, Systems and Technologies 337, https://doi.org/10.1007/978-981-19-9099-1_30

transition to hybrid entrepreneurship and/or full-time entrepreneurship, the risk attitudes of hybrid entrepreneurs, the methods of entry into hybrid entrepreneurship and the identification of determinants for hybrid entrepreneurship and its effects.

30.1 Introduction

The phenomenon of hybrid entrepreneurship, although emerging, has aroused increasing academic interest in recent years [1] given its growth [2] as a result of greater flexibility in the labour market and professional careers [3, 4]. Hybrid entrepreneurs already represent a considerable portion of all entrepreneurial activity [5]. Recent empirical studies have revealed that more than 50% of nascent entrepreneurs, for example in the USA, started their businesses while they were employed [6]. The hybrid entrepreneurship, that is, the process of starting a business while maintaining a job in an existing organization (salaried work), is an increasingly common career transition path [7, 8]. Hybrid entrepreneurs are important to society due to the dynamics generated in the labour market. Its existence reflects changes in market conditions and shows that more and more individuals decide to work in atypical forms of employment and create part-time [9] or temporary jobs [3, 4] or part-time business [10, 11] or second job entrepreneurship [12]. Considered a portfolio of real options in entrepreneurship and paid employment [13], the literature states that uncertainty in entrepreneurship shapes the individual's decision regarding hybrid entry into entrepreneurship or entry into entrepreneurship in full-time [13]. Furthermore, research suggests that the risk and uncertainty associated with entrepreneurial activity impede entry and contribute to high failure rates for new businesses. In this context, as a way to reduce risk and uncertainty, hybrid entrepreneurship can be considered as a means for the entry and survival of the entrepreneur as it allows the realization of his entrepreneurial potential while he is financially and socially guaranteed by an employer [14]. The survival advantage is driven by the learning effect that occurs during hybrid entrepreneurship providing higher levels of enrichment for the entrepreneur as a result of salaried work [14]. Furthermore, entrepreneurial persistence, considered a key factor for the success of hybrid entrepreneurs [15], is higher in individuals with previous experience, as a result of greater entrepreneurial self-efficacy. The probability of becoming autonomous is higher when entrepreneurial self-efficacy is higher and more balanced [16]. In this context, when individuals choose to become full-time entrepreneurs, they tend to register high growth and success rates [17]. The reduction of risk in the transition to full-time entrepreneurship, the increase in income and the possibility for individuals to engage in activities they really enjoy are the economic and social benefits often associated with this type of entrepreneurship [17].

The labour market, currently characterized by uncertainty and volubility, has great and opportune emotional, social and economic challenges. An example of this was the recent pandemic crisis experienced around the world, which triggered the adaptation and/or the creation of new business to compensate the loss of income or lack of employment. In the pandemic context, digital transactions and digital marketing became prominent and the opportunity was not wasted, especially by those who, long time ago, wanted to have their own business without, however, for reasons of security and financial comfort, no longer having a salaried work. So, the main objective of this study was to investigate the current status of hybrid entrepreneurship based on the Preferred Reporting Items for Systematic Reviews and Meta-Analyses (PRISMA) statement.

This article is organized into four sections. In the first section, the literature review is organized, the objective is presented, the methodology used is briefly described and the structure of the article is presented. The second section describes with more detail the methodology used. The third section presents the results including detailed content analysis. Finally, the fourth and last section presents the main conclusions, suggests guidelines for future research and points out the limitations of the research.

30.2 Methods

A systematic literature review was carried out, which enabled data collection and subsequent analysis. For this, some criteria were defined that allow the collection of data in a reliable and replicable way [18], making the literature review process scientific and transparent [19]. The present systematic literature review was based on the PRISMA statement [20]. In this context, the Scopus and WoS databases were screened (Fig. 30.1).

The search was performed in 28 July 28 2022, using the keywords "hybrid entrepreneurship". A total of 76 documents were found (Fig. 30.1). Subsequently, the search was limited to the "Business, Management and Accounting" study area, in Scopus (38 articles), and to the "Business" and "Management" areas, in WoS (32 articles).

Later, 70 documents were screened taking into account the "Article" document type having been accounted 60 articles published (31 in Scopus and 29 in WoS) to date. Of these, 18 articles were removed because they were duplicated in both databases. Therefore, 42 articles were taken into consideration for reading and analysis.

Finally, the entire text was selected, in order to verify the relevance of publications for this research, considering the following criterion: studies that include groups of individuals who combine business activities with salaried work according to the definition proposed [17]. For each article, it was collected information about authorship and publication date, place (country) where the research took place, type of study, sample size, methodology used, objectives of the study and findings, totalling twenty articles.

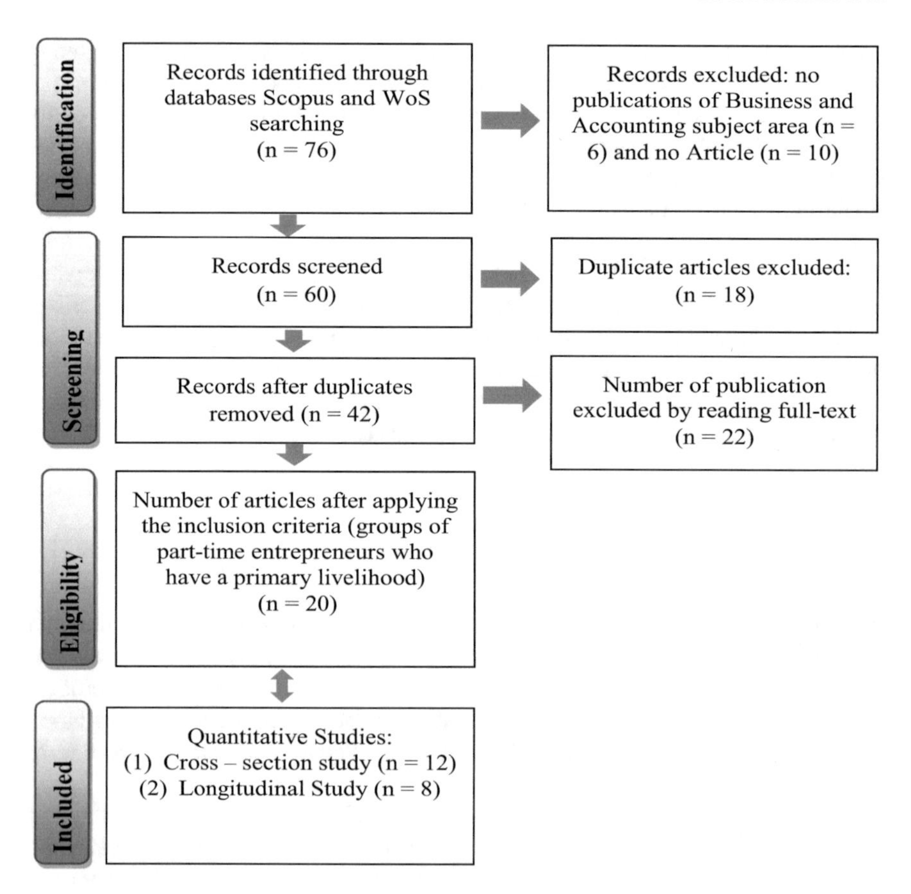

Fig. 30.1 PRISMA flow diagram of data collection process and analysis

30.3 Results and Discussion

Table 30.1 presents all the included publications by ascending order of publication date. The table also contains information about authorship, type of study, country where the research was developed, sample size, objectives and contributions. The selected studies were developed in America (USA, Mexico, Canada), Europe (Sweden, UK, Poland, France, Denmark, Germany, Finland) and Africa (Nigeria, Ghana). There is, also, a study that involved 30 European Union countries allowing a broader view of this subject (Austria, Belgium, Bulgaria, Switzerland, Cyprus, Czech Republic, Germany, Denmark, Estonia, Spain, Finland, France, Greece, Croatia, Hungary, Ireland, Iceland, Italy, Lithuania, Luxembourg, Latvia, Malta, Netherlands, Norway, Poland, Portugal, Romania, Sweden, Slovenia, UK) [21].

As can be seen in Table 30.1, all studies are quantitative, predominantly of the cross-sectional type (60%), with most of them being developed on the European

Table 30.1 General data, objectives and findings of the articles included in the systematic literature review

Author (date) (References)	Type of study Place (country) Sample size Methods	Objectives of the study findings
1. Folta et al. (2010) [17]	– Quantitative (longitudinal study: 1994–2001) – Sweden – Sample: 329,624 individuals (31,7216 salaried work; 5844 hybrid; 6517 self-employed) – Multivariate analysis	To determine the prevalence of hybrid entrepreneurship; to check whether the hybrid entry can influence entry into self-employment; to know the reasons why individuals prefer hybrid entry to full immersion in self-employment and to examine the empirical evidence surrounding these issues Empirical evidence supports the notion that hybrid entrepreneurs are predominantly and systematically different from those who opt for direct entry into self-employment. Furthermore, they found compelling evidence that hybrids are much more likely to enter self-employment than non-hybrids, and that entry into self-employment is significantly influenced by learning (experience) in the hybrid mode

(continued)

Table 30.1 (continued)

Author (date) (References)	Type of study Place (country) Sample size Methods	Objectives of the study findings
2. Raffiee and Feng (2014) [14]	– Quantitative (longitudinal study: 1994–2008) – United States – Sample: 12,909 individuals (salaried work: 5299; hybrid; 1093; self-employed; 6517) – Continuous survival analysis	To explore how hybrid entrepreneurship reduces the risk and uncertainty associated with entrepreneurial activity The results prove that individuals who are risk averse and have low self-evaluation are more likely to engage hybrid entrepreneurship compared to those who are full-time self-employed. Hybrid entrepreneurs who later enter full-time self-employment (i.e. quit daily employment) have much higher survival rates compared to individuals who enter full-time self-employment directly from salaried work; adding support to the theory that survival advantage is driven by a learning effect that occurs during hybrid entrepreneurship. The results, also, show that the decrease in exit risk is stronger for individuals with prior entrepreneurial experience. The results suggest that individual characteristics may play a greater role in determining the process of how entrepreneurial entry occurs, that this process has important implications for the survival of new businesses

(continued)

Table 30.1 (continued)

Author (date) (References)	Type of study Place (country) Sample size Methods	Objectives of the study findings
3. Thorgren, Nordström and Wincent (2014) [1]	– Quantitative (cross-sectional study) – Sweden – Sample: 262 hybrid entrepreneurs – Logistic Regression Analysis	To investigate the reasons for choosing parallel business-employment careers (hybrid entrepreneurship) with a particular focus on passion (i.e. working on something you are passionate about) as the primary reason for this choice The results indicated that (1) the ability to work at something one is passionate about is the main reason to combine the job with a side business; (2) passion is more likely to be the main motive behind the hybrid form among older individuals at the start of the business; (3) passion is less likely to be the main motive behind the hybrid form among individuals who spend more time in the business
4. Thorgren et al. (2016) [22]	– Quantitative (Cross-sectional study) – Sweden – Sample: 256 venture owners The respondents had to meet two criteria: (1) engaging in salaried employment and (2) active engagement in an entrepreneurial venture at the time of data collection – Logistic regression analysis	To examine how age relates to the transition process from a hybrid entrepreneur to a full-time entrepreneur The results reveal a U-shaped association between age and the hybrid entrepreneur's intention to enter full-time entrepreneurship. This means that younger and older hybrid entrepreneurs are more likely to become full-time entrepreneurs

(continued)

Table 30.1 (continued)

Author (date) (References)	Type of study Place (country) Sample size Methods	Objectives of the study findings
5. Block and Landgraf (2016) [23]	– Quantitative (cross-sectional study) – Germany – Sample: 481 participants (379 current part-time entrepreneurs, 82 former part-time entrepreneurs who transitioned into full-time entrepreneurship and 20 former part-time entrepreneurs who abandoned entrepreneurial activity) – Logistic regression analysis	To analyse how the financial and non-financial motives of entrepreneurs influence the propensity of part-time entrepreneurs to become full-time entrepreneurs The results show that the motivation to complement income or the motivation to achieve social recognition is negatively associated with transition behaviour, while the motivation to achieve independence or self-actualization is positively associated with transition behaviour
6. Schulz et al. (2016) [9]	– Quantitative (longitudinal study: 4th trimester 2009-4th trimester 2013) – Mexico – Sample: 212,523 individuals aged between 20 and 65 who are part of households. The individuals who are not able to work (e.g. strong physical or mental disabilities), and the individuals who work less than 30 h in their job were excluded – Multinomial logit analysis	To analyse the impact of introducing one-stop shops for faster and simplified business registration given the need to include hybrid entrepreneurs in public policies According to the results, hybrid entrepreneurs respond better to changes in entry regulations when compared to full-time entrepreneurs While both educated and less educated people respond to reform, the effect is most pronounced in highly educated hybrid entrepreneurs

(continued)

Table 30.1 (continued)

Author (date) (References)	Type of study Place (country) Sample size Methods	Objectives of the study findings
7. Nordström et al. (2016) [24]	– Quantitative (cross-sectional study) – Sweden – Sample: 262 hybrid entrepreneurs – Logistic regression analysis	To examine how entrepreneurial ownership and involvement in entrepreneurial teams influence the passion to engage in entrepreneurship The results show that (1) the longer the individual had the side business, the less likely passion is to be the main motive behind entrepreneurship; (2) passion is less likely to be the main motive behind entrepreneurship among those who are part of an entrepreneurial team and (3) involvement in an entrepreneurial team strengthens the negative association between entrepreneurial ownership and passion for entrepreneurship
8. Schulz et al. (2017) [25]	– Quantitative (longitudinal study: 1991–2008) – United Kingdom – Sample: 6362 employees with ages between 18 e 65 years old (salaried employee and self-employer) – Multinomial logit analysis	Explain why multi-job holders often have higher hourly earnings in their second job compared to their main job. The results show that self-employment as a second job significantly increases the probability of having higher average earnings in second job compared to salaried employment

(continued)

Table 30.1 (continued)

Author (date) (References)	Type of study Place (country) Sample size Methods	Objectives of the study findings
9. Luc et al. (2018) [6]	– Quantitative (cross-sectional study) – Canada (Quebec) – Sample: full-time and part-time employees who participated in the 2015 Quebec Business Index Survey (1787 observations) – Factor and reliability analyses, Probit analysis	To determine whether sociodemographic variables and employees' perceptions of accessibility to resources and work and quality of work favour or hinder the transition from a level of commitment to the entrepreneurial process The results demonstrate that employees' progress on the entrepreneurial ladder is stimulated by gentle support in the form of easy (perceived) access to business advice and also by high (perceived) work autonomy in the employee's salaried work
10. Xi et al. (2018) [26]	– Quantitative (cross-sectional study) – France – Sample: 9032 French hybrid entrepreneurs who started a new company (new venture) or took over an existing company (acquisition of companies) – Regression analysis	To identify determinants of hybrid entrepreneurship related to entry mode decisions, distinguishing between business acquisition and start-up of new ventures; to examine how an individual's work experience, educational level, demographic status, motivation and support received—as well as company characteristics—affect the hybrid entrepreneur's mode of entry into entrepreneurship According to the results, the entry mode of hybrid entrepreneurs is influenced by the individual's personality, human capital, among other factors Educational level and management experience are linked to the creation of new ventures, while worker experience is linked to business acquisition

(continued)

Table 30.1 (continued)

Author (date) (References)	Type of study Place (country) Sample size Methods	Objectives of the study findings
11. Pollack et al. (2019) [8]	– Quantitative (longitudinal study: 20-week study that assessed three time periods, namely initial survey, at 10 weeks and at 20 weeks) – USA – Sample: 29 hybrid entrepreneurs – Confirmatory factor analysis and structural equation modelling	To provide an exploratory view on the longitudinal association between self-efficacy and persistence for hybrid entrepreneurs The results showed that changes in entrepreneurial effectiveness over time affect changes in entrepreneurial persistence of hybrid entrepreneurs
12. Klyver and Lomberg (2020) [27]	– Quantitative (cross-sectional study) – Denmark – Sample: step 1: n = 718 (job search = 164; parallel search = 115; entrepreneurship search = 439); step 2: n = 330 (parallel search = 115; entrepreneurship search = 439). Multinomial probit regression	To check whether the entry into entrepreneurship results from the process of looking for a new job, considering two steps: (1) decision to look for a job, try a start-up, or both and (2) try a start-up According to the results, employment status is important and the "parallel search" for a new job is detrimental to successful entrepreneurial entry

(continued)

Table 30.1 (continued)

Author (date) (References)	Type of study Place (country) Sample size Methods	Objectives of the study findings
13. Kurczewska and Wawrzyniak (2020) [16]	– Quantitative (cross-sectional study) – Poland – Sample: 1600 entrepreneurs – Multivariate logistic regression	To test whether the probability of being linked to a salaried job with the business itself increases with the variety and level of education acquired, the amplitude of professional and management experience and the level of entrepreneurial self-efficacy The results show that the probability of being a hybrid entrepreneur increases with professional experience, while decreasing as the level and diversity of education increases. The results also suggest that hybrid entrepreneurs are an important and discrete population and, therefore, need to be treated separately
14. Adebusuyi and Adebusuyi (2021) [28]	– Quantitative (cross-sectional study) – Nigeria – Sample: 303 bachelor's degree holders teaching in Nigerian public secondary schools in two states of the federation (Ondo and Ekiti states) – Regression analysis (controlling for gender and age)	To investigate how high school teachers, deal with a recessive economy by adopting hybrid entrepreneurship The results showed that (1) teachers had a strong feeling of salary underemployment (reduced pay) and relative deprivation; (2) salary underemployment and relative deprivation have led directly to hybrid entrepreneurship The results suggest that teachers' involvement in hybrid entrepreneurship is motivated by the need to deal with the recessionary economy in Nigeria

(continued)

Table 30.1 (continued)

Author (date) (References)	Type of study Place (country) Sample size Methods	Objectives of the study findings
15. Ardianti and Davidsson (2022) [2]	– Quantitative (longitudinal study: 2011–2017) – United Kingdom – Sample 1: 3847 Hybrid Entrepreneurs; 5149 Individuals in two salaried jobs; 119,446 full-time salaried employed; 1,8496 full-time self-employed – Sample 2: 710 individuals who have moved from full-time salaried employment to hybrid entrepreneurship – Sample 3: 878 individuals who have switched from hybrid entrepreneurship to full-time salaried employment – Sample 4: 111 individuals who have switched from hybrid entrepreneurship to full-time self-employment – Sample 5: 925 individuals who switched from one full-time salaried job to two salaried part-time jobs - Sample 6: 1427 individuals who have switched from part-time salaried employment to full-time salaried employment – Sample 7: 567 individuals who switched from full-time salaried employment to full-time entrepreneurship – Analysis of variance (anova) and the games-howell post-hoc test	To verify whether hybrid entrepreneurs exhibit distinct patterns of psychological well-being (measured through mental pressure, job satisfaction and life satisfaction) compared to full-time salaried employees working two salaried jobs The study shows distinct patterns of psychological well-being of hybrid entrepreneurs compared to full-time entrepreneurs and salaried workers. Furthermore, it shows that the observed patterns can be attributed to personal factors and unique work arrangements (task characteristics)

(continued)

Table 30.1 (continued)

Author (date) (References)	Type of study Place (country) Sample size Methods	Objectives of the study findings
16. Asante et al. (2022) [15]	– Quantitative (cross-sectional study) – Ghana – Sample: 279 hybrid entrepreneurs – Confirmatory factor analysis; structural equation modelling (SEM)	To develop a model in which person-enterprise adequacy, needs-enterprise adequacy and entrepreneurial-skills demand adequacy are associated with entrepreneurial persistence through entrepreneurial self-efficacy According to the results, believing in the ability to perform a task gives the individual an important advantage in any situation. Furthermore, the results suggest that self-efficacy can be modified, harnessed and nurtured. Finally, the results show that the self-efficacy of hybrid entrepreneurs is enhanced when person-enterprise adequacy, needs-enterprise adequacy and entrepreneurial-skills demand adequacy are high
17. Carr et al. (2022) [29]	– Quantitative (Longitudinal Study: 26-week online survey over 17 time points) – United states – Sample: 29 hybrid entrepreneurs – Confirmatory factor analysis and mediation analysis	To develop a new construct based on the theory that hybrid entrepreneurship is associated with what the authors call "work conflict" for the enterprise The authors conclude that an increase in start-up effort over time for hybrid entrepreneurs is related to increased role conflict between work and enterprise, which results in a decrease in job satisfaction and an increase in turnover intentions in relation to the salaried job

(continued)

Table 30.1 (continued)

Author (date) (References)	Type of study Place (country) Sample size Methods	Objectives of the study findings
18. Dvouletý and Bögenhold (2022) [21]	– Quantitative (cross-sectional) – 30 EU countries (Austria, Belgium, Bulgaria, Switzerland, Cyprus, Czech Republic, Germany, Denmark, Estonia, Spain, Finland, France, Greece, Croatia, Hungary, Ireland, Iceland, Italy, Lithuania, Luxembourg, Latvia, Malta, Netherlands, Norway, Poland, Portugal, Romania, Sweden, Slovenia, UK) – Sample: 14,625 (full-time entrepreneurs, hybrid entrepreneurs and full-time salaried workers) – Logistic regression analysis	To identify individual and family characteristics associated with hybrid entrepreneurs The results highlight that hybrid entrepreneurs are a specific subgroup of the population of autonomous individuals. Although hybrid entrepreneurs are also a relatively heterogeneous group, results show that most of them do business in the agricultural sector and live in rural areas. Their entrepreneurial activities appear to be, on average, younger than those run primarily by self-employed workers, which may account for the temporary nature of many hybrid entrepreneurial ventures. On the other hand, hybrid entrepreneurs have, on average, higher levels of education, but this finding concerns mainly those involved in non-agricultural activities. Finally, women are less prone to hybrid entrepreneurship, a fact that becomes more marked when associated with the responsibility of taking care of children
19. Gänser-Stickler et al. (2022) [13]	– Quantitative (longitudinal study: 2006–2019) – United Kingdom – Sample: 6673 individuals, aged 25–59 years old, who have transitioned into hybrid entrepreneurship or full-time entrepreneurship – Logistic regressions analysis: logit model and Heckman probit model	To examine how uncertainty in entrepreneurship and uncertainty in paid employment together influence an individual's choice to enter hybrid entrepreneurship rather than full-time entrepreneurship The results strongly support the hypothesis that uncertainty in paid employment negatively moderates the positive impact of uncertainty in hybrid entrepreneurship

(continued)

Table 30.1 (continued)

Author (date) (References)	Type of study Place (country) Sample size Methods	Objectives of the study findings
20. Viljamaa et al. (2022) [30]	– Quantitative (cross-sectional) – Finland – Sample: 400 part-time entrepreneurs – Multiple regression analysis	To understand the different effects of motives and attitudes on the well-being of part-time entrepreneurs The results show that when controlling for the stability of the financial situation, attitudes of entrepreneurship and self-fulfilment as a motive explain well-being in old age. For the younger age group, financial reasons were also important in explaining well-being, but recognition and independence reasons had a negative effect

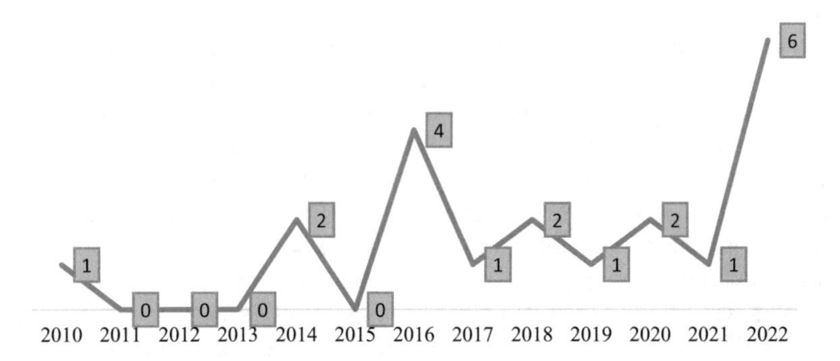

Fig. 30.2 Number of articles published per year

continent (65%), with Sweden standing out as the country where the most studies were carried out. performed and published (25%). Although, in 2022, only the first 7 months were considered for analysis, this already stands out as the year with the highest number of published articles (6 out of 20). From 2010 to date, there has been an average annual growth of 16.1% in the number of published articles (Fig. 30.2).

The publications included in this systematic literature review aimed to explore the reasons (self-actualization, independence, financial stability, recognition, psychological well-being, better performance) that lead individuals to be hybrid entrepreneurs [2, 5, 15, 17, 25], the passion surrounding this career [1, 24], the uncertainty associated with salaried work [13, 28], the role conflict between salaried work and hybrid entrepreneurship [29], the transition to hybrid entrepreneurship and/or full-time entrepreneurship (factors that favour or hinder the transition) [6, 17, 22, 23], risk attitudes of hybrid entrepreneurs [14], entry modes into hybrid entrepreneurship [26, 27] and the determinants of hybrid entrepreneurship as well as its effects (age, gender, rural environment, level of self-efficacy, level of entrepreneurial persistence, academic training, professional experience, opportunity cost, public policies) [8, 9, 15, 16, 21, 22, 26].

In the overwhelming majority of articles, the authors conclude that the group of hybrid entrepreneurs is different from the group of full-time entrepreneurs due to its heterogeneity and, as such, highlight the need to do more research to better understand their specificities [2, 3, 9, 16, 17, 21].

30.4 Conclusion

This research aimed to investigate the current status of hybrid entrepreneurship based on the PRISMA statement. So, a bibliographic search which included all publications in the scientific areas of Business, Management and Accounting and documents of the "Article" type, available in the Scopus and WoS databases was carried out, having been selected twenty articles.

These publications explore the reasons that lead individuals to be hybrid entrepreneurs, the influence of passion around this career, the uncertainty associated with salaried work, the conflict of roles between salaried work and hybrid entrepreneurship, the factors that promote or hamper the transition to hybrid entrepreneurship and/or full-time entrepreneurship, the risk attitudes of hybrid entrepreneurs, the modes of entry into hybrid entrepreneurship and the identification of determinants for hybrid entrepreneurship and its effects.

Some limitations can be pointed out to this investigation, namely publications from other databases, in addition to Scopus and WoS, publications from other scientific areas in addition to the Business, Management and Accounting areas and others beyond the "article" document type were excluded. Also, the employer perspective is poorly researched in studies on hybrid entrepreneurship. So, this should be the object of future research.

Acknowledgements The authors are grateful to the Foundation for Science and Technology (FCT, Portugal) for financial support through national funds FCT/MCTES (PIDDAC) to CIMO (UIDB/00690/2020 and UIDP/00690/2020) and SusTEC (LA/P/0007/2020).

UNIAG, R&D unit funded by the FCT, Portuguese Foundation for the Development of Science and Technology, Ministry of Science, Technology and Higher Education. Project n.o UIDB/04752/2020.

References

1. Thorgren, S., Nordstreom, C., Wincent. J.: Hybrid entrepreneurship: the importance of passion. Baltic J. Manage. **9**(3), 314–329 (2014)
2. Ardianti, R., Obschonka, M., Davidsson, P.: Psychological well-being of hybrid entrepreneurs. J. Bus. Ventur. Insights **17**, e00294 (2022)
3. Solesvik, M.: Hybrid entrepreneurship: definitions, types and directions for future research. Upravlenets **68**(4), 2–13 (2017)
4. Demir, C., Werner, A., Kraus, S., Jones, P.: Hybrid entrepreneurship: a systematic literature review. J. Small Bus. Entrep. **34**(1), 29–52 (2020)
5. Viljamaa, A., Varamaki, E., Joensuu-Salo, S.: Best of both worlds? Persistent hybrid entrepreneurship. J. Enterprising Culture **25**(4), 339–359 (2017)
6. Luc, S., Chirita, G.M., Delvauxm, E., Kepnou, A.K.: Hybrid entrepreneurship: employees climbing the entrepreneurial ladder. Int. Rev. Entrepreneurship **16**(1), 89–114 (2018)
7. Bosma, N.S., Jones, K., Autio, E., Levie, J.: Global entrepreneurship monitor 2007 executive report. Babson College, Babson Park, MA and London Business School, London (2008)
8. Pollack, J.M., Carr, J.C., Michaelis, T.L., Marshall, D.R.: Hybrid entrepreneurs' self-efficacy and persistence change: a longitudinal exploration. J. Bus. Ventur. Insights **12**, e00143 (2019)
9. Schulz, M., Urbig, D., Procher, V.: Hybrid entrepreneurship and public policy: the case of firm entry deregulation. J. Bus. Ventur. **31**(3), 272–286 (2016)
10. Smallbone, D., Welter, F.: The distinctiveness of entrepreneurship in transition economies. Small Bus. Econ. **16**(4), 249–262 (2001)
11. Petrova, K.: Part-time entrepreneurship and financial constraints: evidence from the panel study of entrepreneurial dynamics. Small Bus. Econ. **39**(2), 473–493 (2012)
12. Gruenert, J.: Second job entrepreneurs. OCCUP. Outlook Q. **43**(3), 18–26 (1999)

13. Gänser-Stickler, G.M., Schulz, M., Schwens, C.: Sitting on the fence—untangling the role of uncertainty in entrepreneurship and paid employment for hybrid entry. J. Bus. Venturing **37**(2), 106176 (2022)
14. Raffiee, J., Feng, J.: Should I quit my day job? a hybrid path to entrepreneurship. Acad. Manage. J. **57**(4), 936–963 (2014)
15. Asante, E.A., Danquah, B., Oduro, F., Affum-Oseid, E., Collins, T., Azunu, C., Li, C.: Entrepreneurial career persistence of hybrid entrepreneurs: the opposing moderating roles of wage work-to-entrepreneurship enrichment and entrepreneurship-to-wage work enrichment. J. Vocat. Behav. **132**, 103660 (2022)
16. Kurczewska, A., Mackiewicz, M., Doryń, W., Wawrzyniak, D.: Peculiarity of hybrid entrepreneurs—revisiting Lazear's theory of entrepreneurship. J. Bus. Econ. Manag. **21**(1), 277–300 (2020)
17. Folta, T., Delmar, F., Wennberg, K.: Hybrid entrepreneurship. Manage. Sci. **56**(2), 253–269 (2010)
18. Xiao, Y., Watson, M.: Guidance on conducting a systematic literature review. J. Plan. Educ. Res. **39**(1), 93–112 (2019)
19. Tranfield, D., Denyer, D., Smart, P.: Towards a methodology for developing evidence informed management knowledge by means of systematic review. Brazilian J. Manage. **14**(3), 207–222 (2003)
20. Moher, D., Liberati, A., Tetzlaff, J., Altman, D.G.: The PRISMA group: preferred reporting items for systematic reviews and meta-analyses: the PRISMA statement. PLoS Med. **6**(7), e1000097 (2009)
21. Dvouletý, O., Bögenhold, D.: Exploring individual and family-related characteristics of hybrid entrepreneurs. Article in Press, Entrepreneurship Research Journal (2022)
22. Thorgren, S., Sirén, C., Nordström, C., Wincent, J.: Hybrid entrepreneurs' second-step choice: the nonlinear relationship between age and intention to enter full-time entrepreneurship. J. Bus. Ventur. Insights **5**, 14–18 (2016)
23. Block, J.H., Landgraf, A.: Transition from part-time entrepreneurship to full-time entrepreneurship: the role of financial and non-financial motives. Int. Entrepreneurship Manage. J. **12**(1), 259–282 (2016)
24. Nordström, C., Sirén, C.A., Thorgren, S., Wincent, J.: Passion in hybrid entrepreneurship: the impact of entrepreneurial teams and tenure. Balt. J. Manag. **11**(2), 167–186 (2016)
25. Schulz, M., Urbig, D., Procher, V.: The role of hybrid entrepreneurship in explaining multiple job holders' earnings structure. J. Bus. Ventur. Insights **7**, 9–14 (2017)
26. Xi, G.Q., Block, J., Lasch, R., Robert, F., Thurik, R.: Mode of entry into hybrid entrepreneurship: new venture start-up versus business takeover. Int. Rev. Entrepreneurship **16**(2), 217–240 (2018)
27. Klyver, K., Steffens, P., Lomberg, C.: Having your cake and eating it too? A two-stage model of the impact of employment and parallel job search on hybrid nascent entrepreneurship. J. Bus. Ventur. **35**(5), 106042 (2020)
28. Adebusuyi, A.S., Adebusuyi, O.F.: Predicting hybrid entrepreneurship among secondary school teachers in Nigeria. Afr. J. Econ. Manag. Stud. **12**(4), 516–530 (2021)
29. Carr, J.C., Marshall, D.R., Michaelis, T.L., Pollack, J.M., Sheats, L.: The role of work-to-venture role conflict on hybrid entrepreneurs' transition into entrepreneurship. J. Small Bus. Manage. Article in Press (2022)
30. Viljamaa, A., Joensuu-Salo, S., Kangas, E.: Part-time entrepreneurship in the third age: well-being and motives. Small Enterprise Res. **29**(1), 20–35 (2022)

Chapter 31
Impact of Mobile Apps on Building Customer Relationships and Financial Support for the Football Club: Findings from Ruch Chorzów

Michał Szołtysik and Artur Strzelecki

Abstract Mobile applications have allowed football clubs to conduct marketing and sales activities. The purpose of this paper is to present the mobile application as a tool that allows for building lasting relationships between the football club and its fans, which translates into an increase in sales of the services and products offered, using the example of football club in Poland. Building relationships with fans is part of the process of running a sports club, and social media is the most popular mass medium that makes this possible. An attempt was made to show how they are used for these purposes by Polish football club. An example of a mobile app that directly impacts increasing the revenues of a football club and simultaneously allows building lasting relationships with fans is presented. Based on the research, attention was paid to how the mobile app increased the company's revenue.

31.1 Introduction

On social media, company profiles are no longer owned or set up not only by large corporations and well-established companies but also by local, stationary businesses that realize that the digitization of society is becoming a reality. To reach a wide range of people, they need to switch from the traditional form of advertising in the form of banner ads or advertising in traditional media to an active life in social media [1]. With easy and quick access to various sources of information, customers can choose the most favorable offers, so one of the key elements of doing business is a marketing strategy to encourage potential buyers of the product to take advantage of just their offer [2]. It is difficult, but it can be accomplished with the right marketing strategy. Building relationships with customers, or potential customers, is an integral part of running any business, which can translate into increased interest in the company,

M. Szołtysik · A. Strzelecki (✉)
Department of Informatics, University of Economics in Katowice, 40-287 Katowice, Poland
e-mail: artur.strzelecki@ue.katowice.pl

M. Szołtysik
e-mail: michal.szoltysik@edu.uekat.pl

J. L. Reis et al. (eds.), *Marketing and Smart Technologies*, Smart Innovation, Systems and Technologies 337, https://doi.org/10.1007/978-981-19-9099-1_31

and ultimately increased sales resulting from attracting new customers [3]. Social media is a tool that, among other things, allows you to save financial resources while building lasting ties with the audience [4].

31.2 The Ruch Chorzów Mobile App

The Ruch Chorzów app was created in cooperation between the club and the Fan Foundation. It is available on the App Store and Google Play, so anyone using Android or iOS software can download it to their mobile device. According to the club's owners, the app is just an add-on that will help increase the club's budget revenue, so the club can afford expenses that would not be possible without fan support. What's important about the app is that users don't incur additional costs and don't have to donate their money directly. The main idea behind the app is that each person using it through their daily purchases, which they made before the app came out, supports the club [5].

The application has more than a dozen powerful modules that, when combined, provide various communication opportunities for the users and allow the club to improve its financial situation. Thus, it has become both a social media and a purchasing tool. The Ruch Chorzów application consists of modules like news, galleries, videos, schedule, table, team staff, and match reports. In one place, all users of the application can find all the necessary information on the functioning of the company and the course of football matches. In addition, the application developers decided to provide a board and a messenger through which all club supporters can exchange observations about the activities and post various information. The club's representatives are trying to build a relationship with fans and often interact with them by posting posts, surveys, and encouraging activities.

A key module of the app is the financial module. All app users can support the club financially through daily purchases without spending extra money. Such purchases are possible in two ways, both in stationary form and through online stores. To financially support the club by shopping stationery at local places, it is necessary to declare a payment card in the app. The club has created a network of local partner points that have decided to join the campaign. By paying at them with the payment card registered in the app, part of the money from the transaction goes directly to the club's account. Each partner point voluntarily decides what percentage of the purchases it shares with the club. As part of the business cooperation, the Ruch Chorzów club offers its partners advertising and promoting the local place among fans. The whole system is also based on partnership with payment terminal operators, so the owner of the partner point does not have to transfer the pledged money manually, but everything happens automatically. Online shopping is a bit simpler, as a user wishing to make a purchase and at the same time allocate a portion of the expenses to the club only needs to go through the application to the store website where the user intends to buy something.

The most important principle of the application's financial module is its transparency. The user is informed before starting shopping that the percentage of the transaction made in the store will go to the club's account. A user can check the number of transactions he has already made, for what amounts, and how many of these purchases were transferred to the club. The money that goes to the club's account is always allocated to a specified purpose, visible to all app users. They can see how much has been collected and how much money is missing to complete the goal. The app also included an auction module, where users can bid on unique memorabilia and ventures put up by the club.

Many football clubs choose to develop proprietary mobile apps for the company's customers and business partners [6]. Ruch Chorzów app is different. It is the case of a tool through which the customers of the company, being at the same time customers of other companies or stores without changing their lifestyles, support the club with real money. Such a tool improves communication between the enterprise and the audience and allows for building relationships, but often beyond that, it does not give direct revenues to the budget [7].

An example of such a solution is the Ruch Chorzów app. The app was created and presented at a press conference for the media and other business partners at the end of 2019, and from the first days of its release, it had a considerable interest. Within the first days after the release, the app developers could receive good results, as the app was downloaded more than 7500 times, and as many as 2600 people registered their payment cards.

Figure 31.1 shows an app menu allowing users to select the module they want to use. The Ruch Chorzów application functions as a social medium, a messenger, a place to exchange information, and a financial function that allows purchasing transactions and supporting the club.

31.2.1 Incomes to the Club

Entrepreneurs want to build relationships with their customers on a fair basis, so it was decided to create a fully transparent module so that all users have a full view of the funds credited to the club's account through their transactions. This solution lets the user know how much more money needs to be raised, over what period, and what it will be spent on.

For the money to be credited to the company's account, the user must make a purchase at a stationary store of a partner point. A list of such places can be found in the app. Geolocation helps the user to reach such a store. The only condition for a percentage of the money from such a transaction to go to the enterprise's account is that the purchase must be made through the app.

Using the financial module in the application, the user can select any store on the list of online stores. Once selected, using the application, the store's page is automatically opened, and the user only has to select the goods he is interested in and make a transaction. Before starting such a purchase, the user can see what amount

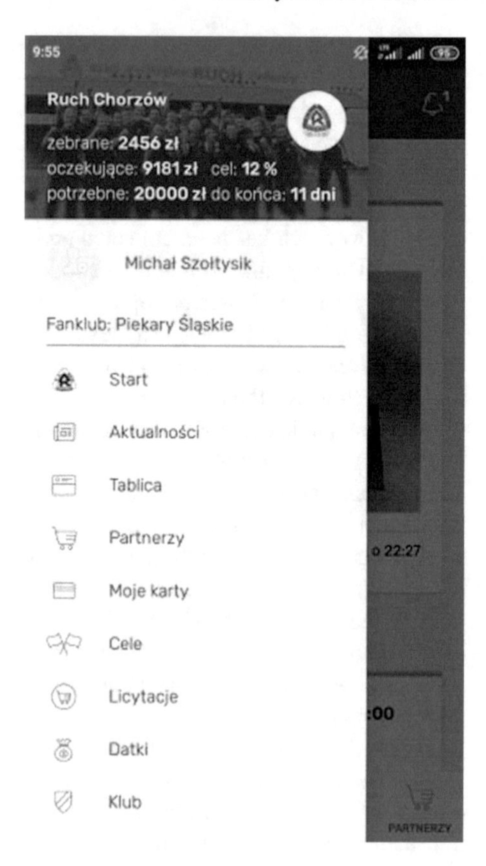

Fig. 31.1 Menu of the Ruch Chorzów application

or percentage of the transaction will go to the club's account. After finalizing such a transaction, the user can see how much money from his purchases has already been transferred. In addition to an aggregate statement of all transactions, each user can check how much money went into the company's account due to his transactions with partner entities cooperating with the football club.

31.2.2 Building Relationships with Fans

The main element of Ruch Chorzów's mobile application is a financial module that allows the company to improve its finances. However, good customer contact is part of the company's long-term strategy, which in the future will enable increasing revenues resulting from the sale of offered products and services in the form of, among others, tickets to football matches [8]. The second important module is the community module, which allows interaction between all users using the application and the club [9]. Easy interface allows any person, regardless of the level of skill in using mobile devices, to easily participate in the fan community. According to statistics kept by the club, the app is used not only by young people but also by older people who are eager to use the tools offered by the app. An example is the community board, with dozens of new posts published daily by the club and supporters.

In Fig. 31.2, we see an excerpt from a board where users post their thoughts, share insights with other app users, and often ask questions to the club owners, who are happy to answer them. This activity allows customers to feel that they are part of the club and are very important to it. Building relationships with football fans is quite specific because they are often hermetic groups with their own representatives, and young people usually follow the voice of their "bosses," so the club must try to avoid conflicts [10]. However, such conflicts do happen, as evidenced by the situation in 2019, when Ruch Chorzów fans dissatisfied with the actions of the management decided to boycott all home matches, and thus, the club lost tens of thousands of zlotys from tickets and gadgets sold on the day of the match [11]. It shows that relationships built over the years, often in one day, can be destroyed and have to be built from scratch. It has been happening since October 2019, and one of the solutions to improve it is the released app. It is intended to help improve communication, understand each other's needs and, above all, allow the company to improve its financial position [12].

31.3 Methodology

The first part of the study evaluates the mobile app's impact on building relationships with the football club's fans. The evaluation was made by providing a questionnaire as the web form to people who have used or are actively using the Ruch Chorzów mobile application. The second part of the study presented the impact of the mobile app on the football club's financial performance. Ruch Chorzów is a joint-stock company operating on the NewConnect stock exchange and is required to make its financial statements publicly available quarterly [13]. The study presented the amounts that went into the club's account from transactions made by app users.

In addition to the financial module, the mobile app is designed to allow club employees and fans to interact more, access information on the club's current activities more quickly, or exchange views between fans who have the app.

This is guaranteed by appropriate features and an intuitive interface that allows all people, regardless of age and IT skills, to use it easily and pleasantly via users' mobile devices.

The research problems that arose during the design phase of the survey were intended to provide answers to questions related to the app, both related to technical aspects and to assess the impact of the app on building relationships with fans. Here are the following questions used in the survey.

1. Usability of Ruch Chorzów app
    - Have users encountered problems downloading and installing the app?

Fig. 31.2 Social board in the Ruch Chorzów app

- Was the operation of the application smooth and intuitive?
- How did the application look graphically and aesthetically?
- Was the application free of errors?

2. Usage of Ruch Chorzów app

   - What influenced you to start using the Ruch Chorzów mobile app?
   - How often do users use the app?
   - What are the benefits to users of using the app?
   - Which of Ruch Chorzów's mobile app activities performed the best, and which the worst?
   - Which activities are most influential in building relationships with supporters, and which are not?

31.4 Results

31.4.1 Usability of Ruch Chorzów App

The survey began with a few questions about the respondents, who were asked to state their age and gender. A total of 155 people participated, the vast majority of whom were men, 138 (89%) of the surveyed group. Thus, only 17 women (11% of all respondents) took part in the survey. Next, the respondents were asked about their age. The results are somewhat surprising because as many as 100 people (64.5% of all respondents) are 30 or older. It is usually considered that only young people use this application, which is not entirely true in this case. The second most common age range in the responses was 26–29. Twenty-six respondents fell into this range (16.8%). The smallest number of respondents, only 12 (7.7%) of all respondents, said they were in the 22–25 age group.

The survey continued with 154 respondents, who were asked to answer questions related to the technical area of the mobile application. Respondents answered several questions intended to show whether the app was designed for people who are not necessarily up on technological innovations. Many football fans are middle-aged people who do not always boast high computer skills, making it important to adapt the app to people of all ages.

The survey began with questions related to the app's smoothness and level of difficulty of use. The mobile app was developed for Android and iOS users, but users typically have more or less powerful smartphones. It is often the case that older phones can't cope with overly demanding apps. Respondents were asked whether they thought the app's performance was smooth and encountered obstacles in the form of stuttering or other performance issues or the phone's memory when working with the app was heavily loaded.

Most respondents felt that using the mobile app was pleasant, did not put much strain on the phone, and could use the app's full potential. One hundred forty-seven respondents (95.5%) said that the application in terms of hardware requirements was adapted very well. Seven respondents, on the other hand, were of a different opinion. According to them, the smoothness of the application was disturbed, making it impossible to enjoy using it fully.

Since the app's audience is of different ages, the developers should adjust its design and operation to make it reasonably easy and possible to use without specialized training. As many as 100 respondents are 30 years old or older, making a question to find out their opinion on whether the app is easy to use. Again, an overwhelming majority of 148 people, which is 96.1% of the surveyed group, said that using the mobile application is easy and intuitive. According to them, there are no significant problems with its operation. Six people had a different opinion (3.9%). In the next part of the survey, respondents were asked about the app's appearance. They were asked their opinion on the app's graphics and aesthetic quality. The majority of respondents believe that the app is easy to read, easy to switch between modules, the graphical elements are well arranged, and they rate the overall aesthetics of the app good or very good.

The most significant number of respondents, 85 (55.2%), believe that the application was executed neatly and aesthetically, is characterized by simplicity and elegance, and, above all, the quality of the graphics stands at a high level. In contrast, 59 respondents gave a good rating (38.3%). On the other hand, only one person (0.6%) believes that the application is not very aesthetically pleasing. This part of the survey evaluated the technical aspects of the Ruch Chorzów mobile app. Most respondents did not encounter any problems while downloading, installing, and using the app. They believe the app has a very intuitive interface, making it easy to use. As for the graphical elements, the majority of respondents also give the app a good rating.

31.4.2 Usage of Ruch Chorzów App

This part of the study consisted of evaluating the mobile app, and more specifically, the activities of the football club with its help, on strengthening ties and building relationships with fans. The app provides several different opportunities that Ruch Chorzów uses to carry out the marketing strategy launched in 2019/2020, including the interaction between club employees and fans or providing a place called the board where fans can post their thoughts and discuss with other users.

The study to assess the impact of the mobile app on building club–fan relations began with a question about why respondents decided to start using a dedicated solution. Respondents were allowed to choose up to three reasons (Table 31.1).

The most common reason for using the mobile app was that the user was a fan of Ruch Chorzów. Such an answer appeared 153 times (which translates into 99.4%). It means that only one respondent did not declare that he is a supporter of this club. Another answer that appeared 71 times was that supporters were encouraged by the club's staff to support the club financially actively. So it follows that belonging to a particular club is one thing, but the club's action is also important, and the club should go out to supporters and encourage them to support by pursuing common goals. The results show that fans are an important part of any football club, as they are strongly attached to the club and do not expect too much from the club's employees. They

Table 31.1 Reasons for using the mobile app

Reason	*N*	%
I am a fan of Ruch Chorzów	153	99.4
I was interested in advertising the club on social media	19	12.3
The popularity of any of the club's players	1	0.6
Encouragement by club employees to support	71	46.1
The overall popularity of the football club	34	22.1

Table 31.2 Benefits of using the mobile application

Benefit	*N*	%
Ability to follow the news	115	74.7
Opportunity to ask questions of footballers	37	24
Chance to win prizes	4	2.6
Opportunity to receive a discount coupon	1	0.6
Opportunity to express your opinion on the operation of the club	36	23.4
Exchanging views with other supporters	35	22.7
Opportunity to raise additional money for the club	135	87.7

often come up with suggestions to support the club without the need for interference from others. However, a mobile application should come with a number of benefits that will make the user want to use it again and again. So respondents were asked what amenities the mobile app brings them and why they use it all the time (Table 31.2).

Respondents were allowed to select up to three main benefits of using the mobile app. The most significant number of respondents, as many as 135 (87.7%), said that the main benefit of using the app is the ability to support the club by gaining additional money, thanks to the financial module, which allows fans to shop with partners by which part of the money from the transaction goes to Ruch Chorzów's account. It shows that the respondents put first not their own benefit but the general good, in this case, the good of the football club, which can maintain liquidity and afford more expenses. In addition to the financial module, which directly supports the club, there are many other functions in the mobile application. Respondents perceive that they can, among other things, keep track of the latest information published on a specially adapted board (115 responses). Users can also exchange views with other fans of the club, as well as express their opinion on the activities carried out by the club. These two opportunities received 35 votes (22.7%) and 36 votes (23.4% of all responses), respectively.

What is important are specific actions that lead to improved relations with the sports club's customers. The respondents were therefore asked which activities carried out in the mobile application by Ruch Chorzów, according to them, most strongly affect the strengthening of ties and building relationships between the club and fans. The results were presented in the form of an aggregate statement. For each activity, an average was drawn, resulting from the number of responses to a given rating (on a Likert scale from 1 to 5) (Table 31.3).

According to those surveyed, three of the seven activities carried out by football clubs are most influential in strengthening ties and building customer relationships. These include "encouraging fans to come to games," "interacting with fans by providing opportunities to ask questions," and "publishing information about football matches. All these activities were rated highest, meaning respondents believe they have a meaningful impact on the relationship between fans and the club.

Table 31.3 Evaluation of the impact of activities in the mobile application on building relationships between the club and fans

Action	Average
Publishing information about football matches	4.02
Publish information about news in the club store	3.86
Interact with fans by providing opportunities to ask questions	4.04
Organization of contests with prizes	3.79
Encouraging fans to come to the game	4.02
Offering discount coupons	3.67
Posting frequency	3.88

31.4.3 Impact of Application on Financial Performance

In financial terms, the app's main idea was to provide functionality allowing fans to earn money for the club. However, this was a rather specific monetization, as the fans did not incur an additional cost from it. Agreements between the football club and business partners who decided to join the campaign allow that a certain percentage of the transactions made by Ruch Chorzów fans owning the mobile application went to the Ruch Chorzów account.

The app went live at the end of 2019, and the ability to use the financial module began with its launch. Ruch Chorzów decided to be fully transparent in what the money that would end up in the club's account would be spent on due to purchases that fans would make from the club's partners. A website was created where current goals and progress in raising money were posted. Once the appropriate amount had been raised, the club would show how the goal had been met by showing photos or receipts for specific services, among other things. As December is the period when no games are played due to the winter break, Ruch Chorzów decided to allocate the first money received to work on renovating the locker room in the club building. Table 31.4 shows the individual collection goals, the date they were completed, and the amounts that went directly into the club's account. The money was earned by app users who made purchase transactions at partners with which Ruch Chorzów cooperates. Each partner has an agreement, which guarantees that the club's account will receive a certain percentage of each transaction or a fixed amount pledged by the partner.

From December 2019 to mid-March 2020, Ruch Chorzów received more than 50 thousand zlotys. This is additional money received due to the involvement of fans who have a mobile app. For a club struggling financially, this is a substantial amount that will relieve the burden on the club's budget. The goals are usually not related to the repayment of specific debts incurred by the club but are things necessary for the operation of the football team. Most are mainly associated with purely sporting issues, such as organizing the team's trips to matches or the physiotherapy equipment needed for the players' wellness.

Table 31.4 Summary of collections and the money raised from them

Purpose of the collection	The amount collected	End date of collection
Renovation of the first team's locker room	PLN 7199.00	31.12.2019
Physiotherapy equipment	PLN 12,137.00	12.01.2020
Subsidized grouping	PLN 15,007.79	04.02.2020
Cup renovation	PLN 7036.67	19.02.2020
First team outings	PLN 8771.37	15.03.2020
Total	PLN 50,151.83	

31.5 Discussion

The survey results made it possible to answer all the research questions posed in the design phase of the study regarding a dedicated app for the Ruch Chorzów football club. Most respondents were in the 30+ age group, meaning that dedicated apps are more likely to be used by fans who identify with the team, go to matches and follow current trends. As for the technical aspects of the mobile app, respondents rated it very well, with almost no technical problems related to downloading or using the app. Users, moreover, were satisfied with the aesthetics of the app's design and all the features it contained.

According to the vast majority, the app was easy to use, intuitive, and had a decent interface that allowed it to be used regardless of the skill level and age of the respondents. In the opinion of respondents, the activities of Ruch Chorzów that were undertaken with the help of the mobile application were usually done well. In some cases, it was even a very good activity of actively participating in fan life by publishing posts daily and encouraging fans to ask questions and express opinions on issues related to the club and sports games.

Social media is becoming more and more a part of the tools supporting the marketing activities of companies in various industries every year. Companies are choosing to set up profiles on social media sites and microblogs, but recently one can also see an increase in accounts set up on content sites such as YouTube. In addition to the free services available on the market, some companies, wanting to get ahead of their competitors, decide on dedicated solutions, which include mobile applications.

Football clubs have the challenge of convincing their new customers, the fans [14]. The sporting aspect is crucial, as the results will determine whether interest in the club will increase, but it is also essential to keep a solid base, which is good contact between the club and the customer. Clubs using tools such as social media can reach a broad audience and try to establish a connection with them, and thus at a later time, build a relationship with them, which can give many benefits, such as increased sales from tickets for football matches or club gadgets. With the use of social media, clubs that do not choose to release own mobile application do not incur any costs related to technical infrastructure or the need to improve technology.

They can allocate these funds to marketing campaigns aimed at potential fans, who, seeing the commitment from the company, will decide to choose their offer.

31.6 Conclusions

In this study, we have presented a mobile app created for a Polish football club. The purpose of the app is to strengthen the fan bond between the club and its fans and create opportunities for financial aid to the club. We have asked app users about app usability, functionality, and reasons for app use. It has been established that the app is used by football club fans and generates income for the club. The possible limitations we see are third-party dependencies for the app to work. A developer company needs to be involved; a payment gate must always work. Despite the pros, the app has generated some costs for the club, such as the cost of developing and maintaining. The possible future work for this app is integrating the ticket sale for matches and connecting to popular social media networks to merge discussion into one place.

References

1. Williams, J., Chinn, S.J.: Meeting relationship-marketing goals through social media: a conceptual model for sport marketers. Int. J. Sport Commun. **3**, 422–437 (2010). https://doi.org/10.1123/ijsc.3.4.422
2. Smoleń, A., Pawlak, Z.: Marketing strategies of professional sports clubs. In: Social Sciences of Sport: Achievements and Perspectives. Wydawnictwo im. Stanisława Podobińskiego Akademii im. Jana Długosza w Częstochowie, pp. 131–146 (2017)
3. Close Scheinbaum, A., Lacey, R., Drumwright, M.: Social responsibility and event-sponsor portfolio fit. Eur. J. Mark. **53**, 138–163 (2019). https://doi.org/10.1108/EJM-05-2018-0318
4. Vale, L., Fernandes, T.: Social media and sports: driving fan engagement with football clubs on Facebook. J. Strateg. Mark. **26**, 37–55 (2018). https://doi.org/10.1080/0965254X.2017.1359655
5. Fenton, A., Cooper-Ryan, A.M., Hardey, M. (Maz), Ahmed, W.: Football fandom as a platform for digital health promotion and behaviour change: a mobile app case study. Int. J. Environ. Res. Public Health **19**, 8417 (2022). https://doi.org/10.3390/ijerph19148417
6. Maderer, D., Holtbrügge, D.: International activities of football clubs, fan attitudes, and brand loyalty. J. Brand Manag. **26**, 410–425 (2019). https://doi.org/10.1057/s41262-018-0136-y
7. Baena, V.: Online and mobile marketing strategies as drivers of brand love in sports teams. Int. J. Sport Mark Spons. **17**, 202–218 (2016). https://doi.org/10.1108/IJSMS-08-2016-015
8. Kim, N., Kim, W.: Do your social media lead you to make social deal purchases? Consumer-generated social referrals for sales via social commerce. Int. J. Inf. Manage. **39**, 38–48 (2018). https://doi.org/10.1016/j.ijinfomgt.2017.10.006
9. Nisar, T.M., Prabhakar, G., Patil, P.P.: Sports clubs use of social media to increase spectator interest. Int. J. Inf. Manage. **43**, 188–195 (2018). https://doi.org/10.1016/j.ijinfomgt.2018.08.003
10. Kossakowski, R., Antonowicz, D., Jakubowska, H.: The reproduction of hegemonic masculinity in football fandom: an analysis of the performance of polish ultras. In: The Palgrave Handbook of Masculinity and Sport, pp. 517–536. Springer International Publishing, Cham (2020)

11. Chynał, P., Cieśliński, W.B., Perechuda, I.: Value of a football club in the context of social media. Inform. Ekon. **4**, 47–57 (2014). https://doi.org/10.15611/ie.2014.4.05
12. Majewski, S., Rapacewicz, A.: The analysis of relationships between sport results and rates of returns of companies involved in sport sponsoring. Ann. Univ. Mariae Curie-Skłodowska, Sect H, Oeconomia **52**, 103–113 (2018). https://doi.org/10.17951/h.2018.52.3.103-113
13. Dziawgo, L.: The use of financial market instruments in supporting professional sports in Poland. J. Phys. Educ. Sport **20**, 2899–2904 (2020). https://doi.org/10.7752/jpes.2020.s5393
14. Krzyżowski, F., Strzelecki, A.: Creating a fan bond with a football club on social media: a case of Polish fans. Soccer Soc. **00**, 1–14 (2022). https://doi.org/10.1080/14660970.2022.2095619

Chapter 32
Understanding Bullying and Cyberbullying Through Video Clips on Social Media Platforms

Janio Jadán-Guerrero, Hugo Arias-Flores, and Patricia Acosta-Vargas

Abstract Bullying and cyberbullying are widely recognized as complex problems that have serious negative repercussions for the health and society of children and adolescents. Bullying can take place in-person at school, home or work, irrespective of age, gender, or role. Cyberbullying can take place on messaging platforms, social media platforms, gaming platforms, or marketing platforms. For example, posting embarrassing photos or videos of someone on social media sending hurtful or spreading lies, as well as abusive or threatening messages, images, or videos through messaging platforms. Offenders impersonate someone and send malicious messages to others in their name or through fake accounts. In this sense, the use of short videos based on the principles of microlearning is scented as a proposal to understand and prevent bullying and cyberbullying. The objective of this article is to analyze how video clips are used on social platforms to raise awareness of bullying and cyberbullying and to propose the creation of new video clips as resources for an awareness-raising MOOC. The methodology followed a phase of analysis of videos on the TikTok and YouTube networks, in order to have a baseline prior to the creation of new video clips by twelve students of the Digital Design and Multimedia career. The content of the videos was evaluated by a psychologist to finally select the appropriate ones for a marketing campaign of an awareness MOOC.

J. Jadán-Guerrero (✉) · H. Arias-Flores
Centro de Investigación en Mecatrónica y Sistemas Interactivos (MIST), Universidad Tecnológica Indoamérica, Av. Machala y Sabanilla, Quito EC170103, Ecuador
e-mail: janiojadan@uti.edu.ec

H. Arias-Flores
e-mail: hugoarias@uti.edu.ec

P. Acosta-Vargas
Facultad de Ingeniería y Ciencias Aplicadas, Carrera de Ingeniería en Producción Industrial, Universidad de Las Américas, Av. de los Granados 12-41 y Colimes, Quito EC170125, Ecuador
e-mail: patricia.acosta@udla.edu.ec

Facultad de Tecnologías de Información, Universidad Latina de Costa Rica, San José, Costa Rica

J. L. Reis et al. (eds.), *Marketing and Smart Technologies*, Smart Innovation, Systems and Technologies 337, https://doi.org/10.1007/978-981-19-9099-1_32

32.1 Introduction

Bullying can happen anywhere and for any number of reasons. Being different in some way, such as belonging to a particular ethnic group, having a disability or thinking differently, can make a person vulnerable to harassment. Bullying can take many different forms, but can often be defined as behavior by an individual or group, usually repeated over time, that intentionally harms another individual or group, either physically or emotionally [1].

Cyberbullying is the use of cell phones, instant messaging, e-mail, chat rooms, or social networking sites such as Facebook, Twitter, and TikTok to harass, threaten, or intimidate someone. In bullying in its traditional form, it is easier to seek refuge, as the act itself depends on the physical proximity the victim has with the bully. The same cannot be said of cyberbullying, as the person is a target no matter where they are, as long as they are connected to the Internet [2].

The use of online platforms and social networks allows for quick and effective interaction between people, making it possible to go beyond the limits of traditional bullying, normally confined to a school campus and an established class schedule, to extend virtually to the homes of the students attacked and bullied, which is known as cyberbullying [3].

Online social marketing campaigns have the potential to contribute to school initiatives on bullying and cyberbullying practices. Operating in digital environments with digital marketing strategies allows you to align with the online practices of young people. One study reveals that extending campaigns beyond the school setting provides an opportunity for youth to engage and revisit campaigns, reinforcing proactive strategies and key messages, which propels youth toward desired behavioral outcomes. Social norms, attitudes, and perceived control were identified as entry points for preventive strategies [4, 5].

Previous studies have been conducted to identify the presence and causes of cyberbullying among students and outline technological tools that help students and parents to control some aspects of cyberbullying [6]. However, it is not enough to make use of technology in a timely manner, it is necessary to conduct social marketing campaigns so that parents, teachers or caregivers are aware of them and even the victims themselves use them to protect themselves.

In the present study, we intend to design learning capsules based on microlearning, a digital marketing strategy for social marketing purposes. Microlearning is understood as a learning perspective oriented toward the fragmentation of didactic content, of short duration, to be viewed anytime, anywhere. The social media platforms are becoming a potential tool in marketing and education, as it allows the delivery of small learning units in a brief period and evaluates them with experts in Psychology and Marketing so that an educational strategy can be implemented through a MOOC course.

32.2 Background

Cyberbullying can include such acts as making threats, sending provocative insults or racial or ethnic slurs, gay bashing, attempting to infect the victim's computer with a virus, and flooding an e-mail inbox with messages.

TikTok, the leading short-form mobile video platform, launches #CreateKindness, a global campaign that reinforces TikTok's commitment to eliminating online bullying and harassment and building a welcoming and supportive community. This campaign kicks off the first installment of a series of creative videos, featuring real and subjective experiences from TikTok creators and animators, to raise awareness about cyberbullying, its effects, and how we can all help prevent it.

Humanity today is experiencing radical changes; technology predominates in this era and makes it possible that thanks to technological advances we can be participants in the digital tools offered by the Internet and the various technological devices. These resources facilitate access to communication and therefore contact with families and unknown people at the national or global level. This is possible thanks to the infinity of social networks that currently exist, such as Facebook, Skype, Instagram, Messenger, Twitter, and the well-known platform that during the COVID-19 pandemic came to revolutionize TikTok adults and children. Technology is not bad, but we must ensure that its use is appropriate, since nowadays problems of different kinds originate such as cyberbullying, mostly originated by children and adolescents, manifesting maladaptive behaviors such as aggression, depression, school violence, low performance in even many are led to suicide [7].

In recent years, research has revealed that one in three teens experience some form of cyberbullying, often resulting in emotional distress, psychosocial trauma, and decreased self-esteem. While researchers agree on the alarming impacts of cyberbullying, a precise consensus on what constitutes cyberbullying is much less clear. It is strongly supported that cyberbullying involves the use of technology to harass or intimidate others, although, there are some discussions as to the nature and regularity of this harassment. Some researchers believe that an action can only be classified as cyberbullying if it involves continuous and repeated attacks [8].

Cyberbullying is defined as an action or decision perceived by another individual that arouses singular or multiple feelings of victimization, shame, or harassment. With the proliferation of the Internet, cybersecurity is becoming a major concern. While Web 2.0 provides easy, interactive, anytime, anywhere access to online communities, it also offers an avenue for cybercrimes such as cyberbullying. Experiences of cyberbullying among young people have been recorded internationally, thus drawing attention to its negative effects [9].

According to the present study, cyberbullying has connections with personal family and social variables, since today it is a very worrying issue in society due to violent behaviors through modern technologies that are increasingly increasing. This phenomenon has resulted in situations of suicide and low self-esteem; although the most serious effects evidenced by the victims are post-traumatic stress and depression. Prevalence studies show that it is a problem of consideration for which different interventions have been conducted to be able to identify as accurately as possible the percentage that exists among the victims, aggressors, and observers in the different modalities, since it is becoming increasingly popular due to the ease of access to them [10].

32.3 Method

The present study applies a social marketing strategy based on microlearning, which is understood as a learning perspective oriented to the fragmentation of didactic contents, of short duration, to be watched at any time and place. The use of short videos in social networks is presented as an alternative to prevent cyberbullying.

Today the social network TikTok has become one of the trends in Social Media and an influential medium in the lives of young people, especially in leisure and entertainment. TikTok has attracted the attention not only of young people, but also of politicians, professionals and individuals, due to the benefits it brings socially and emotionally.

Based on these premises, we selected TikTok as a platform to analyze digital marketing strategies. Twelve students of Digital Design and Multimedia participated in the study, who had to search for three types of videos related to bullying and cyberbullying, one as a victim, another as a victimizer, and the last as a spectator. In total, a database of 19 different videos was created. The next step was to take the best practices from these videos so that the students could create new ones, in which they would be the protagonists. Students worked in pairs and created a script in which they searched for a scenario and volunteer participants. Six video clips were created and then reviewed by a psychology expert and socialized with a social marketing expert. A phase of feedback and improvement of the video was carried out to finally create educational content based on them, which would be used in a MOOC. Figure 32.1 summarizes the process followed for the analysis, creation, evaluation, and application of the video clips.

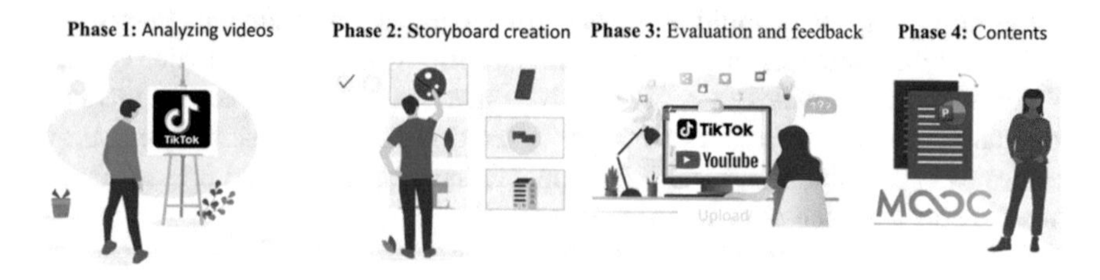

Fig. 32.1 Procedure followed in the development of video clips

Table 32.1 Example of the storyboard of videos

No.	Scenario	Video	Audio	Text
1	**Place**: Hallway **Time**: 10 am **Atmosphere**: Academic (Medium light) **Attire**: Clothing: P1—Victim (blue pants, white sweater, white shoes) P2—Students (various)	**P1** You are in the first person walking through the corridors of the university to go to the terrace	**Music**: Lo-fi **Ambience**: Melancholic **P1:** Sound of footsteps and music on your EarPods **P2:** Various sounds in the hallway	No text
2	**Place**: Terrace **Time**: 10:10 am **Atmosphere**: Cold rainy morning **Clothing**: P1—Victim (Blue pants, white sweater, white shoes) P2—Aggressor (Virtual)	**P1**: He walks toward the edge of the terrace with his hands in his pockets, receives several messages from his aggressor and his mood drops	Music: Lo-fi Ambience: Melancholic **P1**: Sound of footsteps, rain, and music on your EarPods **P2**: Notification sound	The video will have the subtitles with the text of the audios

32.4 Results

Some ideas and inputs were taken to create the educational video clips, for which three working sessions were organized with a psychologist, an expert in audiovisual design and the twelve students of the Digital Design and Multimedia career to coordinate the construction of six educational capsules. The microlearning content of these videos was validated by the psychologist before they were used. Table 32.1 gives an example of the storyboard of the video.

In Table 32.2, we show six video clips developed with the participants.

Analyzing some of the videos presented in Table 32.2. It was possible to identify that the participants are clear about the concepts and differences between bullying and cyberbullying. In all the videos, messages were found that show that cyberbullying can be present in all activities of daily life and that if there is no maturity to know how to deal with it, it can cause psychological and emotional damage.

32.5 Conclusions

Social media is a vital means for information sharing due to its easy access, low cost, and fast dissemination characteristics. However, increases in social media usage have corresponded with a rise in the prevalence of cyberbullying [11, 12].

Microlearning is a didactic strategy that allows rapid and effective interaction between people. The use of short videos based on the principles of microlearning

Table 32.2 Description of video clips

Name	Description	Feedback
Video 1	A video about a foreign student who suffers bullying in high school. It explains the causes and consequences of bullying. In the end talks about some recommendations to prevent the harassment to students	The videos show the problematic of bullying and the bad impacts in the people who suffer it. It analyses how bullying happens and how it has developed with the social media and technology boom. It gives to the person the perspective about what is bullying, but more important how to fight against it with several recommendations. The videos are shocking showing how bullying occurs in the normal daily life. The target looks to be focused in more mature people. However, in several videos in the end; it recommends to the person to not shut up against the aggressions implying that these videos might be also watched by teenagers It looks a good starting point of diffusion to create a digital marketing campaign to fight and create conscience about bullying and how to stop it
Video 2	In the introduction you may see a first person video where a man is suffering harassment through messages. Then, it talks about cyberbullying and how it works in social media, due to the facility of the viralization of the videos, images and messages. Despite the laws and controls of the content in Internet, it is impossible to control all the material there. It explains the difference of cyberbullying and bullying and ends with recommendations about how to afford it	
Video 3	In the beginning it appears the phrase "treat others how you want to be treated." This persuasive catchphrase serves to persuade the viewer about bullying. Moreover, this video focuses in the negative impacts of bullying and cyberbullying. It describes the two ways to commit bullying to a person. First, via direct conflict and fight against the person. And second, via defamation and criticism against the person. In the end it gives some recommendations about how to fight back bullying	
Video 4	It starts with the definition of bullying and then it advances with cyberbullying. The video shows how the kids commit bullying meanwhile it describes, furthermore, how youngsters suffer it in the academic environment in schools and universities	
Video 5	This video illustrates how a person suffers of bullying in the daily normal life of a student. It shows the process of the person toward these hostile attitudes against it. Through images you may see the pain of the person	

is presented as an alternative to prevent cyberbullying. Short, didactically structured content featuring real experiences to raise awareness about cyberbullying, its effects, and how we can all help prevent it, can foster online solidarity to combat cyberbullying [13, 14].

Through the video analysis, it is also evident the need to generate new contents focused on children and adolescents who are in situations of cyberbullying. The limitations of the study could be that the videos were created in Spanish language, and this would limit the use of social marketing campaigns at a global level, concentrating mainly in the Latin American segment.

The future work for this research is to develop an instructional design that articulates the videos created to be used in a MOOC aimed at raising awareness of the problems of bullying and cyberbullying. Subsequently create digital marketing strategies for diffusion in social networks.

Acknowledgements The authors would like to thank the Coorporación Ecuatoriana para el Desarrollo de la Investigación y Academia- CEDIA for their contribution in innovation, through the CEPRA projects, especially the project CEPRA-XVI-2022-04, "Implementación y despliegue de cápsulas de aprendizaje para combatir el bullying y el ciberbullying en niños, niñas y adolescentes"; also the Universidad Tecnológica Indoamérica, Universidad de Cuenca and Universidad del Azuay for the support for the development of this work.

References

1. Acosta-Pérez, P., Cisneros-Bedón, J.: El Fantasma del Acoso Escolar en las Unidades Educativas, CienciAmérica **8**(1) (2019). https://doi.org/10.33210/ca.v8i1.207
2. Mkhize, S., Gopal, N.: Cyberbullying perpetration: Children and youth at risk of victimization during Covid-19 lockdown. Int. J. Criminol. Sociol. **10**, 525–537 (2021)
3. Khlaif, Z.N., Salha, S.: Using TikTok in education: a form of micro-learning or nano-learning? Interdisc. J. Virtual Learn. Med. Sci. (IJVLMS) (2021). https://ijvlms.sums.ac.ir/article_47678.html
4. Zhu C., Huang S., Evans R., Zhang W.: Cyberbullying among adolescents and children: a comprehensive review of the global situation, risk factors, and preventive measures. Front Public Health **9**, 634909 (2021). https://doi.org/10.3389/fpubh.2021.634909. PMID: 33791270; PMCID: PMC8006937
5. Spears, B., Taddeo, C., Barnes, A.: 6-Online social marketing approaches to inform cyber/bullying prevention and intervention: What have we learnt? In: Campbell, M., Bauman, S. (eds.) Reducing Cyberbullying in Schools, pp. 75–94. Academic Press (2018). ISBN 9780128114230, https://doi.org/10.1016/B978-0-12-811423-0.00006-7
6. Jadán-Guerrero,J., Bermeo, A., Cedillo, P., Nunes, I.: Helping kids and teens deal with cyberbullying through informative learning capsules. In: Nunes, I.L. (ed.) Human Factors and Systems Interaction. AHFE (2022) International Conference. AHFE Open Access, vol. 52. AHFE International, USA (2022). https://doi.org/10.54941/ahfe1002176
7. Mishna, F., Birze, A., Greenblatt, A.: Understanding bullying and cyberbullying through an ecological systems framework: the value of qualitative interviewing in a mixed methods approach. Int. J. Bullying Prev. **42**, 1118–1126 (2022). https://doi.org/10.1007/s42380-022-00126-w

8. Gualdo, A.M.G., Hunter, S.C., Durkin, K., Arnaiz, P., Maquilón, J.J.: The emotional impact of cyberbullying: differences in perceptions and experiences as a function of role. Comput. Educ. **82**, 228–235 (2015)
9. Salawu, S., He, Y., Lumsden, J.: BullStop: a mobile app for cyberbullying prevention. In: Proceedings of the 28th International Conference on Computational Linguistics: System Demonstrations, pp. 70–74, Barcelona, Spain (Online). International Committee on Computational Linguistics (ICCL) (2020)
10. Kumar, A., Sachdeva, N.: Cyberbullying detection on social multimedia using soft computing techniques: a meta-analysis. Multimed. Tools Appl. **78**, 23973–24010 (2019). https://doi.org/10.1007/s11042-019-7234-z
11. Emmery, C., Verhoeven, B., De Pauw, G., et al.: Current limitations in cyberbullying detection: on evaluation criteria, reproducibility, and data scarcity. Lang. Resour. Eval. **55**, 597–633 (2021). https://doi.org/10.1007/s10579-020-09509-1
12. Cheng, L., Shu, K., Wu, S., Silva, Y., Hall, D., Liu, H.: Unsupervised Cyberbullying Detection via Time-Informed Gaussian Mixture Model. CIKM '20, 19–23 Oct 2020, Virtual Event, Ireland Association for Computing Machinery, ACM (2020). ISBN 978-1-4503-6859-9/20/10. https://doi.org/10.1145/3340531.3411934
13. Aldosemani, T.: Microlearning for macro-outcomes: students' perceptions of telegram as a microlearning tool. In: Laanpere, M., Väljataga T. (eds.) Digital Turn in Schools—Research, Policy, Practice. Lecture Notes in Educational Technology. Springer Singapore, Singapore, pp. 69–81 (2019). ISBN 978-981-13-7360-2
14. Chen, J., Zhang, Y., Sun, J., Chen, Y., Lin, F., Jin, Q.: Personalized micro-learning support based on process mining. In: Proceedings—2015 7th International Conference on Information Technology in Medicine and Education, ITME 2015, pp. 511–515 (2016). https://doi.org/10.1109/ITME.2015.120

Chapter 33
Gender Bias in Chatbots and Its Programming

Carolina Illescas, Tatiana Ortega, and Janio Jadán-Guerrero

Abstract Nowadays, digital transformation in business seeks to maximize the efficiency of its own processes through automation. These changes will be successful if the process of adaptation and change management is properly carried out. One of the main tools to achieve these transformations is the chatbot. This key technology will help embrace these changes and transformations; however, there have been processes in which their interactions show bias, as inclusive interaction is missing. This research focuses on two main points: first, to identify whether the user is capable of recognizing any bias variables in chatbots and second, to analyze if these variables have an impact when there is an adoption of new technologies in any digital transformation and business goals. The faster and more in depth a company understands the bias and its impact in the target audience, the better automation will be adopted. As a result, chatbots will be properly used to achieve and increase business goals. Some undoubtable chatbot uses in marketing are personalized service, facilitate sales, and trigger client's satisfaction. Most importantly, gathering valuable data for analysis, predictive business models and trends into virtually for any product or service intended to promote. After consulting +150 chatbot users, it can be concluded that it is a proven fact that gender bias is present, also, there is a high trend and acceptance of chatbots with female voices and characteristics, which can be balanced if there is awareness around this topic among developers and change management teams seeking adoption of new technologies.

C. Illescas · T. Ortega
Universidad Galileo, 4A Calle 7a. Avenida, Calle Dr. Eduardo Suger Cofiño, Cdad. de Guatemala 01010, Guatemala
e-mail: carolina.illescas@galileo.edu

T. Ortega
e-mail: tatiana.ortega@galileo.edu

J. Jadán-Guerrero (✉)
Centro de Investigación en Mecatrónica y Sistemas Interactivos (MIST), Universidad Tecnológica Indoamérica, Av. Machala y Sabanilla, Quito EC170103, Ecuador
e-mail: janiojadan@uti.edu.ec

J. L. Reis et al. (eds.), *Marketing and Smart Technologies*, Smart Innovation, Systems and Technologies 337, https://doi.org/10.1007/978-981-19-9099-1_33

33.1 Introduction

Communication technologies have advanced by leaps and bounds in the last decade. In the case of online conversations, new tools have emerged to optimize dialog with users and customer service. Chatbots are resources created in the commercial sphere to improve the provision of uninterrupted attention on websites without or with little dependence on humans.

In this investigation, we explore the impact of gender bias in chatbots and its programming and how that can affect the user's acceptance of automation within the company or impact a business, in a branch like marketing as campaigns around new products or services. Today, there are more bots with associated characteristics to female gender than to male gender or bots without gender associated features, and this can be seen with Siri, Alexa, Waze, Google Maps, and others. There's little information about the variables that have been taken into account to select female voices and names. It is unknown whether the customer perceives better satisfaction when using technology. There is little evidence from studies that identify the biasing and most notable variables in relation to the end user. On the other hand, it is still unknown whether the biased conversation between humans and robots decreases satisfaction and acceptance. [1].

We presume that the adoption level of the use of these new technologies during a process of digital transformation may improve if chatbots without gender bias are implemented. Our objective is, indeed, to validate the impact of inclusive/non-inclusive programming in chatbots guidelines on the level of adoption of new users during a business digital transformation through several steps. First, we proved if the bias in the data is noticeable when a chatbot is programmed and trained. Second, we evaluated what type of discrimination is more sensible for the users and third, we evaluate if a bias or personalization is detected when using chatbots for commercial purposes. Finally, we acknowledge if there are best practices to incorporate in the adoption of bots for digital transformation purposes which may be utilized also for predictive trends and acceptance of marketing strategies utilizing digital media and channels.

A panel of 87 ICT professionals and end users was sampled to ask about gender perceptions of chatbots and quantify the impact on gender bias variables. In a second phase, 60 end users were sampled to evaluate gender bias and efficiency related to male or female bot on commercial tasks. A high percentage of chatbots have characteristics more related to femininity as perceived by users. It is necessary to have more studies in the programming part, since the sample of developers was based on market trends and behavior.

33.2 Related Work

The literature review brings to light studies that show that there is gender bias in chatbots. Most of these studies were conducted on the European continent, however there is no evidence of a specific study for Guatemala or LATAM where the research was conducted.

The first topic is the relevance of the studies found about the bias in technology. It is known, for example, that conceptualizing bias as a programming problem is something wrong, since the data as well as the algorithms and its way of solving queries have been implemented by humans. There is a paradox in taking out of the equation the human intervention in these actions, so an AI may manage it. However, human intervention is still applied in the programming and the training of the bots. In the end, this perpetuates and transmits conscious and unconscious bias [2].

In the same field, it was possible to identify that the predictive models used for the AI training may foment a systematic discrimination toward some groups where it is probable to discuss and find patterns which may help to identify bias and how to minimize its impact [3].

This quality of data used in algorithmic training is also established in other studies. Despite the big data used on chatbots, there are still concepts with bias against minority groups, and it is not always simple to identify. Most programmers are not conscious of these failures since they are not focused on this type of narrative [4].

Other studies have reviewed the quality of the content published in newspapers about chatbots in order to identify the role in the business digital transformation. 74 articles were evaluated from 54 newspapers in Australia. The method was based on a systematic literature where they concluded that chatbots are a key tool in the transformation of several industries. They also identified areas of interest in the perception, acceptance, and use of the chatbot in customer service issues, supporting performance and satisfaction levels [5].

With this basis, it is important to investigate the concept of gender bias and its social perception. There is a study which identifies the effect of the gender bias in the final user experience. Specifically, virtual agents or chatbots with voice functionality were created with specific gender characteristics. The methodology of this study was based on a social experiment with virtual interviews. They put two bots "Kathy and Bill" with the conclusion that the interactions with Kathy were ruder and based on fantasy dialogs. This supports the validation of gender bias and the preference of the users for a specific chatbot with specific gender characteristics [6].

Focusing on the American continent, there are studies that try to identify the injustice in machine learning and how it affects the users. They defined injustice in machine learning as the inequities that technology may take advantage of and its derivatives of bias and algorithms. In this study, they identified seven different types of bias in data that was used, commonly but not consciously, in machine learning. Moreover, there are five biases in the algorithm, which ends up harming the final users. Nonetheless, the users also nourish the bias in the seven variables. One of them, the clearest one, is when people had to choose the variable of the ethnicity in a formulary or in the type of content that is published in social media [7].

As a result of these studies, it is important to recognize that there are clear definitions about what injustice means in the machine learning process, and there is a clear example about how possible it is to merge algorithms, depending on the stage on which it is part. For example: data- pre-process with the user- classification- process with the user, vocabulary- post process with the user.

It is necessary, once the different types of biases were acknowledged, to find studies about the efforts to mitigate and eradicate these biases. Some of them proportioned an insight about experiments with algorithms and databases to eradicate biases in AI. There are several codes to mix variables such as skin color, hair color, and facial traits. This mix concluded in an algorithm which is capable of identifying faces without interaction problems making it possible to recognize any face of any race without complications. After the identification process and the revision of these studies, we found that the training process of the algorithms in machine learning is relevant to eliminate any bias of the data used for programming. The development of AI without any gender, race, and language bias is imperative in the process to adopt new technologies [8].

Our objective is to prove that it is possible to mitigate bias in the predictive models based on data without losing efficiency in the outcomes and prevent algorithms from making decisions based on sensible information. Several variables of sensible information may directly impact in the program and in the interpretation of chatbots who are implemented in the digital innovation process [9].

33.3 Method

This study defines as priority topics the level of impact of the bias in programming over the society to demonstrate if the effect of the gender is linked with the level of satisfaction of the final user, either in commercial transactions, marketing efforts for digital or non-digital products or services or general tasks such as customer service. Also, if the professionals who develop these technologies are aware of these biases when programming data and in the training of machine learning, as well as the algorithmic parameters. Although this paper proposes quantitative research through digital surveys that provided statistical results, it was possible to identify good insights to be considered in automation and digital transformation projects as well as applicability for strategies to launch marketing efforts understanding better the audience and their reaction to their interactions with chatbots enabled to suggest, prompt, or offer products and services.

33.3.1 Participants

The first phase of the investigation was made with users who live in Central America or work for companies that are based in Central America. The population was 87 people who were divided in two different groups. One, composed of ICT developers or people related to this field, and the second, with users who have interacted with chatbots. In the second phase, the universe was of 101 users, six were excluded since they didn't report chatbot interactions or were outside the regions' perimeter, the real sample was 94 unique users.

33.3.2 Instruments

In all cases, two surveys were utilized with open and closed questions, some of them including Likert scale. The two phases were done through digital channels; however, the difference lies in the type of questions and platform used, Google Forms was used for the developers and a Typeform® chatbot for the general Internet user, simulating the interactions which are the subject of this study, including versions incorporating features associated with female and male genders.

33.3.3 Procedure

This research was divided in two phases, phase one required a month to investigate, structure the surveys, and identify the universe to be tested. Phase two lasted a week, since subjects were easier to contact and most of them were already familiar with the chatbot tool.

Phase I: Confident about the data to evaluate, each variable was assigned to each control group, whether the developer group or the final user group. Of all the instruments, at the end, we concluded that the survey was the better one to use in both groups. The first group did the survey in Google Forms, due to the security, anonymity, and because it is simple and easy to use for both sides, the interviewee and the interviewer. The focus on this group was to measure algorithm ethics and the level of conscience about gender bias when programming tools.

The second group used Typeform® with two variables. This platform also gives security in the information and anonymity. However, this survey was under the chatbot scenario. Colors were used during the survey to see if the color influences the perception of the gender and if the answers to check if the answers changed because of this bias. This strategy was implemented to see if there was any bias answer as it was done in the project "Lenguage Masculino" de AI by Amazon [10]. In this group, the perception of the bias and the technology adoption was verified.

Regarding identification with traits associated with the male gender, 11% of the chatbots were associated with a non-binary gender or described themselves as "genderless" and a large majority of the sample, 77%, did not perceive any type of discrimination toward their person by the chatbot. The 11% who perceived some types of discrimination were part of the female gender.

In phase two, the objective was to relate the gender bias, if any, when utilizing chatbots for business purposes and marketing tactics. This time the research group adapted the tool, Typeform®, to neutral colors and font, no avatars were used and the questions were orientated to measure user's satisfaction related to a chatbots gender or bias.

33.4 Results

Both groups have some conscience level about the gender characteristics integrated in the chatbots; however, the technology development professionals in more than 25% of the time these features were derived for the effectiveness perception or of the satisfaction levels aligned with them (Fig. 33.1).

The results show that the final users perceive the name, generally, associated with a gender, but for them this is irrelevant because they already assume it will be feminine. This issue was found in 66.9% of the surveyed population. Regarding identification with traits associated with the male gender, 11% of the chatbots were associated with a non-binary gender or described themselves as "genderless" and a large majority of the sample, 77%, did not perceive any type of discrimination toward themselves by the

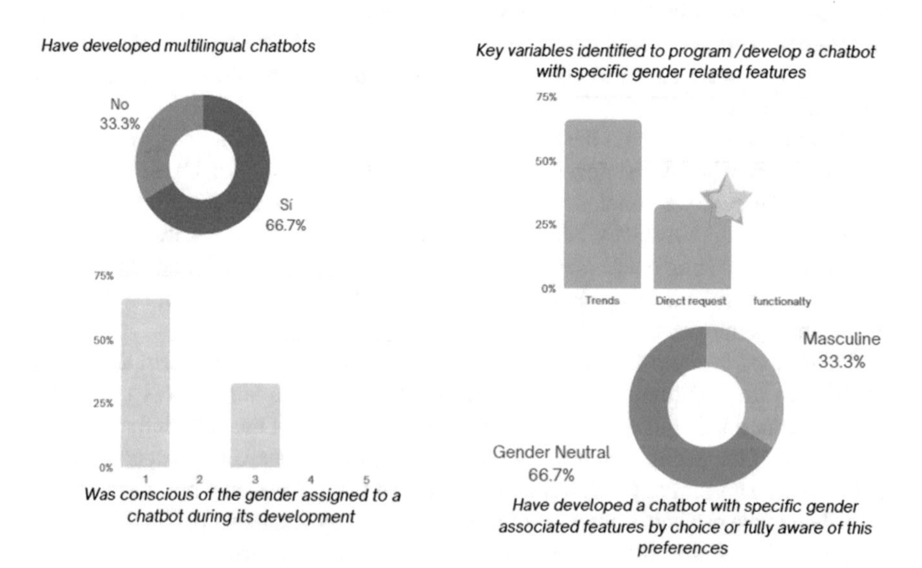

Fig. 33.1 Group 1: developers' key results of gender associated to a chatbot

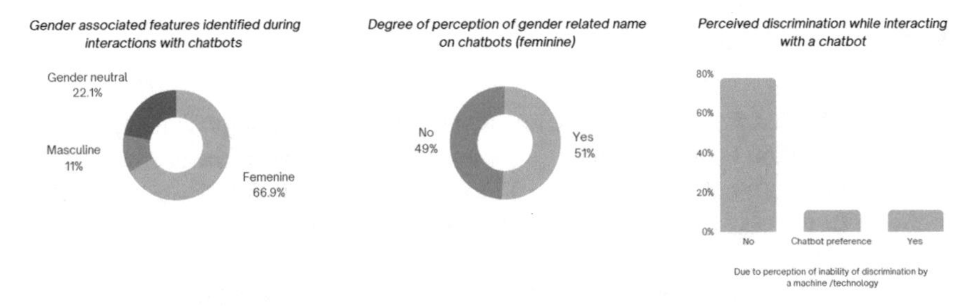

Fig. 33.2 Group 2: end user bias perception and satisfaction

chatbot. The 11% who perceived some type of discrimination identified themselves with the female gender (Fig. 33.2).

Results of the second phase confirm the theory that having female characteristics associated to a chatbot increase clients' satisfaction. The adoption of this technology could be better and faster when associating a female personality to bots. Out of the 94 users, 53% confirmed they have purchased through a bot. Out of the 42 users that responded that have not made a purchase, almost half of them feel a "neutral" satisfaction with the bots' assistance; this is key for business to understand that satisfaction on other areas needs to be improved from neutral to pleased or happy, in all aspects, customer service, sales or just as a companion bot. Finally, and probably most important, for this research, is that a correlation was identified between the bots' gender and the perceived satisfaction of the user who bought using this specific technology. 58% of the users assure to be satisfied and 73% of them recall having a female bot. The users who did not buy though bots recall having a neutral bot or not being able to identify if it was female or male (Fig. 33.3).

Fig. 33.3 Phase II: end user chatbot's gender acknowledgment versus satisfaction when using it for commercial purposes

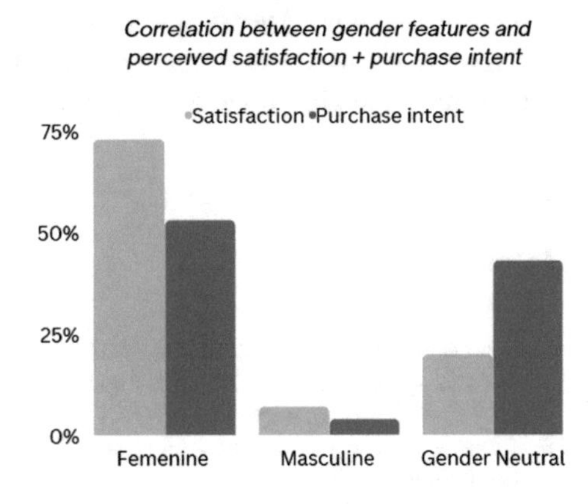

33.5 Conclusions

- During the first stage of investigation, it was possible to distinguish that there is a high perception by the final user about the gender features of the chatbots vs the perception on the developer's side. The people who work in ICT are not always aware of the bias imposed on some features, only if these ones are requested a prior [11].
- Chatbots may help in a positive way to the discrimination problematic in customer support issues and marketing efforts. Not having AI only perpetuates the tendency to discriminate and the users may feel more comfortable using this type of technology.
- A high percentage of chatbots have features more related to femininity as it was perceived by the users. It is necessary to have more studies in the programming side, since the sample of the developers was based on tendencies and market behavior.
- Business should consider the winning insight when automating processes; fixing the data analysis for recommendations along with a female bot could be a tool to achieve marketing objectives, business goals and increase the perception on non-discrimination toward the final user.
- An important insight for marketing and BI is that interests in products or other variables can be gathered when navigating and then use bots to provide options to the users. According to the investigation, people you assured buying with or through a bot did receive those recommendations, however almost half of them were useless options. Theory is that if BI could analyze better and provide clear information, marketing should be able to program along with IT better recommendations to both the one who bought and the ones who didn't.
- In the following steps, it is possible to investigate more in detail what defines the traits of a bot, if in the space where they are being developed, they are using algorithmic ethics, and probably, also, to investigate the level of impact of this implementation in a broader sample of end users.

References

1. Beattie, H., Watkins, L., Robinson, W.H., Rubin, A., Watkins, S.: Measuring and mitigating bias in AI-Chatbots. In: 2022 IEEE International Conference on Assured Autonomy (ICAA), pp. 117–123 (2022). https://doi.org/10.1109/ICAA52185.2022.00023
2. Ajunwa, I.: The Paradox of Automation as Anti-Bias Intervention. Cardozo Law Rev. **41** (forthcoming 2020) (2020)
3. Žliobaitė, I.: A Survey on Measuring Indirect Discrimination in Machine Learning (2015). arXiv pre-print, https://arxiv.org/abs/1511.00148
4. Barocas, S., Selbst A.D.: Big data's disparate impact. Calif. Law Rev. **104** (2016)
5. Miklosik, A., Evans, N., Qureshi, A.M: The use of Chatbots in digital business transformation: a systematic literature review. IEEE Access **9**, 106530–106539 (2017). https://ieeexplore.ieee.org/stamp/stamp.jsp?arnumber=9500127

6. de Angeli, A., Brahnam, S.: Sex Stereotypes and Conversational Agents (2006). https://www.academia.edu/22924095/Sex_stereotypes_and_conversational_agents?sm=b
7. Mehrabi, N., Morstatter, F., Saxena, N., Lerman, K., Galstyan, A.: A survey on bias and fairness in machine learning. ACM Comput. Surv. (2021). https://doi.org/10.1145/3457607
8. Amini, A., Soleimany, A.P, Schwarting, W., Bhatia, S., Rus, D.: Uncovering and mitigating algorithmic bias through learned latent structure. In: Proceedings of the 2019 AAAI/ACM Conference on AI, Ethics, and Society (AIES'19), pp. 289–295 (2019). https://doi.org/10.1145/3306618.3314243. Google Scholar Digital Library
9. Mejía, J., Llop, J., Espinosa, L.F.: Los superhéroes combaten los sesgos de género de los algoritmos. BBVA (2021). https://www.bbva.com/es/los-superheroes-combaten-los-sesgos-de-genero-de-los-algoritmos/
10. Jeffrey, D. (2018) Amazon Scraps Secret AI Recruiting Tool that Showed Bias Against Women. Reuters (2018). https://www.reuters.com/article/us-amazon-com-jobs-automation-insight/amazon-scraps-secret-ai-recruiting-tool-that-showed-bias-against-women-idUSKCN1MK08G
11. Le, Ch., Ma, R., Hannák, A., Wilson, Ch.: Investigating the impact of gender on rank in resume search engines. In: Paper presented at the Annual Conference of the ACM Special Interest Group on Computer Human Interaction, 1–14, 2018. Montreal, Canada

Chapter 34
Production, Exhibition, and Promotion of the Peruvian Web Series: Miitiin, Brigada de Monstruos y Leo en el Espacio

Veruschka Espinoza Zevallos and Yasmin Sayán Casquino

Abstract During the pandemic, many Peruvian productions within the traditional form faced various obstacles in the realization of open signal television series. Nonetheless, new content creators saw the opportunity to explore new methods of production and distribution within the social network (Arboleda in Revista Universidad EAFIT 52:122–125, 2017 [1]). The digital transformation accelerated by the pandemic has provoked content creators to develop their ability to overcome constraints and find new ways to produce. Therefore, the objective of this research is to identify the strategies that have been implemented in the production, exhibition, and promotion of Peruvian web series. It is indispensable to understand the processes, creation, and promotion that can help others to generate more content. To carry out this objective, an interpretative investigation with qualitative data methodology was carried out. The selected categories were digital platforms, production, and exhibition. This study involved 20 interviews with professionals in the audiovisual sector and the team of the web series: Miitiin, Monster Brigade and Leo in Space. From the perspective of the specialists, it has been identified that the most relevant strategy in these processes is to disseminate the project on all possible platforms. In addition, it is important to keep the consumer expectant with different universes that finally help them to reach to the series. Finally, knowing the processes of audiovisual production is of vital importance not only for the learning of the creators but also for the audience.

The current research is the winner of the research contest of the Universidad Peruana de Ciencias Aplicadas (UPC).

V. E. Zevallos · Y. S. Casquino (✉)
Universidad Peruana de Ciencias Aplicadas, Prolongación Primavera 2390, Lima, Perú
e-mail: pcavysay@upc.edu.pe

V. E. Zevallos
e-mail: u201623072@upc.edu.pe

J. L. Reis et al. (eds.), *Marketing and Smart Technologies*, Smart Innovation, Systems and Technologies 337, https://doi.org/10.1007/978-981-19-9099-1_34

34.1 Introduction

The arrival of mobile devices along with the Internet has managed to revolutionize daily tasks, now to entertain ourselves it is enough to make a click from a cell phone, tablet or television [2]. The area of communications is facing new and accelerated technological transformations, such as the incursion of Information and Communication Technologies—ICT [3]. These transformations merge systems for transmission of content, which contrasts with a relevant application of alternative media, and the need to generate changes focused on the requirements promoted by production for the web [3].

Netflix, YouTube, or Vimeo are some of the platforms that are already part of our lives and our consumption habits [4]. Today, the streaming entertainment sector, which is in growth, is demandable and can be customized. It has also accelerated the evolution of streaming to digital applications and platforms [5]. In addition, digital platforms and their large catalog of audiovisual products are easily accessible [4].

Digital content and forms of entertainment are constantly evolving to adapt to this new generation of consumers [6]. The audiovisual world has faced major transformations, due to technological, social, legal factors, and so on [6]. These changes have allowed the development of new series and new formats and created exclusively for online consumption. Likewise, with the advent of the Internet and 2.0 networks, the digital world has been transformed. Thanks to these phenomena, a new distribution possibility has opened up for the audiovisual sector, thereby creating a closer connection with the ultimate goal of the film: the viewer [7]. Today, we are in a media revolution, which shifts the entire culture toward means of production, distribution, and communication mediatized by the computer [8].

In Peru, considering the context we have been going through, a remarkable number of audiovisual productions aimed at the web have emerged: Miitiin, Brigada de Monstruos and Leo en el Espacio are some of those that have been released during the pandemic period. Additionally, one of the web series that had the greatest impact in our country is "Los Cinéfilos", which was produced by Señor Z in 2013. This web series attracted the attention of users who, week after week, anxiously awaited the episodes to follow the adventures of two movie buffs. Thanks to the reception of this series, six seasons with 64 four-minute episodes were completed [9].

As aforementioned, web series have been positioning themselves as one of the main sources of entertainment among users; hence, it is important to investigate and analyze specific cases. On the other hand, production, exhibition, and marketing are key stages for the success of any audiovisual product and especially for web series, since they will surely be the main source of entertainment in the future [1].

Thus, this article aims to identify the strategies that have been implemented in the production, exhibition, and promotion of Peruvian web series. The main categories are digital platforms and production, exhibition, and promotion, since in this way we can delve into all aspects involved in making a web series.

In addition, this article will contribute to the audiovisual sector and the new methods that are emerging to create and exhibit an audiovisual product. To address this issue, we hereby propose a qualitative analysis in which 20 semi-structured interviews will be conducted in order to obtain different points of view.

34.1.1 The Internet as a Protagonist of Audiovisual Evolution

Before taking into consideration all the aspects involved in the evolution of the Internet and web series, it is deemed necessary to know the meaning of some important terms in this article. To begin with, it should be understood that web series are all those audiovisual productions created for broadcasting on the Internet in a serialized way (by means of chapters) [10], which means that web series are a different type of production from traditional ones. These are a new proposal for the consumer with a different content and are the pursuit of new ways to tell stories. According to the report "TV and Media 2015: The empowered TV and media consumer's influence" carried out by Ericsson's ConsumerLab, the consumption of audiovisual and entertainment series on the Internet has increased by more than 120% worldwide since 2010. The Internet is also a computer network that democratizes audiovisual consumption and production. This democratization, coupled with the ease of access to mobile devices with high quality video cameras, opened the possibilities of audiovisual production with alternative formats [1].

The audiovisual industry has been transformed and the consumption of audiovisual productions for the web has increased in an extraordinary way. The Internet has meant that today we can watch a web series whenever and wherever we want [11]. Similarly, the digital transformation has had a global impact on content distribution and consumer habits [12].

To comprehend this process, it is impossible not to talk about the concept of streaming. The author mentions that this term means accessing content before it has been completely downloaded and obviously this fact of playing files in streaming means saving space in memory. Similarly, another of the things it offers is the possibility of live streaming or consumption and publication of live content. Where one or all of the parties broadcast content that has just been recorded so that it can be consumed in the same way [4].

34.1.2 New Features for Web Series

Web series are generally fictional and their episodes have a thematic unity. Their online distribution makes them target an audience very different from the typical television audience [10]. Likewise, web series that have a length between 3 and 6 min, are articulated through the springs of the most classic comedy. Whereas, those that are extended up to 15 min, do so by virtue of a treatment that moves away from comedy or combines other genres [9]. Thanks to the short duration, it is possible for users to watch these series more flexibly as well as frequently. Web series become the emblematic product of digital fiction with the junction of technology, industry and audience. Thanks to their short length and ease of viewing, they offer an effective response to users' interests and demands [9].

Focusing on the narrative and visual characteristics, the web series evidenced a short duration and a linear, agile, and rhythmic montage. Also, they have moderate camera movements and an advantage in the scale of shots, since they tend to be tight or closed frames. This is due to the fact that web series, being an agile product from its production to its display, have to be easy to process and at the same time entertaining [3].

Finally, a factor that must bear in mind in the narratives, since they use digital platforms. There may be many kinds of narratives, but if we do not consider the relationship between the story that is told and how it is told using these platforms. It is evident that the same story can be told and adapted to each of the platforms where it is to be published [13].

34.1.3 *YouTube as an Exhibition Medium*

To first discuss YouTube, we have to see it as a social phenomenon. A social phenomenon, in sociology, is defined as all those events, trends, or reactions that take place within an established human society and are evidenced through collective behavioral modifications. As a result, YouTube has become a trend with an established behavior and the explanation for this is that this platform is part of our lives [14]. Most people browse the Internet on a daily basis and when they wish to watch some audiovisual content or a video of anything, they relate to YouTube as a platform.

YouTube can achieve great things and not only in areas as important as changing the consumption of all mankind, but also in smaller aspects, such as external events related to the Club Media Fest platform. YouTube has been growing over the years and as mentioned before, it is already part of our lives and we consult it daily for any purpose, either for entertainment or to look for information about something. According to a report made in 2019 by Kantar Ibope Media, YouTube is the second most used social network in our country and 40% of the world population uses social networks and invests in them an average of two hours a day. In addition, we can observe the percentages of each social network, with Facebook in first place with 75% (see Fig. 34.1) [15].

Fig. 34.1 Kantar Ibope Media's most used social networks report (2019)

In Latin America, currently, the influence of the Internet, the global screen, the social web as well as collective participation have transformed consumption habits. New filmmakers have taken advantage of this change to rethink the business model and the film production chain [16]. Similarly, with new formats such as web series, filmmakers have taken advantage of this transformation to create their content and broadcast it on various platforms [4].

The exhibition is an important factor in the process of an audiovisual product and today there are several channels where we can exhibit, but currently the two channels with greater relevance, notoriety, and market presence are YouTube and Vimeo [17].

34.1.4 Marketing and Promotion of Web Series

The term digital marketing emerged in the 1990s and gained prominence in the twenty-first century. It occupied an important place in persuading consumer buying behaviors, generating efficiency and proximity. Digital marketing, as a trend, has also caused the evolution of digital devices and all the technology around it [18]. In the audiovisual industry, project marketing begins simultaneously with the product development stage and continues throughout the entire production, distribution, and exhibition process [19].

As it occurs with other audiovisual products, audiences find new universes. These are generated by the fiction that emerges and interconnects users around a common theme. The fan phenomenon also encourages the spreading of these contents, which enjoy good engagement rates on Twitter, Instagram, and Facebook [20]. According to data from the 2020 Social Networks Study (IAB 2020), 87% of Internet users use social networks and the channels with the highest spontaneous awareness are Facebook (94%), Instagram (76%), and Twitter (70%).

Knowing the different aspects that move audiences and generate more interactions on social networks allows us to increase our knowledge of the market they are targeting and the social representations that emerge. In addition, the feedback obtained in these social media help to predict what type of audiovisual product and content marketing strategy will capture and retain a greater number of user-viewers [21].

On the other hand, the distribution of web series is through their official website and social media (YouTube, Facebook, Twitter, etc.). Therefore, the promotion of web series is given through social networks. Facebook is the network that brings together a larger community of users worldwide around the accounts of the brands [8]. To measure the effectiveness, we must bear in mind three types of interactions by the audience: reactions, shares, and comments. In the case of Instagram, publicly measurable interactions are only the "likes" and comments, since it is not allowed to directly share the content of other profiles in the user's feed. Therefore, likes and comments are the measurable interactions for effectiveness [21].

34.2 Methodology

Within the study of the production, exhibition, and promotion of web series, this article focuses particularly on web series exhibited on the YouTube platform. The scope of the study was limited to the Peruvian web series Miitiin, Brigada de Monstruos and Leo en el Espacio.

The choice of these web series as the object of study is justified by the fact that they are Peruvian web series that premiered in 2020, when our country and the world experienced a pandemic. These series become relevant cases, since they were produced and shot in the same year. In addition, it involved innovating and discovering new ways to connect with the technical and artistic team, so that the series are successful. Also, these series, being from independent creators, show that, with a professional team, a low budget and a pandemic, innovative series can continue to be produced. With these web series, we intend to base our research and respond to our objectives, which will be of great help for future audiovisual productions.

Focusing on the object of study, it can be mentioned that the Miitiin team was in talks for the series to be sold to a Peruvian channel. Also, there is a second season ready to be produced. In the case of Brigada de Monstruos, it can be mentioned that it is the first Peruvian web series made with miniatures and has more than five official selections at festivals such as Lift-Off Sessions 2020 in England and Curta-Se Festival Iberoamericano de Cinema de Sergipe 2021 in Brazil. Similarly, Leo in Space was part of the official selection of the Ojo Móvil Fest 2020 Festival, an international film and video festival made with mobile devices.

To address the object of study, an interpretative methodology has been chosen, reviewing indexed research articles related to the topic to be investigated, since this methodology serves to generate evidence of its scientific status [22]. Likewise, several communication journals were reviewed, among them, Comunicación y Sociedad and Latinoamericana de Comunicación. For this study, about 40 articles on the concepts that helped me to strengthen the research were reviewed.

Finally, it is desired to carry out qualitative research, since interviews are a very useful technical instrument in research and serve to collect incommensurable data [23]. Likewise, in this research I do not analyze the reception of the series by the public, but the production and exhibition processes, for example, the ways of producing, how this type of format is exhibited or the strategies that are used. This type of information cannot be measured quantitatively because it is data handled by the series team, and it is essential to carry out qualitative research to obtain it. For this reason, 20 semi-structured interviews were conducted with experts from the audiovisual and marketing sectors, as well as with people involved in the web series Miitiin, Brigada de Monstruos and Leo en el Espacio, in order to have concise information and people with experience in the research topic. Likewise, when interviewing the team of the three web series, there will be a clearer picture of how the process of elaboration and exhibition of the Peruvian web series has been. It is necessary to emphasize that we have the consent of all those interviewed to use the information

Table 34.1 List of interviewees for data collection

Interviewees	Professional profile
José Balado	Director de DOCUPERU
Omar Vite	Experto en marketing digital
Joaquín Sancho	Gestor de proyectos
Christian Yaya	Productor y animador
Gonzalo Benavente	Director de cine
Daniel Rodríguez	Creador y director de Miitiin
Alberto Castro	Productor de Miitiin
Atilio Quesada	Productor y director de Miitiin
Bruno Rosina	Guionista de Miitiin
Ivan Tahara	Director de arte de Miitiin
Julian Amaru	Director de fotografía de Miitiin
Giovanni Rossi	Director de sonido de Miitiin
Franco Parodi	Editor de Miitiin
Marco Castro	Post productor de Miitiin
Valeria Calmet	Asistente de producción y arte de Miitiin
Rogger Vergara	Creador de Brigada de Monstruos
Giovanni Arce	Cocreador y guionista de Brigada de Monstruos
Nataly Vergara	Productora y directora de arte de Brigada de Monstruos
Renato Medina	Creador de Leo en el Espacio
Jorge Cárdenas	Editor y post productor de Leo en el Espacio

for purposes related to the investigation. The following Table 34.1 details the people interviewed and their respective positions.

It should be noted that for this article a specialized expert has validated the instruments and to certify this validation I have a document signed by him.

34.3 Results

As time goes by, forms of entertainment change and with them the way they are produced and consumed. Web series are a clear example, since it is a format that is being produced more frequently not only by independent creators but also by production companies in our country. Despite being a format that is becoming more relevant in these times, the interviewees have innovated with new ways of telling stories and have challenged a global pandemic to continue producing content, since they visualize great potential in web series.

34.3.1 Digital Platforms and Universe of YouTube

As findings of the digital platforms category, Christian Yaya mentions that digital platforms aim to accelerate the processes so that the content exhibited takes off faster and has a larger audience. Thus, digital platforms today are a powerful medium for the audiovisual market. Furthermore, Gonzalo Benavente says that YouTube has become more prominent in the audiovisual world and today is serving as a window to promote an audiovisual product, this is due to the very mechanisms of the platform and its form of consumption. However, Franco Parodi considers that YouTube is not a platform for web series or for a medium/high quality audiovisual product, since these products are drowned by contents that are less demanding at production level and are daily.

On the other hand, Omar Vite mentions that all content should not be deposited on the Internet, but everything has to be interconnected (ecosystem) between everything to be a phenomenon. Nevertheless, Marco Castro assures that if you are not in digital media or social media, your product does not exist; that is why artists disseminate their content in these media and thus remain active. Likewise, Daniel Rodriguez mentions that in each digital platform audiences are different and behave differently; hence, if you want your content to reach more audience you must spread it through all possible platforms and at the same time.

Focusing on Miitiin, Marco Castro assures that in Peru you have to push people everywhere to reach your content; consequently, the main medium was YouTube because it has the most people in the world and it is more accessible to everyone. Then, it was decided to use Facebook and Instagram as support media. However, Franco Parodi mentions that the premiere of Miitiin coincided with the launch of IGTV, and they decided to take advantage of this phenomenon to reach the audience. Thanks to this decision and curiously, Miitiin has had greater receptivity on Instagram, has had greater views and connectivity with the public.

Focusing on Brigada de Monstruos, Rogger Vergara mentions that YouTube is not a platform where users watch fiction. As a result, they decided that the main channel for Brigada de Monstruos was Facebook, since that is where their target audience was focused, which were the fans of role-playing games. The fans of these games were key for the series to have a good reception, since they shared, commented, and made the series known to the entire community. Giovanni Arce mentions that while it is true that the audience was on Facebook, new users who liked the series and shared it were also arriving via YouTube and Instagram.

Focusing on Leo en el Espacio, Renato Medina mentions that YouTube today has become more important to watch quality content and is a great alternative for independent creators. He also states that he was looking for the series to be in the most massive place in order to reach more users. Nonetheless, he also used Facebook to bounce the chapters and Instagram to upload content that would invite people to watch the series.

34.3.2 Production, Exhibition, and Promotion of Peruvian Web Series

As findings of the production, exhibition, and promotion category, José Balado mentions that web series are fundamentally different from traditional series, since they are a new format, a format that is exclusively designed for a digital platform. Likewise, Joaquín Sancho states that web series have their own particularities in terms of length, in addition to the fact that the stories are easier to digest. Atilio Quesada affirms that web series are approached from a different perspective and that is why you can take other licenses. However, Gonzalo Benavente affirms that web series and traditional series are still the same in terms of complexity, since both have to tell a story and tell it well. Nevertheless, web series, being designed in a web format, do not have limitations or parameters to follow, but rather the time and structure are adapted to the story.

Bruno Rosina claims that the difference of a web series is more on the technical side, since the script is written as for any audiovisual product: movie, series, and so on. He also mentions that web series and comedy are closely related, since it is a genre that lends itself to this format. You are able to make a person laugh in 20 s, but making a person laugh throughout a whole film is more complicated. Giovanni Rossi asserts that many times there are series that have been made for the web and are the gateway for them to be sent to other platforms or simply to be sold. He also affirms that the gap between web series and traditional series is getting shorter and shorter, since all the work of the technical team is the same and the demand is the self-same.

Regarding the exhibition, Omar Vite confirms that each project has its own essence and it is necessary to look for what it is about in order to propose a good exhibition strategy. Similarly, Joaquín Sancho mentions that the issue of distribution and exhibition is a decision that is made before producing.

On the other hand, Gonzalo Benavente declares that the level of complexity of an exhibition strategy is developed according to the variables that one incorporates, and the more complex the better. The variables that can be taken into account are marketing, press, target, etc. He also mentions that each product must have a particular exhibition strategy; if a product is similar to the previous one, it is a good reference point, but in reality each story connects with a different public. Although, Ivan Tahara affirms that it depends on what you need, there will be a platform for that and you have to know well what your product needs to decide on which platform to exhibit it. In the following Fig. 34.2, we can observe in a general way what are the distribution and exhibition processes that must be followed for any audiovisual project (see Fig. 34.2).

Focusing on Miitiin, Julian Amaru considers that in this production he did not feel as a director of photography, since the conditions made the web series to be shot with the equipment that the actors themselves had. Given that, he was more as a technical advisor prior to the shootings in which he was not present. With this fact that Julian mentions, we might infer from what Giovanni Rossi said above regarding the differences between web series and traditional series, specifically with

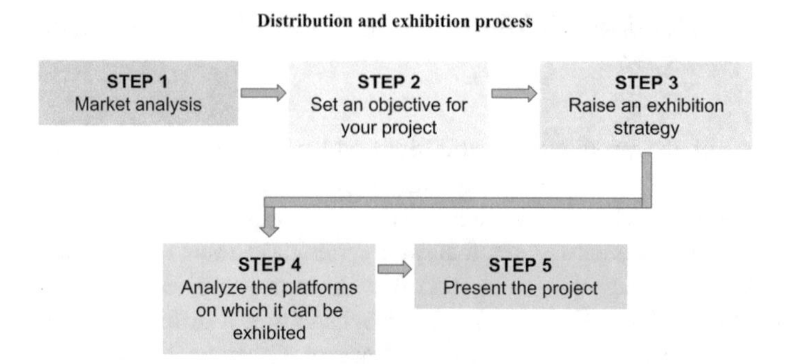

Fig. 34.2 Step by step of the distribution and exhibition process. Own elaboration

the statement "All the work of the technical team is the same and the demand is the same." With Julian on Miitiin, we certainly can affirm that in this case the experience of a cinematographer was different.

Meanwhile, Ivan Tahara mentions that the construction of the characters regarding the art direction was a long process of many video calls, explorations of the actors' homes and the spaces that were going to be used. Ivan explains that although it is true that it has been a new way of working in his area, the art direction process for a web series or a traditional series is the same.

As for Brigada de Monstruos, Rogger Vergara states that it was a great challenge to make the series with the miniature dolls, since he not only had the role of director, but he was in charge of the whole process of making the series. In addition, he indicates that web series have their own particularities but that they undoubtedly take and have as a reference some characteristics of traditional series, such as, for instance, the structure to tell a story. On the other hand, Nataly Vergara points out that the art direction was complex because they had few resources and they had to figure it out. The art direction was based on actions, in other words, on what was happening at the time of the scene. Nataly also adds that because the camera lens was set so close to the stage, everything was seen very large, and thus, they had to be aware of every detail of the framing.

Centering on Leo en el Espacio, Renato Medina mentions that the concept of the series was always thought in the character's personality, in his world. This series was filmed with a single cell phone and everything was handmade by himself, the production was not difficult but he managed with all the objects he had at home to simulate the spacecraft. Undoubtedly, this project was a challenge for him, not only because it was the first time he did a project independently but also because it challenged his creativity and finally the project obtained the expected results.

Regarding the exhibition and promotion, Marco Castro mentions that, in the rush of wanting to be the first to come out, the processes needed to be accelerated and in the exhibition these processes were not addressed as they should have been. Furthermore, he affirms that Miitiin would have been more successful if they had had sponsorship.

Giovanni Rossi mentions that they had "Aj Zombies" as a base, since this series was uploaded to YouTube and was successful. They participated in a contest in Miami and won a prize, thanks to which they were contacted by Canal + France to shoot the film. Bearing this in mind, Alberto Castro mentions that with Miitiin they never thought of winning based on views, the idea was to make a series that would be successful enough to be sold to a channel or platform, something similar to what happened with "Aj Zombies." Bruno Rosina mentions that months later they were approached by a Peruvian audiovisual platform that was interested in the product and that they wanted it for their platform but what they were offering was low. It is understood that if they want to buy it, it is because the web series is good and they are going to make a profit; therefore, he considered that what they were offering was not reasonable.

On the other hand, Daniel Rodriguez mentions that not having a big budget for the series and being a small team, there was not a good management in the screening. He is not a person who knows about the subject and had to take charge of this process with part of the team. Consequently, he believes that in order to analyze the exhibition strategies to be implemented, it is essential to have a person who knows the subject and is 100% in charge of this process.

With regard to Brigada de Monstruos, Rogger Vergara mentions that different strategies were proposed to see where the audience was, since each social network has a different audience. In this case, they did a double strategy: The first was to try to find the target audience in order to make them aware of the series and the second was to promote the voice performances of all the actors who were in the series and the fact that it was the first fantasy web series made with miniatures in Peru. Furthermore, Gionanni Arce mentions that with the series they discovered a niche that had always been there and they didn't know it, many role-playing communities started to watch the series and members of these communities even had companies related to this universe. This is how these two sponsors they had in the series approached them to support them in the way they could. On the other hand, Nataly Vergara explains that one of their strategies was also to make the series transmedia, hence, on the website of La Taberna they uploaded comics where they explained the whole world of dungeons, and this served as support for people who do not necessarily know the world of role-playing games.

Referring to Leo en el Espacio, Renato Medina mentions that for the promotion in the networks they used pieces related to the chapters, and they always invited you not to miss the next chapter. The digital strategy was always aimed at getting the viewers to watch the series. What we did work with a press officer we hired was a campaign that allowed the series to have a press dossier and through this, the series was able to appear in various media such as El Peruano, El Comercio, La República, RPP, allowing it to have a better reception. Moreover, Jorge Cárdenas points out that together with Renato Medina they created Leo's universe so that it would also be visually reflected in the graphic pieces. Jorge affirms that the network strategy was created focusing on the character's personality and the situations he lived in each chapter. He also adds that in the promotion there is a zero point of research, data collection, hypothesis formulation, and preparation of your digital strategy document that is fundamental but that many do not keep in mind.

Finally, most of the interviewees responded to certain concepts that are linked to the profitability of an audiovisual product: budget, distribution, business model, and financing. These concepts are a concern for the audiovisual market, since many times the productions that are made end up not being profitable and are rather a waste of money. In this case, several factors may influence: the product of low-quality, mis-execution of processes, and so forth.

34.4 Discussion and Conclusion

In response to the first objective, how is the production of series content within digital platforms? Franco Parodi mentions that web series and traditional series start from the same thing, if you do not have a solid script nothing will work. Additionally, Rossi also mentions that the differences between web series and traditional series are becoming shorter and shorter for several reasons, among them, the technical team, the requirements, the way they are written, among others. Nonetheless, this differs from author Hernandez who claims that web series are a different type of production than traditional series. As stated by the scriptwriter of Miitiin, Bruno Rosina, in the case of this web series the script was written as for any audiovisual product, it has a beginning, a conflict and a resolution. According to authors Montoya and Garcia, one of the main characteristics that distinguish web series from traditional series is that there is no limit to the number of chapters a web series may have per season.

Based on all these arguments, it can be seen that web series and traditional series have different characteristics that are specific to the format, such as the length; however, they share characteristics with each other, as Julian Amaru mentioned, the web has influenced the traditional series that are now seen on Netflix, for instance, the way they are exhibited. Moreover, another similar characteristic is that all audio-visual products tell a story and for that story to be well told, there must be a good script, a clear objective and, above all, a clear structure.

On the other hand, Cappello mentions that, thanks to the short runtime, it is possible for users to be able to watch these series more flexibly and more frequently. Similarly, Bruno Rosina and Alberto Castro mention that the particularity of web series is their short duration, and that this has made them so popular today, as consumers do not spend more than 20 min watching a video on YouTube. Author Cappello also explains that web series become a dynamic playground where new narrative forms are tested. Rogger Vergara and Renato Medina also mention that these new web formats give you much more freedom to create and narrate new stories. Notwithstanding, Gonzalo differs from this argument and mentions that the formats and narratives in a web series are exhausted if they always follow the same logic, as in the case of Miitiin, if they follow the logic of the zoom meetings, the time will probably come when the series will no longer be able to continue, and they will have to opt for another more innovative proposal.

As the author Rufete states, forms of entertainment are constantly evolving to adapt to this new generation of consumers. In the case of the audiovisual world, the transformation involving various factors has enabled the development of new series and new formats and created exclusively for consumption on the network. Likewise, as Valeria Calmet mentions, web series have particularities such as the audience and the way in which they can be consumed. In addition, Giovanni Arce states that the pandemic allowed the acceleration of this format and that it depends on the type of story you want to tell to find your audience. Regardless, web series fulfill the same objective of any audiovisual product: entertainment. Under these arguments, web series become a new form of entertainment for a specific audience that knows how to adapt to these new ways of consuming an audiovisual product.

An important factor mentioned by the authors Molpeceres and Rodríguez is that the narratives make use of different digital platforms, so it is very important to think from the beginning for which platform the audiovisual product is aimed. However, with Miitiin it was different, since the team produced the series and uploaded it to YouTube because it is the most watched platform in our country. In this case, they did not care about the nature of the product, but rather because they wanted to be the first to be released, they opted for that platform. It should be noted that the option chosen does not deviate from what the authors pointed out, since the format and narrative of Miitiin lends itself to this platform.

Responding to the second objective, what is the strategy for exhibition and promotion of web series within digital platforms, the authors Montemayor and Ortiz indicate that today there are several channels through which you may display your product, but that there are channels with greater relevance, notoriety, and market presence, such as YouTube and Vimeo. Bruno Rosina states that in order to know which platform is relevant to exhibit the product it is necessary to analyze the content; although in Peru local platforms are not being used and those that are easy to access, such as YouTube, are usually used. Also, the strategy used in the exhibition depends on what is the goal you seek to achieve with that product, for example, if what you wish is to generate content to have a large coverage and generate income, it is best to use platforms that are easily accessible, such as YouTube, TikTok, among others.

On the other hand, if you expect your project to be sold to a recognized digital platform, the most important thing is to analyze all the factors that this process includes, highlighting that it is of utmost importance to have someone who knows the subject, to finally present it to a digital platform such as Movistar Play, Claro Play, etc. There is also another option that is widely used in the Peruvian audiovisual market, which is to exhibit an audiovisual production on platforms such as YouTube, in order to achieve a great reach and then be sold to a recognized digital platform in the market. All options are viable, though as I mentioned above, the one you choose will depend solely on what is the goal you want to achieve with that audiovisual product.

According to the Miitiin team, the strategy that was implemented was not the right one. Firstly, because there were several factors at play that affected the decisions that were made, such as, the rush of being the first to come out with a product such as this. Secondly, the small budget and thirdly, the lack of a clear objective as to where the company wanted to go with this product. Atilio Quesada, producer of the series, mentioned that they had a great idea and a great team, however, if they had not been in such a hurry to be the first to come out, probably the decision to release it on YouTube would not have been in their plans. Nonetheless, Alberto Castro, also producer of the series, commented that with Miitiin they never thought of making money based on views, since the idea was to make a series that would be successful enough to be sold to a channel or a bigger platform.

According to the Brigada de Monstruos team, they believe that the strategy was the right one. The first thing they did was to look for role-playing game fans, since the series is 100% inspired by that niche. When they found it, they found out that their audience was on Facebook and that was the platform where they should share the series and all the content, that's how they decided to publish the series on the platform and in a short time it managed to have great reach. The followers of Brigada de Monstruos were so excited about this project that they did not stop supporting the series and it managed to get sponsors.

Meanwhile, Leo en el Espacio sought to implement a strategy with diffusion on all platforms and also through the creator's contacts. Because of its rebound not only on the platforms but also in the media, it was a semi-finalist in the Ojo Movil Fest International Festival, which is a festival created for projects that have been made with cell phones.

We can point out that web series are a growing universe in the audiovisual sector, since in economic terms they are not yet within the metrics of the Peruvian audiovisual market. Additionally, web series are characterized for being a light format to consume, since as I mentioned before, the user can watch the content wherever, whenever, and how he/she desires. On the other hand, at the production level, audiovisual products start from the same point of view. In the specific case of web series and traditional series, web series have their own features of the format; notwithstanding, they still contain characteristics of the series we are accustomed to watch. Web series are becoming a new form of entertainment and thus, more and more are being produced in the Peruvian market.

The commitment and perseverance of each member of the Miitiin team ensured that the series has a structured flow. It is very likely that in the future the team of the series will bear in mind everything that happened with Miitiin, to avoid making the same mistakes in the next projects and to bring them a lot of personal and professional satisfaction.

Regarding the exhibition, we can affirm that this stage is fundamental for the success of the audiovisual product. It is essential to analyze all the factors involved in this process in order to make a good decision. Without a good exhibition strategy your project does not exist, so it is important to analyze the market, innovate, and have mapped out the risks that are taken with the decisions that are chosen.

Brigada de Monstruos and Leo en el Espacio are projects that were created in the pandemic while seeking to be a relief in the face of the crisis we were going through. Also, they knew how to take advantage of digital platforms in relation to the needs of users, and this allowed the series to have great reach and can be seen internationally.

Finally, the limitations that have been encountered in the realization of this article have been the conditioning of the articles related to the research topic and especially of the web series investigated, since being a Peruvian product the environment becomes more complicated. Moreover, the accessibility to the interview-two was a determining factor for this research. The recommendations for content creators and especially for those who explore new formats such as the web series are to be innovative so that their project has greater visibility. In addition, it is important to plan a good advertising and promotion strategy so that the project is able to scale quickly and achieve its objective.

To the Research Direction of the Universidad Peruana de Ciencias Aplicadas for the support provided for the realization of this research article

References

1. Arboleda, A.: La serie web no es el futuro, es el presente. Revista Universidad EAFIT **52**(169), 122–125 (2017)
2. Arango, M.: Análisis de la influencia de internet en las formas de consumir televisión en los jóvenes del grado once de la institución educativa santo Tomás en la ciudad de Cali en el año 2019. Doctoral dissertation, Universidad Santiago de Cali (2019)
3. Urrea, J.E.: Lenguaje y contenido audiovisual de los programas en Internet frente a los programas de televisión convencional. Revista Lasallista de Investigación **11**(2), 36–42 (2014)
4. López Delgado, D.: Estudio de las plataformas de streaming. Trabajo Fin de Grado Inédito, Universidad de Sevilla, Sevilla (2018)
5. Díez de Paz, L.: El streaming como forma de entretenimiento= Streaming as an entertainment source. Trabajo Fin de Grado, Universidad de León, España (2021)
6. Rufete, E.: Análisis del uso de estrategias de crecimiento en Netflix. Tesis de fin de grado, Universidad Miguel Hernández, España (2016)
7. Rodríguez, J.: Instagram como una plataforma de distribución disruptiva para el trabajo audiovisual de estudiantes de pregrado, el caso de Form Factor. Tesis de fin de grado, Universidad de La Sabana, Colombia (2020)
8. Sansó, M.: La producción audiovisual y los nuevos medios: configuraciones narrativas, plataformas y circuitos de exhibición de series web del NEA (2015–2019). Tesis de fn de grado, Universidad Nacional del Nordeste, Argentina (2021)
9. Cappello Flores, G.: Prácticas narrativas en las ficciones seriadas para la web. Una mirada a la producción de cuatro países en Sudamérica. Comunicación y Sociedad, e7122 (2019)
10. Prósper, J., Ramón-Fernández, F.: Propuesta de modelo de evaluación para webseries. La Colmena **105**, 65–76 (2020)
11. Heredia, V.: Revolución Netflix: desafíos para la industria audiovisual. Chasqui: Revista Latinoamericana de Comunicación (135), 275–295 (2016)
12. Del Pino, C., Aguado, E.: Internet, Televisión y Convergencia: nuevas pantallas y plataformas de contenido audiovisual en la era digital. El caso del mercado audiovisual online en España. Observatorio (OBS*) 6(4) (2012)
13. Montoya Bermúdez, D., García Gómez, H.: Estructuras narrativas en relatos cortos y serializados para la web**.** anagramas rumbos sentidos comun. [online]. **15**(29), 103–118 (2016)
14. Gonzáles, M.: YouTube como fenómeno social y su modo de consumo del audiovisual. Trabajo Fin De Grado, Universidad de Sevilla, Sevilla (2020)
15. Asociación de Agencia de medios: Las tendencias en las redes sociales 2019 por Kantar Ibope Media (2020)

16. Sayán Casquino, Y.: Producción, distribución y exhibición del cine desde una nueva mirada: Obra Digit. (12), 27–51 (2017)
17. Montemayor, F.J., Ortiz, M.Á.: El vídeo como soporte en la narrativa digital del Branded Content y los productos audiovisuales en las plataformas online. Poliantea **12**(22), 85–116 (2017)
18. Ancín, J.M.: El plan De marketing digital En La Práctica. ESIC, Madrid (2017)
19. Chumacero, Y.: Marketing de contenidos audiovisuales: Análisis de las herramientas aplicables a la industria cinematográfica peruana. Tesis para optar el título, Universidad de Piura, Perú (2019)
20. Fernández-Gómez, E., Martín-Quevedo, J.: La estrategia de engagement de Netflix España en Twitter. El profesional de la información **27**(6), 1292–1302 (2018)
21. Navío Navarro, M.: Contenidos eficientes en redes sociales: la promoción de series de Netflix. index.Comunicación **11**(1), 239–270 (2021)
22. Barco, B., Carrasco, A.: Explicaciones causales en la investigación cualitativa: elección escolar en Chile. magis, Revista Internacional de Investigación en Educación **11**(22), 113–124 (2018)
23. Torruco-García, U., Varela-Ruiz, M., Martínez-Hernández, M., Díaz-Bravo, L.: La entrevista, recurso flexible y dinámico. Investigación en Educación Médica **2**(7), 162–167 (2013)

Chapter 35
Management Model and Capture of Benefits Integrated into the Practice of Project Management

André Almeida, Carolina Santos, Henrique Mamede, Pedro Malta, and Vitor Santos

Abstract An attempt has been made to address the difficulty of identifying and measuring the benefits derived from investment projects and capturing capital gains for an organization, focusing on developing and implementing a management model and realizing benefits for a leading company in its activity sector. Thus, the objective is to understand how it is possible to achieve the expected benefits of an investment project: A model characterized as generalist was developed (applied to all areas of the company), with the objective of optimizing the realization of benefits, measuring them and thus create value for the organization. Among the methods used, we highlight, in a first phase, the research of some existing Frameworks, which later enabled the development of a proposed framework, validated internally using the existing Business Intelligence platform. Subsequently, based on a satisfaction questionnaire about the framework proposed to users, data related to its development and implementation were collected, with the aim of understanding its acceptance among the users and employees of the company. With the data from this questionnaire, an artifact was developed: a PowerBI dashboard that reflects the benefits identified and captured. In summary, the artifact made it possible to identify, measure, and achieve the benefits generated by the project in question, but also to motivate its use in other existing investment projects, by adapting it to each of the other ones.

A. Almeida · P. Malta (✉) · V. Santos
Nova IMS, Lisboa, Portugal
e-mail: pmalta@novaims.unl.pt

H. Mamede
Universidade Aberta, Lisboa, Portugal

C. Santos
Nova ENSP, Lisboa, Portugal

J. L. Reis et al. (eds.), *Marketing and Smart Technologies*, Smart Innovation, Systems and Technologies 337, https://doi.org/10.1007/978-981-19-9099-1_35

35.1 Introduction

Innovation and sustainable growth are two crucial concepts for the global competitiveness of organizations. In this sense, investing in investment projects is increasingly most vital within each organization, which can generate a set of benefits useful to daily activity.

Santos [1] refers to the management of investment projects as essential to the competitiveness of ever-changing organizations. In this context, it is important to consider the capture and realization of benefits as essential concepts in motivating the implementation of investment projects [2], that is, identifying, capturing, monitoring, measuring, and presenting benefits are steps to consider in the implementation of a project.

The support of these steps by the management of investment projects lacks the use of a Framework [1] capable of measuring and realizing the identified benefits. In addition to being able to assign a value, financial or otherwise, to each objective, it is essential to understand and analyze the impact, even if intangible, for example, on productivity or customer loyalty [3].

With a focus on understanding how to capture the benefits of an investment project, how all its management is carried out and contribute to the existing model in the company, this document exposes the work carried out through the choice of a Business Case of an organization of the Energy sector.

The project aimed to answer the following questions:

- To what extent is project management important to the organization?
- How is the entire management process for the benefits associated with investment projects?
- What benefits did the Business Case bring to the organization?
- How it will be possible to measure these same benefits?
- How will the results/benefits of the Business Case and other projects of the same organization be presented and disseminated?

In this context, where organizations need constant investment projects to update [4], the management of investment projects and benefits are two intrinsically related assumptions [5]. Thus, even considering that a project is characterized as being specific and finite, short or long term, larger or smaller, it is always associated with a stipulated budget [5, 6], and its ultimate objective is to put the organization's strategy into practice and transfer its vision to the real plan [3]. Even the synchronous use of investment project management and inherently of benefits management is more likely to promote the success desired by organizations [7].

35.2 Research Background

At the base of this work is the concept of Investment Project Management. We are talking about "...management of techniques that enable companies to associate their strategy with the successful results of a project." [8] which must be evaluated from different perspectives: cost, time, quality, scope, resource and activity, and also through successful measurement models [9].

This assessment, that is, basically the analysis of its viability, is supported by the correlation between costs and income, referenced in plans, such as financing, investment or exploration. In this sense, analyzing through technical–economic and economic-financial studies is essential to be able to decide whether to invest in a particular Investment Project.

Moreover, according to the Introduction, it is equally crucial to look at benefits management, which "... constitutes an important part of project management," as the latter is "... the means through which organizations/companies achieve their goals." [8].

More focused, he wants the notion of benefit—"... a result whose nature and value is considered an advantage for the organization." [10]-, both their classification are essential, given the subjectivity in question (tangibility and intangibility) and measurement because "... what cannot be measured cannot be managed." [5]. Classifying the benefits is even identifying each one as tangible, intangible, financial, or non-financial [6].

There are several models to consider supporting benefits management: Ward's Cran-field Process Model [8]; Thorp's Benefits Realization Approach [11]; the Active Benefits Realization [12]; or even Kaplan and Norton's Balanced Scorecard [13]. All of them with specific focuses, yet promoters of the knowledge of managing benefits through the act of executing each one of them [14], useful in conflict management [4], crucial in the pre-definition of benefits [5], important in assessing the success of implementation [15], or even in identifying the organizational value of each benefit [16].

The implementation of investment projects and the inherent management of benefits is based on a Business Case of the organization, which must be classified in different perspectives: financial, management, commercial, and strategic [17]. The objective is to identify investment reasons, benefits and costs, terms to consider and associated risks, in pursuit of the feasibility of the investment project [18], in order to expose the results as a form of table that summarizes information providing the indicators in question, with its final output being a data analysis [19].

35.3 The Solution

The approach we followed includes the creation of a conceptual framework and, based on it, an implementation of a tool for the proof-of-concept. This has been realized as project steps, with the creation of the framework being step 1 and the tool development as step 2.

In this chapter, we describe the proposed framework, the project developed using the framework, and an evaluation of the result.

35.3.1 Proposed Framework—Step 1

A conceptual framework relates concepts, empirical investigations, and relevant theories to advance and systematize knowledge about related concepts or issues.

The proposed framework was developed based on the study of several articles, as already mentioned, and its structure is presented in Table 35.1. It consists of a phased approach to the problem.

The first phase is called pre-project implementation: It consists of the entire definition of the vision and business objectives that arise from that same vision. Project managers also define all KPI's. Then, still in the pre-implementation phase of the project, the Business Case is created and developed, which will be essential to assess the reliability of the methodological framework. The Business Cases provide information regarding benefits, costs and investment details allocated to projects, risk analysis, among many other aspects.

The second phase is related to the implementation of the project. At this stage the assumptions will be validated, and updated if necessary, according to the framework. It will be during the implementation of the project that all initiatives will be monitored and changed in order to achieve all pre-defined benefits.

Finally, in the post-implementation phase of the project, the main focus is managing and monitoring the benefits arising from the project, as well as analyzing and reviewing the project results.

The last point relates with the reporting and dissemination of results.

In summary, this framework applies to the case study that was chosen, as will be seen in the following section, but due to its structure it can be applied to any type of investment project, from any area within an organization, being for generic application.

35.3.2 Tool Development—Step 2

The developed tool consists on a business intelligence application; under an Oracle platform already implemented in the organization for two years. BI-SCADA included developing a Data Warehouse (DW) and ETL Processes, which group data from different sources.

In order to provide reliable information for analysis, management and decision-making support within the organization, a Framework was created (Table 35.1) to determine the scope of all defined benefits before the implementation of an investment

Table 35.1 Proposed framework

		Stages	Highlights
Pre-project implementation	1	Set vision and objectives	I.I Define vision and derive business objectives from the vision
			1.2 Define KPI's and CSF's
	2	Business case development	2.1. Identify benefits
			2.2. Estimate costs/Investment details
			2 3. Risk assessment
			2.4. Assessment the feseability of the project
	3	Benefits identification	3.1. Define which benefits we want to achieve and create essential changes to reach these objectives
			3.2. Link benefits to the business objectives, and ensure that they are realistic, specific, measurable, actual and linked to the strategic outcomes
			3.3. Identification of metrics and targets
			3.4. This stage include benefits and changes description associated to the project
			3.5. This stage include the date of realization of benefits as well as the used resources
Project implementation	4	Benefits management	4.1. Validate the approaches, and if not valid, update the benefits (turn to stage n 3)
			4.2. All management and follow-up of the changes/initiatives that be done internally, in order to achieve all the benefits
After project implementation	5	Benefits realization	5.l. Monitorization and benefits measurement
			5.2. Review and analyze of the results as well as identification of measures to adjust to targets

(continued)

Table 35.1 (continued)

		Stages	Highlights
	6	Benefits report	6.1. Reporting and dissemination of results

project, ensuring that the management of benefits is processed in the best way: there is a first step, pre-implementation of the project, which defines the vision and business objectives that arise from this same vision and the managers project also define all KPI's; a second step, which consisted of project implementation, in which the assumptions were validated and updated; and finally, post-implementation of the project, focusing on the management and monitoring of benefits as well as the analysis and review of project results, including the reporting and dissemination of results.

A satisfaction questionnaire was applied to assess the effectiveness of the realization model and capture benefits in the business as mentioned above intelligence application. The questionnaire was applied in two phases: at the beginning of the use of the application (January 2021) and in the post-implementation phase of the project (June 2021). The main purpose was to understand whether the pre-established benefits for this project were achieved or not, and for those that were not, to identify those reasons.

The satisfaction questionnaire also has the following objectives:

- understand the evolution of this application since the first completion of the questionnaire;
- understand if its users are satisfied with the application and if they are well using it;
- and understand if there were problems adapting to the platform by end users, to promote measures in the future that can solve this type of problem.

The satisfaction questionnaire has 44 questions, divided into eight distinct groups to achieve the defined goals: user profile, user experience, BI system functionalities, performance and availability, integrity and security of the data extracted from the program, application and evolutionary maintenance, user training for a better use of the tool and, finally, the global appreciation.

Finally, a dashboard was developed to present the results obtained, both from the questionnaire and from other investment projects. This dashboard focuses on the standardization and centralization of all information (costs, benefits, investment, benefits catalog, status of investment projects) allocated to investment projects in all areas of the company: This artifact allows project managers to use all the information associated with investment projects, eliminating their dispersion across the company's various media.

35.4 Evaluation

For the evaluation purpose of the proposed artifact, a questionnaire has been applied. In a universe of 246 users, a response rate of approximately 38.6% was obtained: 95 users of the platform answered the questionnaire. These users are divided into these 10 departments, most of them aged between 31 and 40 years old (34.74%), or even 30 years old or less (30.53%).

For the rest of the questionnaire, the organization defined a set of targets to be addressed by each question:

- user experience—5.3;
- functionalities—4.5;
- performance and availability—4.3;
- integrity and security—unassigned;
- application and evolutionary maintenance—unassigned;
- user training—5.5;
- overall appreciation—5.5.

Only two groups reached the values defined by the target and only in the second pass of the questionnaire: Performance and Availability (4.47) and Training (5.68). The remaining groups fell short on the defined targets. Below is a summary Table 35.2.

From the analysis of these averages by group, we got evidence that contributed to the results, as presented below:

- user experience: SCADA is unappealing, not very user-friendly and graphs, reports and tables could have a better design;
- functionalities: 20 of the 95 respondents chose the option "I completely disagree" in the question about the use of the help menu frequently, lowering the averages of this group (2.18 in the 1st time and 2.37 in the second);

Table 35.2 Summary averages of responses in each group

Group	Average questionnaire 1st time	Average questionnaire 2nd time
User experience	4,31	4,71
Functionalities	3,81	4,09
Performance and availability	4,49	4,87
Integrity and security	5,69	5,86
Maintenance	4,05	4,59
Formation	5,06	5,68
Global appreciation	4,58	4,97

- performance and availability: The question about the speed of production of results by the system had an interesting rise from 3.94 (1st time of the questionnaire) to 4.47 (2nd time), certainly contributing to the reach of target with the 2nd pass of the questionnaire;
- integrity and security: Although without a defined target, it is possible to verify high averages in both sections of the questionnaire (5.69 and 5.86), reflecting the satisfaction of users in this area;
- maintenance: Also without a defined target, it is possible to verify lower values (4.05 and 4.59) than those of the previous topic, which may indicate lower user satisfaction in this topic;
- training: In this context, the average of 5.68 in the question about the usefulness of available training is highlighted, which is very close to the top value of the scale (between 1 and 7);
- global appreciation: The averages of this scope are below the target, showing that the application in general will have to be improved—the answers to the question about the "user-friendly" characteristic (4.3) stand out.

35.4.1 Results Discussion

In a first approach, the fact that the application is considered not very "user-friendly," compared to the target defined for the user experience group. It suggests some negative impact on productivity: Users spend more time using the application than expected. One of the reasons for this evidence is the lack of speed in finding the information necessary for the execution of activities: It increases time wastage and yes it contributes to reducing productivity rates.

From the analysis, it is important to highlight the integrity and security, as well as the training available, of this platform. Integrity and security really support the application, as there is trust in using the software. The group of training questions turned out to be the one that had averages closest to the target: To improve this scope, it will have to be felt, namely the monitoring of users and the materialization of platforms that allow to clarify doubts and provide clarifications on the app.

On the other hand, the averages of applicational and evolutionary maintenance were among the lowest compared to the values obtained in other areas. This is a topic to bet on, in the sense that it will certainly have an impact on productivity, as already mentioned in the scope of application use.

Following these ideas, the development of the dashboard allowed the dissemination of the results of the questionnaire made and contributed to internal communication, also an objective of the organization's management. This was the most useful result for the organization, since it includes a list of all projects carried out internally, with an indication of investment values (initial investment, financial benefits, fixed costs and variable costs).

The dashboard is supported in four main parts:

- a "Benefits Tree" that consists of exposing the benefits in the form of a tree, being possible to consult each benefit by itself regardless of the level and ascending or descending dependence it has, associated with a set of three filters—organization area, area products and associated metrics;
- a "Project/Product Status," depending on the benefit target, to consult the implementation status of each project (pre, during or post-implementation), with associated filters regarding the forecast of costs and income, in addition to the benefits actually generated;
- "Captured Values" area supported by a balanced scorecard perspective, to identify which organizational unit is generating more revenue, and which one needs some type of development and investment, supported by an improved cost structure, giving as yes, a better financial view of the impacts of each project;
- • "Continuous improvement initiatives in progress" area, which is a list of initiatives to improve ongoing investment projects (includes expected start and end dates), with reference to the internal person in charge, being possible to filter by area and by project.

This artifact promotes the simplification and streamlining of the entire process of filling in the "Business Case sheets" by project managers, thus enabling them to allocate part of their time to other activities within the company. Likewise, it centralizes and standardizes all information associated with investment projects, as well as eliminating the dispersion of information about them, due to their more interactive use.

35.5 Conclusion

As mentioned at the beginning of the paper, "the difficulty of identifying and measuring the benefits derived from investment projects and capture of capital gains for the organization" was the motto for this exhibition, which began with the identification of the issues inherent to the concept of "benefit."

After framing and theoretical delimitation, by bibliographic review of concepts and frameworks in this area, a Business Case was used to explore the benefits of an organization's investment projects.

Regarding the necessary steps to methodologically support the study carried out, it was possible to identify benefits inherent to the chosen Business Case, promoting the identification of information, namely financial, on the evolution of the investment project in question.

With the construction of the dashboard, a set of data/information was evidenced, which promotes evidence of the evolution of the pre-implementation phase to the post-implementation phase, highlighting a set of pertinent information for the different users. This was undoubtedly the most useful and most transversal result, in the sense that it allowed the adaptation of the layout of this Business Case to others that the organization was also interested in exploring and identifying inherent benefits.

This work was carried out in a very specific context, so it was difficult to sustain its relevance given the almost non-existence of scientific publications in this area. This limitation, however, provides motivation in the dissemination of this type of projects in scientific forums, due to the opportunity to obtain feed-back.

Even so, issues such as the complexity of the organizational database or the use of little-known software to develop the dashboard were also limitations for carrying out the work.

In this sense, it would be very interesting to carry out a broader analysis of this organization, through a comparative study of more projects, both in terms of the benefits of each one, and in terms of the impact of each one on the results, for example financial, of the organization. It is clear to dimension this study for the Energy sector, it could bring useful inputs to the development of strategies by the sector, either by the organizations of the same, or by the tutelage in the regulation for which it is responsible.

References

1. Santos, C.: Fatores de sucesso da gestão de projetos da saúde pública. Doutoral dissertion, Universidade Nova de Lisboa (2019). Retrieved from https://1library.org/document/yd2g3o6q-fatores-de-sucesso-da-gestao-projetos-saude-publica.html
2. Williams, T., Vo, H., Bourne, M., Bourne, P., Cooke-Davies, T., Kirkham, R., Valette, J.: A cross-national comparison of public project benefits management practices-the effectiveness of benefits management frameworks in application. Prod Plan Control Manage Oper (n.d.). Retrieved 26 Nov 2020 from https://doi.org/10.1080/09537287.2019.1668980
3. Capenda, A.: Proposta de Implementação duma Metodologia de Gestão de Projeto de Tecnologia de Informação para o Ministério da Justiça e dos Direitos Humanos de Angola. Trabalho de Projeto de Mestrado. Instituto Superior de Estatística e Gestão de Informação, Universidade Nova de Lisboa, Lisboa (2015)
4. Serra, C.E.M., Kunc, M.: Benefits realization management and its influence on project success and on the execution of business strategies. Int. J. Proj. Manage. **33**(1) (2015)
5. Badewi, A.: The impact of project management (PM) and benefits management (BM) practices on project success: towards developing a project benefits governance framework. Int. J. Proj. Manage. **34**(4), 761–778 (2016)
6. State of New South Wales through Department of Finance, S. and I.: No Title. In: Benefits Realization Management Framework. NSW Govern, p. 6 of 40 (2015)
7. Aladwani, A.M.: An integrated performance model of information systems projects. J. Manag. Inf. Syst. **19**(1), 185–210 (2002)
8. Drabikova, E., Svetlik, J.: Improving management of the company through cranfield process model. MM Sci. J. **2018**(March), 2153–2157 (2018)
9. Mladen Radujkovic, M.S.: Project management success factors. In: Creative Construction Conference (2017)
10. Gomes, J.: Gestão de Benefícios numa Empresa de GeoEngenharia. Master Thesis in Management (2011)
11. Pereira, R.A.B.: Gestão de Benefícios em Programas de Projetos de I & D em Colaboração Universidade-Indústria. Master thesis, Universidade do Minho (2016).

Retrieved from https://repositorium.sdum.uminho.pt/bitstream/1822/47128/1/Rita%20de%20Andrade%20Brites%20Pereira.pdf.pdf.
12. Dejahang, D.F.: Benefits Management and its Applicability (2016). Retrieved from https://www.slideshare.net/DrFereidounDejahang/benefits-management-and-itsapplicability-69111829.
13. Kaplan, R.S.: Conceptual foundations of the balanced scorecard. In: Handbooks of management accounting research, vol. 3, pp. 1253–1269 (2009)
14. Nomakuchi, T., Takahashi, M.: A study about project management for industry-University Cooperation Dilemma. Procedia Comput. Sci. **64**, 47–54 (2015)
15. Rocco, S.T., Plakhotnik, S.M.: Literature reviews, conceptual frameworks, and theoretical frameworks: Terms, functions, and distinctions. Hum. Resour. Dev. Rev. **8**(1), 120–130 (2009)
16. Management, A. for P.: What is a Business Case? (2021). Retrieved from https://www.apm.org.uk/resources/what-is-project-management/what-is-abusiness-case
17. Marques, E.: Business Case: O que é, quando utilizar e como montar um (2017). Retrieved from https://www.gp4us.com.br/business-case-preliminar/
18. Almeida, G.B.: DashBoard para Internet das Coisas com Dados Abertos. Instituto Federal de Educação, Ciência e Tecnologia do Rio Grande do Sul (IFRS) (2013). Retrieved from http://atom.poa.ifrs.edu.br/uploads/r/bibliotecaclovis-vergara-marques4/8/7/a/87ada6afc512a9800b8470540807550d42a18b00df3d777ee90363657df12495/Artigo-Final-Gutierre.pdf
19. Mendonça, R.H., Tavares, P.F.R., Malta, R.D.F.B., de Brito, M.F.: Planejamento Logístico Dashboard Para Apoio À Tomada De Decisão Relacionada a Escolha De Frota—Estudo De Caso, pp 300–311 (2020)

Chapter 36
The Influence of Social Media on Voters' Decision-Making Process in Portugal: A Case Study

Jorge Esparteiro Garcia, Eduardo González Vega, Patrícia Purificação, and Manuel José Fonseca

Abstract Nowadays, social media are inevitably part of people's daily lives. Thus, political communication should also go through digital communication channels, particularly on social media. In such channels, it is important to define a digital marketing and communication strategy to attract new voters and consecutively more votes. As in offline communication channels and also in digital communication, one of the indispensable points in political communication is the candidate's image. This image must show its own style and differentiate the candidate from his opponents. The main objective of this study is to understand the influence of social media on Portuguese voters' decision-making process. Throughout the study, different research questions were also analyzed to access which social media are the most used to follow the online political campaign and which criteria influence the voting decision-making process. To achieve this purpose, exploratory research was carried out through questionnaire surveys. Three surveys were conducted based on the Portuguese presidential elections of January 24, 2021. The surveys were distributed before, during, and after the end of the electoral campaign, and 106 people were questioned and answered all 3 surveys. With the results of this study, it was possible to conclude that only 11% of respondents changed their voting intention due to the political communication made by political parties on social media during this electoral campaign. The social media most used by respondents was Facebook, which is also the one they consider the safest and most trustworthy to follow political communication in online media.

J. E. Garcia (✉)
ADiT-LAB, Instituto Politécnico de Viana Do Castelo, Viana Do Castelo, Portugal
e-mail: jorgegarcia@esce.ipvc.pt

INESC TEC, Porto, Portugal

E. G. Vega
Universidad Camilo José Cela - Facultad de Comunicación Y Humanidades, Madrid, Spain

P. Purificação · M. J. Fonseca
Instituto Politécnico de Viana Do Castelo, Viana Do Castelo, Portugal

M. J. Fonseca
UNIAG—Unidade de Investigação Aplicada em Gestão, Viana Do Castelo, Portugal

J. L. Reis et al. (eds.), *Marketing and Smart Technologies*, Smart Innovation, Systems and Technologies 337, https://doi.org/10.1007/978-981-19-9099-1_36

36.1 Introduction

It is from the marketing that it is possible to create, communicate, and confer value to the target audience, customers, or consumers of some product or service [1]. It is also through marketing that a continuous development and exchange of ideas, goods, and services is achieved to meet the needs of consumers and companies [2]. In politics, it does not work differently, political marketing is characterized by the creation and maintenance of the relationship between politicians and voters, in which the idea of mutual exchange is implicit [3]. According to [4], the political market is based on the central principle of marketing, the exchange between buyer and seller, since voters give their votes to politicians in exchange for the promise of seeing their interests defended and provided for. There are within political communication four basic elements, the sender, the message, the means used/channels of communication and the receiver. In the work *Comunicación política en campañas electorales* [5], one can see the importance of the feedback between the sender and the receiver, that is, between the candidate and the voter, this part of political communication is fundamental, because it is at this moment that the voter can express the consternation and feel that they are being heard and that they have a chance to be resolved in the future, this way the candidate gains supporters. Durán et al. [6] consider that strategy is the path to follow to achieve the main objective of the campaign, in other words, to make the candidate win the elections. The strategy mainly involves the definition of the electoral space, the message, the communication, the image of the candidate and his opponents and the campaign calendar. The candidate's image, positioning, and message are three fundamental strategies in both traditional and online political communication [5]. The image of the candidate is important because it can dictate whether the candidate wins or loses the elections. It should include factors such as personal qualities, convictions, credibility, history, and the ability to communicate and create empathy with the other. The positioning, on the other hand, concerns the way in which the candidate manages to interact, to reach the mind or emotion of the voters. Finally, the message concerns everything the candidate speaks publicly during the campaign. It can be conveyed in a speech or through symbols, metaphors, images, colors, or shapes, all procedures previously planned and used for the candidate to gain new votes and keep current supporters [5]. According to [7], the integration of the Internet in the political environment benefits the development of horizontal networks of interactive communication that connects locally and globally. Finally, the characterization of the voters, they are the main key of the political landscape; without these elements, there are no elections, and hence, there is such a great concern regarding the large abstention present in recent years of political elections [8]. With this study, it is intended to better understand political marketing and the influence of social media in the decision-making process of voters in Portugal. That is, to understand if the communication that is made online, such as publications of images, videos, interviews, online debates have any influence in the final decision of the Portuguese at the time of voting. The study will focus on social media such as Facebook, Instagram, Twitter, and WhatsApp on political communication and the communication strategies and political content that are implemented by political parties on these platforms. And it aims to understand and study the influence that social media have on Portuguese voters' decision making, as well as to understand the use of social networks in the political environment. In addition to the general objective, the following research questions were identified: Q1: Do young people consume more political communication on social media?; Q2: During the election campaign, which social media did respondents prefer and considered safe?; Q3: Are voters who always vote for the same party less influenced by political communication on social media?; Q4: Does the monthly income of voters make a difference regarding the influence that political communication has on their voting intention?; Q5: Are people influenced differently by different social media?; Q6: Did the voters who changed their voting intention during the political campaign show at the beginning of the study that they already knew which candidate they were going to vote for?; Q7: Is there a difference by gender regarding the influence of political communication on social media on voting decision making? This research paper is structured in five parts. After the introduction, the second part refers to the theoretical framework. This section defines the key concepts that underpin the study. In Sect. 36.3, it is characterized by the methodological approach adopted. In Sect. 36.4, it is proceeded with the presentation and analysis of the results. The fifth part presents the conclusions and the practical implications of the study, as well as its main limitations and lines for future research.

36.2 Theoretical Background

36.2.1 Marketing

In recent years, the concept of marketing has undergone several changes, becoming an increasingly varied and less universal definition among marketers [9].

According to [1], managers, academics, and management students know dozens of definitions of marketing. All of them have contributed to the understanding and practical use of an area of management that has exceeded the boundaries of the functional area normally referred to as the commercial function in companies. In addition, still be able to make profit, both for the company and for the surrounding market [10]. Marketing is a continuous method of development and exchange of ideas, goods, and services that aim to meet the needs of consumers and organizations [2]. It is a set of means available to a company to sell its products or services, and it can be said that marketing adds value.

36.2.2 Political Marketing

Political marketing can be characterized as the creation and maintenance of the relationship with voters, in which the idea of mutual exchange is implicit; that is, the voter expects and believes that the promises made by candidates are fulfilled, thus satisfying both parties, on the one hand the candidate wins and on the other hand, voters expect to see fulfilled the promises made during the campaign. In the same way that a person exchanges a part of the money he has for a piece of clothing, a sofa or an appliance, the voter exchanges his vote for the expectation of having a better government [3]. In the work, Conference report [11], the author lists the main aspects of political marketing, a focus on exchange relationships, a long-term perspective, voter orientation, and mutual benefits for all parties involved. According to the Journal of Marketing Management [4], political marketing is defined as the party's or candidate's use of opinion research and analysis of the environment to produce and promote a competitive offering that will help realize organizational objectives and satisfy constituency groups in exchange for their votes. According to [12], political marketing is related to the development of propaganda and the media. According to the publication *Comunicación política en campañas electorales* [5], political communication can be characterized by all the actions and activities practiced transmitting a message to the receiver, in this case, the voters. The political campaign should be directed toward a strategy to capture new votes/voters, all communication should be carried out with a view to winning the election and all members of the campaign, such as the presidential candidate or the head of campaign, for example, should remember this main objective and not allow themselves to be distracted by other concerns [5].

36.2.3 Digital Marketing and Social Media

The most recent advances in the field of marketing are mainly due to rapid technological advancement. Some authors argue that digital marketing is a restructuring of traditional marketing [13]. In the work *Changing from Traditional to Digital*, digital marketing does not appear as a replacement for traditional marketing, but rather to complement it, with the most facilitated and fast interactions through the internet [14]. Digital marketing is defined as marketing that uses strategies with some digital component in the marketing mix—differentiating itself from commercial marketing, because its function becomes to disseminate the image of an organization or person through the internet [15, 16]. Social media are virtual platforms where people can relate to each other, either by sending messages or sharing content [17].

36.2.4 Online Political Communication

Currently, people use new technologies and consequently social media through all day. Thus, it is important that political communication also goes through these new sharing channels. According to [7], the integration of the Internet into the political environment benefits the development of horizontal networks of interactive communication. Furthermore, there remains an antagonism within online political communication; for one viewpoint, there are the digital dynamics identical to Habermas' description of the public sphere [18]. On the other, it represents the public space, i.e., the unspecified sociocultural sphere [19]. According to the study Internet and Elections [20], the Orkut communities in Brazil are organized into different virtual groups that defend different parties, with nuclei called "I love …", "I hate …" or "I will vote for …". Frequently online democracy is present, and there has been a huge increase in the presence of parties on the Internet. In this regard, [21] underline the "me too" phenomenon as the cause of this increasing use of social media by politicians or political parties, with the premise that "if others have it, I have to have it too" [22].

36.3 Methods

The objective of this paper was to study the influence of social media in the decision-making process of voters in Portugal. That is, to understand how the communication made by politicians or political parties influenced the Portuguese over 18 years old who exercised their right to vote. The study was based on the presidential elections of January 24, 2021, which elected the President of the Republic of Portugal. It was used a non-probability convenience sample [23, 24] and composed of voters aged enough to exercise their right to vote. Three questionnaire surveys were created with the aim of analyzing voting intention and its variability throughout the campaign (before, during, and after the elections). The first form corresponds to the pre-election campaign period and was carried out in person, between January 1st and 9th, at the School of Business Sciences of the Polytechnic Institute of Viana do Castelo, for the more controlled group, and in the municipality of Vila Nova de Cerveira, for the more heterogeneous group. The central themes of the questions aimed at collecting sociodemographic data, and the access to and trust in online political information. The second questionnaire survey referred to the period during the election campaign, which ran from January 10 to January 22, 2021. Questions were asked related to the opinion of voters during the campaign and whether they were attentive to the information disseminated. An attempt was also made to understand how they had access to the campaign contents. The survey was answered between January 17th and 22nd. The third questionnaire survey was carried out after the election campaign period and was administered between January 25th and 29th. The selected questions aimed to analyze how many respondents were influenced by political communication on social media, as well as to assess which digital communication channels and social media were used by candidates to influence voters. The three questionnaire surveys were divided into six different parts, which categorized the information collected. Thus, in the first survey (administered in the pre-election campaign period), sociodemographic data were collected (part I), as well as the information collected about social media and the politicians/political parties (part II) that the respondents follow or have more affinity with. In the second survey (administered during the political campaign), the questions focused on political campaigns (part III) and social media communication (part IV). Finally, the third survey (administered in the post-election period) addressed the campaign (part V) and the voting decision (part VI). The structure of the surveys involved the use of closed questions with direct answers, as well as questions with 5-point Likert scales. In total, 106 responses were collected, which were repeated for all three surveys. The participants, divided into two groups, were the same answering the three surveys.

36.4 Presentation and Results Analysis

36.4.1 Questionnaire Survey 1—During the Pre-election Campaign Period

The first part of the analysis of this study consists of the characterization of the sample of individuals interviewed, namely their age, gender, location, education, occupation, as well as financial condition. Regarding age, most of the respondents are between 18 and 24 years old and between 35 and 44, each corresponding to 25.5%. As for the gender, there are more female respondents. In fact, 58.5% of the respondents answered female (62 answers) while only 41.5% answered male (44 answers). In terms of place of residence, the greatest number of respondents belong to the district of Viana do Castelo, corresponding to 75.5%, followed by Vila Real (8.5%) and Braga (7.5%). Lastly, Porto with 6.6% and Leiria and Coimbra with only one respondent (0.9% each). In this sample, there are only responses from respondents residing in the north of Portugal, given that the questionnaire survey was administered in person. Concerning the respondents' academic qualifications, 48.1% have secondary schooling and 35.8% have a bachelor's degree. In terms of professional occupation, 64.2% are employees, 18.9% are students, and 8.5% are self-employed. Regarding the financial background, most of the respondents are in a remuneration bracket between 10,000 and 20,000€ per year (44.3%), followed by incomes below 10,000€/year (43.4%). In the second part of the first questionnaire, it is intended to verify if those involved in the study had access to the Internet, if they usually vote, what their interest in politics is, and how this corresponds with their relationship with social media. The trust and influence that politicians/political parties have on the interviewed individuals were also analyzed. The media most used to obtain information on political issues are television (35.4%), followed by paper newspapers/magazines (13.7%), digital newspapers/magazines with 13.1%, Facebook also with 13.1%, and radio with 12.7%. It was possible to verify that a large part of the interviewees does not resort to social media much to obtain political information, since excluding Facebook, the other social media options showed little participation: Instagram with 6.2%, Twitter with 3.8%, and WhatsApp with 1.4%. When respondents were asked whether they regularly follow any politician or political party on social media, 69.8% assumed not to follow any politician, 15.1% follow one politician/political party, 10.4% follow 2–5 politicians, about 1.9% follow between 5 and 10 politicians, and 2.8% follow more than 10 politicians/political parties on social media. Regarding the trust in the communication made by politicians/political parties in the media, 32.1% assume themselves as indifferent, 30.2% trust partially. When this communication is made in social media, most respondents assume indifference (42.5%), 26.4% partially trust, 9.4% do not trust. As for the influence of communication made by politicians/political parties on the voting intention of respondents, 46.2% assume that it is partially influenced, 17% indifferent, 13.2% assume no influence whatsoever. In terms of loyalty to political parties, 51.9% occasionally vote for the same party, 6.6% never vote for the same political party, and only 2.8% of respondents said they always vote for the same political party. When asked if they had already decided who to vote for in the 2021 presidential elections, 69.8% of respondents said yes, 25.5% had not yet decided, and 4.7% did not intend to vote.

36.4.2 Questionnaire Survey 2—During the Election Campaign Period

During this period, it was intended to verify if the participants of the study were following the political campaigns and through which media they were doing so, as well as following politicians/political parties on social media. Thus, 40.4% were following political campaigns through television, 14% through radio, and 13.6% through digital newspapers/magazines. As for social media, it was found that 13.2% of respondents were following through Facebook, 3.6% through Instagram, 2.8% through Twitter, and 0.8% through WhatsApp. However, 76.4% of the respondents were not following any campaign, and of the 23.6% who were doing so, 36.8% were following through television and 16.9% through paper newspapers/magazines. Some 15.4% were following via digital newspapers/magazines and on social media in a very residual way, with Facebook accounting for 15.4%, then Instagram with 3.1% and Twitter and WhatsApp with 1.5% each. Specifically for the presidential elections, only 21.7% of respondents followed the campaign on social media, with Facebook being the most used, corresponding to a percentage of 64.3%, then Twitter with 17.9%, Instagram with 10.7%. Finally, WhatsApp with only 3.6% and another social media, Telegram, was also pointed out with 3.6%. Regarding the results concerning the perceived influence of communication actions on social media on voting intention,

39.1% of respondents consider themselves partially influenced, 26.1% indifferent and 8.7 do not consider themselves influenced. Still, while in Questionnaire 1 only 69.8% had made their decision, during the campaign they were 85.8%. Those respondents who had not yet decided decreased from 25.5% to 8.5%, and those who did not intend to vote went from 4.7% to 5.7%.

36.4.3 Questionnaire Survey 3—During the Post-election Period

With this survey by questionnaire, it was intended to measure only three dimensions: which communication channels did the respondents follow during the electoral campaign, how many candidates saw communication elements, and which were the most relevant for the change of opinion or voting decision. As for the communication channels that respondents followed during the campaign, television was in the lead, with 38.3%. As for social media, Facebook represents 12% of the answers, Instagram represents 5.6%, Twitter 4.9%, and WhatsApp only 1.9%. Regarding the communication elements seen by the respondents, 55.7% saw from 2 to 5 politicians/political parties, 19.8% saw from 5 to 10, 16% saw none, and 8.5% saw only one politician/political party. Before the election campaign, 64.2% of the respondents said they were not undecided and only 35.8% said they were undecided. In relation to the specific communication of the candidates, it was intended to know if the communication of any politician/party made the respondents change their opinion, to which 67.9% answered "no", and 32.1% answered "yes". Objectively, to understand if the way they voted had changed during the election campaign, 77.4% answered "no", 20.8% answered "yes", and 1.9% of the respondents "did not go to vote". Those who answered "yes" were asked if their vote had been changed based on the candidates' communication on social media, to which 54.5% answered "yes" and 45.5% answered "no". Additionally, it was asked which contents the respondent saw most frequently on social media, with the answers being 29% for debates between candidates, 22.6% for campaign videos, 19.4% corresponds to news about the candidate's campaign, 16.1% for photographs, and 12.9% for video interviews with the candidate. The contents perceived as most influential in decision making were debates between candidates (29.2%), campaign videos (25%), news about the candidate's campaign (20.8%), interview with the candidate on video (16.7%) and photographs (8.3%). As for the reasons why, the respondents decided or changed their vote, it is possible to see that the way the candidate behaved in the debates between candidates was the most chosen (35.3%), followed by the statement: "I had a wrong idea about the candidate's ideas" (23.5%) and "I had the opportunity to get to know the candidate better" (23.5%). Last was the way the candidate responded in the interviews (17.6%). As for the answers alluding to the initial research questions, a brief systematization follows. Based on the analysis of the results obtained, it can be stated that research question 1 (Q1), "Do young people consume more political communication on social media?" has a positive answer. It is possible to make this statement because it is verified that of the 27 young people interviewed (100%), aged between 18 and 24 years, 17 say they consume political communication on social networks, corresponding to approximately 63%, while the remaining participants of the study (79 participants = 100%), aged between 25 and 65 + years, only 38% approximately consume politics on social media. Regarding question 2 (Q2: During the election campaign, which social media did respondents prefer and considered safe?), it can be said that respondents expressed their preference for Facebook (64.3%), followed by Twitter (17.9%) and only then Instagram (10.7%), WhatsApp (3.6%) and Telegram (3.6%). In response to research question 3 (Q3: Are voters who always vote for the same party less influenced by political communication on social media?), it was found it to be true, as of the 20 people who said they "always" or " often" vote for the same party (18.8%), only two were influenced during the online political campaign. Regarding research question 4 (Q4: Does the monthly income of voters make a difference regarding the influence that political communication has on their voting intention?), the answer is also positive: of the respondents with lower income (93 respondents), 12% changed their voting intention due to political communication on social media. As for the respondents with a higher income (13 individuals), only 8% changed their voting intention. As for the research question 5 (Q5: Are people influenced differently by different social networks?), respondents feel more affinity for Facebook, because it is where they watch more online political communication. "Did the voters who changed their voting intention during the political campaign show at the beginning of the study that they already knew which candidate they were going to vote for?" (Q6). It was possible to verify that out of 106 respondents, only 12 people (100%) changed their voting intention based on the candidates' political communication on social media. Of these people, 7 (58%) said at the beginning of the study that they already knew which candidate they were going to vote for, and only 5 (42%) said they were still undecided. Thus, it is possible to see that the majority of voters who changed their voting intention during the political campaign, at the beginning of the study already had their voting intention set. Finally, regarding the question (Q7: Is there a difference by gender regarding the influence of political communication on social media on voting decision-making process?), 58% of women respondents assumed the influence of online political communication on their voting decision, against 42% of men.

36.5 Conclusions and Practical Implications

The present study, whose purpose was to understand the influence of social media in the political field, allowed the identification of different strategies to be used during political communication and the importance that marketing has in multiple dimensions of this activity sector. It was also possible to analyze how fundamental it is, nowadays, to be present in online platforms, to be able to reach a larger number of people. This was developed through an exploratory study to assess some of the aspects related to the initial research questions. Throughout the study, a several research questions were identified regarding age, income, loyalty to political parties, social networks used in the online political campaign and voting intention. The study focused on the Portuguese presidential elections held on January 24, 2021. Three questionnaire surveys were conducted, distributed before, during, and after the end of the electoral campaign, and 106 individuals were surveyed and responded to all 3 surveys. A quantitative exploratory methodological approach was implemented with a non-probabilistic convenience sample. The results and conclusions of the empirical work indicated that only 11% of the interviewees changed their voting intention due to political communication on social media. Young people, aged 18 to 25, consume more political communication on social media than other age groups. It can also be stated that voters who usually always vote for the same party are less influenced by online political campaigns. As for the voting intention and the gender of each participant of the study, between the people who changed their vote during the political campaign on social media, 58% were female, and 42% were male, thus concluding that the female participants of the study were more influenced by the political communication of the parties on social media. Overall, it can be concluded that there were few people interviewed who changed their voting intention during the political campaign, as out of 106 respondents in the study, only 12 changed their vote. The online content most viewed and that most changed the voting intention of the voters surveyed in this study were debates between candidates, campaign videos, and news about the candidate's campaign. The main reason for the voter to have changed their vote was the way the candidate behaved in the debates between candidates, followed by trust, responsibility, and security as other characteristics to have changed their vote during the campaign. In the future studies, we suggest the use of a larger sample size. In addition, in the future research, as this is a longitudinal study, we suggest the use of inferential statistics for the population. This may be considered a limitation of the present study, given that the results are only applicable to the sample under analysis. Another aspect to consider in the future studies might be the diversification of the type of elections to be investigated: presidential elections are more focused on the personal side of candidates than on the parties that support them.

References

1. Pires, A.: Marketing—Conceitos, Técnicas e Problemas de Gestão (3a Edição). Lisboa, São Paulo: Editorial Verbo (2002)
2. Kotler, P.: Marketing, Edição Compacta, 1st edn. Editora Atlas, São Paulo (1996)
3. Kotler, P.: A generic concept of marketing. J. Market., (1975)
4. Wring, D.: Reconciling marketing with political science: theories of political marketing. J. Market. Manag., (1997)
5. Democrática, Á.: Comunicación política campañas electorales. Int. IDEA (2006)
6. Durán Barba, J., Nogueira, F., Izurieta, R., Perina, R.M., Christopher, A., Garnett, J., Vega, H.: Estrategias de Comunicación para Gobiernos. Washington, Quito-Ecuador (2001)

7. Castells, M.: Communication, power and counter-power in the network society. Int. J. Commun. (2007)
8. Fernandes, A.M.: Mulheres na Política—Retratos na Primeira Pessoa. Safaa Dib, Lisboa (2017)
9. Keefe, L.M.: What is the meaning of "marketing"? AMA's Marketing News (2004)
10. Kotler, P., Keller, K.L.: Marketing Management (14a edição). Prentice Hall (2012)
11. Henneberg, S.C.M.: Conference report: second conference on political marketing: judge institute of management studies. University of Cambridge March 1996 **12**(8), 777–783 (1996)
12. Figueiredo, R.: O que é marketing político. São Paulo, Brasiliense (1994)
13. Çizmeci, F., Ercan, T.: The effect of digital marketing communication tools in the creation brand awareness by housing companies (2015)
14. Kotler, P., Kartajaya, H., Setiawan, I.: Marketing 4.0—Mudança do tradicional para o digital. Lisboa, Conjuntura Atual Editora (2017)
15. Garcia, J.E., Rodrigues, P., Simões, J., Serra da Fonseca, M.J.: Gamification strategies for social media. In: Remondes, J., Teixeira, S. (eds.) Implementing Automation Initiatives in Companies to Create Better-Connected Experiences, pp. 137–159. IGI Global (2022). https://doi.org/10.4018/978-1-6684-5538-8.ch007
16. Rodrigues, M.I., Fonseca, M.J., Garcia, J.E.: The use of CRM in marketing and communication strategies in Portuguese non-profit organizations. In: Andrade, J., Ruão, T. (eds.), Navigating Digital Communication and Challenges for Organizations, pp. 223–244. IGI Global (2022). https://doi.org/10.4018/978-1-7998-9790-3.ch013
17. Marques, V.: Mkt Digital 360. Lisboa, Actual Editora (2018)
18. Habermas, J.: A transformação estrutural da esfera pública : investigações sobre uma categoria da sociedade burguesa. Lisboa, Fundação Calouste Gulbenkian (2012)
19. Dahlgren, P.: Participation and alternative democracy: social media and their contingencies. Livros LabCom (2014)
20. Chaia, V. (2007). Internet e eleições: as comunidades políticas no orkut nas eleições de 2006. Logos Comunicação e Universidade.
21. Gibson, R., Nixon, P., Ward, S.: Political Parties and the Internet—Net Gain? Routledge, Londres (2003)
22. Silva, R.J.E.S.D.: Comunicação política no Facebook: um estudo exploratório de sete políticos portugueses (Doctoral dissertation, ES Comunicação Social) (2013)
23. Henn, M., Weinstein, M., Foard, N.: A Short Introduction to Social Research. SAGE Publications, London (2006)
24. Malhotra, N.K.: Pesquisa de marketing: uma pesquisa aplicada. Porto Alegre, Bookman (2006)

Chapter 37
How Endorser Promotes Emotional Responses in Video Ads

Luísa Augusto, Sara Santos, and Pedro Manuel do Espírito Santo

Abstract The importance of advertising on social networks has been growing, and the use of videos has become a more present reality. Understanding the narrative is not always understood by consumers. Closely associated with advertisements on YouTube are influencer marketing and the use of influencers by brands to produce content and narratives about brands, generating greater trust, engagement, and more positive emotional responses. In this sense, understanding how consumers see the narrative and how the video ad generates emotional responses for consumers is still an unexplained issue. This study seeks to fill this gap and studies how video ad elements contribute to positive emotional responses. Through a cross-sectional investigation, data were collected from a group of consumers of a telecommunications brand that were analyzed through structural equation modeling. Therefore, this investigation understands that the endorser is a constituent element of the narrative that promotes narrative transportation and emotional responses.

37.1 Introduction

YouTube has become one of the most used social media in brand advertising campaigns [1], which is why it has aroused the interest of academia [2]. The authors are unanimous in considering that the way YouTube disseminates information associated with entertainment generates more positive attitudes toward advertisements and their respective brands [3]. Dehghani et al. [2] understand that factors such as

L. Augusto (✉) · S. Santos
CISED—Instituto Politécnico de Viseu, 3504-510 Viseu, Portugal
e-mail: laugusto@esev.ipv.pt

S. Santos
e-mail: ssantos@esev.ipv.pt

P. M. do Espírito Santo
ESTGA, Universidade de Aveiro, 3750-127 Águeda, Portugal
e-mail: pedroes@ua.pt; pedro.m.santo@ipleiria.pt

ESTG- Instituto Politécnico de Leiria, 2411-901 Leiria, Portugal

J. L. Reis et al. (eds.), *Marketing and Smart Technologies*, Smart Innovation, Systems and Technologies 337, https://doi.org/10.1007/978-981-19-9099-1_37

entertainment and information add value to the ad, affecting brand recognition and purchase intention.

Closely related with announcements on YouTube are influencer marketing and the use of endorsers by brands to produce content and narratives about brands, generating greater trust, engagement, and more positive emotional responses on the part of audiences concerning the ad and brands [4, 5]. Stories and narratives, or the process of telling the brand or product's story, make advertisements more persuasive and effective [6].

The literature review suggests that the fit of celebrities to the target audience, the sharing of values [7], a certain familiarity [8], as well as a transparent, altruistic behavior perceived in the influencers' messages, generates greater trust in audiences, giving greater value to the advertisement [9], facilitating transport and immersion in the narrative. In turn, immersion in narrative positively impacts consumers' emotional responses [10].

This study aims to analyze the role of endorser in brand narratives and, consequently, in emotional responses, in particular in the sympathy of audiences toward brands. It is crucial to understand whether celebrities' adjustment, familiarity, and altruism influence audiences' narratives and emotional responses.

37.2 Theoretical Framework

37.2.1 Endorser Fit, Familiarity, and Altruism

Currently, brands have opted for advertising strategies that include using celebrities who fit the advertised product. Previous studies demonstrate that celebrity endorsement advertising responses can be transferred to the brand, increasing campaign effectiveness [11].

A celebrity endorser is defined as "an individual who enjoys public recognition and who uses this recognition on behalf of a customer good by appearing with it in an advertisement" [12].

Several recent studies [13] have confirmed that celebrity endorsements substantially increase advertising effectiveness. The celebrity's image is transferred to the advertised product [14]. With the advancement of technologies, not only "traditional celebrities" (actors, athletes, supermodels, etc.) but also social media influencers started to endorse brands [15]. The high number of followers on social networks of celebrities positively impacts advertising campaigns [16]. However, not all celebrities can represent every type of product or brand.

When brands look for celebrities for their advertising campaigns, it is essential to have a brand-celebrity fit and a product-celebrity fit. This fit is the similarity between the product and the celebrity endorser [11]. Many studies [17, 18] confirmed that endorser-brand fit is essential in advertising because images of the endorser are transported to the product. When consumers identify with the celebrity, they respond more to the brand endorser, leading to increased purchase intention [19].

So, the product-endorser fit is related to the effectiveness of the ad [20]. When endorsers advertise products, they do not fit in, they become less effective and credible [3]. Likewise, endorsers are less committed to each brand when they advertise several different brands and, therefore, less credible [21].

Likewise, familiarity with the endorser affects the endorser's communication and decision-making process [22]. Brand familiarity refers to "the number of product-related experiences accumulated by the consumer" [23]. Other authors define it as "knowledge of the source through exposure" [24].

When familiarity with the person, it is recognized by a large part of the public who have a positive or negative image of it. The more familiar and attractive the celebrity is to the consumer, the more positive effects it will have on the brand [25].

Studies in the area of endorsement show that familiarity with the celebrity is more persuasive than with unknown endorsers [20]. This familiarity also attracts and keeps consumers' attention to the brand [26].

On another side, altruism is also a critical endorser characteristic. When people point out altruistic intentions to endorsers, attitudes are more favorable [17].

Endorser effectiveness increases when consumers recognize altruistic motivations such as sincere commitment [27].

37.2.2 Narrative Transportation

Narratives in advertisements in video format have been increasingly used, becoming a useful creative technique for different and varied types of products [28]. For Brechman and Purvis [29], transport to the narrative is "an intrinsic orientation toward greater absorption into an imaginative world created by a story and its characters," so the story or narrative must be memorable and engaging, that encourage viewers to integrate their thoughts and experiences [30].

Casaló et al. [7] mention that influencers must publish original content that is perceived as authentic that matches their lifestyle and personality.

Glaser and Reisinger [6] emphasize that the ad story must be linked and have a strong connection with the product, as it helps viewers understand and interpret the story's meaning, being more easily involved in the story.

Grigsby et al. [31] add the role and importance of images in activating consumers' imagination and transporting them to the narrative.

Advertisements should privilege an authentic and engaging story [28]. An engaging story influences a good ad [6]. According to Green and Brock [32], the transport to the narrative involves images, affection, and attention of the spectators.

However, communication managers must create only the appropriate level of transport for the narrative so that viewers do not overstep their limits [10].

The narrative carryover effect is more substantial when the story is generated by the user rather than the professional. Furthermore, it is received by one story receiver at a time [33].

37.2.3 Emotional Responses: Sympathy

Advertising narratives play a crucial role in immersing and engaging consumers in these brand narratives and stories [8]. They are a powerful communication tool and can lead to changes in attitude, as they usually use emotional appeals and a certain emotional intensity [29].

The experience of immersion or transport to the narrative, leading the viewer, through an imaginary mental process, to stay away from real life or even forget reality, can have positive effects on the viewers' attitude [32], giving rise to affective, evaluative and behavioral responses from consumers [34]. Transport leads the viewer to experience the message without feeling pressured, generating increased appreciation and trust [29].

Chang [35] considers that narrative processing leads to strong emotional responses, and Escalas [36] adds that it reduces levels of critical thinking, which has implications for brand evaluation.

The brand's history and transport to the narrative can positively influence the formation of the brand's image [37], and the higher the level of transport, the more intense the affective responses [36]. Escalas and Stern [36] analyzed sympathy and empathy responses as different, albeit related, emotional responses of consumers concerning advertisements, considering that transport and immersion in the narrative generate more sympathy and empathy toward the advertisement and the brand, and in turn, these generate a more positive attitude.

Based on the literature, we formulated the following:

H1: Endorser familiarity has positive effects on narrative transportation;
H2: Endorser fit has positive effects on narrative transportation;
H3: Endorser altruism has positive effects on narrative transportation;
H4: Narrative transportation has positive effects on emotional responses;
H5: Endorser familiarity has positive effects on emotional responses;
H6: Endorser fit has positive effects on emotional responses;
H7: Endorser altruism has positive effects on emotional responses.

37.3 Methodology

In order to validate the proposed research model, we used a video ad, available at YouTube. (https://www.youtube.com/watch?app=desktop&v=5tFYfASqMCo), that we presented before the consumers answered the questionnaire.

The selection of the video ad took into account the advertised brand (NOS—one of the largest national telecom operators) as well as the chosen influencer (the 3rd with the most notoriety, according to a Marketest study [38]).

A cross-sectional research was conducted. The data was collected through a self-administered questionnaire and had two phases. First, consumers showed a video, in which the main character is a well-known Portuguese blogger. In the second phase,

Table 37.1 Survey instrument

Construct	Items	Item description
NAT	NAT1	While I was watching the ad, I thought of nothing else
	NAT2	This ad attracted me from start to finish
	NAT3	While watching the ad, I imagined myself in the scenes of the video
EFA	EFA1	This character is very familiar to me
	EFA2	I am informed about this character
	EFA3	I know a lot about this character
EFI	EFI1	This character adds value to this brand
	EFI2	There is a logical connection between this brand and the character
	EFI3	It makes sense for this character to be associated with this brand
EAL	EAL1	I think the main reason why the influencer is involved with NOS is because she believes she deserves their support
	EAL2	I think that influencer is passionate about the NOS brand
	EAL3	I think the influencer would agree to do this video at any price
ERE	ERE1	In the ad presented, I understood the characters' feelings
	ERE2	In the ad presented, I understood the characters' concerns
	ERE3	In the ad presented, I understood the motivation of this ad
	ERE4	In the ad presented, I recognize a sympathetic message

NAT: Narrative Transportation; EFI = Endorser Fit; EFA = Endorser Familiarity; EAL = Endorser Altruism; ERE = Emotional Responses

we presented a questionnaire to analyze the constructs. To measure the constructs, we adapted items from the literature. Thus, we used Silva and Topolinski [39] items for narrative transportation. The items used for endorser altruism, endorser fit, and endorser familiarity were adapted from Carlson et al. [19] (Table 37.1). The items to measure emotional responses were adapted from Escalas et al. [36].

All the items used in this study were measured using a 5-point Likert scales, ranging from strongly disagree (1) to strongly agree (5).

37.3.1 Sample

Previous to data collection, we conducted a pretest with 15 YouTube users. Thus, the data were collected from a non-probabilistic sample of YouTube users who were contacted through social networks to answer the questionnaire. The sample is composed of 286 YouTube users whose characterization is presented in Table 37.2.

Table 37.2 Demographic profiles (N = 286)

Variables	Category	N	%
Gender	Male	197	31.1
	Female	89	68.9
Age	≤20	47	16.4
	20–29	147	51.4
	30–39	37	12.9
	40–49	38	13.3
	≥50	17	5.9
Education	Primary school studies	31	10.8
	Secondary school studies	135	47.2
	High school	120	42.0

37.4 Results

Given the sample size, we considered that the conditions were met for data analysis through structural equation modeling (SEM), which is a multivariate analysis technique that allows testing dependency relationships between concepts. To this end, we used AMOS v.25 software [40].

In a preliminary analysis of the data, we verified that the Variance Inflation Factor (VIF) value met the criterion for no multicollinearity (VIF <5). We also analyzed skewness (Sk) and kurtosis (Ku) and concluded that the items do not deviate from normality since Sk <3 and Ku <7 as suggested in the literature [40]. Additionally, we calculated the Kaiser–Meyer–Olkin (KMO) value and performed Bartlett's test of sphericity. We obtained KMO = 0.866, and Bartlett's test of sphericity is significant for $p < 0.05$.

The data analysis was developed in two phases: First phase we analyzed the validity of the measurement model, and in the second phase, we analyzed the causal effects between the variables under study by validating the hypotheses under study.

37.4.1 Measurement Model

The first phase of this study consisted of analyzing the measurement model. Although the items were adapted from other studies and showed consistency and reliability, we performed a confirmatory factor analysis through maximum likelihood estimation. According to the literature [41], the measures model has an acceptable fit ($\chi 2 = 147.52$; df = 80; GFI = 0.936; NFI = 0.955; IFI = 0.979; TLI = 0.972; CFI = 0.979; RMSEA = 0.054).

Table 37.3 shows the values obtained for the measurement model and shows that the average variance extracted (AVE) values are between 0.677 and 0.878, and the composite reliability (CF) ranges between 0.855 and 0.956. These values are above the recommended threshold values (AVE >0.5; HR >0.7). In addition, the analysis of the measurement model of the items under study verified that the standardized factor loadings were above the value of 0.7 (λ >0.5; p <0.05) [41, 42]. The items under study have a coefficient R^2 >0.2.

Table 37.4 presents the Fornell and Larcker [43] criterion to assess discriminant validity and verify if the construct is distinct from each other. In this sense, we verified that the values of the average variance extracted from all constructs are higher than the square of the interconstruct correlations. These values show discriminant validity.

Table 37.3 Measurement model (N = 286)

Latent variable	Items Code	λ	*t*-values	R^2	CR	AVE
Narrative transportation					0.875	0.700
	NAT1	0.845	(a)	0.714		
	NAT2	0.863	16.542	0.744		
	NAT3	0.800	15.253	0.641		
Endorser fit					0.855	0.747
	EFI1	0.844	(a)	0.713		
	EFI2	0.884	14.004	0.782		
Endorser familiarity					0.956	0.878
	EFA1	0.938	(a)	0.879		
	EFA2	0.952	31.910	0.907		
	EFA3	0.921	28.645	0.848		
Endorser altruism					0.862	0.677
	EAL1	0.921	(a)	0.847		
	EAL2	0.812	16.793	0.659		
	EAL3	0.723	14.238	0.523		
Emotional responses					0.908	0.711
	ERE1	0.834	(a)	0.696		
	ERE2	0.859	17.567	0.738		
	ERE3	0.805	15.945	0.648		
	ERE4	0.874	18.002	0.763		

NAT: Narrative Transportation; EFI = Endorser Fit; EFA = Endorser Familiarity; EAL = Endorser Altruism; ERE = Emotional Responses; λ = Standardized loadings; AVE = Average Variance Extracted; CR = Composite Reliability; (a) the path is fixed to 1.0 to set the construct metric; Model fit: $\chi 2$ = 147.52; df = 80; Good fit index (GFI) = 0.936; Normal fit index (NFI) = 0.955; Incremental fit index (IFI) = 0.979; Tucker-Lewis index (TLI) = 0.972; Comparative fit index (CFI) = 0.979; Root mean square error approximation (RMSEA) = 0.054

Table 37.4 Discriminant validity: Fornell and Larcker criterion

Construct	NAT	EFI	EFA	EAL	ERE
Narrative transportation (NAT)	**0.700**				
Endorser fit (EFI)	0.287*	**0.747**			
Endorser familiarity (EFA)	0.104*	0.173*	**0.878**		
Endorser altruism (EAL)	0.174*	0.401*	0.229*	**0.677**	
Emotional responses (ERE)	0.389*	0.283*	0.154*	0.364*	**0.711**

NAT: Narrative Transportation; EFI = Endorser Fit; EFA = Endorser Familiarity; EAL = Endorser Altruism; ERE = Emotional Responses; * All correlations are significant (p <0.01)

Table 37.5 Structural theory results

Hyp	Path	β	t values	p-values	Hypothesis
H1	EFA→NAT	0.095	1.431	0.152	Not supported
H2	EFI→NAT	0.436	4.991	< 0.01	Supported
H3	EAL→NAT	0.096	1.111	0.267	Not supported
H4	NAT→ERE	0.426	6.478	< 0.01	Supported
H5	EFA→ERE	0.061	1.095	0.274	Not supported
H6	EFI→ERE	0.046	0.596	0.551	Not supported
H7	EAL→ERE	0.367	4.941	< 0.01	Supported

NAT: Narrative Transportation; EFI=Endorser Fit; EFA=Endorser Familiarity; EAL=Endorser Altruism; ERE=Emotional Responses; β = Path loadings; Model fit: $\chi 2 = 147.52$; df = 80; Good fit index (GFI) = 0.936; Normal fit index (NFI) = 0.955; Incremental fit index (IFI) = 0.979; Tucker-Lewis index (TLI) = 0.972; Comparative fit index (CFI) = 0.979; Root mean square error approximation (RMSEA) = 0.054

37.4.2 Structural Model

After confirm, the validity of the measurement model was analyzed and the results of the structural model in order to test the hypotheses under study. In Table 37.5, we present the relationships between the constructs and the results suggest that the structural model presents a good fit ($\chi 2 = 147.52$; df = 80; GFI = 0.936; NFI = 0.955; IFI = 0.979; TLI = 0.972; CFI = 0.979; RMSEA = 0.054).

37.5 Discussion

In the context of communication and publication, this study examined the effects of endorsement on narrative carry and emotional responses. From the results, it was found that the R^2 values for narrative transport are equal to 0.305 and the R^2 value for emotional responses is equal to 0.536. Thus, it is considered that the model presented explains, through the hypotheses under study, the dependent variables.

Hypothesis H1 proposed to analyze the direct effects between endorser familiarity and narrative transportation. The results obtained do not allow us to conclude that there are relevant effects between these two constructs ($\beta_{EFA} \rightarrow {}_{NAT} = 0.095$; $t = 1.431$; $p > 0.0$). Thus, it is considered that the hypothesis is not confirmed by this study.

From the results, hypothesis H2 reveals that there is statistical support to confirm that adjusting the characteristics of the endorser as a character in the advertising video promotes greater narrative transport ($\beta_{EFI} \rightarrow {}_{NAT} = 0.436$; $t = 4.991$; $p < 0.01$). In the literature, we found that, when characters with high notoriety have a high fit with consumers, this increases the effectiveness of ads through narrative transportation. Thus, it is considered that hypothesis H2 is confirmed by this study.

In the analysis of the effects of endorser altruism on narrative transportation, no significant effects were found ($\beta_{EAL} \rightarrow {}_{NAT} = 0.096$; $t = 1.111$; $p > 0.05$). Thus, it is considered that hypothesis H3 is not confirmed by this study.

According to the literature, narrative transportation leads the viewer to experience the message without feeling pressured and emotional responses appeared [36]. In this point of view, this research confirms this theory and H4 is supported by the results ($\beta_{NAT} \rightarrow {}_{ERE} = 0.096$; $t = 1.111$; $p > 0.05$.

When compared to ads that do not use endorsers, the use of endorsers generates more positive responses to the ad as sympathy [14, 21]. Therefore, we analyzed the direct effects of endorser familiarity on emotional responses generated by the video ad. Thus, we tested hypothesis H5 and the results do not show that there is direct impact of endorser familiarity on emotional responses ($\beta_{EFA} \rightarrow {}_{ERE} = 0.061$; $t = 1095$; $p > 0.05$). Hypothesis H6 was also not supported by this study ($\beta_{EFI} \rightarrow {}_{ERE} = 0.046$; $t = 0.596$; $p > 0.05$).

This result may be associated with the fact that some consumers did not recognize the character in the video which may have led to some responses being biased. Hypothesis H5 was not supported by this study.

Finally, this study assessed the direct effects of endorser altruism on emotional responses and found statistical evidence to confirm hypothesis H7 ($\beta_{EAL} \rightarrow {}_{ERE} = 0.367$; $t = 4941$; $p > 0.05$).

37.6 Conclusions

The model used obtained a good fit and confirmed the importance of the endorser in advertising. In detail, this study confirms the relationship between endorsement (fit and altruism) and narrative transport and emotional responses.

In the literature, it is considered very important for brands to include characters known to consumers in their ads so that their message is more easily understood and interpreted. In this sense, consumers analyze the video as their real life and the message that the brand intends to spread has a greater efficiency.

This study shows that not all endorser characteristics are equally relevant, since, while the altruism of the endorser proved to be very relevant for consumers to have positive emotional responses, the consumer's adjustment to the endorser promotes greater immersion in the video scenes and, consequently, more meaningful emotional responses. Thus, we consider that this study is very relevant, and we believe that brands should consider including endorsers in their ads.

Therefore, we suggest that brand managers take the findings of this study and recognize in their ads that endorser altruism and fit have benefits for their consumers' perception. Although the study found valid results, these were subject to some limitations.

Among the limitations found, it was identified that the sample consisting of young individuals with high school education may have limited the results. More educated individuals have greater skills. Another limitation of the work was the fact that the study analyzed only one video. Therefore, it is proposed to carry out studies with different samples in advertising and countries. In these studies, new variables can also be included, such as the counter arguing and video creativity.

Acknowledgements This work is funded by National Funds through the FCT—Foundation for Science and Technology, I.P., within the scope of the project Refª UIDB/05583/2020. Furthermore, we would like to thank the Research Centre in Digital Services (CISeD) and the Polytechnic of Viseu for their support.

References

1. Kim, J.: The institutionalization of YouTube: from user-generated content to professionally generated content. Media Cult. Soc. **34**, 53–67 (2012)
2. Dehghani, M., Niaki, M.K., Ramezani, I., Sali, R.: Evaluating the influence of YouTube advertising for attraction of young customers. Comput. Hum. Behav. **59**, 165–172 (2016)
3. Lee, J., Lee, M.: Factors influencing the intention to watch online video advertising. Cyberpsychol. Behav. Soc. Netw. **14**, 619–624 (2011)
4. Lou, C., Tan, S.-S., Chen, X.: Investigating consumer engagement with influencer- vs. brand-promoted ads: the roles of source and disclosure. J. Interact. Advertising **19**, 169–186 (2019)
5. Augusto, L., Santos, S., Santo, P.E.: Youtube as a means of immersing consumers in brand narratives. RISTI—Revista Iberica de Sistemas e Tecnologias de Informacao **2021**, 518–530 (2021)
6. Glaser, M., Reisinger, H.: Don't lose your product in story translation: how product-story link in narrative advertisements increases persuasion. J. Advert. **51**, 188–205 (2022)
7. Casaló, L.V., Flavián, C., Ibáñez-Sánchez, S.: Influencers on Instagram: antecedents and consequences of opinion leadership. J. Bus. Res. **117**, 510–519 (2020)
8. Augusto, L., Santos, S., Santo, P.E.: Factors influencing the credibility of tourist advertisements shared on Youtube. RISTI—Revista Iberica de Sistemas e Tecnologias de Informacao **2021**, 478–492 (2021)
9. Phua, J., Jin, S.V., Kim, J.: Pro-veganism on Instagram. Online Inf. Rev. **44**, 685–704 (2020)
10. Dessart, L.: Do ads that tell a story always perform better? The role of character identification and character type in storytelling ads. Int. J. Res. Mark. **35**, 289–304 (2018)
11. Bergkvist, L., Zhou, K.Q.: Celebrity endorsements: a literature review and research agenda. Int. J. Advert. **35**, 642–663 (2016)

12. McCracken, G.: Who is the celebrity endorser? cultural foundations of the endorsement process. J. Consum. Res. **16**, 310–321 (1989)
13. Bergkvist, L., Hjalmarson, H., Mägi, A.W.: A new model of how celebrity endorsements work: attitude toward the endorsement as a mediator of celebrity source and endorsement effects. Int. J. Advert. **35**, 171–184 (2016)
14. Santos, S., Santo, P., Ferreira, S.: Storytelling in Advertising: From Narrative to Brand Distinctiveness. In: Martins, N., Brandão, D. (eds.) Advances in Design and Digital Communication, pp. 516–527. Springer International Publishing, Cham (2021)
15. Marwick, A.E.: Instafame: luxury selfies in the attention economy. Publ. Cult. **27**, 137–160 (2015)
16. Boerman, S.C.: The effects of the standardized instagram disclosure for micro- and meso-influencers. Comput. Hum. Behav. **103**, 199–207 (2020)
17. Park, H.J., Lin, L.M.: Exploring attitude–behavior gap in sustainable consumption: comparison of recycled and upcycled fashion products. J. Bus. Res. **117**, 623–628 (2020)
18. Yoo, J.-W., Jin, Y.J.: Reverse transfer effect of celebrity-product congruence on the celebrity's perceived credibility. J. Promot. Manag. **21**, 666–684 (2015)
19. Carlson, B.D., Donavan, D.T., Deitz, G.D., Bauer, B.C., Lala, V.: A customer-focused approach to improve celebrity endorser effectiveness. J. Bus. Res. **109**, 221–235 (2020)
20. Augusto, L., Santos, S., do Espírito Santo, P.M.: Endorser altruism effects on narrative transportation in video ads. In: Smart Innovation, Systems and Technologies, pp. 711–721 (Year)
21. Santos, S., Santo, P.E., Ferreira, S.: Brands Should Be Distinct! the contribution of ad's narrative and joy to distinctiveness. In: Martins, N., Brandão, D., Moreira da Silva, F. (eds.) Perspectives on Design and Digital Communication II: Research, Innovations and Best Practices, pp. 319–332. Springer International Publishing, Cham (2021)
22. Tian, S., Tao, W., Hong, C., Tsai, W.-H.S.: Meaning transfer in celebrity endorsement and co-branding: meaning valence, association type, and brand awareness. Int. J. Advert. **41**, 1017–1037 (2022)
23. Alba, J.W., Hutchinson, J.W.: Dimensions of consumer expertise. J. Consum. Res. **13**, 411–454 (1987)
24. Erdogan, B.Z., Baker, M., Tagg, S.: Selecting celebrity endorsers: the practitioner's perspective. J. Advert. Res. **41**, 39–48 (2001)
25. Swartz, T.A.: Relationship between source expertise and source similarity in an advertising context. J. Advert. **13**, 49–54 (1984)
26. Premeaux, S.R.: The attitudes of middle class versus upper class male and female consumers regarding the effectiveness of celebrity endorsers. J. Promot. Manag. **15**, 2–21 (2009)
27. Samman, E., Auliffe, E.M., MacLachlan, M.: The role of celebrity in endorsing poverty reduction through international aid. Int. J. Nonprofit Voluntary Sect. Market. **14**, 137–148 (2009)
28. Yang, K.C.C., Kang, Y.: Predicting the relationships between narrative transportation, consumer brand experience, love and loyalty in video storytelling advertising. J. Creative Commun. **16**, 7–26 (2021)
29. Brechman, J.M., Purvis, S.C.: Narrative, transportation and advertising. Int. J. Advert. **34**, 366–381 (2015)
30. Singh, S., Sonnenburg, S.: Brand performances in social media. J. Interact. Mark. **26**, 189–197 (2012)
31. Grigsby, J.L., Jewell, R.D., Zamudio, C.: A Picture's worth a thousand words: using depicted movement in picture-based ads to increase narrative transportation. J. Advertising 1–19 (2022)
32. Green, M., Brock, T.: The role of transportation in the persuasiveness of public narrative. J. Pers. Soc. Psychol. **79**, 701–721 (2000)
33. van Laer, T., Feiereisen, S., Visconti, L.M.: Storytelling in the digital era: a meta-analysis of relevant moderators of the narrative transportation effect. J. Bus. Res. **96**, 135–146 (2019)
34. Nguyen, T.-T., Grohmann, B.: The influence of passion/determination and external disadvantage on consumer responses to brand biographies. J. Brand Manag. **27**, 452–465 (2020)

35. Chang, C.: Narrative ads and narrative processing. In: Thorson, E., Rodgers, S. (eds) Advertising Theory, pp. 241–254. New York, Routledge (2012)
36. Escalas, J.E., Stern, B.B.: Sympathy and empathy: emotional responses to advertising dramas. J. Consum. Res. **29**, 566–578 (2003)
37. Ryu, K., Lehto, X.Y., Gordon, S.E., Fu, X.: Effect of a brand story structure on narrative transportation and perceived brand image of luxury hotels. Tour. Manage. **71**, 348–363 (2019)
38. Marques, O.M.: Os influenciadores com maior notoriedade em Portugal. Meios and Publicidade (2022). https://www.meiosepublicidade.pt/2022/05/os-influenciadores-com-maior-notoriedade-em-portugal/
39. Silva, R.R., Topolinski, S.: My username is IN! The influence of inward vs. outward wandering usernames on judgments of online seller trustworthiness. Psychol. Market., **35**, 307–319 (2018)
40. Hair, J., Black, W., Babin, B., Anderson, R.: Multivariate Data Analysis, Global Annabel Ainscow, New York (2018)
41. Bagozzi, R.P., Yi, Y.: On the evaluation of structural equation models. J. Acad. Mark. Sci. **16**, 74–94 (1988)
42. Hair, J., Risher, J., Sarstedt, M., Ringle, C.: When to use and how to report the results of PLS-SEM. Euro. Bus. Rev., **31** (2018)
43. Fornell, C., Larcker, D.F.: Evaluating structural equation models with unobservable variables and measurement error. J. Mark. Res. **18**, 39–50 (1981)

Chapter 38
The Impact of Surprise Elements on Customer Satisfaction

Márcia Martins, Mafalda Teles Roxo, and Pedro Quelhas Brito

Abstract This study intends to understand whether hotels should choose to surprise through a discount or a surprise gift. The experiment consisted in identifying whether there were differences in satisfaction and delight, according to the associated treatment (no surprise, surprise discount, or gift). With this purpose, a fictional hotel website was created for participants to simulate a reservation. Through the analysis of the experiment, the impact of surprise on customer satisfaction was confirmed. It was also found that, in the hospitality industry, a gift has a higher impact on satisfaction than a discount. When analyzing the guest delight, the results differ from what is stipulated in the literature (which points to the significant impact of surprise in this measure). It was concluded that between the two promotion tools, only the gift can significantly increase customer delight. This study demonstrates the importance of understanding the concept of surprise according to different industries. It also points to the importance of identifying the best methods to surprise customers, as different methods may lead to different results.

38.1 Introduction

The hospitality research focusing on the customer experience has been increasing over the last years since tourism activity is managed having into consideration

M. Martins (✉) · M. T. Roxo · P. Q. Brito
Faculdade de Economia, Universidade Do Porto, Porto, Portugal
e-mail: marcialemos1999@gmail.com

M. T. Roxo
e-mail: mafalda.t.roxo@inesctec.pt

P. Q. Brito
e-mail: pbrito@fep.up.pt

M. T. Roxo · P. Q. Brito
LIAAD/INESC TEC, Porto, Portugal

M. T. Roxo
Universidade da Maia, Maia, Portugal

J. L. Reis et al. (eds.), *Marketing and Smart Technologies*, Smart Innovation, Systems and Technologies 337, https://doi.org/10.1007/978-981-19-9099-1_38

that customers are looking for positive and pleasant experiences [6, 21, 43]. One of the company's main priorities and challenges is to create a unique and memorable customer experience [30]. Using surprise enables a lasting relationship with customers since it makes them more satisfied [53]. To achieve customers' loyalty, it is essential to understand what is the "uniqueness" that customers are looking for and, on the other hand, what is the exclusive factor that the company can offer to the market [34]. It is no longer enough for the company's strategy to focus only on the price or product/service differentiation [37]. Therefore, it is necessary to comprehend the impact of using different w ays of surprise. Nowadays, people feel increasingly bored and unenthusiastic about the purchase process and surprise can prevent this situation. The exploration of emotions by organizations enables them to achieve a competitive advantage, given that such strategy is hardly imitated [46]. Surprise, both negative and positive, provokes emotions in human beings [38] that last in their memory and, therefore, it is used to define their satisfaction [52]. This research aims to understand how hotels can use surprise, more precisely find if there are differences between surprising with a gift or a discount. The focus of this research is to answer the following research question: "How can hotels use the element of surprise to increase customer satisfaction and delight: discount or gift?" After gathering the theoretical background, a questionnaire was carried out to better understand which gift is preferred by generation Z to be used on the experiment, subsequently performed. The data was analyzed, and at the end, it was defined the theoretical contributes and managerial implications, as well as limitations and direction for future research.

38.2 Theoretical Background

38.2.1 Surprise

According to Teigen and Keren [50], a surprise is the difference between the expected result and what really happened. Therefore, it is present when a stimulus does not meet expectations [25]. Since it is unlikely to happen, this stimulus is considered unexpected and surprising [32]. Stimuli that are not part of the individual schema draw the customers' attention, making them stop doing their ongoing activity, once they try to match this stimulus with the familiar ones [8]. Besides, they become more conscious about the stimulus, having deeper memory retention, because the cognitive process is more profound. Surprise can be used by an organization in several ways, allowing to achieve success in the respective market [7], since it can cause positive emotions next to the customer [20].

38.2.2 Customer Satisfaction

Satisfaction is present when consuming a product or service, a pleasurable level of fulfillment is reached [42]. Customer satisfaction defines the long-term customer behavior, causing attitude change, repeat purchase, and brand loyalty [41]. It is determined through cognitive and affective processes, having into consideration the pre-consumption mood and expectations [35, 55]. People's mood affects their satisfaction. If they are in a negative mood, they will make a worse evaluation than it would be expected [35]. Contrarily, if they are in a good mood, they are more propitious to consider forgivable possible mistakes [24].

Surprise is a crucial factor to understand and achieve customer satisfaction [27, 47]. Clearly, when the element of surprise is present, the customer gives a better evaluation to the same product [27]. There is also a bigger impact on the attitude toward the brand and on their word of mouth when the customer is loyal compared to the non-customers [23]. Indeed, positive surprise generates satisfaction, regardless of the pre-consumption mood of the individual [27].

38.2.3 Customer Delight

Several researchers claim that it is no longer enough to just satisfy the customer. In order to have loyal and committed customers, companies need to go beyond satisfaction. To achieve competitive advantage, companies have to delight their customers [47, 48, 54]. Satisfaction is present when a company meets the needs of its customers. When a company not only satisfies its customers but also surprisingly goes beyond their expectation, it delights them [10, 42]. In fact, it has been increasing the number of researches that demonstrate that satisfaction does not always lead to significant favorable behaviors [4]. Indeed, not all satisfied customers are loyal [26] and, according to Bartl et al. [5], satisfaction is not that relevant when it comes to repurchasing intentions. Otherwise, delight results in more relevant outcomes, such as a bigger impact on customer memory [47]. Other differences are that delighted customers are more loyal and committed, having higher willingness to pay and repurchase [3]. Therefore, after companies guarantee the satisfaction of their customers, they must switch their focus to customer delight [16]. In the service experience, surprise potentiates delight, which consequently will help achieve long-term success [1]. Magnini et al. [31], throughout the analysis of hotels' online posts from TripAdvisor, discovered that surprise is crucial to obtain customer delight. Besides, they concluded that in the case of guests who affirmed being surprised and delighted, there was a perfect correlation with loyalty. According to this research, very satisfied in comparison to satisfied guests are 24% more probable to recommend the hotel and share positive word-of-mouth, while when delighted they are 40% more likely to have such behavior [31].

38.2.4 Sales Promotion

Sales promotion increases sales in the short run. It can be addressed to customers, members of the distribution channel or to employers [2]. The promotion tools chosen to this study approach are the discount and gift. The most used sales promotion technique is price discounts [14, 28] because it allows companies to gain new customers and increase sales [19]. Indeed, the discount tool may cause a positive effect on consumer perceptions, increasing the perception of value of the offer [12]. The customer may also receive a free gift/premium when buying a product or service. This technique is highly valuable from the customers' perspective [14] since they appreciate the experience of accepting and using a gift, as well as the value of the desirable premium [17]. It can change customers' attitudes and preferences, developing a good opinion about the brand [11]. When offering gifts of small value, there is the risk of reducing the purchasing intention of the customer, who may consider the offer an opportunist marketing strategy [17]. When the surprise element is not present, by comparing these two promotion techniques it is possible to conclude that offering a gift is more effective in the communicating the companies' values and culture. In fact, the consumer's evaluation of the gift depends on the perceived price and not on the cost price for the company. When using the discount, it is noted that there is vulnerability in relation to the negative perceptions of quality [14]. Wu et al. [57] analyzed the behavior of loyal restaurant customers, comparing the impact of discount surprise (10%) with rewards surprises (free dessert). It was concluded that, in dissatisfied clients, surprise rewards cause more delight and satisfaction than a discount. In contrast, when trying to surprise a satisfied client, it is not relevant the type of promotion chosen between these two.

38.2.5 Relevant Aspects When Using the Surprise Element

Before satisfyingly surprising the customer, companies must ensure that the current product/service (which does not yet contain the element of surprise) already satisfies the essential needs of the customer [36]. After ensuring that the minimum expected is respected, it is important to find the best way to include the surprising element [34]. To comprehend what surprises their customers, managers need to understand their target groups and how they react to surprise. For that, it is crucial using relevant and current information [20, 23]. It also should be considered that it is not appropriate to try to surprise all market segments [34]. Before including the surprise element, it is necessary to reflect about the aspects that contribute to the company uniqueness. They must be aligned with the values, vision, and mission since the WOW effect, to be successful, must be based on the organizational culture [34] and brand identity [9]. The use of surprise is unlikely to influence the customers' feelings if the company's employees are not motivated. Regarding this referred to the concept of internal marketing particularly the importance of satisfaction and well-being of employees in the marketing and sales department, as they represent the company [34]. When using surprise, it should be considered that it will have an impact on the company's reputation, but also that it will send the message that customers are a priority. This may positively influence their behavior and make them choose the company in question or even become loyal to it [34].

38.2.6 Research Hypotheses

With the purpose of answering the research question (How can hotels use the element of surprise to increase customer satisfaction and delight: discount or gift?) several hypotheses were defined, regarding each one of the following topics:

H1: Satisfaction:
H1.1 A surprise gift (versus without surprise) leads to a higher customer satisfaction.
H1.2 A surprise discount (versus without surprise) leads to a higher customer satisfaction.
H1.3 There are not relevant differences on customer satisfaction between surprising with a gift or a discount.
H2: Delight:
H2.1 A surprise gift (versus without surprise) leads to a higher customer delight.
H2.2 A surprise discount (versus without surprise) leads to a higher customer delight.
H2.3: Customer delight is higher when receiving a gift, instead of a discount.
H3: Acknowledgment:
H3.1: When the customers are surprised with a gift (versus without surprise), they acknowledge more the effort of the hotel.
H3.2: When the customers are surprised with a discount (versus without surprise), they more the effort of the hotel.
H3: Acknowledgements
H3.3: The customers acknowledge more the effort of the hotel when surprised with a discount instead of a gift.
H4: Expectancy Disconfirmation:
H4.1: The expectations of the customers who receive a surprise gift are most exceeded comparing to the ones "without surprise."
H4.2: The expectations of the customers who receive a surprise discount are most exceeded comparing to the ones "without surprise."
H4.3: The expectations of the customers who receive a gift surprise are more exceeded than the ones who are surprised with a discount.
H5: Surprise:
H5.1: When customers are surprised with a gift (versus without surprise), surprise is felt more intensely.

Table 38.1 Hypotheses

Constructs	Hypotheses
H1: Customer satisfaction	H1.1: surprise gift > without surprise
	H1.2: surprise discount > without surprise
	H1.3: surprise gift = surprise discount
H2: Customer delight	H2.1: surprise gift > without surprise
	H2.2: surprise discount > without surprise
	H2.3: surprise gift > surprise discount
H3: Acknowledgment	H3.1: surprise gift > without surprise
	H3.2: surprise discount > without surprise
	H3.3: surprise discount > surprise gift
H4: Expectancy disconfirmation	H4.1: surprise gift > without surprise
	H4.2: surprise discount > without surprise
	H4.3: surprise gift > surprise discount
H5: Surprise	H5.1: surprise gift > without surprise
	H5.2: surprise discount > without surprise
	H5.3: surprise gift > surprise discount

H5.2: When customers are surprised with a discount (versus without surprise), surprise is felt more intensely.
H5.3: When surprising with a gift, the surprise is higher than with a discount (Table 38.1).

38.3 Methodology

This study intends to understand whether hotels should choose to surprise through a discount or a surprise gift. The experiment consisted in identifying whether there were differences in satisfaction and delight, according to the associated treatment (no surprise, surprise discount, or gift). With this purpose, a fictional hotel website was created for participants to simulate a reservation. Through the analysis of the experiment, the impact of surprise on customer satisfaction was confirmed. It was also found that, in the hospitality industry, a gift has a higher impact on satisfaction than a discount. When analyzing the guest delight, the results differ from what is stipulated in the literature (which points to the significant impact of surprise in this measure). It was concluded that between the two promotion tools, only the gift can significantly increase customer delight. This study demonstrates the importance of understanding the concept of surprise according to different industries. It also points to the importance of identifying the best methods to surprise customers, as different methods may lead to different results, with the specific goal to find whether hotels should choose to surprise through a discount or a surprise gift. The experiment consisted in identifying whether there were differences in satisfaction and delight, according to the associated treatment (no surprise, surprise discount, or gift). With this purpose, a fictional hotel website was created for participants to simulate a reservation. Through the analysis of the experiment, the impact of surprise on customer satisfaction was confirmed. It was also found that, in the hospitality industry, a gift has a higher impact on satisfaction than a discount. When analyzing the guest delight, the results differ from what is stipulated in the literature (which points to the significant impact of surprise in this measure). It was concluded that between the two promotion tools, only the gift can significantly increase customer delight. This study demonstrates the importance of understanding the concept of surprise according to different industries. It also points to the importance of identifying the best methods to surprise customers, as different methods may lead to different results. This study intends to understand whether hotels should choose to surprise through a discount or a surprise gift. The experiment consisted in identifying whether there were differences in satisfaction and delight, according to the associated treatment (no surprise, surprise discount, or gift). With this purpose, a fictional hotel website was created for participants to simulate a reservation. Through the analysis of the experiment, the impact of surprise on customer satisfaction was confirmed. It was also found that, in the hospitality industry, a gift has a higher impact on satisfaction than a discount. When analyzing the guest delight, the results differ from what is stipulated in the literature (which points to the significant impact of surprise in this measure). It was concluded that between the two promotion tools, only the gift can significantly increase customer delight. This study demonstrates the importance of understanding the concept of surprise according to different industries. It also points to the importance of identifying the best methods to surprise customers, as different methods may lead to different results articles, and websites. The primary data was collected through an experimental study. For the data collection, it was used a convenience sample (composed by college students), it was decided to use this sample since guests and consumers are often segmented according to their age because individuals with similar age have similar experiences and needs [22]. All the individuals that contributed to the data collection belong to Generation Z, as they are born between the mid-1990s and late 2000s. It is important for the hospitality industry to better understand this group since they are becoming a bigger influence on their family travel decisions and are seen as the next generation of travelers. When they travel, around half of them (52.7%) stays in hotels, indeed most of them have already stayed in hotels. This preference for hotels by Gen Z is mainly because these establishments are able to offer them everything they may need when compared to other types of accommodation [56].

38.3.1 Experimental Design

To respond to the research question, it is necessary to prove if the presence of the surprise element (discount or gift) causes an impact on customer satisfaction and delight. Despite it is not possible to prove causality, through experimentation, cause-and-effect relationships can be inferred [33]. For this reason, when studying the element of surprise, experimental studies are essential [52]. The main idea of this experiment was to see if there was any difference in the satisfaction and delight of the participant when they received a surprise. It was developed a fictitious hotel website for this purpose, so participants could simulate the purchase of a one-night stay. When the participants concluded the reservation, some of them were surprised with a pop-up notification informing they had just received a surprise (a gift or a discount). It was necessary to define that the discount and the gift used in this study would have the same value. The gift chosen was "A credit of 15€ to use at the customer choice (hotel bar, fridge minibar of your bedroom, hotel restaurant, game room)." It was decided to use in the experiment the one regarding a friend's trip since all members of the group belong to the generation Z (population under study). The discount surprise used was of 15€ (to be deducted on the client bill, paid on the hotel reception). At the end of the online reservation, participants completed a survey to measure their satisfaction and delight. Posteriorly, this data was analyzed to see if the answers differ according to the treatment of the group (no surprise vs gift surprise vs discount surprise). It was decided to do all this process online.

The dependent variables demonstrate how the independent variables affect the test units. In this study, they are customer satisfaction, customer delight, acknowledgment, expectancy disconfirmation, and surprise (these were measured using a questionnaire). The independent variables (without surprise; surprise discount; surprise gift) are changed by the researcher to measure and compare their impact.

The treatment levels are the following:

- ***No Surprise***: "normal reservation," there aren't any surprises offered.
- ***Surprise Gift***: after booking on the fictitious website, it pops up a notification offering a surprise gift (credit of 15€ to use at the customer choice on the hotel services).
- ***Discount Gift***: after booking on the fictitious website, it pops up a notification offering a surprise discount (15€ discount to deduct on the bill to be paid on the hotel reception).

In order to control the extraneous variables, it was used randomization and design control. The participants' assignment was random since depending on which group were the responses to the questionnaire being collected, the website settings would change. For example, when collecting answers regarding the gift surprise treatment, on the website settings the corresponding pop-up notification would be selected (appearing at the end of the participant reservation). After a specific questionnaire (gift surprise, discount surprise, or no surprise) having reached 50 additional answers comparing to the previous number of answers, the website would switch to another condition. So, the links (both questionnaire and the hotel website), even though those already sent to participants, were adjusted to the treatment selected). In this study, the experimental design applied was the true experimental (the participants are randomly attributed to groups, and the treatment of the group is also chosen randomly) using posttest-only control (the measurements are only collected once. It was not taken any pretest measures, only after the exposition of the treatment [13].

38.3.2 Choosing and Assignment of the Participants

There were three groups: one for the discount surprise treatment, another for the gift surprise treatment, and the other one was the control group (that didn't receive any surprise). It was collected the participation of 400 participants (at least 132 students per condition). The invitation of university students to take part in the study was done using social media platforms, institutional email, and by QR Codes placed in the university.

38.3.3 Experimental Design Results

Sample Description

In this study participated 402 students, from several universities in Portugal, mostly from University of Minho and University of Porto. The average age of the experiment students is 21.39 years. It was guaranteed the homogeneity of the groups in terms of demographic characteristics.

38.3.4 Manipulation Check

It was analyzed through the presence of two questions on the questionnaire. On them it was asked to rank, on a 5-point Likert scale, the agreement with the sentences: "I felt surprised" and "I felt astonished." Several ANOVA and multiple comparison tests were performed. Since in each case $p \leq 0.05$ ($p = <0.001$), the hypothesis H0 is rejected (the average of the dependent variable has the same value in each group), and therefore, it is possible to conclude that at least two averages are different from each

Table 38.2 Cronbach's alpha of each construct

Constructs	Cronbach's alpha
Customer satisfaction	
Customer delight	0.91
Acknowledgment	0.918
Expectancy disconfirmation	0.873
Surprise	0.858

other. Consequently, it is necessary to do multiple comparisons (pair by pair) to understand the existing differences between the respective combinations. In conclusion, as it was hoped there are significant differences in having surprise (gift or discount) or not (without surprise), both on the "surprised" and the "astonished." Due to the similar nature of discount and gift (both are surprises), it was not found significant differences between them. In summarizing, the manipulation was successful.

38.3.5 Internal Consistency

It was decided to use internal consistency methods to evaluate the several constructs defined according to the literature (customer satisfaction, customer delight, acknowledgment, expectancy disconfirmation, and surprise). Table 38.2 full version is in appendix 2. (the construct "Customer Satisfaction" wasn't included on this analysis since it is composed by only one item). From the analysis, it is noticed that the scale items that measure the same construct are highly intercorrelated. It wasn't necessary to eliminate any measure since all the constructs were acceptable (in all $\alpha \geq 0.7$). Of the four analyzed constructs, two had a good internal consistency (Expectations: $\alpha = 0.873$ and Surprise: $\alpha = 0.858$). The other two had an excellent internal consistency (Customer Delight: $\alpha = 0.91$ and Acknowledgment: $\alpha = 0.918$). All measures were maintained on this construct since if one was deleted Cronbach's alpha would decrease. On the customer delight construct, since it was composed by three parts, it was also possible to conclude that if one of the items of this construct were deleted, Cronbach's alpha would decrease, and therefore the three items must be kept together.

38.3.6 Questionnaire and Hypotheses Research Analysis

Customer Satisfaction

H1.1 A surprise gift (versus without surprise) leads to a higher customer satisfaction.

By considering Table 38.3, it is possible to conclude that the satisfaction has a higher value when the participant was surprised with a gift ($M = 5.85$). When the surprise is not present, as expected, the customer satisfaction is lower ($M = 4.85$). To understand if this difference was statistically significant, an ANOVA analysis was made (the assumptions are guaranteed ($p > 0.05$), and therefore the variances are equal). Since $p < 0.05$, it indicates that the differences in the customer satisfaction mean between groups are statistically significant. By doing a multiple comparison test, Table 38.4 was created. To sum up, from a statistical point of view, the customer satisfaction varies significantly according to the treatment.

The customer satisfaction is average—when it is offered, a surprise gift is higher ($M = 5.85$) than when the surprise element is not present ($M = 4.85$). As claimed by the multiple comparison tests, the difference between these averages is statistically significant. Thereby, H1.1 is confirmed.

H1.2 A surprise discount (vs without surprise) leads to a higher customer satisfaction.

Table 38.3 Mean values of each construct according to each treatment level

Treatment levels				
Means values		Discount	Gift	Without surprise
	Customer satisfaction	5.3	5.85	4.85
	Customer delight	3.89	4.01	3.71
	Acknowledgment	3.44	3.3	2.93
	Expectancy disconfirmation	5.49	5.39	4.99
	Surprise	3.62	3.67	2.77

Table 38.4 Conclusions from the multiple comparison tests

Multiple comparison tests						
Dependent variable	Combinations		Tukey HSD	Bonferroni	Games-Howell	Conclusion
Customer satisfaction	Without surprise	Discount	≠	≠	–	≠
		Gift	≠	≠	–	≠
	Discount	Gift	≠	≠	–	≠
Customer delight	Without surprise	Discount	=	–	–	=
		Gift	≠	–	–	≠
	Discount	Gift	=	–	–	=
Acknowledgment	Without surprise	Discount	≠	≠	–	≠
		Gift	≠	≠	–	≠
	Discount	Gift	=	=	–	=
Expectancy disconfirmation	Without surprise	Discount	≠	≠	–	≠
		Gift	≠	≠	–	≠
	Discount	Gift	=	=	–	=
Surprise	Without surprise	Discount	–	–	≠	≠
		Gift	–	–	≠	≠
	Discount	Gift	–	–	=	=

There are significant differences between not offering a surprise ($M = 4.85$) and giving a discount surprise ($M = 5.30$). By comparing these two treatments, it is possible to see that discount creates a higher customer satisfaction. H1.2 is supported.

H1.3 There are not relevant differences on customer satisfaction between surprising with a gift or a discount.

H1.3 is not supported, customer satisfaction when surprised with a gift ($M = 5.85$) is statistically significantly higher than when surprised with a discount ($M = 5.30$).

Customer Delight

Starting by looking at the construct "Customer delight," it is possible to note that the one with the higher score was the gift ($M = 4.01$), followed by the discount ($M = 3.89$). Robustness of means was used instead of ANOVA since there is no equality of variances. The means of the groups are statistically different ($p < 0.05$: H0 is rejected). According to Games-Howell Test (inequality of the variances), there is only one difference between the conditions (without surprise versus discount). There is no impact on the customer delight by offering a discount surprise or not since the means of these two groups are considered to be equal. The data is contradictory considering that there are no significant differences between surprising with a gift or discount. It is inconsistent since comparing the condition without surprise and discount the averages is statistically similar, and comparing it to the gift are different, so it wouldn't be expected the equality between surprise discount and gift. At this point, it must be considered the following statement: "In mathematics, if A = B and B = C, then A = C. However, in statistics, when A = B and B = C, A is not the same as C because all these results are probable outcomes based on statistics." The inconsistent results must be due to inadequate statistical power, caused, for example by the small size of the used sample [29]. It is also useful to analyze the means of the homogenous subsets, due to the odd results prevenient from the multiple comparison test [15]. Indeed, considering the similar averages of each condition regarding the "customer delight" construct, there are two homogeneous groups: One composed by the discount and the without surprise, and the other gift and discount. The condition surprise discount has a similar average to each one of the other groups, although gift and without surprise had differences statistically significant. Overall, the conclusion is that it is preferable to surprise through a gift that doesn't use surprise.

H2.1 A surprise gift (vs without surprise) leads to a higher customer delight.

Analyzing the research hypostasis of the delight construct, it begins by supporting H2.1. Indeed, the surprise gift ($M = 4.01$) leads to a higher customer delight than when it is not offered any surprise ($M = 3.71$), being this difference statically significant (Games-Howell' test).

H2.2 A surprise discount (vs without surprise) leads to a higher customer delight.

However, offering a surprise discount ($M = 3.89$) increases customer delight (mean of without surprise = 3.71); according to the multiple comparison tests, this difference is not relevant, and therefore H2.2 is rejected.

H2.3: Customer delight is higher when receiving a gift, instead of a discount.

H2.3 is also not supported; despite the surprise gift generating a higher customer delight (M = 4.01) comparing to discount (M = 3.89), this difference is not statistically relevant.

Acknowledgment

H3.1: When the customers are surprised with a gift (versus without surprise), they acknowledge more the effort of the hotel.

Since there is variances equality, it is possible to use the ANOVA. It indicates that there are significant differences between the groups, indeed it can be seen by the comparisons (Table 38.4). According to this statistic, by using surprise, regardless of the nature (gift or discount) there is a significant impact on the guests, by making them feel special and unique. When surprised with a gift (M = 3.30), they acknowledge more the effort of the hotel (without surprise mean = 2.93). These averages are statistically significant differences, and therefore H3.1 is supported.

H3.2: When the customers are surprised with a discount (vs without surprise), they acknowledge more the effort of the hotel.

H3.2 is confirmed since the treatment level where participants received a discount (M = 3.44) resulted in a significant higher score of this measure (without surprise = 2.93). Indeed, when it is offered a surprise discount, the customer recognizes more the intention of the hotel, feeling more unique and special (M = 3.44), in comparison with the other groups.

H3.3: The customers acknowledge more the effort of the hotel when surprised with a discount instead of by a gift.

H3.3 was not supported since there are no significant differences on the customer acknowledgment between surprising with a discount or gift. Besides, the score of this construct is smaller than in the discount treatment (M = 3.44) than the gift (M = 3.30).

Expectancy Disconfirmation

H4.1: The expectations of the customers who receive a surprise gift are most exceeded comparing to the ones "without surprise."

Due to the equality of the sample variances, it is possible to analyze the values of the ANOVA analysis, which indicates that the means in each group are statistically different. Considering Table 38.4, it is seen that surprise, regardless of which of the two methods it is used, affects the participants. H4.1 is supported since the presence of a surprise gift (M = 5.49) results in a expectancy disconfirmation higher than the treatment level without surprise (M = 4.99), being this difference statistically significant.

H4.2: The expectations of the customers who receive a surprise discount are most exceeded comparing to the ones "without surprise."

H4.2 is also supported since there is a significant increase of the expectancy disconfirmation when the customer is surprised with a discount (M = 5.49), comparing to when it is not (M = 4.99). The surprise discount was the one in which expectations were most exceeded (M = 5.49), followed by the gift (M = 5.39).

H4.3: The expectations of the customers who receive a gift surprise are more exceeded than the ones who are surprised with a discount.

H4.3 is rejected, despite the treatment of discount presenting a higher score on the expectancy disconfirmation measure ($M = 5.49$); the difference between this promotion tool and the surprise gift is not relevant.

Surprise

H5.1: When customers are surprised with a gift (vs without surprise), surprise is felt more intensely.

There is a difference between those who received a surprise and those who do not, being slightly higher on the gift treatment ($M = 3.67$). Both methods provoked equally surprise in the student. H5.1 is confirmed since the treatment level gift ($M = 3.67$) has a significant higher score of this measure than without surprise ($M = 2.77$).

H5.2: When customers are surprised with a discount (vs without surprise), surprise is felt more intensely.

H5.2 is also supported since the treatment level discount ($M = 3.62$) has a significant higher score of this measure than without surprise ($M = 2.77$).

H5.3: When surprising with a gift, the surprise is higher than with a discount.

H5.3 is not supported since there are no significant differences between these two levels.

Answering the Research Question Figure

Order of preference when implementing a surprise:

Customer Satisfaction: 1° Gift- > 2° Discount- > 3° Without Surprise.

Customer Delight: 1° Gift- > 2° Discount or Without Surprise.

Acknowledgment, Expectancy disconfirmation, Surprise: 1° Gift or Discount- >

2° Without Surprise

Answering the research question, between the two options, it must be chosen a gift. Regarding the acknowledgment of the hotel effort, the expectancy disconfirmation, and the surprise the impact is the same. On the other hand, since using this promotion tool, results in a higher customer satisfaction and delight, comparing to the discount.

38.4 Discussion

38.4.1 Final Considerations

This study's purpose was to better understand generation Z's perception of surprise, as well as recognize how can hotels achieve customer satisfaction and delight. The experiment's results allow to answer the research question (How can hotels use the element of surprise to increase customer satisfaction and delight: discount or gift?).

The goal was to see if, according to the treatment level associated (without surprise, surprise gift, or surprise discount), there would be significant differences between these two measures. Since H1.1 e H1.2 were supported, it is confirmed the impact of surprise, whether by a discount or gift, on satisfaction. In fact, surprise has the ability to increase satisfaction [27, 47, 52]. According to this research, that effect is also present in the hospitality industry. In different research, it was analyzed the behavior of satisfied and loyal restaurant customers, by comparing the impact of surprise when using each of the two promotion tools. No relevant differences were found in customer satisfaction between surprising with a gift or a discount when the initial cumulative satisfaction is high. On the other hand, when this cumulative satisfaction is low, this indifference is not confirmed. Indeed, in that condition, using a surprise gift instead of a discount has a higher impact (positive) on customer satisfaction [57]. The conclusions from this hotel experiment indicate that offering a gift surprise generates higher customer satisfaction than a discount does. To accomplish customer commitment and loyalty, companies need to go beyond satisfaction, focusing on delighting them [47, 48, 54]. It was already proven that surprise increases customer delight [1, 31]. However, the experiment results indicate that such effect depends on the promotion tool chosen. More precisely, a gift surprise increases customer delight; however, the same is not confirmed when using a discount. Regarding the other three components (acknowledgment, expectancy disconfirmation, and surprise), it is not relevant which one of the two promotion tools is used, both statistically increase similarly customer satisfaction and delight. Portuguese hotels must consider implementing surprise techniques. Indeed, when surprised, customers give a better evaluation of the same product [27] and besides generate a higher impact on the attitude toward the brand and word of mouth [23]. The experiment results confirm that using surprise leads to higher customer satisfaction. A customer delight increase is also notable when offering a gift. On the other hand, when surprised with a discount, no significant differences were found. Generation Z acknowledges the hotel effort similarly, regardless of having received a surprise discount or gift. Likewise, on the expectancy disconfirmation and the intensity of surprise were not found any relevant differences. Answering the research question: between the two promotion tools analyzed, the best way for Portuguese hotels to increase customer satisfaction and delight is by surprising them with the offer of a gift.

38.4.2 Theoretical Contributes and Managerial Implications

The hospitality research focusing on the customer experience has been increasing over the last years since the key is to create positive experiences for the customers [6, 21, 43], which can be achieved by using surprise [34]. Many studies show the several effects of surprise, especially in customer satisfaction and delight. However, it cannot be found any research indicating specifically the best way for hotels to surprise their guests, considering that their main intention is to increase customer satisfaction and delight. The developed study helped fill a part of a research gap previously pointed.

Companies must also consider the resulting costs in surprising their customer [34]. The entire guest's process and steps during their stay must be analyzed to identify how the hotel can surprise them in a personalized way, even in the most banal steps and small details. Surprising efficiently implies using wisely the company's exclusive factor, in alignment with the organizational values, vision, mission and culture [34] and brand identity [9]. Part of the customers, search for more information about the company after being surprised and want to tell others about it. Since these attitudes are also present online, companies need to update and invest in the website and social media. When the element of surprise is present, the customer gives a better evaluation to the same product [27]. Taking this into consideration, hotels can also try to get more guest reviews on their websites or on reservation platforms, like booking. On the other hand, there are situations when surprise may be a double-edged sword, for example, if customers view this action as an opportunist marketing strategy [17, 49]. This research helped to understand that a hotel must choose to surprise their customers through the offer of a gift (in this case, a voucher of 15€ to use in the hotel according to the customer preference). To use a gift instead of a discount, does not provoke significant changes regarding the acknowledgment of the hotel effort, the expectancy disconfirmation, and the intensity of the surprise. However, it does significantly generate higher customer satisfaction and delight. This research also alerts to the limitations associated with using this strategy. A surprise is something unexpected that includes the novelty factor. If customers were previously surprised with the same or similar stimulus (either by the company or competitors), it won't generate surprise again [23, 53].

38.5 Conclusions

This study contributed to the research fields that approach the impact of surprise and customer behavior on the hospitality industry. In fact, it is necessary more information about how hotels can use the surprise element to achieve the company objectives. According to this research results, if a hotel intends to increase customer satisfaction and delight, a surprise gift must be offer to the guests instead of a discount. This choice is mainly due to the higher ability of this tool to increase customer satisfaction. In addition, there are significant differences in customer delight between not surprising or surprising with a gift, the same is not confirmed when it's given a discount. Both promotion tools increase the components evaluated, excluding the customer delight. Indeed, H1.1 and H2.2 indicate that the surprise gift and discount significantly increase customer satisfaction. The hypotheses measuring the positive impact of each technique in other components were all supported (acknowledgment of the hotel effort: H3.1 + H3.2: expectancy disconfirmation: H4.1 + H.4.2 and surprise: H.5.1 + H5.2). The impact of the gift and the discount on these three components are similar. H2.2 compares the impact on customer delight in surprising through a surprise discount or not surprising. This hypothesis was rejected, indicating that there are no significant differences in this component between the conditions. As it was confirmed H2.1, the gift increases customer delight in the defined scenario. On the other hand, the rejection of H2.3 supports that, regarding this measure, it is indifferent to use the discount or gift. This study contains some limitations. First, these research conclusions are based on hypothetical online booking; as any laboratory experiment, it does not represent real life in all its elements and distraction. Developing this study in a real setting would make it more meaningful. The self-selection of the sample used may also have contributed to the existence of bias in this study since it reflects only the perceptions of a specific group of individuals [33]. For causal research, such as this one, convenience samples are not suggested. However, considering that this is a pilot study in the hospitality field, and this type of sample is acceptable. Indeed, they stimulate the development of ideas, perceptions, and hypotheses [39]. In this experiment, it was compared the impact of a discount of 15€ and a voucher to use at the hotel. It could be analyzed the results of offering others amount of discount, as well as other gifts. This research is restricted to the perception and reaction to surprise by Portuguese university students who belong to generation Z. Students' perception regarding services can be different from the general consumers [40]. In fact, age is an influencing element. If the study participants were older, most likely different conclusions would be found. Indeed, in older individuals a surprise may not have such impact on their satisfaction and delight [45]. It is also known that reactions to surprise differ between cultures [51] and in other experiments could be used other participants' profiles to generalize the findings to other populations. Another issue to be considered is that hospitality industry is unpredictable. Guest experience depends on countless factors [18] and decides the emotional responses to surprises may differ having into account individual personality traits and emotional instability [44]. Another research limitation was that the measurement of surprise and respective outputs could have been affected by the use of self-reporting scales [23]. Another limitations are due to the fact that the increase in the participant satisfaction and delight may have been caused by the fact of receiving a gift or discount, regardless of the presence of surprise. Future research should control the surprise element. In other words, understanding if the customer satisfaction and delight change between receiving these promotional tools with the presence or not of surprise. From this research, it was concluded that the impact of the surprise discount and gifts in restaurants is different in hotels. It would also be interesting to carry out the same study in another economic sector. In the experiment, the discount and the gift had different effects, demonstrating the need of comparing different surprise techniques in the future.

Acknowledgement This work is financed by National Funds through the Portuguese funding agency, FCT—Fundação para a Ciência e a Tecnologia—within project LA/P/0063/2020.

References

1. Ariffin, A.A.M., Omar, N.B.: Surprise, hospitality, and customer delight in the context of hotel services. Tour. Hosp. Manag. **12**, 127–142 (2016). https://doi.org/10.1108/S1871-317320160000012010
2. Bandyopadhyay N, Sivakumaran B, Patro S, Kumar RS (2021) Immediate or delayed! Whether various types of consumer sales promotions drive impulse buying?: an empirical investigation. J. Retail. Consum. Serv. 61. https://doi.org/10.1016/j.jretconser.2021.102532
3. Barnes, D., Beauchamp, M., Webster, C.: To delight, or not to delight? This is the question service firms must address. J. Mark. Theory Pract. **18**, 275–284 (2010). https://doi.org/10.2753/MTP1069-6679180305
4. Barnes, D.C., Collier, J.E., Howe, V., Douglas Hoffman, K.: Multiple paths to customer delight: the impact of effort, expertise and tangibles on joy and surprise. J Serv Mark **30**, 277–289 (2016). https://doi.org/10.1108/JSM-05-2015-0172
5. Bartl, C., Gouthier, M. H. J., Lenker, M.: Delighting consumers click by click: antecedents and effects of delight online. J. Serv. Res. **16**(3), 386–399 (2013). https://doi.org/10.1177/1094670513479168
6. Bazerman, M.H., Tenbrunsel, A.E., Wade-Benzoni, K.: Negotiating with yourself and losing: making decisions with competing internal preferences. Acad. Manag. Rev. **23**, 225–241 (1998). https://doi.org/10.2307/259372
7. Becattini N, Borgianni Y, Cascini G, Rotini F (2020) Investigating users' reactions to surprising products. Des. Stud. 69. https://doi.org/10.1016/j.destud.2020.05.003
8. Berlyne, D.E.: Motivational Problems Raised by Exploratory and Epistemic Behavior. In: Koch, S. (ed.) Psychology: A Study of a Science, pp. 284–364. McGraw-Hill, New York (1962)
9. Bronner, F., de Hoog, R.: Conspicuous consumption and the rising importance of experiential purchases. Int. J. Mark. Res. **60**, 88–103 (2018). https://doi.org/10.1177/1470785317744667
10. Chandler, C.C.: Specific retroactive interference in modified recognition tests: evidence for an unknown cause of interference. J. Exp. Psychol. Learn. Mem. Cogn. **15**, 256–265 (1989). https://doi.org/10.1037//0278-7393.15.2.256
11. Chang, C.: Effectiveness of promotional premiums: the moderating role of affective state in different contexts. Psychol. Mark. **26**, 175–194 (2009). https://doi.org/10.1002/mar.20266
12. Compeau, L.D., Grewal, D.: Comparative price advertising: an integrative review. J. Public Policy Mark. **17**, 257–273 (1998). https://doi.org/10.1177/074391569801700209
13. Cooper DR, Schindler PS (2013) Business Research Methods, 12th ed. New York
14. Darke, P.R., Chung, C.M.Y.: Effects of pricing and promotion on consumer perceptions: it depends on how you frame it. J. Retail. **81**, 35–47 (2005). https://doi.org/10.1016/j.jretai.2005.01.002
15. Field A (2009) Descobrindo a estatística usando o SPSS, 2nd ed. Artmed
16. Finn, A.: Customer delight: distinct construct or zone of nonlinear response to customer satisfaction? J. Serv. Res. **15**, 99–110 (2012). https://doi.org/10.1177/1094670511425698
17. Foubert, B., Breugelmans, E., Gedenk, K., Rolef, C.: Something free or something off? a comparative study of the purchase effects of premiums and price cuts. J. Retail. **94**, 5–20 (2018). https://doi.org/10.1016/j.jretai.2017.11.001
18. Grace, D., O'Cass, A.: Examining service experiences and post-consumption evaluations. J. Serv. Mark. **18**, 450–461 (2004). https://doi.org/10.1108/08876040410557230
19. Grewal, D., Monroe, K.B., Krishnan, R.: The effects of price-comparison advertising on buyers' perceptions of acquisition value, transaction value, and behavioral intentions. J. Mark. **62**, 46–59 (1998). https://doi.org/10.2307/1252160
20. Gupta A, Eilert M, Gentry JW (2020) Can I surprise myself? a conceptual framework of surprise self-gifting among consumers. J. Retail. Consum. Serv., 54. https://doi.org/10.1016/j.jretconser.2018.11.017
21. Hirschman, E.C., Holbrook, M.B.: Hedonic consumption: emerging concepts, methods and porpositions. J Mark **46**, 92–101 (1982)

22. Hoyer, W.D., MacInnis, D.J., Pieters, R.: Consumer Behavior, 7th edn. South-Western, Cengage Learning, Mason, OH (2018)
23. Hutter, K., Hoffmann, S.: Surprise, surprise. ambient media as promotion tool for retailers. J. Retail. **90**, 93–110 (2014). https://doi.org/10.1016/j.jretai.2013.08.001
24. Isen, A.M.: An influence of positive affect on decision making in complex situations: theoretical issues with practical implications. J. Consum. Psychol. **11**, 75–85 (2001). https://doi.org/10.1207/153276601750408311
25. Izard, C.E.: Surprise-Startle. In: Human Emotions, 1st edn., pp. 277–284. Springer, Boston (1977)
26. Jones, T.O., Sasser, W.E.: Why satisfied customers defect. Harv. Bus. Rev. **73**, 88–91 (1995). https://doi.org/10.1061/(asce)0742-597x(1996)12:6(11.2)
27. Kim, M.G., Mattila, A.S.: The impact of mood states and surprise cues on satisfaction. Int. J. Hosp. Manag. **29**, 432–436 (2010). https://doi.org/10.1016/j.ijhm.2009.10.022
28. Kim EL, Tanford S (2021) The windfall gain effect: Using a surprise discount to stimulate add-on purchases. Int. J. Hosp. Manag. 95. https://doi.org/10.1016/j.ijhm.2021.102918
29. Lee, S., Lee, D.K.: What is the proper way to apply the multiple comparison test? Korean J. Anesth. **71**, 354–360 (2018)
30. Lemon, K.N., Verhoef, P.C.: Understanding customer experience throughout the customer journey. J. Mark. **80**, 69–96 (2016). https://doi.org/10.1509/jm.15.0420
31. Magnini, V.P., Crotts, J.C., Zehrer, A.: Understanding customer delight: an application of travel blog analysis. J. Travel. Res. **50**, 535–545 (2011). https://doi.org/10.1177/0047287510379162
32. Maguire, R., Maguire, P., Keane, M.T.: Making sense of surprise: an investigation of the factors influencing surprise judgments. J. Exp. Psychol. Learn. Mem. Cogn. **37**, 176–186 (2011). https://doi.org/10.1037/a0021609
33. Malhotra, N.K., Nunan, D., Birks, D.F.: Marketing Research: An Applied Approach, 5th edn. Trans-Atlantic Publications Inc., Harlow, Essex (2017)
34. Marcus P, Reunanen T, Arndt B (2018) Finding the wow-factor to enhance business. In: Advances in Human Factors, Business Management and Leadership. Springer International Publishing AG 2018, pp. 96–105
35. Mattila, A., Wirtz, J.: The role of preconsumption affect in postpurchase evaluation of services. Psychol Mark **17**, 587–605 (2000)
36. Michelli J (2012) The Zappos Experience 5 Principles to Inspire, Engage and WOW
37. Millard, N.: Learning from the “wow” factor—how to engage customers through the design of effective affective customer experiences. BT Technol. J. **24**, 11–16 (2006). https://doi.org/10.1007/s10550-006-0016-y
38. Noordewier, M.K., Topolinski, S., Van, D.E.: The temporal dynamics of surprise. Soc. Personal. Psychol. Compass. **10**, 136–149 (2016)
39. Nunan D, Birks DF, Malhotra NK (2020) Marketing Research: Applied Insight. Harlow
40. Ok, C., Shanklin, C.W., Back, K.J.: Generalizing survey results from student samples: implications from service recovery research. J. Qual. Assur. Hosp. Tour. **8**, 1–23 (2008). https://doi.org/10.1080/15280080802103037
41. Oliver, R.L.: A cognitive model of the antecedents and consequences of satisfaction decisions. J. Mark. Res. **17**, 460–469 (1980). https://doi.org/10.2307/3150499
42. Oliver, R.L., Rust, R.T., Varki, S.: Customer delight: foundations, findings, and managerial insight. New Zealand J. Retail. **73**, 311–336 (1997)
43. Pizam, A.: Creating memorable experiences. Int. J. Hosp. Manag. **29**, 343 (2010). https://doi.org/10.1016/j.ijhm.2010.04.003
44. R. MR, Costa PT (2003) Personality in Adulthood. In: A Five-Factor Theory Perspective, 2nd ed. The Guilford Press, New York
45. Rimé, B., Finkenauer, C., Luminet, O., Zech, E., Philippot, P.: Social sharing of emotion: new evidence and new questions. Eur. Rev. Soc. Psychol. **9**, 145–189 (1998). https://doi.org/10.1080/14792779843000072
46. Robinette, S., Brand, C., Lenz, V.: Emotion Marketing- The Hallmark Way of Winning Customers for Life. McGraw-Hill (2001)

47. Rust, R.T., Oliver, R.L.: Should we delight the customer? J. Acad. Mark. Sci. **28**, 86–94 (2000)
48. Schneider, B., Bowen, D.E., David, E.: Understanding customer delight and outrage. Sloan Manage. Rev. **41**, 35–45 (1999). https://doi.org/10.1108/14637150310496758
49. Shibly, S.A., Chatterjee, S.: Surprise rewards and brand evaluations: the role of intrinsic motivation and reward format. J. Bus. Res. **113**, 39–48 (2020). https://doi.org/10.1016/j.jbusres.2020.03.009
50. Teigen, K. H., Keren, G.: Surprises: low probabilities or high contrasts? Cogn. **87**, 55–71 (2003). https://doi.org/10.1016/s0010-0277(02)00201-9
51. Valenzuela, A., Mellers, B., Strebel, J.: Pleasurable surprises: a cross-cultural study of consumer responses to unexpected incentives. J. Consum. Res. **36**, 792–805 (2010). https://doi.org/10.1086/605592
52. Vanhamme, J.: The link between surprise and satisfaction: an exploratory research on how best to measure surprise. J. Mark. Manag. **16**, 565–582 (2000). https://doi.org/10.1362/026725700785045949
53. Vanhamme, J., Snelders, D.: The role of surprise in satisfaction judgements. J. Consum. Satisf. Dissatisfaction Complain Behav. **14**, 27–45 (2001)
54. Wang, X.: The effect of unrelated supporting service quality on consumer delight, satisfaction, and repurchase intentions. J. Serv. Res. **14**, 149–163 (2011). https://doi.org/10.1177/1094670511400722
55. Westbrook, R.A.: Product/consumption-based affective responses and postpurchase processes. J. Mark. Res. **24**, 258–270 (1987). https://doi.org/10.2307/3151636
56. Wiastuti, R.D., Lestari, N.S., Ngatemin, B.M., Masatip, A.: The generation z characteristics and hotel choices. African J. Hosp. Tour. Leis. **9**, 1–14 (2020)
57. Wu, L., Mattila, A.S., Hanks, L.: Investigating the impact of surprise rewards on consumer responses. Int. J. Hosp. Manag. **50**, 27–35 (2015). https://doi.org/10.1016/j.ijhm.2015.07.004

Chapter 39
The Social and Financial Impact of Influencers on Brands and Consumers

Inês Melo and José Luís Reis

Abstract With the growth and development of Instagram, it is more possible to see brands using this platform, not only to show and promote their products and services but also to collaborate with various influencers to increase visibility with their target audiences. These partnerships are based on hiring people with some influence on Instagram, with intrinsic and brand-like characteristics, for them to disclose the products and services to their followers. This research seeks to understand the social and financial impact of influencers on brands from the point of view of influencers and consumers. A mixed, quantitative and qualitative methodology was used to elaborate this work. The quantitative data were obtained through a questionnaire to Instagram users. The results were analysed to understand the perception of digital influencers and consumers, in relation to the social and financial impact of marketing strategies on social networks, by the influencers in the dissemination of products and services of brands. Qualitative data were obtained through interviews with nine influencers. The results obtained allowed us to realize that there are several characteristics inherent to the influencers and their work that positively influence the consumer's purchasing decision, namely the proximity between influencer and follower, opinion about brand or product, recognition/credibility of the influencer and brand, number of followers of the influencer and constant communication. It was also possible to conclude that recognition/credibility positively affects the collaboration between brands and the influencer and the financial impact of brands.

I. Melo · J. L. Reis (✉)
University of Maia - ISMAI, Avenida Carlos de Oliveira Campos, Maia, Portugal
e-mail: jreis@ismai.pt

J. L. Reis
LIACC - University of Porto, FEUP, Rua Dr. Roberto Frias, Porto, Portugal

J. L. Reis et al. (eds.), *Marketing and Smart Technologies*, Smart Innovation, Systems and Technologies 337, https://doi.org/10.1007/978-981-19-9099-1_39

39.1 Introduction

The development of digital marketing is associated with partnerships between brands and influencers in digital platforms. The use of influencers in marketing strategies by companies is increasing, which makes the development of this research relevant to better understand this phenomenon, being this a current theme that aims to understand the importance of influencers for brand strategies and how influencers are accepted by consumers. This research aims to understand the work of influencers from different points of view, analysing the repercussion that influencers can exert on brands and their marketing strategies, how consumers receive and recognize influencers, as well as they recognize the work they do with brands.

This work, in the first part, presents information about the evolution of marketing, the decision-making process of buying consumers, the concept of social networks and influencers, and how these are the main elements of influencer marketing. In the second part, the research methodologies are presented, including de hypotheses, the conceptual model and the data collection method. This is followed by an analysis of the results of the questionnaires and interviews and the validation of the hypotheses. Finally, the conclusions, limitations and future work are presented.

39.2 From Traditional Marketing to Digital Marketing

Marketing, since the 50 s, has been failing to focus so much on the product (the so-called Marketing 1.0) to focus on the consumer (Marketing 2.0). However, there are currently more and more transformations [1].

Marketing involves dealing with a market that is constantly changing, and it is necessary to always understand the evolution of the market. It is this constant advance that also causes marketing to change and adapt more and more [2]. Marketing, in general, is the discipline capable of dealing with the needs of consumers and their wills, which in turn will bring return to brands, whether financial as social.

Traditional marketing strategies were to communicate with their customers on a programmed schedule at specific times, whether festive seasons or promotions. However, it is verified that, with the introduction of the Internet and the passage of these strategies to the digital world, this communication has become continuous and creating possibilities for multidirectional interaction between brands and consumers [3]. The communication model has gone from “one-to-many” to “many-to-many”, now allowing for greater interaction between businesses and consumers.

In this way, digital marketing can be defined as the development of marketing strategies, in order to promote products or services, through digital channels and electronic devices, such as computers, mobile phones and tablets [4].

With the evolution of the Internet, people now have the possibility to choose what they want to see, where and when they want to do it. They also end up becoming own content producers by being able to share their opinions and ideas, interacting with others on a channel or platform, now more digital [5].

39.3 Consumer Buying Process

Consumers are influenced by three factors when they want to make purchasing decisions [2]. The first is the communications that marketing departments create for the products; second, the opinions of friends and family; and third, their own knowledge and feelings in relation to a particular product and brands. In view of the whole decision-making model, it is increasingly becoming more and more that both the recognition of needs and the information collection are increasingly influenced by information available on the Internet and on digital platforms. With the increasing use of the mobile network, consumers are now connected and with access to crowds of people who can help in the decision-making process of purchasing a product or service. Given this, most decision-making becomes social decisions now, since consumers currently communicate with each other and can influence each other on a brand or product [2].

The new generation of people draws inspiration from digital personalities and sees them as a reference to their decision-making regarding any product or service. Young people aspire to be like these figures since they believe they are ordinary people, so they accept their words and opt for products referred to them [6]. In addition to the opinions of nearby groups, young people are also attracted by current trends in areas such as sports, music, art and fashion. Many of these trends end up being currently on digital platforms and addressed not only by other young people but also by these new opinion leaders and influencers [2].

39.4 Social Media

One of the points of digital marketing goes through social networks and marketing strategies using these platforms to promote products and services to increase the visibility of brands. [6].

Social networks are increasingly present in the marketing communication of companies that have begun to realize, as Campbell and Marks say, "that sales and promotional messages are not usually the kind of messages consumers want to see" on these platforms, but rather more informal content [7].

Social networks become increasingly a space for the introduction of marketing strategies, being the main means of communication in digital marketing to serve customers, clarify doubts and generate traffic to some website and a great complement to other existing digital media [4]. They need to be managed effectively to produce long-term results, becoming a way to create more loyal relationships and dialogues with the target audience of each company [8].

Social networks seem to facilitate and enable communication between a greater number of people, attracting proximity between brands and their consumers while also providing greater competition between companies and to capture more attention from customers, brands now have more and more to compete while maintaining credibility and responding to the needs of their target audience [9].

39.4.1 Instagram

Instagram is an Internet-based application and a social network that aims to share images, with also integrated features such as IGTV and Instagram Stories [10]. It turns out to be a creative tool where there are millions of people connected to exchange information and opinions on various subjects [15].

With the increasing development of this digital platform, several brands and companies began to take advantage of its functionalities and use it as one of the tools in their digital marketing strategies to disseminate projects, products and even show their entire commercial identity and entrepreneurship [6]. It was on Instagram that the "micro-celebrities" were set off, now referred to as influencers, where they see their work recognized [13].

39.4.2 Influencers

As the number of Instagram and Facebook users began to emerge, "professionals who use social networks to influence the decision-making of their followers" have begun to emerge as well [4].

There are several terms to designate the user who produces content on digital platforms, from digital influencers, online opinion former, content producer, creator to youtuber, blogger, vlogger, etc. This user shares on social media platforms or even on their own digital properties such as websites and blogs and may or may not receive money [14].

Currently, the significant growth of influencers that can be found on social networks such as Instagram, Facebook, Snapchat, YouTube and TikTok is noticeable [15]. The influencers are recognized as experts in this field, share content that is seen by thousands and even millions of people, generating with them an engaging community with very large interaction and engagement rates [16].

These opinion leaders become "content creators with a large number of followers on digital platforms who are able to influence others through their behaviours and attitudes" [17]. They are chosen and selected by various brands to disseminate various products, through their opinions.

39.4.3 Influencer Marketing

As the number of influencers and the number of users of the various social networks grows, companies begin to realize that the latter joining the influencers and become a great platform to activate their marketing strategies. Thus, a new marketing term is born, “Influencer Marketing”.

Influencer marketing is the art and science of engaging people who are influential online to share branded messages with their audience in the form of sponsored content [17]. It is a technique that is growing, since it benefits from lower costs, faster and more specificity in relation to its targets.

Brands must be able to identify opinion leaders, based on various characteristics, such as the number of followers or traffic they generate on a web page, so that they can help organizations connect naturally and spontaneously with their customers [9, 16]. They must be confident and convey this since doing so also promotes confidence in the brand [18].

39.5 Research Methodology

In this section, the methods used in this research are identified and described, which is divided into three parts; in the first part, the objectives of the research, in which the problem and the general and specific objectives to be achieved will be evaluated; in the second part, the research hypotheses and the conceptual model are presented; and in the third part, the methodology that will respond to the problem will be selected and, finally, the techniques of data collection appropriate to meet the objectives.

39.5.1 Objectives

The overall objective of this work is to understand how the work of influencers on the social network Instagram influences the perception about brands, as well as the perception that consumers, have about influencers. Regarding the specific objectives of the research, it focuses on four distinct objectives to answer the question set out in the general objective:

- Realize the importance of the work of influencers with brands;
- Understand consumers’ perception of influencers and their influence on purchasing decisions;
- Analyse the size of brands and the work of influencers on the social network *Instagram*;
- Understand the work of influencers and their perception of working with brands.

39.5.2 Research Hypotheses, Propositions and Conceptual Model

Seven research hypotheses were formulated that aim to respond to the specific objectives formulated in the previous point, especially the first and second objectives—to understand the importance of the work of influencers with brands; and understand consumers' perception of influencers and their influence on purchasing decisions.

The following research hypotheses and proposition have been formulated following the objectives proposed for the elaboration of the research:

H1: The proximity between follower and influencer has a positive influence on the decision to purchase a product.

H2: An influencer's opinion of a brand or product is a positive factor in consumers' purchasing decision.

H3: The recognition and credibility of an influencer have a positive impact on consumers' purchasing decisions.

H4: The number of followers of an influencer is a key factor in the decision to purchase a product.

H5: The constant communication between influencer and follower makes an impact on the purchase decision greater.

H6: Brand recognition is positively influenced through collaboration with influencers.

P1: Brand and influencer recognition has a positive factor in the financial impact of brands.

According to the literature review that served as the basis for the research, the conceptual model that supports the research was constructed in accordance with the objectives and hypotheses defined—see Fig. 39.1. This model serves to understand the relationship between the various constructs selected for the investigation.

As seen in Figure 39.1, it is intended to understand the relationship between proximity and purchase decision; how opinion of a product or brand influences the purchasing decision; how the recognition/credibility of an influencer is important for purchasing decision-making; how the number of followers can be an important factor in the purchase decision; and whether constant communication by influencers positively interferes with consumers' purchasing decisions. Next, it is intended to understand if the collaboration between brands and influencers brings a good recognition to them and whether this recognition / credibility has a positive financial impact.

39.5.3 Data Collection Method

This research presents a mixed methodology since it presents a quantitative analysis, collected through an online questionnaire to consumers, and a qualitative analysis,

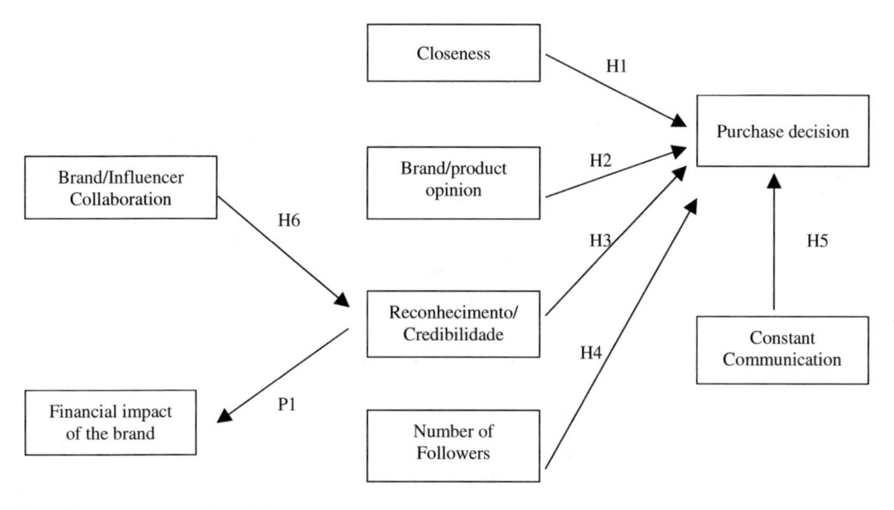

Fig. 39.1 Conceptual model

collected from an e-mail interview to some influencers. Using the mixed methodology, it is intended to give preference to data validation and its compliance with the object under study [12]. Since we cannot say that any methodology is perfect, both presenting advantages and disadvantages, the use of both allows a better interpretation of the data, combining different approaches to better study the same phenomenon.

The sample used for this investigation is a random sample for convenience, in two categories of individuals. On the one hand, we have a set of individuals/consumers, who are users of social networks, who answered a questionnaire, which obtained 305 valid answers. On the other hand, we have a set of 9 digital influencers who responded to an interview via email.

39.6 Analysis of Interviews

The aim of the interview is to understand the point of view of digital influencers in what is related to working with brands on Instagram, as well as the perception of the impact they can have on consumers' purchasing decisions. The collection of influencers for the sample went through research by the works carried out between influencers and brands that the researcher knows, as well as their acceptance for the interview.

The interviews were conducted with a pre-defined script and included 9 participants that answered everything through email, given the situation of COVID-19.

It was possible to understand that the influencers give great importance to the identification with the brand or product and to the interest and curiosity of themselves or their followers when they decide to accept some proposal of partnership with a brand. These influencers also claim that they work with brands often sporadically, but one can count a gap between 5 and 23 brands with which the interviewees are working. It was also possible to realize that the work between brands and influencers is beneficial for brands since, according to the interviewees, the greater reach of their accounts and the identification with their followers causes the visibility of the brand to increase.

When asked if they believed that their followers only buy products because they mention it, the interviewees responded positively, although they stated that it sometimes depends on the type of product and the type of followers.

Regarding the work with brands and the perception that followers may have, the interviewees unanimously believe that the more solid and long the relationship with a brand, the more their followers believe not only in the products, but in their honesty.

39.7 Analysis of the Questionnaire Results

The launch of the questionnaire was made through the social network Facebook, obtaining a total of 312 responses to the questionnaire. This was limited to individuals aged 18 years or older, of both genders, who use the Internet and social networks, specifically Instagram and were considered 305 valid responses for this study. In this questionnaire, it was chosen to provide the answers to the questions by completing a 5-point Likert scale, 1 meaning "Totally Disagree" and 5 means "Totally Agree", to obtain more credible results and a better statistical evaluation.

Regarding responses to the frequency of use of Instagram, it is noticed that 95.4% of respondents visit this social network every day and 70.8% of the respondents are between 1 and 5 h a day on the social network. To understand if respondents consider it important that brands are present on the social network Instagram, 51.5% fully agree with the statement.

Regarding the importance of proximity between influencer and follower, 41.3% of respondents agree that the closer the proximity, the greater the recognition of a need and the search for information about a product.

When confronted with the statement "I only follow influencers with a large community of followers", to perceive their position on the number of followers of an influencer, 39.3% responded totally disagreed, showing that a larger number of followers is not the only factor in the purchasing decision. As for the belief that the more an influencer interacts with its followers, the greater the power of influence in the purchasing decision, 43.6% partially agreed on—see Table 39.1.

39.7.1 Exploratory Factor Analysis

To perceive the veracity of the information obtained with the respondents' answers, an exploratory factor analysis was performed on constructs that have a correlation between them. In this analysis, four instruments were performed to test: the first,

Table 39.1 Frequency of influence of interaction with followers in purchasing decision

Q14: The more you interact with an influencer with your followers, the more power of influence you have in deciding to purchase a product		
	Frequency	Percentage (%)
I totally disagree	8	2.6
I partially disagree	24	7.9
Indifferent	53	17.4
I partially agree	133	43.6
I totally agree	87	28.5

Bartlett's field test, which intends to test whether the first hypothesis of the correlation matrix is identity or not; the second test is the KMO test (Kaiser–Meyer–Olkin) that aims to evaluate the intensity and variance of the analysis and its variables; in the third phase, in order to evaluate the quality and internal consistency of the factors, Cronbach's alpha is used; and finally, in order to interpret the factors obtained, the indicator of total variance is used.

After verifying each of the variables, it was possible to perform factor analysis at a total of four constructs, and three of them, because they are identity variables, have no correlation to be verified in factor analysis.

Thus, it was possible to verify the results obtained in each construct and it is concluded that:

- All variables are one-dimensional and extracted 1 component in each analysis.
- In the Bartlett scouting test, all variables achieved a result of 0.000, meaning that there is a correlation between them.
- In the KMO test, the variables obtained values between 0.500 and 0.714, which demonstrates results between admissible and good, respectively.
- Referring to the percentage of variance, values between 48.157 and 83.427% can be observed.
- Finally, verifying the consistency and quality of the variables, the values reached in Cronbach's alpha are between 0.451 and 0.811.

39.8 Validation of the Hypotheses and Propositions

Pearson's correlation analysis was used to validate the hypotheses and the proposition created for this study since it provides proof of the relationship between the constructs in each hypothesis.

Thus, it can be concluded that all the hypotheses and the proposition proposed in this investigation are valid, within their scales and values, as can be seen in summary in Table 39.2.

In conclusion, it is possible to perceive that for the validation of all hypotheses and propositions, the two methods chosen for this analysis resulted in conclusions

Table 39.2 Summary of hypothesis validation

Hypotheses/Propositions	Correlation	Validity of hypotheses
H1	0,246	Valid
H2	0,499	Valid
H3	0,479	Valid
H4	0,327	Valid
H5	0,247	Valid
H6	0,320	Valid
P1	In interviews	Valid

that allowed to validate all hypotheses, both from the point of view of the influencers and from the point of view of consumers.

39.9 Conclusions, Limitations and Future Work

This research lies in understanding the work of digital influencers present in social networks, particularly Instagram, for brands as well as for consumers. With the main objective of understanding how the work of influencers, on Instagram, achieves the perception of brands as well as the view of consumers towards this work, it was possible to ascertain, throughout the investigation, that the reach that these influencers have in their digital accounts allows to increase the visibility of the brand, by making known to their consumers the products that later also end up being purchased, indirectly influencing the sales of the brand with which they collaborate. Given this, it becomes important for those responsible for brand marketing, when developing marketing strategies, bet on working with influencers with characteristics that meet the brand's vision. In the view of consumers, it was possible to realize that many characteristics that influencers value—such as proximity, recognition and credibility, constant communication, opinion they transmit and the number of followers—result in a positive influence at different levels in their purchasing decisions.

Regarding the size of the brands on Instagram and the work of the influencers in this social network, it was possible to understand that, from the point of view of consumers, they believe that the presence of brands in this social network is important. From the point of view of influencers, Instagram turns out to be the favourite social network for collaboration with brands since it has several tools and at the same time achieves greater and better communication with its followers.

In relation to the work of influencers with brands, they consider that all the partnerships they carry out end up being beneficial for both sides once the visibility and reach of the two parties are increased. In addition, it was also possible to perceive, through the view of the influenced interviewees, that even if the types of collaborations vary, the longer a partnership with a brand, the better the results, since these partnerships demonstrate greater honesty on the part of the influencers and increase the interaction with their followers.

Finally, regarding the limitations of this study and in order to have guidelines for future studies, it is important to highlight that the results obtained are limited, and the global population cannot be generalized since the sample obtained is not probabilistic. Thus, for future work in this area, even with inspiration from the results of this research, it will be necessary a study with a more comprehensive population to obtain a more generalized result and closer to reality. Another limitation found was the lack of collaboration on the part of brands to answer some questions, which would be used to understand their point of view in this investigation, resulting only in the analysis of the point of view of influencers and consumers, thus being important in future studies that the view of the brands be studied.

References

1. Kotler, P., Kartajaya, H., Setiawan, I.: *Marketing 3.0.* Actual Editora (2011)
2. Kotler, P., Kartajaya, H., Setiawan, I.: Marketing 4.0: do tradicional para o digital. Actual Editora (2017)
3. Teixeira, C.: *Influência das Redes Sociais na construção de uma marca. Estudo de Caso: Officelink* (2017)
4. Faustino, P.: *Marketing digital na prática: como criar do zero uma estratégia de marketing digital para promover negócios ou produtos.* Marcador (2019)
5. Silva, C.R.M. da, Tessarolo, F.M.: Influenciadores Digitais e as Redes Sociais Enquanto Plataformas de Mídia. *XXXIX Congresso Brasileiro de Ciências Da Comunicação*, 14 (2016)
6. Oliveira, M., Barbosa, R., Sousa, A.: *The Use of Influencers in Social Media Marketing.* In: Rocha, Á., Reis, J.L., Peter, M.K., Bogdanović, Z. (eds.), pp. 112–124. Springer, Singapore (2020)
7. Couto, A.F., Brito, P.Q. de.: *Tactical Approaches to Disclose Influencers' Advertising Partners.* In: Rocha, Á., Reis, J.L., Peter, M.K., Bogdanović, Z. (eds.), pp. 88–100. Springer Singapore (2020)
8. Costa, A.C.F.: *Impacto das Redes Sociais no Marketing.* 57. (2013) https://eg.sib.uc.pt/bitstream/10316/24613/1/Relatório_CarolinaCosta.FEUC.pdf
9. Soares, D.: *A análise comportamental dos pais e jovens perante presença de microinfluenciadores na rede social Instagram.* Instituto Português de Administração e Marketing do Porto (2020)
10. Kristian, H., Journal, I., Kristian, H., Bunawan, S.G., Wang, G., Sfenrianto, S., Marketing, A.D.: Social user behavior analysis of purchasing decisions in instagram online store. Int. J. Emerg. Trends Eng. Res., 8– 2(February) (2020)
11. Bessa, A.R.: *Influenciadores em Redes Sociais Digitais: Uma análise aplicada ao Instagram.* Universiade de São Paulo—Escola de Comunicações e Artes (2018)
12. Portelada, B.: *Os Influenciadores Digitais e a Relação com a Tomada de Decisão de Compra de seus Seguidores* [Instituto Português de Administração e Marketing do Porto] (2019). http://hdl.handle.net/10400.26/33784
13. Korotina, A., Jargalsaikhan, T.: Attitudes towards Instagram micro-celebrities and their influence on consumers' purchasing decisions. *Master Thesis in Business Administration*, 69 (2016). http://www.divaportal.org/smash/get/diva2:950526/FULLTEXT01.pdf
14. Terra, C.F.: Marcas e Influenciadores: quem precisa de quem? *X Simpósio Nacional Da ABCiber* (2017)

15. Oliveira, J. De, Neto, Á.C.J.: Marketing Digital como influenciador de moda: relatos sobre o comportamento de compra com base nos digital influencers. South Am. Develop. Soc. J., **4**(11), 254 (2018). https://doi.org/10.24325/issn.2446-5763.v4i11p254–265
16. Castelló, A.: El Marketing de Influencia: Un Caso Práctico, pp. 49–65. Tendencias Publicitarias en Iberoamérica. Diálogo de Saberes y Experiencias. Alicante, Revista Mediterránea de Comunicación, Julio (2016)
17. Silva, S., de Brito, P.Q.: *The Characteristics of Digital Influencers and Their Ethically Questionable Attitudes*. In: Rocha, Á., Reis, J.L., Peter, M.K., Bogdanović, Z. (eds.), pp. 101–111. Springer Singapore (2020)
18. Glucksman, M.: The rise of social media influencer marketing on lifestyle branding: a case study of Lucie Fink. Elon J. Undergraduate Res. Commun. **8**(2), 77–87 (2017)

Chapter 40
Has It Ever Been the Fashion Blog's Dusk? A Thematic Analysis-Based Research on the Anguishes in the Post-transition from Text Blog Writers to the Ready-Made Scroll, Scroll, Scroll Instagram Images

Maria Inês Pimenta and José Paulo Marques dos Santos

Abstract The first decade of the current century witnessed the emergence of fashion blogs. Although initially bloggers were sawn as interlopers by established fashion journalists, the power of the large audiences of the former imposed their acceptance in the fashion industry, pushing changes in the status quo. The early 2010s were bloggers' golden age, but then social networks emerged with hordes of ad-lib content co-creators, culminating in the indistinguishable Instagram influencers and Instagrammers. At the beginning of the 2020s, it is questionable if there is still a place for fashion bloggers and their blogs. We conducted semi-structured interviews with fashion consumers in Portugal, aiming to capture their perspective, rationale and prognoses for this sector of the digital marketing arena. Data is analysed under the thematic analysis method. Results are interpreted under the modernity / postmodernity paradigm. Although the media differ, fashion journalists and professional fashion bloggers do not differ in their essence. They inscribe in the modernity paradigm. Their creations are well-structured and receive important investments in their production. Authors speak backed by authority, aligning with the Foucauldian "regimes of truth". Influencers and Instagrammers inscribe in the postmodernity paradigm. Fragmented lives consume (and produce) fragments of everything. The subject is decentred and is comfortable with the hype(r)-reality, and the producers are the images and the products of their authors. It is unlike that there will be a place for bloggers again as postmodernist fluidisation is underway… until one rupture relieves the tensions.

M. I. Pimenta · J. P. Marques dos Santos (✉)
Universidade Europeia, Lisboa, Portugal
e-mail: jpsantos@umaia.pt

J. P. Marques dos Santos
University of Maia, Maia, Portugal

J. L. Reis et al. (eds.), *Marketing and Smart Technologies*, Smart Innovation, Systems and Technologies 337, https://doi.org/10.1007/978-981-19-9099-1_40

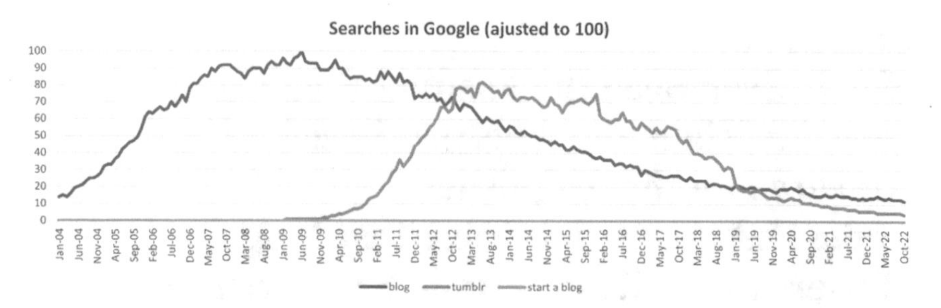

Fig. 40.1 Searches (adjusted to 100 for the higher value) in Google of the terms "blog", "tumblr" and "start a blog" since January 2004. Because of its low magnitude, the line corresponding to the term "star a blog" overlaps the abscissa, and it is detailed in Fig. 40.2. *Source* Google Trends

40.1 Introduction

In a recent article in Forbes, Mimi Polner and Cassie Bottorff explain "how to start a blog and make money in 2022".[1] Eight years before, Onur Kabadayi announced in The Guardian that "Bloggings is dead, long live blogging".[2] He explains:

> Blogs haven't disappeared – they have simply morphed into a mature part of the publishing ecosystem. The loss of casual bloggers has shaken things out, with more committed and skilled writers sticking it out. Far from killing the blog dream, this has increased the quality of the blogosphere as a whole.

He concludes by stating that "bloggers should be excited, not downhearted. (…) best years are ahead of it". However, by the end of 2017, Farah Mohammed sums up in the JSTOR Daily article "The Rise and Fall of the Blog"[3]:

> Today, writers lament the irrelevance of blogs not just because there's too many of them; but because not enough people are engaging with even the more popular ones. Blogs are still important to those invested in their specific subjects, but not to a more general audience, who are more likely to turn to Twitter or Facebook for a quick news fix or take on current events.

According to Google Trends, the search terms "blog", "tumblr" and "start a blog" have been evolving, as depicted in Figs. 40.1 and 40.2, since January 2004.

It seems that opinions are contradictory, although the data, which, however, refers to searches in Google and is not a direct measure of blogs' interest, suggests that the golden era of bloggers and blogs is gone, and they are leaving their dusk today. Pedroni et al. [9] recognise that the golden era is past, and bloggers have entered the silver age. They split the bloggers into three groups: at the bottom, the amateurs, which are the vast majority, the lovers, who maintain their blogs due to their

[1] https://www.forbes.com/advisor/business/start-A-blog/.

[2] https://www.theguardian.com/media-network/media-network-blog/2014/jul/16/blogging-dead-bloggers-digital-content.

[3] https://daily.jstor.org/the-rise-and-fall-of-the-blog/.

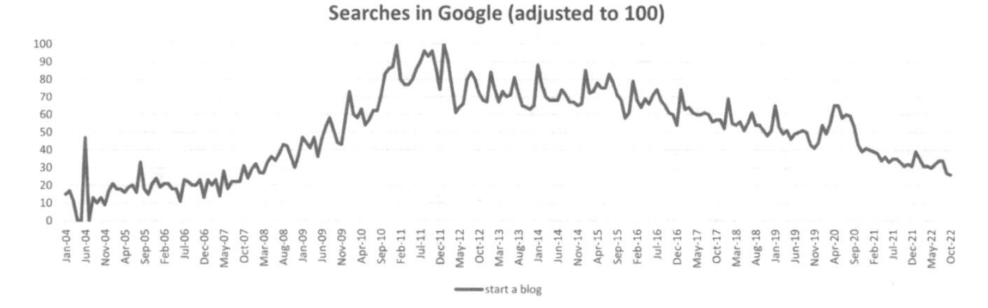

Fig. 40.2 Searches (adjusted to 100 for the higher value) in Google of the term "start a blog" since January 2004. *Source* Google Trends

passion for fashion; and, on top, the professionals, who have an international reputation and accordingly structure. Professional bloggers have an established position, mediating companies and brands, and fashion consumers, and influencing the latter decision-making process [11]. Nonetheless, in 2014, Robin Givhan, in her article "The Golden Era of 'Fashion Blogging' Is Over" in the Cut, questioned[4]:

> Fashion followers can thank bloggers for making fashion coverage more democratic (…) But, now that so many bloggers have been embraced by the industry (…) is there still an opportunity for new voices at shows? And if so, what kind of voices can still flourish?

Hence, these are the same questions that guide the present research. We conducted semi-structured interviews with fashion consumers in Portugal, aiming to understand more specifically:

- How do consumers see fashion blogs' past, present and future in Portugal?
- Is there still an opportunity for new voices (blogs)?
- Have we really entered a new stage in the history of this relatively recent social phenomenon (fashion blogs)?
- What pros and cons have social networks brought to fashion blogs?
- Has social media come to revolutionise the way fashion blogs are perceived?
- Have digital fashion influencers replaced fashion bloggers?

40.2 Method

40.2.1 Participants

This study contains five face-to-face semi-structured interviews with five women of different ages, ranging from 19 to 53, to have a longitudinal perspective. The chosen participants are female because it is the gender that follows more fashion blogs. Fictitious names are used to ensure the participants' privacy, and their characterisation is shown in Table 40.1.

[4] https://www.thecut.com/2014/04/golden-era-of-fashion-blogging-is-over.html.

Table 40.1 List of participants and their characterisation

Fictitious name	Age	Marital status	Children	Occupation	Field
Sophia	19	Single	No	Student	Architecture
Rachel	25	Single	No	Working-student	Researcher (PhD) in the environmental area
Maria	26	Single	No	Worker	Doctor
Clare	37	Married	3	Worker	Fashion
Amelia	53	Married	2	Worker	Teacher

Sophia is 19 years old and lives in the North of Portugal. She entered university this year, where she is studying Architecture. She is single and has no children. However, the area of fashion is a personal interest of hers, and it was this interest that led her to research and follow fashion blogs.

Rachel lives in the North of Portugal and is 25 years old. She is a student-worker, doing a PhD and working in the environmental area. She started following fashion blogs in her adolescence, where she decided to create her blog. The fashion area continues to be a personal interest until today. She is single and has no children.

At 26 years old, Maria is single and has no children. She is a dedicated doctor who lives in Northern Portugal. The area of fashion is a personal interest of hers, which began in her adolescence when she built her blog because of her curiosity in the area of blogs. Nowadays, she still keeps up with it.

Clare is 37 years old, married and has three children. Fashion has always been an area of her personal and professional interest. She works in a multinational company, where she employs specialised positions in the fashion area. She lives in the North of Portugal.

Amelia is married, has two children, lives in the North of Portugal and is 53 years old. Fashion is like a hobby; she used to follow trends through fashion magazines. She is a textile engineer, and at the moment, she works as a teacher.

40.2.2 Semi-Structured Interview

The semi-structured interview was prepared with the following objectives in mind:

- Cross-referencing and relating information.
- To understand the perspectives of female blog consumers towards the present, past and future of fashion blogs;
- To understand the importance that social networks have for the communication strategies of bloggers;
- To understand whether social media has revolutionised how blogs are perceived;
- Understand the role of digital fashion influencers vis-à-vis fashion bloggers;
- To achieve an overview of how fashion blogs are positioned in Portugal.

The semi-structured interview is composed of seven questions. There is a progression, starting with general questions to understand the interviewees' opinions about fashion blogs, followed by questions related to social networks, the digital world and the same blogs. The emergence of digital fashion influencers is a pivotal point in understanding the future of fashion bloggers. The following questions focus on these topics. One question, particularly, is where fashion bloggers' or digital fashion influencers' pictures are shown. First, the interviewee has to identify if they knew the person and if the same is a fashion blogger or a digital influencer. In this question, the aim is to understand if the interviewees can, in practice, distinguish what a digital influencer is and what is a fashion blogger. Next, the question focuses on four sentences where the interviewees have to say their degree of agreement with the sentence and explain the reason. We intend to demystify what people think about certain subjects through these questions. The last question focuses on the main reason for the interview.

In detail:

Question 1: Tell me a bit about how you started following Portuguese fashion blogs and the reasons and motivations

Themes to explore:

- Reasons and motivations to follow;
- Whether the environment in which the interviewee is inserted is favourable or not for her to get to know blogs;
- Means of discovery of blogs;
- The method you use when you want to visit the blogs you follow.

Question 2: What characteristics come to mind when you think of a "fashion blogger"?

Themes to explore:

- To understand the adjectives and memories that interviewees have about "fashion bloggers";
- To understand at what moments the interviewee associates the "fashion bloggers".

Question 3: What is the difference between a fashion blogger's social media and the blog?

Themes to explore:

- Which social networks do you use the most to follow the bloggers you like the most;
- What is the opinion of the interviewees about: whether social networks have come to replace blogs;
- If you think social media has come to transfer more merit to people who own a fashion blog;
- What are your views on Instagram as a content viewing tool;

- What's your opinion on fashion vlogs.

Question 4: Tell me about your opinion on fashion digital influencers

Themes to explore:

- What distinction do you make between a digital fashion influencer and a fashion blogger;
- Whether they understand the difference between the two.

Question 5: Can you identify digital influencers or bloggers from the fashion/lifestyle sector on this list? I will show you the pictures one by one, and you will tell me if you know who it is and if you think it is active

Question 6: I'm going to say a few sentences, and tell me how much you agree or disagree and why?

Themes to explore:

- I met the bloggers I follow through social media (Instagram; YouTube; Facebook; etc.);
- Whenever I want to check blog X, I go directly to his blog address;
- I view vlogs (YouTube) more compared to blogs;
- I rarely go to a blogger's blog; however, I regularly follow social media.

Question 7: What are your expectations for the future of fashion blogging?

Themes to explore:

- Whether blogs should adopt specific changes to grow the blog and which ones;
- If there is still room for new fashion bloggers in Portugal;
- If fashion bloggers spread a style.

40.2.3 Data Analysis with Thematic Analysis

The method of analysis of the semi-structured interviews is thematic analysis, which contains six phases in sequence [2]: (1) familiarisation with the data, (2) creation of initial codes; (3) search for themes; (4) review of themes; (5) definition and naming of themes; and (6) report writing. According to the developers, "Thematic analysis is a method for identifying, analysing and reporting patterns (themes) within data. It minimally organises and describes your data set in detail." [2], p. 79. In parallel, for [1], it is essential to follow three guidelines: describe, compare and relate. First, it is necessary to describe in detail the context of the study and all its pertinent details. Next, comparing to understand the differences in characteristics and boundaries between the themes is essential. Finally, by asking questions, one must relate categories and themes through other details.

The first stage focuses on familiarisation with the data to become aware of the vastness and depth of the content. Before reaching the next stage (coding), it is essential to read all the data so that the analyst is exposed to all ideas and possibly identifies appropriate patterns early on. The transcription of verbal data is done at this point, which helps this familiarisation process. All oral, as well as non-verbal utterances are reported in detail.

The second stage focuses on coding. In this stage, one must have present all the data, analyse it separately and compare it, which allows for identifying recurring aspects that help form patterns that lead to the themes. The coding applied is of the open type.

The third phase focuses on the search for themes. Since the data have been previously coded, it is necessary to transform the data summarised in the codes into possible themes. In this stage, some techniques are used to group the themes, such as thematic maps and visual representations.

The fourth stage is the review of the themes against the codes they contain. At this stage, it is important to go back and revisit the coherence of the content of the codes of each theme again. The aim is to refine, "purify" and decide which themes will be kept and which will be discarded or merged. Scrutiny takes place at two levels: at level 1, the themes must be coherent with the content of the codes, and at level 2, the themes must reliably represent the overall data set, producing a thematic map of the analysis.

The fifth phase concerns the definition and naming of themes. The multiple but coherent facets within a theme converge into a name. The names of each theme are precise and solid, translating to the reader the transparent idea of the theme.

The last phase, the sixth, is the production of the analysis report. This report considers all the data for each theme summarised in a memo. After the necessary polishing, the grouping of the memes is the theory that emerges from the analysis, which is grounded in the data. Several snippets of data are presented to give evidence of this foundation.

40.3 Data Analysis and Results

This section presents the data analysis sequence and results, which integrates the six phases of the thematic analysis.

40.3.1 Data Familiarisation

At this stage, the transcripts of the five interviews conducted were made. In addition, all notes and details observed during the interview were reviewed.

40.3.2 Coding

Codes were created for all interviews. In quantitative terms, Sophia's interview generated 48 codes, Rachel's 133, Maria's 71, Clare's 146 and Amelia's 39. The following paragraphs reproduce the most relevant ones for each participant:

Sophia #16: The blog ends up being more tiring (compared to social networks): "(…) in the blog itself it is more descriptive, it is more a text where they write what they do, and it ends up being a bit more tiring (…)";

Rachel #22: To consume content more instantly, Instagram is the best fit: "(…) you have Instagram for that, on Instagram, you easily see all the pictures of all the looks she shared that week in a matter of seconds, so you don't have to go to the blog".

Maria #22: Fashion bloggers currently use social media more than the blog itself: "(…) the older ones that I used to follow have turned to social media a lot, especially Instagram, and I think it's an easier and more accessible way to reach people than the blog itself (…)";

Clare #43: The content in a fashion blog is boring because nobody wants to read: "(…) in a blog the content was boring, it was boring because nobody wanted to read. In these areas that I am telling you about, fashion (…)";

Amelia #13: Instagram is easier to access and faster: "(…) I go there because I think it's easier to access. Being on Instagram is faster".

40.3.3 Search for Themes

Themes emerge from coherent sets of codes. Some of the themes found:

She got to know fashion blogs through her interest in the fashion field:

- She got to know fashion blogs through her taste in the fashion area (Sophia; #1);
- She started following fashion blogs because of her interest in fashion (Clare; #1);
- She began following blogs because of her interest in the area (Maria; #1).

When you really want to visit a specific blog, you go directly to its blog address:

- Go directly to the address of a blog that interests you to follow it (Maria; #10);
- I had saved in thematic favourites the blogs I followed most so that it would be easier to follow (Clare, #17);
- She usually goes to the blog via her blog address to see more complete articles (Sophia; #7);
- Some blogs directly visit the blog address when you want to check their blog (Sophia; #40);
- Only enter their blog if they offer an exciting tip (Amelia, #11);
- You only go to the blog if the content interests you (Amelia, #34).

40.3.4 Review and Refinement of the Themes

During the iterative process of reviewing and refining the themes, which contemplated the two levels, i.e. scrutinising the coherence between codes under the same theme and ensuring that the thematic map reproduces the data set, there was an intermediate convergence to twelve themes. Still, in the end, it reduces to seven:

- How the participants met the fashion blogs, and how did they use or are using
  - How did you meet the fashion blogs
  - Motivations to start following the fashion blogs
  - When you started following fashion blogs
  - Shares, or not, the consumption of fashion blogs with someone else
  - Has, or does not have, an own blog
  - How sees blogs and why
- What represents Instagram to the consumers
- What fashion blogs represent to the consumers
- What fashion vlogs represent to consumers
- Participants' opinions about the digital influencers
- Differences between fashion blogs and social networks
- The future of fashion blogs.

40.3.5 Naming the Themes

The definition and nomination of themes were carried out, taking into account several aspects, such as the research questions, the problem spectrum, the division between social networks and blogs, and the temporal space between opinions. The final convergence is for seven themes (cf. Section 3.4). However, the first theme, which focuses on the magnitude of fashion blogs, how participants started to follow them, and their motivations, is unfolded into six subthemes.

40.3.6 Final Reporting

Clare, Maria and Sophia state that their interest in the fashion area was a strong point for them to get to know fashion blogs. Amelia and Clare, the oldest participants, aged 37 and 53, looked at fashion blogs as a follow-up to fashion magazines because before the existence of fashion blogs, they followed fashion in magazines, and it was through these magazines that they got to know some fashion blogs and started to follow them. Clare "(…) because in the old days we used to buy a lot of magazines to see what other people were wearing and suddenly we could see that quickly on the internet, right, and it was a bit of that (…)". Most participants state that one of how they discovered fashion blogs were through Google searches. Maria says that she started by searching on Google "fashion blogs", and Clare goes further as she states that she met bloggers through this search on Google and then met other people through these people she found on Google.

It is unanimous that the main reason for following fashion blogs is to be inspired by looks and to follow fashion trends. Maria started following fashion blogs to see what the bloggers wore and the shops where they bought clothes. She also found it interesting when they published looks with cheaper pieces to recreate. Rachel, 26 years old, agrees and admits that she has already met new shops through blogs. For example, she got to know Primark through a blogger many years ago. Clare remembers that when she started following blogs, she saved the photos in an image bank for later inspiration "(…) looks to copy, trainers that I didn't know the brand (…) lots of things that I have saved (…)".

Despite significant age differences, the participants started following fashion blogs at more or less the same time. Maria and Rachel comment that they began following fashion blogs when they were more or less 15/16/17 years old, i.e. in adolescence. Clare, 37 years old, comments that she started following fashion blogs more or less ten years ago.

Rachel, a researcher in the environmental area, is the only one who does not interact much with her environment regarding her taste for fashion blogs. However, she states that it is only from her side that she tries to make her friends aware of fashion blogs "In the beginning it was more like a desire of her own (…) it was very much "mine", it was a personal taste (…)". The remaining participants affirmed that they shared this taste with their environment. Sophia, 19 years old, says that she has already met fashion blogs through friends, while Amelia comments that she shares this taste with friends and with her sisters "(…) we share tips: "(…) 'follow this one is interesting, follow that one and I will also tell her: 'look this one is cool', and we exchange ideas (…)".

Maria and Rachel created a blog about fashion and other themes. Rachel still has the blog active, while Maria's is no longer in action. However, Rachel comments that she no longer dedicates much time to the blog: "Yes, without a doubt, also because my free time is completely different and, well, having a blog demands work, it requires you to leave home thinking what you are going to photograph, where, who is going to photograph you. Maybe you want to go out with your friends, with your boyfriend, but knowing that you have to photograph limits you a little bit and then it is much more work (…)". Rachel thinks that her blog's followers don't miss her blog posts. Over the years, Rachel has changed the content of her blog and bet on Instagram, too: "I tend to change my content on the blog myself. While on Instagram, maybe I bet much more on photographing clothes, making more short videos, on the blog maybe I focus more on content posts, more informative, maybe with more text, more boring (…)".

When they want to enter a specific fashion blog, they usually go straight to its blog address. Amelia comments that she only really goes to a blog if the content is exciting. Sophia goes along the same lines, saying that she usually goes when she wants to see a complete article. Clare had saved the blogs she most followed in her favourites, so it was easier to go there.

All participants admit that, now, they no longer open fashion blogs. Clare justifies this attitude by the new technologies, more precisely by the arrival of Instagram and the speed that this network offers, and being so, "(…) one ends up forgetting the blog". Amelia agrees. At this time, she follows the blogs mainly through Instagram. Years ago, she went daily or weekly to the blogs. Sophia says the same and confesses that she follows the blogs that appear in her Instagram feed: "(…) even if I'm not looking for it, it appears in the feed (…)".

You can see the demand for quicker, ready-made content. Rachel even says "You consume content very instantly". Clare concludes that what is fast is good, and Amelia agrees. After all, she thinks the blog is not so used because people want everything faster. Ready-made, at a glance.

Instagram is, without a doubt, the most used social network of the five participants. For them, Instagram is the favourite social network to follow fashion blogs, with advantages over these, even. At 19 years, Sophia follows the bloggers daily and in more detail on Instagram because it is easier to follow the instant and daily content. Maria shares the same opinion: "(…) and I think it's an easier and more accessible way to reach people than the blog". Rachel and Maria comment that on Instagram, they can see everything about the blogger X in seconds, thus not needing to resort to her blog. Maria says: "(…) and there it is, allowing Instagram to make the description of the photos and so everything they did on the blog is much more accessible". Clare even says she prefers using Instagram compared to blogs and other social networks. For Rachel, Instagram offers better accessibility than blogs. There is no need to read this because the content you seek is on the social networks of bloggers and is easier to access.

Instagram is a more instant social network, where the more personal and daily contact is more intense. Clare praises Instagram, seeing it as an instant social network, fast and easy to use and search. Maria compares bloggers' interaction with the public through blogs and through Instagram, where she says that Instagram fosters more shares than the blog. She continues by stating that they don't "have to spend as much time" on a post on Instagram. Rachel comments "The Instagram itself is also made in a way that gives you what you seek (…)".

Rachel says that "(…) it is one of my favourite social networks (…)". Amelia agrees commenting "(…) Instagram, I've been a fan for a long time (…)". However, she also admits that she already liked Instagram more before because she does not like the advertising part of the network.

For the participants, Instagram allows bloggers to forward the Instagram audience to the blog and quickly reach a good range of followers. Some participants say they met bloggers through Instagram. Rachel says, currently "(…) I go to Instagram. I don't go to the search engine looking for 'blogs', no". Sophia also agrees, saying she has met more bloggers through social networks than blogs. Maria mentions that instastories are an Instagram tool where bloggers can take advantage and captivate and refer people to their blog. For Maria "(…) anyone who invests a bit in this can get a range of people who will follow (…) with the social networks and, especially with Instagram, they will be able to captivate the same number and even a larger number of people (…)".

The tools that Instagram offers its users are an asset to a good performance on this platform, such as IGTV, the instastory and the use of hashtags. Rachel comments "(…) on Instagram you can also take advantage of the videos, the instastories; ah! those masks now of Snapchat, you can draw much more attention and people in the background look at that (…)". Maria, Clare and Rachel agree that videos on Instagram are an exciting tool. Maria comments that, with IGTV, the videos on Instagram have started to be longer. Clare agrees, saying that they have the perfect length, so they are not dull "(…) although now with IGTV, they are not boring because if you are in someone's profile, you watch that minute and then if you want, you can keep watching; otherwise, you may leave, which makes all the difference (…)".

The participants' opinion on blogs is diverse. For them, a fashion blog is a personal fashion diary, very descriptive, where they show people who dress very well. Rachel comments "(…) maybe a style icon, maybe what few people use is not (…), but on the other hand, maybe I value more those bloggers who use pieces that I could easily also use (…) it is me wanting to use what they have (…)". Sophia says that she looks at fashion bloggers as fashion journalists who keep a diary of what they wear and states that a fashion blog is more descriptive than social networks. Clare says that when she thinks of fashion bloggers, she automatically thinks of fashion and comments "The blog was a kind of diary, which at the beginning was a kind of personal diary of the person who wrote about it. People would throw up everything they wanted in there (?)". For Clare, it no longer makes sense to use blogs in the fashion area, and the blog platform is no longer the same as it was a few years ago.

Currently, getting to know new bloggers is done through social media. Sophia admits that she has met more fashion bloggers through social networks than from any other source. Likewise, Maria admits that she has met more bloggers through Instagram than any other source. However, it wasn't like this in the old days. Clare agrees. Rachel agrees, saying "(…) nowadays I take my mobile phone, I go to Instagram. I don't go to the search engine to look for 'blogs', no".

A fashion blog is considered something with text and images, a personal diary that is massive. Rachel comments that the blog has text and pictures and doesn't go much beyond that "(…), so it is easy you get bored there (…)". Clare even comments "(…) it no longer works as I was saying a moment ago, having a diary to go and write in there, that's a nuisance".

When we talk about fashion vlogs, opinions diverge. The youngest participants, Sophia, Rachel and Maria, admit they are interested in this content. The older participants, Clare and Amelia, say they have no interest in this type of content. However, Rachel and Maria are neither assiduous consumers of this media type either. Sophia states "(…) I see more fashion vlogs (…) it is more a message passed more quickly (…) because it is visual than reading a text (…) I can see it in more places, and it is not so boring (…)". On the other hand, Rachel says that she watches more fashion vlogs than blogs. For Maria, "(…) while I stopped going to blogs a lot, YouTube is something that I still watch because Instagram also doesn't have videos with a level (…)".

Amelia comments that she doesn't view fashion vlogs and has no interest in discovering YouTube. Clare says that YouTube has never interested her and that she rarely goes there. Only when someone sends her a video to go and watch. She thinks that YouTube is more geared towards her children or people who like technology. For Clare, "(…) I think it's boring (…)". Rachel, while stating that she consumes more fashion vlogs compared to blogs, also says that YouTube does not interest her much. In her perspective, fashion vlogs are directed to a lower age group. She also thinks that people her age have the same opinion. She also comments that "(…) for me, spending 10/15 min watching a video about the whole day of a person without any kind of specific interest for me is not enough (…)".

Rachel quickly says she watches a fashion vlog but also easily gets bored. She thinks the younger generation has more patience to watch endless videos. Sophia believes that fashion vlogs have replaced fashion blogs. However, YouTube is not as instantaneous as Instagram. In this context, Maria thinks fashion vlogs are more dynamic than blogs but that Instagram will overtake YouTube and videos.

Digital influencers are a relatively new concept. To be a digital fashion influencer, it is not necessary to have a fashion blog. Sophia and Rachel are of this opinion. Rachel says that they can be concepts that come together, or not. Sophia says that "(…) because they now through Instagram, which in this case is the best known social network, they can do everything, photos, videos, pass information (…)".

All participants say that digital influencers are a "new way" to promote fashion, and brands are increasingly betting on them as a form of advertising. Sophia even says that digital influencers are the most recent era. Rachel comments that brands increasingly bet on digital influencers for their campaigns: "(…) it's simply an update, it's not futility". Clare agrees, saying that brands have started to enter this market and invite digital fashion influencers to give their faces and get to know their products because this works in promoting a brand.

Only Rachel, 25, says that what digital fashion influencers communicate is authentic and credible. It should be noted that this participant has a fashion blog and relies on social networks to disseminate its content. The same defends that she more easily buys something recommended by a digital influencer because she finds it more accurate: "(…) the way she does it creates much more empathy (…)". She reflects that digital fashion influencers should not be called as such, but rather as content creators: "(…) even I think it makes much more sense to say that you are a content creator than to dominate yourself as an influencer because this is a bit relative, who says you are an influencer or not are your metrics, not you (…)". Clare is peremptory when she says that what digital influencers communicate is neither authentic nor credible. For her, being careful and suspicious of what they say is necessary because they are always trying to sell something. They create expectations of something that does not exist: "(…) In this area of fashion, there is always a product to be sold. Nothing works as we see in the photographs (…) again turned to the commercial area, because everything they are talking about and everything they are wearing, everything they are doing and eating is for profit (…) that is, nothing is real (…)". For her, digital influencers often put their faces to brands they don't even identify with. Maria has the same opinion. However, she is not so drastic. She only states that the content that digital influencers share is not exciting and that not all of them do a good job.

Clare thinks that the world of digital influencers is already saturated. Clare feels that the amount of digital influencers out there is getting boring and that people don't want to see unboxing anymore. As a result, she believes that people will start not trusting them. Maria also thinks that there are too many fashion influencers, but, on the other hand: "(…) I think there is room for everyone, and then one also doesn't get tired of always seeing the same thing (…)". For Rachel, it is as legitimate for an actor as a fashion influencer to advertise a brand. As such, she thinks that digital fashion influencers have a future.

Most participants in the study value a person who has a blog more than someone who is just a digital fashion influencer. Amelia thinks a blog is more structured and thought out than a digital fashion influencer. Sophia is of the same opinion, saying that fashion bloggers work harder than digital fashion influencers. However, she does not take the spotlight away from female fashion digital influencers. Rachel has a divided opinion on this issue: "I would not say believe more. I think having a blog has an increased weight. You complement Instagram with a blog. You know that the person spends some time on that (…). It probably gives you more maturity, but it's a bit relative. But, I think it always has a bit of influence".

The participants say that social networks have replaced fashion blogs. Sophia admits that social networks are replacing fashion blogs because they contain a bit of everything. Maria agrees because she says that bloggers manage to pass all their content to Instagram "(…), so I think that very difficultly the blog will be able to overcome no matter how many changes it makes, it will not overcome the social networks". Since Instagram's new features, Clare believes that blogs will start to be forgotten. Rachel thinks that, given the excess of content that blogs present, social networks have come to replace them: "I think it no longer makes sense for you to put a photo on Instagram. Maybe it can make sense, but you put a photo on Instagram and say to see more photos go to the blog… like no, that photo is good, like there is no need to go to the blog (…)".

Sophia says that for fashion blogs to survive, they should make some changes, captivating the public through vlogs, for example. Clare thinks fashion blogs should be more irreverent, and Rachel says they should reformulate their concept. Maria feels the blog platform should not adopt changes, and Amelia agrees: "now I go to Pipoca's blog and the text is well written. I won't say that she should be more succinct. I don't know. I think it is the speed of access that is easier. I also can't say that changes are needed. I'm not going to say that they are bad (…)". In this context, Amelia comments that there is still room for new fashion bloggers, whereas Clare shares the same opinion, but only if the new fashion bloggers bring something new. Rachel thinks the same: "it's in that sense, I think that nowadays having a blog forces you to stand out for the difference, don't you, to have contents that you may not find so easily elsewhere (?)". Maria says there is no room for new fashion bloggers: "I don't think it's worth it (…)".

Maria, Amelia and Sophia say that fashion blogs are likely to end. Rachel doesn't think so: "I don't think they will end. I think they will reformulate the concept. I think they will be seen more as a search engine than as a way of consuming instant content". Clare can not predict the future of fashion blogs: "(…) I don't think I can predict what will happen in 10 years. Just like I couldn't predict what would happen ten years from now, which is where I am now, and it's amazing what's changed. And it's very fast, very fast".

Table 40.2 shows that the five Portuguese participants have a heterogeneous knowledge, recognising some fashion bloggers to the detriment of others. The blogger "Pipoca mais Doce" is the blogger that all participants know, and all of them say that she has her blog. It should be noted that "Pipoca mais Doce" is one of the oldest bloggers, having had her blog since 2004. Only Clare, 37 years old, recognises "Carlota's Blog". This blog is for mothers and is owned by Fernanda Velaz. It is interesting to note that Clare, with three children, is the only one who knows this blog. Bárbara Inês is a 25-years-old fashion blogger. Except for the older participants, Amelia and Clare, all participants know Bárbara Inês. However, Maria is unsure if she has a blog or not: "I think Barbara Ines is just an influencer. At least, I only know her from there". Concerning Anita Costa & The Blog, everyone knows them except 53-years-old Amelia. Clare comments that Anita Costa "(…) sells a product. She looks good, she does gym classes with Nilton Bala, so everyone wants classes with a PT like hers (…)". Finally, most of the participants recognised the blogger "Stylista", a 34-year-old fashion blogger. Only 19-year-old Sophia and Maria do not know her. Clare comments: "(…) is Maria Guedes, has a brand and has a blog…".

Table 40.2 Recognition of fashion bloggers or fashion influencers

Blogger	Sophia (19)	Rachel (25)	Maria (26)	Clare (37)	Amelia (53)
Pipoca mais Doce	Knows blog and blogger	Knows blog and blogger	Knows blog and blogger	Knows blog and blogger	Knows blog and blogger
Blog da Carlota	Does not know	Does not know	Does not know	Knows blog and blogger	Does not know
Bárbara Inês	Knows blog and blogger	Knows blog and blogger	Knows the blogger only	Does not know	Does not know
Anita and The Blog	Knows the blogger only	Knows blog and blogger	Knows blog and blogger	Knows blog and blogger	Does not know
Stylista	Does not know	Knows blog and blogger	Does not know	Knows blog and blogger	Knows blog and blogger
Maria Vaidosa	Knows vlog & influencer	Knows vlog & influencer	Knows vlog & influencer	Does not know	Knows vlog & influencer
Catarina Filipe	Does not know	Knows vlog & influencer	Does not know	Does not know	Does not know
Bárbara Corby	Does not know	Knows vlog & influencer	Does not know	Does not know	Does not know
Mafalda Patrício	Does not know	Does not know	Knows influencer	Knows influencer	Does not know

Opinions are controversial when talking about "Maria Vaidosa", Mafalda Sampaio, the author of this character. "Maria Vaidosa" is a fashion digital influencer and YouTuber and does not have a blog. Only Clare does not know this digital influencer. All the others do. However, she is not sure if she has a blog or not. Sophia says: "(…) this is Mafalda Sampaio, Maria Vaidosa, this is more of an influencer and makes some videos (…)". The YouTuber and digital influencer Catarina Filipe is not "famous" among the participants. Only Rachel knows her and knows that she makes vlogs. The same goes for Barbara Corby, a digital influencer and YouTuber. Only Clare, 37 years old, and Maria, 26 years old, know who is Mafalda Patrício, a digital fashion influencer.

In sum, the oldest blogger is the one all participants recognise and know that she has a blog. A more specific blog, such as "Carlota's Blog", is recognised by only one participant, coincidentally being 37 years old and having children. With surprise, it is found that the two YouTubers and digital fashion influencers are almost unknown to most. Only Rachel knows them, with Rachel being one of the biggest YouTube consumers in the sample. "Maria Vaidosa" is identified as a digital influencer and someone "who makes videos". However, the participants were unable to remember if she had a blog or not. Finally, the characters who have a blog are those the participants most recognise, which is paradoxical given the consensus that blogs and bloggers are on the decline. However, these blogs and bloggers are followed more through Instagram nowadays.

40.3.7 Results of the Analysis Centred Around the Research Questions

How Do Participants see Fashion Blogs' Past, Present and Future In Portugal? The participants distinguish between the past, the present and the future of fashion blogs. In the past, people viewed blogs as a complement to fashion magazines, as a new way to visualise fashion. It is noticeable that female participants started looking at blogs as a platform of interest to follow fashion trends. They looked at bloggers as fashion journalists.

Clare: "The blog was a kind of diary, which at the beginning was a kind of personal diary of who wrote about the theme (…) People threw up everything in there (…). Some people followed that because they had common interests, but in the meantime, it started to become segmented. Some people talked about everything, and then they started to segment (…)".

Sophia: "(…) it's more like a diary of what you use".

Today, female consumers already look at fashion blogs differently. Most participants state that they no longer open a blog, following it through the bloggers' social networks. They complain that blogs are massive; i.e. there is a lot of text and images and not something so personal. Instead, they have started to look at social networks as something more appealing, where they follow bloggers more daily and intimately. In addition, participants say that they know more bloggers through social networks.

Rachel: "Yes, it is. I agree, ah, because nowadays I pick up my phone, I go to Instagram. I don't go to the search engine looking for "blogs" no".

Clare: "ah, and in a very clever way, it no longer works as I was saying earlier, having a diary to go and write in, that's a hassle".

Amelia: "Initially, I would go. I would go to the blog a few years ago. I was interested, and I would see either daily or weekly. I would go to the person's blog and see things back (…) nowadays it's more on Instagram only".

In future, participants think that blogs will never return to what they were. Given the limited use of the blog platform by the participants, they believe that if fashion blogs do not change their form, they will be forgotten, and few people will consume them.

Maria: "(…) I think that blogs are tending to go increasingly into disuse (…)".

Federica: "I don't think it will grow. I think it will stop. I think it has to be. It tends to stop (…)".

Rachel: "But I feel a lot of difference between a few years ago and now [when asked what the difference is between using blogs in the past and now]. For example, I come from work on the train, and I'm on Instagram scrolling. I see a photo I like. So I enter the page, scroll, scroll, scroll, instastories… I'm much more consuming that, than " let me see this link here (…)".

Rachel: "I think it's that way, they won't stop existing, but it will be different. Basically, I think the blog will be almost a complement to Instagram and not the opposite".

Is there Still an Opportunity for New Voices (Blogs)?

The participants are somewhat divided on this question. However, it is noticeable that they think there is no longer an opportunity for new blogs. However, they say there might be an opportunity, but only if they are different and bring something new.

Rachel: "Yes, in that sense, I think that nowadays, having a blog forces you to stand out for your difference. You have contents that you may not find so easily elsewhere (…)".

Maria: "I don't think it's worth it (…)".

Have We Really Entered a New Phase in the History of this Relatively Recent Social Phenomenon (Fashion Blogs)?

Yes, at this point, we have entered a new phase in the history of fashion blogs. Participants no longer look at blogs like they did when they started following them. Fashion blogs are in a declining era of usage. Participants state that they no longer open fashion blogs but rather follow them on social networks and do not miss opening the blog platform itself.

Clare thinks that not all blog themes have this outcome, but she feels that when we talk about fashion, it no longer makes sense to open a blog. She says that when it comes to cooking blogs, she still opens them because she is interested in reading their content, while when we talk about fashion, she is no longer interested in reading the content.

Fashion vlogs are seen as a genre that is replacing fashion blogs. However, older participants do not share this perspective, as it is something that has never caught their attention. They think this type of content is more suitable for younger age groups.

Rachel: "I think it's more for younger girls. I think as I connected with blogs eight years ago, the younger generation connects with vlogs".

Amelia: "I haven't discovered YouTube yet [laughs]. But that's enough too, so much".

Those who consume fashion vlogs comment that they use fashion vlogs more than blogs, as they deliver a quicker message, are more visual and are not as dull.

What Pros and Cons have Social Networks Brought to Fashion Blogs?
Social media has brought both pros and cons. However, social media is the cause of blogging's demise. With the emergence of social networks, participants began to look at blogs differently. On the one hand, social networks brought visibility to blogs, as participants admitted that they got to know new bloggers through social networks.

Maria: "(…) I met most of the bloggers I currently follow through Instagram. (…) because with social networks, especially Instagram, they are managing to captivate the same number, and even more people (…)".

They offered bloggers the possibility to create more varied, dynamic and personal content, being able to pass their content to platforms other than the blog.

Maria: "(…) even those who started on blogs are currently on Instagram. So it's easier to follow (…) the content ends up being more varied on social networks, and you follow it more closely".

However, as cons, the blog platform lost preponderance, as social media have replaced it.

Has Social Media Come to Revolutionise the Way Fashion Blogs are Perceived?
Yes, social networks have revolutionised how fashion blogs are seen and used. However, for the participants, the functionalities of social networks replace fashion blogs, contributing to their obsolescence. Through the emergence of social networks, consumers have become used to wanting everything faster, more dynamic, current and instant content, and they realise that the blog platform can no longer offer such capabilities.

Clare: "Yes, yes. Precisely due to the excess of content".

Through social networks, the participants realised that people could publish all the content on social networks that they post on the blog, but better. So social networks can be a kind of introduction or summary. For those who want to see more, then open the blog.

Rachel: "(…) for example, if you go to Double Trouble's blog, you go to the comments box, and there are zero comments, nobody writes anymore, they make comments saying "ai! loved the look! ". Nobody does that anymore. And as a consumer, it is much more intuitive to make a like or a comment on social networks (…) you on a blog you have no way of making a quick appreciation, let's say (…)".

Federica: "(…) it's the speed of information. I think that we now want everything very fast, and sometimes we go to the blog, and things appear that don't interest us either, you know?".

Maria: "(…) I don't think it's worth it. But, of course, if [the new blog] starts with a social network, it may have some visualization. But, even so, I think that nowadays it is more worth investing in a social network than in a blog".

Today, female consumers look at blogging as uninteresting, dull and massive.

Have Digital Fashion Influencers Replaced Fashion Bloggers?
The participants admit that having a blog is not mandatory for a digital fashion influencer. Digital fashion influencers are a "new way" to promote fashion, where brands bet on them as a form of advertising. However, participants also think digital influencers are more commercial, where you can see that they are trying to sell something. In addition, most participants say that what digital fashion influencers communicate is neither authentic nor credible.

Clare: "(…) and now it's all fakes, it's all 'let's say this cream is wonderful for wrinkles' and the brand is giving them I don't know how much [money] and everyone will buy the damn cream. No wonder now brands are starting to herd influencers to introduce a product".

Given the growth of social media and, as such, the consequent increase of digital influencers, these have come to replace fashion bloggers. However, participants value someone who has a blog more than someone who is only a digital influencer. The underlying reason is that fashion bloggers dedicate more time to blogging than digital influencers dedicate to social media, i.e. bloggers master more topics and have more depth.

Sophia: "(…) they are people who are working for that and end up informing themselves a little bit (…) which does not take away, in this case, the highlight of the influencers who several people also follow for some reason (…)".

Rachel: "believing more, I wouldn't say. I think that without a doubt having a blog has another weight (…) if you complement your Instagram with a blog, you know that the person spends some time on that (…) it gives you a greater maturity. But it's a bit relative. I think it always has a bit of influence".

40.3.8 Summary of Results

Table 40.3 presents a summary of the results that are derived from the data analysis process.

Table 40.3 Summary of the discussion of results

Result	Support in data
In the past, fashion blogs were the complements of fashion magazines	Blogs were and are seen as platforms of interest for following fashion trends One looked at bloggers as fashion journalists The blogs were like a diary with the garments that the bloggers wore Most people know about fashion blogs through Google searches or fashion magazines
At present, people no longer open fashion blogs directly. Eventually, they follow them, but on social networks	The participants perceive blogs as massive, being too much text and images. They are not something personal but rather a manufactured product, over-produced Social media is something more accessible and appealing, and intimate because there's no time for big productions. So it's ready-made but genuine They have started following bloggers on social media as it is easier The participants are now getting to know more bloggers through social media
In the future, bloggers will start to be "forgotten"	The participants admit that they use the blog platform very little They believe that blogs will cease to be used gradually Social media, specifically Instagram and YouTube, will be used even more Participants now follow bloggers on social media rather than on the blog platform
There is no opportunity for new fashion blogs	There may be an opportunity. However, only if they excel at making a difference
We have entered a new phase of the fashion blogging phenomenon	Increasingly the blog platform is less and less used People follow bloggers on social media and don't feel the need to open the blog platform The blog platform can still make sense when we talk about topics other than fashion. However, when the subject is only fashion, it no longer makes sense to use the blog platform

40.4 Discussion and Conclusions

The present study characterises the Portuguese panorama in digital marketing for the fashion market based on digital influencers. The social network with greater expressiveness in Portugal is Instagram. This recent digital phenomenon, which is impactful in terms of marketing, is the proliferation of influencers and their growing presence in the daily lives of the general public.

The main conclusion is that fashion blogs in Portugal are no longer in their golden age. On the contrary, they are in the dusk. Fashion blogs were a platform recognised by all participants as important for direct contact with the fashion segment. It was very much through fashion blogs that it was possible to know the trends first-hand, the shops where you could buy a particular piece of clothing, or suggestions on how to wear or combine elements. However, the relevance of fashion blogs has been decreasing, being seen today as a complement to the information consumed through social networks and not the opposite. Finally, and in line with the previous statement, it was also possible to verify that, despite not being central to the communication between the influencers and their audience, the existence of a blog allows them to gain credibility and notoriety.

Somewhat paradoxically, the panel of participants believes that there are still opportunities for new voices if, however, they bring something new and innovative. Regardless of age, from 19 to 53, there is nostalgia. Blogs and bloggers are in their twilight years, and the feeling is anguish. The substitutes, the digital influencers and their ready-made content seem not to convince. They are being followed because it's quick, ready-to-consume, and, today, there is no time to be wasted. Open Instagram… scroll, scroll, scroll… click… print screen… scroll, scroll, scroll. But on the other hand, the works of digital influencers seem to be more authentic, genuine and intimate even. The frugal means of production, often completely amateurish, put a stamp of authenticity that is appreciated, which, however, contrasts with the highly produced articles, with professional means, that were posted on blogs. After all, not everything used to be the best, nor is everything now mere pastiche. Is there a balance?

Although there is a common ground, communication via digital media, the confrontation between text blogs and Instagram photos is nothing more than a particularisation of the transition from the dusk of the Modern world to the dawn of Postmodernity. Firat and Venkatesh [5] found five conditions of Postmodernity: hyperreality, fragmentation, reversal of consumption and production, decentring of the subject, and paradoxical juxtaposition. The postmodernity framework may interpret the transition from text blogs to Instagram photos [4, 6]. Text blog productions, much created under journalistic paramount, i.e. rigorous text writing, alluring images, and graphic composition, fit into the modernity order. Bloggers are few but are committed to their mission and authorities on the matter. Invest time and resources in their productions, which are a reference for pertinent social groups. Fashion bloggers author the metanarratives, the stories that form one's beliefs and that people share to keep the community aggregated. Fashion blogs help people perceive and interpret the environment without much anguish as they reproduce the rules. Fashion bloggers and blogs are players and products in the Foucauldian "regimes of truth" [6, 8, 10]. But then comes the "unchartered" Instagram digital influencer. There is no need to write a text structured as an inverted triangle. In fact, the text is minimal, if there is text at all. The photo or the 6-s video takes the spotlight, and everything comes to the screen at once: fashion, cuisine, body fit, travelling, childcare, parties, Netflix promos, pets, friends and so on. Fragmented lives consume fragments of everything. Consumers, like the one who is scrolling, reverted to be producers. In Marketing jargon, they are the content's co-creators. The hype(r)-reality is omnipresent in the smiles, outfits, decors and props, all elements that contribute to building personal sovereignty. Paradoxically, everything goes with everything, but not in the blog text, just in the Instagram selfie photos, where the producers are the images and the products, their authors. The subject is decentred [3], as their life flow is determined by searching for the perfect shot or shooting that will permit the most ravishing post. In the words of the study participants, what was massive, time-consuming, personal-reflective, overproduced and boring, now is intimate, genuine, live, dynamic and "ready".

The claim that bloggers and blogs could have an opportunity nowadays must be interpreted with scepticism. It may be due to nostalgia because of the loss of references in Postmodernity [3]. Losing references causes vertigo and melancholy for the "good old days", but no one would honestly choose to return to the massive, overproduced, boring texts. Holt [7] theorised how Mountain Dew became an icon and how icons are reborn during crisis situations to give people grounded frames of reference. Thus, it is not expectable that there will be an opportunity for bloggers and their blogs again unless an identity crisis emerges. Until then, fashion bloggers are in the dusk heading to darkness while swift ready-made images scrolling shine on smartphones' screens illuminating Instagramers. One may expect that fragments become atoms, lives will be even fluid and ungrounded, the pastiche will dominate more and more… until one rupture relieves the tensions.

References

1. Bazeley, P.: Analysing qualitative data: More than 'identifying themes'. Malaysian J. Qualit. Res., **2**(1), 6–22 (2009). https://www.qramalaysia.org/journals-vol2
2. Braun, V., Clarke, V.: Using thematic analysis in psychology. Qual. Res. Psychol. **3**(2), 77–101 (2006). https://doi.org/10.1191/1478088706qp063oa
3. Brown, S.: Postmodern marketing? Eur. J. Mark. **27**(4), 19–34 (1993). https://doi.org/10.1108/03090569310038094
4. Cova, B.: The postmodern explained to managers: implications for marketing. Bus. Horiz. **39**(6), 15–23 (1996). https://doi.org/10.1016/S0007-6813(96)90032-4
5. Firat, A.F., Venkatesh, A.: Postmodernity: the age of marketing. Int. J. Res. Mark. **10**(3), 227–249 (1993). https://doi.org/10.1016/0167-8116(93)90009-n
6. Fuat Firat, A., Dholakia, N., Venkatesh, A.: Marketing in a postmodern world. Eur. J. Mark. **29**(1), 40–56 (1995). https://doi.org/10.1108/03090569510075334
7. Holt, D.B.: What becomes an icon most? Harvard Bus. Rev., **81**(3), 43–49 (2003). https://hbr.org/2003/03/what-becomes-an-icon-most
8. Lorenzini, D.: What is a "Regime of Truth"? Le foucaldien, **1**(1), 1–5 (2015). https://doi.org/10.16995/lefou.2
9. Pedroni, M., Sádaba, T., SanMiguel, P.: Is the golden era of fashion blogs over? An analysis of the Italian and Spanish fields of fashion blogging. In: Mora, E., Pedroni, M. (eds.) *Fashion Tales. Feeding the Imaginary*, pp. 105–124. Peter Lang (2017). https://doi.org/10.3726/b11234
10. Reyna, S.P., Schiller, N.G.: The pursuit of knowledge and regimes of truth. Identities **4**(3–4), 333–341 (1998). https://doi.org/10.1080/1070289X.1998.9962594
11. Sádaba, T., SanMiguel, P.: Fashion blog's engagement in the customer decision making process. In: Vecchi, A., Buckley, C. (Eds.), *Handbook of Research on Global Fashion Management and Merchandising*, pp. 211–230. IGI Global (2016). https://doi.org/10.4018/978-1-5225-0110-7

Part IX
Web Marketing, E-Commerce and V-Commerce

Chapter 41
Impact of E-commerce on Corporate Sustainability—Case Study

Agostinho Sousa Pinto, Marta Guerra-Mota, and Inês Dias

Abstract Sustainability (environmental, social, or economic) has become an increasingly present theme in our daily lives. This issue has been mobilising society and organisations to opt for more sustainable, ecological, fair and transparent behaviours. The E-commerce (EC) has experienced exponential growth in the last year and has become an activity that impacts environmental, economic and social sustainability. Through a case study methodological approach, the impacts of e-commerce on sustainability were analysed, considering the activity of a multinational company. We conclude that the impacts of e-commerce on sustainability can be both positive and negative and that they are related to several dimensions that are interdependent, such as packaging, logistic and product development. Companies have a self-interest to get compromised with sustainable development, as confirmed by the company under study, as they can shape, guide and report on their strategies, goals and activities, capitalising on various benefits. On the other hand, e-shoppers are more sustainability-conscious and demanding on issues such as transport (environmentally friendly delivery options) and packaging (sustainable packaging solutions). The study also concluded that the impacts of e-commerce on sustainability will be more positive the more committed companies are, with defined plans, goals and metrics.

A. S. Pinto (✉)
CEOS.PP, ISCAP - Polytechnic of Porto, R. Jaime Lopes Amorim, São Mamede de Infesta, Portugal
e-mail: apinto@iscap.ipp.pt

M. Guerra-Mota
Universidade da Maia - ISMAI, Avenida Carlos de Oliveira Campos, Maia, Portugal
e-mail: marta.guerra@umaia.pt

I. Dias
ISCAP - Polytechnic of Porto, R. Jaime Lopes Amorim, São Mamede de Infesta, Portugal

J. L. Reis et al. (eds.), *Marketing and Smart Technologies*, Smart Innovation, Systems and Technologies 337, https://doi.org/10.1007/978-981-19-9099-1_41

41.1 Introduction

Electronic commerce (EC) represents all transactions for the purchase or sale of products or services that occur using the Internet, the Web and applications and browsers. These transactions may occur through devices such as computers, mobile phones and tablets, to carry out business transactions, regardless of the payment system chosen and the product or service contracted. So, the EC consists of commercial transactions supported digitally, between individuals and organisations and between organisations among themselves. Thus, EC has several benefits for organisations, consumers and society [1].

The constant improvement of e-commerce is an effective way to support businesses and provide services with higher quality [2]. EC with social connections implies consumer-led online transactions, being a necessary aspect of business policy and a powerful accelerator of the economy [3].

Although the Internet allows the saving of certain resources, the truth is that e-commerce does not only have positive impacts. For every potential positive impact, there will also be a negative impact. The ease and convenience potentiate a higher number of purchases and, therefore, the Internet has increased the production of various products on a global scale [4].

The report "2021 European E-commerce", carried out by the University of Applied Sciences of Amsterdam and Centre for Market Insights (CMIHvA), shows that in Europe, in all 27 member states of the European Union and 10 neighbouring digital economies, the turnover of the e-commerce sector had a growth rate of 10% compared to the previous year. By 2022, the trend is expected to continue—the study predicts a 12% growth and a turnover of 844 billion euros [5].

Despite most companies progressively employing sustainability plans, e-tailers in their online communication end up focusing on item bundling and promotional dynamics, concentrating on stimulating consumers to buy and less on raising shoppers' awareness about the effect of accepting the natural world or supporting sustainable consumption and development models [6].

In this sense, sustainable consumption and development models have three fundamental components: environmental protection, economic growth and social equity. According to the United Nations Commission on Environment and Development, equity, growth and environmental maintenance are simultaneously possible and each country is capable of reaching its full economic potential while increasing its resource base [7].

This paper intends to analyse the impacts of e-commerce on sustainability through a case study methodological approach. Therefore, it is intended to understand, through the theoretical framework, several concepts, as well as, to find an analysis and discussion of the obtained results and respective conclusions.

For this reason, the following chapter presents a literature review of the topics in focus, with the purpose of providing knowledge that allows establishing the outline of the research to be conducted.

The third chapter explains the methodological option adopted, as well as the procedure used in the data collection and analysis phase.

The fourth chapter sets out the results obtained from the collection and analysis of the information from the semi-structured interviews conducted with 8 employees of the company under study, with an average duration of 30 min, as well as a discussion of the results obtained by comparison with the compiled theory and the general and specific objectives defined.

Finally, the last chapter points out some of the conclusions of the case study and presents the contributions, limitations and relevance for future research on the theme of this dissertation.

41.2 Literature Review

The literature review aims to understand the state of the art of the concepts of e-commerce and sustainability in its three scopes. In this sense, this exhibition reflects the main conclusions drawn from the scientific production associated with the theme of the impacts of e-commerce on environmental, economic and social sustainability.

Thus, the literature review assumed the reading and analysis of research related to the research question, involving the interpretation of the collected data. The databases used to collect the scientific documents were as follows: B-On (Online Knowledge Library), Google Scholar (Google Scholar), SCOPUS, RECIP (Scientific Repository of the Polytechnic Institute of Porto), Elsevier, Science Direct and Web of Science.

The literature review was conducted using the following search terms "sustainability" or "environmental sustainability" or "social sustainability" or "economic sustainability" and "e-commerce" or "digital" or "innovation" in Portuguese and English. Comprising the period between 2000 and 2021, privileging the most recent results since 2017. Additionally, reports from several identities, which due to their relevance, were crucial to these themes and were considered.

Relevant documents were identified through a careful analysis of the abstract, table of contents, keywords and introduction of each article. Simultaneously, the search was restricted to those scientific documents that provided the full content, as well as their thematic applicability for this purpose. Thus, after exporting all results, we proceeded to refine them; i.e. we sought to analyse only documents that were related to e-commerce in retail, as well as data relating to Portugal, since the company under study is national. The Portuguese company chosen is a market leader, listed on the stock exchange, operating in the food and non-food retail sector.

41.2.1 Sustainability in Companies

The concept "sustainable" means to be bearable, capable of being maintained and preserved, if certain conditions are considered. In this sense, sustainability is a continuous, long-term process, which under certain circumstances maintains a stable system [8].

For the Business Council for Sustainable Development Portugal (BCSD) sustainability is a key theme for the competitiveness of companies and for their short-, medium- and long-term strategies, as it is increasingly a requirement of the various stakeholders and is related to the concept of sustainable development [9].

The "Our Common Future" report of the "Brundtland Commission" presented, in 1987, the definition of sustainable development—" it is the development that meets the needs of the present without compromising the ability of future generations to meet their own needs, ensuring the balance between economic growth, environmental care and social welfare". On the other hand, sustainable development is defined by Furtado [8] as a way of achieving better conditions, with quality consumption through harmonious, long-term relationships that contribute to the growth and development of the human community, with equity and guaranteeing the physical and biological quality of the ecological systems, which provide and guarantee the means for the sustainability of human society itself.

In this sense, sustainability, in a holistic approach, respects the environmental, social and economic dimensions. In this way, all must be considered to ensure lasting prosperity.

For companies, sustainability means having a business and quality management model, in the long term, keeping the company competitive, with guaranteed access to goods and services, through the preservation, conservation and replacement of resources and services provided by the economic, natural, human and social capital [8]. Thus, organisations should be concerned with social issues internal and external to the place where they operate.

41.2.2 Sustainability and E-commerce

Dynamic Parcel Distribution's (DPD) e-shopper Barometer 2021 revealed that some 82% of regular online consumers believe brands need to be environmentally responsible, but only half ensure they buy environmentally friendly products, and only 43% are willing to pay more for environmentally friendly products and services [10].

The study published, March 2022, shows that e-shoppers are increasingly aware of sustainability in e-commerce. About 65% of online shoppers consider that choosing eco-friendly delivery alternatives is important when shopping online [10].

Even so, the percentage increases when the issue relates to using low-emission vehicles, as 70% of e-shoppers say they are more likely to choose an online shop if it offers more environmentally friendly delivery options [10].

The speed and convenience of e-commerce have benefited from marked growth in recent years; however, this expansion has given rise to a range of issues related to environmental, economic and social sustainability.

41.2.3 Environmental Sustainability and E-commerce

The environmental dimension of sustainability is to ensure that natural resources are conserved and managed through actions that minimise negative impacts on air, water and soil, preserve biodiversity, protect and improve the quality of the environment and promote responsible production and consumption [9]. Taking into account the literature review on e-commerce, transport and packaging are the most relevant themes to explore in the environmental dimension of sustainability for e-commerce.

The Dynamic Parcel Distribution (DPD) e-shopper Barometer 2021 for Portugal revealed that about 82% of regular online consumers consider that brands have to be environmentally responsible, but only half guarantee that they buy environmentally friendly products, and only 43% are willing to pay more for environmentally friendly products and services [10].

The study published, March 2022, shows that e-shoppers are increasingly aware of sustainability in e-commerce. About 65% of online shoppers consider that choosing eco-friendly delivery alternatives is important when shopping online. Yet, the percentage increases when the issue is related to using low-emission vehicles, as 70% of e-shoppers state that they are more likely to choose an online shop if it offers more environmentally friendly delivery options [10].

From the consulted studies, the packaging and transport issues are the most relevant, as the reduction of packaging and the alternatives of less polluting delivery vehicles are two of the challenges for brands and transport companies to improve the environmental sustainability of e-commerce [11].

On average, 24% of the volume of an e-commerce order package is empty, so adopting new tools for optimising packaging, materials and handling technologies will significantly increase efficiency, sustainability and productivity across supply chains.

The sustainable use of packaging consists in checking whether the material chosen for it is the most appropriate one, based on the purpose for which it is intended, ascertaining the stages in which environmental impacts can be minimised, with a view to reusing and recycling the packaging material.

For 43% of Europeans, one of the greatest indicators of a brand's sustainability is the use of recycled packaging. The survey by Smurfit Kappa—Europe's leading corrugated cardboard packaging manufacturer—conducted in 2021, revealed that for 40% of Europeans the symbols appearing on packaging are one of the main sources of information they have about a company's environmental responsibility. The study involved 5028 European consumers from France, Germany, Italy, Poland and Spain and highlights the three main factors that influence the perception of sustainable packaging by consumers: packaging can be recycled (43%), packaging is biodegradable (35%) and packaging is made of recycled materials (35%). Packaging thus becomes, in e-commerce, a tangible and visual proof of the commitment that brands have to sustainability [12].

The mission of logistics is to obtain the right goods or services, at the right place and time, and in the desired condition, at the lowest cost and with the highest return on investment. Economic and technological conditions act as a major lever of logistics development.

The requirements of fast deliveries tend to create situations where various means of transport deliver multiple orders without having their transport capacity fully filled. At the same time, e-commerce, due to its globalisation capacity, boosts faster means of transport which can represent high fuel and energy consumption and pollutant gas emissions.

The growth of e-commerce thus promotes innovation and sustainability in logistics. In short, the optimisation of processes, materials, new propulsion techniques and intelligent facilities are the great potential for logistics to become more environmentally friendly. It is important to modernise all touch points in supply chains, from a digital or consumer journey, to transport and delivery at the destination. Those who adopt new technologies and improve the workforce faster will have a competitive advantage in the market [13].

41.2.4 Social Sustainability and E-commerce

The social dimension of sustainability indicates that human rights and equal opportunities of all individuals in society should be respected, contributing to the promotion of a fairer society, with social inclusion and equitable distribution of goods and focusing on the elimination of poverty [9].

Because of globalisation, producers in developing countries have been integrated into global supply chains and, at this moment, they have significant potential for improvement in the field of social sustainable development, as they are associated with practices such as child labour, excessive overtime, pollution, among others [14].

For companies, social sustainability is about how they contribute to the well-being of society, the context in which they operate and their employees. This way of being of companies should be transversal to the various stakeholders, such as, for example, employees, suppliers, clients and service providers.

Social sustainability in e-commerce is reflected in the design of digital assets that must be inclusive to avoid inadvertent exclusion of certain communities. At the same time, it is related to supply chains that must be fairer in both supply and distribution.

Inclusive innovation is a concept intrinsic to this dimension as it relates to the ability of businesses to create shared prosperity and enable access to quality goods and services at low prices, creating livelihood opportunities for excluded populations [14].

41.2.5 Economic Sustainability and E-commerce

The economic dimension refers to prosperity at different levels of society and the efficiency of economic activity, including the viability of organisations and their activities in generating wealth and promoting decent employment [9].

Sustainable attitudes benefit efficiency in management as they limit the consumption of materials, energy and water. There are advantages between sustainability and economy that influence the profit as also the financial and economic results of a company.

Allied to the economic sustainability of companies, the storage of certain products happens both to achieve advantages in the purchase price and to mitigate possible price increases. At the same time, there is also the advantage of transport costs that become lower when compared to the multiple transport of small quantities [15].

Without hurting the company's profit margins, inventories offer a quick response to demand needs, building customer loyalty and offering more competitive prices. However, inventories do not only have advantages for companies. Over time, this asset deteriorates and loses value. When there are risks of obsolescence and product expiration, companies feel the need to dispose of products quickly through the practice of significant discounts that reduce profit margins and contribute to the company's liquidity problems [18]. According to Cláudia Silva [15], inventories are an important item in financial statements and an important component in the management of any company.

41.3 Methodological Approach

The case study is a method that seeks to understand complex phenomena while preserving the holistic characteristics of certain events. In this case, specifically, the case study aims to understand the impacts of e-commerce on sustainability that are dependent on several circumstances, which must be analysed within a context, i.e. the application of sustainable practices to e-commerce activity varies from business context to business context. Therefore, for us to be able to pose casual questions and conduct holistic and in-depth research, this phenomenon must be studied within the context in which it occurs.

As it is intended to establish a cause-and-effect relationship between the impacts of e-commerce on sustainability and the practices of the company under analysis this is an explanatory case study. Given the objective of this case study, the main data collection method consists of semi-structured interviews, as this format allows the flexibility to ask extra scripted questions whenever necessary [16]. Eight interviews involved respondents aged between 30 and 55 and with a range of time in the company between 2 and 15 years.

The interviewees occupy different functions. The choice focused on areas such as product development, packaging, as well as quality area managers with an active role in supplier relations. In the e-commerce area, it was possible to interview digital development coordinators and brand managers. The interviews were conducted with the aim of collecting data about the employees' perception of social, environmental and economic sustainability associated with e-commerce.

The interviews lasted on average 30 min, in an informal environment and the interviews were audio recorded, with prior permission. The collection of information during the interviews is considered the criterion of theoretical saturation, which determines that the interviews should end as soon as the information obtained becomes repetitive, no longer adds new elements, contents, properties and connections between categories [17].

After the interviews had been conducted, the transcription verbatim was carried out using the recordings. Verbatim transcription consists of reducing the entire interview to writing in the interviewee's words and is, according to Merriam (2009), the best basis for further analysis.

After transcription, the interviews were analysed according to the principles of Grounded Theory, which aims to create a theory based on systematic data collection and analysis [18] i.e. existing data from the literature review (codified information) were compared with the collected data (uncoded information) and a conceptual framework was created with the aim of enriching the existing theory.

Grounded Theory goes beyond the mere verification of pre-existing theories or hypotheses and is embodied in the creation of theories about the social phenomena under study and is currently the most widely used method in qualitative research. This method provides for the creation of categories that emerge from the data collected, through a comparative stance between the existing codes and the information to be codified. The conceptual structure created allows for an enrichment of the theory, as connections are created between the different elements of the theory through the categories obtained [10].

41.4 Presentation, Analysis and Discussion of Results

According to the "Collaborative Report on Sustainability and e-Commerce" developed by E-commerce Europe, there is an opportunity to incorporate digitalisation as a key component of sustainability policies and vice versa. The e-commerce sector is a bridge between the digitalisation of society and its transition to a more sustainable economy. This is a constantly and rapidly evolving sector, shaped by new technologies, demands and capable of leading innovative sustainable solutions [19].

E-commerce should be seen as an opportunity to structurally change retail and consumer practices. Yet, the end-consumer remains crucial for the sustainable transition of e-commerce. In recent years, consumers have changed their behaviour and expectations to the extent that they are making more sustainable choices.

Online shopping can allow shoppers to benefit from greater transparency about product information, as certain digital tools can empower users. However, any information aimed at empowering consumers needs to be accessible, understandable, but also comprehensive.

The results obtained during the interviews are in line with the aforementioned information. The interviewees also considered that information, in e-commerce, is very important and should contain all the data about the product, namely origin, technical specifications, recyclability, certifications (if applicable), among others to combine information with the purchase decision process. Additionally, participants also listed several e-commerce features that could assist in communicating this information, such as banners, pop-ups, search bar and filters and product multi-photography with zoom to detailed and important information on the packaging.

This view agrees with Jones [6] who noted that the sustainable commitments of e-commerce can be enhanced by providing information on sustainably manufactured items and product life cycle effects at the point of sale online. E-commerce Europe warns that there is still no standardisation in the way sustainability information is presented, stressing that it is important that this basis exists but is flexible.

E-commerce packaging needs to be functional, protect the product and allow brands to implement their marketing strategies and, therefore, represents a challenge for e-commerce sustainability [19]. In this sense, the interviewees revealed several movements that the company under study has already made and is currently making—always starting from the assumption that the ideal is not to have packaging, opting to print all the important information directly on the product, and if packaging is needed opt for recycled, reusable and recyclable materials, with raw materials from sustainably managed sources.

The central part of e-commerce is logistics. It is interdependent on various themes such as mobility, urban planning, but also accessibility and is therefore shaped by various developments such as consumer expectations (e.g. delivery time), the growth of omnichannel commerce and transport infrastructure. Thus, the issues around this theme focus on pollutant gas emissions and urban congestion.

Interviewees indicate that the logistics chain of the company under study is extensive and that, therefore, fleet electrification, efficient transport and stock management may be ways to improve the company's ecological footprint, as well as, the construction of more sustainable last mile solutions, such as collection in physical shops or electronic lockers.

As of January 1, 2026, in Portugal, all national SMEs will have to comply with the European Commission Directive and will have to start reporting non-financial information in a more integrated, targeted, reliable and accessible way in order to support and encourage sustainable decision-making [20].

The data collected from this dissertation indicates that the company has well-defined metrics, communicated to the market and with concrete deadlines, such as elimination of PVC by the end of 2022; elimination of EPS by 2023; making all packaging recyclable, reusable and compostable by 2025 and use of at least 30% recycled raw material.

Sustainability is one of today's most pressing issues. Companies must rethink the way they produce, how they trade and how they sell to the final consumer, as they are part of a social, environmental and economic complex where collectively and individually they can help in the fight against climate change and resource rarefaction [19]. For João Meneses, secretary-general of the BCSD Portugal, there are several reasons for companies to be more sustainable, such as cost reduction, value chain optimisation, better risk management, differentiation of competitors from consumers and investment and business opportunities.

The Secretary General of BCSD 2020 noted that "sustainability is going to be a factor of positive differentiation and increasing competitiveness. Not only for consumers, but also for employees, as a factor in attracting talent".

According to the interviews carried out, the organisation under analysis has sustainability as a priority, defining objectives with concrete deadlines and communicated transparently to the market. Even so, for the interviewees, a different importance is given to each of the aspects of sustainability. Environmental sustainability was the most mentioned by most of the interviewees and the one that allowed collecting the most examples. The social dimension was presented by most interviewees as the area of sustainability that can still be further explored. Economic sustainability did not meet consensus, since for some it is considered the most important for the company, while for others it is the one that has the least positive influence on the other areas of sustainability.

Economic sustainability represents the sustained economic growth of a company, with respect for natural resources, progressive reduction of the environmental footprint of products (full cycle) and equitably distributed wealth [21]. During the interviews, economic sustainability was associated with the cost of sustainable raw materials and, consequently, the margin of the products that impact the selling price to the final consumer.

Environmental sustainability represents the minimisation of negative environmental impacts arising from business activity—with the aim of creating positive impacts. To this end, it relies on a product life cycle approach, internalising good practices at all levels of the company's internal structure [21]. The data collected during the interviews demonstrates several actions in this area: mapping of all packaging, products and their components to ensure that it is known what needs to be changed, seeking to have recyclable, reusable or compostable packaging and products with at least 30% of recycled raw materials, from sustainably managed sources and with certifications.

According to the United Nations Global Compact, "directly or indirectly, companies affect what happens to employees, workers in the value chain, customers and local communities, and it is important to manage these impacts proactively"; therefore, the social aspect of sustainability mirrors the sustainable relationship that the company should promote in the human dimension—internally, but also for the benefit of society and the local community where it operates. According to the interviews carried out, in this field, the company tries to audit suppliers to ensure that there are no situations of forced labour, child labour or others, assessing the conditions of the

Table 41.1 Summary table on the impact of e-commerce on the sustainability of the company (according to the interviewees)

Sustainability topic	Overall view	E-commerce topic
Environmental sustainability	Information, in e-commerce, is important and should contain all the data about the product. Several e-commerce features could assist in communicating this information, such as banners, pop-ups, search bar and filters. E-commerce packaging needs to be functional, protect the product and allow brands to implement their marketing strategies. The ideal is not to have packaging, opting to print all the information directly on the product, and if packaging is needed opt for recycled, reusable and recyclable materials (from sustainably managed sources)	Content marketing, logistics, packaging
Economic sustainability	The company has well-defined metrics, communicated to the market and with concrete deadlines	Reporting non-financial information
Social sustainability	The company tries to audit suppliers to ensure that there are no situations of forced labour, child labour or others, assessing the conditions of the factories and workers, while at the same time working internally on issues related to diversity and inclusion	Suppliers and employees

factories and workers, while at the same time working internally on issues related to diversity and inclusion (Table 41.1).

41.5 Conclusion

In short, it was possible to conclude that the retail company under study actively acts on the three aspects of sustainability although it does so with different intensities. Environmental sustainability was the aspect that received the greatest focus during the interviews. At the same time, some of the practices that the company under study applies were identified, which have an impact on e-commerce and which contribute to the company's sustainability, as indicated by the following examples: supplier audits; definition of requirements and specifications for products and packaging; fostering recycling, reuse and durability of products; elimination of EPS, black packaging and PVC; betting on certifications and upcycling; making packaging recyclable, reusable and compostable; identifying the origin of products and raw materials; reduction of waste; use of recycled raw materials; reduction of packaging; detailed analysis of packaging and product components and optimisation of transport and stocks.

Through the interviews carried out, some limitations and opportunities for improvement were identified for the organisation that may impact on the sustainability of e-commerce. On the one hand, we highlight the following limitations: higher costs of sustainable raw materials; lack of dedicated recycling flows for certain raw materials; legal issues that impose certain forms of packaging for food vs. non-food e-commerce; extensive logistics chain. On the other hand, we highlight the following opportunities for improvement: lack of information regarding the sustainability of products on e-commerce platforms; greater focus on social sustainability, namely diversity and inclusion; use of various features to improve communication in digital commerce such as pop-ups, banners, filters, search bars, photographs and descriptions; innovating and seeking differentiating solutions; applying recyclability iconography in e-commerce on product pages; informing through e-commerce what the availability of the product is in physical shops to avoid unnecessary visits; electrification of the fleet and developing more sustainable last mile solutions.

E-commerce has both positive and negative impacts on the sustainability of companies. However, there are several steps prior to the e-commerce platform and transaction that need to be ensured to make e-commerce even more sustainable, of which we highlight the following: product development, packaging and the logistics chain. With the results obtained it was possible to enrich the existing theory. E-commerce is interdependent on several processes, and thus, for it to be sustainable, each of the parties also needs to be so. There is complicity between several aspects that influence e-commerce such as, for example, product and packaging development (design, materials, etc.), distribution, supply, among others.

Acknowledgements This work is financed by Portuguese national funds through FCT—Fundação para a Ciência e Tecnologia, under the project UIDB/05422/2020.

References

1. Laudon, T.C.: E-commerce 2019, Business, Technology, Society (2020)
2. Drumea, M.C.: Stopping forced labor. Economics, Management, and Financial Markets, pp. 839–842 (2011)
3. Ionescu, L.: Audit committees as a governance device. Economics, Management, and Financial Markets, pp. 127–132 (2014)
4. Tiwari, S.: E-Commerce: Prospect or Threat for Environment, Junho (2011)

5. Lone, S., Harboul, N., Weltevreden, J.: European E-commerce report, setembro 2021. [Online]. Available: https://ecommerce-europe.eu/wp-content/uploads/2021/09/2021-European-E-commerce-Report-LIGHT-VERSION.pdf
6. Jones, H.D.: E-retailers and Environmental Sustainability (2014)
7. Shah, M.: Sustainable development. In: Encyclopedia of Ecology, pp. 3443–3446 (2008)
8. Furtado, J.S.: Sustentabilidade empresarial: guia de práticas econômicas, ambi-entais e sociais (2005)
9. BCSD Portugal: O que é a Sustentabilidade? (2022) [Online]. Available: https://bcsdportugal.org/sustentabilidade/. [Acedido em 14 julho 2022]
10. DPD, "European E-shoppers in 2021.," 2021. [Online]. Available: https://www.dpd.com/wp-content/uploads/sites/77/2022/02/European-E-Shoppers-in-2021.pdf. [Acedido em 14 julho 2022].
11. Serrão, M.: O impacto do e-commerce na sustentabilidade, 13 julho 2020 [Online]. Available: https://expresso.pt/opiniao/2020-07-13-O-impacto-do-e-commerce-na-sustentabilidade. [Acedido em 14 julho 2022]
12. Correia, R.: Embalagem sustentável influencia 44% dos con-sumidores a comprar moda online, 5 novembro 2021. [Online]. Available: https://www.distribuicaohoje.com/destaques/embalagem-sustentavel-influencia44-dos-consumidores-a-comprar-moda. [Acedido em 14 julho 2022]
13. Santos, M.: O crescimento do e-commerce continua a promover a agenda da inovação e sustentabilidade na logística, 22 outubro 2020. [Online]. Available: https://ecommercenews.pt/o-crescimento-do-e-commerce-continua-apromover-a-agenda-da-inovacao-e-sustentabilidade-na-logistic/. [Acedido em 14 julho 2022]
14. Thorlakson, L.E.F.: Companies' contributions to sustainability through global supply chains, pp. 2072–2077 (2018)
15. Silva, C.A.: A gestão dos resultados e os inventários, p. 5 (2018)
16. Patton, M.: Qualitative evaluation and research methods (1990)
17. Taylor, D.M.: Introduction to Qualitative Research (2016)
18. Glaser, S.A.: The Discovery of Grounded Theory (1967)
19. Ecommerce Europe: Collaborative Report on Sustainability and ecommerce, junho 2021 [Online]. Available: https://ecommerce-europe.eu/wp-content/uploads/2021/06/Collaborative-Report-on-Sustainability-and-e-Commerce-June-2021-2nd-edition.pdf. [Acedido em 14 julho 2022]
20. Comissão Europeia, 24 fevereiro 2022 [Online]. Available: https://www.consilium.europa.eu/pt/press/press-releases/2022/02/24/council-adopts-position-on-the-corporate-sustainability-reporting-directive-csrd/. [Acedido em 14 julho 2022]
21. PHC Software: Que indicadores medem a sustentabilidade da sua empresa? 15 dezembro 2017 [Online]. Available: https://phcsoftware.com/business-atspeed/indicadores-de-sustentabilidade-empresa/. [Acedido em 14 julho 2022]

Chapter 42
Social Commerce—When Social Media Meets E-commerce: A Swiss Consumer Study

Marc K. Peter, Alain Neher, and Cécile Zachlod

Abstract Social commerce was established with the growing commercial use of social media platforms, centered around e-commerce features and actors (buyers, influencers, and sellers) which provide user-generated content through recommendations on products and services. Important factors are both the utilization of social media platforms as trusted sources by users and the trusted content from users posting their experiences and opinions. This study focuses on Swiss consumers (9 out of 10 Swiss residents are active online) and their usage of social media, e-commerce, and social commerce as a source of trusted product information and platforms to share their own product experiences. While social media usage in Switzerland is high (almost all online users are using some form of social media), only one-third are using social media (primarily YouTube) as a source of information that guides decision-making. Merely one in ten respondents post their experiences—satisfaction or recommendations for products and services—on social media at least once a month. Our research data show that in Switzerland, social commerce is an under-developed marketing concept as to available e-commerce features and the degree of trusted user-generated content about products and services on social media platforms.

42.1 Introduction

Social commerce as a marketing concept was established with the growing commercial use of social media platforms. It is part of an extended customer journey that covers components such as need recognition, information seeking, purchase intention, participation, information sharing and brand loyalty building, connecting both networks of sellers and networks of buyers [1, 2]. Social commerce is a Web 2.0 phenomenon allowing interactions on social media platforms based on

M. K. Peter (✉) · C. Zachlod
School of Business, University of Applied Sciences and Arts Northwestern Switzerland FHNW, Riggenbachstrasse 16, 4600 Olten, Switzerland
e-mail: marc.peter@fhnw.ch

M. K. Peter · A. Neher
School of Business, Charles Sturt University, Panorama Ave, Bathurst, NSW 2795, Australia

J. L. Reis et al. (eds.), *Marketing and Smart Technologies*, Smart Innovation, Systems and Technologies 337, https://doi.org/10.1007/978-981-19-9099-1_42

user-generated content. Social commerce further enables sharing of content as part of primarily web-based social communities, allowing actors to participate in the marketing, selling, buying, and sharing of products and services on marketplaces [2–4].

In Switzerland, with its 8.7 million residents [5], 89% are online, that is, are using the Internet. While practically all younger residents are online (99% in the age groups of 14–39), older residents are not using the internet at the same frequency on a regular basis (only 79% of the age group 60–69, and 53% of the age group 70+) [6]. Swiss households spend about CHF 5,870 per month on housing and energy, food, insurance, and consumer products [7], making Swiss consumers in theory a lucrative market for social commerce.

An important factor in consumers' use of social commerce is trust [8–10], which is based on quality-assured shared information (i.e., user-generated content), and familiarity and endorsement by other users. This implies that both content from trusted users might be consumed to inform purchase decisions, and users might post their experiences on social media, as members play a crucial role in social commerce [8]. Hence, this research project aims at identifying if (and which) social media platforms in Switzerland are being used as trusted sources, and to which degree users are posting their experiences on social media. Hence, this research project is designed to understand the degree to which social media and e-commerce platforms among Swiss Internet users are being used, and to identify the degree to which social commerce activities with a focus on user-generated content about recommendations and product/services feedback are prevalent to inform purchase decisions via social media marketplaces.

42.2 Social Commerce

42.2.1 Background

The term social commerce has been slowly established as it started to emerge about two decades ago and was defined in the literature in the past decade [1–4, 9, 10]. Back in the late 1990s, the two e-commerce giants Amazon and eBay laid the foundation for social commerce by providing product and performance ratings for sellers [11, 12]. The term itself first appeared in 2005, used by tech blogger Steve Rubel and venture capitalist David Beisel, as well as by Yahoo, who used it to describe a new area (Shoposphere) on websites that allowed consumers to browse product listings, exchange information and comment on products and services [8–10, 12, 13].

For a while, the term social commerce was defined very broadly as an e-commerce form in which the focus lay on the active participation of users and the personal relationship between buyers and sellers. Central elements were customer involvement in the design, sales, and marketing via purchase recommendations or comments from other customers [2, 4, 13]. Due to developments in social media platforms and enhanced e-commerce features, the term social commerce started to emerge in marketing practice.

In the literature, social commerce is thus referred to as a new delivery platform of e-commerce that supports social interaction, based on user-generated-content to assist their purchase decision-making process [8–10], leading to value co-creation for e-commerce on social media [14], based on interpersonal relations that integrate customers in the e-commerce process [10]. Hence, social commerce is a social, creative, and collaborative approach, based on interactions and the posting of experiences and opinions which add value to the participating actors [15, 16].

A famous case study often referred to is the WeChat app, which is described as an "all-in-one app." Users can communicate with each other through various features such as chat or video, participate in large group chats or play games. Driven by its integrated payment service, users can order products and services online and offline (e.g., taxis and food) and pay via WeChat. Its user-generated content, including product and service recommendations, drives e-commerce activity on the platform [12]. Western social media platforms (as presented below) are also developing into e-commerce platforms with new sales features and integrated shop systems. Operators such as Facebook, Instagram, Pinterest, and TikTok are slowly starting to embrace social commerce.

42.2.2 Social Commerce Features on Major Social Media Platforms

Ongoing developments on social media platforms have shown various social commerce features evolve. This section provides an overview of enhancements on Facebook, Instagram, TikTok, Pinterest, and YouTube, and their increased collaboration with e-commerce providers.

In 2014, Facebook introduced the "buy button" that appears at the bottom of ads and posts. Furthermore, products can be marked in posts and stories (shoppable content) and users will be redirected to the product page of the online shop with a single click where a purchase can be completed. In 2020, Facebook Shops was launched along with other new features, replacing the Facebook Pages shop (as well as the Instagram Profile shop). However, it is not possible to buy directly in the shop on the business profile of sellers located in the EU and Switzerland, and therefore, the user will continue to be forwarded to the online shop. For Switzerland, it is still unknown when the so-called direct checkout (or checkout button) will be introduced with which users in the USA can remain on Facebook and complete the purchase (including payment) directly on the platform [17–20].

Instagram, like Facebook, has a buy button and the ability to tag products in posts, reels, and stories. Instagram is currently testing an affiliate feature for creators with which influencers (content creators) can also tag products of their partners in their posts, reels, or stories [21]. In addition, the swipe-up feature has been replaced by a sticker (shop link). However, this still leads users to the external online shop. In almost all views, products are displayed to the user in thumbnails with the title and price. More details such as a description or information about the shop can be displayed in the individual product display. Users can also send a direct message to the shop. The direct checkout feature is already available in the USA, but the launch date in Switzerland is not yet known [20–26].

TikTok will most likely also expand its shopping features. So far, a clickable link can be added to one's own profile, including swipe-ups in videos. TikTok Shopping with the associated features, such as the shop tab in the profile or product links in videos for organizations and influencers, and live stream shopping, is currently being used by selected retailers and influencers in the USA, the UK, and Canada. Since the Chinese counterpart Douyin already provides extensive shopping features as well as direct purchase options, it can be assumed that a feature like Meta's direct checkout will certainly be soon provided to the marketplace [25–27].

The pinboard network Pinterest has developed some advertising and shopping features in recent years. With the so-called product pins, information such as product descriptions, availability, and price can be added. By linking the products to the online shop, the information will be automatically updated if changes occur. With the shop-the-look feature, several products can be marked on one pin. As with the other platforms, users are directed to the external online shops for the actual purchase. Currently, Pinterest also tests a direct payment/purchase option with selected organizations. In addition, Pinterest offers recommendation features with which, for example, product pins can be searched for that are visually similar to the content of a selected pin [27, 32].

So far, YouTube has only offered common shopping options with external links, for example, with a link to the online shop being included in the video description, or clickable links in the video advertising in campaigns such as True View. In the future, YouTube will also offer livestream shopping, which may be used for live auctions. Livestream shopping has been popular in China for a long time and was already available on the Taobao and Tmall platforms (both owned by Alibaba) back in 2016. Again, the YouTube social commerce features are not yet available in Switzerland [12, 27].

Most social media platforms have partnerships with e-commerce store providers like Shopify or BigCommerce, allowing product catalogs to be synchronized and used across platforms. This promises further potential in terms of targeting, ad creation, or tracking [14, 26]. Furthermore, the mobile-first approach is clearly noticeable among all platforms, whether this might be in the planned use of augmented reality (AR) on Pinterest or in the announcement of the new Facebook/Instagram shop [20]. The associated consumer behavior of social media as an everyday companion to users will drive platforms to improve usability, but also to further stimulate the development of user-generated content as part of digital marketing strategies [20, 25, 28].

42.3 Research Methodology

To measure the degree to which social media, e-commerce, and social commerce are prevalent among Swiss consumers, 1,008 telephone interviews were conducted in the German-speaking and French-speaking parts of Switzerland [29] using the computer-assisted telephone interview (CATI) method [30].

Telephone interviews were conducted between January 14 and February 19, 2022, with consumers aged 18 and over (population). The basis for the sample was (1) a random selection of landline telephone numbers from the public telephone directory (for 80% of the sample); and (2) mobile phone numbers selected according to a random system (random digit dialing; for 20% of the sample). This mix addresses the availability gap of consumers on landline numbers. The contacted consumer, either within the household or on the mobile phone, was identified against a quota, that is, subgroups. Quotas were set for the language region (German-speaking and French-speaking parts of Switzerland), three age groups (18–39, 40–64, 65+), and gender (male/female). The subgroups prevent groups from being over-represented in the sample (e.g., being easier to reach or more willing to participate) and guarantee a structurally equal image of the population.

The population represented in the sample thus includes the linguistically assimilated part of around 6.8 million inhabitants of the German-speaking and French-speaking regions of Switzerland who are at least 18 years old [31]. The confidence interval of the total sample with 1,008 interviews is ±3.2% with a certainty of 95% (50/50 distribution). The survey shows a structurally identical image of the population. Hence, the results can be extrapolated to the population, considering the confidence interval. It is a proportional (unweighted) sample. Table 42.1 provides a demographical overview of the distribution of the quota population subgroups in comparison with the achieved proportion in the sample (deviations of ≤1% arise due to rounding to whole percentage numbers). The response rate of the CATI survey was 7.3% (152,820 calls less 139,059 unavailable contacts = 13,761 contacts, of

Table 42.1 Comparison of population versus sample

	Actual share [31] (%)	Proportional sample ($n = 1008$) (%)
German-speaking Swiss consumers	74	74
French-speaking Swiss consumers	26	26
Men	49	50
Women	51	50
Age group 18–39	35	35
Age group 40–64	43	43
Age group 65+	23	22

which 1,008 interviews were completed). Interview data were analyzed in SPSS, and a research report was prepared.

42.4 Results

42.4.1 Social Media Usage Among Swiss Consumers

Figure 42.1 presents the use of social media platforms in Switzerland. Only one in 20 respondents stated that they did not use social media platforms. For almost nine out of ten consumers (87%), WhatsApp is the most frequently used social media platform, followed by YouTube (80%). Facebook or Facebook Messenger is used by almost every second respondent at least once a month and is thus the third most frequently used social media platform.

Women and men differ in their usage of social media platforms to some degree: Women use WhatsApp (90% vs. 84%), Facebook (52% vs. 44%), Instagram (38%

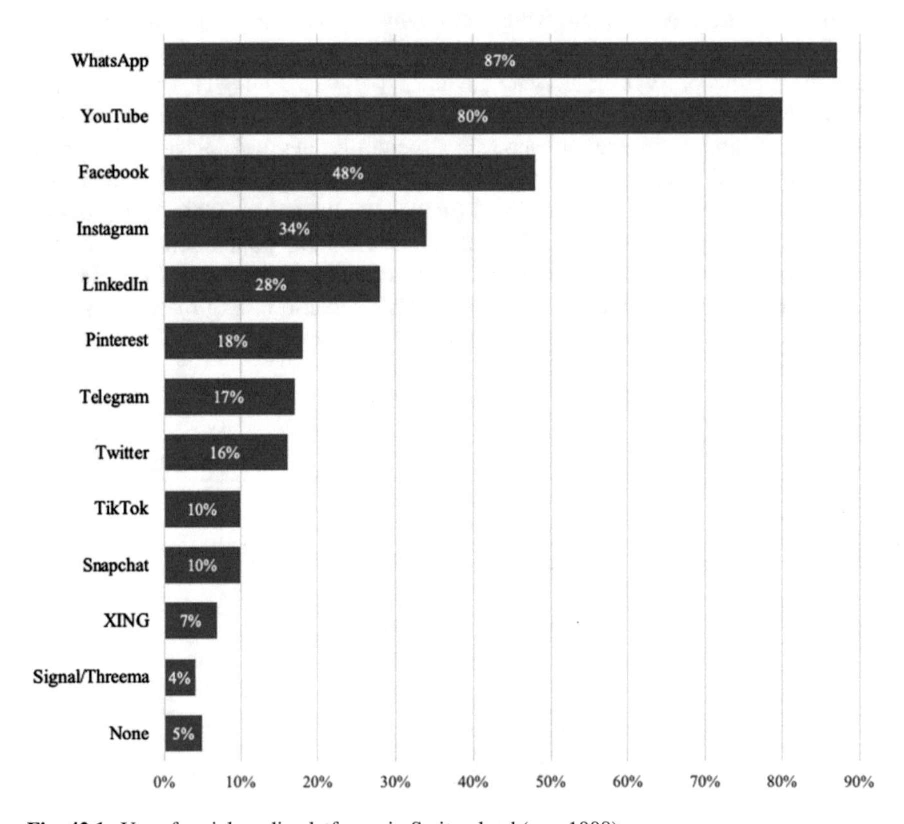

Fig. 42.1 Use of social media platforms in Switzerland ($n = 1008$)

vs. 30%), and Pinterest (26% vs. 10%) more than men and stated more frequently motives such as they want to know what their friends are doing (53% vs. 44%) and they want to share what they are currently doing or thinking (8% vs. 4%). Men, on the other hand, use YouTube (83% vs. 77%), LinkedIn (35% vs. 22%), Telegram (21% vs. 14%), Twitter (20% vs. 11%), and Xing (9% vs. 5%) more frequently, mentioning business purposes significantly more often than women (23% vs. 12%).

Respondents in the youngest age group (18–39 years) are the most active and diversified on social media. On average, they named 5.1 different platforms that they visit at least once a month and 2.5 different reasons for doing so. The middle (40–64 years) and oldest (65+) age groups named 3.9 and 2.7 different platforms used and 1.8 and 1.6 different motives, respectively. The two most frequently mentioned usage motives are "write text messages/organize free time" (50%) and "find out what my friends are up to/stay in touch" (48%).

TikTok (86%), YouTube (84%), Pinterest (83%), Xing (70%), and Twitter (71%) are mostly used passively; that is, contributions are only read or viewed. The remaining platforms are actively used by a majority; that is, consumers also post comments or media: WhatsApp (93%), Telegram (66%), Facebook (62%), Snapchat (59%), Instagram (54%), and LinkedIn (51%).

Study results further reveal that consumers spend 1.7 hours (h) a day on their mobile phones, about half of which (0.9 h) are on social media. Here, too, there are differences according to age: The youngest age group (18–39 years) spends an average of 2.5 h on mobile phones, of which 1.4 h on social media. The 40–64-year-olds spend 1.4 h a day on their mobile phones, including 0.7 h on social media. The oldest age group (65+) reported 0.8 h on mobile phones and 0.4 h per day on social media platforms.

42.4.2 E-commerce Usage Among Swiss Consumers

Results show that more than four-fifths (84%) of the surveyed population shop online. Over a quarter (27%) do it less than once a month, more than a fifth (22%) once a month, and another fifth (20%) two to three times a month. One in ten respondents (10%) shop online once a week, and the remaining respondents (4%) several times a week or even daily.

Overall, women shop online less often than men: Every fifth woman (20%) and only about every tenth (12%) man stated that they did not shop online at all. Almost two-thirds of men (63%) and half of the women (50%) shop online at least once a month. In the oldest age group (65+), almost a third (31%) said they did not shop online at all. About another quarter (28%) shop online less than once a month and around two-fifths (41%) at least once a month. In contrast, the youngest age group (18–39 years) shop online most frequently: Three-fifth (60%) shop online at least once a month, about one-third (30%) less often, and almost one in ten (9%) do not shop online at all.

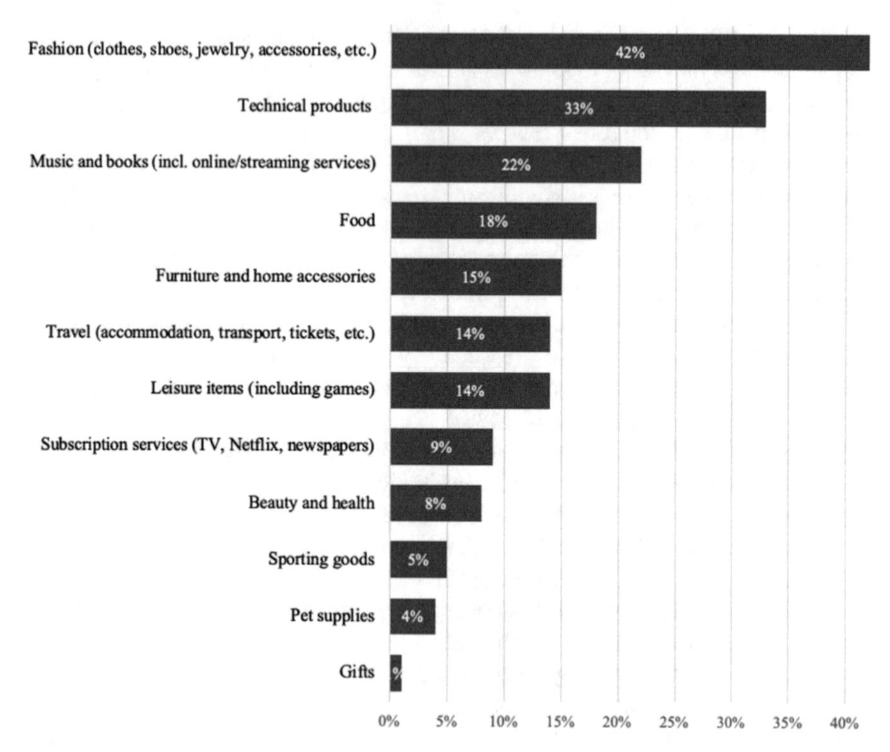

Fig. 42.2 Purchase categories of Swiss online consumers (n = 844/filter: consumer purchases products/services at least sometimes online)

Figure 42.2 shows that fashion items such as clothing, shoes, or jewelry are bought most frequently online (42%), followed by technical products (33%) as well as music and books (including streaming services: 22%). While women are more likely to buy fashion items (54%), men are more likely to buy technical products (47%). The younger and middle-aged respondents (18–39 and 40–64 years) on average indicated more categories (2.0) than the oldest group (65+: 1.7).

42.4.3 Social Commerce: When Social Media Meets E-commerce

Regarding social commerce and user-generated content, just under a quarter (24%) of respondents said that YouTube helps them make purchasing decisions. These respondents mainly represented the youngest age group (18–39 years: 32%). The middle (40–64 years: 22%) and oldest (65+ years: 12%) age groups are less inspired by YouTube. Facebook (9%), Instagram (6%), and Pinterest (2%) were also mentioned as a source of content driving purchasing decisions, however, significantly less than

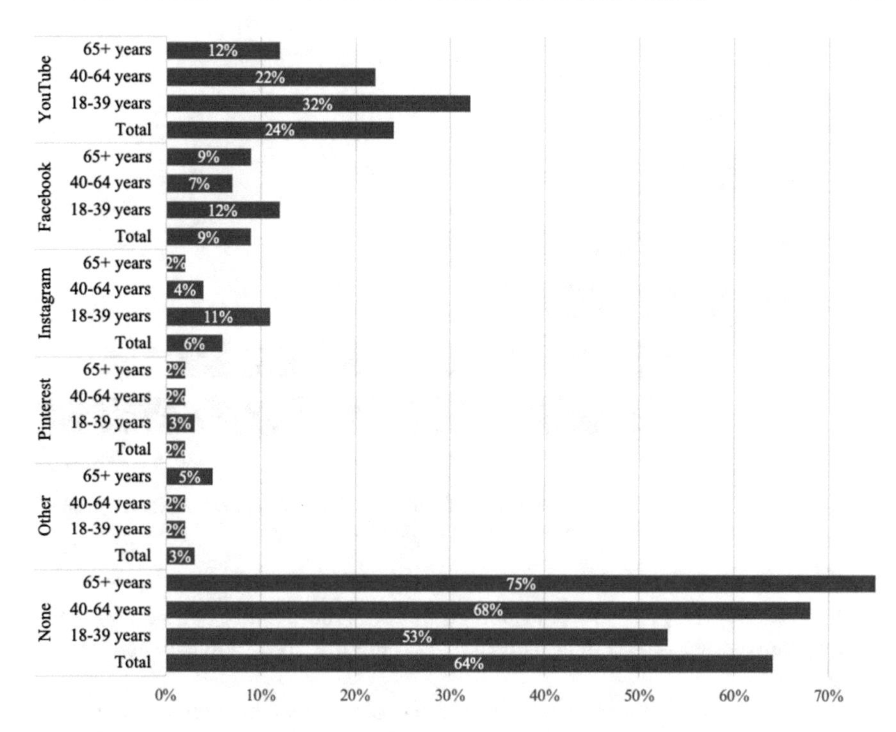

Fig. 42.3 Social media platforms that inform purchase decisions ($n = 900$/filter: consumers who utilize social media to obtain information on products and services)

YouTube (Fig. 42.3). Almost two-thirds (64%) of the interviewees believe that social media platforms do not help them with their purchasing decisions.

The results further show that almost two-thirds of the interviewees (61%) believe that social media platforms are not important to become aware of a product or service based on user-generated content (mean value 2.4 on a scale of 1–6). The youngest age group (18–39 years) estimates the importance to be the highest (2.6), followed by the middle age group (40–64 years: 2.3) and the oldest age group (65+: 2.1). Furthermore, social media is given the highest importance as a source of information for purchasing decisions by respondents in the lowest education category (compulsory schooling: 2.8); in contrast, respondents in the middle (vocational/technical degree) and highest (university) educational categories achieve an average value of 2.3 each.

The value of experts/influencers on social media platforms, who can guide or influence purchasing decisions, was also analyzed. Almost two-thirds (63%) of the respondents rated the value as rather low or very low (mean value 2.2). Social media influencers received the highest rating from the 18–39-year-old group (2.4). The 40–64-year-olds and those surveyed over 65 rated the value of influencers at 2.1.

As shown in Fig. 42.4, almost two-fifths (39%) of the interviewees stated that they did not use any information from social media platforms to subsequently buy products or services. Almost 30% use user-generated content at least once a month

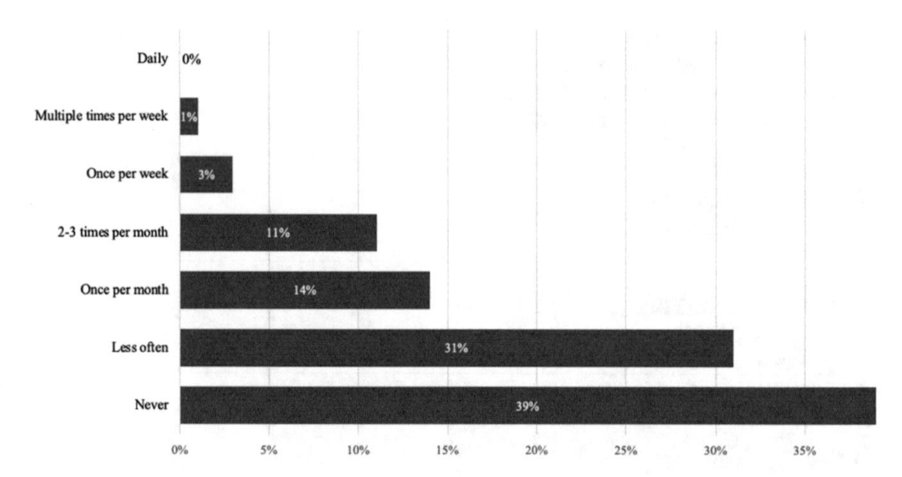

Fig. 42.4 Usage of social media content to inform purchasing decisions ($n = 900$/filter: consumers who utilize social media to obtain information on products and services)

(30%) and another 31% less than once a month. Additionally, women answered "never" more often (44%) than men (35%). There are no differences between the age groups.

Lastly, Fig. 42.5 illustrates the results of consumers posting content about purchases. Almost two-thirds (63%) of the respondents do not post their experiences and satisfaction with or recommendations for products and services on social media at all. Only about one in ten of the respondents (11%) does it once a month and around a quarter (26%) less often. Again, men reported a higher frequency (13%

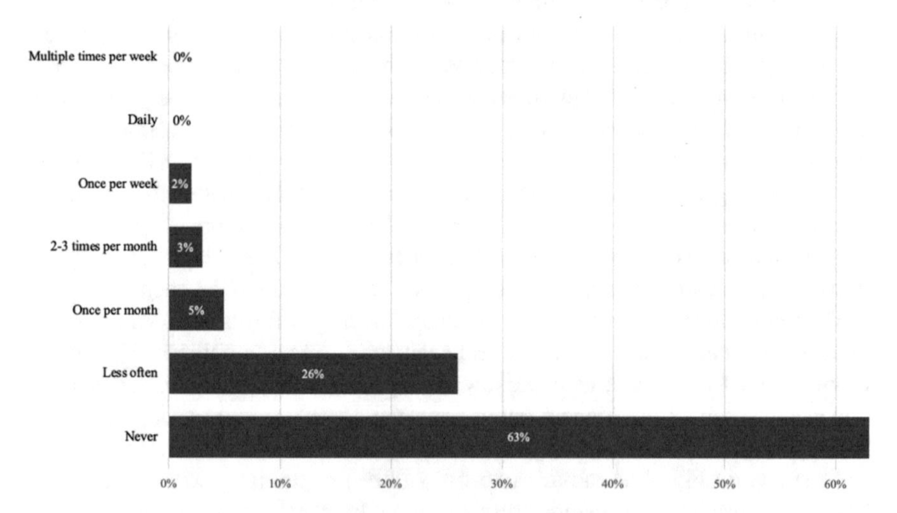

Fig. 42.5 Consumers posting content about purchases (products and services) ($n = 900$/filter: consumers who utilize social media to obtain information on products and services)

at least once a month) than women (8%). Somewhat surprising, the oldest age group stands out with the highest frequency of user-generated content: Almost a fifth (17%) of the over 65-year-olds who use social media stated that they post their experiences with products and services at least once a month. Among the 40–64-year-olds, this is only about one in ten (11%), and among the 18–39-year-olds, it is only about one in fourteen (7%).

42.5 Discussion and Conclusion

Research results show that the use of social commerce is still a young topic in Switzerland. While social media is highly prevalent among Swiss consumers (87% are on WhatsApp, 80% on YouTube, 48% on Facebook, and 34% on Instagram), only YouTube (for 24% of Swiss consumers) provides content that influences purchase decision-making. This might raise the question if these platforms are regarded as trusted sources as suggested in the literature [8–10], which is seen as an important success factor for social commerce. For Swiss consumers, social media certainly is an important use case of digital technology and is part of their daily lives. About half of the survey respondents use social media to write text messages, organize their free time and find out what their friends are doing. The respondents spend almost one hour a day on social media via their mobile phones. This appears to be a lot as it corresponds to 6% of the daily 16-h awake time of a human being (assuming eight hours of sleep and rest).

However, social commerce not only lacks e-commerce features on social media platforms in Switzerland (and the EU), but it is also hindered by current consumer behavior and the earlier mentioned potential lack of trust: Almost two-thirds (64%) of those surveyed believe that social media platforms do not help them with purchasing decisions. Additionally, almost two-thirds (61%) think that social media platforms are not important to become aware of a product or service. This is surprising, as social media seems to be important to Swiss consumers based on their usage, and advertising is common on all platforms. Could it be that the content driving commercial consumer decisions is perceived subconsciously (i.e., users are unaware of their influence on decision-making)?

Finally, only around one in ten respondents (11%) post their experiences and satisfaction with or recommendations for products and services on social media at least once a month. This could indicate that social media in the consumer's customer journey is still a pure information channel, but not yet a commercial information exchange channel that triggers user-generated content about products and services among friends and colleagues in their networks. More active participation might drive trust perceptions among the participating users on social media platforms.

How do organizations deal with this challenge? Do they react to the posts and do customers expect feedback from organizations following the posting of their product recommendations? Data from this research project showed empirical evidence that both the posting frequency/volume of experiences and opinions [15, 16], and the utilization of social media platforms as trusted sources [8–10] are low. As a contribution to theory, the study provides a national consumer data set that allows for conducting cross-cultural studies as recommended in the literature [14]. For business practice, our research shows that social commerce features offered in Switzerland are still in development compared to countries with more advanced social media platforms. Additionally, social media platforms as a source of trusted information are not yet widely used by Swiss consumers, while at the same time, their participation to provide recommendations might need stimuli from actors participating in social commerce. According to the literature [15], managers first need to identify their existing e-commerce and social networking applications and capabilities as part of their social commerce strategies. Next, with the ongoing developments and the availability of e-commerce features on social media platforms in Switzerland, it is expected that the concept of social commerce will gain increasing impetus. While organizations nowadays can no longer be inactive on social media, social commerce is still a young and slowly developing field in the Swiss digitization and marketing landscape.

42.6 Limitations and Future Research

While this study provides a more refined understanding of the use of social media, e-commerce, and social commerce based on recommendations and feedback on products and services (i.e., user-generated-content) with Swiss consumers, it does not provide the reasons (drivers) for the lack of activity and interest in opinions from influencers and does not explain the differences in male and female online usage. In addition, the age group under 18-year-olds (i.e., 6–18-years-olds) was not surveyed. Lastly, as it was a quantitative study, qualitative research instruments such as consumer observation or the collection of user information would allow a deeper understanding of behaviors and motivations. We recommend addressing these gaps in future research.

References

1. Afrasiabi Rad, A., Benyoucef, M.: A model for understanding social commerce. J. Inf. Syst. Appl. Res. **4**(2), 63 (2011)
2. Zhang, K.Z., Benyoucef, M.: Consumer behavior in social commerce: a literature review. Decis. Support Syst. **86**, 95–108 (2016)
3. Zhou, L., Ping, Z., Zimmermann, H.-D.: Social commerce research: an integrated view. Electron. Commerce Res. Appl. **12**, 2, 61–68 (2013)
4. Liang, T.P., Turban, E.: Introduction to the special issue social commerce: a research framework for social commerce. Int. J. Electron. Commer. **16**(2), 5–14 (2011)
5. Bundesamt für Statistik: Bevölkerungsentwicklung und natürliche Bevölkerungsbewegung im Jahr 2021: Provisorische Ergebnisse. Neuenburg, Schweiz. https://www.bfs.admin.ch/news/de/2022-0464 (2022)

6. Bundesamt für Statistik: Internetnutzung 1997–2021. Neuenburg, Schweiz. https://www.bfs.admin.ch/bfs/de/home/statistiken/kultur-medien-informationsgesellschaft-sport/informationsgesellschaft/gesamtindikatoren/haushalte-bevoelkerung/internetnutzung.assetdetail.22404521.html (2022)
7. Bundesamt für Statistik: Konsum und Sparen. Neuenburg, Schweiz. https://www.bfs.admin.ch/bfs/de/home/statistiken/querschnittsthemen/wohlfahrtsmessung/wohlfahrt/wohnsituation/konsum-sparen.html (2020)
8. Cheng, X., Gu, Y., Shen, J.: An integrated view of particularized trust in social commerce: an empirical investigation. Int. J. Inf. Manage. **45**, 1–12 (2019)
9. Han, H., Xu, H., Chen, H.: Social commerce: a systematic review and data synthesis. Electron. Commer. Res. Appl. **30**, 38–50 (2018)
10. Wang, X., Wang, H., Zhang, C.: A literature review of social commerce research from a systems thinking perspective. Systems **10**(3), 56 (2022)
11. Friedrich, T.: Analyzing the factors that influence consumers' adoption of social commerce: a literature review. In: Proceeding of Twenty-First Americas Conference on Information Systems, Fajardo, pp. 1–16 (2015)
12. Ghani, L., Hofreiter, S.: Wie Social Commerce die Welt des Online-Handels verändert. In: Gutting, D., Tang, M., Hofreiter, S. (Hrsg.), Innovation und Kreativität in Chinas Wirtschaft, pp. 353–377. Springer Gabler, Wiesbaden, Deutschland (2021). https://doi.org/10.1007/978-3-658-34039-1_14
13. Zhang, P., Wang, C.: The evolution of social commerce: an examination from the people, business, technology, and information perspective. Commun. Assoc. Inf. Syst. **31**(5), 105–127 (2012)
14. Pouti, N., Taghavifard, M.T., Taghva, M.R., Fathian, M.: A comprehensive literature review of acceptance and usage studies in the social commerce field. Int. J. Electron. Commerce Stud. **11**(2), 119–166 (2020)
15. Huang, Z., Benyoucef, M.: From e-commerce to social commerce: a close look at design features. Electron. Commer. Res. Appl. **12**(4), 246–259 (2013)
16. Terra, L., Casais, B.: Moments of truth in social commerce costumer journey: a literature review. In: Martínez-López, F.J., D'Alessandro, S. (eds.), Advances in Digital Marketing and eCommerce, DMEC 2021, pp. 236–242. Springer (2021). https://doi.org/10.1007/978-3-030-76520-0_24
17. Facebook: Facebook Shops, ein neues Online-Shopping-Erlebnis. https://www.facebook.com/business/news/announcing-facebook-shops (2020)
18. Facebook: Shops auf Facebook und Instagram. https://www.facebook.com/business/tools/shops/tips (2022)
19. Facebook: Infos zu Shops. https://www.facebook.com/business/help/2343035149322466?id=1077620002609475 (2022)
20. Zumstein, D., Bärtschi, D.: Social commerce report 2021: Potenzialanalyse des Direct Checkout auf Facebook und Instagram für Onlinehändler. ZHAW School of Management and Law (2021). https://doi.org/10.21256/zhaw-3131
21. Instagram: Affiliate. https://creators.instagram.com/earn-money/affiliate (2022)
22. Facebook: Wie funktionieren Shops auf Instagram? https://www.facebook.com/help/instagram/240116693975803/?cms_platform=iphone-app&helpref=platform_switcher (2022)
23. Facebook: Einen Merchandise-Launch auf Instagram erstellen. https://www.facebook.com/business/help/237743385002947?helpref=search&sr=4&query=Swip-up%20Instagram (2022)
24. Facebook: Tag Products in Live on Instagram. https://www.facebook.com/business/help/3051388324964403?helpref=search&sr=3&query=Swip-up%20Instagram&locale=en_US (2022)
25. Heinemann, G.: Der neue online-handel. Springer Gabler, Wiesbaden (2021). https://doi.org/10.1007/978-3-658-32314-1_4
26. Erbslöh, B.: Suxeedo. 5 wichtige Social Commerce Trends für 2022—welche Entwicklungen sind zu erwarten? https://suxeedo.de/magazine/social/social-commerce-trends/ (2021)

27. Kästner, A.: Internet World. Die Zukunft des Social Commerce: Was kommt jetzt? https://www.internetworld.de/social-media-marketing/zukunft-social-commerce-kommt-jetzt-2731881.html?ganzseitig=1 (2022)
28. Peter, M.K., Dalla Vecchia, M.: The digital marketing toolkit: a literature review for the identification of digital marketing channels and platforms. In: Dornberger, R. (ed.) New Trends in Business Information Systems and Technology, pp. 251–265. Springer, Berlin (2021)
29. Peter, M.K., Niedermann, A., Zachlod, C., Lindeque, J., Mändli Lerch, K., Dalla Vecchia, M., Gnocchi, A.: Social Commerce Schweiz 2022: Social Media und E-Commerce in der Schweizer Bevölkerung. FHNW Hochschule für Wirtschaft, Olten, August (2022)
30. Breen, D., Donnelly, R.D., Chalmers, J.: Finding out what the customer wants: computer-assisted telephone interviewing (CATI)? Int. J. Health Care Qual. Assur. **5**(3), 19–21 (1992). https://doi.org/10.1108/09526869210014926
31. Bundesamt für Statistik: STATPOP 2020, Statistik der Bevölkerung und der Haushalte. https://www.bfs.admin.ch/bfs/de/home/statistiken/bevoelkerung/erhebungen/statpop.html (2020)
32. TikTok: How to Add Product Links to TikTok Videos. https://ads.tiktok.com/help/article?aid=10005125 (2022).

Chapter 43
Web Marketing Trends—Case Study of Trigénius

Patrícia Duarte and Madalena Abreu

Abstract Digital marketing is nowadays the most common and acclaimed used method for marketers to survive in the markets. In fact, digital marketing has broken out after the twenty-century revolution on technology, creating a new era for the business in general. Digital market encompasses a set of strategies and tools using Internet and electronic devices, making it possible for companies to connect with prospective and actual customers. Also, these new tools and strategies have created a new world for communicating in a personalized way with customers and on a global scale. Besides, in this highly competitive world, customer's loyalty has become a business survival concern, and companies do strive for improving its relationships with customers on a permanent basis. And so digital marketing is being used as a communication strategy to foster customer loyalty. To better recognize this trend, this paper explores a real situation. A case study is employed as the methodology, and the use of communication tools in the digital environment by the enterprise Trigénius is explored. After the work developed in order to increase loyalty specially within digital marketing channels, and the implementation of several operations within different web marketing tools, it was possible to see the increase of loyalty by the company actual customers.

43.1 Introduction

The role of marketing has changed dramatically over the years. The development of marketing is inseparable from technology development and the era of Internet pushed and allowed marketing to become absolutely market driven. In fact, strategic decision concerning business opportunities and markets in general have been vital for marketing performance, and the technological enhancements made a revolution in the

---

P. Duarte
Instituto Politécnico de Coimbra, Coimbra Business School | ISCAC-IPC, Coimbra, Portugal

M. Abreu (✉)
Business School Research Centre, Coimbra, Portugal
e-mail: mabreu@iscac.pt

J. L. Reis et al. (eds.), *Marketing and Smart Technologies*, Smart Innovation, Systems and Technologies 337, https://doi.org/10.1007/978-981-19-9099-1_43

arena for finding and managing information about customers, products, marketplaces and the overall environment [4].

The Internet itself is constantly developing, and with the constant improvements in technologies easily available in the market, it has become the primary business playground and also the most popular communication channel [11]. What is more, internet does play a vital role in several spheres, from personal life to commercial ventures and has brought many advantages. These benefits are seen both from companies and customers, such as the elimination of geographic barriers, the ability to reach more and diversified publics, the customization of commercial relationships, building relationships on a twenty four hours a day and seven days a week basis, within a list of profits and surplus for the society at large [22].

Alongside with technology massive advances, Internet transformed dramatically the life of each consumer daily, causing a massive impact in terms of consumer behavior, communication channels and frequency. This turmoil within the context of this digital world has an enormous impact within the world business reality. What is more, as digital marketing became more sophisticated, businesses do follow on a regular base all the changes in the digital world, adopting approaches and strategies for gaining sustainable competitive advantages [11].

Currently, to capture the attention of online consumers, marketing communication must follow the last novelties and fashion; communication channels and messages need to be informative, personalized, unexpected and fun, on an innovative and changing scenario. This means that information without engagement and entertainment is just not enough [11].

Furthermore, in a world characterized by high competitiveness in the market, loyalty is a requirement in every business. There are several precedents that help to achieve loyalty; however, communication presents itself as a key factor, fundamental in building lasting relationships, winning over customers and ultimately, their loyalty. Relationships have not always been valued in business, nevertheless, if we study the evolution of marketing, they began to be increasingly recognized by marketing professionals, and by all the other individuals who maintain a business, and a new concept emerged, the relationship or relationship marketing.

In this study, the focus will lie on communication to achieve loyalty. Communication is essential for all companies and organizations that intend to make themselves known, including their products and services, to their target audience and all stakeholders.

Having in mind these challenges and particular situation, the aim of the research is to understand how Internet marketing can be used for a better communication with its customers. And thus foster its relationship, and finally customer's loyalty. In other words: "How can digital marketing, as a communication strategy, help customer loyalty? Besides this broad goal, this paper does want to have a deeper overview of digital marketing tools; to be able to point out digital marketing importance and its impact on customer relationship; and to realize how digital marketing contributes to customer loyalty.

For the present study, these main goal and objectives are addressed analyzing a real situation to better understand the issues raised above. In other words, a real situation is explored: the case study of the enterprise called Trigénius.

And so, the main objectives are to promote the company's brand, and its products or services through various digital media, and seek to improve and maintain its relationship with its customers.

43.2 Literature Review

43.2.1 Digital Marketing at a Glance

While looking at the main question that gives the reason for this inquire, the first topic to be raised lies in analyzing the digital marketing trends. In fact, to conduct marketing effectively, companies need to track trends in markets and also trends in marketing [21].

Web marketing, Internet marketing, e-marketing or online marketing are concepts that emerged on a global scale in the beginning of the 90s of last century. The age of information, globalization and the outburst of new technologies is now "on the headlights of every house or window". And text-based websites, which offered information about products, were just the beginning of this novelty [4].

Within this new era, the concept and operationalization of digital marketing became pervasive. Digital marketing does incorporate channels, strategies and tools as the Internet, mobile phones, display advertising and other digital medium, being this simply an umbrella word for the marketing of products using digital technologies.

In Sristy and Rungta [22] words, digital marketing is a vast concept for the use of modern technologies and Internet connections to reach marketing objectives, including thus the promotion of goods and services and the delivery of messages to buyers through various channels. And as already pointed out, this vast spectrum of functions does happen through different ways, for example, through social media, search engine, banner ads on specific websites, e-mail and app development [23].

According to [20, p. 7], web marketing is "about delivering useful content at just the precise moment a buyer needs it". This means web marketing is considered as a subset of digital marketing. There can be in fact several benefits of web marketing, even though, one must be careful because the excessive or wrong use of Internet in marketing can have its losses, for example, an excessive amount of Ad clutter, the unserious recognition, causing harm by negative feedback, the absence of trust and others disadvantages can arise unexpectedly [22].

An effective web marketing plan must thus consider the way potential buyers talk and the possible real words they would use. All these efforts and are compulsory for planning effective search engine marketing strategies, and by doing it, then it is possible to build a positive online relationship with buyers [20].

There are diverse trends of web marketing, such as search engine marketing, social media marketing, content marketing, e-mail marketing and video marketing, which will be explored below. These tools will be described within this paper as they were central in the case study analysis.

43.2.2 Relationship Marketing and Loyalty

Moreover, for the present study, the concepts of relationship marketing and loyalty are also important to bring up as they drive and foster the accomplishment of business success. One must also remember the importance of digital marketing on the advancement of relationship marketing, i.e., digital marketing strategies and tools are nowadays a vital aid to relationship marketing success.

The expression "relationship marketing" first appeared in the service marketing literature in 1983, an expression coined by Leonard Berry, being relationship marketing conceptualized as "attracting, sustaining and—in multiservice organizations—improving customer relationships" [7]. Nowadays, this new approach focuses on building stable and lasting relationships, where the business model becomes customer-centric, and supported by technological development, information management and customer service. Ultimately, as Chaffey and Ellis-Chadwick [8] point out, building long-term relationships with customers is essential for any sustainable business, and this applies equally to the online elements of a business. In other words, digital marketing appears to expand these relationships, using technology to attract a greater number of customers.

Besides, customer loyalty is inevitable for all companies that want to remain competitive and become a reference for consumers. And by retaining its consumers, the company wins, since customers will repeat their purchases with regularly, and they will pursue the relationship with the same company, brand or shop outlet [10].

43.2.3 SEM and SEO

Search engine marketing (SEM) is a collection of activities that include search engine optimization, social media marketing and other search engine-related activities, with the goal of increasing visibility in search engines, whether through the acquisition of more free or paid traffic [23].

Search engine optimization (SEO) can be defined as the process of increasing a website's visibility and ranking in search engine result pages, which means the higher you rank on the result pages, more often you will appear in the search results list, receiving more traffic from search engine users [27]. This traffic can, later, be converted into customers [23].

To achieve SEO, one can use the so-called organic SEO to reach top ranking search results without using any paid means. And so, the paid SEO is employed when a website owner achieves a high position in search results, by purchasing an advertising space in search engine results, for example, through Google Ads.

To use search engine marketing successfully, the choice of keywords is a very important and valuable process, which requires a special attention to the intent of user's request.

This process is carried out by identifying popular keywords that correspond to the website's content, based on relevance and potential conversion rate. This identification can be made by using keyword suggestion tools, available on most search engines or through information about popular keywords. By doing so, website developers can build their website with keywords that have a higher value and return over investment (ROI), allowing the website developer to successfully position their site at the top of the search engines results. However, keywords must be monitored and updated on a regular basis [13, 23].

On the realm of SEM, it is also important to note the Digital Display Advertising.

Digital Display Advertising is a subset of SEM efforts. It is the application of diverse display advertising formats to target a potential audience, such as text, image, video ads, banner and others. The message can be personalized according to customer's interests, content topics or the position of customer's buying cycle [4]. Over the past years, advertising has emerged as a primary source of revenue for many websites [6].

According to the same author, there is a growing preference for display ads, mainly image.

43.2.4 Social Media Marketing

Social media has transformed the behavior of consumers. They have provided new ways to look for, evaluate, choose and buy goods or services [1]. Besides this, consumers can interact with brands and express their opinions about brand experiences. On the other side, organizations can invest in their online presence on platforms and develop more aimed campaigns, communicate with consumers, gain insights about how their costumers perceive their brand [25].

Social media marketing is a tool businesses can use to increase website traffic, brand awareness [22] and create a strong relationship with their clients [15]. It allows businesses to promote their websites, products or services through different platforms or tools, for example, Facebook, Instagram, LinkedIn, among others, and reach a larger community that may not be available in traditional channels. As mentioned before, the communication through these channels must be informative, interesting, and fun, in order to catch the attention of the consumer [11, 18]. In fact, if the content created is good and well created, it can reach hundreds of new visitors, and boost traffic [26]. Sedej [21] adds by saying that social media is becoming an increasingly important tool to reach the rest of external and internal audience, besides customers.

Nowadays, the word-of-mouth has a big impact on social media. In the platforms is possible, for customers, to talk about their brand experiences, reaching a great number or people online and influencing other potential customers' purchase decisions or perceptions about brands, products or services. On the other side, companies can see or hear the feedbacks and recalibrate their social media marketing strategies accordingly [25].

While talking about social media, nowadays, influencers have been a vital topic for the success of marketing under these devices.

As mentioned before, word-of-mouth has an enormous impact on social media. Therefore, some well-known businesses use celebrities or well-known opinion leaders to influence consumers. However, smaller brands will most likely use micro-influences, meaning, influencers who are not so well-known as celebrities, but have strong and enthusiastic followers. These micro-influencers are considered more trustworthy and authentic, who can encourage others to see the content they create e engage with them. Even though, influencer marketing is not new, it is believed that is has a lot of potential to develop further [3].

43.2.5 Content Marketing

Content marketing is considered the key for a successful online marketing. It is a strategy focused on the creation and distribution of valuable, relevant and consistent content, that aims to attract and retain a defined audience, and ultimately, drive a costumer to an action, for example, a purchase [9].

The content is what keeps social media "working", so it is important to create quality content and think about the variety of content that can be made, always having in mind the topics that costumers care about, and be careful with the SEO, keywords, that customers can search into search engines. Also, it is important for businesses to engage with people who share and comment, as it is an opportunity to connect with potential customers. The content created can go from images, newsletters, eBooks, to videos and posts in social media or blogs [5].

As the world is dominated by social media, all businesses must bet on content marketing, as it can define the success of a company. The creation of informative and quality content will keep the attention of readers and will help building trust toward the company and assist in brand recognition, authority, trustworthiness and authenticity [14, 16, 17].

In the realm of content marketing, video marketing has been conquering a privileged position.

Video marketing has gained prominence in marketing communication strategies as they are not only a source of information, but also a learning tool [21]. According to Ansari et al. [2], videos are considered to be one of the most effective marketing techniques.

Before, video was mostly used by big brands and businesses with large marketing budgets. However, today, is it being used by all companies, even small and medium-sized enterprises. This type of content presents opportunities for businesses, as it is able to provoke emotions on customers or potential and appeal to their needs [21].

According to research made by Sedej [21], who pretended to find manager's views on awareness and understanding of trends in marketing, focusing on video marketing, he found out that most of managers agree that video marketing is important and that it, in fact, can increase the probability of business success.

Currently, successful marketing is about focusing the strategy on customer's experiences and maintaining long-lasting relationships with them. In consideration of the modern business environment, where creativity and intelligence are a requirement, videos foster relationship building, since they grab attention, have high information value and are entertaining. Besides, it can increase the understanding of products or services within the target audience, increase in sales, web traffic, brand awareness and others [21].

In the future, video marketing is expected to continue on the rise, as the technology development and the trend of overall digital transformations persists [21].

43.2.6 E-mail Marketing

E-mail marketing is a digital channel used to address customers or potential customers. At business with enterprises, within B2B or B2C, newsletters are very common to see. It is way to increase brand awareness and a sense of community building. It also recalls the existence of a company, provides information, raises the credibility of the brand, leads readers to make a purchase and allows to receive feedback [12].

Newsletters should always and only be sent to people who have signed up for subscription, if not it is sent as spam. Beyond that, all of the information provided should be useful to consumers, otherwise, people may unsubscribe the newsletter [12].

E-mail platforms are very useful for businesses since it allows to send personalized newsletters, in a way that it can be sent to a segmented database, according to profile of each customer or potential customer [19].

43.3 Methodology

This article follows the case study methodology, being the research question "How can web and digital help Trigénius enterprise on a daily basis?", being thus the object of study developed in a real context. And so, more specifically, the present case study addresses the web marketing operations of the enterprise Trigénius, falling thus the analysis on a single case; and consisting on a descriptive research.

Following Yin [28], the case study method can be used while researching and analyzing events in their own context, in real time.

So, this method was chosen for the present research, as this is a real situation, with the description of daily operations of an enterprise.

For this analysis, the main sources of information were the analysis of Trigénius documents, and direct observation while developing the operational function as a manager of the web operations of this business.

43.4 Case Study—Trigénius

43.4.1 Brief Description of Trigénius

Trigénius is a leading company in technological solutions, standing out for the implementation of business management software. Founded in 1998 by 3 young entrepreneurs, and based in the city of Fátima, Trigénius is a company for the creation, development and maintenance of technological solutions, standing out in the implementation of business management software.

This company has currently two branches, in the cities of Lisbon and Ourém and has partnerships with its main hardware and software suppliers that is: Primavera, PHC and SAP. With around 70 employees, this company has the necessary skills and competencies for the implementation, development and support of more complex information systems. Its customer typology is mostly the business-to-business (B2B) market, with a total of 90% of its all operations in the overall market, and the remaining 10% share of the B2C (business-to-consumer) market.

The company's mission is to provide adequate solutions to the specific technological needs of each company, seeking to guarantee the quality and stability of the relationship [24].

Trigénius pursues to be a reference company in the market of software development and implementation, within the information technology sector, both locally and nationally [24].

This company has a full team working solely on the development of branding strategies, from the adaptation of a company's visual identity to the implementation of an online store or the development of a website from scratch, being thus present on the first pages of Google.

43.4.2 The Study Case Operations

The work performed on web marketing operations consists in a range of different activities related to Internet marketing, such as social media management content marketing, e-mail marketing and inorganic SEO (paid media, e.g., Google Ads).

In the beginning of each month, a social network content plan is created for the coming month.

This plan includes the choice of "special days" such as "International fun day at work", "Thank you day", "Christmas" or others such as "world backup day" and "data privacy"; moreover, these 'offers' are related with products of technology or other areas, where Trigénius does its business. Besides this, testimonials from customers are published monthly, also like promotion of products or services, or informative issues, such as registration for events. All this content is mainly presented by images and text and is published on social networks such as Facebook, Instagram and LinkedIn, and sometimes, on the company's website.

When it comes to e-mail marketing, the communication sent is mainly useful information from partners, information about Trigénius events and communications regarding updates, legal information and company's products or services. Every single newsletter is sent to a segmented database, based on different needs and the customers history.

As already said, communication must be useful, entertaining and valuable. Having this in mind, and as an example, when Vodafone was the target of a cyber-attack, there were several communications and complaints from companies: They wanted solutions to provide them from frauds and that would help them secure their business. However, in Trigénius' case, it was decided to do something simple but meaningful, and not to be just another company promoting cybersecurity products.

Also, on "Thank you Day", the department of marketing and communication did something a little different. A newsletter was created and sent to customers and partners saying, "The word that best defines our feeling for you is: Thank you" and then, we wrote things the company was thankful for, such as "trust", "encouragement", "challenges", "preference", "for making us grow", "dedication", "commitment", "partnership" and others. This newsletter ended up having a very positive and unexpected feedback from customers and partners.

According to the literature review, a content marketing strategy is important to attract clients, or followers. To achieve this goal, Trigénius developed especially its search engine marketing (SEM), developing asset of activities in social media marketing and thus increasing its visibility.

And so Trigénius developed a program on inorganic SEO.

One of the tasks performed was the creation of ad campaigns for a client. First, it was necessary to set the goal the company wants to achieve, then the work went on using the segmentation of ads which are based on locations, ages and interests. And so, the choice of keywords was based on the services that the client wanted to enlarge giving it more visibility. For this, keyword research was made, to discover words that could achieve better results and a higher probability of being searched. To do this, a keyword planner was used, which shows search estimates and cost. Then, the customization of specific ads, adding titles, text, images, phone numbers, if necessary and an URL was added to each ad, to determine a landing page, which people were directed to when they click on the ad. After every month, the data was extracted, such as campaign impressions, clicks, CTR (the number of clicks that campaign or ads receive divided by the number of times it is shown), average CPC (the average amount charged for a click on the ad or campaign) and total cost. Later, when conversions were made from Google Ads, a record was made with the keywords that gives rise to the conversion and keywords that generate most conversions. And so, adjustments could be made to the campaign, and if necessary, there was the need to delete keywords that do not give the expected results but cost money. In addition, when necessary, new keywords were added to the ads or improvements were made by using the quality index that provides information regarding certain aspects that can be improved when compared to other advertisers. When search results that had nothing to do with the ads were observed, negative keywords were added, which means, a word or phrase that prevents the ads from being triggered.

43.5 Results and Discussion

According to the literature review, a content marketing strategy is important to attract actual clients, or followers.

To achieve this goal, Trigénius intended to show up as a reliable and professional organization by sharing testimonies on social networks and newsletters. Moreover, all these messages did highlight a link that redirected the viewer to a case study in the company's website. By promoting services or products with a link to the respective webpage, Trigénius contributed to the increase of website traffic and brand awareness. On the other side, by sharing content like "Thank you day", the company shared content that, in the present and in the future, will contribute to build and maintain relationships with customers or partners, and at the same time, their importance was being acknowledged by companies, as Sedej [21] stressed for companies while using strategically social media with customers and other important people around the company.

Regarding e-mail marketing, Trigénius sent a lot of relationships communication, useful for customer's consumption experience, important to increase the strength of preference for a brand and strengthen ties between businesses and customers and raise credibility as stressed by Sristy and Rungta [22]. Besides this, the platforms that exist today are very helpful since they allow to personalize messages by allowing to send newsletters to a specific database, with different targets.

Finally, in terms of SEO, the ads that were created and displayed in search results can have a great impact on the increase of brand awareness and recognition. Knowing the importance of good content, the titles, texts or imagens displayed on the ads were eye-catching and interesting so that consumers or potential consumers could be curious or interested enough to open the ads, that are then redirected to a landing page, the same conclusions according to Hudák et al. [12].

Moreover, and on a global perspective, Trigénius failed on some issues.

The most important aspect while addressing the mistakes of this company in web marketing, Trigénius did not plan and develop a proper marketing and communication strategy. On this topic, the biggest error begun with the absence of a vital step for a business strategy: Objectives were not established. Most of the communication that was sent, like newsletters, were made according to the company's daily needs; in other words, communication was developed "now", and that also happened in the case of the promotion of a product or service or even another information of interest.

Besides this, video marketing was something that should be already "on the way" but nothing has been done on this account. However, its importance is acknowledged, especially when a business wants to obtain a greater engagement from followers and boost results.

43.6 Conclusion

This study presents a case study of the enterprise Trigénius and its efforts to apply digital marketing, especially within the web marketing channel, as a communication strategy, to enlarge its customer loyalty.

For this purpose, the literature on web marketing was conducted, and different strategies and technologies were briefly analyzed. As a conclusion of the literature, one can stress that with the rapid evolution of internet and its impact on consumers' behavior and businesses everyday life, Internet marketing has become crucial for business and people's life in general. Nowadays, to be successful, besides being constantly on websites or platforms, businesses must deliver the right content, to the right audience on the right time, with the right tools and platforms. It is vital for companies to have constant information about new digital tools and being aware of new trends, that are constantly emerging and consequently, companies must rethink their business strategies. And offer new communication channels and messages to its customers.

Having analyzed Trigénius use of a content marketing strategy, it was possible to realize a set of good examples that were a real good example for maintaining and improving good relationship with its customers. On the other side, the company failed on its digital communication strategy by not having a structure plan for this area.

The present paper, within a single case, does show the importance of digital marketing tools for promoting customer relationship.

Future research can develop in two directions: This enterprise can be analyzed with a special focus in its customers. Trigénius does have a special relation with some of its clients: So, it could be very helpful to learn from them what Trigénius should develop, writing content marketing, for instance, and answer to clients in a better way.

Moreover, in the scope of digital marketing, other B2B enterprises could be also analyzed, and the different tools and digital communication strategies could be compared.

References

1. Alves, H., Fernandes, C., Raposo, M.: Social media marketing: a literature review and implications: implications of social media marketing. Psychol. Mark. **33**(12), 1029–1038 (2016)
2. Ansari, S., Ansari, G., Ghori, M., Kazi, A.G.: The impact of brand awareness and social media content marketing on consumer purchase decision. J. Publ. Value Admin. Insights **2**, 5–10 (2019)
3. Appel, G., Grewal, L., Hadi, R., Stephen, A.T.: The future of social media in marketing. J. Acad. Mark. Sci. **48**(1), 79–95 (2020)
4. Bala, M., Verma, D.: A critical review of digital marketing. Int. J. Manage. IT Eng. **8**(10), 321–339 (2018)
5. Baltes, L.: P content marketing—the fundamental tool of digital marketing. Bulletin Transilvania Univ. Brasov Ser. V Econ. Sci. **8**(2), 111–118 (2015)
6. Barford, P., Canadi, I., Krushevskaja, D., Ma, Q., Muthukrishnan, S.: Adscape: harvesting and analyzing online display ads. In: Proceedings of the 23rd International Conference on World Wide Web, pp. 597–608 (2014)
7. Berry, L.: Relationship marketing of services—perspectives from 1983 and 2000. J. Relation. Market. **1**(1), 59–70 (2002)
8. Chaffey, D., Ellis-Chadwick, F.: Digital marketing: strategy, implementation & practice, 7.ª ed. Pearson UK (2019)
9. Content Marketing Institute: B2C Content Marketing: Benchmarks, Budgets, and Trends (2020)
10. Dutra, K., Rangel, L.: O marketing como ferramenta para fidelização dos clientes. Revista Eletrônica da Faculdade Metodista Granbery **1**, 1–12 (2006)
11. Grubor, A., Jaksa, O.: Internet marketing as a business necessity. Interdisciplinary Des. Complex Syst. **16**(2), 265–274 (2018)
12. Hudák, M., Kianičková, E., Madleňák, R.: The importance of E-mail marketing in E-commerce. Proc. Eng. **192**, 342–347 (2017)
13. Iskandar, M.S., Komara, D.: Application marketing strategy search engine optimization (SEO). IOP Conf. Ser. Mat. Sci. Eng. **407**, 012011 (2018)
14. Jacinto, D.F.A.: Marketing de conteúdo integrado no Instagram: Um estudo sobre o envolvimento da geração milénio com a qualidade e o valor percebido da informação, Universidade Nova de Lisboa (2018)
15. Jan, A., Khan, M.F.: Social media is nothing but a public relation tool. Int. J. Bus. Manage. **2**(12), 272–277 (2014)
16. Le, D.: Content Marketing [Tese de Bacharelado, University of Applied Sciences] (2013)
17. Lieb, R.: Content Marketing: Think Like a Publisher—How to Use Content to Market Online and in Social Media, 1st ed. Que Publishing (2011)
18. Ryan, D.: Understanding digital marketing: marketing strategies for engaging the digital generation, 3.ª ed. Kogan Page Limited (2014)
19. Samantaray, A., Pradhan, B.B.: Importance of E-mail marketing. PalArch's J. Archaeol. Egypt/Egyptol. **17**(6), 5219–5227 (2020)
20. Scott, D.M.: The new rules of marketing and PR: how to use social media, online video, mobile applications, blogs, news releases, and viral marketing to reach buyers directly, 3rd ed. Wiley (2015)
21. Sedej, T.: The role of video marketing in the modern business environment: a view of top management of SMEs. J. Int. Bus. Entrep. Dev. **12**(1), 37–48 (2019)
22. Sristy, A., Rungta, S.: Digital marketing VS internet-529.pdf. Int. J. Novel Res. Market. Manage. Econ. **3**(1), 29–33 (2016)
23. Terrance, A.R., Shrivastava, S., Mishra, A.: Importance of Search Engine Marketing in the Digital World **14**, 155–158 (2018)
24. Trigénius. https://trigenius.pt/empresa (2022)

25. Vinerean, S.: Importance of strategic social media marketing. Expert J. Market. **5**(1), 28–35 (2017)
26. Weinberg, T.: The new community rules: marketing on the social web. Develop. Learn. Org. Int. J. **25**(3) (2009)
27. Yasmin, A., Tasneem, S., Fatema, K.: Effectiveness of digital marketing in the challenging age: an empirical study. Int. J. Manage. Sci. Bus. Admin. **1**, 69–80 (2015)
28. Yin, R.: Estudo de caso: planejamento e métodos (2ª Edição). Porto Alegre: Bookman (2001)

Chapter 44
Augmented Reality: Toward a Research Agenda for Studying the Impact of Its Presence Dimensions on Consumer Behavior

Virginie Lavoye

Abstract Augmented reality (AR) virtual try-ons (VTO) have emerged as an important decision-making tool because of the highly realistic experience. For instance, AR enables users to virtually try-on sunglasses by placing the virtual product on their face. Research increasingly emphasizes the importance of spatial presence in the realistic AR experience. However, prior research on AR presence remains scant and overlooks social and self-presence. To fill this gap, we review literature on presence in the context of prior immersive technologies and propose a future research agenda on the impact of AR presence dimensions on product-relevant outcomes. This article starts by presenting AR spatial presence definition and proposing definitions for AR social and self-presence by drawing parallels between AR apps features and the presence dimensions of prior immersive shopping technologies. Thereafter, our review uncovers how each presence dimension leads to positive consumer outcomes. Then, we propose a research agenda for future studies of AR presence in marketing that outlines the need for a multidimensional perspective of presence to help uncover their unique impact on consumer responses. In addition, future research should investigate which contextual factors (marketing channels, for instance, in store and online as well as the types of products displayed in AR for instance makeup and sunglasses) might explain differences in the outcomes of presence. Our study has several limitations as it only considers the type of presence dimensions relevant to current AR-VTO experiences.

44.1 Introduction and Research Aim

The key advantage of augmented reality (AR) service is the highly contextual and realistic information [11, 13]. For instance, Sephora AR mirror is an augmented service that enables consumers to try-on the company's entire online assortment without needing to go to the physical stores [6]. Moreover, Sephora's color match

V. Lavoye (✉)
LUT Business School, Lappeenranta, Finland
e-mail: virginie.lavoye@lut.fi

J. L. Reis et al. (eds.), *Marketing and Smart Technologies*, Smart Innovation, Systems and Technologies 337, https://doi.org/10.1007/978-981-19-9099-1_44

helps customers find the right color shade for their skin tone [6]. Overall, such service augmentation strategy focuses on services that are typically available in stores [13]. The digitalization of physical aspects of services poses novel challenges to firms and marketers, for instance, whether the experience is realistic, and the products are tangible enough to attract consumer engagement [10]. AR-based virtual try-ons (VTOs) provide a tangible service experience by replacing tangible service elements with superimposed digital content on the real environment [10]. Tangible virtual experiences rely on presence, the psychological state in which consumers perceive a virtual object to be real [18].

For consumers, the potential benefits of VTO include being able to try the products wherever and whenever they want, and without size restrictions [6]. While for firms, AR service augmentation can free employee input and replace the need for employees to bring boxes and advice consumers on best fitting products for instance. Overall, AR service augmentation has the potential to save time and money for consumers and service providers [10]. However, determining whether to use AR is a difficult decision for any business, and 52% of retailers are not ready to use AR as part of their service experience [4]. Notably, one key issue is that AR remains expensive to develop and the possible marketing-relevant outcomes remain unclear. Thus, providing clearer description of the mechanisms that enable AR to enhance consumer outcomes is highly important and timely.

The optimal AR experience should deliver a realistic experience of the product, the virtual self, and the social context [5]. Despite preliminary studies on spatial presence in AR [13, 22, 26], little is known about the holistic presence dimensions (spatial, social, and self-presence) in AR and their specific impact on consumer outcomes. Therefore, we aim to focus on presence dimensions because research in prior immersive shopping technology asserts their role in enhancing marketing-relevant outcomes. In addition, focusing on presence would enable to propose guidelines to marketers and developers to improve consumer experience.

This article starts by presenting AR spatial presence definition and proposing definitions for AR social and self-presence by drawing parallels between AR apps and the presence dimensions of prior immersive shopping technologies. By studying the impact of these presence dimensions on product-relevant outcomes, we uncover the psychological mechanisms that enable the persuasive impact of presence dimensions.

We address two research questions in line with this aim:

RQ1: What is the definition of our three AR presence dimensions?

RQ2: How does each presence dimension influence consumers' responses?

Then, we are able to propose a research agenda for future studies of AR in marketing that outlines the need for a (1) multidimensional perspective of presence to unravel their unique impact on consumer outcomes, as well as boundary conditions such as (2) the type of consumer experience they deliver in different marketing channels including in retail and online and that (3) different product types may require different combination of presence.

44.2 Background

Presence refers to the psychological state in which consumers perceive a virtual object to be real [18]. Presence dimensions can be facilitated by a range of immersive technologies including AR, virtual reality (VR), e-commerce website, or virtual worlds [18, 23]. Presence dimensions vary in intensity and types between technologies, however, we emphasize similarities between prior presence dimensions and AR presence dimensions and propose to study whether and to what extent the outcomes of AR presence will be similar to prior presence outcomes. In addition, there are studies on AR spatial presence that we also include into the analysis. This study starts by defining AR spatial presence and proposes definition for AR self-presence and social presence by drawing parallels between presence experience in AR-VTO apps with the presence experience in prior immersive shopping technologies.

Based on spatial presence in virtual reality context, Hilken et al. [13] developed AR spatial presence. AR superimposes virtual object on the real world in real time [2] thus, studying spatial presence in AR entails that the object "is here" rather than the user being transported as in virtual reality research [13]. Specifically, when using IKEA or makeup AR apps, AR spatial presence involves that the location of the product appears to be in one's living room or on their body [13]. In addition, spatial presence also entails that the product can be moved around in the real world [13]. To sum up, spatial presence is defined as the sense that the object is embedded in the real environment and embodied on consumers [13].

Self-presence in video game occurs when players get a sense of physical resemblance and identification with their virtual self [25]. When users feel they are physically similar, they often relate with the virtual self personally [24] and experience self-presence [3]. AR superimposes virtual object on one's virtual body or self. For instance, L'Oréal Makeup Genius displays a virtual lipstick on a live feed of a consumer's face. Thus, AR users may experience self-presence because the virtual self can be considered highly physically similar (not perfectly similar because a virtual lipstick is superimposed) and enables users to identify with the virtual self. Therefore, we propose that AR self-presence refers to the sense that one's virtual representation is oneself in the real world [21] and is conceptualized as physical similarity and identification with the virtual self [21].

Based on social presence on e-commerce website, social presence occurs when consumers get a sense of human contact when they interact with technology at the company's frontline [8]. AR apps convey highly contextual information that help match the characteristic of a try-on technology with the actual try-on experience and address needs for consumers as if it was a salesperson in a store [13]. Thus, we propose that AR social presence refers to the sense that the AR app is a social actor [8] and is conceptualized as a sense of human contact in the online environment [8].

Overall, we define AR presence experience as follows: AR gives a sense that the offering is located in the physical environment and can be interacted with (i.e., spatial presence), involves a sense of self in the experience (i.e., self-presence), and the AR app gives a sense of human warmth similar to a salesperson in a store (i.e., social presence).

44.3 Methodology

We follow the recommended steps for literature review from Xiao and Watson [27]. Based on our definitions and conceptualizations of AR presence dimensions presented above, we keep only the studies on immersive shopping technologies that discuss the impact of similar presence features (e.g., object presence is similar to AR spatial presence, while game character identification is similar to AR self-presence) on product-relevant affect, cognition, and behavioral intentions. This approach to literature review based on technologies' effects on users has been used in previous reviews [16]. We repeat the selection process three times, once for each presence dimensions.

First, we searched in title and abstract for terms such spatial presence, physical presence and augmented reality or online shopping on Web of Science. We identified 14 relevant studies that we checked for eligibility and included in the review. Second, we searched in title and abstract for self-presence combined with game character on Web of Science. We identified 7 relevant studies that we checked for eligibility and included in the review. Third, social presence was combined with purchase, retail, shopping, or consumer on Web of Science. We find 14 relevant studies that we checked for eligibility and included in the review.

44.4 Results

This section is a short presentation of the effects of presence dimensions on consumers' responses. First, spatial presence delivers highly contextual information about the product, and thus, it enhances decision comfort [13]. Second, self-presence increases the sense that the situation is self-involving, and it enhances self-efficacy and loyalty [15]. In addition, self-presence increases product diagnosticity when the product directly involves one's body or identity [24]. Third, social presence increases consumers' sense of closeness with the seller, the AR app gives virtual proximity to the social actor as a seller in a store [24]. Social presence enhances trust [8] and results in positive product attitude [9]. To sum up, we show that immersive shopping technologies can decrease the physical, personal, and social intangibility inherent to buyer–seller relationships.

44.5 Future Research Directions

From prior literature, we find that each presence dimensions has a unique role in influencing positive consumer outcomes. In addition, literature on AR suggests that AR delivers an optimal realistic product experience [10, 13], thus we ask:

FRQ1: Whether and to what extent can the unique role of each AR presence dimensions enhance consumer outcomes?

Moreover, when people experience high self-presence online, they are more comfortable to disclose personal information, as long as they are not identifiable [14]. Thus, self-presence in store might have a negative effect with people feeling too self-cautious to look at themselves, in a virtual mirror, around strangers. We propose that different uses of AR will explain contextual differences in presence outcomes impact consumers thus, we ask:

FRQ2: What are the optimal AR presence dimensions for different shopping experiences such as in retail and online?

Spatial presence was found to have a positive effect on purchase intentions in sunglasses AR apps [26] but not in the makeup app [22]. When consumers play an exergame, self-presence influences behavioral intentions, while spatial and social presence do not [3]. Such contradictory findings reveal the need for additional research thus, we ask:

FRQ3: Whether and to what extent would the impact of AR presence dimension on consumer outcomes be influenced by product types?

44.6 Implications for Theory and Practice

First, our multidimensional approach of presence confirms the importance of presence dimensions and their potential to benefit both consumers and firms [13]. We suggest that holistic view of presence dimensions enables to distinguish their effects on consumer outcomes. For instance, spatial and social presence increase attitude certainty for sunglasses AR-VTO, while self-presence does not [17]. Therefore, the authors suggest that a firm that aims to enhance decision-making should focus on spatial and social presence in the fashion accessories context. Enhancing spatial presence can include making the virtual product more realistic and improve the interaction with the product [13]. While social presence can be enhanced by implementing an AR recommendation system enabled by artificial intelligence technology (e.g., Ray-Ban matches glasses shape to user's face shape) or as an add-on outside of the app (e.g., Nordstrom proposes to book a virtual call with a stylist).

Second, our study proposes that boundary conditions (e.g., different touchpoints such as offline and online or differences in the type of products that AR displays) should be researched to provide guidelines to firms and marketers on the contextual elements that explains that each AR presence dimension does not always lead to increased marketing-relevant outcomes. For instance, high self-presence in public might not be beneficial. People dislike seeing personalized advertisement in public [12]. In addition, people prefer to explore styles by watching influencers they can identify with rather than with AR technology [7]. Therefore, identification is appreciated as long as people are not identifiable thus, physical similarity creates privacy issues in this context. This exemplifies that the highest presence dimension is not always beneficial and depends on the context.

Third, we find that presence dimensions (spatial, self, and social) are highly interrelated [20] thus, studying one presence dimension at the time would still capture other dimensions. For instance, a highly realistic embodied experience with a product enhances spatial presence in AR [13], however, a highly embodied experience is often part of the conceptualization of self-presence [1]. Studying three dimensions help attribute the outcomes of presence to its specific enabler and provide more consistent ground for recommendations to marketers. We also encourage authors to be more consistent in the conceptualization of presence dimensions.

44.7 Conclusion

This study is a short version of our review on presence dimensions and a call to research presence in AR in a holistic manner that considers the impacts of spatial presence, self-presence, and social presence. Moreover, we show that presence dimensions trigger different mechanisms that lead to positive consumer behavior. Therefore, understanding the effect of each presence dimension can inform marketers and app developers on the elements of the experience to implement in priority to reach the firm's strategic goals. Finally, contextual differences, such as whether the technology is used in store or online and what type of products is displayed in the AR-VTO, may explain differences in the outcomes of presence and should be investigated further.

This study has several limitations that are avenue for future research. First, presence is a psychological state thus, it depends more on users' perception rather than on specific technological features. For instance, immersion is a strong predictor of social presence, however, increasing immersion does not always lead to higher social presence [20]. Therefore, technological features do not linearly translate into presence and in turn, our recommendations are not based on specific technological features. Second, based on our definition of AR social presence, we study strictly the computer as social actor (CASA) definition in which consumers perceive a sense of human touch in the virtual experience [8]. We do not discuss social presence defined as the presence of another embodied or disembodied real (vs. imagined) social actor and co-presence as the sense of "being together" in the virtual environment [19]. However, we can foresee that development in AR and VR technology, as well as the multiverse will make this dimension of presence highly relevant and timely. Thus, future research should improve our multidimensional perspective with new dimensions.

References

1. Allen, J.J., Anderson, C.A.: Does avatar identification make unjustified video game violence more morally consequential? Media Psychol. **24**(2), 236–258 (2021)
2. Azuma, R.: A survey of augmented reality. Presence Teleoperators Virtual Environ. **6**(4), 355–385 (1997)
3. Behm-Morawitz, E.: Mirrored selves: the influence of self-presence in a virtual world on health, appearance, and well-being. Comput. Human Beh. Elsevier Ltd **29**(1), 119–128 (2013)
4. Chandukala, S.R., Reddy, S.K., Tan, Y.-C.: How augmented reality can—and can't—help your brand. Harvard Bus. Rev. (2022). Available at: https://hbr.org/2022/03/how-augmented-reality-can-and-cant-help-your-brand
5. Chylinski, M., Heller, J., Hilken, T., Keeling, D.I., Mahr, D., de Ruyter, K.: Augmented reality marketing: a technology-enabled approach to situated customer experience. Australas. Mark. J. **28**(4), 374–384 (2020)
6. DeNisco Rayome, A.: How Sephora is leveraging AR and AI to transform retail and help customers buy cosmetics. TechRepublic, available at: https://www.techrepublic.com/article/how-sephora-is-leveraging-ar-and-ai-to-transform-retail-and-help-customers-buy-cosmetics/
7. El-Shamandi Ahmed, K., Ambika, A., Belk, R.: Augmented reality magic mirror in the service sector: experiential consumption and the self. J. Service Manage. (2022). Available at: https://doi.org/10.1108/JOSM-12-2021-0484
8. Gefen, Straub: Managing user trust in B2C e-services. E-Serv. J. **2**(2), 7
9. Hassanein, K., Head, M.: The impact of infusing social presence in the web interface: an investigation across product types. Int. J. Electron. Commer. **10**(2), 31–55 (2005)
10. Heller, J., Chylinski, M., de Ruyter, K., Keeling, D.I., Hilken, T., Mahr, D.: Tangible service automation: decomposing the technology-enabled engagement process (TEEP) for augmented reality. J. Serv. Res. **24**(1), 84–103 (2021)
11. Heller, J., Chylinski, M., de Ruyter, K., Mahr, D., Keeling, D.I.: Let me imagine that for you: transforming the retail frontline through augmenting customer mental imagery ability. J. Retail. **95**(2), 94–114 (2019)
12. Hess, N.J., Kelley, C.M., Scott, M.L., Mende, M., Schumann, J.H.: Getting personal in public!? How consumers respond to public personalized advertising in retail stores. J. Retail. New York Univ. **96**(3), 344–361 (2020)
13. Hilken, T., de Ruyter, K., Chylinski, M., Mahr, D., Keeling, D.I.: Augmenting the eye of the beholder: exploring the strategic potential of augmented reality to enhance online service experiences. J. Acad. Mark. Sci. **45**(6), 884–905 (2017)
14. Hooi, R., Cho, H.: Avatar-driven self-disclosure: the virtual me is the actual me. Comput. Hum. Behav. **39**, 20–28 (2014)
15. Hooi, R., Cho, H.: Virtual world continuance intention. Telematics Inform. Elsevier Ltd. **34**(8), 1454–1464 (2017)
16. Javornik, A.: Augmented reality: research agenda for studying the impact of its media characteristics on consumer behaviour. J. Retail. Consumer Services Elsevier **30**, 252–261 (2016)
17. Lavoye, V., Tarkiainen, A.: Toward an improved understanding of AR-based presence dimensions and their impact on attitude certainty. Eur. Market. Acad. (EMAC) 50th, p. 94505 (2021)
18. Lee, K.M.: Presence, explicated. Commun. Theory **14**(1), 27–50 (2004)
19. Mennecke, B.E., Triplett, J.L., Hassall, L.M., Conde, Z.J., Heer, R.: An examination of a theory of embodied social presence in virtual worlds. Decis. Sci. **42**(2), 413–450 (2011)
20. Oh, C.S., Bailenson, J.N., Welch, G.F.: A systematic review of social presence: definition, antecedents, and implications. Frontiers Robot. AI **5**(OCT), 1–35 (2018)
21. Seo, Y., Kim, M., Jung, Y., Lee, D.: Avatar face recognition and self-presence. Comput. Human Beh. Elsevier Ltd. **69**, 120–127 (2017)

22. Smink, A.R., van Reijmersdal, E.A., van Noort, G., Neijens, P.C.: Shopping in augmented reality: the effects of spatial presence, personalization and intrusiveness on app and brand responses. J. Bus. Res. Elsevier **118**, 474–485 (2020)
23. Steuer, J.: Defining virtual reality: dimensions determining telepresence. J. Commun. **42**(4), 73–93 (1992)
24. Suh, K.-S., Kim, H., Suh, E.K.: What if your avatar looks like you? Dual-congruity perspectives for avatar use. MIS Q. **35**(3), 711–729 (2011)
25. Teng, C.I. (2021). How can avatar's item customizability impact gamer loyalty? Telematics and Inform. Elsevier Ltd. **62**(July 2020), 101626
26. Verhagen, T., Vonkeman, C., Feldberg, F., Verhagen, P.: Present it like it is here: creating local presence to improve online product experiences. Comput. Human Beh. Elsevier Ltd **39**, 270–280 (2014)
27. Xiao, Y., Watson, M.: Guidance on conducting a systematic literature review. J. Plan. Educ. Res. **39**(1), 93–112 (2019)

Chapter 45
Online Purchasing Behavior of Portuguese Consumers of Garment and Beauty Products During the COVID-19 Pandemic

Maria I. B. Ribeiro, Isabel M. Lopes, José A. M. Victor, and António J. G. Fernandes

Abstract This research aimed to understand the digital behavior of Portuguese consumers of garment and/or beauty products purchases during the COVID-19 pandemic; to verify if gender is a differentiator factor of the intention of repeating the purchase at the same store/brand and its frequency; and to correlate the intention of repeating the online purchase with age, education level, time of Internet use and frequency of online purchase. For this, a cross-sectional research based on a non-probabilistic snow ball type sample of 1521 consumers was carried out. Most consumers were female students with secondary education from the North of Portugal. They have been using the Internet for more than 5 years, and in the last 6 months, have shopped online garment and/or beauty products for 2–5 times. Among those who shopped online in the last 6 months, the most searched store was Zara. The main reason given by consumers for choosing the online purchase was comfort. The most used device to make the purchase was the mobile phone and the most payment methods used were the ATM and credit card. It was found that the intention of repeating the online purchase at the same store/brand was statically higher for women than for men. Also, women were the ones who most often made online purchases of this type of products. Furthermore, age was negatively correlated with the intention of repeating the online purchase at the same store/brand. On

M. I. B. Ribeiro · A. J. G. Fernandes
Centro de Investigação de Montanha (CIMO), Instituto Politécnico de Bragança, Campus Santa Apolónia, 5300-253 Bragança, Portugal

Laboratório Associado Para a Sustentabilidade e Tecnologia em Regiões de Montanha (SusTEC), Instituto Politécnico de Bragança, Campus de Santa Apolónia, 5300-253 Bragança, Portugal

I. M. Lopes (✉)
Unidade de Pesquisa Aplicada Em Gestão, Instituto Politécnico de Bragança, Bragança, Portugal
e-mail: isalopes@ipb.pt

Centro ALGORITMI da Universidade Do Minho, Bragança, Portugal

J. A. M. Victor
Instituto Politécnico da Maia, Maia, Portugal

Universidade da Maia, Maia, Portugal

J. L. Reis et al. (eds.), *Marketing and Smart Technologies*, Smart Innovation, Systems and Technologies 337, https://doi.org/10.1007/978-981-19-9099-1_45

the contrary, the education level and the number of years of Internet use were not correlated with the same variable.

45.1 Introduction

The persistent increase in the use of the Internet and information and communication technologies and their advancement in the last two decades has allowed people to have greater space–time flexibility, contributing to the growth of online shopping [1, 2], an increasingly common phenomenon these days [2]. The COVID-19 pandemic outbreak has altered consumer attitudes, behaviors and purchasing habits worldwide, causing online shopping to increase substantially [2–7]. On the one hand, this increase resulted from preventive health behavior adopted by consumers to limit the risk of infection by avoiding crowds in stores [8], and on the other hand, from measures to combat the pandemic enacted by governments around the world which resulted in numerous restrictions on citizens' daily routines, including social distancing [5]. Since March 2020, society around the world has been forced to change its routine, namely in terms of mobility, shopping and work [9]. In this context, during the pandemic, online purchases, in all product groups, increased, with products that previously had greater demand online being those that recorded the highest growth rates [5]. Although most online shoppers prefer to buy garment products, skin care products and cosmetics [10], during the pandemic, in addition to garment, food and hygiene, there was a great demand for equipment, electronic and technological devices, books and among others [2]. If before the pandemic, regardless of the advantages associated with online shopping, traditional shopping was, for many consumers, the preference [2] during the pandemic everything changed. Consumers have literally changed their shopping channel from offline to online [11]. Consumers who shopped offline and online before the pandemic began to shop completely online; consumers who had never shopped online before COVID-19 began doing so [12, 13]. Many people, who had never or rarely shopped online before COVID-19, shopped online for the first time and/or more frequently [6]. Therefore, the increase in online shopping was positively influenced by fear of illness, security concerns and the adoption of full-time teleworking [6, 14]. However, and as a result of the lay-off adopted by several companies and the increase in the unemployment rate, online shopping was negatively affected by the decrease or loss of household income [6]. The changes in the markets that occurred in 2020, as a consequence of the COVID-19 outbreak, were drastic and unprecedented, for companies, but especially for consumers [15]. A study developed in Hanoi, Vietnam, aimed to explore the factors associated with more frequent online shopping for five types of products, namely food, health products, garment, electronic devices and books [6]. The results showed that around 80% of consumers shopped online more often than before the COVID-19 outbreak. Compared with men, women were more likely to buy online more often certain product categories, and as a consequence, spend more time shopping online from home, which, in the authors' opinion, highlights not only the burden on women, in terms of providing care to family members during the period of social distancing, but also the important role of women in confronting and managing the consequences of the pandemic [6]. Interestingly, in Portugal, in 2021, the proportion of women in telework (22.7%) exceeded the proportion of men (17.5%) [16].

Among product types, the frequency of online purchases of food and garment was much more common when compared to online purchases of books, health products and electronic and technological devices [6]. Similar results were obtained by other research [5] in which women choose to purchase clothes, cosmetics, pharmaceuticals and household items in their online purchases, while the products most purchased by men were books, CDs and DVDs and technological devices. A trend that was already present before COVID-19 and continued during the pandemic outbreak, although online purchases of these products have registered substantial growth, fundamentally as a result of social distancing and the fear of infection. Other researchers conclude that gender decreased income and pleasure in shopping are statistically significant predictors of a higher frequency of online purchases of certain products, among which, garment and beauty products (cosmetics) [6].

A large number of studies prior to the COVID-19 pandemic consistently report that people with higher educational levels are more likely to engage in online shopping [6]. In addition, variables related to Internet use, including the number of years of Internet use, are predictors of online shopping behavior [6].

In Portugal, in 2021, 40.4% of people aged between16 and 74 years old placed orders online [16]. The proportion of Internet users who placed online orders were significantly higher in women, in the 25–34 age group and in the case of users who had higher education or were students and in families with higher income. In 2020, the proportion of users for products or services ordered shows a similar pattern, maintaining the predominance of garment, footwear and fashion accessories (69.0% in 2021 and 60.4% in 2020), take-away or home delivery meals (46.0% in 2021 and 38.2% in 2020) and movies, series or sports programs (34.9% in 2021 and 34.3% in 2020) [16]. In Portugal, the growth rate of online shopping, in the period during the COVID-19 pandemic compared to the period before the pandemic, was 1125%. The results reveal that the COVID-19 pandemic had a significant impact on the growth of online shopping in Portugal [17].

Globally, consumers appreciate the experiences they have with the store/brand and, above all, in a digital environment, they highly value the aspect of trust with the virtual seller, which can translate into satisfaction and continued purchase in the future [18]. In this sense, creating close ties between the consumer and the store/brand is extremely important for any business. Associated with this close relationship, there is the concept of the intention of repeating the purchase, which is based on the idea that the consumer will buy a product/service from the same store/brand again, taking into account their current situation and circumstances [19]. Some researchers define the intention of repeating the purchase as the decision, on the part of the consumer, to relate in the future with an activity performed by the service/product provider and the way in which this activity will elapse [20]. Although, there are many marketing activities aimed at attracting new customers, it is extremely important that companies focus on activities that lead to repeat the purchase by current customers [21].

In view of the above, the objectives of this research are: to describe the consumers' digital behavior in the context of COVID-19 pandemic outbreak; to verify whether gender is a differentiator factor of the intention of repeating the online purchase of garment and/or beauty products and the frequency of online purchases of this type of products; and to analyze whether the intention of repeating the online purchase is associated with age, education level, time of Internet use and frequency of online purchases of garment and/or beauty products.

This article is organized into four sections. In the Sect. 45.1, the literature review is organized, the objective is presented, and the structure of the article is presented. Section 45.2 describes the methodology used. Section 45.3 presents the results and discusses it. Finally, the Sect. 45.4 presents the main conclusions, suggests guidelines for future research and points out the limitations of the research.

45.2 Methods

To achieve the objectives of the research, a quantitative and cross-sectional study was developed based on a non-probabilistic sample of snow ball type constituted by 1521 consumers. In this context, the following hypotheses were formulated:

1. H_1: The intention of repeating the purchase is the same for female and male consumers.
   H_{a1}: The intention of repeating the purchase is higher in the female gender.
2. H_2: The frequency of online purchases of garment and beauty products is gender-independent.
   H_{a2}: Women buy garment and/or beauty products more often than men.
3. H_3: Age is independent of repeating the online purchase of garment and beauty products.
   Ha3: The manifestation of the intention of repeating the online purchase is greater in younger consumers.
4. H_4: The intention of repeating the online purchase and the education level is independent.
   H_{a4}: The intention of repeating the online purchase is stronger in consumers who have a higher education level.
5. H_5: The intention of repeating the online purchase and the time of Internet use (years) is independent.
   H_{a5}: The intention of repeating the online purchase is higher in consumers who have been using the Internet for a longer time (years).
6. H_6: The intention of repeating the online purchase does not depend on the frequency of online purchases.
   H_{a6}: The probability of repeating the online purchase is greater the more frequent the online purchases are.

In this study, consumers were asked about their digital behavior when shopping for garment and/or beauty products in the last 6 months, in the midst of the

COVID-19 pandemic outbreak. To this end, a questionnaire with two sections was adapted [22]. The first section included questions that allowed the sociodemographic characterization of the respondent, namely nationality, district of residence, age, gender, professional situation and education level. Nationality was a screening question, that is, if participants answered a nationality other than Portuguese, they could not continue to answer the questionnaire. On the other hand, only consumers with 18 years old or older were considered. The second part included questions about the respondent's digital behavior, namely, how long they have been using the Internet (years), whether they have made online purchases of garment and/or beauty products in the last 6 months, reasons that justify not having made online purchases during the COVID-19 pandemic, reasons that justify the online purchase, the frequency of online purchases of garment and/or beauty products in the last 6 months, the store/brand where they bought the last garment and/or beauty product online, the device used to make the last online purchase of garment and/or beauty products, the payment method used in the last online purchase of garment and/or beauty products and the intention of repeating the online purchase at the same store/brand. The items used to construct the variable "intention of repeating the online purchase at the same store/brand" were adapted [19]. In fact, instead a 7-point Likert scale [19], these items were measured using a 5-point Likert scale ranging from 1 (strongly disagree) to 5 (strongly agree).

The questionnaire was posted on social media and sent by email to avoid personal contacts during the COVID-19 pandemic. Data collection was carried out from November 2021 to January 2022.

During the preparation of this research, all the ethical requirements in force, set out in the General Data Protection Regulation in Law No. 58/2019, were complied with. That is, respondents, when approached, were informed about the objectives and scope of the work. They were also assured that their participation in this study would be voluntary and confidential and that the data would be treated in aggregate way, thus ensuring anonymity. In addition, respondents had at their disposal a consent form in which they declared that they understood the objectives and all associated and necessary procedures for the development of this study, giving their consent for the processing of their data, in the defined manner.

To edit and process the data, the IBM SPSS Statistics software version 28.0 was used. The results will be presented in tables or graphs since the study is a quantitative one. Regarding the statistical treatment of the data, absolute (n) and relative (%) frequencies were calculated for nominal and ordinal variables [23, 24]. For the ordinal and quantitative variables, measures of central tendency were calculated, namely mean, median, mode and also, measures of dispersion, such as standard deviation (SD), maximum and minimum [23, 24]. The t-Student test was used to compare two independent samples [23, 24]. This test allows testing the null hypothesis that the means are equal (H_0: $\mu_{male} = \mu_{female}$) against the alternative hypothesis that the means are different (H_1: $\mu_{male} \neq \mu_{female}$), where μ is the mean. The r-Pearson correlation test was used to study the correlation between two variables [24]. This test provides a correlation coefficient (r), corresponding to the level of intensity of the correlation between the variables that varies between -1 and 1. If $r = -1$, it is a perfect inverse or negative correlation. If $r = 1$, it is a perfect direct or positive correlation. The closer the r value is to 1 or -1, the stronger the correlation is. Finally, if the coefficient is closer to 0, the weaker is the correlation between the variables. The r-Pearson test allows testing the null hypothesis H_0*: The variables are not correlated* against the alternative hypothesis H_1*: The variables are correlated.* Finally, to analyze the internal consistency of the items that make up the variable "intention of repeating the online purchase of garment and/or beauty products at the same store/brand", the Cronbach Alpha (α) was used. The value must be positive, ranging from 0 to 1. Values greater than 0.9 mean that the consistency is very good; between 0.8 and 0.9 mean it is good; between 0.7 and 0.8 correspond to reasonable; between 0.6 and 0.7 to weak; and values below 0.6 are not admissible [25]. The significance level used was 5%. For a given significance level, the general decision rule is to reject the null hypothesis ($H_{0)}$), if probability of significance or p-value (p) $\leq$ significance level. The smallest value of the significance level from which H_0 is rejected is called probability of significance (p). So, if this probability is low because H_0 cannot be true, and therefore must be rejected [23].

45.3 Results and Discussion

The majority of respondents were female (63.3%), which have completed the secondary education or equivalent (60.9%), were students (58.1%) and live in the North of Portugal (55.0%), namely in the Bragança (36.7%) and Porto (18.3%) districts (Table 45.1).

Respondents were aged between 18 and 89 years old. The average age recorded was 25.9 years old (SD = 11.4). The median and mode were 20 and 18 years, respectively.

As given in Table 45.2, only 17.2% of consumers did not make any purchases online in the last 6 months. In fact, the majority used the Internet between 5 and 15 years and 52.0% purchased garment and/or beauty products, in the last 6 months, 2–5 times.

Of the 262 respondents who did not make any online purchase in the last 6 months, the main reason given was "there was no need to buy any garment and/or beauty products" (69.2%), followed by "I don't feel safe about buying online" and "I like to try the product to reduce the risk of uncertainty", both with values above 10.0% of the responses (Fig. 45.1).

Interestingly, although online shopping has increased during the pandemic, it seems that confidence in online shopping has not changed, as consumers still feel insecure as a result of their past experiences and during the pandemic [2]. In addition, many consumers felt and expressed the concern regarding the change in prices during the pandemic. In fact, in the opinion of consumers, many companies took advantage of it to raise prices in an attempt to minimize the impact of the crisis generated by COVID-19 pandemic outbreak. Among those consumers who shopped online in the last 6 months ($n = 1259$), most opted for stores/brands, such as Zara (18.0%),

Table 45.1 Sociodemographic characterization ($n = 1521$)

Variables	Groups	n	%
Gender	Female	963	63.3
	Male	556	36.6
	Not binary	2	0.1
Education level	1st cycle (4 years)	4	0.3
	2nd cycle (6 years)	16	1.1
	3th cycle (9 years)	61	4.0
	Secondary education or equivalent (12 years)	927	60.9
	Higher education	513	33.7
Professional situation	Employed	556	36.5
	Unemployed	55	3.6
	Retired	16	1.1
	Student	883	58.1
	Worker-student	10	0.7
	Domestic	1	0.1
Residence district	Aveiro	91	6.0
	Braga	110	7.2
	Bragança	558	36.7
	Castelo Branco	32	2.1
	Coimbra	63	4.1
	Faro	52	3.4
	Guarda	14	0.9
	Leiria	38	2.5
	Lisboa	126	8.3
	Porto	278	18.3
	Santarém	58	3.8
	Setúbal	26	1.7
	Vila Real	62	4.1
	Others	13	0.9

Pull and Bear (14.0%), Shein (8.6%), Bershka (7.8%) and Stradivarius (5.5%), as shown in Fig. 45.2. These stores are, especially, directed to the youth market. A study developed in Portugal also found that Zara was the most searched store/brand in an online purchase made by respondents in the last 6 months [22].

The most used device to make the last online purchase was the mobile phone/smartphone (55.5%), as shown in Fig. 45.3. Similar results were obtained by another research [17]. In fact, regarding the devices used to make online purchases, the results were similar before and during the COVID-19 pandemic outbreak. The smartphone, the laptop/desktop computer and the tablet were the preferred devices

Table 45.2 Internet use, purchases and frequency of online purchases

Variables	Groups	n	%
Number of years of Internet use ($n = 1521$)	<5 years	33	2.2
	5–10 years	534	35.1
	10–15 years	586	38.5
	15–20 years	289	19.0
	>20 years	79	5.2
Purchase of garment and/or beauty products in the last 6 months ($n = 1521$)	Yes	1259	82.8
	No	262	17.2
Frequency of purchase of garment and/or beauty products in the last 6 months ($n = 1259$)	1 time	333	26.4
	2–5 times	655	52.0
	6–9 times	4	0.3
	>9 times	267	21.2

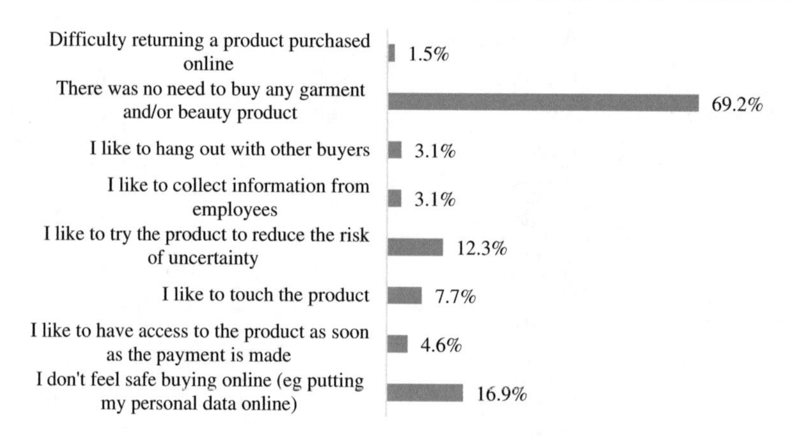

Fig. 45.1 Reasons for not buying garment and/or beauty products in the last 6 months ($n = 262$)

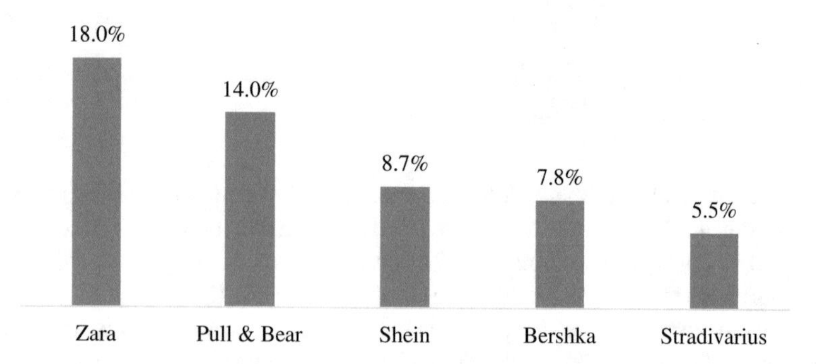

Fig. 45.2 Top 5 most searched stores/brands in online shopping in the last 6 months ($n = 1259$)

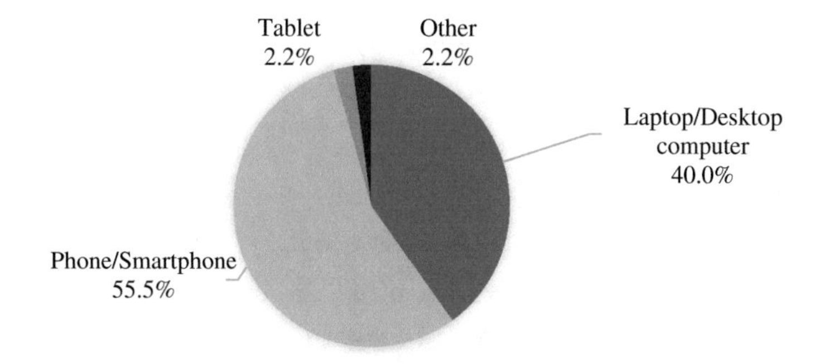

Fig. 45.3 Device used to make the last online purchase ($n = 1259$)

that were used by Portuguese consumers to make their purchases online. In the study developed by other research in Portugal, the device used to make the last online purchase was the laptop/desktop computer [22].

As shown in Fig. 45.4, and with values above 20.0%, the main reason for choosing the online purchase was comfort, that is, being able to make the purchase from home (27.3%), followed by convenience (26.9%), speed and simplicity of the purchase process (24.9%) and greater variety of products (24.5%). In other research developed in Portugal, the choice of respondents was comfort, convenience and low price [22]. The results obtained by another research [10] show that most consumers who shop online do so because "they don't need to leave the house to purchase the products they like in a quickly and convenient way". Furthermore, the variety and discounts offered online are very appealing.

Among the methods of payment used, the ATM (40.6%) and credit card (21.5%) stood out (Fig. 45.5). However, according to the results of other research [17], the COVID-19 pandemic changed the preferences of the Portuguese consumers on the method of payment for online purchases. Apparently, before the pandemic, Portuguese consumers used the ATM as their main payment method. However, during the pandemic, the results were completely different. In fact, MBWay has increased exponentially, surpassing the ATM reference that remained in second position as the most used payment method for online purchases during the COVID-19 pandemic outbreak [17].

The variable "intention of repeating the online purchase at the same store/brand" resulted from the aggregation of 3 items, namely (1) "I plan to continue using this website to buy products", (2) "This website is my first choice for online shopping in the future" and (3) "It is likely that I will continue to buy products from this website in the future". Table 45.3 gives the mean ($\overline{X}$), the standard deviation (SD) of the items, as well as the mean ($\overline{X}$), the standard deviation (SD) and the Cronbach Alpha (α) of the variable "Intention of repeating the online purchase at the same store/brand".

As given in Table 45.3, the Cronbach Alpha was 0.893, which means that the internal consistency is good. The item that most contributed to the intention of

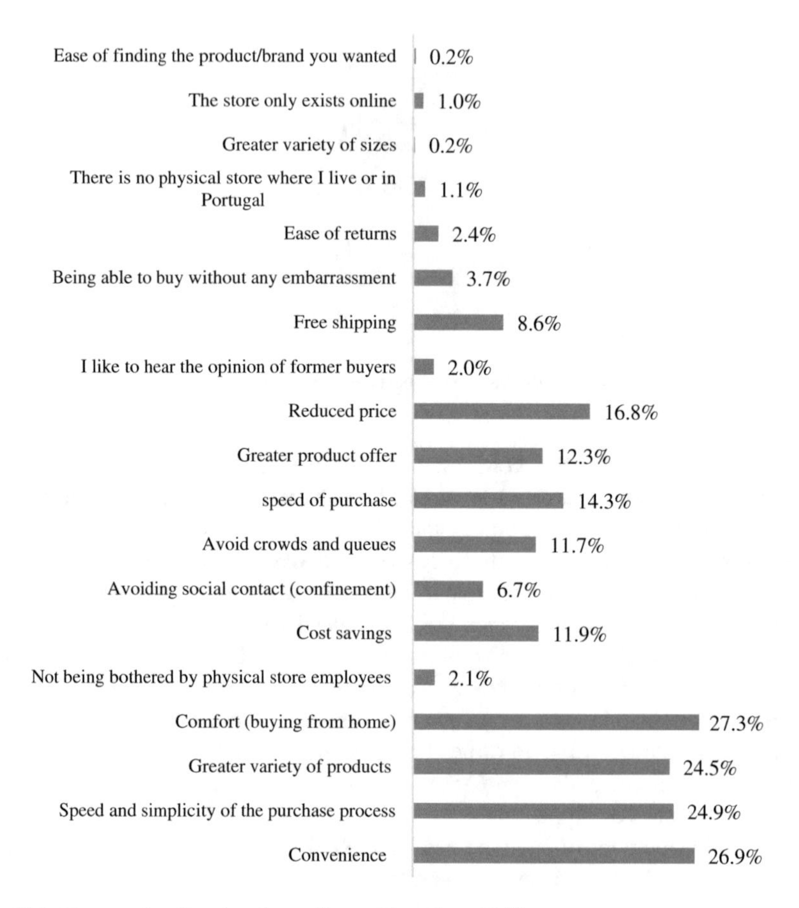

Fig. 45.4 Reasons for choosing the online purchase ($n = 1259$)

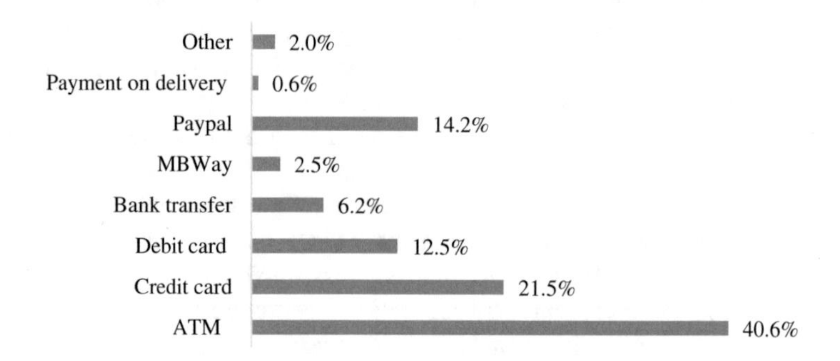

Fig. 45.5 Payment method used in the last online purchase ($n = 1259$)

Table 45.3 Mean, standard deviation and Cronbach Alpha of the intention of repeating online purchase at the same store/brand

		$\overline{X}$	SD	α
Items	1. I plan to continue to use this website to purchase products	3.45	1.1	
	2. This website is my first choice for online shopping in the future	3.17	1.9	
	3. It is likely that I will continue to purchase products from this website in the future	3.48	1.3	
Variable	Intention of repeating the online purchase at the same store/brand	3.36	1.1	0.893

repeating the online purchase was "It is likely that I will continue to purchase products from this website in the future" ($\overline{X} = 3.48$; SD = 1.3) and the lower score was "This website is my first choice for online shopping in the future" ($\overline{X} = 3.17$; SD = 1.9).

Testing the formulated hypothesis H_1, the results show that gender is a differentiator factor of the intention of repeating the online purchase at the same store/brand ($p = 0.000$), with this intention being manifested, above all, by women (Table 45.4). In addition, women are also the ones who most frequently shop online garment and/or beauty products ($p = 0.000$). In this sense, the null hypothesis H_1 and H_2 is rejected, concluding that there are statistically significant differences between male and female individuals with regard to the intention of repeating the online purchase at the same store/brand and the frequency of online purchases of garment and/or beauty products made in the last 6 months. These results are consistent with the findings of other researchers [26, 27]. According to these researchers, women more often bought clothes online before and during the COVID-19 pandemic outbreak than men.

The results of the tests of hypotheses H_3, H_4, H_5 and H_6 are given in Table 45.5. As can be seen, age is statistically correlated with the intention of repeating the online purchase at the same store/brand. This correlation, although weak, is negative ($r = -0.102$; $p = 0.000 < 0.05$). That is, the younger the consumers, the greater the intention of repeating the online purchase at the same store/brand. In the opinion of some researchers [28], the beauty and personal care sector has predominantly young and digitally connected customers who recognize the distinct advantages of online shopping, namely, competitive prices, convenient delivery, shopping 24 h a day, unlimited variety, and most importantly, access to information.

Table 45.4 t-Student test results to compare two independent samples

Variable	Groups	$\overline{X}$	SD	p
Intention of repeating the online purchase at the same store/brand	Female	3.5	1.05	0.000*
	Male	3.2	1.20	
Frequency of online purchases of garment and/or beauty products in the last 6 months	Female	2.3	1.04	0.000*
	Male	1.9	0.99	

*The means are statistically different at the significance level of 5%

Table 45.5 Results of the r-Pearson correlation test

Variables	Statistics	(1)	(2)	(3)	(4)	(5)
(1) Intention of repeating the online purchase at the same store/brand	r	1				
	p					
(2) Age	r	−0.102**	1			
	p	0.000				
(3) Education level	r	0.031	0.173**	1		
	p	0.277	0.000			
(4) Number of years of internet use	r	−0.052	0.443**	0.203**	1	
	p	0.066	0.000	0.000		
(5) Frequency of online shopping of garment and/or beauty products	r	0.156**	−0.002	0.068*	0.009	1
	p	0.000	0.946	0.016	0.760	

*The variables are statistically correlated at the significance level of 5%
**The variables are statistically correlated at the significance level of 1%

According to the literature, the purchasing behavior of young adults is strongly oriented toward online stores. The COVID-19 pandemic may have contributed significantly to this trend, although there are differences between consumers in developing and developed countries [9]. According to other researchers [5], since the beginning of the COVID-19 pandemic, the products with the highest demand in the pre-COVID-19 period were those that registered the highest relative growth during the pandemic. The main reasons for this are the fact that online shopping is a hybrid between the advantages of online commerce (simple and uncomplicated) and the measures related to COVID-19. On the other hand, the higher the frequency of online purchases, the greater the intention of repeating the online purchase at the same store/brand ($r = 0.156$; $p = 0.000 < 0.01$). Thus, the null hypotheses H_3 and H_6 are rejected, concluding that purchase intention is correlated with age and frequency of online purchases.

Furthermore, the results showed that the education level ($r = 0.031$; $p = 0.277 > 0.05$), as well as the number of years of Internet use ($r = -0.052$; $p = 0.066 > 0.05$) are not correlated with the intention of repeating the purchase at the same store/brand. Therefore, it can be concluded that there are not enough data to reject the null hypotheses H_4 and H_5. As this research was carried out only online, it is normal that the answers are mainly from people who already have a great affinity of using the Internet. As for the educational level, similar results were found in a study developed before the COVID-19 pandemic [29]. However, these results contradict the findings of other research [30].

According to other researchers, the number of years of Internet use is a predictor of online shopping behaviors [31]. However, in the present research, it was found that there is no correlation between the number of years of Internet use and the intention of repeating the online purchase at the same store/brand.

45.4 Conclusion

This research aimed to know the digital behavior of Portuguese consumers in online purchases of garment and/or beauty products, during the COVID-19 pandemic outbreak and to verify if gender was a differentiator factor of the intention of repeating the online purchase and of the frequency of online purchases. In addition, it was intended to analyze whether the intention of repeating online purchase at the same store/brand was associated with age, education level, time of Internet use and frequency of online purchases.

In this context, a quantitative and cross-sectional study was developed that involved the collection of primary data, in a non-random sample of 1521 individuals. Respondents were aged between 18 and 89 years old. Most were female, had secondary education or equivalent qualifications, were students and resided in the North of Portugal.

Most of the consumers have been using the Internet for more than 5 years and, in the last 6 months, they have shopped online garment and/or beauty products between 2 and 5 times. A small minority of consumers did not make any online purchase and the main reason given was that they did not need to purchase such products.

Among those who shopped online, in the last 6 months, the most searched store was Zara. The main reason given by consumers for choosing the online purchase was comfort (shopping from home). The most used instrument to make the last online purchase was the mobile phone and the payment methods used, by the majority, were the ATM and credit card.

The results also showed that gender is a differentiator factor of the intention of repeating the online purchase of garment and/or beauty products at the same store/brand. In fact, this intention was higher for women than for men. Women were also the ones who most often made online purchases of this type of products. Furthermore, it was found that age was negatively correlated with the intention of repeating online purchases at the same store/brand and that the education level as well as the number of years of Internet use were not correlated with the intention of repeating the online purchase at the same store/brand.

This study is relevant because it was developed during the COVID-19 pandemic outbreak, when consumer behavior changed and online shopping increased substantially. In this sense, it is crucial to know the behavior of consumers in order to act according to their needs. Therefore, this research can be used by companies that sell garment and beauty products online and are interested in increasing their customers' satisfaction in order to promote their proximity and guarantee the repetition of online purchase at the same store/brand even in crisis periods.

In this study, a non-probabilistic sampling was used. Despite being easier to operationalize, has as limitation the fact that the results cannot be generalized to the entire population with statistical precision. For future research, it is suggested to study a specific brand of garment, based on a probabilistic sample, to analyze whether, after the COVID-19 pandemic outbreak, the frequency of online purchases and the proximity of the consumer to the store/brand are maintained since there are studies [9, 14, 17] which show the intention of many consumers to return to pre-COVID-19 purchase behavior, that is, to go back to using the usual and traditional face-to-face purchase. An intention that is manifested to a greater extent by consumers in developed economies.

Acknowledgements The authors are grateful to the Foundation for Science and Technology (FCT, Portugal) for financial support through national funds FCT/MCTES (PIDDAC) to CIMO (UIDB/00690/2020 and UIDP/00690/2020) and SusTEC (LA/P/0007/2020).

UNIAG, R&D unit funded by the FCT, Portuguese Foundation for the Development of Science and Technology, Ministry of Science, Technology and Higher Education. Project no. UIDB/04752/2020.

References

1. Andreev, P., Salomon, I., Pliskin, N.: Review: state of teleactivities. Transp. Res. Part C Emerging Technol. **18**, 3–20 (2010)
2. Ismajli, A., Mustafa, A., Velijaj. F., Dobrunaj. L.: The impact of COVID-19 on consumer behaviour and online shopping: the case study in the developing country. Corporate Gov. Org. Beh. Rev. **6**(3), 34–43 (2022)
3. Sharma, A.J.: Changing consumer behaviours towards online shopping—an impact of COVID-19. Acad. Market. Stud. J. **24**(3), 1–10 (2020)
4. Zhang, H.-Z., Wang, Q.-Y.: E-commerce live broadcast influencing factors on impulsive clothing purchases. In: 14th Textile bioengineering and informatics symposium proceedings (TBIS 2021), pp. 275–280 (2021)
5. Ludin, D., Wellbrock, W., Gerstlberger, W., Müller, E., Nolle., S., Ruchti. S.: Changes in shopping behavior under the aspect of sustainability and the COVID-19 pandemic. Eur. J. Sustain. Devel. **11**(2), 142–154 (2022)
6. Nguyen, M.H., Armoogum, J., Thi, B.: Factors affecting the growth of E-shopping over the COVID-19 Era in Hanoi, Vietnam. Sustainability **13**(16), 9205 (2021)
7. Nagpal, S., Gupta, G.: Impact of pandemic communication on brand-specific outcomes: testing the moderating role of brand attitude and product category. J. Creative Commun. Article in Press (2022)
8. Dryhurst, S., Schneider, C.R., Kerr, J., Freeman, A.L.J., Recchia, G., van der Bles, A.M., Spiegelhalter, D., van der Linden, S.: Risk perceptions of COVID-19 around the world. J. Risk Res. **23**(7–8), 994–1006 (2020)
9. Rossolov, A., Aloshynskyi, Y., Lobashov, O.: How COVID-19 has influenced the purchase patterns of young adults in developed and developing economies: factor analysis of shopping behavior roots. Sustainability **14**(2), 1–24 (2022)
10. Yan, X.: Research on consumers' attitudes towards online and offline shopping. In: Oladokun, S.O., Lu, S. (eds.), E3S Web of Conferences, vol. 218, id 01018. ISEES, Chongqing, China (2020)
11. Sheth, J.: Impact of COVID-19 on consumer behavior: will the old habits return or die? J. Bus. Res. **117**, 280–283 (2020)
12. Watanabe, T., Omori, Y.: Online consumption during the COVID-19 crisis: evidence from Japan. Covid Econ. **32**, 218–252 (2020)
13. Youn, S-Y., Lee, J.E., Ha-Brookshire, J.: Fashion consumers' channel switching behavior during the COVID-19: protection motivation theory in the extended planned behavior framework. Clothing Textiles Res. J. **39**(2), 139–156 (2021)

14. Wang, X., Kim, W., Holguín-Veras, J., Schmid, J.: Adoption of delivery services in light of the COVID pandemic: who and how long? Transp. Res. Part A Policy Practice **154**, 270–286 (2021)
15. Veselovská, L., Závadský, J., Bartková, L.: Consumer behaviour changes during times of the COVID-19 pandemic: an empirical study on Slovak consumers. Ekonomie a Manage. **24**(2), 136–152 (2021)
16. INE: Inquérito à utilização de tecnologias da informação e da comunicação pelas famílias—2021. Serviço de Comunicação e Imagem, Lisboa (2021)
17. Pires, S.: E-commerce e pandemia—Comportamento de compra online antes e durante a pandemia COVID-19. Dissertação de mestrado em Data-Driven Marketing com especialização em Marketing Intelligence. NOVA Information, Management School, Instituto Superior de Estatística e Gestão de Informação, Universidade Nova de Lisboa, Lisboa (2021)
18. García-Salirrosas, E.E., Acevedo-Duque, Á., Marin Chaves, V., Mejía Henao, P.A., Olaya Molano, J.C.: Purchase intention and satisfaction of online shop users in developing countries during the COVID-19 pandemic. Sustainability **14**(10), 1–14 (2022)
19. Hellier, P.K., Geursen, G.M., Carr, R.A., Rickard, J.A.: Customer repurchase intention: a general structural equation model. Eur. J. Mark. **37**(11–12), 1762–1800 (2003)
20. Hume, M., Mort, G., Winzar, H.: Exploring repurchase intention in a performing arts context: who comes? and why do they come back? Int. J. Nonprofit Voluntary Sector Market. **12**(2), 135–148 (2007)
21. Spreng, R., Harrell, G., Mackoy, R.: Service recovery: impact on satisfaction and intentions. J. Serv. Mark. **9**(1), 15–23 (1995)
22. Monteiro, N.: As motivações para a compra online: Comportamento de compra do consumidor digital. Dissertação de Mestrado em Marketing Digital. Instituto Superior de Contabilidade e Administração, Porto (2018)
23. Marôco, J.: Análise Estatística com o SPSS Statistics. Report Number, Pero Pinheiro (2021)
24. Pestana, M., Gageiro, J.: Análise de Dados para Ciências Sociais: A complementaridade do SPSS. Edições Sílabo, Lisboa (2014)
25. Nunnally, J., Bernstein, I.: Psychometric Theory. McGraw-Hill, New York (1994)
26. Zhen, F., Cao, X., Mokhtarian, P.L., Xi, G.: Associations between online purchasing and store purchasing for four types of products in Nanjing, China. Transp. Res. Record **2566**, 93–101 (2016)
27. Saphores, J.-D., Xu, L.: E-shopping changes and the state of E-grocery shopping in the US evidence from national travel and time use surveys. Res. Transp. Econ. **87**(2), 100864 (2020)
28. Prasad, A., Krithika. R., Susshruthi, G.: A study of digital shopping behaviour of women with respect to beauty and personal care products. In: Conference Proceedings of 10th International Conference on Digital Strategies for Organizational Success. Prestige Institute of Management, Gwalior, India (2019)
29. Loo, B.P.Y., Wang, B.: Factors associated with home-based e-working and e-shopping in Nanjing, China. Transportation **45**, 365–384 (2018)
30. Shi, K., De Vos, J., Yang, Y., Witlox, F.: Does E-shopping replace shopping trips? Empirical evidence from Chengdu, China. Transp. Res. Part A Policy Practice **122**(C), 21–33 (2019)
31. Farag, S., Schwanen, T., Dijst, M., Faber, J.: Shopping online and/or in-store? A structural equation model of the relationships between e-shopping and in-store shopping. Transp. Res. Part A Policy Practice **41**(2), 125–141 (2007)

Chapter 46
Systematic Literature Review—Factors of Loyalty and Acceptance in Voice Commerce

Matilde Vieira, Victor Santos, and Lara Mendes Bacalhau

Abstract Voice commerce is a new trend in e-commerce and it is boosted by the growing number of equipment connected to the Internet with voice recognition capability. There is an increasing number of users who use their smartphones to perform voice searches when shopping in online stores, especially inside their homes. If voice commerce has gained over the last years importance in marketing and in the e-commerce world, it is necessary to understand what factors are affecting this loyalty and acceptance over voice commerce. The purpose of this scoping review was therefore to provide a broad synopsis of pertinent studies in a structured and comprehensive way over two years (2021 and 2022). In doing so, the findings from 19 studies were analyzed and suggestions were made for future studies. The results not only synthesize existing empirical evidence of the factors of acceptance and loyalty, but also identify some knowledge gaps in the literature to guide future studies.

46.1 Introduction

We live in a new era where there is a rapid growth of technological adaptation. And with this transformation, e-commerce uses different digital marketing channels to give visibility and target to reach products and services. Marketing as we know it has changed and evolved significantly to increasingly digital media and, as proof of this, we have the new trend of voice commerce. A recent study [1] found that consumers prefer products recommended by virtual assistants when shopping online. That is the main reason for this research. Brands must understand the customers' needs and desires, principal the factors of loyalty and acceptance, about voice commerce and the help of the different types of virtual assistants. By 2024, it is estimated that consumers will use voice assistants on more than 8 billion devices [2]. From

M. Vieira · V. Santos (✉) · L. M. Bacalhau
Polytechnic of Coimbra, Coimbra Business School | ISCAC, Coimbra, Portugal
e-mail: vsantos@iscac.pt

L. M. Bacalhau
e-mail: lmendes@iscac.pt

J. L. Reis et al. (eds.), *Marketing and Smart Technologies*, Smart Innovation, Systems and Technologies 337, https://doi.org/10.1007/978-981-19-9099-1_46

a marketer's perspective, voice technology opens a new world of marketing channels that enable brands to reach customers through a range of voice-enabled digital devices, such as smartphones, smart speakers, smart televisions, and cars [3].

Brands that have found their voice and use it not just to improve customer experience, but to build brand awareness too, will have a distinct competitive advantage [4]. Businesses have more opportunities to generate meaningful data and information from these interactions by working to build predictive models that allow their business divisions to develop the insights they need around voice behaviors.

The more significant inflows of data and information provided by conversational technologies create lower-cost automated solutions, like search engine optimization (SEO) built on the insights culled from exploring the predictive data generated by voice-powered communications platforms.

Voice search is driven by asking questions, which means benefits for the brands that can harness this technology to anticipate and answer customer queries. Implementing voice interfaces allows brands to deliver interactive and personalized experiences while still allowing customers to maintain social distance and choose how much they want to connect.

The global digital transformation of the last couple of years has reinforced customers' connection to smart speakers and assistive voice-powered technologies such as Alexa and Siri [5].

46.1.1 Voice Commerce

Voice commerce is a form of digital commerce where purchases are made through voice [6]. It exists because of natural language processing (NLP). This is a form of artificial intelligence (AI) that focuses on analyzing human language to draw insights, create advertisements, help the consumer text (suggesting words), and more. It takes many forms, but at its core, the technology helps a machine understand, and even communicate with, human speech. The first step in NLP depends on the application of the system. Voice-based systems like Alexa or Google Assistant need to translate words into text. That is done (usually) using systems that use mathematical models to determine what the customer said and translate that into text usable by the NLP system that tries to break each word down into its part of speech [7]. To conduct commerce via voice, simply initiate an interaction with the voice assistant, whether via smartphone, smart electronics, or an appropriate mobile device such as Amazon's Alexa [6].

46.1.2 Virtual Assistants

The beginning of voice commerce started with browsing using voice instead of tipping. The first interaction between a machine and humankind started with the voice. With the recent advantages in technology interaction through AI, consumers have gained new interaction habits with virtual assistants [8]. Virtual assistants are advancement of recent technology that is driven by data embedded in Internet of Things (IoT) devices or applications, such as browser, voicebot, or Alexa, which can communicate with consumers and respond to their queries through voice commands [9].

Digital voice assistants conduct the implementation of voice commerce, facilitating the purchase journey for consumers. They offer convenience value unmatched by most technology systems, which allow consumers to perform tasks with minimal effort (reading, typing, and holding devices), twenty-four hours, and seven days a week. Given the benefits of digital voice assistants in eliminating the need to interact with a physical user interface and enabling multitasking during interactions, consumers are expected to experience a hassle-free smart shopping experience [10].

46.1.3 Loyalty in e-Commerce

Loyalty to e-commerce through voice improves the consumer experience with a much easier and more convenient purchase journey, creates more personalized experiences according to consumer preferences and discovers better opportunities for products and marks.

Talking is the most natural form of communication, and it is the fastest form of communication than any other type of communication. Voice communication with intelligent assistants means that nothing is paused in the consumer's life, being able to carry out several activities at the same time [11].

46.2 Methodology

This study contains a review to map the existing literature in the context of voice commerce loyalty, and what are the reasons for this loyalty, in a structured and comprehensive way. The literature review used in this study is based on the principles and guidelines of the PRISMA-ScR method. Students and schools widely use this method to identify any gaps in a topic that has not been sufficiently explored. The literary review follows the five phases of Arksey and O'Malley (2005).

46.2.1 Phase 1: Formulating the Research Question

The central question that guided this review was: What is the factors that lead to loyalty and acceptance in the use of voice commerce and the help of the virtual assistants.

This question led to several sub-questions:

Sub-question 1: Is voice commerce relevant in online shopping and in new digital media?

Sub-question 2: Consumer loyalty is enhanced by virtual assistants/how important are virtual assistants in the consumer journey?

Sub-question 3: Is voice commerce important to consumers in digital environments?

Sub-question 4: Do loyalty factors increase the importance of voice commerce?

Sub-question 5: Is voice commerce the key to brand loyalty?

46.2.2 Phase 2: Identifying the Relevant Studies

The key concepts of loyalty and acceptance in voice commerce guided the search strategy. The following databases were used to find relevant studies: Web of Science, Scopus, and ScienceDirect. These databases were selected because they cover management and marketing topics and represent academic journals that would consider publishing studies on voice commerce.

An advanced search mode was firstly used to query these databases with the following Boolean search commands:

Loyalty AND Voice AND Commerce

Artificial Intelligence AND Voice Commerce

Customer OR Experience AND "Voice Commerce".

For a better result, there were some differences in each database. In Web of Science, the queries used were Loyalty AND voice AND commerce; Costumer OR experience AND voice commerce and "AI" AND "voice commerce". In ScienceDirect, the queries used were Loyalty AND voice AND commerce; Artificial intelligence AND voice commerce and "Consumer Experience" AND "Voice Commerce". For the database Scopus, the queries were Loyalty AND voice AND commerce; AI AND "voice commerce" and Costumer OR experience AND "voice commerce".

46.2.3 Phase 3: Inclusion and Exclusion Criteria

A methodological protocol guided the study's inclusion and exclusion criteria. The publications had to adhere to the following inclusion criteria to be selected:

- Since Voice Commerce is a new trend were only accepted papers from 2021 and 2022;
- Research or conceptual study;

- Published in English;
- Addressed the loyalty in voice commerce;
- Subject areas business, management, and accounting;
- Referred to loyalty or acceptance in voice commerce in the title and/or abstract and/or body of the paper.

In addition, it was also important that the studies had a clear and focused research question/hypothesis, appropriate use of methods, and a clear description of results to ensure the credibility of the empirical evidence reported.

Gray literature, such as working papers, theses/dissertations, reports, and white papers, was excluded from the study since the focus was on traditional academic peer-reviewed research on the topic and more conclusive evidence on the topic was required.

As depicted in Fig. 46.1, out of 465 registers, only 403 were screened because of duplicated and ineligible. Only 36 reports were sought for retrieval, because the title 367 was excluded. 26 reports were assessed for eligibility with title and abstract, only 10 were excluded because the abstract did not answer the main question of this research. Each study was independently evaluated by the author, using a dichotomous scale to determine the presence or absence of each criterion. Full papers were obtained at this stage to make a more detailed assessment and were automatically retrieved with EndNote reference management software. The 7 excluded studies failed on one or more of the inclusion criteria. The final sample comprised of 19 journal articles.

46.2.4 Phase 4: Extracting and Analyzing the Relevant Data

The content of the documents in the final sample was extracted systematically in accordance with a summary table that covered the authors, year of publication, journal, study objectives, methodology, and research approach. The data was then grouped according to the sub-questions of this study. It was also necessary to hand search the literature to ensure the credibility of the process.

46.2.5 Phase 5: Compiling, Summarizing, and Reporting

First comprised of a quantitative process to quantify and categorize the raw data. Second, the data was qualitatively interpreted, using a deductive approach to provide meaning following the sub-questions of this study. In this regard, it was important to map what the author currently knows about the voice commerce and the loyalty and acceptance of this new trend and then to report on the insights gained for future studies.

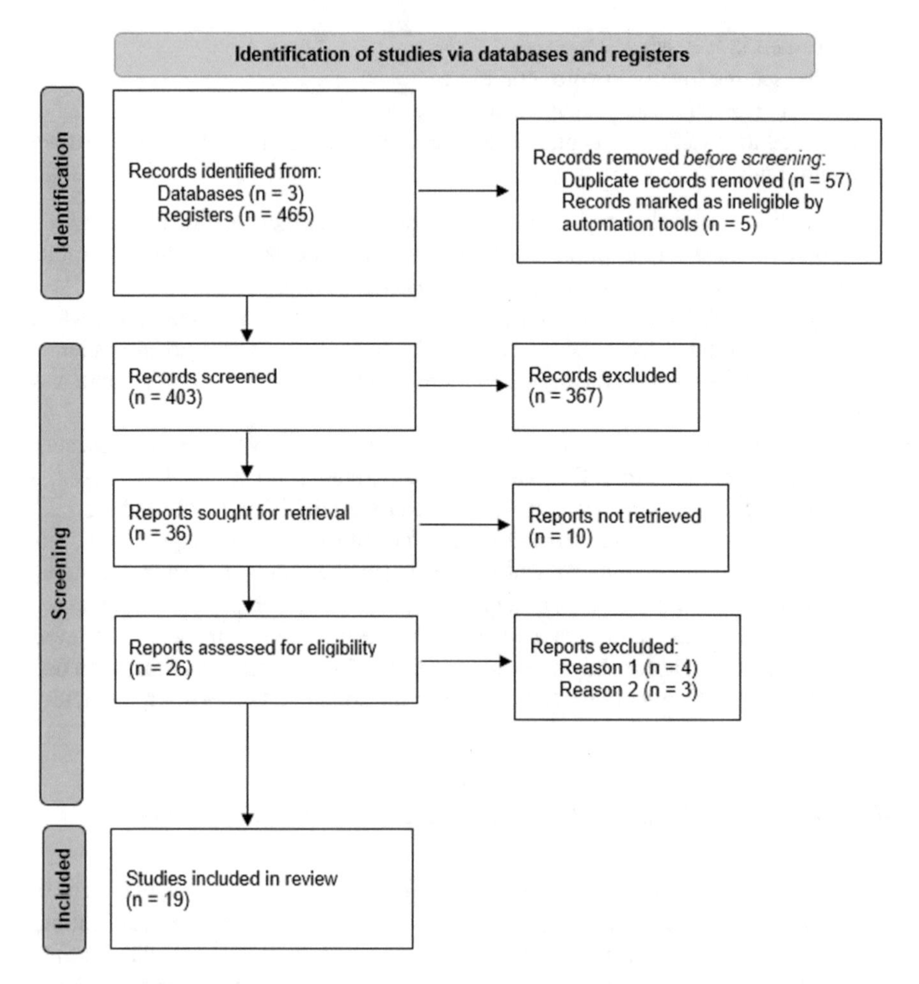

Fig. 46.1 PRISMA flow diagram for the scoping review process to identify relevant studies

46.3 Results

In this scoping review, 19 primary studies ($n = 19$) dealt with voice commerce and what makes a customer accept and loyal to this new trend. After an initial scholarly focus on what voice commerce is and entails, the results show that there has recently been a gradual shift to studies that also investigate what makes a customer loyal to this new e-commerce.

For analysis, the studies were grouped under the sub-questions of this study to answer the central question of what is known about voice commerce. The results are now explained in terms of the key characteristics of the selected studies which evidence the causal connection between voice commerce and the factors that make a customer loyal. The results can be further explained as follows.

46.3.1 Key Characteristics of Studies on the Topic

Table 46.1 depict the characteristics of the range of methodologies and research approaches adopted by the studies in the sample. The results show that 19 studies adopted a quantitative, qualitative, and conceptual model research approach to determine the characteristics or factors that make a customer loyal to voice commerce and with virtual assistants. Advanced statistical methods were used to evaluate proposed models; experiments were conducted with consumers of voice commerce and virtual assistants. The quantitative studies also show evidence of rigorous statistical analysis, which can be attributed to the complexity and the little information that exists about the topic.

There are six quantitative studies (2, 5, 7, 15, and 19) that used interviews, questionnaires, case studies, and focus groups. One of these approaches (17) involved two studies where the first was interviews and the second study examines a questionnaire. There are three studies (1, 9, and 10) that involve both qualitative and quantitative research.

Five of these studies (3, 4, 8, 13, and 14), involve a conceptual model, which is the model of an application that the designers want users to understand. Although there is another set of five studies that use conceptual model and quantitative approaches (6); proposed model and quantitative approaches (11); systematic review (16); a novel with a structural equation modeling (18); and a contextualized theory with a quantitative review (12).

Table 46.1 also depicts the journals where the studies were published. The journals that have more studies are the International Journal of Innovation and Technology Management (with two papers), Journal of Business Research (with two papers), Computers in Human Behavior (with 2 papers), Journal of Retailing and Consumer Services (with two papers), and Technological Forecasting and Social Change (with three papers). There are more eight journals that each has only one study with the characteristics requested for this research.

The studies in this sample confirm the factors that affect the acceptance and loyalty to the usage of voice commerce and the help of its virtual assistants.

46.4 Discussion

This scoping review synthesizes the factors that transform a customer's acceptance and loyal to what is known in voice commerce over the past two years. In doing so, the study provides more insights into the extent of the current body of knowledge on this topic, as depicted in Table 46.1. Although the causal connection between voice commerce and several virtual assistants has been acknowledged in earlier literature, the studies included in the final sample verified this connection with empirical data, using a wide range of methodologies. By strategically focusing on relevant factors for customers to become loyal and accept more easily this new trend—voice commerce,

Table 46.1 General description of the key characteristics and findings of the studies included in the review

Study number	Authors/date	Journal name	Study objective	Methodology and research approach
1	Rabassa et al. (2022) [7]	Technological forecasting and social change	Consumers' perception of conversational commerce and product choice offers delivered by voice assistants	Qualitative and quantitative approaches
2	Aw et al. (2022) [8]	Technological forecasting and social change	Validating human-like attributes and contextual factors as the antecedents to continuance usage of digital voice assistants to shop	Quantitative approaches
3	Zaharia and Würfel (2021) [10]	Advances in intelligent systems and computing book series	Customers' willingness to use smart speakers for online shopping (voice commerce)	Conceptual model
4	Balakrishnan and Dwivedi (2021) [11]	Annals of operations research	Investigate the role of technology attitude and AI attributes in enhancing purchase intention through digital assistants	Conceptual model
5	Maroufkhani et al. (2022) [12]	Technological forecasting and social change	Identify the mechanism through which the users of voice assistants might develop reuse intention and loyalty toward a specific service provider brand	Quantitative approaches
6	Lee et al. (2022) [13]	International journal of human–computer interaction	Key factors that influence intention to use voice assistants	Conceptual model and quantitative approaches

(continued)

Table 46.1 (continued)

Study number	Authors/date	Journal name	Study objective	Methodology and research approach
7	Hu et al. (2022) [14]	Computers in human behavior	Examine how the power experience in human-AI interaction impacts voice shopping	Quantitative approaches
8	Klaus and Zaichkowsky (2022) [15]	Journal of retailing and consumer services	View of consumer choice based on AI and inherent convenience addiction to smart speakers	Conceptual model
9	Chung et al. (2022) [16]	International journal of innovation and technology management	Describe factors influencing user acceptance to voice commerce	Qualitative and quantitative approaches
10	Ramadan (2021) [17]	Journal of retailing and consumer services	Amazon's captive relationship strategy on shoppers	Qualitative and quantitative approaches
11	Ashrafi and Easmin (2022) [18]	International journal of innovation and technology management	Attitude and trust on the behavioral intention of adopting artificial intelligence	Proposed model and quantitative approaches
12	Bawack et al. (2021) [19]	International journal of information management	Explores how personality, trust, privacy concerns, and prior experiences affect customer experience performance perceptions and the combinations of these factors that lead to high customer experience performance	Contextualized theory and quantitative approaches

(continued)

Table 46.1 (continued)

Study number	Authors/date	Journal name	Study objective	Methodology and research approach
13	Canziani and MacSween (2021) [20]	Computers in human behavior	Explores how voice-activated smart home devices (SHDs) like Amazon Alexa and Google Home influence consumers' retail information seeking and ordering behaviors	Conceptual model
14	Hasan et al. (2021) [21]	Journal of business research	Determine the influence of consumer trust, interaction, perceived risk, and novelty value on brand loyalty for AI supported devices	Conceptual model
15	Hsu and Lin (2021) [22]	Service industries journal	Factors that increase smart speakers usage stickiness	Quantitative approaches
16	Lim et al. (2022) [23]	Psychology and marketing	Review the performance and intellectual structure of conversational commerce	Systematic review
17	McLean et al. (2021) [24]	Journal of business research	Uncovers the key drivers of consumer brand engagement through voice assistants	Study 1, through a set of in-depth exploratory interviews and Study 2 examines a questionnaire
18	Pal et al. (2022) [25]	Heliyon	The effect of trust on the behavioral intention of users toward voice assistants	Novel research framework and structural equation modeling
19	Shao and Kwon (2021) [26]	Hum behav and emerg Tech	Understand why users interact with a voice assistant system	Quantitative approaches

six factors are making the difference in the last two years (2021 and 2022), they are: Expected performance; Hedonic Motivation; Perceived Value; Previous experience with Voice Commerce; Power Experience; Perceived Risk; Social Presence; Convenience and Animation.

46.4.1 Expected Performance

The performance expectation is related to the fact that virtual assistants are easy to use. Many consumers consider the idea of using voice commands in online shopping without a completely new visual display. This can be a big initial barrier to use. Especially more complex online shopping tasks, such as product comparisons, can be a cognitive burden for consumers. When technology is considered easy to use, it reduces the effort required to use it, thus having a positive effect on performance expectations. Effort expectancy also influences hedonic motivation, because if technology is seen as complex and challenging, it is likely to generate frustration for the user and counteract the enjoyment of its use [12].

Intelligence has become an important identity for any AI-powered system [13].

46.4.2 Hedonic Motivation

In a voice commerce context, hedonic motivation is defined as the extent to which a consumer perceives their journey as fun, exciting, and enjoyable [12]. Hedonic value is one of the most crucial factors in customer perceived value. Hedonic shoppers consider sites that not only offer secure transactions, confidentiality, privacy, cooperative interactions, and quick access to vast amounts of information, but also intrinsic emotional and aesthetic value that enrich the desire to shop online.

In addition to the hedonic value, there is anthropomorphization in robotics, which is the tendency of humans to perceive robots as human when presented with visual, auditory, or tactile stimuli. Anthropology satisfies two essential human needs: the desire for social bonding and the desire to control and understand the environment. Studies have supported that anthropomorphic traits can induce high-level trust through an emotional connection with the object, which sustains a stronger relationship with nonhuman objects [14].

46.4.3 Perceived Value

Literature suggests that users typically believe that machines' performance is more precise and accurate than humans' and these beliefs include the common stereotypes of machines such that they are more accurate, objective, and precise than human beings in performing tasks [15]. Perceived value is linked to the performance expectation and costs required to acquire a virtual assistant. The more the perceived value is positive, the easier it is to accept the technology and the willingness to make online purchases. The experience in the use of devices and virtual assistants can demonstrate stronger intentions in the use of voice commerce to buy a product [12].

46.4.4 Previous Experience with Voice Commerce

The more users interact with virtual assistants, the more their perceived intelligence can increase as they interact with digital assistants. The intelligence theory of the naive suggests that people's experience in processing information can increase their perceived understanding. From this rational view, we propose that users with more experience with digital assistants can improve their ability to perceive the intelligence of digital assistants [13].

46.4.5 Power Experience

Power is a fundamental concept in social interactions, within the context of relationships with other people. When people feel they can control or influence the actions of others, the experience of power occurs. In back-and-forth interactions between people and computers, some experience may occur that originally pertains to interpersonal interaction, such as the experience of power. The experience of power is not only pervasive during social interactions between humans but, can also be perceived in voice interactions between users and AI assistants. When consumers experience power over AI assistants, they may feel that their AI assistants are under their control, which is unlikely. For example, users cannot control which products or brands AI assistants recommend [16].

46.4.6 Perceived Risk

Perceived privacy risk can be defined as the fear of inappropriate use of consumer data and violation of their privacy by online companies. Perceived risk is a major obstacle for voice shopping because, in this context, the tasks of searching for information and evaluating product alternatives are delegated to AI assistants [17]. In this sense, AI is likely to recommend a product or service that benefits companies but harms users' interests. In addition, the perceived risk may also involve the security of voice printing payment and the leakage of consumer privacy in voice purchases [16].

Consumers are more concerned about the privacy issues of virtual assistants when they interact with lesser known brands and experience less value delivered to them due to the constant threat of loss of information and privacy [14].

46.4.7 Social Presence

Social presence can be described as the ability of virtual assistants to expressively portray themselves, socially and emotionally, as human beings in facilitated communications. This social presence is vital in defining the concept of messaging and will, therefore, impact social interactions between consumers. Social responses are enriched in the interaction involving digital devices that embrace human aspects. Furthermore, an analysis of Amazon user reviews argued that a voice assistant is much more enjoyable when the social aspects of voice assistants appeal to them. Therefore, it is correct to assume that the greater the degree of social presence felt by the consumer over the voice assistant, especially the voicebot, the richer the experience he obtains when using these devices [14].

There is an advanced level of social presence compared to traditional online shopping websites, making it reasonable to conclude that they will affect the behavioral intention and resistance to using voice commerce [18].

46.4.8 Convenience

The consumer using the voice assistant has significantly less information than in the store, but it saves not only physical time, but also a cognitive time in the process [12].

Digital voice assistants carry out the implementation of voice commerce, facilitating the purchase journey for consumers. They offer convenience value unmatched by most technology systems, which allow consumers to perform tasks with minimal effort and within less time [10].

The usefulness is indisputable because: (1) consumers may not be able to identify all options; (2) even if all choices are accessible, consumers do not have the time or ability to consider all choices, evaluating each one in terms of their needs and inclinations; and (3) not all decisions are high-involvement for consumers, who make many low-involvement decisions without realizing the risk of a bad choice [12].

Given the benefits of digital voice assistants in eliminating the need to interact with a physical user interface and enabling multitasking during interactions, consumers are expected to experience a hassle-free smart shopping experience [10].

46.4.9 *Animation*

While the literature on animacy has been derived from a psychology background, its application in recent years is predominantly used in information systems research. The growth of AI-oriented applications is increasing day by day, and so does the involvement of humanistic characteristics and animatic involvement. Digital assistants are one such example, in which the developers use multi-functional AI attributes with human embedded features to create a perceived animatic environment. One good example is Amazon Alexa, a well-known digital assistant, which creates human-like voices and intelligent algorithms to recreate a human–human experience rather than a human–machine experience [13].

46.5 Example of a Success Story—Alexa (Amazon)

After reading the articles selected for this study, the loyalty factors on the part of consumers are aggregated in a specific example that can be found in the market. Amazon's virtual assistant, Alexa, manages to gather several loyalty factors, being an excellent success story. Alexa strives to differentiate itself through a set of functional, relational, emotional, and experiential benefits that it promises to deliver to the consumer. These self-service technologies allow customers to use services independently, without the assistance of employees. While the proliferation of virtual assistants has mostly focused on home use, Amazon recently launched a set of Alexa-enabled mobile devices that should be instantly accessible and used anywhere. Amazon offers a seamless omnichannel experience (offline and online usage).

Customer experience is becoming personalized at the persona level using service preferences, location data, content consumed, purchase history, and communication preferences. The relationship between Alexa and its users is largely based on emotional interactions that are characterized by a need for connection. This enhanced social experience is triggered by the personification of the virtual assistant and the unique perceived personality, which amplifies the overall impact of the Amazon experience on the consumer journey.

Amazon is now able to track the entire consumer journey through the ecosystem of virtual assistants they have built since 2014 [19].

46.6 Conclusion

The purpose of this scoping review was to summarize empirical evidence of the factors that consumers take into count to accept and became loyal to voice commerce and virtual assistants in all its forms, and to identify some knowledge gaps for future studies on the topic. The study thus provides more insights into the extent of verified knowledge and which areas to still pay attention to.

This study also has some practical implications. Greater awareness of more conclusive evidence of the main factors affecting this new e-commerce trend. In this regard, the results provide more insights into what voice commerce consumers expect from brands with which they interact on digital platforms; and how brands can benefit when these consumer expectations are met. Customers' perception of a brand as attractive, motivated by meeting self-definition needs, contributes both to brand loyalty and serves as a willingness to discard any unpleasant information acquired about the brand. Individuals tend to find a brand attractive if it offers them the opportunity to satisfy one of three essential needs: self-continuity, self-distinction, or self-improvement.

This scoping review had some limitations. It is acknowledged that the inclusion of gray literature and other databases could have contributed to a larger sample. The main limitation was the few studies that could be included and analyzed and the few that talk more specifically about the voicebot and browser assistants, the new studies are more focused on the IoT virtual assistants like Alexa.

The study also allowed the identification of knowledge gaps that can guide the future research agenda in this area. This study, therefore, has several research implications for further studies.

While studies on the acceptance of virtual assistants and voice commerce seem to mount, some knowledge gaps in this area remain. For example, further research is needed into voice branding, being necessary to consider how the brand will speak, with what sound and signature and how are the additional contact; positioning and differentiation, and how to know how to position and differentiate the brand through the use of voice and related services to complement brand products; communication, as it is a new, highly customizable channel, and communication must be done according to the right context and the new content; interaction with customers, brands must know how to integrate all existing services and digital voice assistant creators must promote trustworthiness, responsibly address privacy issues, and create a trusting climate for their customers to interact and remain loyal to their companies.

Voice commerce remains a vibrant new field with increased scholarly interest. This scoping study, therefore, provides some clarification of the key role of virtual assistants and voice branding when it comes to online consumer behavior. However, there is a need for further studies to advance content marketing's effect in the literature.

46.7 Declaration of Conflicting Interests

The authors declares that there are no possible conflicts of interest in this article regarding research, authorship, and/or publication.

Funding The authors will apply for financial support for the research, authorship, and/or publication of this article.

References

1. Zoovu, Reseachscape: The Consumer Report Humanizing Digital 2020 (2021). https://zoovu.com/resources/humanizing-digital-2020, last accessed 2022/08/23
2. Forbes Technology Council: The psychology of voice technology: building a better voice assistant for everyone (2021). https://www.forbes.com/sites/forbestechcouncil/2021/12/09/the-psychology-of-voice-technology-building-a-better-voice-assistant-for-everyone/?sh=ad68de82c276, last accessed 2022/08/23
3. Forbes Communications Council: Is voice the next big thing to transform consumer behavior (2020). https://www.forbes.com/sites/forbescommunicationscouncil/2020/03/23/is-voice-the-next-big-thing-to-transform-consumer-behavior/?sh=2549306e224f, last accessed 2022/08/23
4. Aithority: What is voice commerce and why is it so important in 2021 (2021). https://aithority.com/machine-learning/what-is-voice-commerce-and-why-is-it-so-important-in-2021/, last accessed 2022/10/05
5. Blutag: What is Voice Commerce and Why Does it matter in 2022 (2022). https://blu.ai/what-is-voice-commerce-and-why-does-it-matter-in-2022, last accessed 2022/10/05
6. Blog Mailmunch: What is voice commerce and how can it add value to your business (2022). https://www.mailmunch.com/blog/voice-commerce-can-add-value-business, last accessed 2022/08/23
7. Forbes Technology Council: What is natural language processing and what is it used for (2018). https://www.forbes.com/sites/forbestechcouncil/2018/07/02/what-is-natural-language-processing-and-what-is-it-used-for/?sh=536b06d75d71, last accessed 2022/08/23
8. Personal Technology: Google's duplex uses A.I. to mimic humans (sometimes) (2019). https://www.nytimes.com/2019/05/22/technology/personaltech/ai-google-duplex.html?auth=login-google, last accessed 2022/08/23
9. Rabassa, V., Sabri, O., Spaletta, C.: Conversational commerce: do biased choices offered by voice assistants' technology constrain its appropriation? Technol. Forecast. Social Change (174) (2022)
10. Aw, E.C.-X., Tan, G.W.-H., Cham, T.-H., Raman, R., Ooi, K.-B.: Alexa, what's on my shopping list? Transforming customer experience with digital voice assistants. Technol. Forecast. Social Change (180) (2022)
11. Blog Whiplash: Is voice commerce the future of ecommerce (2022). https://whiplash.com/blog/voice-commerce-future-ecommerce/, last accessed 2022/08/23
12. Zaharia, S., Würfel, M.: Voice commerce—studying the acceptance of smart speakers. Polish Academy of Sciences, Poland (2021)
13. Balakrishnan, J., Dwivedi, Y.K.: Conversational commerce: entering the next stage of AI-powered digital assistants. Annals Oper. Res. (35) (2021)
14. Maroufkhani, P., Asadi, S., Ghobakhloo, M., Jannesari, M.T., WanIsmail, W.K.: How do interactive voice assistants build brands' loyalty? Technol. Forecast. Social Change (183) (2022)
15. Lee, S., Oh, J., Moon, W.-K.: Adopting voice assistants in online shopping: examining the role of social presence, performance risk, and machine heuristic. Int. J. Human–Comput. Interact. 1–15 (2022)
16. Hu, P., Lu, Y., Wang, B.: Experiencing power over AI: the fit effect of perceived power and desire for power on consumers' choice for voice shopping. Comput. Human Beh. (128) (2022)
17. Klaus, P., Zaichkowsky, J.L.: The convenience of shopping via voice AI: introducing AIDM. J. Retail. Consumer Serv. (65) (2022)
18. Chung, D., Kim, H., Ahn, S.: An integrated study of user acceptance and resistance on voice commerce. Int. J. Innov. Technol. Manage. (41) (2022)
19. Ramadan, B.: "Alexafying" shoppers: the examination of Amazon's captive relationship strategy. J. Retail. Consumer Serv. (62) (2021)
20. Ashrafi, D.M., Easmin, R.: Okay Google, good to talk to you... Examining the determinants affecting users' behavioral intention for adopting voice assistants: does technology self-efficacy matter? Int. J. Innov. Technol. Manage. (30) (2022)

21. Bawack, R.E., Wamba, S.F., Carillo, K.D.A.: Exploring the role of personality, trust, and privacy in customer experience performance during voice shopping: evidence from SEM and fuzzy set qualitative comparative analysis. Int. J. Inf. Manage. (58) (2021)
22. Hasan, R., Shams, R., Rahman, M.: Consumer trust and perceived risk for voice-controlled artificial intelligence: the case of Siri. J. Bus. Res. (131) (2021)
23. Forbes Business Council: Transforming Customer Service Starts With Voice AI (2022). https://www.forbes.com/sites/forbesbusinesscouncil/2022/07/29/transforming-customer-service-starts-with-voice-ai/?sh=33b12b424836, last accessed 2022/08/23
24. Hsu, C.L., Lin, J.C.C.: Factors affecting customers' intention to voice shopping over smart speaker. Service Ind. J. (21) (2021)
25. Lim, W.M., Kumar, S., Verma, S., Chaturvedi, R.: Alexa, what do we know about conversational commerce? Insights from a systematic literature review. Psychol. Market. (39) (2022)
26. McLean, G., Osei-Frimpong, K., Barhorst, J.: Alexa, do voice assistants influence consumer brand engagement?—examining the role of AI powered voice assistants in influencing consumer brand engagement. J. Bus. Res. (124) (2021)
27. Canziani, B., MacSween, S.: Consumer acceptance of voice-activated smart home devices for product information seeking and online ordering. Comput. Human Beh. (119) (2021); J. Bus. Res. (131) (2021)
28. Pal, D., Roy, P., Arpnikanondt, C., Thapliyal, H.: The effect of trust and its antecedents towards determining users' behavioral intention with voice-based consumer electronic devices. Heliyon (8) (2022)
29. Shao, C., Kwon, K.H.: Hello Alexa! Exploring effects of motivational factors and social presence on satisfaction with artificial intelligence-enabled gadgets. Human Beh. Emerg. Tech. (3) (2021)

Chapter 47
Analyzing Driving Factors of User-Generated Content on YouTube and Its Influence on Consumers Perceived Value

Ana Torres, Pedro Pilar, José Duarte Santos, Inês Veiga Pereira, and Paulo Botelho Pires

Abstract Companies are increasingly focusing on audiovisual content as part of their strategy, and YouTube being a massive video hosting platform that makes content sharing possible has been the most successful platform for reaching their consumers products and services. Researches have proven that user-generated content impacts brand engagement, loyalty and firm revenue. Therefore, it is necessary to determine what factors stimulate the creation of consumers perceived value from user-generated content, on social media. We analyze the driving factors of user-generated content on YouTube and its influence on consumers perceived value. The sample data consists of 282 YouTube users' responses collected through an electronic survey. This research contributes toward the digital content marketing literature by complementing existing research exploring consumer behavior on social media, assessing the driving factors of user-generated content and its impacts on customer perceived value. The study findings provide academic contributions and several challenges for firm and user-generated content, on actions they can tackle. Finally, based on the study limitations, we discuss future research in generated content in social media, providing insights for future research directions.

A. Torres (✉)
University of Aveiro, Campus Universitário de Santiago, 3810–193 Aveiro, Portugal
e-mail: ana.torres@ua.pt

INESC TEC, Campus da FEUP, Rua Dr. Roberto Frias, S/N, 4200 – 465 Porto, Portugal

J. D. Santos · I. V. Pereira · P. B. Pires
CEOS.PP, Rua Jaime Lopes Amorim, S/N, 4465 – 004 Matosinhos, Portugal
e-mail: jdsantos@iscap.ipp.pt

I. V. Pereira
e-mail: ipereira@iscap.ipp.pt

P. B. Pires
e-mail: paulopires@pbs.up.pt

P. Pilar · J. D. Santos · I. V. Pereira
Accounting and Business School of the Polytechnic of Porto, (ISCAP/P.PORTO), Rua Jaime Lopes Amorim, S/N, 4465 – 004 Matosinhos, Portugal
e-mail: p.pilar87@gmail.com

J. L. Reis et al. (eds.), *Marketing and Smart Technologies*, Smart Innovation, Systems and Technologies 337, https://doi.org/10.1007/978-981-19-9099-1_47

47.1 Introduction

Currently, many people depend on social media for their social lives. Social media are needed for people to interact with each other and for firms to promote their products. Thus, generated content in social media has become an important part of marketing.

According to Digital, in 2020, the Internet has reached 60% of the world's population, and social media has reached 50% [1]. The number of online video (including short videos) reached millions of Internet users worldwide. People spend on average two and a half hours a day on social media and are willing to receive and publish content through social media. During the COVID-19 pandemic government lockdowns, consumers were (and some still) confined to their homes, prompting them to spend more time on platforms such as Twitch, YouTube, which caused ratings to spike across platforms.

According to Miller [2], YouTube is a video sharing platform, public and free, with social networking features where countless virtual communities have the power to transmit and share knowledge in the widest areas of interest. YouTube is characterized as a content community, that is, a website that allows users to organize and share specific types of content with other users. YouTube is also considered a user-generated media (UGM), as it allows the possibility to publish and create content, and in addition permits to comment and share, on other platforms, material essentially created by its users and members (e.g., videos, music, text, photography, etc.) [3].

Scholars have confirmed that generated content on social media has a significant impact on consumer loyalty, engagement, purchase intent, and other consumer behaviors. This impact will change the business paradigms of industries and companies [4]. Therefore, companies are increasingly valuing social-media-generated content, and it is even more common in the experience goods industry (e.g., tourism, literature, films and music).

While, firm-generated content (FGC)—for instance, firm-created videos, comments, tweets—is produced under professional supervision and authority, user-generated content (UGC)—e.g., user-created videos, comments, original tweets—is freer because it focuses on user perspectives and is created by users rather than marketing professionals [5, 6].

To date, most studies have focused on how FGC and UGC affect other research subjects. For example, Nagoya et al. [7] argue that FGC and UGC positively impact "values". Radovic et al. [8] found that content on social media (whether FGC or UGC) significantly impacts teenagers' attitudes and value systems. Qi and Tan [9] suggested that the frequency of FGC-UGC interactions is a crucial factor in minimizing negative UGC communication; Irelli and Chaerudin [10] confirmed that consumers' perceptions of FGC and UGC have a positive impact on purchase intention. Ma and Gu [11] show that some attributes of firm-generated video have a significant impact on user-generated video, from e-Sports platform on China.

However, there are some issues in the UGC research field that scholars have addressed less. (i) Most researchers have studied UGC and FGC in the form of images or text but have rarely addressing UGC perceptual dimensions and their related attributes. (ii) There are already sufficient papers focusing on social media platforms such as Twitter, Facebook and YouTube, studying online videos including the following features: like, share, reply and view. These attributes represent users' perceptions of videos. But relatively few studies examine UGC reliability, usefulness on YouTube social media platform. (iii) Few studies have explored the impact of UGC dimensions on consumer value creation perception. In order to fill these gaps, our work identifies the relevant UGC measures dimensions and explores their impact on users perceived value creation.

This study provides some academic and managerial contributions. First, make available a set of validated metrics that can be used to develop an Index to rating reliable, useful and valuable UGC, on social media platforms. The phenomena of "fake news" is a growing problem on the Internet, increasing biased information exponentially, as the network expands. Therefore, an index that can provide accurate ratings of UGC and should be useful for social media users. Second, for managers, this study provides interesting insights into the use of the best ratings of user-generated content to increase the production of firm-generated content.

The remainder of this paper is organized as follows. In the next section, we present earlier papers dealing with user-generated content in social media platforms. Sections methodology, results and discussion describe the main research questions and the proposed hypothesis, the methodology conducted to collect data which are further evaluated, present the findings and analyze the results obtained. Finally, based on the main findings of the study, a conclusion and global remarks are presented, highlighting practical implications for online managers and study limitations and suggestions for future research directions are presented.

47.2 Existing Studies of UGC and Background

Extant research focus on content marketing on social media aiming to understand how to use content marketing to influence consumers. Content marketing includes two hot topics: FGC and UGC, which have been studied by scholars. User-generated content (UGC) refers to all types of communication generated by and between consumers on social media [12]. While, firm-generated content (FGC) is referred as social media communication generated on the firm's side [13].

Many scholars are interested in studying the topic of UGC influencing user behavior, and some scholars have studied it from a marketing perspective. Previous research has demonstrated the multiple beneficial marketing effects of UGC and its ability to influence brand image [14]. UGC can influence users' brand engagement [15, 16] and the effect is continuous [17] but subjected to professionalism [18]. The impact of UGC on users can also have an impact on other user behaviors. An increase in user engagement can stimulate user-generated content creation intentions [19, 20]. Additionally, UGC can directly influence users' purchase intentions [21–23]. UGC not only affects users' behavior, but also their psychology. UGC also affects consumer satisfaction after the purchase is completed. Consumer satisfaction refers to the degree of consumers' psychological feelings and expectations regarding a particular product or service [24, 25]. For example, UGC influences consumers' expectations, which indirectly affects consumer satisfaction [26] and contains information that helps companies understand user satisfaction [27]. Satisfaction also mediated the relationship between UGC and brand value [28]. Given all the benefits mentioned above, it is logical that companies want to promote their brand image by choosing UGC [29, 30].

With the potential marketing benefits of UGC, many researchers have taken an interest in determining the motivations and driving factors of UGC. For instance, Hennig-Thurau et al. [31] found that the main reasons for eWOM are users' need for social interaction, desire for economic benefits, concern for others and willingness for self-improvement; Bazi et al. [32] advocate that the motivations of customers' engagement on social media are grouped into several dimensions: perceived content relevancy, brand-customer relationship, hedonic, aesthetic, socio-psychological, brand equity and technology factors; Nikolinakou and Phua [33] indicate that conservation, self-enhancement, openness to change and self-transcendence are significant drivers of brand-related UGC. Antón et al. [34] showed that the intention to create UGC depends on on-site experience and satisfaction. Wang and Li [35] confirmed the impact of trust on UGC production using the self-determination theory. Based on the findings of Verma et al. [36] in Facebook brand communities, they show that FGV can provide value to users in four areas: engagement, perceived usefulness, popularity and brand identity.

Although scholars have exhaustively studied psychological motivations to stimulate UGC, little research has been conducted on the topic of specific attributes of UGC influencing consumer's value perceptions and behaviors toward user-generated-content. For instance, Mir and Rehman [37] found that reliable UGC depending on unbiased, honest, truthful, dependable and useful UGC, related to good, valuable relevant and convenient source of product information is important factors affecting consumer attitudes and intentions toward user-generated product content on YouTube. Kim et al. [38] advocate that user perception of the quality, value and utility of user-generated content is dependent on interesting, enjoyable, having acceptable standard of quality and feeling good when using the UGC.

47.3 Methodology

47.3.1 Research Questions and Conceptual Model

The two research questions of this research are:

RQ1: What are the relevant dimensions of UGC perceived by YouTube consumers?

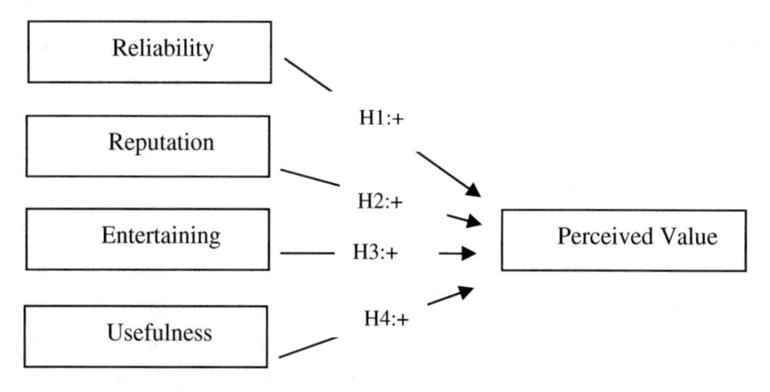

Fig. 47.1 Conceptual model

RQ2: Can the UGC dimensions be used to predict the consumer's perceived value from the YouTube use-experience?

According to the objective of the study, a model (Fig. 47.1) was developed to explain the relationship between the dimensions of user-generated content (UGC) and consumers' perceived value, from the YouTube use-experience.

Four research hypotheses are presented, one for each dimensions of the UGC construct supported in the literature:

H1: Reliability (RE) of UGC is positively related to the perceived value (PV).

H2: Reputation (RP) of UGC is positively related to the perceived value (PV).

H3: Entertaining (EN) of UGC is positively related to the perceived value (PV).

H4: Usefulness (US) of UGC is positively related to the perceived value (PV).

47.3.2 Questionnaire Structure

As a research tool, a questionnaire was created that enabled the collection of data from our sample, allowing us, later through statistical treatment, to test the research hypotheses and thus understand whether the dimensions of UGC have a positive impact on consumer perceived value. At the beginning of the questionnaire, individuals were informed about the research goal. They were asked about the product/service category mostly used on YouTube and invited to answer the questions accordingly. The questions were grouped into four groups:

- A screening question to identify who are the YouTube users/consumers;
- Questions about to the use-experience of YouTube (e.g., product/service categories; frequency of use);
- Questions, in a 5-point Likert scale (totally agree/disagree), to assess the UGC item dimensions and consumer perceived value, from the YouTube use-experience, respectively, adapted from the literature scale measures (described in Table 47.3).

- Sociodemographic characterization (e.g., age, gender, education level, average net monthly income and professional status) of the respondent.

47.3.3 Data Collection

After conducting a pre-test with faculty staff and master's students, the data were collected through surveys provided via the web through a Google Form, and disseminated through email (personal) using different social networks (e.g., Facebook, WhatsApp and LinkedIn). This was a convenience sampling procedure, which cannot offer the guarantee of a sample representative of the entire population, and one must proceed with caution when generalizing the results [39]. A total of 307 responses were collected. After excluding 25 responses from people who have never used the YouTube, only 282 remaining responses were taken into account for further data analysis. Of the 282 respondents, 55.7% of participants spend less than one hour/day on YouTube, 22% spend until 2 h/day, 9.9% do it at least 3 h/day, 6.4% until 4 h/day and 6.0% spend more than 4 h/day.

The data were analyzed using the software SPSS [40]. A factorial analysis was conducted to identify the relevant measures dimensions of UGC on YouTube from the perspective of consumer perceived value. A multiple linear regression model was used to study the relationship between the various dimensions of UGC and the consumer perceived value.

47.3.4 Sample Characteristics

The age of the respondents varies between 15 and 65 years (mostly aged between 26 and35 years, 39.75%). The majority of respondents are female (53.9%), and 46.1% are male. Regarding the level of education, 24.5% of the respondents were postgraduate and had a master's/MBA/other higher graduate level. The remaining respondents have a Bachelor's degree level (28.0%) and 30.5% have a high-school education level. Regarding their professional situation, 81.2% are employed, 7.1% are working students, 6.4% are students and 4.3% are unemployed. With regard to the average net monthly income, 59.6% of respondents reported between €500 and €1500, 17.0% do not know or prefer not to answer, 12.4% earn between 1501€ and 2500€, 7,4% earn between 0€ and 500€ and 3.6% earns an average net monthly income between 2501€ and 3000€ or more. Table 47.1 illustrates those demographic characteristics.

The category of products and services mostly searched on YouTube (Table 47.2) is music (62.4%), followed by learning (6.7%), sports (5.7%), beauty and fashion and movies (with 4.6% each), stand-up comedy (3.9%), science and technology (2.5%), cars and vehicles (2.1%). The least searched categories (with $\leq$1% each) are

Table 47.1 Sample: demographic characteristics (% of total answers)

Characteristics	Number of respondents	% Total
Gender:		
Male	130	46.1
Female	152	53.9
Age:		
[15, 25]	50	17.8
[26, 35]	112	39.7
[36, 45]	77	27.3
[46, 55]	36	12.8
[56, 65+]	7	2.5
Education level:		
<High school	43	15.3
High school	86	30.5
Bachelor's degree	79	28.0
Postgraduate	69	24.5
Other	5	1.8
Income (monthly):		
<500€	21	7.4
[500, 1500]	168	59.6
[1501, 2500]	35	12.4
[2501, 3000+]	10	3.6
NR	48	17.0
Professional status:		
Student	18	6.4
Working student	20	7.1
Worker	229	81.2
Unemployed	12	4.3
Retired	3	1.1

videogames, traveling and events, cooking, how´s do and do it yourself, influencers and celebrities and pets/animals.

47.4 Results and Discussion

Following are presented and discussed the results obtained from the empirical study in order to answer the two main objectives of this research: (i) What are the relevant dimensions of UGC perceived by YouTube users/consumers? and (ii) Can the UGC

Table 47.2 Sample: YouTube use by product/services category (% of total answers)

Products/services category	Number of respondents	% total
Music	176	62.4
Learning	19	6.7
Sports	16	5.7
Beauty and fashion	13	4.6
Movies	13	4.6
Stand-up comedy	11	3.9
Science and technology	7	2.5
Cars	6	2.1
Videogames	5	1.8
Traveling and events	5	1.8
Cooking	4	1.4
How´s do and do it yourself	3	1.1
Influencers and celebrities	3	1.1
Pets and animals	1	0.4

dimensions be used to predict the consumer's perceived value from the YouTube use-experience?

Results of Factor Analysis presented in Table 47.3, identify the significant measures to assess UGC in YouTube, in terms of the users´ perceived value.

We analyze first the individual item reliability, and we verify that the factor loadings are all greater than 0.7, satisfying the reliability condition (>0.5). Contributions of two items were below the cut off value (0.5), therefore, CH2 and PRO4 were excluded from the analysis. To ensure the reliability of the constructs, we examined the Cronbach's alpha (α), and acceptable levels (>0.7) [39, 41, 42] were obtained for all variables. As showed in Table 47.3, the reliabilities of constructs (i.e., ranging from 0.85 to 0.94) are all above the recommended values. From these results, we verify that all constructs satisfy the internal consistency and reliability conditions.

The results of the multiple linear regression carried out, having as the dependent variable the Perceived Value (PV) and as independent variables the extracted factors Reliability (RE), Reputation (RP), Entertaining (EN) and Usefulness (US), allowed to support only H2, H3 and H4 (Table 47.4, Fig. 47.2). When examining the hypothesized relationships proposed, all received empirical support, except H1, and globally, the model result explains a significant amount of variance (69%). Specifically, the results presented in Table 47.4 showed that RP has a significant positive effect on PV ($\beta = 0.143$; $p = 0.001$), (H2 supported); EN has a significant positive effect on PV ($\beta = 0.257$; $p = 0.000$), (H3 supported), and US has a significant positive effect on PV ($\beta = 0.498$; $p = 0.000$), (H4 supported). From this results, we verify that

Table 47.3 Measurement scales and reliability analysis

Construct/items loadings	Adapted from
UGC-reputation (α = 0.944, KMO = 0.871) I believe that this content producer on YouTube: • Has an excellent public image *(loading: 0.938)* • Has an excellent reputation *(loading: 0.935)* • Is trustworthy *(loading: 0.910)* • Is a celebrity *(loading: 0.872)* • Is a public person *(loading: 0.864)*	Kim et al. [38]
UGC-reliability (α = 0.915, KMO = 0.868) User-generated content on YouTube is: • Honest *(loading: 0.925)* • Reliable *(loading: 0.920)* • Truthful *(loading: 0.899)* • Dependable *(loading: 0.852)* • Unbiased *(loading: 0.718)*	Mir and Rehman [37]
UGC-entertaining (α = 0.852, KMO = 0.730) User-generated content on YouTube is: • Enjoyable *(loading: 0.888)* • Pleasing *(loading: 0.884)* • Exciting *(loading: 0.864)*	Kim et al. [38]
UGC-usefulness (α = 0.915, KMO = 0.881) User-generated content on YouTube: Is useful (loading: 0.885) • Is a convenient source of product information *(loading: 0.873)* • Supplies relevant product information *(loading: 0.860)* • Is valuable *(loading: 0.826)* • Is good *(loading: 0.792)* • Makes information immediately access *(loading: 0.791)*	Mir and Rehman [37]

(continued)

all the standardized coefficients signs show the expected direction, confirming the subjacent value perception of user-generated content theory.

From Table 47.4, we can see that the p-value of Usefulness is 0.000 (<0.001), indicating that it is a significant independent variable. If there is a 100% increase in usefulness, the perceived value of UGC will increase by 50%. This verifies our hypothesis H4, indicating that usefulness strongly stimulate UGC perceived value. We also found that Entertaining and Reputation, at the significance level <0.05, with p-values of 0.000 and 0.001 for EN and RP, respectively. This means that 100%

Table 47.3 (continued)

Construct/items loadings	Adapted from
Perceived value (α = 0.934, KMO = 0.901) User-generated content on YouTube: • Is interesting *(loading: 0.904)* • I enjoy to use it *(loading: 0.893)* • Has an acceptable quality standard *(loading: 0.892)* • I feel good when I use it *(loading: 0.881)* • Performs consistently *(loading: 0.859)* • Delivers clear information *(loading: 0.771)*	Kim et al. [38]

Table 47.4 Regression analysis results (adjusted $R^2 = 0.687$; *p*-value 0.000)

Factors	Standardized coefficient	p-value	Decision
Reliability	0.059	0.197	H1 not supported
Reputation	0.143	0.001**	H2 supported
Entertaining	0.257	0.000***	H3 supported
Usefulness	0.498	0.000***	H4 supported

Statistical significance level <0.05, with p-values of 0.000*** and 0.001**

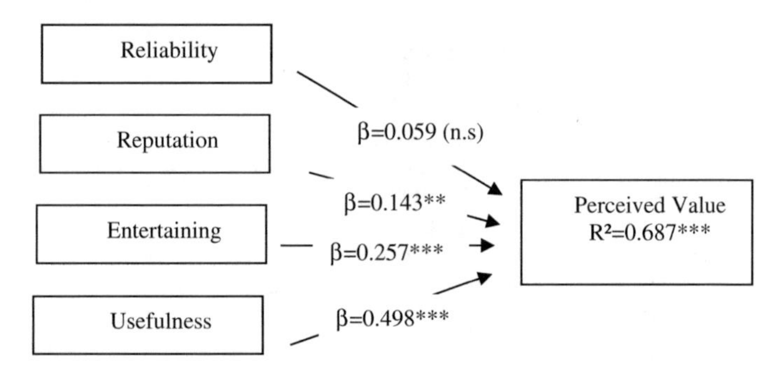

Fig. 47.2 Empirical model with main results

increase in the entertaining results in a 25.7% increase in the PV, and 100% increase in reputation, results in a 14.3% increase in PV of UGC. This confirms our hypothesis H2 and H3 that EN and RP can impact PV of UGC. However, it is surprising that Reliability, which also represent UGC did not influence PV at the significance level of 0.05, with p-value of 0.197. This indicates that the amount of RE of UGC achieved does not indicate whether it can arouse the PV. This may be because when users collect videos on YouTube, they focus more on the usefulness and entertaining of the video than on the content reliability of the video itself, thus failing to elicit inherent perceived value.

Following, the conclusions of the study are presented. Finally, the managerial contributions of the study and directions for further research are provided.

47.5 Conclusion

Addressing the research questions of this study, findings indicate that reliability, reputation, entertaining and usefulness are the relevant antecedent dimensions of UGC, in terms of YouTube consumers' perceptions. When analyzed the interplay between the antecedent dimensions of UGC and the perceived value outcomes of the YouTube consumers, the study provides interesting conclusions. First, the study results show that usefulness (US) strongly impacts UGC PV outcomes. This result indicates that, the impact of US that may lead to PV of UGC which will be carried out by perceptions of "useful, convenient and relevant information source, easily available on YouTube".

Second, entertaining (EN) plays an important role in determining UGC PV outcomes. This also suggests that results may be context-dependent: For instance, music and films are the most popular categories in YouTube from the users' experience. Also, it has been referred the increasing popularity of video in several contexts, such as learning, do it yourself, cooking and traveling.

Third, reputation (RP) of the producer of UGC significantly interplays with UGC PV outcomes. In the perceptions of YouTube users, reputation of the producer of UGC is also an indicator at same point of reliability, in terms of excellent reputation, outstanding public image and trustworthy person.

Finally, the amount of reliability of UGC achieved does not indicate whether it can arouse the PV. This may be because when users collect videos on YouTube, they focus more on the usefulness and entertaining of the video than on the content reliability of the video producer itself, thus failing to elicit inherent perceived value. Taken altogether, a major conclusion from the study findings encourages firms and producers of generated content to tackle efforts in increasing users' perceptions of usefulness, entertaining and reputation of UGC in order to create value.

The study findings provide some academic contributions. Make available a set of validated metrics that can be used to develop an Index to rating reliable and valuable UGC, on social media platforms. As we mention earlier, the phenomena of "fake news" is a growing problem on the Internet, increasing bias information exponentially, as the Networks expands. Therefore, an Index that can provide accurate ratings of UGC, should be useful for Internet and social media users.

The study also offers managerial contributions, on actions and challenges to tackle. This study provides interesting insights into the use of firm-generated content to increase the production of user-generated content. For instance, for managers the UGC Index ratings could be beneficial to stimulate UGC authors with best ratings to produce FGC. Also, UGC should increase positive e-Wom not only of most popular

UGC, but as well, reliable and useful UGC, on YouTube and other social media. In addition, best ratings of UGC on social media, should be used by companies to stimulating FGC. This is a win–win strategy: If companies could encourage users to post valuable user-generated content on social media, merchants can stimulate consumption to some extent. Therefore, best UGC ratings could be beneficial for stimulating FGC, as companies wish to promote their brand image by choosing reliable and valuable UGC. Given all the benefits mentioned above, it is logical that companies want to promote their brand image by choosing valuable UGC.

Some limitations of this study allow suggesting some directions for further research.

First, the sample of the study typically comprises YouTube users. Future research should include other social media, blogs and online communities, in order to develop a comparative cross-sector study. Secondly, the sample is cross-section and the use of online panel data in further studies could increase the predictive accuracy of the research model. Finally, in future research should be taken into account other explanatory variables of OGC, such as: social influence, helpful, aesthetics or altruism in sharing content which are referred to play potential roles on consumers' perceptions of UGC which lead to positive impact on satisfaction, value, engagement and purchase intention.

Acknowledgments This work is financed by portuguese national funds through FCT—Fundação para a Ciência e Tecnologia, under the project UIDB/05422/2020.

The authors gratefully acknowledge the anonymous respondents that helped in the data gathering process in this study.

References

1. Wearesocial. Digital in 2020 (2020). Available from: https://wearesocial.com/digital-2020
2. Miller, K.: Playing Along: Digital Games, YouTube, and Virtual Performance. Oxford University Press, Oxford (2012)
3. Dao, W.V.-T., Le, A.N., Chen, J.M.-S.: Social media advertising value: the case of transitional economies in Southeast Asia. Int. J. Advert. **33**(2), 271–295 (2014)
4. Corbo, L, Pirolo, L, Rodrigues, V.: Business model adaptation in response to an exogenous shock: an empirical analysis of the Portuguese footwear industry. Int. J. Eng. Bus. Manage. (2018)
5. Colicev, A., Kumar, A., O'Connor, P.: Modeling the relationship between firm and user generated content and the stages of the marketing funnel. Int. J. Res. Market. **36**(1), 100–116 (2019)
6. Schivinski, B., Dabrowski, D.: The effect of social media communication on consumer perceptions of brands. J. Market. Commun. **22**(2), 189–214 (2016)
7. Nagoya, R., Innocentius Bernarto, F.A.: Do private universities still need social media? Firm generated and user generated content in social media. Psychol. Educ. J. **58**(2), 6953–6964 (2021)
8. Radovic, V., Ljajic, S., Dojcinovic, M.: Media paradigms and programmed code of values orientation in adolescence. Tem. J. **10**(2), 682–691 (2021)
9. Qi, J., Tan, Y. (eds.): User-generated content (UGC) encountered enterprise-generated content (EGC): quantifying the impact of EGC on the propagation of negative UGC. PACIS (2014)

10. Irelli, R.S., Chaerudin, R.: Brand-generated content (BGC) and consumer-generated advertising (CGA) on Instagram: the influence of perceptions on purchase intention. KnE Social Sci. 882–902 (2020)
11. Ma, Z., Gu, B.: The influence of firm-generated video on user-generated video: evidence from China. Int. J. Eng. Bus. Manage. **14**, 1–17 (2022)
12. Lopes, A.R, Porto, I., BJAoSMJ, C.: Digital content marketing: conceptual review and recommendations for practitioners (2022)
13. Poulis, A., Rizomyliotis, I., Kjit, K., et al.: Do firms still need to be social? Firm generated content in social media (2018)
14. Dolan, R., Seo, Y., Kemper, J.J.T.M.: Complaining practices on social media in tourism: a value co-creation and co-destruction perspective. Tour. Manage. **7**, 35–45 (2029)
15. Chu, S.-C., Lien, C.-H., Cao, Y.J.I., Joa: Electronic word-of-mouth (eWOM) on WeChat: examining the influence of sense of belonging, need for self-enhancement, and consumer engagement on Chinese travellers' eWOM. Int. J. Adv. **38**(1), 26–49 (2019)
16. Yang, M., Ren, Y.: Adomavicius, GJISR. Understanding user-generated content and customer engagement on Facebook business. Inf. Syst. Res. **30**(3), 839–855 (2019)
17. Han, W., McCabe, S., Wang, Y., et al.: Evaluating user-generated content in social media: an effective approach to encourage greater pro-environmental behavior in tourism? J. Sustain. Tour. **26**(4), 600–614 (2018)
18. Khurana, S., Qiu, L., Kumar, S.J.I.S.R.: When a doctor knows, it shows: an empirical analysis of doctors' responses in a Q&A forum of an online healthcare portal. Inf. Syst. Res. **30**(3), 872–891 (2019)
19. Asmussen, B., Harridge-March, S., Occhiocupo, N., et al.: The multi-layered nature of the internet-based democratization of brand management. J. Bus. Res. **66**(9), 1473–1483 (2013)
20. Van Doorn, J., Lemon, K.N., Mittal, V., et al.: Customer engagement behavior: theoretical foundations and research directions. J. Serv. Res. **13**(3), 253–266 (2010)
21. Menon, B.: Determinants of online purchase intention, towards firm generated content in Facebook. Management **2**(2), 47–56 (2017)
22. Abdelkader, A.A., Ebrahim, R.: Decomposing customer engagement effect between marketer- and user-generated content and repurchase intention in the online airline service community. Int. J. Online Market. **11**(4), 1–22 (2021)
23. Demba, D., Chiliya, N., Chuchu, T., et al.: How user-generated content advertising influences consumer attitudes, trust and purchase intention of products and services. Communicare J. Commun. Sci. South Africa **38**(1), 136–149 (2019)
24. Abdullah, N.N., Prabhu, M., Othman, M.B.: Analysing driving factors of customer satisfaction among telecommunication service providers in Kurdistan region. Int. J. Eng. Bus. Manag. 14 (2022)
25. Hu, M., Chaudhry, P.E., Chaudhry, S.S.: Linking customized logistics service in online retailing with E-satisfaction and E-loyalty. Int. J. Eng. Bus. Manag. 14 (2022)
26. Narangajavana Kaosiri, Y., Callarisa Fiol, L.J., Moliner Tena, M.A., et al.: User-generated content sources in social media: a new approach to explore tourist satisfaction. J. Trav. Res. **58**(2), 253–265 (2019)
27. Jia, S.: Tourism. Measuring tourists' meal experience by mining online user generated content about restaurants. Scand. J. Hospital. Tour. **19**(4–5), 371–389 (2019)
28. Colicev, A., O'Connor, P.: How social media impacts brand value: The mediating role of customer satisfaction. Multidisc. Bus. Rev. **13**(1), 82–96 (2020)
29. Colicev, A., Malshe, A., Pauwels, K., et al.: Improving consumer mindset metrics and shareholder value through social media: the different roles of owned and earned media. J. Market. **82**(1), 37–56 (2018)
30. Kübler, R.V., Colicev, A., Pauwels, K.: Social media's impact on the consumer mindset: when to use which sentiment extraction tool? J. Interact. Market. **50**, 136–155 (2020)
31. Hennig-Thurau, T., Gwinner, K.P., Walsh, G., et al.: Electronic word-of-mouth via consumer-opinion platforms: what motivates consumers to articulate themselves on the internet? J. Interactive Market. **18**(1), 38–52 (2004)

32. Bazi, S., Filieri, R., Gorton, M.: Customers' motivation to engage with luxury brands on social media. J. Bus. Res. **112**, 223–235 (2020)
33. Nikolinakou, A., Phua, J.: Do human values matter for promoting brands on social media? How social media users' values influence valuable brand-related activities such as sharing, content creation, and reviews. J. Consumer Behav. **19**(1), 13–23 (2020)
34. Antón, C., Camarero, C., Garrido, M.-J.: What to do after visiting a museum? From post-consumption evaluation to intensification and online content generation. J. Trav. Res. **58**(6), 1052–1063 (2019)
35. Wang, X., Li, Y.: How trust and need satisfaction motivate producing user-generated content. J. Comput. Inf. Syst. **57**(1), 49–57 (2017)
36. Verma, R., Jahn, B., Kunz, W.: How to transform consumers into fans of your brand. J. Serv. Manag. **23**, 344–361 (2012)
37. Mir, I.A., Rehman, K.U.: Factors affecting consumer attitudes and intentions toward user-generated product content on YouTube. Manag. Market. **8**, 637–654 (2013)
38. Kim, C., Jin, M.-H., Kim, J., Shin, N.: User perception of the quality, value, and utility of user-generated content. J. Electron. Commer. Res. **13**(4), 305–319 (2012)
39. Hair, J., Black, B., Babin, B., Anderson, R., Tatham, R.: Multivariate Data Analysis. 6th ed., NJ: Pearson (2006)
40. IBM Corp.: IBM SPSS Statistics for Windows, Version 24.0 (2016)
41. Cronbach, L.J.: Coefficient alpha and the internal structure of tests. Psychometrika **16**(3), 297–334 (1951)
42. Gefen, D., Straub, D., Boudreau, M.: Structural equation modeling and regression: guidelines for research practice. Commun. Assoc. Inf. Syst. **7**, No. (August), 1–78 (2000)

Chapter 48
Exploratory Analysis of Financial Literacy and Digital Financial Literacy: Portuguese Case

Ana Paula Quelhas, Isabel N. Clímaco, and Manuela Larguinho

Abstract In last few years, the fast development of digital technology and e-commerce has changed the spending and saving behavior, particularly after the COVID-19 pandemic. There has been a strong evidence in the literature that establishes a relationship between digital literacy and financial literacy. According to this, the study intends to present a descriptive analysis about financial literacy and digital financial literacy among Portuguese adults. The data were collected using a questionnaire made available through the Google Forms platform. The questionnaire consists of two main sections of questions, and there were received 311 valid responses. The sample is distributed almost identically between both genders, but more than a half belongs to the age cohort between 36 and 66 years old. Respondents are highly educated, about 70% holding an academic degree, and about a half are in intermediate income levels. The results allow us to identify, generally speaking, a good level of financial literacy, both in objective terms and in terms of perception. Moreover, a vast majority of respondents use mobile phones to make payments (65.9%), and use digital means in their financial transactions. In addition, differences in actual financial literacy level are significant for gender, while no significant differences were found between genders, both in terms of perceived financial literacy levels and in the use of mobile phones and other digital devices.

A. P. Quelhas
Polytechnic Institute of Coimbra, Coimbra Business School, Quinta Agrícola - Bencanta, 3045-231 Coimbra, Portugal

I. N. Clímaco (✉)
Centro de Estudos e Investigação em Saúde da Universidade de Coimbra, Coimbra, Portugal
e-mail: iclimaco@iscac.pt

M. Larguinho
Centro de Investigação em Matemática e Aplicações da Universidade de Évora, Évora, Portugal

J. L. Reis et al. (eds.), *Marketing and Smart Technologies*, Smart Innovation, Systems and Technologies 337, https://doi.org/10.1007/978-981-19-9099-1_48

48.1 Introduction

The development of digital technology and e-commerce has changed the spending and saving behavior, particularly after the COVID-19 pandemic. There has been found an association between digital literacy and financial literacy [12] showing that digital financial literacy (DFL) must be the new focus for researchers and policy makers. In 2016, G20 leaders focused on DFL more closely and endorsed the high-level principles for Digital Financial Inclusion which includes a principle that must "Strengthen Digital and Financial Literacy and Awareness" (GPFI, 2016). However, DFL is a complex and multi-dimensional concept and there is still no standardized and consensual definition of DFL. Nevertheless, some approaches refer to DFL as the developing financial education contents and competences specifically targeted to face the challenges and risks created by financial products' digitalization [10]. Some authors proposed a more detailed measure by considering four dimensions of DFL knowledge of digital financial products and services, awareness of digital financial risks, knowledge of digital financial risk control, and knowledge of consumer rights and redress procedures [9]. Moreover, DFL can also be affected by social characteristics or social-economic standing such as age, income, and level of education [15]. Agarwal et al. [1], for instance, found that digital payment may be caused overspending in India. It is recognized that the current development of information technology has brought enormous changes in all sectors and may also affect and change the behavior of consumers in their spending and saving patterns. In line with Lusardi and Mitchell [6] study, the lack of financial literacy may cause less saving and more spending in the future. Yet, the rapid development of digital products in the financial sector has not been accompanied by an increase in public literacy in the field of digital finance, as known as DFL. Therefore, an adequate understanding of digital finance is essential since, like other technological innovations, technological innovations in the financial sector not only offer benefits in the form of convenience, speed, and economy for users but also have potential risks [10, 15]. Prete [12] found that across countries, on average, the percentage of the population who are expertise in digital technologies (31%) is much lower than the percentage of financially literate people (55%). Apparently, DFL is higher in Nordic countries and in the Netherlands and lower in Eastern Europe [11]. This emphasizes the importance of considering the socio-demographic characteristics of the population to be studied. For instance, the use of digital financial technology is affected by Internet and mobile phones uses, and the young generation shows a heavy use of digital technology which is of particular interest given that young cohorts are better educated than older ones. On the other hand, it is well identified the aging of population process in Europe and in other developed countries. Gender issues are also often scrutinized in financial literacy studies [2, 17].

The main purpose of this work is to explore the levels of financial literacy (actual and perceived) and DFL among Portuguese adults, for male and female. This study could potentially advance the research on the subject of DFL, because most of the literature tends to mainly focus on financial literacy (actual and/or perceived).

This paper is structured as follows. Section 48.2 provides a synopsis of relevant literature. Section 48.3 describes the study design and data analysis process. Section 48.4 provides the results and discussion. Finally, Sect. 48.5 presents the main conclusions of the study.

48.2 Literature Review

48.2.1 Actual Financial Literacy

There is not a consistent definition of financial literacy in the literature. The first studies on the topic date back to the 60s, which were conducted among college students [3]. Lusardi and Mitchell [6] stated that financial literacy is the "knowledge of basic financial concepts and ability to do simple calculations". Some other studies linked financial literacy to financial capability. The former is related with financial knowledge, while the second refers to the ability or the opportunity to act. However, financial knowledge may be useless if it is not reflected on financial behavior [16]. Warmath and Zimmerman [19] consider financial literacy as a combination of three levels of knowledge: financial skills, self-efficacy, and explicit knowledge.

Rahim et al. [13] also distinguish between the possession of financial information (knowledge and skills) and the capacity to apply financial knowledge when it deals to make sound financial decisions.

In the context of the present study, this dimension of financial literacy is named as actual or objective financial literacy [3, 15].

Financial literacy is often measured using the "big three" questions proposed by Lusardi and Mitchell [6]. The first question is about the amount of interest in a capitalization process, the second question focuses on inflation and real value of capital over time; and the third question is related with risk of different financial assets. In the context of this study, two more other questions are considered, respectively, about the concepts of spread and Euribor.

48.2.2 Perceived Financial Literacy

Also related with financial behavior, subjective or perceived financial literacy involves the self-assessment of a specific financial topic [3, 15]. Frequently, people may think what they know is really true. Subjective financial knowledge results the one's beliefs about a topic.

Setiawan et al. [15] pointed out that subjective financial knowledge positively influences consumers' behavior and decision-making. Nevertheless, this relationship is not always clear and evident.

In our work, this dimension of literacy is evaluated through a self-assessment question of financial knowledge, and also through a set of understandings about specific financial topics, using a 5-points Likert scale.

48.2.3 Digital Financial Literacy

More recently, another dimension of financial literacy has been assuming a relevant role in the literature—the DFL. This dimension encompasses both financial and digital financial literacies [8, 13, 18].

The authors referred above to analyze DFL taking into consideration five dimensions: (i) knowledge; (ii) practical know-how; (iii) self-defense; (iv) basic digital knowledge; and (v) decision-making related to financial digital activities.

According to Rahim et al. [13], DFL is still a novel concept, possibly hard to evaluate. In the present work, we use a set of questions concerning the use of digital devices in the financial decision-making process.

48.3 Study Design and Data Analysis

The data in this research were obtained using an online structured questionnaire, using the convenience sampling method. Before the final version of the questionnaire, we performed a pre-test, with five volunteers, which allowed us to build the final and definitive version of the questionnaire, with the reformulation of some questions, making them simpler and clearer. The questionnaire is composed of two main sections. The first section includes several demographic characteristics, namely age, gender, education level, and monthly net income.

The second section consists of three groups of questions: (i) the first is dedicated to actual financial literacy, where five objective questions are included which are related to concepts of inflation, compound interest, risk diversification, and spread and Euribor rates (see, for example, [5–8]), (ii) the second group includes questions that pretend to measure Portuguese's subjective perceptions of financial literacy using a questionnaire adapted from Swiecka et al. [17]. The respondents were required to rate the level of agreement with the statement related to financial behavior, using a 5-point Likert scale (1—strongly disagree, 5—strongly agree). (iii) The third group of questions is dedicated to DFL, adapted from Lyons and Kass-Hanna [8].

The questionnaire was carried out between March 10 and May 30, 2022 and yielded 311 valid responses. All respondents are 18 years old or over.

According to the above, gender can affect knowledge and perception of financial literacy. In order to verify whether or not there are differences, we propose to test the following research hypotheses:

H1: Gender has influence in the actual financial literacy.

and

H2: Gender affects perceived financial literacy.

We are also interested in verifying if the is there is an association between gender and two variables related to DFL. So, we propose to test the following research hypotheses:

H3: Gender is associated with the use of the mobile phone to pay for goods and services.

and

H4: Gender is associated with the use of digital media to carry out your financial transactions/operations.

In order to verify the research hypothesis, H1 and H2, we performed independent-samples t-tests, and in order to verify the research hypothesis, H3 and H4, we performed the Pearson chi-square test of independence. We also used descriptive statistics.

We used the statistical software IBM SPSS version 27 for the analysis of data results.

48.4 Results and Discussion

As already mentioned, in this research, 311 respondents' volunteers participated. Sociodemographic characteristics based on gender, age, education level, and monthly net income are presented in Table 48.1.

As can be seen from Table 48.1, our sample constituted of 52.1% females and 47.9% males; 55.3% are in age class 36–66 (the minimum age is 18, the maximum is 77, and the average is 39.7 years old); around 70% have a higher education level; 19.9% do not have own income, 31.8% have a net income between 705.01€ and 1200€, and only 0.6% of respondents have a monthly net income superior to 5000€.

48.4.1 Actual Financial Literacy

As previously said, one of the main objectives of this research is to analyze the level of financial literacy of Portuguese adults and their perception about this. Therefore, in the second section of the questionnaire, we started with the following question: "How do you assess your financial knowledge?": 10.6% of respondents consider that they are below average, 59.2% are identical to the average, and 30.2% think they are above average.

Concerning the actual financial literacy (AFL) level, we considered five objective questions (see Table 48.2), as already mentioned. We have considered two methods to assess knowledge about financial literacy. The first, inspired by the work of Rajdev

Table 48.1 Characteristics of respondents

Variables	Descriptions	N of respondents	% of respondents
Gender	Female	162	52.1
	Male	149	47.9
Age	18–35	131	42.1
	36–66	172	55.3
	Age 67 and over	8	2.6
Education	Until basic level	4	1.3
	Secondary	88	28.3
	Bachelor	149	47.9
	Master/PhD	70	22.5
Income	Without own income	62	19.9
	Until 705€	18	5.8
	705.01€–1200€	99	31.8
	1200.01€–2000€	84	27.0
	2000.01€–3000€	31	10.0
	3000.01€–5000€	15	4.8
	More than 5000€	2	0.6

et al. [14], evaluates the performance of the Portuguese in the 5 questions, and the results are presented in Table 48.2. The second method, based on studies by Lyons and Kass-Hanna [8] and Rahim et al. [13], analyzes the AFL scores. According to Lyons and Kass-Hanna [8], AFL scores can range from 0 (i.e., zero correct responses) to 5 (the maximum number of correct responses). Table 48.3 presents these results.

Table 48.2 gives the results of the performance for AFL, indicating the percentage of respondents who answered each question correctly, organized in descending order. The questions related to risk diversification, inflation, and compound interest rates have an elevated rate of correct answers, all above 80%. Participants performed poorly on questions related to spread and Euribor rates, only 37.6% know what the spread is, and only 27.7% know what the Euribor rate means.

Table 48.3 presents the distribution of AFL scores of the respondents. All participants answered at least one question correctly. Only 7.7% of the respondents answered all the questions correctly. The mean score is 3.1801 which reveals a relatively good financial literacy level.

Results in Table 48.4 give that the AFL score of females is 3.006 and for males is 3.369, and this difference is statistically significant ($p\text{-value} = 0.01 < 0.05$).

Table 48.2 Actual financial literacy ranks

Rank	Question	Answer	% of the correct answers
1	Where do you think it's safest to invest your money?	In more than one type of business/Investment	86.8
2	Assume that you have a savings account that earns interest at the rate of 1% per year and that the inflation rate is 2% per year. In a year from now, with the money from this account, you will be able to purchase a number of goods:	Smaller than what I would get today	85.2
3	Suppose you have €100 invested in a savings account and the interest rate is 2% per year. After 2 years, you will have in this account:	More than 102€	80.7
4	What is spread?	It is the variable accrual that your bank establishes against an interest rate of reference	37.6
5	In most home loans, the interest paid to the bank is calculated based on a rate indexed to a reference rate, which is the "Euribor". This rate	It results from loans made by a set of European banks	27.7

Table 48.3 Actual financial literacy scores

Score	0	1	2	3	4	5
No. of respondents	0	18	49	127	93	24
% of respondents	0	5.8	15.8	40.8	29.9	7.7
Mean						3.1801
Std. deviation						0.9836

Table 48.4 Result of t-test to examine gender differences in actual financial literacy scores

Gender	*n*	Mean	Std. deviation	t-test	df	p-value
Female	162	3.006	0.981	−3.303	309	0.01
Male	149	3.369	0:954			

Table 48.5 Perception of financial literacy scores

Statement	Mean	Std. deviation
I often spend more money than I intended	2.75	1.20
I often buy things without evaluating or thinking about whether or not I can afford them	1.59	0.86
I think I can manage my money	4.39	0.92
At the moment, I have some financial concerns	3.04	1.30
I always know how much money I have in my wallet and in my current account	4.16	1.00
Overall	3.185	1.136

Table 48.6 Result of t-test to examine gender differences in perceived financial literacy scores

Gender	*n*	Mean	Std. deviation	t-test	df	p-value
Female	162	3.154	0.544	−0.999	309	0.319
Male	149	3.219	0:594			

48.4.2 Perceived Financial Literacy

In order to measure Portuguese's subjective perceptions of financial literacy, the respondents were required to rate the level of agreement with the statement related to financial behavior, using a 5-point Likert scale (1—strongly disagree, 5—strongly agree). The sentences and the results are presented in Table 48.5.

Table 48.5 illustrates the subjective perceptions of participants on financial literacy. The results reveal that the participants have a fairly good perception, nevertheless, some respondents show some doubt about the reasonability of their spending behavior.

Table 48.6 gives the result of t-test of perceived financial literacy scores by gender. We can conclude that there is no significant effect of gender on the perceived financial literary (p-value $= 0.319 > 0.05$).

48.4.3 Digital Financial Literacy

Finally, we analyze the participants' behavior regarding the use of digital media in financial transactions, the statements and the results are presented in Table 48.7. By analyzing this table, we found that the vast majority of respondents use mobile phones to make payments (65.9%), and use digital means in their financial transactions (88.1%).

The results of the Pearson chi-square independence test for hypotheses H3 and H4 are presented in Table 48.8. These results showed that there is not a significant

Table 48.7 Use of digital media in financial transactions

Statement	Descriptions	No. of respondents	% of respondents
Do you use your mobile phone to pay for goods and services? (grocery shopping, restaurant, gasoline, etc.…)?	Yes	205	65.9
	No	106	34.1
Do you use digital media to carry out your financial transactions/operations?	Yes	274	88.1
	No	37	11.9

relationship between gender and the use of the mobile phone to pay for goods and services (p-value = 0.956 > 0.05), and there is not a significant relationship between gender and the use of digital media to carry out your financial transactions/operations. (p-value = 0.799 > 0.05).

In the following analysis, we only considered respondents who use digital media in their financial transactions (274 participants). The results are presented in Table 48.9. These results show that the majority use the smartphone (52.6%), carry out their transactions without any help (81.8%), and 54.7% cancel or reverse a transaction/financial transaction without any help.

Table 48.8 Results of χ^2-test to examine hypotheses H3 and H4

Hypothesis	χ^2-test	p-value
H3	0.003	0.956
H4	0.065	0.799

Table 48.9 Digital financial literacy

Variable	Description	No. of respondents	% of respondents
What digital media do you use to carry out your financial transactions (choose only the option you usually use)?	Smartphone	144	52.6
	Desktop	40	14.6
	Laptop	85	31.0
	Tablet	5	1.8
Performs financial transactions/operations in digital media:	Without help	224	81.8
	With the help of your bank	34	12.4
	With the help of family or friends	16	5.8
Can you, if necessary, cancel or reverse a transaction/financial transaction that you have initiated (in digital media)?	Yes	150	54.7
	Yes, but with help	86	31.4
	No	38	13.9

48.5 Conclusions

Most of prior studies in the literature are mainly focused on financial literacy and on its different measures. However, the holding of financial information is not sufficient to ensure that individuals will made sound financial decisions. Additionally, the rapid development of digital technology requires that economic agents also hold digital literacy, in order to have an adequate understanding of digital finance. To analyze the level of financial of the Portuguese adults, we considered two perspectives: actual financial literacy (or objective) and perceived financial literacy (or subjective). Concerning the first approach, all the respondents answered at least at one question correctly. They perform poorly on the question related to spread and Euribor (only 37.6% know what the spread is, and only 27.7% know what the Euribor rate means). Concerning the second approach, the highest perception scores are about money management and about liquidity. Nevertheless, respondents reveal a less consistent perception about their ability to afford their spending levels when compared with the other statements. Regarding the use of digital media in financial transactions, our results show that a vast majority use mobile phones to make payments (65.9%), and use digital means in their financial transactions (88.1%). When we consider only respondents who use digital media in their financial transactions, the majority use the smartphone (52.6%) and carry out their transactions without any help.

In our research, hypothesis H1 is validated, while hypotheses H2, H3, and H4 are not validated. In fact, gender is relevant for objective financial literacy, but it is not relevant for subjective (or perceived) financial literacy. This result is not in line with prior research because some references pointed out that male use to be more confident and willing to take risks than female [4].

Further research could focus on the determinants of financial literacy and of digital financial literacy within the Portuguese context.

References

1. Agarwal, S., Qian, W., Yeung, B.Y., Zou, X.: Mobile wallet and entrepreneurial growth. AEA Papers Proc. **109**, 48–53 (2019)
2. Antonijević, M., Ljumović, I., Ivamović, D.: Is there a gender gap in financial inclusion worldwide? J. Women's Entrepreneurship Educ. **1–2**, 79–96 (2022)
3. Goyal, K., Kumar, S.: Financial literacy: a systematic review and bibliometric analysis. Int. J. Consum. Stud. **45**(1), 50–105 (2021)
4. Hohl, L., Bican, P.M., Guderian, C.C., Riar, F.J.: Gender diversity effects in investment decisions. J. Entrepreneurship **30**(1), 134–152 (2021)
5. Klapper, L., Lusardi, A., Van Oudheusden, P.: Financial literacy worldwide: insights from the standard & poor's rating services global financial literacy survey. International Bank for Reconstruction and Development & World Bank, 16 (2015). Available online: https://gflec.org/wp-content/uploads/2015/11/3313-Finlit_Report_FINAL-5.11
6. Lusardi, A., Mitchell, O.S.: Planning and financial literacy: how do women fare? Am. Econ. Rev. **98**(2), 413–417 (2008)

7. Lusardi, A., Mitchell, O.S.: Financial literacy and retirement planning in the United States. J. Pension Econ. Fin. **10**(4), 509–525 (2011)
8. Lyons, A.C., Kass-Hanna, J.: A methodological overview to defining and measuring "digital" financial literacy. Fin. Plan. Rev. **4**, e1113 (2021)
9. Morgan, P.J., Huang, B., Trinh, L.Q.: The need to promote digital financial literacy for the digital age. In the Digital Age (2019)
10. OECD: G20/OECD INFE Policy Guidance on Digitalization and Financial Literacy. OECD Publishing, Paris (2018)
11. Organization for Economic Cooperation and Development (OECD): skills matter: additional results from the survey of adult skills. In: OECD Skills Studies, OECD Publishing, Paris (2019)
12. Prete, A.L.: Digital and financial literacy as determinants of digital payments and personal finance. Econ. Lett. **213**, 110378 (2022)
13. Rahim, N.M., Ali, N., Adnan, M.F.: Students' financial literacy: digital financial literacy perspective. J. Fin. Bank. Rev. **6**(4), 18–25 (2022)
14. Rajdev, A., Modhvadiya, T., Sudra, P.: An analysis of digital financial literacy among college students. Pacific Bus. Rev. Int. **13**(5), 32–40 (2020)
15. Setiawan, M., Effendi, N., Santoso, T., Dewi, V.I., Sapulette, M.S.: Digital financial literacy, current behavior of saving and spending and its future foresight. Econ. Innov. New Technol. **31**(4), 320–338 (2022)
16. Sherraden, M.S.: Building blocks of financial capability. In: Birkenmaier, J., Curley, J., Sherraden, M. (eds.), Financial Education and Capability: Research, Education, Policy, and Practice, pp. 3–43. Oxford University Press, Oxford, UK (2013)
17. Swiecka, B., Yeşildağ, E., Özen, E., Grima, S.: Financial literacy: the case of Poland. Sustainability **12**(2), 700 (2020)
18. Tony, N., Desai, K.: Impact of digital financial literacy on digital financial inclusion. Int. J. Sci. Technol. Res. **9**(1), 1911–1915 (2020)
19. Warmath, D., Zimmerman, D.: Financial literacy as more than knowledge: the development of a formative scale through the lens of bloom's domains of knowledge. J. Consum. Aff. **53**(4), 1602–1629 (2019)

Chapter 49
The Influence of Instagram on Consumer Behavior and Purchase of Home Decor Items in Brazil

Manuel Sousa Pereira, Silvia Faria, António Cardoso, Eulália Sabino, Jéssica Fonseca, and Renan Soler

Abstract Internet has changed the way people and brands communicate, produce, share content on social media and buy products or services. Several interaction platforms emerged, making it easy for individuals to exchange information. Facebook and Instagram are the most used networks all over the world; Instagram, in particular, has become one of the social networks with the higher number of followers/users, across different ages. The Internet has promoted changes not only in terms of communication, but also in terms of consumer buying behavior. Individuals started buying more online, so it is now essential to clearly understand the consumers' behavior and perceptions. To buy, they no longer use just websites or marketplaces; social networks that have been updated and allow them to complete a purchase are being widely used. Instagram, due to the exponential growth in the number of followers, is now used as a source of inspiration for the purchase of various items, highlighting decorative items, the subject of this study. We aim to understand the importance of social networks on purchasing decisions, focusing on Instagram commerce, more specifically on buying intention on decoration products in Brazil. The study uses quantitative and descriptive methods. Data collection was carried out through a questionnaire, built on the basis of the literature review carried out. According to the data obtained, it was observed that most women, aged between 25 and 34 who shop online, do access

M. S. Pereira (✉) · E. Sabino · J. Fonseca · R. Soler
Instituto Politécnico de Viana do Castelo, Viana do Castelo, Portugal
e-mail: pereiramanuel@esce.ipvc.pt

E. Sabino
e-mail: eulaliasabino@ipvc.pt

J. Fonseca
e-mail: j.fonseca@ipvc.pt

R. Soler
e-mail: renansoler@ipvc.pt

S. Faria
Universidade Portucalense, Porto, Portugal

A. Cardoso
Universidade Fernando Pessoa, Porto, Portugal
e-mail: ajcaro@ufp.edu.pt

J. L. Reis et al. (eds.), *Marketing and Smart Technologies*, Smart Innovation, Systems and Technologies 337, https://doi.org/10.1007/978-981-19-9099-1_49

Instagram several times a day, being influenced by the content that is shared on this social network. The study proves the growing importance of Instagram for brands, specifically those that offer articles and services in the area of home decoration.

49.1 Introduction and Theoretical Framework

With recent technological advancement, the Internet has changed the way people communicate with each other, how they produce, share and consume content, reducing borders, distances, overcoming barriers and bringing people together. These changes were extended to brands, in terms of their communication, the goods or services they offer to the market, the management of their distribution channels and, also, pricing policies and sales promotions. Through smartphones and other mobile devices, data transmission has become faster and more accessible, facilitating communication worldwide.

With the growing popularity of Internet access, several social networks have emerged that have facilitated the exchange and consumption of information, as well as products and services. A kind of democratic, accessible environment was created, capable of promoting changes in various segments of society, allowing everyone to create and share content. Thus, the democratization and popularization of access to the Internet and social networks changed the way brands communicate and, also, the entire communication process and consumer behavior itself. We are witnessing a transformation of traditional business models, and sales and promotion can now be done through social networks; this has brought several changes to society, including the possibility of self-expression and socialization through technological devices [1].

These changes have made it possible for any company to increase its competitiveness by adapting business models to the current digital and online scenario. Brands that developed marketing strategies are focusing content on its target and on the social network, in which it is most connected, betting on technology, get satisfied and delighted consumers [2].

One of the most used networks today, with more than 1 billion users (1221 in 2021), is Instagram [3], which is a social network for sharing photos and videos that enables different forms of interactions between users. The platform has become a diversified, interactive, personalized environment; with a wide variety of profiles available on it, such as fashion profiles, commercials, news, entertainment, information, nutrition, pets, among others, has been capturing the attention of a growing number of individuals. Instagram is a social network in which several companies, of different segments and sizes, have profiles, in order to be closer to their customers and potential customers. The objective is to make the company/brand, its products and services known and, in this way, attract the attention of consumers and lead them to buy [3]. People stay connected, all the times, through social networks, and therefore, they are up to date on special offers, products' features and promotional campaigns [4]. Thus, social networks have become a large-scale communication network that promotes interactivity between different peoples of the world; people share opinions, photos, reflections, moments in their personal and professional lives, achievements, problems, outbursts [5]. They are easily accessible, relatively inexpensive and extremely desirable and easy to use; people who do not have profiles on social networks are considered invisible and the same for brands [6, 7].

Studies done in 2021 considered Facebook, YouTube, WhatsApp and Instagram as the networks with the highest number of active followers in the world [3]. From approximately 8 billion people in the planet, 5.22 billion use mobile phones (66%), which exceeds Internet users through computers (4.66 billion, corresponding to 59.5%). Active users of social networks total 4.2 billion (53.6% of the world population); 4.15 billion of them connect to social networks from a mobile phone (98.8%), and the average global time spent on social networks, in 2020, was 2h25m; in 2021, it was 2 h 30 m [8]. On average, social media users have 8.4 accounts across platforms around the world [3].

Instagram is generally known for being a photo and video sharing social network, which was created by a Brazilian and an American in 2010 and sold to Facebook in 2012 (Meta, now), in a billion-dollar transaction: US$ 1 billion [9, 10]. Currently, Instagram is recognized as a social network that supports and sustains the social behaviors of its users: parallel communication, interaction, leisure and entertainment. It is an interesting space, full of inspiration, opinions and experiences of other people [6, 10].

This increase in popularity and, consequently, in the number of users made this social network widely used for commercial purposes. Brands realized its potential and started to consider it as an important marketing tool. The companies created commercial profiles and started to make content available for free. Using the various features available on the platform, managers began to get essential data about their customers (current and potential). With these data, they are now able to create and understand commercial profiles and, in this way, adopt marketing strategies that allow a proper answer, in line with the needs and desires of consumers [11].

With the technological revolution (the information age), a new type of consumer has emerged: an individual who has quick access to various information about products, services and companies in general, having ample bargaining power: the possibility to choose the brand(s), store(s), or any other means/touching points that make the purchase process easier and offer a continuous positive experience. This requires marketers to know the main trends in consumer behavior, devoting full attention to them [12].

At least two issues can be identified that make it possible to distinguish online consumers from offline consumers. First, online consumers need to interact with technology to buy the goods and services they need. The physical store environment is replaced by an electronic shopping environment or, in other words, by the use of information systems (IS). This gives rise to technical and human–computer interaction issues [2]. Second, a greater degree of trust is needed in an online shopping environment than in a physical store. It is commonly accepted that trust is an important issue; it reduces any feeling of uncertainty, doubt and risk that arise when (1) the store is unknown, (2) the brand is unknown and (3) the quality of the product is unknown. If in the physical store, contact with products and employees can decrease any discomfort and/or mistrust, when buying online, doubts and insecurities increase, and therefore, the need for the selling entity to gain trust and reduce any perceived risk in the mind of the consumer is high. AI, Augmented Reality, Robots, among others, need to be used by brands in order to exceed expectations and ensure a good shopping experience [2].

Actual clients are being Blended—ROPO: research online, purchase offline and vis-a-vis [13]—therefore, adopting an omnichannel strategy is crucial for any company survival and competitiveness [14]. E-commerce (e.g., own site, market-places, dropshipping) and Digital Marketing are other touching points for clients to get information on products, services, prices and to decide for the better alternatives (what to buy and where to buy), ending the buying process [14, 15]. Managing all the touching points in a consensual way (omnichannel strategy) allows the possibility for customers to make purchases anywhere in the world—any virtual store on the Internet, some social media networks allowing to purchase goods or services—without leaving home, just by using a smartphone [14, 16]; this makes the purchase journey easier and leads to a more confident consumer, believing that he/she made the right choice [7].

Social media gave consumers new opportunities and make them engage in social interaction on the Internet. Consumers use social media and this is something that brands no longer ignore; in fact, social media development has led to the development of e-commerce into social commerce [17]: social media improves the social interaction of consumers, promoting trust and repurchase intention [17], and the completion of the purchase can take place offline or online (e.g., brand website, marketplace, social network). Instagram is one of the most used social media networks all over the world [3] and is part of the new means of social commerce for many companies to sell goods and services [18, 19]. Thus, understanding the growing acceptance and use of Instagram, we aimed at knowing if customers use this social media network to get information and buy home decor products and services.

49.2 Methods

The following research question was defined to meet the general purpose of the study: does the social network Instagram impact on consumer buying behavior and, in particular, does it influence the process of home decor products purchasing in Brazil?

As specific objectives, the research aims to: (1) identify the Instagram's frequency of use; (2) analyze the influence of Instagram on consumption; (3) identify the frequency of online purchases; (4) find out if respondents follow stores and home decor brands on Instagram; (5) try to evaluate where do consumers prefer to buy home decoration items—physical or virtual stores; (6) know the advantages of Instagram in the decision-making process to buy home decor items.

In order to verify the influence of Instagram on consumer purchase behavior of home decor products, descriptive research was carried out that analyzes, reports and correlates the facts that the researchers decided to be important indicators; specifically, data collection was made through a questionnaire, built from the literature review and intending to assess which factors impact the consumer's decision to buy online, focusing on Instagram. The questionnaire was pre-tested with ten participants to analyze possible changes. As there were no problems, the questionnaire was made available online (Google Forms), in December 2021 and January 2022. A filter question was used—age of the respondent—to ensure answers from individuals aged 18 or above.

Survey respondents who are social media users and buy home decor products were targeted among the authors' network contacts. To be considered, individuals needed to (1) be 18 or above, (2) use social networks and (3) use to buy home decor products. At the beginning of the questionnaire, participants were clearly informed about the nature of the study and were required to consent participating and data treatment. Complete confidentiality was guaranteed.

We used a non-probability sample. Two hundred and three people were participated in the questionnaire; 193 of the respondents claimed to have an Instagram account. About 67.9% of the respondents are female and 32.1% male. The predominant age group was 26–34 years old (37.8%), followed by 35–44 years old (26.4%), 18–24 years old (17.6%), 45–54 years old (11.9%), and from 55 to 64 years old, we had 6.3% of the respondents.

The majority of participants have higher education (74.6%) and only 1% has only elementary school. This was not a surprise, since participants were individuals belonging to the researchers' network. With regard to earnings, we obtained a slightly dispersed distribution, with 30.1% of the participants indicating a monthly salary above $5000 reais and only 4.8% reporting an income below $1000 reais.

49.3 Results and Discussion

Looking carefully at the average of the answers to the different questions, it was possible to reach an interesting and enlightening interpretation about the impact of social networks, with a focus on Instagram, in the decision process and consumer purchase of home decor items in Brazil.

When asked about the number of times they use Instagram; 50% of individuals reported that they use the network several times a day and 34.7% indicated that they access it daily. It was also possible to verify that 14% of individuals reported that they are on Instagram for more than 7 h a week, 14.5% of individuals claim to be on the network between 5 and 6 h a week, 17.1% connect between 3 and 4 h, 36.8% stay a total of 1 to 2 h per week and 17.6% indicated that they stay for less than 1 h. These data seem to corroborate the average daily hours indicated in recent studies (an average of 2:30 min spent on NET, daily [8]) and highlight the importance for companies to include social media (particularly, Instagram) in their marketing and communication plans as an important touching point to share information, understand consumers trends and even impact on consumers' choices and intention to buy.

When asked whether they consider Instagram to influence them to consume, 71% of respondents said yes. The average of the responses obtained indicates that 58.5% of respondents follow some profiles of furniture stores and home decor on Instagram; of these, 43.5% follow these profiles daily, 39.1% follow them weekly and only 13.9% indicated that they rarely do it. Our study seems to reveal that a considerable number of individuals use Instagram and like to access information about home decor and products; 71% of the people analyzed consider that they are influenced to consume through what they follow on Instagram. Then, focusing on the existing home decoration profiles on Instagram, research showed that they undoubtedly influence 51.3% of the people interviewed; 26.9% of individuals reveal some degree of influence and 21.8% consider that they do not feel influenced by the profiles and respective contents shared. In other words, most participants report that they are impacted by what is shared on this social network.

We asked participants how many of them use to make purchases online: 85.3% have already made purchases online. Of these 85.3%, 40.3% use to shop monthly, 40.3% shop weekly, 7.3% do so every fortnight and 5.8% sporadically. About 6.3% indicated that they rarely buy online. When asked whether they prefer to buy home decor items in online stores and/or physical stores, we found that 76.4% choose to buy in physical stores and only 23.6% prefer online stores. Nevertheless, all confirmed doing offline and online researches before deciding where to buy (ROPO). This is in line with the current client definition—Blended [13, 14].

Most people interviewed agree that Instagram is useful to search for information about brands, products and services they are considering to buy; 49% fully agree on the importance of this social network to gather diverse information and make comparisons, 21% agree and only 1% claimed to disagree. When asked whether or not the fact that a brand has a presence on Instagram is relevant, 22.5% of respondents said that they prefer brands with a profile on the network, and 21.5% do not have an opinion; however, 56% of the participants indicated that it was indifferent to them if a brand does not have a profile on Instagram. In other words, not having an Instagram profile does not make the 56% of the individuals that constitute our sample to give up buying the brand's products, online (e.g., e-commerce) or offline (e.g., physical store).

Even if the objective at first is not to buy, 42.4% of respondents follow the profiles of their favorite brands, the ones that they feel match their lifestyle. Photos of decor items and environments with planned furniture are great sources of inspiration for consumers; they are also a way for companies to make their products known, capture the customers' attention and reinforce brand awareness. Therefore, we can say that according to data obtained in this study, it is important that brands invest in high-quality photos and valuable content for their target, allowing or not the purchase, but creating desire through scenarios considered attractive. This seems to be even more important in terms of home decor items: Instagram's creative content influences the vast majority of respondents (37% of individuals strongly agree, 28.6% agree and only 5.2% strongly disagree). That said, the social network really inspires those who like home decor items.

49.4 Conclusion

The main objective of this research was to analyze the influence of Instagram on the consumer's decision to buy home decor items. The data obtained allowed to trace a profile of Instagram users in Brazil focusing on home decor products: gender, average age, income level, frequency with which they access the network, online shopping habits vs. decision process and purchase intention. According to the sample obtained, we can say that most users are female, mostly aged between 25 and 34 years old, having a high education level, accepting or being predisposed to make online purchases, connecting to Instagram several times a day and considering that they are influenced by the content shared on the network.

After analyzing the answers to each question, it appears that a significant number of participants give Instagram a high utility in terms of searching and gathering information about brands, products and trends. In terms of decoration, 45% of the participants stated that they use the network as a source of inspiration, trying to replicate decoration trends at home. The importance of the quality and creativity of the content shared by decoration brands on the social network in increasing the purchase intention was generally accepted.

Instagram is associated with a greater interaction between individuals and companies, greater visibility for brands, lower costs (e.g., fewer trips to physical stores) and less time spent. The interaction on Instagram manages to inspire and impact the consumption of household items through quality content. Many of the participants reported feeling inspired and motivated to buy goods and/or invest in the decoration of their homes by accessing this network.

Despite the recognition of Instagram's ability to capture, retain attention and drive consumers to purchase, it was found that there is still a certain resistance from consumers to buy home decor items online, preferring to do so in physical stores. However, participants prove the concept of Blended consumer, as most indicated that they do online research before deciding to buy.

The main limitations of the study are the fact that it is restricted to a single country and that the sample used has a number of participations below the desired and is the result of the researchers' contact network (non-probabilistic), which prevents the extrapolation of the results. It is recommended to replicate this study with a larger sample and to extend the research target to other countries and types of products. It would also be interesting to develop a qualitative study from a business perspective (e.g., market experts, using the Delphi model) in order to also understand the companies' perspective about Instagram importance to reach customers, provide leads and turn them into actual clients.

References

1. Recuero, R.: Redes Sociais na Internet. Ecompós **2**(2003) (2009). https://doi.org/10.1590/S0102-88392005000300006
2. Cruz, W.L.M.: Crescimento do e-commerce no Brasil: desenvolvimento, serviços logísticos e o impulso da pandemia de Covid-19. GeoTextos **17**(1) (2021). https://doi.org/10.9771/geo.v17i1.44572
3. https://www.linka.com.br/analytics/relatorio-global-do-digital-2021, retrieved at 12th September 2022
4. Felipe, B., Lins, E.: A evolução da Internet : uma perspectiva histórica. Caderno ASLEGIS **48**, 11–45 (2013)
5. Colares, D.M., Braga, M.M.S.: Redes sociais e o marketing de relacionamento: o uso da Facebook na relação editora/leitor. Intercom – Sociedade Brasileira de Estudos Interdisciplinares Da Comunicação XVII Congresso de Ciências Da Comunicação Na Região Nordeste – Natal-- RN – 2 a 4/07/2015, 12 (2015). https://www.portalintercom.org.br/anais/nordeste2015/resumos/R47-2030-1.pdf
6. Kotler, P., Kartajaya, H.,Setiawn, I.: Marketing 5.0: Technology for Humanity (2021). ISBN-10:1119668514
7. Eliab, J., Santos, D., Silva, D.A.: O Instagram Como Meio De Influência Na Decisão De Compra Do Consumidor. Universidade Federal Rural Do Semi-Árido Pró-Reitoria De Graduação Departamento De Ciências Exatas, Tecnológicas e Humanas-Dceth Bacharelado Em Sistemas De Informação Marketing Digital (2020)
8. https://www.terravista.pt/dados-redes-sociais/, retrieved at the 13th September 2022
9. G1: Instagram faz 10 anos como uma das maiores redes sociais do mundo e de olho no TikTok, para não envelhecer | Tecnologia | G1 6, 1–13 (2020). https://g1.globo.com/economia/tecnologia/noticia/2020/10/06/instagram-faz-10-anos-como-uma-das-maiores-redes-sociais-do-mundo-e-de-olho-no-tiktok-para-nao-envelhecer.ghtml
10. Liao, S.H., Widowati, R., Cheng, C.J.: Investigating Taiwan Instagram users' behaviors for social media and social commerce development. Entertainment Comput. **40**, 100461 (2022). https://doi.org/10.1016/J.ENTCOM.2021.100461
11. Sampaio, V.C.F., Tavares, C.V.C.C.: Marketing digital: o poder de influência das redes sociais na decisão de compra do consumidor universitário da cidade de Juazeiro do Norte-CE. Revista Científica Semana Acadêmica, 1–26 (2017)
12. Lourenço, P., Rodrigues, E.C.C., Lima, C.M.: A influência do Instagram no comportamento do consumidor online. R. Adm. Faces J.Belo Horizonte **19**, 89–102 (2020)
13. Faria, S., Ferreira, P., Carvalho V, Assunção, J.: Satisfaction, commitment and loyalty in online and offline retail in Portugal. Eur. J. Bus. Social Sci. **Vl.2**(7), 49–66 (2013). http://www.ejbss.com/recent.aspxISSN:2235-767X
14. Albertin, A.L.: Comércio eletrônico: modelo, aspectos e contribuições de sua aplicação. Revista de Administração de Empresas **40**, 108–115 (2000)
15. Bravo, R.A.G.: E-commerce : a influência da Confiança na Intenção de Compra Online. 197. Dissertação em Publicidade e Marketing, Escola Superior de Comunicação Social (2017), https://repositorio.ipl.pt/bitstream/10400.21/8475/1/Disserta%C3%A7%C3%A3o%20Rafaela%20Bravo.pdf, retrieved at 27th July 2022
16. Neslin, S.A.: The omnichannel continuum: integrating online and offline channels along the customer journey. J. Retail. Vl. **98**(1), 111–132 (2022). https://doi.org/10.1016/j.jretai.2022.02.003
17. Hajli, M.N.: A study of the impact of social media on consumers. Int. J. Market Res. Vl **56**(3), 387–404 (2014). https://doi.org/10.2501/IJMR-2014-025

18. Herzallah, D., Muñoz-Leiva, F., Liébana-Cabanillas, F.: Selling on Instagram: factor that determine the adoption of Instagram commerce. Int. J. Human-Comput. Interaction **Vl.38** (11), 1004.1022 (2021). https://doi.org/10.1080/10447318.2021.1976514
19. Staniewski, M., Awruk, K.: The influence of Instagram on mental well-being and purchasing decisions in a pandemic. Technol. Forecast. Soc. Chang. **174**, 121287 (2022). https://doi.org/10.1016/j.techfore.2021.121287

Chapter 50
Study of the Online Fashion Consumer Shopping Journey and the Effects of Digital Communication Media: Case Study MO Online

Isabel Valente and Mafalda Nogueira

Abstract The following investigation aims to study the consumer behavior, more specifically of digital communication media and Online Consumer Shopping Journey. In this exploratory-natured study of quantitative methodology, the collection of data was obtained through a questionnaire where 166 valid answers were processed. This study resulted in important contributions, namely the construction and validation of the assessment instrument that combines the constructs of online shopping journey and digital media. In addition, it contributed as a step to solving a need of the MO Online, the company in which the internship took place—to know more about e-commerce—and where and how the brand's potential consumers were present in an online environment. The obtained findings helped to build a digital communication plan for the internationalization of MO's e-commerce, according to the profile and digital behavior of its target audience. It was found that digital media shape the online shopping journey. This is a study that can serve as a premise for future investigations but, above all, it is focused on a real need of the fashion retail market—knowing how to reach its potential consumers more effectively. It also has value on a practical level, as it presents very relevant insights for fashion marketing managers, providing them with relevant information that they can adopt daily on their digital platforms.

50.1 Introduction

With the proliferation of internet use among a large part of the population around the world, the creation and growth of the digital environment have become the necessity of the moment for developed and emerging economies [45].

I. Valente (✉) · M. Nogueira
IPAM, Porto, Portugal
e-mail: isabelmmvalente@gmail.com

M. Nogueira
e-mail: mafalda.nogueira@ipam.com

J. L. Reis et al. (eds.), *Marketing and Smart Technologies*, Smart Innovation, Systems and Technologies 337, https://doi.org/10.1007/978-981-19-9099-1_50

The role of the environment is particularly important for e-commerce companies, both in the industrial goods market and, above all, in the consumer goods market, such as fashion retail [12].

In the modern digital world, the digitalization of retail has phenomenally changed the reasons for shopping, making it more convenient and accessible [12]. Online shoppers have a more favorable perception of convenience (in terms of access, search, transaction and purchase) than physical store shoppers [3, 19].

The pandemic experienced in the last two years also changed the consumption paradigm through the closing of physical stores for months. According to the Global Consumer Insights Pulse Survey, conducted in 22 countries during the first quarter of 2021, most consumers are currently adopting a more digital lifestyle compared to the pre-COVID period [11].

Fashion products represent an important category in e-commerce and have witnessed the development of a new form of online marketing [12, 29].

Omnichannel marketing recognizes that consumers typically go through the purchase journey in each transaction across stores, computers, smartphones, tablets and social media. These different channels and touchpoints are used constantly, interchangeably and simultaneously by customers and the company [44].

Brands should therefore begin to consider communication congruence and customer experience when switching from one channel to another and whether they are maintaining brand uniformity and consistency through channel-specific marketing attributes in order to increase the value in the purchase process [35].

In the case under study, the possibility of expansion, through digital media and points of the purchase journey, of a nationally recognized fashion brand will be evaluated.

The main topics present in this report dedicated to fashion e-commerce explore some research gaps in a professional context, such as: the role of technology and touchpoints in the online shopping journey and the personalization of the communication mix.

50.2 Literature Review

50.2.1 Online Consumer Journey and Touchpoints

Digitization has influenced both companies and customers, improving the digital aspects of their assets, processes and value chains [7, 50].

This behavior is largely due to the maturation of technology in recent years. The drivers of this maturation include increased quality and ease of internet, more advanced digital content, more convenient payment methods and carefully customized customer journeys. All these factors contribute to more satisfying e-commerce experiences for customers [16]. Logistics, which is the backbone of e-commerce, has also improved [50].

Digitization, in general, offers companies attractive strategic opportunities. For the first time, almost all companies can reach global markets with digital e-commerce technology platforms [27].

Rapid growth and technological innovation have given consumers the opportunity to interact with businesses through multiple digital channels. The use of advanced technologies can be an influencing factor of the consumer experience throughout your shopping journey [25]. In this context, the development of e-commerce, in particular that of fashion retail, is an advantage in that it internationalizes the business, since companies have the facility to reach customers worldwide [12]. That said, organizations integrate a variety of technologies into their multichannel strategies through the combination of offline and online resources, in order to add value to consumers and thereby attract and retain more consumers [26, 46].

Making an online purchase has advantages, one of which is the fact that it is easier and more convenient for individuals to choose the product they want, saving time and avoiding queues in physical stores [19]. Still, as advantages, it is observed that certain products are cheaper when purchased online and the consumer can perform this process anywhere in the world. Supporting this perspective, Ramachandran et al. [38] report that online shopping improves the effectiveness of the business, strengthens the growth of the brand and its notoriety and accelerates the internationalization of operations, expanding the company's distribution network.

Online consumer behavior on the shopping journey is influenced by demographic factors and characteristics [30], risk perceptions, i.e., the risk and insecurity of the consumer in sharing their personal data online and by trust [13, 37]. Furthermore, the involvement and interaction of the consumer throughout the purchase journey [23] and the technology used by it are also factors that influence the behavior of the online consumer [43].

Thus, it can be confirmed that the online shopping journey is not linear. Contrary to the "traditional" journey, consumers can interact with different touchpoints at different stages of the shopping journey [26, 44].

The online-focused shopping journey can be defined, according to Wolny and Charoensuksai [46], as a description of the consumer experience and their interaction with different touchpoints that present it to the product and/or service. Wright [47] reinforces that the online consumer journey translates into the consumer experience inserted in digital touchpoints, when he browses the internet and finds the website or other digital platform of a company, makes a purchase, processes the monetary transaction and ends in the post-purchase experience. According to Lemon and Verhoef [25], touchpoints translate into the variety of individual contact points between the company and the consumer throughout the purchase journey.

Following the same logic, Kotler et al. [22] describe touchpoints as an aspect, in which companies must invest to ensure that consumers experience positive experience and a shopping journey consistent with what the brand intends to communicate. Supporting this conclusion, Wright [47] adds that touchpoints are fundamental to measure and parameterize the consumer journey depending on the business practiced by the company. In this way, digital touchpoints have been added to describe the individual digital environment used by the consumer, such as the brand's website.

However, the term digital channel corresponds to the group of touchpoints used by the consumer to interact with the brand. In this sense, the touchpoints of greatest relevance and use by organizations are: inter alia, the website, the mobile application, social media, e-mail marketing through newsletters, digital advertising, among others [41].

Thus, within the consumer's shopping journey, different touchpoints can be identified. Four types of touchpoints can be observed: brand ownership, partner property, customer property and social touchpoints. Thus, the consumer can interact with each of these categories of touchpoints throughout the shopping experience since, depending on the nature of the product or service and the purchase journey itself, the impact and importance of each category of touchpoints may vary at each stage of the purchase journey [2, 15, 25].

Since touchpoints are an integral part of the digital communication strategy and, consequently, in the digital media used, in the next section presented, digital marketing communication will be explored.

50.2.2 Digital Marketing Communication

Within the communication mix, we will delve only the digital media, since it is part of the core concepts of the research carried out.

As mentioned above, in the age of omnichannel behavior, the shopping journey varies throughout its process, fruit of the different channels and digital media that the consumer has at his disposal [35].

The introduction and use of technology have transformed the dynamics of interpersonal communication, both in everyday life and in the professional sphere. Digital marketing and the emergence of new tools and media are two of the indicators that show this transformation fueled by the development of the Internet [8].

Digital communication can be defined as the form of communication that allows promoting products and services through online channels, which involves marketing activities such as image sharing or the publication of online information [48].

Kim and Ko [21] present the elements that guide marketing and communication activities in digital channels, which are personalization, proximity, interaction, personalization and entertainment. These channels appear to have an extremely positive impact on brand awareness, brand membership, brand loyalty and perceived quality [9]. These platforms can drive business sales and profit, being a powerful tool to create brand-level value [10]. The same authors reinforce that communication through these vehicles allows a large-scale reach at a lower cost compared to traditional means of communication, besides allowing to attract and attract new consumers, as well as retain them.

The use of digital channels in value creation should be strategic, requiring planning to maintain a high interaction with consumers, in order to allow easy further monitoring [4, 18].

Currently, there are several channels, with exponentially different potentialities, within the widely growing communication tool, which is digital communication. In this case under study, we will explore the role that digital media play on the different phases of online consumer journey.

50.3 Conceptual Model

This research aims to investigate the consumer behavior, regarding the online shopping journey and digital communication. Thus, it was possible to understand and create, in a more personalized way to the case in question, which digital communication strategy is most appropriate for the growth of MO Online e-commerce in the Iberian Peninsula market.

This objective was born from the following research question, proposed by the company:

What communication strategy should we adopt for growth and entry into an international market of MO Online, as a fashion retail e-commerce brand?

After explaining the conceptual model of this study, we intend to theoretically frame the scales used in the survey carried out on the brand's consumers (Table 50.1).

50.4 Methodological Decisions

50.4.1 Methodological Approach Case Study

To explore the research goal, the adopted methodology consists of a case study of the MO Online company. Thus, a case study is a method that allows the realization of an adequate study for an entity. In this context, the case study basically corresponds to a survey that can be used as a means of identifying the key issues of a company that deserve further investigation [49].

Considering what has been presented, it is essential for this study to use quantitative techniques, namely the questionnaire survey.

50.4.2 Data Collection

The technique of data collection through a questionnaire survey can be defined as a particular questioning about a given situation where individuals are included. Thus, the questionnaire survey translates into a rigorous data collection technique, both in terms of the questions asked and answers, as well as their order, in order to ensure

Table 50.1 Theoretical framework of the questionnaire

Constructs	Questions	Question type	Reference authors
A —online consumer shopping journey			
Recognition of need	1. Digital channels lead me to realize the need to buy clothes and accessories	Scales 1 to 5, in which 1 represents "totally disagree" and 5 represents "totally agree"	[9]
Search for information	2. I increasingly use digital media to search for information about clothing and accessories 3. Digital channels offer me a good amount of information about clothing and accessories 4. Digital media reduce my time to search for information about clothing and accessories 5. Digital media help me reduce visiting time to physical clothing and accessories stores 6. Digital means of clothing and accessories are interactive by nature 7. Digital media offer me more relevant information about clothing and accessories 8. Digital media are easy to use when I want to find information about clothing and accessories		

(continued)

Table 50.1 (continued)

Constructs	Questions	Question type	Reference authors
Alternatives evaluation	9. I find it easy to find expert advice on clothing and accessories in digital media 10. Comments made by other consumers about clothing and accessories in digital media help in my product evaluation 11. I feel positive about the clothing and accessories brand when I have the most positive reviews in digital media 12. I consider it important to agree on the recommendations of clothing and accessories brands in digital media 13. Clothing brands and accessories are easily comparable in digital media		

(continued)

Table 50.1 (continued)

Constructs	Questions	Question type	Reference authors
Moment of purchase	14. I find it easy to decide to purchase a clothing and/or accessories product 15. I find it easy to select a brand of clothing and accessories 16. I intend to join online forums, blogs and social media groups about clothing and accessories 17. Subscribe to newsletters for clothing and accessories brands 18. I participate in online discussions about clothing and accessories 19. I buy clothing and accessories based on the information contained in the digital media 20. I buy clothes and accessories online 21. I find it easy to make available online payment methods 22. My decision to purchase clothes and accessories can be anticipated 23. The clothing and accessories brand that selecionei, pode mudar após a pesquisa online		

(continued)

Table 50.1 (continued)

Constructs	Questions	Question type	Reference authors
Post-purchase experience	24. I give feedback on factors of satisfaction and dissatisfaction in the online purchase of clothing and accessories 25. I consider the use of digital media to purchase clothes and accessories a good idea 26. I recommend buying clothes and accessories online, through digital channels, to friends and family		
B—digital communication			
Digital communication media	27. Which of the digital media do you consider triggering the need to purchase clothing and accessories? 28. Which of the digital media do you use to search for information about clothing and accessories? 29. Which digital media do you use to compare clothing and accessories purchase alternatives? 30. Which of the digital media do you use when you intend to buy clothing and accessories? 31. Which of the digital media do you use after purchasing clothing and accessories (e.g., to exchange and/or make complaints)?	Options—social media, electronic word-of-mouth, websites, paid media advertising, e-mail, blogs and forums	Adaptado [9, 18]

the comparability of respondents' answers [6]. In this sense, the objective of this technique is to obtain information about a certain population to be investigated in a systematic and orderly way [31].

That said, for the present study, the survey was carried out among online shoppers, in various groups, forums and social media, obtaining a total of 166 valid responses. Thus, the sampling process applied to the study is randomly probabilistic.

50.5 Results

50.5.1 Characterization of Respondents

Starting with the characterization of the sample, in relation to the gender of the respondents, it was verified that 86% of the individuals are female and the remaining 14% are male. These data are somewhat expected, since the brand under study is a fashion brand that is usually more acquired by female individuals. It should be noted that this characterization, as well as that of the other variables, was performed considering the sample considered valid, with a total frequency of 166 individuals.

Most of the sample, representing 59% of the total, is in the age range between 28 and 42 years. The remaining sample is between 18 and 27 years old (33%), between 43 and 57 years (7%) and only 1% are over 57 years old.

Regarding educational qualifications, 10% of the individuals under analysis completed secondary education and 90% completed some university education degree (including bachelor's, master's and/or doctorate).

Finally, regarding occupation, most of the sample (88%) performs work, which is being therefore considered an active population, and consequently, consumers and/or potential consumers of the specific products: 65% of the individuals are employees, 13% are student workers, 10% are self-employed, 9% are students and the remaining 2% are unemployed.

In a total, the sample consists essentially of women aged between 28 and 42 years, who have finished higher education and are active workers.

50.5.2 Instrument Validation

In this validation, questions that did not meet the recommended criteria and that, therefore, could negatively influence the validity of the scale were excluded. This variation may be related to the fact that the scale has been adapted from a scale applied to populations other than the population of the Iberian Peninsula, and it is not advisable to use it in this population.

Originally, the instrument was composed of 25 items, and through factor analysis and after being removed the items whose commonalities ditches were less than 0.5, a scale with four factors was obtained, composed of the 18 items that presented factor loadings greater than 0.5. The factors were characteristic of:

- Factor 1—this factor includes five questions that refer to the Information Search phase.
- Factor 2—this factor includes three questions that refer to the Alternative Evaluation phase.
- Factor 3—this factor includes five questions that refer to the Moment of Purchase phase.
- Factor 4—this factor includes two questions that refer to the Post-Purchase Experience phase.

After this change, it turns out that the Online Consumer Purchase Journey scale has a KMO value of 0.810, which corresponds to a value within the recommended parameters. Therefore, it is considered that the factorial recommendation for this construct is good. Regarding Bartlett's measurability test, it was observed that the variables correlate, and the significance level is less than 0.05 (sig = <0.001) in all items. It is also important to analyze that, according to the commonalities presented, all values are higher than 0.5 in all variables, ranging from 0.512 to 0.819. This means that these statements have a strong relationship with each other.

Next, the matrix of rotating components was analyzed, which allows us to understand the weight of each variable compared to the construct analyzed. The rotating component matrix shows the optimal solution of the instrument after all the seat changes have already been made. Thus, through this method, results with values higher than 1 and items with factor loadings greater than 0.5 were presented. Thus, it can be concluded that the values are higher than 0.5, ranging from 0.530 to 0.870.

Therefore, after these changes, it is possible to calculate the construct of the Online Purchase Journey, to be used in the validation tests of the hypotheses of the conceptual model, grouping it into four factors and reinforcing the value obtained from the total explained variance of 64.175%.

The reliability of the scales used was then analyzed using Cronbach's alpha calculation. When the value of this index is greater than 0.7, it means that there is appropriate reliability. A value of 0.870 is observed, which indicates internal consistency and a high robustness of the scale under study for the research sample.

50.5.3 Consumer Behavior Analysis: Online Shopping Journey and Digital Media

Regarding the survey, it was found that the average number of respondents' responses focused on level 4 (I agree), which indicates that, overall, respondents show that the various phases of the Online Consumer Day have positive characteristics. In this sense, the item that presented an average of the most positive responses was the item "I buy clothes and accessories online" (M = 4.482; SD = 0.703). On the other hand, there is one of the items that contradicts the global scenario, demonstrating more "neutral" responses from the Likert scale, which is the "I buy clothing and accessories based on the information contained in digital media" (M = 3.42; SD = 1.05), demonstrating that individuals do not agree to disagree that the information on online channels influences their purchase of clothing and accessories.

Regarding the Information Search phase, respondents on average tend to agree that they use digital media to search for information on clothing and accessories ($x = 4.46$) and that, during the search, these channels offer a good amount of information ($x = 4.25$). They also recognize the convenience of digital media, agreeing that they reduce the time of searching for information ($x = 4.00$), which help reduce the time of visiting physical stores ($x = 4.28$). However, when the accessibility and relevance aspect are seen, there is no longer so much consistency in agreeing that digital media are interactive ($x = 3.64$).

Moving on to the Alternative Evaluation phase, on average, respondents agree that the comments ($x = 4.18$) and recommendations ($x = 3.93$) made by other consumers in digital media help the evaluation of the product and that they feel more positive about the brand when there are more positive comments in the same media ($x = 4.03$).

At the time of purchase, consumers find online payment arrangements ($x = 4.43$) easy. There is also a tendency to agree that the purchase decision can be anticipated by digital media ($x = 3.78$) and that the brand chosen to buy may change after an online search ($x = 3.89$). It is also verified that consumers do not consider that the information contained in digital media is linked to their buying behavior ($x = 3.42$) but that, despite this, there is a positive trend regarding the purchase of clothes and accessories online ($x = 4.48$). Finally, in the Post-Purchase phase, consumers consider that the use of digital means to buy clothes and accessories was a good idea ($x = 4.34$) and, in general, recommend the purchase of these items via digital means to family and friends ($x = 4.07$).

50.5.4 Descriptive Statistics: Digital Media Communication Preferences

To study the preferences regarding the use of the various digital media, at each stage of the online consumer journey, several specific questions were asked, and the descriptive statistics of which are broken down in the Table 50.2.

It should be stressed that several options could be selected on all questions, since during the journey, the respondent can use various means.

Considering the answers, it can be observed that:

- In the need recognition phase, respondents preferably use social media (80.1%), digital advertising (49.4%) and electronic word-of-mouth (34.3%). The least used medium is blogs and forums (4.8%).

Table 50.2 Descriptive statistics of the digital communication media

	N	Mean	Standard deviation	Selection frequency
1. *Which of the digital media do you consider triggering the need to purchase clothing and accessories?*				
1.1. Social media	166	1.20	0.400	133 (80.1%)
1.2. Electronic word-of-mouth		1.66	0.476	57 (34.3%)
1.3. Websites		1.67	0.472	55 (33.1%)
1.4. Digital advertising/paid media		1.51	0.501	82 (49.4%)
1.5. E-mail		1.77	0.425	39 (23.5%)
1.6. Blogs and forums		1.95	0.215	8 (4.8%)
1.7. None of the previous options		1.95	0.227	9 (5.4%)
2. *Which of the digital media do you use to search for information about clothing and accessories?*				
2.1. Social media	166	1.50	0.502	83 (50%)
2.2. Electronic word-of-mouth		1.64	0.482	60 (36.1%)
2.3. Websites		1.23	0.421	128 (77.1%)
2.4. Digital advertising/paid media		1.88	0.327	20 (12%)
2.5. E-mail		1.92	0.269	13 (7.8%)
2.6. Blogs and forums		1.90	0.304	17 (10.2%)
2.7. None of the previous options		1.97	0.171	5 (3%)
3. *Which digital media do you use to compare clothing and accessories purchase alternatives?*				
3.1. Social media	166	1.64	0.482	60 (36.1%)
3.2. Electronic word-of-mouth		1.73	0.446	45 (27.1%)
3.3. Websites		1.25	0.436	124 (74.7%)
3.4. Digital advertising/paid media		1.92	0.269	13 (7.8%)
3.5. E-mail		1.99	0.109	2 (1.2%)
3.6. Blogs and forums		1.93	0.260	12 (7.2%)
3.7. None of the previous options		1.95	0.227	9 (5.4%)
4. *Which of the digital media do you use when you intend to buy clothing and accessories?*				
4.1. Social media	166	1.71	0.455	48 (28.9%)
4.2. Electronic word-of-mouth		1.86	0.347	23 (13.9%)
4.3. Websites		1.10	0.296	150 (90.4%)
4.4. Digital advertising/paid media		1.95	0.227	9 (5.4%)

(continued)

Table 50.2 (continued)

	N	Mean	Standard deviation	Selection frequency
4.5. E-mail		1.97	0.171	5 (3%)
4.6. Blogs and forums		1.98	0.154	4 (2.4%)
4.7. None of the previous options		1.99	0.078	1 (0.6%)
5. *Which of the digital media do you use after purchasing clothing and accessories (e.g., to exchange and/or make complaints)?*				
5.1. Social media	166	1.79	0.409	35 (21.1%)
5.2. Electronic word-of-mouth		1.89	0.319	19 (11.4%)
5.3. Websites		1.30	0.458	117 (70.5%)
5.4. Digital advertising/paid media		1.95	0.215	8 (4.8%)
5.5. E-mail		1.89	0.319	19 (11.4%)
5.6. Blogs and forums		1.99	0.109	2 (1.2%)
5.7. None of the previous options		1.85	0.359	25 (15.1%)

- In the Information Search phase, respondents preferably use brand websites (77.1%) on social media (50%) and electronic word-of-mouth (36.1%). The least used means at this stage are e-mail/newsletter (7.8%).
- In the Alternative Evaluation phase, respondents preferably use the websites of the various brands (74.7%), social media (36.1%) and electronic word-of-mouth (27.1%). The least used means at this stage ar e-mail/newsletter (1.2%).
- In the buying phase, respondents preferably use the websites of the various brands (90.4%) and social media (28.9%). The least used means at this stage are blogs and forums (2.4%).
- In the Post-Purchase phase, respondents preferably use the websites of the various brands (70.5%) and social media (21.1%). The least used means at this stage are blogs and forums (1.2%).

50.6 Discussion and Conclusions

50.6.1 Discussion of Results

Through the analysis of our conceptual model and, more specifically, the problem presented by the company, it can be partially verified that, for the study public, digital media have an impact on the various phases of the Online Fashion Shopping Journey. This validation was based on descriptive statistics of the results of the questionnaire, more specifically the questions of the construct of the Online Consumer Shopping

Journey, which indicate that respondents consider that online media are beneficial in the various phases of the journey.

Most respondents consider that digital media offer a lot of product information and that, combined with the ease of access to this information, they also reduce the time of searching for information and the time spent in physical stores.

These data follow the same results cited in the literature, which argue that the shopping journey is increasingly complex due to the emergence and expansion of the online "universe" [26, 46]. These channels allow all consumers to have access to all the information so that their choice can be informed. Despite this, it simultaneously reduces the time spent on searching since, instead of the consumer having to visit several stores, visit several sites in a few seconds and get the same (or even more) information. This potential helps not only consumers but also brands to ensure that there is greater coherence in their communication and that it is updated in real time, since online communication (on the social media and website of each fashion brand, for example) can be more easily managed and has more space versus in-store communication [32]. In addition, there is also the buying behavior already mentioned in the literature review of consumers present in various channels, who search for information online before wanting to see the product offline [26, 35].

Following the same line of thought and going against conclusions from studies such as those of Gupta and Harris [14] and Stephen [40], respondents consider that the opinions and comments of other consumers and consensus on the recommendations given help in the evaluation of the product and that positive comments in the online media help in the time of purchase decision.

Although most respondents present some difficulty in deciding to purchase a clothing and accessories product online, they state that, increasingly, the purchase of this product and comparison between the various options are based on the information contained in the media and that these same means can anticipate the purchase and even that the selected brand may change after a more in-depth online search.

These responses support the fact that, increasingly, e-commerce is here to stay. Although some studies have reported that the growth of e-commerce has been related to the confinement that has happened in recent years, the most recent evidence has been proving that growth has continued even after that [1, 28, 39]. Consumers continue to want to buy online and recognize the convenience associated with this type of purchase, and they will only cross-reference this online purchase with offline channels [19, 34]. As a result, digital media will be important touchpoints in the consumer journey and the consumer's experience in making them could shape their purchasing decision [2, 5].

The exception to this positive behavior is the case of a specific digital media, blogs and forums, where a large part of respondents shows no interest in either joining or having an active role in participating in discussions. This finding is contrary to what is expected [12, 20], raises the possibility that consumers in the Iberian Peninsula behave differently and are not so interested in sharing and/or reading more extensive articles, preferring media such as social media, as can be seen next. It should be said that this possibility, to be confirmed, will require future studies with a larger sample and collected at random.

However, it is clear from the results obtained that the purchase of clothing and accessories through online channels is carried out and that, after the purchase, most respondents advise family and friends to make online purchases of these products.

Moving on, it has proven that there are preferential digital media, according to the stage of the online fashion consumer shopping journey. In the initial phase of the journey, particularly in the phase of recognition of the need, the most used means are social media and online advertising. Also, in pre-purchase, in times of search for information and evaluation of alternatives, social media remain one of the preferred means, but websites and electronic word-of-mouth gain a relevant weight in decision-making.

These findings demonstrate the importance of a good communication strategy, adjusted to the target audience and making use of the means where the public is in the different stages of purchase [43]. In the case of fashion retail brands, more specifically in the case under study (which is a project focused on online and e-commerce), social media have a relevant weight at the beginning of the shopping journey, and therefore, brands must bet on properly feeding these networks, organically and pays, ensuring relevant content and a coherent message that meets the needs of potential consumers [17, 24, 33].

After the decision is made, the preferred means change, since people buy mainly on the websites of fashion brands, through existing e-commerce platforms. The same behavior occurs at the post-purchase time. Still, as alternative means to the brand's website, people highlight social media, as a buying and post-purchase platform (to make complaints, for example) which, once again, proves to be one of the digital media that should be worked on the most since its use is transversal during the journey [5, 17, 36].

Corroborating the behavior verified previously, blogs and forums as well as e-mail have a reduced expression at all stages of the online shopping journey. According to Omar and Atteya, [32], regarding e-mail, this behavior would already be expected. These results may be due to several factors such as the need for more recent evidence on the subject or the change in the paradigm of content consumption, favoring shorter content, especially when we are analyzing younger generations, such as the study in question [42].

50.6.2 Contributions of the Study

In practical terms, it is important to highlight the relevance of this analysis for MO and for companies inserted mainly in the fashion retail sector, especially those that aim to grow in the digital and e-commerce environment. This relevance is high in view of the exponential growth of this channel in recent years, largely due to the pandemic experienced. This paradigm shift in consumption has intensified the investment of brands in their digital footprint and, consequently, both in the use of various digital media and in providing a satisfactory and safe experience to consumers—that is, this study has leveraged the data-based construction of a consistent and coherent marketing strategy across digital channels, by responding to "how" and "when" to reach consumers during the various stages of their journey.

Considering all the results presented in the section above, it can be concluded that, to build an appropriate communication strategy for the growth of a fashion retail e-commerce brand and entry into an international market, in this particular case study, the following parameters should be taken into account:

- Use the different digital media throughout the shopping journey as they are valued by respondents and enhance the best demand for information, anticipate online shopping and reduce the time of searching for products and clothing and accessories brands.
- Valuing comments made by other consumers since they broadly influence the choice of product and brand.
- Consider that blogs and forums and newsletters did not show much interest and interaction on the part of respondents.
- Consider social media, digital advertising and electronic word-of-mouth as relevant digital communication media for the communication of clothing and accessories.

50.6.3 Limitations and Suggestions for Future Research

Understanding this research needs the recognition of the main limitations identified, so that these can become opportunities for future research.

One of the limitations of this study was the sample size—166 individuals. If this same study was carried out in a larger sample and ideally in a random probabilistic way, this would make it possible to prove more solidly the validity of the results obtained.

Another chance for improvement will be to extend and replicate the same research to other fashion retail companies or even other sectors of activity, since this research was focused on the consumer of a specific brand.

Another limitation is that the sample analyzed is mostly female. Although the general consumer of this type of products is feminine, nowadays men have shown a greater interest in the purchase of textile fashion, and thus, it would be relevant for other studies to investigate the experience of male consumers in this sector.

Furthermore, as a limitation, it is important to highlight the difficulty in capturing responses from Spanish consumers. The small number of consumers living in Spain does not translate the effort made to disseminate the questionnaire since the effort was superior to that of disclosure in Portuguese consumers, but its return was much lower.

For future investigations, it would be equally interesting to analyze the present data using structural equations, allowing a more in-depth and integrated analysis of the complex relationships between the multiple independent variables and the dependent variable of the conceptual model of analysis.

References

1. Alberto, P.: CTT e-Commerce Report 2020 (2021). https://www.ctt.pt/grupo-ctt/media/noticias/e-commerce-cresce-46-em-2020-para-4-4-mil-milhoes-de-euros
2. Baxendale, S., Macdonald, E.K., Wilson, H.N.: The impact of different touchpoints on brand consideration. J. Retail. **91**(2), 235–253 (2015). https://doi.org/10.1016/j.jretai.2014.12.008
3. Beauchamp, M., Ponder, N.: In-store and online shoppers. Mark. Manag. J. **20**(1), 49–65 (2010)
4. Blanchflower, T.M., Watchravesringkan, K.T.: Exploring the impact of social networking sites on brand equity. Proc. Market. Manage. Assoc. 20–24 (2014)
5. Cain, P.M.: Modelling short-and long-term marketing effects in the consumer purchase journey. Int. J. Res. Mark. **39**(1), 96–116 (2022). https://doi.org/10.1016/j.ijresmar.2021.06.006
6. Carmo, H., Ferreira, M.M.: Inquéritos por entrevista e por questionário. In: Metodologia da investigação - Guia para Auto-aprendizagem, 2° edição, pp. 135–165. Universidade Aberta (2008)
7. Chailom, P.: Antecedents and consequences of e-marketing strategy: evidence from e-commerce business in Thailand. Int. J. Bus. Strategy **12**, 75–87 (2012)
8. Cizmeci, F.: The effect of digital marketing communication tools to create brand awareness by housing companies. MEGARON/Yıldız Tech. Univ. Faculty Arch. E-J. **10**(2), 149–161 (2015). https://doi.org/10.5505/megaron.2015.73745
9. Dahiya, R., Gayatri, G.: A research paper on digital marketing communication and consumer buying decision process: an empirical study in the Indian passenger car market. J. Glob. Mark. **31**(2), 73–95 (2018). https://doi.org/10.1080/08911762.2017.1365991
10. Dewhirst, T., Davis, B.: Brand strategy and integrated marketing communication (IMC): a case study of player's cigarette brand marketing. J. Advert. **34**(4), 81–92 (2005). https://doi.org/10.1080/00913367.2005.10639211
11. Durand-Hayes, S., Doornbosch, H., Cheng, M., Palmer, M.: PwC's March 2021 Global Consumer Insights Pulse Survey. PwC Global (2021). https://www.pwc.com/gx/en/industries/consumer-markets/consumer-insights-survey.html
12. Guercini, S., Bernal, P.M., Prentice, C.: New marketing in fashion e-commerce. J. Glob. Fash. Market. **9**(1), 1–8 (2018). https://doi.org/10.1080/20932685.2018.1407018
13. Gull, M., Pervaiz, A.: Customer behavior analysis towards online shopping using data mining. In: 5th International Multi-Topic ICT Conference: Technologies For Future Generations, vol. 21(14), pp. 952–956 (2018). https://doi.org/10.1109/IMTIC.2018.8467262
14. Gupta, P., Harris, J.: How e-WOM recommendations influence product consideration and quality of choice: a motivation to process information perspective. J. Bus. Res. **63**(9), 1041–1049 (2010)
15. Haan, E., Wiesel, T., Pauwels, K.: The effectiveness of different forms of online advertising for purchase conversion in a multiple-channel attribution framework. Int. J. Res. Market. **33**(3), 491–507 (2016). https://doi.org/10.1016/j.ijresmar.2015.12.001
16. Harrysson, H., Hesseborn, L., Weber, I.: Leadership and brand alignment. In: LBMG Strategic Brand Management—Masters Paper Series (2019)
17. Hudson, S., Huang, L., Roth, M.S., Madden, T.J.: The influence of social media interactions on consumer-brand relationships: a three-country study of brand perceptions and marketing behaviors. Int. J. Res. Market. **33**(1), 27–41 (2016). https://doi.org/10.1016/j.ijresmar.2015.06.004
18. Jagtap, S.: A study on impact of digital marketing in customer purchase in Chennai. J. Contemporary Issues Bus. Gov. **2**(10), 332–338 (2016). https://doi.org/10.47750/cibg.2020.26.02.136
19. Kavitha, T.: Consumer buying behavior of online shopping—a study. Int. J. Res. Manage. Bus. Stud. (2017)
20. Keng, C.J., Ting, H.Y.: The acceptance of blogs: using a customer experiential value perspective. Internet Res. **19**(5), 479–495 (2009)
21. Kim, A.J., Ko, E.: Do social media marketing activities enhance customer equity? An empirical study of luxury fashion brand. J. Bus. Res. **65**(10), 1480–1486 (2012)

22. Kotler, P., Kartajaya, H., Setiawan, I.: Marketing 4.0-Moving from Traditional to Digital (2017)
23. Laurent, G., Kapferer, J.N.: Measuring consumer involvement profiles. J. Market. Res. **22**(1), 41–53 (1985). https://doi.org/10.2307/3151549
24. Lee, Y.K., Kim, S., Chung, N., Ahn, K., Lee., J.W.: When social media met commerce: a model of perceived customer value in group-buying. J. Serv. Market. **30**(4), 398–410 (2016). https://doi.org/10.1108/JSM-04-2014-0129
25. Lemon, K.N., Verhoef, P.C.: Understanding customer experience throughout the customer journey. J. Market. **80**(6), 69–96 (2016). https://doi.org/10.1509/jm.15.0420
26. Lynch, S., Barnes, L.: Omnichannel fashion retailing: examining the customer decision-making journey. J. Fashion Market. Manage. Int. J. **24**(3), 471–493 (2020). https://doi.org/10.1108/JFMM-09-2019-0192
27. Mäki, M., Toivola, T.: Global market entry for Finnish SME e-Commerce companies. Technol. Innov. Manage. Rev. **11**(1) (2021)
28. Marketeer: E-Commerce cresce 58% no primeiro trimestre. Marketeer (2021). https://marketeer.sapo.pt/e-commerce-cresce-58-no-primeiro-trimestre
29. Mir-Bernal, P., Guercini, S., Sádaba, T.: The role of e-commerce in the internationalization of Spanish luxury fashion multi-brand retailers. J. Glob. Fash. Market. **9**(1), 59–72 (2018). https://doi.org/10.1080/20932685.2017.1399080
30. Mittal, V.: Satisfaction, repurchase intent, and repurchase behavior: investigating the moderating effect of customer characteristics. J. Market. Res. **38**(1), 131–142 (2001). https://doi.org/10.1509/jmkr.38.1.131.18832
31. Oliveira, E.R., Ferreira, P.: Métodos de Investigação-Da interrogação à descoberta científica. S.A. (2014)
32. Omar, A.M., Atteya, N.: The impact of digital marketing on consumer buying decision process in the Egyptian market. Int. J. Bus. Manage. **15**(7), 120 (2020). https://doi.org/10.5539/ijbm.v15n7p120
33. Öztamur, D., Karakadılar, İS.: Exploring the role of social media for SMEs: as a new marketing strategy tool for the firm performance perspective. Proc. Soc. Behav. Sci. **150**, 511–520 (2014). https://doi.org/10.1016/j.sbspro.2014.09.067
34. Park, Chung-Hoon Kim, Y.-G.: Identifying key factors affecting consumer purchase behavior in an online shopping context. Int. J. Retail Distribution Manage. **31**(1), 16–29 (2003)
35. Paz, M.D.R., Delgado, F.J.: Consumer experience and omnichannel behavior in various sales atmospheres. Front. Psychol. 11 (2020)
36. Pop, R.A., Săplăcan, Z., Dabija, D.C., Alt, M.A.: The impact of social media influencers on travel decisions: the role of trust in consumer decision journey. Curr. Issue Tour. **25**(5), 823–843 (2022). https://doi.org/10.1080/13683500.2021.1895729
37. Qi, X., Chan, J.H., Hu, J., Li, Y.: Motivations for selecting cross-border e-commerce as a foreign market entry mode. Ind. Market. Manage. **89**(October 2019), 50–60 (2020). https://doi.org/10.1016/j.indmarman.2020.01.009
38. Ramachandran, K.K., Karthick, K.K., Kumar, M.S.: Online shopping in the UK. Int. Bus. Econ. Res. **10**(12), 23–36 (2011). https://doi.org/10.19030/iber.v10i12.6647
39. Rosário, A., Raimundo, R.: Consumer marketing strategy and e-commerce in the last decade: a literature review. J. Theoret. Appl. Electron. Commerce Res. **16**, 3003–3024 (2021). https://doi.org/10.3390/jtaer16070164
40. Stephen, A.T.: The role of digital and social media marketing in consumer behavior. Curr. Opin. Psychol. **10**, 17–21 (2016). https://doi.org/10.1016/j.copsyc.2015.10.016
41. Straker, K., Wrigley, C., Rosemann, M.: Typologies and touchpoints: designing multi-channel digital strategies. J. Res. Interact. Market. **9**(2), 110–128 (2015). https://doi.org/10.1108/JRIM-06-2014-0039
42. Tellis, G.J., MacInnis, D.J., Tirunillai, S., Zhang, Y.: What drives virality (sharing) of online digital content? The critical role of information, emotion, and brand prominence. J. Mark. **83**(4), 1–20 (2019). https://doi.org/10.1177/0022242919841034
43. Thaichon, P., Liyanaarachchi, G., Quach, S., Weaven, S., Bu, Y.: Online relationship marketing: evolution and theoretical insights into online relationship marketing. Mark. Intell. Plan. **38**(6), 676–698 (2020). https://doi.org/10.1108/MIP-04-2019-0232

44. Verhoef, P., Kannan, P., Inman, J.: From multi-channel retailing to omni-channel retailing: introduction to the special issue on multi-channel retailing. J. Retail. **91**(2), 164–181 (2015)
45. Vieira, V.A., de Almeida, M.I.S., Agnihotri, R., da Silva, N.S.D.A.C., Arunachalam, S.: In pursuit of an effective B2B digital marketing strategy in an emerging market. J. Acad. Mark. Sci. **47**(6), 1085–1108 (2019). https://doi.org/10.1007/s11747-019-00687-1
46. Wolny, J., Charoensuksai, N.: Mapping customer journeys in multichannel decision-making. J. Direct Data Digit. Mark. Pract. **15**(4), 317–326 (2014)
47. Wright, S.J.: Customers buying and interacting with the help of technology. In: Digitizing the Customer Journey, pp. 21–31. Bluetrees GmbH (2019)
48. Yazdanparast, A., Mathew, J., Muniz, F.: Consumer based brand equity in the 21st century: an examination of the role of social media marketing. Young Consumers **17**, 243–255 (2016)
49. Yin, R.: How to do better case studies. In: Applied Social Research Methods, pp. 234–260. SAGE Publications Inc. (2009)
50. Yu, Y., Wang, X., Zhong, R.Y., Huang, G.Q.: E-commerce logistics in supply chain management: Implementations and future perspective in furniture industry. Ind. Manag. Data Syst. **117**, 2263–2286 (2017)

Chapter 51
Optimizing Marketing Through Web Scraping

Diego Albuja, Laura Guerra, Dulce Rivero, and Santiago Quishpe

Abstract The objective of this research was to develop an indexed database on electrical appliances marketed in the city of Ibarra, Ecuador, using web scraping in order to optimize the search time of consumers for these products. It is an applied research, where the survey technique was applied to determine the marketers and the products to be considered. The results of the survey led to directing the study in the marketers Todohogar, Almacenes Japon and La Ganga and to the products: televisions, kitchens and refrigerators. The web scraping process contemplated three phases: search and inspection of the web page, data extraction and cleaning and transformation and storage of final data, for which the Python program and the Request and Beautiful Soup libraries were used. The Data Access Object (DAO) design pattern was used to separate the logic layer from data access, and the connection to the MySQL database was defined with the Python programming language using the mysql.connector library. For the automatic and indexed storage of the data extracted and refined from the pages of the marketers, some databases called product table, characteristic table, brand table were built. As a result, there is a web page that makes use of the web scraping program developed to show the user the summarized and updated information of the three products offered by each of the marketers analyzed.

D. Albuja · L. Guerra (✉) · D. Rivero · S. Quishpe
Pontifical Catholic University of Ecuador Ibarra Headquarters La Victoria,
100150 Ibarra, Ecuador
e-mail: lrguerra@pucesi.edu.ec

D. Albuja
e-mail: dfalbuja@pucesi.edu.ec

D. Rivero
e-mail: dmrivero@pucesi.edu.ec

S. Quishpe
e-mail: squishpe@pucesi.edu.ec

J. L. Reis et al. (eds.), *Marketing and Smart Technologies*, Smart Innovation, Systems and Technologies 337, https://doi.org/10.1007/978-981-19-9099-1_51

51.1 Introduction

The progress and ease of publishing information on the Internet have had an impact on the fact that this network has increased its content substantially. Additionally, the current era is characterized by a consumer of data/information/services/products, more digitally educated, and therefore, more likely to use the internet as a means of contact and knowledge. On the other hand, if one considers the non-renewable and generally scarce resource that most individuals have that is time, the "search" in digital pages can do something long, tedious and that does not always lead to the best or accurate information. In view of this situation, the web scraping technique emerged, also known as Content Scraping, Screen Scraping, Web Harvesting or Web Data Extraction [1], which consists of extracting data from web pages quickly, automatically and efficiently to present it in a better structured and more user-friendly format. Web pages are made up of HTML elements and data within these elements. These elements can be made up of texts, figures, videos, labels and others, that is, information that is not always useful for all people [2]. The extraction of the data is carried out simulating human exploration through computer programs that increase efficiency and optimize search time, while allowing the extracted information to be indexed and ordered in databases for later consultation [1]. This technique is used by a large number of very popular applications and websites, such as the vast majority of search engines (Google, Yahoo, others) [3]. Its field of application is very broad and can be summarized in security and defense; content change tracking; content analysis focused on any discipline [4]. From the commercial point of view, current businesses try to have the greatest visibility on the internet, betting on the investment of resources for the positioning of online businesses; on the other side of the coin, this situation can represent for the consumer, a strenuous task of searching for information. Hence, this study aimed to develop an indexed database of household appliance products offered by marketers in the city of Ibarra through the technique of web scraping, in order to make more efficient and productive the time that the user spends buying such items on the network.

51.2 Literature Review

According to [5], there are different types of web scarping, from the simplest such as copy-pasting (copy and paste) to more complex such as HTML parsing (syntactic analysis of the HTML file), DOM parsing (defines the document structure and content), HTTP programming (based on the HTTP protocol that consists of making requests to the remote web server and analyzing the response obtained), among others. However, regardless of the type of web scraping, there are some fundamental steps to execute them. In the opinion of [6], web scraping includes three basic phases:

Search phase: which implies determining which will be the web pages that present the data of interest and that must be analyzed.

Extraction phase: It consists of extracting the data of interest from each of the pages considered.

Transformation phase: After the data is extracted, it is cleaned and transformed to a certain structure to be stored in a database.

In the context of the use of web scraping, various projects are generated, intertwined with different fields of knowledge. In the field of finance, specifically in relation to the stock exchange, Lin and Yang [7] used web scraping to periodically read the securities of the stock market of the territory of Taiwan. The application captures the price of a company's shares every second, which is useful for users who carry out daily transactions in this market. The scraping algorithm (web scraping) first extracts the name of the company and second the data such as: the price, trend and fluctuations of the shares with the use of the Beautiful Soup library provided by Python; later, this data is exported to a file of Excel for the user who can interpret them and make decisions based on this information. Cárdenas Rubio et al. [8] focused on using web scraping, using data mining techniques and regular expression analysis, in order to collect data to establish fluctuations in housing prices and rentals to determine the strength of Colombia's economy and financial system. In the project proposed by Manrique Andrade and Sanchez Rubio [9], it is mentioned that in Peru, the e-commerce business model is growing, so their proposal aims to streamline the digital purchasing process and thus enhance people's quality of life, through the comparison of prices of the different products offered by various supermarket chains. Cauna Huanca [1] generated the proposal of a website, in which indexed information is stored that has been extracted with the web scraping technique using the Beautiful Soup Python library; The information extracted is related to tourist packages offered by the agencies in a region of Peru. With reference to the legal framework and the legal implications of the use of web scraping, Sanabria De Luque [10] contrasts the Colombian legislation with the legislation of the European Union and that of the USA to conclude that the technique of extracting data from the web could violate the rights of third parties due to the load that the requests could generate to the website servers. In addition, the website may have quite specific prohibitions to limit the use of web scraping, such as the use of the captcha resource. But, the author finally determined that there is currently no legal prohibition on the use of the technique to extract data from a web page that could prevent its use, since the information that is collected in them is of a public nature and their use does not violate regulations.

51.3 Materials and Methods

The study is framed in an applied research, where the survey technique was used to determine the marketers with which web scraping would be applied according to the opinion of the members of the Cantonal Sports League of Antonio Ante. The marketers Todohogar, Almacenes Japón and La Ganga were selected as study centers; in the same way, it was established that the items to be considered would

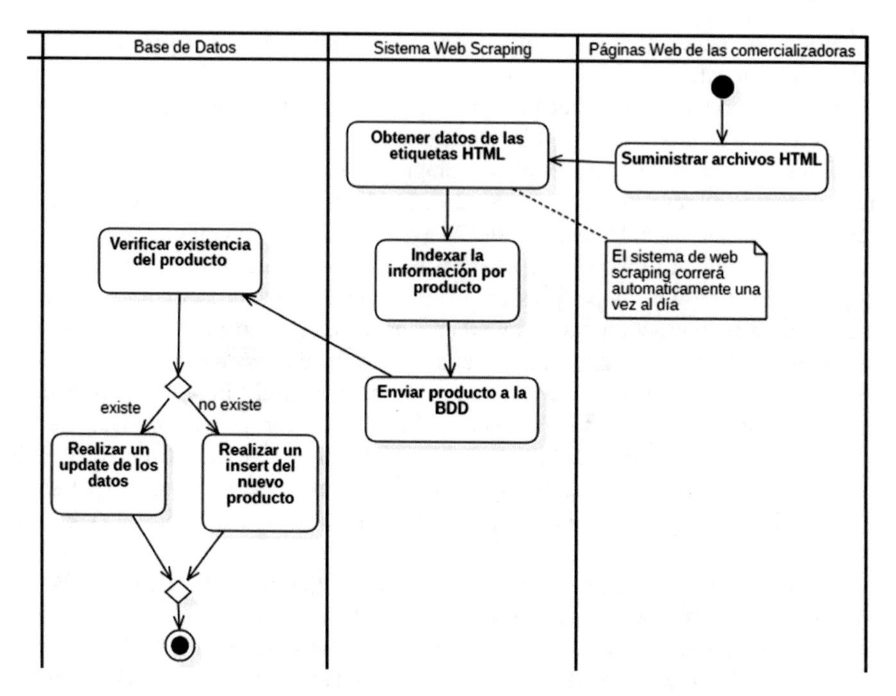

Fig. 51.1 Activity diagram

be televisions, kitchens and refrigerators. The specific URLs of each page where the HTML documents were extracted are as follows:

Todohogar [11]:

https://www.todohogar.com/309-televisores?q=Local-Todohogar+La+Plaza

Almacenes Japón [12]:

https://almacenesjapon.com/55-tv-y-video?q=Subcategorías-Televisores/Disponibilidad-En+stock

La Ganga [13]:

https://www.almaceneslaganga.com/pedidos-en-linea/efectivo/Televisores

The guidelines proposed by Persson [6] were followed to achieve the objective of the research, summarized in the activity diagram (Fig. 51.1), where the interaction of the pages of the marketers, the web scraping software and the database is visualized.

51.3.1 *Web Scraping Process*

To implement the web scraping technique, certain steps were followed that are general and were repeated for all the web pages of the different marketers. Once the Requests and bs4 libraries have been installed in Python, the sequence to follow to do the web scraping included the following phases:

Phase 1. Search and inspection of the data source: The first step was to carefully observe and analyze the website in which the scraping will be applied (web scraping) because the HTML structure that manages the site must be understood in order to be able to extract the relevant data. This stage began by opening the web page from which the information was to be extracted in the browser and sending an HTTP request to its URL. The server responded to the request by returning the structure and HTML content of the page. This task was accomplished with the Python Requests library. For example, for the marketer Todohogar, Fig. 51.2 shows that on the left there are the brands of the products handled by the marketer and on the right there are the products structured in blocks, where product details are shown at the top, and below is the description of the product and its price. The page also presents by default the products of all the branches. In this study, the option of the Ibarra location was selected. Finally, in the final section, no pagination was found, that is, all the products are presented on the same page.

After the general analysis, the website was inspected with the developer's tools, to understand how the information was distributed in the HTML structure of the site and to be able to select the relevant elements for extraction. Figure 51.3 shows a page element on the left and the browser developer tools on the right showing their structures and HTML tags.

In the HTML structure of the Todohogar products in Fig. 51.4, it was found that the information of each product is arranged in a div tag, within which there is a separation generated in two blocks. The first block is the image and the link of the product and the second is the description (brand, model, etc.) and the price of the product.

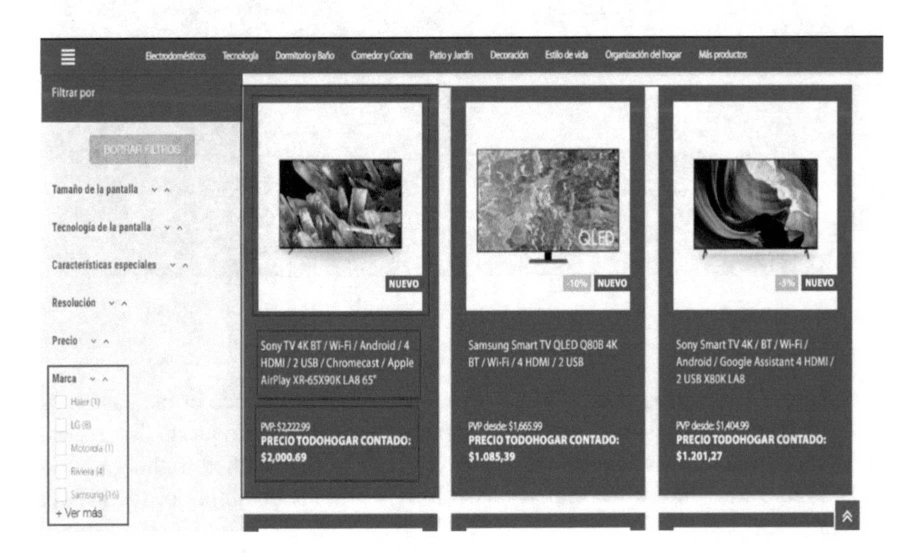

Fig. 51.2 General view of the page—Todohogar

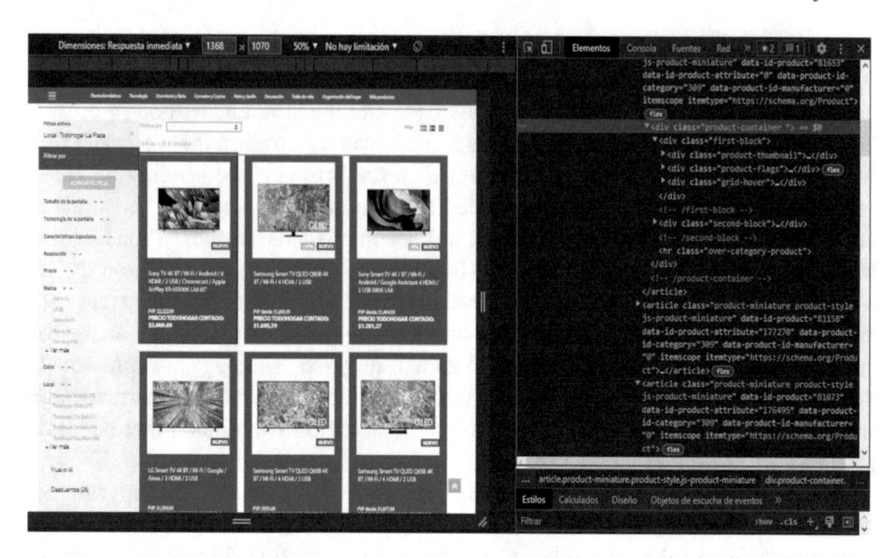

Fig. 51.3 HTML element and structure

```
▼<div class="product-container ">
  ▼<div class="first-block">
    ▼<div class="product-thumbnail">
      ▶<a href="https://www.todohogar.com/televisores/81653-sony-tv-4k-bt-wi-fi-android-4-hdmi-2-usb-chromecast-apple-airplay-xr-65x90k-la8-65-4548736127791.html" class="product-cover-link">…</a>
      <div class="wishlist-product-actions"> </div>
    </div>
    ▶<div class="product-flags">…</div> flex
    ▶<div class="grid-hover">…</div>
  </div>
  <!-- /first-block -->
  ▼<div class="second-block">
    ▼<h5 class="product-name" itemprop="name">
      ▼<a href="https://www.todohogar.com/televisores/81653-sony-tv-4k-bt-wi-fi-android-4-hdmi-2-usb-chromecast-apple-airplay-xr-65x90k-la8-65-4548736127791.html" itemprop="url" title="Sony TV 4K BT / Wi-Fi / Android / 4 HDMI / 2 USB / Chromecast / Apple AirPlay XR-65X90K LA8 65""> == $0
        "Sony TV 4K BT / Wi-Fi / Android / 4 HDMI / 2 USB / Chromecast / Apple AirPlay XR-65X90K LA8 65""
      </a>
    </h5>
    ▼<div class="second-block-wrapper"> flex
      ▼<div class="informations-section">
        ▼<div class="price-and-status d-flex flex-wrap align-items-center"> flex
          ▼<div class="product-price-and-shipping d-flex flex-wrap align-items-center" itemprop="offers" itemscope itemtype="https://schema.org/Offer"> flex
            ▼<div class="first-prices d-flex flex-wrap align-items-center"> flex
              <span class="price product-price pvp" itemprop="price" style="display:inline-block;width:100%;">PVP: $2,222.99</span>
              <span class="price product-price" itemprop="price" content="2000.69" style="display:block;">Precio Todohogar Contado: $2,000.69</span>
```

Fig. 51.4 HTML products structure—Todohogar

Phase II. Extracting the HTML document: The HTML code of the page with which Python could interact was used; for this function, the Requests library was used. This library generates a request with the HTTP GET method to the specified URL. Figure 51.5 shows the import of the library and the obtaining of the HTML content of the page.

The HTML document obtained from the request was transformed into a format that Python can understand, that is, a parsing was performed. For this, the Beautiful Soup library was used, which creates an object of its Python type. Figure 51.6 shows

```
import requests
from bs4 import BeautifulSoup
import logging


# Script para hacer la petición HTTP y generar la sopa
def generarSopa(url):
    try:
        # Hacer la petición a la URL
        res = requests.get(url)
```

Fig. 51.5 Request of the HTML document to the URL

the import of the library and the subsequent creation of the so-called "soup" which would be the HTML content of the page, but already parsed to an element of type Beautiful Soup.

Finally, once the relevant tags for data extraction were known and having obtained the HTML document of the request and having parsed it, we proceeded to search for the elements to extract the necessary information. For this purpose, the Beautiful Soup library presents its own methods to locate an element; in this study, the following were used:

Search by tag name: soup.findAll('tag name').

Search by id: soup.findAll (id = 'id').

Search by class name: soup.findAll (class_ = 'class name').

Search by CSS selector: soup.select(label.CSSClassName).

Search by specific attribute: soup.findAll('a', attrs = {"attribute": "text"}).

To extract a specific piece of data from an HTML tag, it was interacted with as a dictionary: element['attribute'] or if what you want to obtain is the text: element.text.

```
from bs4 import BeautifulSoup
import logging


# Script para hacer la petición HTTP y generar la sopa
def generarSopa(url):
    try:
        # Hacer la petición a la URL
        res = requests.get(url)
        estado = res.status_code
        print(f"Estado del la petición: {estado}")
        # Realizar el parsing del documento HTML extraído con el request
        # para que sea estructurado y manejado por la librería como elementos interactuables
        if estado < 400:
            soup = BeautifulSoup(res.text, 'html.parser')
```

Fig. 51.6 HTML content parsing

```python
class Producto(ABC):
    @abstractmethod
    def __init__(self, comercializadora, marca, modelo, precio, descripcion, imagen, enlace):
        # Atributos del producto
        self.comercializadora = comercializadora
        self.categoria = ''
        self.marca = marca
        self.modelo = modelo
        self.precio = precio
        self.descripcion = descripcion
        self.imagen = imagen
        self.enlace = enlace
```

Fig. 51.7 Product class

Phase III. Transformation: data cleaning and storage: For the development of this research, it was necessary to determine the following data for each product: brand, model, price, description, image, link, product characteristics (screen size, number of burners, among others). However, not all the data obtained by the scraping process was ready to be stored, since there were some that had to be processed in order to generate the desired information. The data that was generally separated was the image, the link and the price of the product. For this reason, the text strings that were extracted from the HTML tags went through a debugging in order to specifically obtain each piece of data. This process was supported by the language's own functions to work with strings and arrays such as: split(), strip(), upper(), lower(), append(), which is allowed to manipulate the text strings. With the specific data already refined, an object of the "Product" class was generated (Fig. 51.7) to which said data was assigned as attributes. The purpose of this class was to generate an object so that any product of any marketer could adapt to a specific model, and with this, the products are standardized to handle them in an orderly and efficient manner when entering their information into the database.

In order to have the code clean, orderly and also reusable, the Data Access Object (DAO) design pattern was used, which allowed separating the logic layer from data access. The DAO pattern was precisely reflected in a class defined with the same name (Fig. 51.8). Within this class were the methods for performing the information management actions or CRUD (create, read, update, delete) and the connection to the MySQL database was defined with the Python programming language with the use of the library mysql.connector.

Each product generated in the previous processes, at the time of storage, is compared taking as reference the "model" of the product in order to identify if said model of the same marketer already exists in the database. If the product does not exist, it is inserted, and if it exists, an update is performed to update attributes that may have changed, such as: price, link or description.

```
import mysql.connector
from mysql.connector import Error
from datetime import date


class DAO():
    def __init__(self):
        conexion = None

    # Conectar a la BDD
    def conectar(self): ...

    # Terminar conexión a la BDD

    def desconectar(self): ...

# ======================== SENTENCIAS TABLA : COMERCIALIZADORA================

    # Select where nombre de comercializadora

    def selectWhereComercializadora(self, nombre): ...

# ======================== SENTENCIAS TABLA : MARCA===========================

    # Select where el nombre de la marca

    def selectWhereMarca(self, nombre): ...

    # Insert de la marca, comprobando si ya no existe

    def insertMarca(self, marca): ...
```

Fig. 51.8 DAO class

51.3.2 Design of the Topic Base

In the design of the physical model presented in Fig. 51.9, the structure and distribution of the relational database are illustrated from the product information extracted from the marketing pages through web scraping.

51.4 Results: Development of Data Extraction Software

The web scraping process was standardized for each marketer and each product, but at the time of developing the specific code for each stage and for each data, particularities were found. For example, in the marketer Todohogar, the most notable characteristics were the following: when searching by "id" for the label of the brands handled by the marketer, it was found that it varied with each update of the page, so

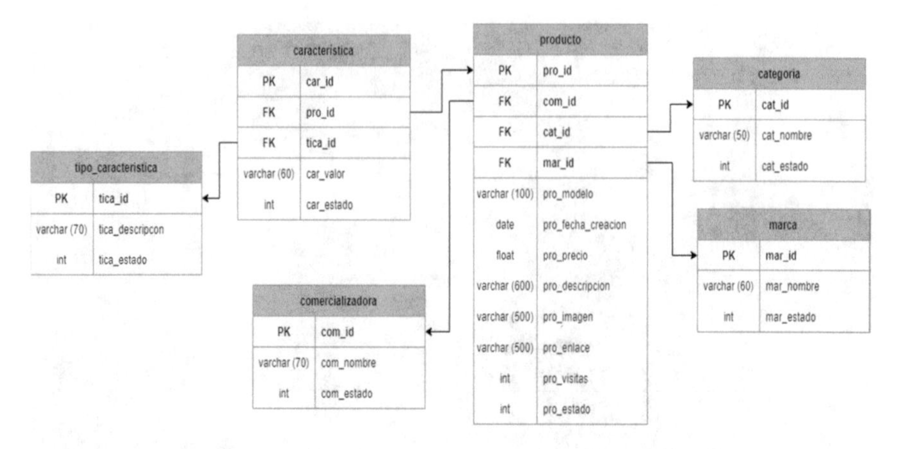

Fig. 51.9 Physical layout of the database

each time it became obsolete. Each time a new request was generated, that is why it was necessary to choose to find the label that had the text "Brand" through iteration through each element of the left sidebar in order to obtain the brands that are handled (Fig. 51.10).

Because the page did not present the information (brand, model, etc.), in different labels, but presented them in a single label, often delimiting each characteristic by a slash (/), as can be seen in Fig. 51.11. And, for this same characteristic and in order not to repeat processes at the same time in the case of a characteristic in the model extraction method, the screen size was also extracted.

A series of lines of code had to be generated to extract generic text strings that appeared throughout all the products and that were not useful for the present project. In the case of televisions, these chains or generic words were: HDMI, 4 K Chromecast, etc. This type of data was refined, and from the chain presented, the brand, model and specific characteristics of each type of product were extracted in the case of televisions, for example, the screen size. All these were achieved with the programming language's own functions for managing arrays and text strings.

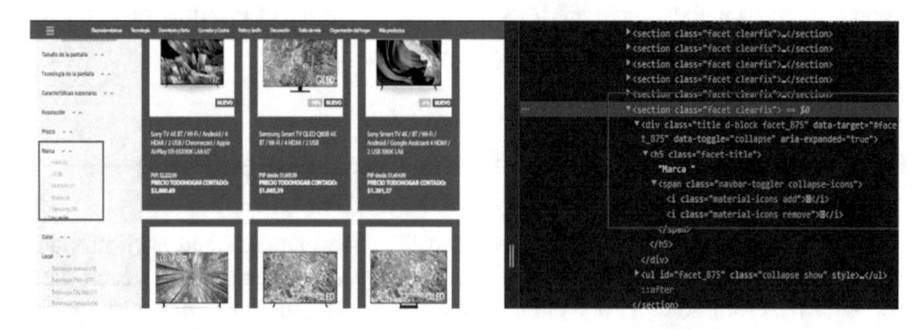

Fig. 51.10 Brand block—Todohogar

Fig. 51.11 Product (TV)—Todohogar

51.4.1 Web Scraping Software

This process was similar for all the marketers, in relation to the framework, because there was previously an analysis work to be able to standardize the process in a sequence of steps to maintain uniformity at the time of programming, as shown in Figs. 51.12 and 51.13, where the abstract parent class "Scraper" is observed from which each class to perform the scraping of each product of each marketer inherited its attributes and methods. Figure 51.14 shows the daughter class "ScraperTV" that has inherited the structure, and it is here where the value of the attributes is defined and the specific code for extracting each data that refers to a particular characteristic of the product is generated.

The result obtained at the end of this first process, which is scraping, was the description of the product in an unprocessed string where the information is found that later has to be cleaned to obtain the data that is really relevant. Figure 51.14 shows the response of the code by terminal of this stage.

Data cleansing process: The cleaning process is closely linked to the scraping process, so the class structure is the same. Broadly speaking, the methods have the same purpose, but the coding and the steps to achieve it were different for each commercial house and to a certain extent for each product as well. For this, the data type (in a text string) was transformed so that it was similar to the one that the database would receive, for example, the data such as the price was transformed to float, which is of decimal numeric type. Finally, once the text string of the consulted product was processed, its attributes or data were known and stored in the database. Figure 51.15 shows the result of this stage.

Storage process: For the storage of the data extracted and refined from the pages of the marketers, some databases called product table, characteristic table, brand table were built. The purpose of these tables was to carry out the storage process of

```
class Scraper(ABC):
    @abstractmethod
    def __init__(self, url):
        self.url = url
        self.soup = None

    @abstractmethod
    def encontrarProductos(self):
        pass

    @abstractmethod
    def obtenerMarcas(self):
        pass

    @abstractmethod
    def extraerMarca(self, hilera):
        pass

    @abstractmethod
    def extraerModelo(self, hilera, marca):
        pass

    @abstractmethod
    def extraerPrecio(self, bloque):
        pass

    @abstractmethod
    def extraerImagen(self, bloque):
        pass

    @abstractmethod
    def extraerEnlace(self, bloque):
```

Fig. 51.12 Scraper abstract class: data extraction

the extracted, refined and indexed data automatically. Figure 51.16 shows the table of the searched products, where their general data is specified. Figure 51.17 shows the characteristics table, where the qualities that each product can present are reflected, such as: screen size, number of burners. Similarly, Fig. 51.18 presents the brands of the products handled by the marketers.

51.4.2 Web System—Appliance Search Optimization

The user interested in purchasing any of the electrical appliances considered in this study will be able to access the web page where the results of the search process through web scraping are presented, as shown in Fig. 51.19.

```
class ScraperTV(Scraper):
    def __init__(self, url):
        super().__init__(url)
        try:
            self.soup = peticion.generarSopa(self.url)
        except Exception:
            logging.exception(
                "Ha ocurrido un error en la petición al generar la sopa en el ScraperTV")

    # Método para encontrar la sección HTML de cada producto
    def encontrarProductos(self): ...

    # Método para obtener las marcas que maneja la comercializadora
    def obtenerMarcas(self): ...

    # Método para extraer la marca de la TV
    def extraerMarca(self, hilera, marcas): ...

    # Método para extraer el modelo
    def extraerModelo(self, hilera, marca): ...

    # Método para extraer el precio
    def extraerPrecio(self, bloque): ...

    # Método para extraer el link de la imagen
    def extraerImagen(self, bloque): ...

    # Método para extraer el link del producto
    def extraerEnlace(self, bloque): ...

    # Método para generar los productos: Televisores
    def generarProductos(self): ...
```

Fig. 51.13 ScraperTV child class

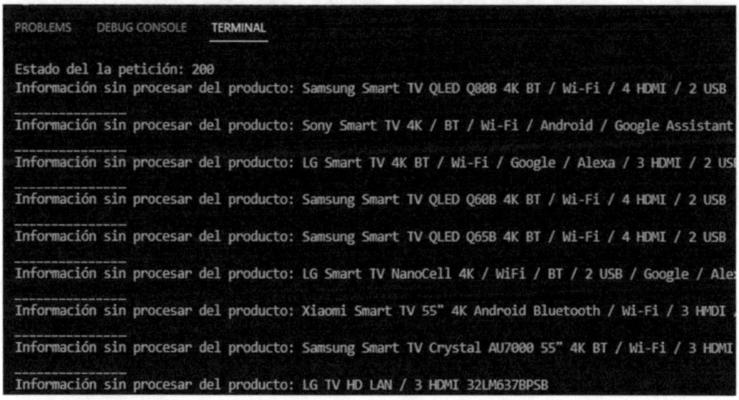

```
PROBLEMS    DEBUG CONSOLE    TERMINAL

Estado del la petición: 200
Información sin procesar del producto: Samsung Smart TV QLED Q80B 4K BT / Wi-Fi / 4 HDMI / 2 USB
----------------
Información sin procesar del producto: Sony Smart TV 4K / BT / Wi-Fi / Android / Google Assistant
----------------
Información sin procesar del producto: LG Smart TV 4K BT / Wi-Fi / Google / Alexa / 3 HDMI / 2 US
----------------
Información sin procesar del producto: Samsung Smart TV QLED Q60B 4K BT / Wi-Fi / 4 HDMI / 2 USB
----------------
Información sin procesar del producto: Samsung Smart TV QLED Q65B 4K BT / Wi-Fi / 4 HDMI / 2 USB
----------------
Información sin procesar del producto: LG Smart TV NanoCell 4K / WiFi / BT / 2 USB / Google / Ale
----------------
Información sin procesar del producto: Xiaomi Smart TV 55" 4K Android Bluetooth / Wi-Fi / 3 HMDI
----------------
Información sin procesar del producto: Samsung Smart TV Crystal AU7000 55" 4K BT / Wi-Fi / 3 HDMI
----------------
Información sin procesar del producto: LG TV HD LAN / 3 HDMI 32LM637BPSB
```

Fig. 51.14 Terminal printout view of raw product data

51.5 Conclusions

One of the main challenges of the web scraping technique is the variety that each website presents, since it is very different from one another, and finding an HTML structure that is repeated is certainly complex. Each web page is unique and needs

Fig. 51.15 Terminal printout view of the cleaned data

to be treated individually if you want to get the most out of it and obtain the relevant data for the project. It must be taken into account that in some products, certain information is not present (brand, model, screen size, etc.) that is why, despite the fact that the code fulfills its purpose when linked to third-party pages (marketers), there may be problems obtaining certain data either because they are omitted or do not have the same distribution as the others. Websites are subject to constant changes, which is why for the durability of web scraping software to be prolonged, you must be willing to provide constant maintenance and be attentive to possible damage that may occur due to changes in the websites where it is used. extract the data. There are certain pages that present captchas or considerations regarding the extraction of your data, which reflect their opposition to their data being used by third parties. Despite

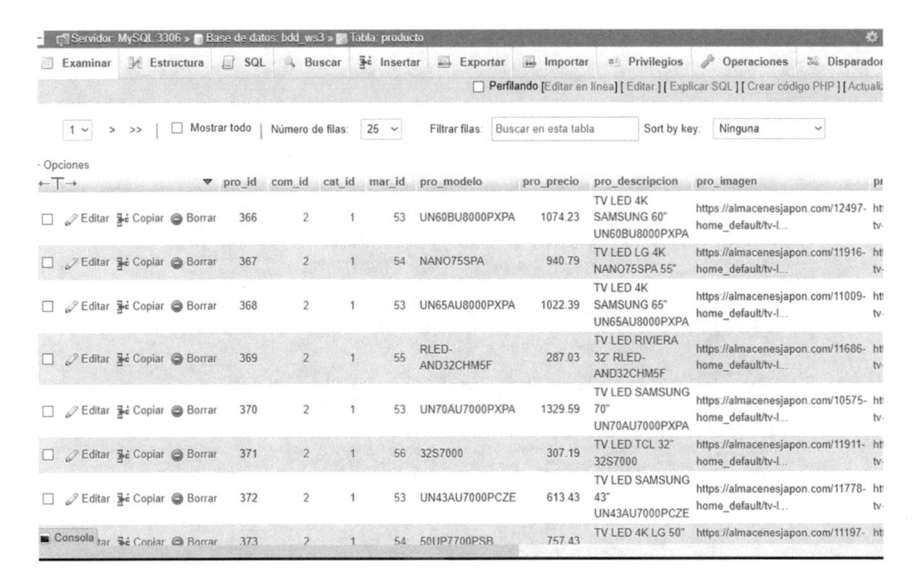

Servidor: MySQL:3306 » Base de datos: bdd_ws3 » Tabla: producto

Examinar | Estructura | SQL | Buscar | Insertar | Exportar | Importar | Privilegios | Operaciones | Disparador

Perfilando [Editar en línea] [Editar] [Explicar SQL] [Crear código PHP] [Actuali

1 > >> | Mostrar todo | Número de filas: 25 Filtrar filas: Buscar en esta tabla Sort by key: Ninguna

Opciones

	pro_id	com_id	cat_id	mar_id	pro_modelo	pro_precio	pro_descripcion	pro_imagen
Editar Copiar Borrar	366	2	1	53	UN60BU8000PXPA	1074.23	TV LED 4K SAMSUNG 60" UN60BU8000PXPA	https://almacenesjapon.com/12497-home_default/tv-l...
Editar Copiar Borrar	367	2	1	54	NANO75SPA	940.79	TV LED LG 4K NANO75SPA 55"	https://almacenesjapon.com/11916-home_default/tv-l...
Editar Copiar Borrar	368	2	1	53	UN65AU8000PXPA	1022.39	TV LED 4K SAMSUNG 65" UN65AU8000PXPA	https://almacenesjapon.com/11009-home_default/tv-l...
Editar Copiar Borrar	369	2	1	55	RLED-AND32CHM5F	287.03	TV LED RIVIERA 32" RLED-AND32CHM5F	https://almacenesjapon.com/11686-home_default/tv-l...
Editar Copiar Borrar	370	2	1	53	UN70AU7000PXPA	1329.59	TV LED SAMSUNG 70" UN70AU7000PXPA	https://almacenesjapon.com/10575-home_default/tv-l...
Editar Copiar Borrar	371	2	1	56	32S7000	307.19	TV LED TCL 32" 32S7000	https://almacenesjapon.com/11911-home_default/tv-l...
Editar Copiar Borrar	372	2	1	53	UN43AU7000PCZE	613.43	TV LED SAMSUNG 43" UN43AU7000PCZE	https://almacenesjapon.com/11778-home_default/tv-l...
Editar Copiar Borrar	373	2	1	54	50UP7700PSB	757.43	TV LED 4K LG 50"	https://almacenesjapon.com/11197-

Consola

Fig. 51.16 Products table

+ Opciones

	car_id	pro_id	tica_id	car_valor	car_estado
Editar Copiar Borrar	325	366	1	60	1
Editar Copiar Borrar	326	367	1	55	1
Editar Copiar Borrar	327	368	1	65	1
Editar Copiar Borrar	328	369	1	32	1
Editar Copiar Borrar	329	370	1	70	1
Editar Copiar Borrar	330	371	1	32	1
Editar Copiar Borrar	331	372	1	43	1
Editar Copiar Borrar	332	373	1	50	1
Editar Copiar Borrar	333	374	1	50	1
Editar Copiar Borrar	334	375	1	32	1
Editar Copiar Borrar	335	376	1	55	1
Editar Copiar Borrar	336	377	1	43	1
Editar Copiar Borrar	337	378	1	50	1
Editar Copiar Borrar	338	379	1	70	1
Editar Copiar Borrar	339	380	1	55	1

Fig. 51.17 Characteristic table

+ Opciones

←T→	mar_id	mar_nombre	mar_estado
Editar Copiar Borrar	53	SAMSUNG	1
Editar Copiar Borrar	54	LG	1
Editar Copiar Borrar	55	RIVIERA	1
Editar Copiar Borrar	56	TCL	1
Editar Copiar Borrar	57	INDURAMA	1
Editar Copiar Borrar	58	DAEWOO	1
Editar Copiar Borrar	59	JVC	1
Editar Copiar Borrar	60	NANOCELL	1
Editar Copiar Borrar	61	SONY	1
Editar Copiar Borrar	62	AIWA	1
Editar Copiar Borrar	63	XIAOMI	1
Editar Copiar Borrar	64	HAIER	1
Editar Copiar Borrar	65	MOTOROLA	1

Fig. 51.18 Brand table

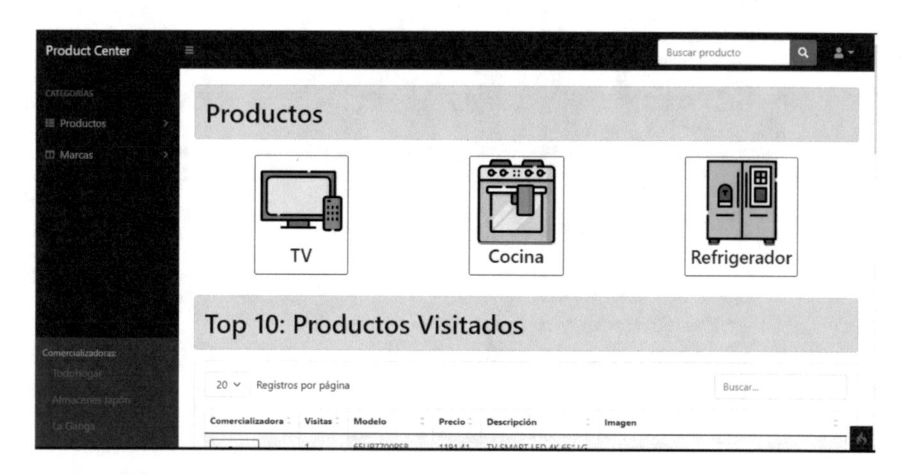

Fig. 51.19 Web system—appliance search optimization

the fact that the legislation is very ambiguous regarding the use of data obtained by the web scraping technique, the developer must appeal to their professional ethics and respect the wishes of the owners of the information that is intended to be extracted.

References

1. Cauna Huanca, G.J.: Indexación de sitios web para optimizar la búsqueda de paquetes turísticos de la región de Puno basado en Web Scraping. Universidad Nacional del Altiplano (2021). http://repositorio.unap.edu.pe/handle/UNAP/16821
2. Uzun, E.: A novel web scraping approach using the additional information. IEEE Access **8**(1), 61726–61740 (2020). https://doi.org/10.1109/ACCESS.2020.2984503
3. Matta, P., Sharma, N., Sharma, D., Pant, B. and Sharma, S.: Web scraping: applications and scraping tools. Int. J. Adv. Trends Comput. Sci. Eng. **9**(5) (2020). https://doi.org/10.30534/ijatcse/2020/185952020
4. Khder, M: Web scraping or web crawling: state of art, techniques, approaches and application. Int. J. Adv. Soft Comput. Appl. **13**, 145–168 (2021). https://doi.org/10.15849/IJASCA.211128.11
5. Valle, G.: Aplicación de técnicas de web scraping y procesamiento del lenguaje natural para la extracción y evaluación de información de una página web de empleo. (2021). https://repositorio.comillas.edu/xmlui/bitstream/handle/11531/55160/TFM%20-%20Valle%20Gutierrez%20Guillermo.pdf?sequence=1
6. Persson, E.: Evaluating Tools and Techniques for Web Scraping. Dissertation (2019). https://www.diva-portal.org/smash/get/diva2:1415998/FULLTEXT01.p
7. Lin, C.F., Yang, S.C.: Taiwan stock tape reading periodically using web scraping technology with GUI. Appl. Syst. Innov. **5**(1), 28 (2022). https://doi.org/10.3390/asi5010028
8. Cárdenas Rubio, J.A., Chaux Guzmán, F.J., Otero, J.: Una base de datos de precios y características de vivienda en Colombia con información de Internet. Revista de Economía del Rosario **22**(1), 25 (2019). https://doi.org/10.12804/revistas.urosario.edu.co/economia/a.7768
9. Manrique Andrade, S.R., Sanchez Rubio, C.J.: Compras eficientes—supermercados. Universidad de Lima (2019) https://hdl.handle.net/20.500.12724/12357
10. Sanabria De Luque, J.J.: Sector privado y libre competencia: implicaciones jurídicas del web scraping. Universidad Externado de Colombia (2021). https://bdigital.uexternado.edu.co/handle/001/4246
11. Todohogar: Todohogar.com. https://www.todohogar.com/
12. Almacenes Japón: Almacenesjapon.com. https://almacenesjapon.com/
13. Almacenes La Ganga: Almaceneslaganga.com. https://www.almaceneslaganga.com/pedidos-en-linea/

Chapter 52
Online Shopping Experience on Satisfaction and Loyalty on Luxury Brand Websites

Ricardo Oliveira, Inês Veiga Pereira, José Duarte Santos, Ana Torres, and Paulo Botelho Pires

Abstract The internet massification and e-commerce growth that have been driven by "millennials" and the coronavirus pandemic cannot remain indifferent to luxury brands. These brands have had to adapt to e-commerce and develop an online shopping experience which satisfies its customers, so that they repeat purchase. Therefore, the main objective of this research is to understand the main impacts of shopping experience on luxury brand websites on satisfaction and loyalty. A model which analyzes the relationship between the three constructs was developed and information was gathered through an online survey, from which resulted 356 valid answers. Through the analysis of data collected and using a structural equation model, using SmartPLS software, we realized that online shopping experience is positively related to satisfaction. Loyalty, in turn, is positively affected by brand satisfaction. This study makes an important contribution to luxury brands and to people in charge of marketing and online platforms selling luxury goods. It helps brands understand that enhancing online shopping experience can positively impact satisfaction and loyalty levels.

R. Oliveira · I. V. Pereira · J. D. Santos (✉)
Accounting and Business School of the Polytechnic of Porto (ISCAP/P.PORTO), Rua Jaime Lopes Amorim, S/N, 4465-004 Matosinhos, Portugal
e-mail: jdsantos@iscap.ipp.pt

I. V. Pereira
e-mail: ipereira@iscap.ipp.pt

I. V. Pereira · J. D. Santos · P. B. Pires
Centre for Organizational and Social Studies of the Polytechnic of Porto (CEOS.PP), Rua Jaime Lopes Amorim, S/N, 4465-004 Matosinhos, Portugal
e-mail: paulo.botelho.pires@gmail.com

A. Torres
University of Aveiro, Campus Universitário de Santiago, 3810-193 Aveiro, Portugal
e-mail: ana.torres@ua.pt

INESC TEC, Campus da FEUP, Rua Dr. Roberto Frias, S/N, 4200-465 Porto, Portugal

J. L. Reis et al. (eds.), *Marketing and Smart Technologies*, Smart Innovation, Systems and Technologies 337, https://doi.org/10.1007/978-981-19-9099-1_52

52.1 Introduction

E-commerce is one of the trends in various economic sectors. However, the shift from a traditional commerce environment to an E-commerce environment, in a business-to-consumer (B2C) model, can be delicate especially for a luxury brand.

The choice of this theme arises in the current context of Internet massification and growing adherence to e-commerce that has been driven by "millennials" and more recently by the coronavirus pandemic. Luxury brands cannot remain indifferent, or else, they may be letting a new distribution channel and a new generation of consumer's "escape". The digital environment brings luxury closer to customers with purchasing power, but geographically distant and with little motivation to go to a physical store of the brand, regardless the generation they belong to. The presence of luxury brands in the digital world is crucial to continue to be relevant and desired.

However, there are values inherent to luxury brands, which at first glance do not seem to be in line with what the digital environment provides. In the online shopping process, when consumers recognize the need for a product or service, they go online and search for information related to their need. Online shopping process, information search, interest in several goods or services online is directly related to online shopping experience [1, 2], which states that online shopping satisfaction reflects the post-purchase judgment, analyzing the expectations generated by the online system. The probability to repeat purchase increases when consumers feel more satisfied with the brand [3] and [4], which adds that the quality of online service has a positive impact on users' satisfaction.

The big challenge is to have access to new customers, educate a new generation of consumers, expand horizons, reach destinations that would not be reached in the traditional format, be aligned with new times and new codes of consumption and, doing all this, keeping the essence of luxury of personalization, customization, exclusivity, incredible stories, the full experience that thrills and delights. Okonkwo [5] refers several challenges that luxury brands face when entering the online market. Translating brand identity, personality, and image on a luxury brand website, as some of the atmosphere of the luxury store is lost, is a challenge.

Therefore, in the online growing business of luxury brands, the shopping experience may have serious reflections on satisfaction and loyalty. The online shopping process, the navigation through the various products, the search for information and interest in various goods and services in the electronic media are directly related to the shopping experience. So, this paper aims to answer an important literature gap, and it becomes important to analyze the impact of the shopping experience in luxury brands websites, in particular, on consumer satisfaction and its impact on loyalty.

This paper starts with literature review, which sustains the hypothesis and the research model. Later, the methodology is explained and results are analyzed and discussed. It ends with conclusion and some future research ideas.

52.2 Literature Review and Hypothesis Development

Luxury concept is very complex, and it is associated to a specific context and it is very common that people's opinion about it is ambivalent and varied [6]. Luxury represents the concept of total excellence: products and services that reach the heights of the extraordinary; demand for what can be detachable and what is not necessary; impulsive purchase, the pleasure or desire to have it [7].

The constant evolution of the meaning of luxury contributes to its unstable character, causing the perception of existence of different levels and categories of luxury, particularly when applied to products. Definitions about luxury brands can be derived from consumer perspective or product perspective [8]. From a consumer perspective, a luxury brand is identified by its psychological value, functional value, and important function as a status symbol. On the other hand, from a product perspective, a luxury brand is defined in terms of its excellent quality, high value, disparity, exclusivity, and handmade or individualized manufacturing.

Digital brings luxury closer to a customer with purchasing power, but geographically distant and unwilling to go to a physical store of the brand regardless of the generation to which they belong. The same author also states that the presence of luxury brands in the digital world is crucial to continue to be relevant and desired.

Shopping experience was defined as being a set of consumer attitudes or behaviors from the moment they visit a certain website to purchase something planned, for recreational purposes or when looking for promotions [9].

The online shopping process, when consumers recognize the need for some product or service, they access the Internet and search for information related to their needs. The online shopping process, the navigation through the various products, the search for information and the interest in various goods and services in the electronic medium are directly related to the shopping experience [9].

A company's main goal should be to make its customers satisfied. This satisfaction is influenced by perceptions about buying experience and other factors [3].

Satisfaction is defined as a person's analogy of the performance of a product in relation to his or her expectations. When that performance does not meet expectations, the customer is disappointed. If the performance meets expectations, the customer is satisfied. When the performance of the product exceeds expectations, the customer is delighted [10].

So, there is positive relationship between the dimensions of e-service quality (design, trust, security/privacy, and customer service) and customer satisfaction [3]. E-service quality dimensions have a positive impact on customer satisfaction. The same authors state that cyber-consumers tend to focus on the site interface and the entire surrounding experience to form their satisfaction judgments.

Based on several studies [4, 11, 12], the following hypothesis was formulated:

H1: Online shopping experience positively affects satisfaction to a certain luxury brand.

The probability of repeated purchase is extremely high with satisfied customers [3]. Brand satisfaction has a significantly positive influence on brand loyalty [13].

Satisfaction is an antecedent of brand loyalty, that is, increases in satisfaction lead to increases in brand loyalty [14].

Loyalty gives the customer a high level of preference, decreasing the likelihood of switching, increasing willingness to pay more, and also the possibility to recommend the product or service or even the company itself [14].

Customer satisfaction is also important in the online environment, as it creates customer confidence in buying online [15]. According to the same author, satisfaction is thus an essential element in building strong long-term customer relationships and plays an essential role in creating online loyalty. If a customer is satisfied, he will have less intention to switch to another brand online [16].

Satisfaction affects customer loyalty and means that satisfied customers are more willing to share their opinions, so word-of-mouth tends to increase [17]. This is advantageous, because satisfied customers have a higher usage intention and are more likely to repurchase, than an unsatisfied customer.

Several authors have established the connection between satisfaction and brand loyalty [3, 13–17], therefore:

H2: Loyalty to a particular luxury brand is positively affected by brand satisfaction.

52.3 Research Methodology

In this research, a quantitative methodology was implemented, through an online distributed questionnaire, and considered the most adequate as it allows relating facts and variables [18]. In this scope, the Structural Equation Model (SEM) was implemented, which allows to evaluate the fit of the theoretical model to the correlational structure between the various variables [19]. The structural equation model was specified following Hair et al. (2010) instructions, performing all the phases. The theoretical model was represented by a path diagram, which shows the relationships between constructs (as in Fig. 52.1.)

After specifying the structural model, it was chosen the estimation methodology, so in this case, the partial least squares (PLS-SEM) method was considered more adequate as the model has at least one construct which is an antecedent and consequent. Calculations were performed using Smart PLS software version 3.3.1 for Windows.

In order to reach desired results, a questionnaire was used. The questionnaire was divided into four parts: the first includes questions for sociodemographic characterization; the second aims to filter the respondent purchases luxury brands; third, the

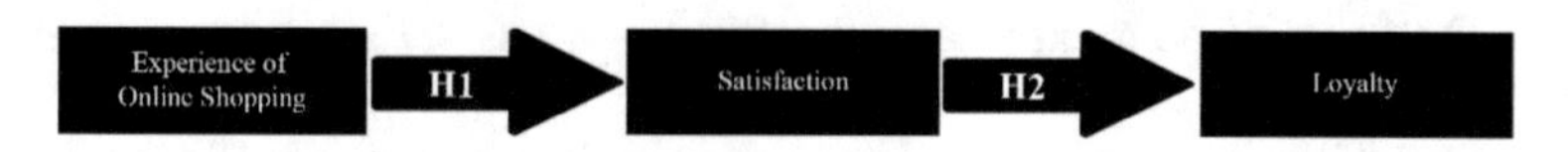

Fig. 52.1 Conceptual model

Table 52.1 Scales used in the questionnaire

Construct	Adapted from
Online shopping experience	Scheinbaum and Kukar-Kinney (2009)
Satisfaction	Mehta and Tariq (2020), Gallart-Camahort et al. (2021)
Loyalty	Anderson and Srinivasan (2003), Gallart-Camahort et al. (2021)

respondent identifies the last luxury website where he/she purchased and the final part where constructs are assessed. The questionnaire was spread by the authors online contacts using the snowball strategy.

The constructs were assessed using scales previously tested in the literature, as shown in Table 52.1. These scales were adapted to Portuguese, as the questionnaire was distributed in Portugal.

52.4 Analysis and Results

52.4.1 Sociodemographic Characterization

For the sample demographic characterization, the following variables were considered: age, gender, nationality, academic background, and finally, professional situation.

As it can be seen in Table 52.2, regarding the age of the respondents, it appears that the youngest individual is 17 years old and the oldest is 67 years old. The average age of the respondents to the questionnaire is 39 years old. About 18.82% were between 31 and 35 years old; 13.20% was being between 36 and 40 years old; and 19.94% was being between 41 and 45 years old.

Table 52.2 Respondents according to age

Groups	Number	Percentage (%)
<25 years old (yo)	16	4.49
26 30 anos	51	14.33
31 aos 35 anos	67	18.82
36 aos 40 anos	47	13.20
41 aos 45 anos	71	19.94
46 aos 50 anos	48	13.48
51 aos 55 anos	23	6.46
Mais de 56 anos	33	9.27
Total	356	100.00

Table 52.3 Respondents according to gender

Gender	Number	Percentage (%)
Female	171	48.03
Male	185	51.97
Total	356	100.00

Regarding gender, the elements of this sample are mostly male. Of the 356 respondents, 185 (51.97%) are male, and 185 (48.03%) are female (Table 52.3). Most of the respondents are Portuguese (97.47%).

Most respondents have completed higher education (university degree). One hundred and fifty-five (43.54%) of the individuals have completed higher education and 106 (29.78%) of the individuals have completed high school.

Of the 356 respondents, 308 (86.52%) are employed, 8 (2.25%) are full-time students, 23 (6.46%) are workers/students, and finally, 17 (4.78%) are self-employed (Table 52.4).

Table 52.5, presented below, shows the luxury brands most mentioned by respondents, where they usually shop online. It should be noted that it was an open question, so 45 different brands were obtained, and all the answers obtained were valid for the study. This is because, as previously mentioned, the concept of luxury is very complex and it is associated with a specific context, being common that a person's opinion about the subject is ambivalent and varied [10].

The brand most mentioned by respondents where they usually shop online was Farfetch with 61 answers (17.13%). There were 45 different answers in an open

Table 52.4 Respondents according to professional situation

Professional situation	Number	Percentage (%)
Full-time student	8	2.25
Student and worker	23	6.46
Employed	308	86.52
Self-employed	17	4.78
Total	356	100.00

Table 52.5 Respondents according to luxury brands they usually shop online

Brand	Number	Percentage (%)
Farfetch	61	17.13
Polo Ralph Lauren	36	10.11
Gucci	22	6.18
Tommy Hilfigher	19	5.34
Cartier	16	4.49
Prada	16	4.49
Outras	186	52.25
Total	356	100.00

question. With 36 mentions, was Polo Ralph Lauren, equivalent to 10.11% of the total respondents. This was followed by Gucci with 6.18% (22) of respondents (see Table 52.5).

52.4.2 *Validation of the Measurement Model*

Factor validity occurs when the items are reflective of the factor they are intended to measure. Most items in this investigation relate to their constructs with full reliability, since they have values greater than 0.700, as it can been in Table 52.6.

Cronbach's alpha of approximately 0.900 is considered excellent and 0.800 is considered very good [20]. Smart PLS suggests a threshold of 0.700 for Cronbach's alpha. All variables in this study are consistent and have excellent values since they all have values greater than 0.900.

Composite reliability (CR), which is calculated from the results of the AFC, is a good indicator of construct reliability when it has a value equal to or greater than 0.700 [21]. This is true for all the variables under study.

Table 52.6 Outer loadings, Cronbach alpha, composite reliability, and AVE

	Outer loadings	Cronbach's alpha	Composite reliability	Average variance extracted (AVE)
EXP1	0.885	0.940	0.953	0.772
EXP2	0.920			
EXP3	0.904			
EXP4	0.705			
EXP5	0.920			
EXP6	0.918			
SAT1	0.898	0.935	0.949	0.758
SAT2	0.936			
SAT3	0.921			
SAT4	0.837			
SAT5	0.805			
SAT6	0.817			
LEAL1	0.923	0.975	0.979	0.887
LEAL2	0.940			
LEAL3	0.954			
LEAL4	0.954			
LEAL5	0.956			
LEAL6	0.923			

Table 52.7 Discriminant validity

	EXP	SAT	LOY
EXP	0.879		
SAT	0.476	0.871	
LOY	0.120	0.418	0.942

Fig. 52.2 Structural model via SmartPLS

Finally, the average variance extracted (AVE) shows values considered acceptable. All of them with values higher than 0.500 are between 0.709 and 0.862.

Discriminant validity was tested through the square root of the AVE values and the correlation between the constructs. It can be concluded that the AVE values are higher than the correlation values between the constructs, and therefore, discriminant validity is supported, meaning that the variables measure different aspects and do not overlap (Table 52.7).

52.4.3 Validation of the Structural Model

The structural model has an acceptable adjustment quality; since the recommended value is SRMR ≤ 0.08 [22], we have a value of 0.067.

The results show that there is significance in the prediction of the constructs since the R^2 is greater than 0.100. A Q^2 greater than zero shows that the model has predictive relevance, a situation verified for all constructs (see Fig. 52.2).

52.4.4 Measurement Analysis of the Total and Specific Indirect Effects

There is a positive relationship between all variables, which solidifies the main focus of this study.

The mediating effect of satisfaction between online shopping experience and loyalty is also significant. In this case, P-value < 0.05, so the relationship is significant.

Table 52.8 Comparison by gender

	β Female	β Male	Difference B female–B male	Values of *P* (female vs male)
Hypothesis 1 EXP > SAT	0.389	0.565	−0.176	0.070
Hypothesis 2 SAT > LOY	0.265	0.533	−0.269	0.048

52.4.5 *Multi-Group Analysis*

52.4.5.1 Comparison of Results by Gender

Starting with the analysis of the gender difference, the first group, representing the male gender, is made up of 185 people, while the second group, referring to the female gender, is made up of 171 people.

As can be seen in Table 52.8, there were significant differences between the genders. The relationship between online shopping experience and satisfaction (hypothesis 1) can show significant differences. This shows that for female respondents, the impact of the online shopping experience on satisfaction is lower than for male respondents. The same can be said for the relationship between satisfaction and loyalty (hypothesis 2).

52.4.5.2 Comparison of Results by Age

For the analysis by age, two groups were established. A first group is made up of people aged up to 40 years old and is made up of 181 answers. The second group is made up of people over 40 years old and is made up of 175 responses.

As can be seen in Table 52.9, there was not a significant difference in the relationship between online shopping experience > satisfaction (hypothesis 1); it is shown that for respondents up to 40 years old, the impact of online shopping experience on luxury brand image is greater than for respondents over 40 years old. The same happens when the relationship between satisfaction and loyalty is concerned. Despite the impact of satisfaction on loyalty is higher for people over 40, it is not statistically different.

Table 52.9 Comparison by age

	β ≤40	β >40	Difference = < 40– > 40 – B Male	Values of *P* (≤40 vs >40)
Hypothesis 1 EXP > SAT	0.383	0.538	−0.155	0.129
Hypothesis 2 SAT > LOY	0.392	0.471	−0.079	0.546

Table 52.10 Comparison by brand and multi-brand

	β Brand	β Multi-brand	Difference brand–multi-brand	Values of *P* (brand vs multi-brand)
Hypothesis 1 EXP > SAT	0.476	0.525	−0.049	0.712
Hypothesis 2 SAT > LOY	0.463	0.239	0.224	0.171

52.4.5.3 Comparison of Results by Multi-Brand Store or Brand Store

Regarding the analysis of online multi-brand store or brand store purchases, two groups were built from the analysis of all the answers.

The first group, composed of 274 answers, is inserted all respondents who say they buy in brand stores (e.g., Cartier, Gucci, Prada, Polo Ralp Lauren, etc.). The second group is made up of 82 responses, where respondents say they shop at multi-brand online stores (e.g., Farfetch, Lyst, Mytheresa, Mr. Porter, and Browns Fashion) (Table 52.10).

As can be seen in the table, there were no significant differences in the relationship between online shopping experience and satisfaction (hypothesis 1). However, there were no significant differences between respondents buying in branded stores and those buying in multi-brand stores in the relationship between image > satisfaction > loyalty (hypothesis 2); for respondents buying in branded stores, the impact of brand image on satisfaction is higher than for respondents buying in multi-brand stores.

52.5 Discussion

From the results of the study, we can conclude that there is a positive relationship of online shopping experience in relation to satisfaction in the case of luxury brands. Consumers shop online for various reasons, such as time saving, convenience, to avoid crowds, for the anonymity, to look for the best prices or to seek entertainment while shopping [14]. All these advantages associated with online shopping experience can have a moderating effect on satisfaction [14]. A company should have as its main goal to make its customers satisfied. This satisfaction is influenced by perceptions about the shopping experience and other factors. So, this research conclusion in consistent with previous research, as online shopping experience affects satisfaction also in the case of luxury brands.

It was also concluded that loyalty to a particular luxury brand is positively affected by brand satisfaction. This conclusion is consistent with previous studies [15] who state that satisfaction is an essential element in building strong long-term customer relationships and plays a key role in creating online loyalty. Also consistent with other research [13] which states that satisfaction with the brand has a significantly positive influence on brand loyalty. Satisfaction is an antecedent of brand loyalty, that is, increases in satisfaction lead to increases in brand loyalty.

52.6 Conclusion

This research had a focus to understand the main impacts of shopping experience in luxury brand websites on satisfaction and loyalty.

In the analysis of the relationships between the variables, it was possible to realize that all of them have a positive value. Thus, from the statistical analysis of the data collected, it was found that the two hypotheses were confirmed.

Therefore, through the results, it was shown that online shopping experience impacts on satisfaction. Likewise, it was shown that satisfaction impacts loyalty.

It was found that most relationships have a greater impact on men than on women, and there is a significant difference between genders when satisfaction impact on loyalty is concerned.

It was also concluded that for respondents up to the age of 40, the impact of the former online shopping experience on luxury brand image is greater than for respondents over the age of 40; the same between satisfaction and loyalty although differences were not significant.

It is concluded that for respondents who shop in a branded store, the impact of experience on satisfaction is lower than for respondents who shop in a multi-brand store.

52.6.1 Research Implications

This study helps to understand the various impacts of shopping experience on luxury brand websites. Therefore, it can make an important contribution, both theoretically and practically, to the luxury brands themselves and to other entities with responsibilities in the management of marketing and online sales platforms of luxury goods. In what concerns online sales platform management, it is believed that this work is useful for a better understanding and conceptual knowledge of the buying experience in luxury brands' websites. There are several details associated with the online shopping experience that can positively impact the customer. From the online buying process to the final use of the purchased good, there is a series of steps that should be valued according to the objectives and values of the brand so that the online buying experience is positive.

In relation to marketing, this study offers important contributions to its development and to a greater conceptual understanding of satisfaction and brand loyalty. The different relationship established between the different constructs also allows important conclusions to be drawn for the marketing teams of luxury brands.

And finally, it is thought that it may be useful for brands to understand that enhancing the online shopping experience can positively impact their satisfaction and loyalty levels.

52.6.2 Limitations and Future Research

The first limitation that can be pointed out is the fact that the sample is a convenience sample, resulting from the fact that the questionnaire was distributed randomly. A survey based on a different sample could guarantee the representativeness of the sample and, thus, produce different results.

This sample has imbalances in educational levels and professional status. This means that the conclusions may be predominantly influenced by a specific sociocultural group, which may condition the final conclusions.

Although the sample is quite comprehensive, both in terms of number and characterization, it does not allow us to generalize. Therefore, it would be interesting to invest in a more representative sample to better define the conclusions.

In future studies, other dependent variables can also be evaluated, which could increase the predictive power of advertising on social networks. Looking for other antecedents that may impact loyalty to luxury brands could also be interesting.

Finally, it may also be interesting to compare the results of this research with studies conducted in other countries. That is, compare countries where shopping habits on luxury brand websites are much higher and e-business is even more evolved with countries where exactly the opposite happens.

Acknowledgements This work is financed by portuguese national funds through FCT—Fundação para a Ciência e Tecnologia, under the project UIDB/05422/2020.

References

1. Scheinbaum, A.C., Kukar-Kinney, M.: Beyond buying: motivations behind consumers' online shopping cart use the uses and gratifications of the consumers' online shopping cart use. J. Bus. Res. **63**(10) (2009)
2. Verhagena, T., Frans, F., Van Den Hooff, B., Meents, S., Merikivi, J.: Satisfaction with virtual worlds: an integrated model of experiential value. Inf. Manage. **48**(6), 201–207 (2011)
3. Mansano, A.T.R., Gorni, P.M: Satisfação do consumidor com o comércio eletrônico: estudo de caso de uma fabricante de tapetes. **1**(1), 12–22 (2014)
4. Bressolles, G., Durrieu, F., Giraud, M.: The impact of electronic service quality's dimensions on customer satisfaction and buying impulse. J. Cust. Behav. **6**(1), 37–56 (2007)
5. Okonkwo, U.: Sustaining the luxury brand on the Internet. J. Brand Manag. **16**(5–6), 302–310 (2009)
6. Dubois, B., Laurent, G., Czellar, S.: Consumer rapport to luxury: analyzing complex and ambivalent attitude. In: Consumer Research Working, **736** (2001)

7. Ferreira, P.S.B.: Como Marcas de Luxo Se Comportam Nos Meios Digitais. Relatório de Estágio (2017)
8. Fionda, A.M., Moore, C.M.: The anatomy of the luxury fashion brand. J. Brand Manag. **16**(5), 347–363 (2009)
9. Close, A.G., Kukar-Kinney, M.: Beyond buying: motivations behind consumers' online shopping cart use. J. Bus. Res. **63**(9–10), 986–992 (2010)
10. Kotler, P., Keller, K.L.: Marketing Management—Global Edition, 15th edn. Pearson (2015)
11. Puccinelli, N.M., Goodstein, R.C., Grewal, D., Price, R., Raghubir, P., Stewart, D.: Customer experience management in retailing: understanding the buying process. J. Retail. **85**(1), 15–30 (2009)
12. Wolfinbarger, M., Gilly; M.C.: ETailQ: Dimensionalizing, measuring and predicting etail quality. J. Retail. **79**(3), 183–98 (2003)
13. Şahin, A., Zehir, C., Kitapçi, H.: The effects of brand experiences, trust and satisfaction on building brand loyalty; an empirical research on global brands. Procedia. Soc. Behav. Sci. **24**, 1288–1301 (2011)
14. Ladhari, R., Souiden, N., Ladhari, I.: Determinants of loyalty and recommendation: the role of perceived service quality, emotional satisfaction and image. J. Fin. Serv. Mark. **16**(2), 111–124 (2011)
15. Kim, A.J., Ko, E.: Impacts of luxury fashion brand's social media marketing on customer relationship and purchase intention. J. Glob. Fash. Market. **1**(3), 164–171 (2010)
16. Tsai, H.T., Huang, H.C.: Determinants of E-repurchase intentions: an integrative model of quadruple retention drivers. Inf. Manage. **44**(3), 231–239 (2007)
17. Anderson, R.E., Srinivasan, S.S.: E-satisfaction and E-loyalty: a contingency framework. Psychol. Mark. **20**(2), 123–138 (2003)
18. Augusto, A.: Metodologias Quantitativas/Metodologias Qualitativas: Mais Do Que Uma Questão de Preferência. In: Forum Sociológico, pp. 73–77. CESNOVA (2014)
19. Marôco, J.: Análise Estatística Com o SPSS Statistics, 8th edn. Reporter Number (2021)
20. Kline, R.B.: Principles and Practice of Structural Equation Modeling, 4th edn. Guilford Publications (2015)
21. Marôco, J.: Análise de Equações Estruturais: Fundamentos Teóricos, Software e Aplicações. 3nd edn. Reporter Number (2021)
22. Hair, J.F., Sarstedt, M., Ringle, C.M., Mena, J.A.: An assessment of the use of partial least squares structural equation modeling in marketing research. J. Acad. Mark. Sci. **40**(3), 414–433 (2012)

Chapter 53
The Impact of Food Delivery Applications on the Restaurant Industry: The Perception of Restaurant Managers in the Metropolitan Area of Porto

Jorge Boabaid and Sandra Marnoto

Abstract This work investigates the use of Food Delivery Applications (FDAs) by restaurants. It aims to understand the main perceived impacts, from the perspective of restaurant managers, of this technological innovation on the ecosystem, on the relationship between restaurants and their customers and on the value proposition of restaurants. Recently, an expressive number of studies investigated the motivations for the use of FDAs by users. However, the literature has devoted little attention to the perception of restaurant managers regarding the impact of these platforms on the restaurant ecosystem. To achieve the proposed objectives, we used qualitative research, based on semi-structured interviews (case study) with entrepreneurs and restaurant managers. The results of this investigation suggest that FDAs represent a new channel for restaurants and can be considered a facilitator to leverage sales. However, from the perspective of restaurant managers, there is a trade-off; the practices, policies, and strategies implemented by these platforms cause (1) loss of control over restaurant sales (dependence), (2) changing customer behavior and expectations, (3) greater difficulty in managing and retaining consumers, and (4) negatively affect the value proposition of restaurants. To practitioners, we suggest that managers and restaurant owners are aware of the implications of these platforms for their businesses. Ideally, we recommend that they seek differentiation through a "unique" value proposition, avoiding exclusive dependence on this channel, working on strategies to improve customer relationships and loyalty.

Sandra Marnoto—F01B-4E84-5653.

J. Boabaid
MIETE, Engineering University of Porto, Porto, Portugal

S. Marnoto (✉)
UNICES, University of Maia, Av. C. Oliveira Campos, 4475-690 Avioso S.Pedro, Portugal
e-mail: smarnoto@umaia.pt

J. L. Reis et al. (eds.), *Marketing and Smart Technologies*, Smart Innovation, Systems and Technologies 337, https://doi.org/10.1007/978-981-19-9099-1_53

53.1 Introduction

The growth and diffusion of e-commerce (electronic commerce) made it possible for consumers to buy almost every product from the comfort of their homes, at any time and in a convenient way [34]. It was in this context that the Online Food Delivery (OFD) industry emerged and grew.

Food Delivery Applications (FDAs), that operate in that industry, are seen as marketplaces where sellers come together to sell their products or services to a customer base. The role of the marketplace owner is to bring together sellers and customers and manage sales through a technological platform. Sellers have a place to gain visibility and sell their products, and the FDAs' owners earn a commission from each sale [26]. Through a technological interface, the consumer swiftly receives the desired food directly from the restaurants' own channels or through one of many intermediary technological platforms [27, 28]. The use of Food Delivery Applications (FDAs) is especially appealing for the new generation of consumers, who prefer to eat at the comfort of their home without cooking, and enjoying a quick delivery and an easy payment. Therefore, these consumers prefer speed and convenience to having a meal at a restaurant [7, 30].

The topic of technological platforms that intertwine restaurants, couriers, and consumers [28] attracts the attention of many researchers. Most studies focus on the motivations of users [4, 6, 34]. However, just a few studies seek to understand the motivations and implications in the context of restaurants [5, 7, 24]. What is the trade-off between brand promotion, number of sales, margins, operation control, and customer relationships? What are the impacts for restaurants? These, among other issues, are mostly overlooked.

The objective of this study is to understand the perception of managers regarding the implications of these technological innovations (FDAs), in a traditional sector such as it is the case of restaurants. More specifically how (a) FDAs' customer acquisition strategies and (b) promotional strategies and practices influence the restaurant industry, the restaurant–customer relationship, and the value proposition of restaurants, in the managers' opinion.

53.2 Reviewing the Literature

In the preparation for this research, a literature review was conducted using SCOPUS as a source of scientific information and using the keywords "FOOD" and "DELIVERY" (please see Appendix 1 for further information). Although the catering industry is a topic widely explored by the literature, in many other contexts, the specific treatment of this technological innovation can still be considered a raising topic with most of the existing works on Food Delivery Applications (FDAs) and Online Food Delivery (OFD) studying the motivations of users to use this type of platform.

The literature can be divided according to the main agents on focus: (1) users; (2) platforms (FDAs); and (3) restaurants. Regarding (1) users, the main topics are motivations for use, behavioral analysis, experience, and technological acceptance (e.g. [6, 33]). The articles dealing with (2) platforms mainly analyze the contents, features, profitability, and pricing strategies adopted by FDAs [9, 27, 29, 32]. Studies related to (3) restaurants can be considered initial and exploratory and aim to discuss the benefits and points of attention of the use of OFD systems by these businesses [5, 7, 24].

53.2.1 Food Delivery Applications—FDAs

Until recently, food delivery services were offered exclusively by restaurants, most times accepting orders by telephone and having a complete structure to attend orders, prepare meals, process payments, and make deliveries. With the development of digital platforms (FDAs), restaurants have gained a new option of delivery, which allows them to focus only on the operational processes of cooking and serving customers in loco [27]. Cho et al. [6] say that before the existence of the FDAs, the delivery service was extremely limited, being present only in some types of restaurants, such as pizzerias.

Furunes and Mkono [9] describe the process of how FDAs work: Customers open the "app", choose one or more products from the menus available from various restaurants, and complete the purchase online. The restaurant then accepts the order, and the customer can track the order's progress through the app. The restaurant makes the meal and provides the order to the courier, who delivers the prepaid food to the customer. Besides, FDAs use listings (such as traditional real estate listings) and recommendation algorithms for users, facilitating customers' decision-making [28].

All economic agents involved in the platform obtain benefits. Restaurants can focus on the operational process and do not need to invest in physical spaces to serve more meals [24, 8] and increase sales [6]. Customers have several ordering options, digital payment methods, access to reviews and opinions [27]. FDAs obtain financial gains through transaction commissions, which vary between 15 and 30% of the value of each order [9, 12, 27].

On the users' side, the ease of use and convenience brought by these applications were the main reasons for the popularization of online food ordering systems [27]. With the resulting increase in orders placed through this system, FDAs became the most popular method, among restaurants, to get new customers and increase restaurants' revenue [6]. Restaurants began to view FDAs with great interest as they provide an additional sales channel and have the potential to reduce customer acquisition cost, which typically represents 30–40% of revenue [33]. The popularization of FDAs was so significant that new business models were designed exclusively for these channels, such as Cloud Kitchens, also known as Delivery Kitchens, that are, in fact, virtual restaurants. These work as a food preparation center for restaurants, operating in the delivery-only model, without face-to-face customer service [23]. The model allows the reduction of some costs—such as those related to space, which can be smaller and less luxurious, and the reduction of the number of employees as they are not needed to attend tables—thus gaining a competitive advantage [30].

As of March 2020, the impossibility of serving customers in person—due to the confinements of the COVID-19 pandemic—was a catalyst for the FDAs' industry that achieves exponential growth, as consumers and restaurants felt encouraged to adopt these online channels. FDAs made it possible to overcome the barrier of confinements for restaurants, that otherwise could only serve customers through the takeaway system. During these periods, many customers preferred not to choose the takeaway option for fear of contagion, thus giving greater traction for FDAs, and pushing traditional restaurants to FDAs [36].

53.2.2 The Practices and Strategies Implemented by the FDAs

The adoption of technologies that support the e-commerce, such as the FDAs, has created a large number of opportunities for entrepreneurs to create new businesses and to expand and grow their operations. However, according to [1] and Mehendale [22], in the case of catering, the industry appears to have already reached a point of market saturation, with many competing platforms, with few differences between value propositions, using venture capital (investment groups) to finance unusual promotions in the restaurant industry, with the ultimate goal of acquiring new customers. According to [5], such strategies ended up creating a dystopia in the relationship between restaurants and consumers, with all the benefits going to FDAs and consumers.

On the restaurants' side, the margins dropped significantly, due to the high commission rates, making some restaurants feel pressured to increase prices on online channels, thus losing competitiveness [24]. Cho et al. [6] argue that presently the restaurants that decide to enter the FDAs market will find a scenario of intense competition, needing to invest money in marketing on the platforms, so that their virtual store appears among the first options of the application. He et al. [12] complement saying that those who hold power are the platforms. If any FDA decides to change the delivery policies (for example, increasing commissions), it will directly affect the restaurants' profits.

According to Furunes and Mkono [9], these models of action based on technological innovations do not always imply improvements in the services provided. Depending on the demand and availability of the couriers linked to the FDAs which are having, at the time of an order, as well as the policy of incentives that are offered to these workers, the waiting time of consumers will be affected. In other words, platforms hold most power and are important determinants of the performance on the OFD market.

Chhonker and Narang [5] highlight two points that can negatively impact restaurants: (1) mass customer acquisition strategies through discounts and (2) technological dependence. In line with that study, Kapoor and Vij [19] and Chandra and Cassandra [4] argue that, when placing orders, one of the most relevant factors for customers is the existence of discounts, vouchers, cashback receipts, among other similar possibilities. The significance of discounts is so strong that [30] suggest that giving discounts is the main promotional activity that restaurants can use to succeed in FDAs. Regardless of the point of analysis (FDAs, restaurant, and final consumer), price guidance is always present, namely the recommendation for restaurants to lower prices to be successful on the platforms [10]. This context creates a cycle, where end customers (users) of FDA always expect to save when they make purchases through these channels [19].

According to Mehendale [22], the explanation for these practices and strategies may be the aggressive competition for customer acquisition among FDAs. In addition to competitiveness, the need for large amounts of capital to invest in technological infrastructures makes these businesses attract financial investors, who want a quick payback of their investment, thus promoting the practice of aggressive market growth strategies.

53.2.3 The Context of Restaurants

High competition is not new for the catering industry, which is characterized by working with extremely small net profit margins. According to Chhonker and Narang [5], what is new is that new restaurants, opening businesses today, will end up being technologically dependent on FDAs. This dependence results in high commissions plus the logistics fees that can have a major impact on restaurant revenue, severely declining profit margins. Since this is a sector with many inexperienced entrepreneurs, most of them are attracted to the advantages offered by FDAs, but many of those end up losing control of their operations to the delivery apps, leading up to unfeasibility. This causes restaurants to lose confidence in new technologies.

In contrast, Gunden et al. [10] argue that FDAs can help managers with obtaining data that restaurants did not have before.

53.2.4 Customers' Relationships

In an environment of extreme competition and challenges, as in restaurant industry, being able to satisfy customers is not enough. The key to survival and success is gaining customer loyalty [31]. With a wide range of restaurants offering a variety of options to choose from, it has become a challenge for restaurants to retain customers in the long term. Users become loyal to food delivery applications rather than restaurants [7].

Customer relationship and loyalty can be measured by repurchase intention, long-term continuity of the relationship, positive referral to new customers (word-of-mouth), and most importantly, behavioral loyalty, which ultimately contributes to the company's overall profitability, providing the basis for an healthy competitive advantage. In addition, creating and maintaining customer loyalty reduce marketing costs and increase profits [2, 11, 16, 17]. In the restaurant industry, there are five other dimensions that can influence customer relationship and loyalty: food quality, service quality, price, location, and environment [14, 16]. According to Kivela et al. [21], restaurants are extremely dependent on loyal customers, as they not only bring in recurring revenue, but also provide predictability, security, and pleasure for those involved in the business. Suhartanto et al. [31] analyzed the quality of electronic service, quality of food, perceived value, and customer satisfaction and found that all dimensions were significant for online customer loyalty.

53.2.5 Value Proposition and Value Perceived by Customers

According to Osterwalder [25], the value proposition refers to the value that a particular company creates for its customers and partners. It is a global view of the offerings, products, and services that together represent value for a specific customer. It describes how the company differentiates itself from its competitors and it is the reason customers buy from the company.

In the context of restaurants, the literature does not deal with the concept of value proposition, but with the value perceived by customers (VPC). Customer-perceived value is defined as the result of an assessment of the relative rewards and sacrifices associated with an offer. Hence, it is a critical factor in evaluating and making decisions regarding restaurants. Because it has a personal nature, the perceived value is subjective [20].

Jin et al. [18] say that brand image is also a vital concept in consumer behavior in relation to the perceived value of offers and subsequent customer behavior. In addition to influencing decision-making in the pre-consumption phase, VPC also influences customer satisfaction and behavioral intentions in the post-purchase phase. Other dimensions that also interfere with VPC are food quality, environment, price, and customer relationship dimensions, such as satisfaction (meeting the need) and customer loyalty [15, 17, 18]. This suggests a strong link between the value proposition and customer relationships, and because of this, a catering strategy focused on creating more value and building quality customer relationships is effective in earning more profits and increasing the competitive advantages.

The literature suggests that managers should make sure to create value-centric strategies when planning their menu items, service quality, and customer relationship management. In addition to the functional value that customers perceive from restaurant food and services, it is important for managers to implement certain tactics that can support a value-based marketing strategy, such as ensuring that employees know the products being served and that they have the necessary social skills, which increase the value perceived by customers [15, 18, 20].

In the context of FDAs, in the customer perception of the restaurant, there is a very significant link between the quality of the food (restaurant), the electronic service provided (FDAs), and the service provided by the courier. All these factors affect the gastronomic experience. Therefore, we may say that the service provided by platforms is an extension of the restaurant service [31]. Furunes and Mkono [9] report that there are problems perceived by customers regarding the lack of professionalism of the courier and that this problem directly affects the customer experience. In this sense, Das and Ghose [7], point out that quality problems during the delivery phase happen frequently and that all quality problems result in a negative customer experience, which directly or indirectly damages the restaurants' image.

53.3 Methodology

Due to the exploratory nature of this study, as well as the research objectives, the methodology chosen for this work was the analysis of case studies. Four restaurants with different characteristics and from different segments were chosen to be studied in the research. The UberEats platform, being the most popular, was used to initially prospect restaurants and characterized them. Based on these definitions, the companies were listed, categorized, and then, selected, following the requirements: (1) to be listed in a FDA; (2) to be located in the metropolitan region of Porto, Portugal; (3) to select at least one restaurant that works exclusively on FDAs; (4) to select at least one restaurant that is a "Delivery-Only" company. Then, attempts were made to contact them by email and Instagram messages. Without much difficulty, four managers agreed to participate in the research.

The data collection instrument chosen was the elaboration of semi-structured interviews, for which an interview guide was prepared. This way we ensured that the most important issues for this research were addressed from the restaurant managers' perspective. The reason for following a semi-structured interview was that it allows the interaction with the interviewee to be more natural and less biased and that previously unplanned points could find space to be discussed [35]. During the interviews, we tried to create a conversation type of environment, always being careful not to direct the answers and giving priority to the interviewees' interventions. All interviews were conducted by videoconference, using the Google Meet software—this method was preferred because it was easy to record the conversation and subsequently transcript and analyze it. Also, this option was due to the health crisis experienced during the period of the interviews. Respondents were asked, whenever possible, to give concrete examples for each of the 20 questions. All four interviews took place in April 2021 and lasted an average of 1 h.

The interview guide was divided into five groups, based on the literature review: characterization and contextualization of the restaurant; perception about FDAs' practices and strategies; perceived impacts and changes in customer relationships;

perceived impacts and changes in the value proposition and financial structure of the restaurant; and finally, perceptions about the future. In each of these groups of questions, the following information was gathered:

I. Characterization and contextualization of the restaurant—in this group of introductory questions, we obtained information regarding the company's activity, the time in operation, the number of employees, the number and with which FDAs it works with, among other information regarding the history of the restaurant in the context of the platforms.
II. Perception about FDAs' practices and strategies—at this point, we listened to the managers' perceptions of customer acquisition practices and the nuances of the relationships with each FDA.
III. Perceived impacts and changes in customer relationships—this section allowed us to understand how these managers perceived changes in customer profiles, changes in habits, and the impact of FDAs on long-term relationships with customers.
IV. Perceived impacts and changes in the value proposition and financial structure of the restaurant—here, we aimed to understand how the performance in the FDAs was perceived to change the product and service offered by the restaurants, being it related to market positioning, to competition, or to quality of the products, and also, to understand the changes in the cost and price structure, that the managers believed to have influenced the value perceived by customers (VPC).
V. Perceptions about the future—through open questions, the managers were given the possibility of reflecting and sharing with us what they thought the next steps of the restaurant within the context of delivery apps should/would be.

The main method for analyzing the data obtained in case studies is the "Identification of Patterns", which according to Yin [35] allows certain patterns to be identified within each of the cases studied and then compare them with the literature. Thus, the data were analyzed using qualitative analytical techniques, including coding (using Taguette software), thematic development, and comparison of patterns with existing literature [3]. Descriptive and interpretive tables were then constructed, regarding the restaurant category, and in terms of perceived impacts. In order to organize and facilitate the analysis of the results, tags were used to identify patterns found in the perceptions of the managers.

A cross-case analysis was also performed, but it is not included in this short paper.

53.4 Data Analysis and Empirical Results

The cases are divided into four parts: (a) the description of the case; (b) the perceived impact in the restaurant caused by the FDAs' strategies and practices; (c) the perceived effect on customer relationships; and (d) the perceived influence on the value proposition and VPC.

53.4.1 Restaurant "A"

(a) This restaurant operates only by delivery. It belongs to a Brazilian franchised chain, being the first unit opened abroad. Their main product is fried chicken, and according to the owner–manager, they seek differentiation through superior quality and larger portions than the competition. The restaurant has started operating in March 2020, corresponding to the period of the first COVID-19 lockdown. It is located in Porto, and it has three employees, including the manager, his wife, and a kitchen assistant. It also works with drivers that deliver orders made through direct channels and has contracts with three FDAs. It operates with UberEats, Glovo, and TakeAway. Initially, the restaurant has worked exclusively with UberEats, in an agreement that included slightly lower commissions (27% vs. 30%) and helps in the area of marketing. Not much later, the manager decided to terminate this exclusivity deal, as he understood that the FDAs' promises were not being fulfilled. Today, the commission charged by FDAs varies between 10 and 30%, depending on the platform. The manager reported that the main motivations for using the platforms are the increase in sales and the promotion of the brand in such a popular channels. In an attempt to maximize the net margin by reducing commissions, the restaurant also has its own delivery channel, with two couriers making deliveries. In addition, the restaurant tries to transfer customers from the FDAs to its direct channel, by sending pamphlets and a having a different menu, with lower prices. About 80% of the restaurant's total revenue comes from FDAs and 20% from direct sales, which makes the restaurant significantly dependent on the FDAs, in the perspective of the interviewee.

(b) The owner–manager of this restaurant believes that, today, platforms control the market. They are in a comfortable position in relation to restaurants, mainly due to the COVID-19 pandemic, which caused an increase in the use of FDAs by customers, as well as in the number of restaurants working with FDAs, generating an aggressive competition. He believes that this situation results in opportunistic behaviors on the part of the FDAs, which do not provide the level of service expected by restaurants. He reported, for example, great difficulty in getting in touch with the technical support teams and with the commercial sector of the platforms, which significantly affects his sales. For instance, the manager could not get in contact with UberEats to renew the contract for days, and due to the lack of response, the restaurant was unable to operate on this channel. Furthermore, the manager believes that there is market manipulation on the part of platforms. He explained that he receives orders at unusual times and frequencies, suggesting that the FDAs manipulate the positioning of the restaurant in the listings provided to costumers, making him feel that he does not have control of the restaurant in the "apps". Despite having no experience in the period prior to the pandemic, the manager noticed a drop in the average sales value (average ticket) and explained that over time, customers' preferences moved toward cheaper items and that this greatly affected the restaurant by decreasing margins. He also indicated that there is a very strong use of discounts on the platforms, and it is even common for him to resort to competition to guide the discounts given. He perceives discount practices as very aggressive, where only the restaurant waives its revenue share, and that such actions have an insignificant practical result in the acquisition of new customers—he recalled the case, in which he placed 100 hamburgers for 1 euro and sold out in less than ten orders. He thinks that there is a significant portion of customers who are completely dependent on the discounts offered, reporting some cases in which customers send him messages through social networks, asking about the discount of the day, and when the answer is negative, they do not place any order. He believes that the increase in the number of competitors working exclusively through these platforms has transformed it in an extremely competitive environment, fostering a state of cannibalism among restaurants. However, he wondered: "Is it bad with Uber? Yes, but it is worse without Uber". Regarding the future, due to relationship problems and dissatisfaction with the FDAs, he plans to soon also work with table and terrace service (in loco).

(c) Since the beginning of the operation, he has seen that the profile of customers has not changed: public between 18 and 30 years old, composed mostly of students. He reported that it is very difficult to manage customer relationships because the only information he has access to is the customer's first name. Despite all difficulties, he thinks that there still is customer loyalty. "There is an important portion of customers that order several times, who then go to our WhatsApp. But, in general it is a great challenge", he said.

(d) This is a business model and value proposition designed to prioritize the delivery system (delivery-only). The manager is satisfied with the presentation of the menus, the photos, and the level of differentiation that the restaurant has in the "apps". The thinks that the negative impacts on the VPC result from the frequent problems the restaurant faces with couriers, such as theft, undelivered orders, delays, among other problems, which affect the customer's perception of the value of the company, also generating restaurant evaluation problems on the platforms and reducing the perceived value of the brand.

53.4.2 Restaurant "B"

(a) Based on a multichannel strategy, the Restaurant "B" sells meals and hamburgers with a typical bread from Madeira Island. Located in Porto, the business started in January 2018, with only the physical channel. According to the manager–owner, the restaurant operates now with six platforms, with UberEats and Glovo being the main ones. She reported to not having any exclusivity agreement, and therefore, the commissions charged by the "apps" are 30%. She started working with the FDAs in the middle of the first year in operation, and since then, she has witnessed the significant growth of the online channel, which now represents 50% of the total sales revenue. She confessed that she cannot see the restaurant surviving without the FDAs. According to the manager, the main reason for the restaurant's presence in the FDAs is the increase in sales. In addition, the back office of the UberEats platform is another important motivation. It reports important market information, sales history, and suggestions for actions for general improvements. The manager felt the impact of the pandemic on in loco sales, as the restaurant is located next to many commercial activities that had to close. She reports that the platforms represented the only sales channel for the restaurant during the confinement and acknowledges that this has taken away control over the sales made.

(b) The interviewee believes that there are abusive behaviors and control of restaurant disclosures by the platforms, depending on whether or not they participate in the promotions suggested by the FDAs. She says that to be at the top of the listings in FDAs and increase sales, she needs to engage in aggressive discounting practices. She understands that this makes customers more price sensitive and uses examples, saying that when they offer a "simple" free delivery promotion, sales can increase by 50%. Despite being dissatisfied with some nuances of the use of delivery apps, she understands that online channels are the future of the business, so she plans to open four more virtual restaurants (ghost kitchen) soon.

(c) According to the manager, the profile of customers remains the same: adults, between 30 and 50 years old, belonging to the middle and upper-middle classes. Recently, she has noticed a change in the behavior of these customers, who, according to her, are looking for products with more discounts, but she says that she does not know if it is because of the crisis or if it is for another reason. In addition, she perceives some difficulties in managing customers through the platforms, where she is unable to respond to comments and evaluations. This situation makes her feel that the customers are now very far away. She also realized that the number of competing restaurants is growing, and with that, the difficulty in retaining customers has increased. According to her, customers today have many options, which is "an advantage for the customer".

(d) In general, despite this perception regarding the difficulties in managing and retaining customers, the interviewee suggests that it is a trade-off that is worth making. During the interviews, the manager revealed that she will open four more virtual stores (ghost kitchens), that is, it will slightly change the restaurant's value proposition, which includes face-to-face service. The main idea is to reach a larger geographical area, but without radical changes in brand positioning and differentiation. In the manager's view, it is natural that the products sold via delivery lose quality, but it is the dissatisfaction with the delivery service of the "third party" couriers, linked to the "apps", which most affects the perception of value of the products. She reported several negative customer experiences, such as exacerbated delays due to the collection of multiple orders by a single driver. According to her, this type of experience is then reflected on the restaurant's evaluation in the "app" and invariably ends up changing the perception of potential customers about the product.

53.4.3 Restaurant "C"

(a) Restaurant "C" is an "all you can eat" buffet business model, and it sells "complete" and healthy meals to the vegan niche. According to owner–manager, it is a franchise with ten restaurants in the Porto district, of which the interviewee owns two units, one located in Gaia and another in Matosinhos. They both opened in the first half of 2019. In total, the two units employ nine employees. Regarding the FDAs, he has an exclusive agreement with UberEats. This agreement was made by the franchise owner and allows the chain to pay a lower commission rate: 20%. The main objective of working with FDAs is to promote the brand, and secondly, to increase sales. Currently, the FDAs represent 5–7% of total restaurant turnover. Additionally, he notes that the platforms allow restaurants to reach customers who would never have known about them. It is worth noting that, as the restaurants would not be able to bear the costs of operation, the manager preferred to keep the restaurants closed during the confinement phase.

(b) The manager considers that there is a very aggressive discount practice, leveraged by large international restaurant chains. He reported that FDAs encourage restaurants to carry out promotions, but that entering in these discount cycles and seeking to compete with these large networks are not sustainable. He realizes that there is a relationship between the discounts offered and the positioning of restaurants in the listings of the "apps". He affirmed that putting a discount on top of the commissions charged is "impossible", as it would "crush" the margins, which are already very small. On the other hand, he understands that the characteristics of his restaurants, which work with customers focused on sustainability, within a vegan niche, mean that he does not have to worry about price wars in relation to the competition. Regarding the future, he does not consider and does not see any advantage in increasing the participation of restaurants on the platforms. However, he recognized that online channels tend to grow and said that attention must be paid to the "leonine" practices of the platforms.

(c) Despite the niche strategy designed by the brand, the manager reported that they lose control of customer service on FDAs, as they need an intermediary for any issue that needs to be dealt with. He said that at certain times, there is a significant demand on the platforms, and as he only has two kitchen employees, he ends up having problems in meeting the demand of "online" orders and the demands of "offline" customers. In the manager's view, because the restaurant operates in a niche where the consumer is very loyal, he thinks that he has no difficulties with customer loyalty. Moreover, he thinks that the FDAs help to retain customers, due to the promotion of the brand: "I learned that customers are loyal in both channels, I see that they repeat orders, and we end up seeing the same behavior that we have physically".

(d) According to the manager, there is a significant loss in the value proposition of his restaurants when suppling an FDA order. He reports that the value proposition changes completely when operating in the FDAs, as it needs to significantly limit the offer, not being able to offer the buffet service, and ending up having an à-la-carte model of action in online channels. Issues such as commoditization and integrated experience arose during the interview. He understands that the value of the product is affected by FDAs, as it is impossible to maintain the quality of the food. He also cited frequent problems with "third party" couriers, who sometimes deliver multiple orders, from multiple restaurants, taking a long time and ending up affecting the product, and consequently the value perceived by customers.

53.4.4 Restaurant "D"

(a) The main business of Restaurant "D" is the sale of hamburgers, through a multi-channel model. It is located in Porto's downtown, and it has been in operation since March 2019. The restaurant has four employees, although that number is variable, according to customer demand. The owner–manager explained the restaurant's nuances. According to the manager, they are operating with the UberEats and Glovo platforms, but they also explore the direct sales channel for delivery. He said that he had analyzed proposals with the FDAs to work in exclusively, but had decided not to move forward because he had realized that he would lose sales. Because the restaurant has its own couriers, the commissions paid to FDAs are between 15 and 30%. He started operating with the platforms six months after the beginning of operations, and today, sales on the platforms correspond to 65% of total sales, which according to the manager, was a significant increase in sales. The manager understands that the participation in the FDAs is vital to his business, and despite acknowledging dependence on FDAs, he admitted that he does not renounce the advantages offered by platforms. He perceives a lot of value in the back-office system offered by UberEats as he obtains information that he would never get with his own systems. In his own words, FDAs are "a necessary evil".

(b) The manager perceives that there is a great challenge imposed on restaurants, due to the high commissions charged by FDAs, which he cannot significantly pass on to customers, thus losing net margin. Regarding discount practices, the manager said: "In the past, you would go to a pizzeria and you would not go to the counter and ask for a discount. Today, the platforms, in addition to encouraging and highlighting the restaurants that offer discounts, have a filter for discounts, and people go straight to this option". To escape the discount practices suggested by the FDAs, the entrepreneur looks for alternatives to invest in marketing, such as, for example, investing in alternative channels. He admitted that discounts increase sales, but that practice is not sustainable. So, sometimes, he tries to balance it with campaigns on social networks. However, he confessed that he is afraid of being forgotten by customers if he stops using the FDAs. For the future expansion of the restaurant, he is thinking of working on a delivery-only model, focusing exclusively on virtual kitchens, without table service. Not only through the FDAs, but also focusing on direct channels, as he understands that if he worked only with FDAs, he would be too dependent, forcing him to achieve a very significant sales volume to have a satisfactory financial return.

(c) The restaurant faces challenges both in managing customers and in retaining them. The combination of a location where many tourists stay, with the use of FDAs, makes the difficulties harder. The only information he has about the customer is the address and telephone number, that the platform provides. The way the restaurant relates to customers has changed completely. "It is necessary to offer excellent service and there is no margin for error, largely due to the evaluations, feedbacks and ranking that the FDAs propose", he said. According to the manager, with only one negative note, the restaurant drops considerably in the ranking of the FDAs. As it is located in a very touristic geographical area of the town, he understands that there is no way to circumvent customer loyalty problems. However, he still implements strategies to try to attract customers to the restaurant's direct channel. In each FDAs' delivery, he sends a pamphlet publicizing a menu with lower prices, to encourage direct orders.

(d) The manager reveals that he realizes that it is difficult to differentiate the restaurant in the FDAs. Despite including elements in the value proposition, such as: buying fresh products and using organic products, the manager finds it very difficult to make the customer understand these differences. In the manager's opinion, FDAs' customers fail to experience an important aspect of the restaurant. "Our service is very personalized, nothing mechanical, we try to create a connection with the customer, we want the whole process to be remarkable… and, because we do not have control of the couriers linked to the applications, we are not sure how this interaction with the customer was, but we believe that it is worse." That, in the manager's view, lowers the value perceived by customers.

53.5 Conclusions

The main objective of this work was to understand the impacts of FDAs on the restaurant industry, from the perspective of restaurant managers. The study focused on the perception of restaurant managers regarding the impacts caused by FDAs' practices and strategies on the restaurant industry, on the restaurant–customer relationship, and on the value proposition of restaurants. By doing so, this research contributes to the blossoming literature on Food Delivery Applications.

The main practices and strategies used by FDAs that are described in the literature were identified in the research. The study case of four restaurants, using semi-structured interviews, enabled us to corroborate a few elements described in the literature, such as the technological dependency, and the aggressive practices of discounts [5].

The analysis of the four cases allowed us to confirm the heterogeneity of the existing situations in the context of ADFs. In the case of restaurants, this heterogeneity exists not only in the segments, but in the different forms of action and thinking among managers. These different aspects enrich our knowledge of the object of study.

53.5.1 Research Conclusions

1. Impact on the catering industry

As expected, FDAs are a reality in the market and should not lose strength anytime soon [19, 30]. Although decreasing the margins of restaurants due to the promotion activities proposed by the platforms, these latter have the predicted impact of increasing sales and also are the most import channel of distribution (except for Restaurant "C" which serves a particular niche). The restaurants in this study prefer not to work in exclusivity with a FDA because although the commission could be lower, it does not compensate for having less distribution channels (less sales), which, in addition, are also considered displays for the restaurants. Further, it does not compensate for taking the risk of something going wrong with that platform (as it has happened with Restaurant "A").

2. On the relationship between FDAs, restaurants, and consumers

Based on the perception of managers, the study identified the practices and strategies adopted in the sector in response to the emergence of platforms and their impact. Thus, the study identified that the platforms:

a. create sales dependency for restaurants [5], which, combined with the perception of manipulation of the business positioning in the "apps" listings [6], results in a feeling of loss of control of the operation by the managers,
b. encourage the practice of discounts [4, 5, 19, 30] which ends up fostering aggressive competition [1, 5, 22], leading to changes in customer expectations and behavior [19].

3. About restaurant–consumer relationship management

In most cases, the extensive use of platforms by restaurants also has a negative impact on managing customer relationships. The existence of an intermediary (FDAs) makes it more difficult for restaurants to manage customers; and, the large number of competitors and options in the "apps" makes customer loyalty more difficult to achieve [1, 5, 22]. Therefore, some restaurants (e.g., Restaurants "A" and "D") try to lure customers to their direct channels.

4. About the value proposition and the value perceived by customers

There is a dystopia between the value proposition planned by restaurants and the value perceived by customers (VPC). In the perception of managers, FDAs have a negative impact through poor delivery experiences, in some cases limiting offers, causing the overall value perceived by consumers to be lower, even when there is no change in the value proposition of the restaurants [7, 9, 31]. As a result, this dystopia might influence restaurants to change their value proposition, such Restaurant "B" is considering.

These four conclusions and their inter-relations are illustrated in Fig. 53.1.

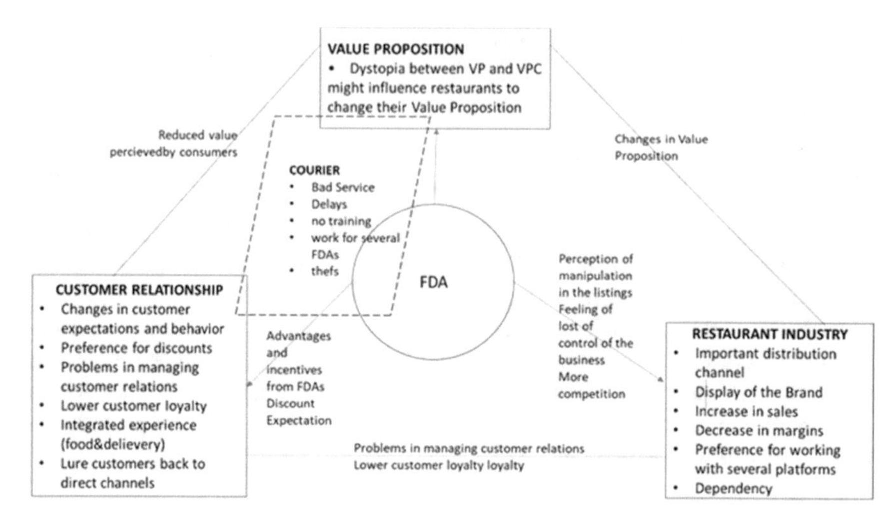

Fig. 53.1 Generalization of perceived impacts on the ecosystem

Considering that FDAs already have and tend to have significant effects on the national economy, bringing attention to these impacts can be useful for all stakeholders. If the trend found in this exploratory study is confirmed, the problem will possibly demand that the Portuguese authorities intervene in the market, to curb destructive practices. This would be in line with [5], who recommend that there should be a market regulatory body to help restaurant managers in order to minimize FDAs' potential negative impacts.

Specifically for restaurants, as Gunden et al. [10] suggest, those who intend to grow and continue to operate in the context of FDAs should seek differentiation and clear communication of the value proposition, to minimize negative impacts on the VPC. The case of Restaurant "C" is a good example of these differentiation strategies.

For FDAs, this study seems to suggest the need to create a value proposition that is more like a win–win strategy for both the platform and the restaurant. A good example is UberEats, which, according to Restaurants "B" and "D", in spite of the negative impacts of some of its practices, offers restaurants' valuable information that is facilitated by the platform's back office.

Due to length limitations, this short paper does not discuss the results of the cross-case analysis that was performed and which has also brought to light some interesting results.

As limitations and suggestions for future research, we believe that doing the research during the pandemic period, which accelerated the adoption of FDAs, might have somehow biased some managers' opinions. Therefore, it will be interesting to study how the perceptions on the use and the impacts of delivery apps in the post-pandemic period unfold. In fact, a longitudinal study, to understand the perceived impact of FDAs on the Portuguese economy, as well as what the long-term impacts of the extensive use of this new technology are and will be are of upmost importance.

It is also worth mentioning other categories of platforms not analyzed in the work, due to the lockdown, such as Zomato and Fork which also have components and strategies similar to FDAs and could also improve our knowledge about this theme.

Appendix 1—Method Used for Literature Review

Following the recommendations of Pinto Ferreira et al., (2020) a systematic and rigorous approach was used to review the literature, which consisted of using the terms "Delivery" and "Food" on March 4, 2021. The combination resulted in 8649 articles, and due to the high number of studies, the criterion of limiting the "research area" to "Business, Management and Accounting" was used, resulting in 144 documents. As this is a recent innovation, which gained popularity from 2016 onward [13], the results were limited between 2016 and 2021, resulting in 104 documents. After that, only Portuguese and English language works were limited, finally resulting in 102 documents.

After reading the title and abstract (abstract) of the 102 documents, 43 documents were selected. Using credentials from the university, EndNote and Google, all documents were found for full reading. Of these, 25 articles were included in the literature review, and another 5 were included after searching on Google Scholar with the terms "Food Delivery Application".

The main inclusion criterion was that the article approached the perspective of one of the three agents involved: The platform, the user or the restaurant. Articles that did not meet the criteria were excluded. The final string used in Scopus was: [KEY (delivery and food)] AND [LIMIT-TO (SUBJAREA, "BUSI")] AND [LIMIT-TO (PUBYEAR, 2021) OR LIMIT-TO (PUBYEAR, 2020) OR LIMITTO (PUBYEAR, 2019) OR LIMIT-TO (PUBYEAR, 2018) OR LIMIT-TO (PUBYEAR, 2017) OR LIMIT-TO (PUBYEAR, 2016)] AND [LIMIT-TO (LANGUAGE, "English")].

After identifying the literature gap, articles were also included in order to contextualize the concepts related to the research objectives, namely, (1) customer relationship and (2) value proposition. These articles were also selected from the Scopus database, with the following string: KEY (restaurant AND "customer loyalty" OR "relationship" OR "value proposition") AND [LIMIT-TO (SUBJAREA, "BUSI") OR LIMIT-TO (SUBJAREA, "SOCI")] AND [LIMIT-TO (LANGUAGE, "English")]. We obtained 88 documents, listed in order of relevance. The following inclusion criterion was defined: the article deals with the analysis of the restaurant industry as a central theme, and as variables include at least one of the subjects: relationship with consumers, consumer loyalty or value proposition. After reading the abstract of the 88 articles, 12 were selected for full investigation, and after reading these articles, 10 were finally included in the literature review.

References

1. Bhotvawala, M.A., Bidichandani, N., Balihallimath, H., Khond, M.P.: Growth of food tech: A comparative study of aggregator food delivery services in India. In: Proceedings of the International Conference on Industrial Engineering and Operations Management, pp. 140–149 (2016)
2. Bowden, J.: Customer engagement: a framework for assessing customer-brand relationships: the case of the restaurant industry. J. Hosp. Leis. Mark. **18**(6), 574–596 (2009). https://doi.org/10.1080/19368620903024983
3. Braun, V. and Clarke, V.: Using thematic analysis in psychology. Qual. Res. Psychol. **3**(2), 77–101 (2006). ISSN1478-0887 Available from: http://eprints.uwe.ac.uk/11735
4. Chandra, Y.U., Cassandra, C.: Stimulus factors of order online food delivery. In: 2019 International Conference on Information Management and Technology (ICIMTech), vol. 1, pp. 330–333 (2019). https://doi.org/10.1109/icimtech.2019.8843715
5. Chhonker, M.S., Narang, R.: Understanding food delivery-apps and FHRAI guidelines: A restaurant entrepreneur's perspective. In: GIT2020 Conference Proceedings, January, pp. 338–344 (2020)
6. Cho, M., Bonn, M.A., Li, J.: Differences in perceptions about food delivery apps between single-person and multi-person households. Int. J. Hosp. Manag. **77**(June), 108–116 (2019). https://doi.org/10.1016/j.ijhm.2018.06.019
7. Das, S., Ghose, D.: Influence of online food delivery apps on the operations of 50 the restaurant business. Int. J. Sci. Technol. Res. **8**(12), 1372–1377 (2019)
8. Edelman, B.G. and Geradin, D.: Efficiencies and regulatory shortcuts: How should we regulate companies like Airbnb and Uber? (November 24, 2015). Stanford Technol. Law Rev. 19 (2016), 293–328 (2015)
9. Furunes, T., Mkono, M.: Service-delivery success and failure under the sharing economy. Int. J. Contemp. Hosp. Manag. **31**(8), 3352–3370 (2019). https://doi.org/10.1108/IJCHM-06-2018-0532
10. Gunden, N., Morosan, C., DeFranco, A.L.: Consumers' persuasion in online food delivery systems. J. Hosp. Tour. Technol. **11**(3), 495–509(2020). https://doi.org/10.1108/JHTT-10-2019-0126
11. Han, H., Ryu, K.: The roles of the physical environment, price perception, and customer satisfaction in determining customer loyalty in the restaurant industry. J. Hosp. Tour. Res. **33**(4), 487–510 (2009). https://doi.org/10.1177/1096348009344212
12. He, Z., Han, G., Cheng, T.C.E., Fan, B., Dong, J.: Evolutionary food quality and location strategies for restaurants in competitive online-to-offline food ordering and delivery markets: an agent-based approach. Int. J. Prod. Econ. **215**, 61–72 (2019). https://doi.org/10.1016/j.ijpe.2018.05.00851
13. Hirschberg, C., Rajko, A., Schumacher, T. and Wrulich, M: The changing market for food delivery. McKinsey & Company (2016)
14. Hyun, S.S.: Predictors of relationship quality and loyalty in the restaurant industry chain. Cornell Hosp. Q. **51**(2), 251–267 (2010). https://doi.org/10.1177/1938965510363264
15. Itani, O.S., Kassar, A.N., Loureiro, S.M.C.: Value get, value give: the relationships perceived among value, relationship quality, customer engagement, and value consciousness. Int. J. Hosp. Manage. **80**(March 2018), 78–90 (2019). https://doi.org/10.1016/j.ijhm.2019.01.014
16. Jani, D., Han, H.: Investigating the key factors affecting behavioral intentions: evidence from a full-service restaurant setting. Int. J. Contemp. Hosp. Manag. **23**(7), 1000–1018 (2011). https://doi.org/10.1108/09596111111167579
17. Jin, N., Line, N.D., Goh, B.: Experiential value, relationship quality, and customer loyalty in full-service restaurants: the moderating role of gender. J. Hosp. Market. Manag. **22**(7), 679–700 (2013). https://doi.org/10.1080/19368623.2013.723799
18. Jin, N., Lee, S., Huffman, L.: Impact of restaurant experience on brand image and customer loyalty: moderating role of dining motivation. J. Travel Tour. Mark. **29**(6), 532–551 (2012). https://doi.org/10.1080/10548408.2012.701552

19. Kapoor, A.P., Vij, M.: Technology at the dinner table: ordering food online through mobile apps. J. Retail. Cons. Serv. 43 (September 2017), 342–351 (2018). https://doi.org/10.1016/j.jretconser.2018.04.001
20. Kim, W., Han, H.: Determinants of restaurant customers' loyalty intentions: a mediating effect of relationship quality. J. Qual. Assur. Hosp. Tour. **9**(3), 219–239 (2008). https://doi.org/10.1080/15280080802412727
21. Kivela, J., Inbakaran, R., Reece, J.: Consumer research in the restaurant environment. Part 3: Analysis, findings and conclusions. Int. J. Contemp. Hosp. Manage. **12**(1), 13–30 (2000). https://doi.org/10.1108/09596110010304984
22. Mehendale, S.: Food—Delivery Start-Ups: In Search of the Core. Scanned by CamScanner, Oct 2016 (2020)
23. Muller, C. (Boston US of HA): Restaurant delivery: are the "ODP" the industry's "OTA"? Boston Hosp. Rev. (2018)
24. Niu, B., Li, Q., Mu, Z., Chen, L., Ji, P.: Platform logistics or self-logistics? Restaurants' cooperation with online food-delivery platform profitability and sustainability. Int. J. Prod. Econ. 234(May 2020), 108064 (2021). https://doi.org/10.1016/j.ijpe.2021.108064
25. Osterwalder, A.: The business model ontology: A proposition in a design science approach. THESE. Présentée à l'Ecole des Hautes Etudes Commerciales de l'Université de Lausanne, Pour l'obtention du grade de Docteur en Informatique de Gestion (2004)
26. Pavlou, PA, Gefen, D.: Building effective online marketplaces with institution-based trust. Inf. Syst. Res. **15**(1) (2004). https://doi.org/10.1287/isre.1040.0015
27. Pigatto, G., Machado, JG de C.F., Negreti, A. dos S., Machado, L.M.: Brazil. Electron. Libr. 34(1), 1–5 (2017)
28. Ray, A., Dhir, A., Bala, P.K., Kaur, P.: Why do people use food delivery apps (FDA)? A uses and gratification theory perspective. J. Retail. Consum. Serv. **51**(March), 221–230 (2019). https://doi.org/10.1016/j.jretconser.2019.05.025
29. Seghezzi, A., Mangiaracina, R.: On-demand food delivery: investigating the economic performances. Int. J. Retail Distrib. Manage. (2020). https://doi.org/10.1108/IJRDM-02-2020-0043
30. Senthil, M., Gayathri, N., Chandrasekar, K.S.: Changing paradigms of Indian foodtech landscape-Impact of online food delivery aggregators. Int. J. Food Syst. Dyn. **11**(2), 139–152 (2020). https://doi.org/10.18461/ijfsd.v11i2.46
31. Suhartanto, D., Helmi Ali, M., Tan, K.H., Sjahroeddin, F., Kusdibyo, L.: Loyalty toward online food delivery service: the role of e-service quality and food quality. J. Foodserv. Bus. Res. **22**(1), 81–97 (2019). https://doi.org/10.1080/15378020.2018.1546076
32. Tong, T., Dai, H., Xiao, Q., Yan, N.: Will dynamic pricing outperform? Theoretical analysis and empirical evidence from O_2O on-demand food service market. Int. J. Prod. Econ. **219**(July 2019), 375–385 (2020). https://doi.org/10.1016/j.ijpe.2019.07.010
33. Ye-Eun, S., Jeon, S.-H., Jeon, M.-S.: The effect of mobile food delivery application usage factors on customer satisfaction and intention to reuse. Culinary Sci. Hosp. Res. **23**(1), 37–47 (2017). https://doi.org/10.20878/cshr.2017.23.1.005
34. Yeo, V.C.S., Goh, S.K., Rezaei, S.: Consumer experiences, attitude and behavioral intention toward online food delivery (OFD) services. J. Retail. Cons. Serv. 35(December 2016), 150–162 (2017). https://doi.org/10.1016/j.jretconser.2016.12.013
35. Yin, R.K.: Case study research: design and methods. SAGE Publications, London (1994)
36. Zhao, Y., Bacao, F.: What factors determining customer continuingly using food delivery apps during 2019 novel coronavirus pandemic period? Int. J. Hosp. Manag. **91**(May), 102683 (2020). https://doi.org/10.1016/j.ijhm.2020.102683

Chapter 54
Exploring the Impact of Esthetics and Demographic Variables in Digital Marketing Campaigns

Erika Yang

Abstract This paper will examine a survey conducted on participants to aim to understand how digital marketing campaigns and a company's branding efforts are affected by esthetics and demographics. The esthetic variables that will be examined include color, font, and staging and the participant's perception of them. The demographic variables that will be included are gender, age, and employment from participants. Discussions of these survey results will use statistical tests such as *t* tests and correlation tests in order to determine which esthetic and demographic variables are the most important to consumers. From these results, companies can understand which variables are the most significant to implement in their digital marketing campaigning.

54.1 Introduction

As the world becomes more digitally connected, companies are adapting to a new digital sphere [5]. When implemented successfully, a social media or digital campaign becomes a coordinated marketing effort that reinforces a company's business goals and leverages consumers to increase brand awareness, generate sales, and cultivate consumer loyalty [12]. In modern day, new media avenues such as social media channels and Website advertisements have allowed for companies to utilize their resources to successfully transform the perceptions and attitudes of consumers [11].

Since the early 2000s, social media marketing has opened up two-way communication between brand and audience by increasing brand visibility and providing transparency [13]. Social media platforms have also given businesses full control over who they wish to target with their advertising, resulting in an increase of business sales and brand awareness [6]. Real-world interactions have moved toward the virtual world, and now, more than ever, consumers are provided with a plethora of

E. Yang (✉)
Annenberg School for Communication and Journalism, University of Southern California, Los Angeles, CA 90007, USA
e-mail: erikayan@usc.edu

J. L. Reis et al. (eds.), *Marketing and Smart Technologies*, Smart Innovation, Systems and Technologies 337, https://doi.org/10.1007/978-981-19-9099-1_54

information at their fingertips and are curious to see how different businesses market themselves on these various platforms [11].

54.2 Theoretical Framework

Previously, companies may have employed the usage of email blasts, television and radio advertisements, and other marketing mechanisms to reach their audiences [2]. These forms of marketing were successful during a time when traditional media was prominent and catered to consumers with longer attention spans [1]. Now with the digital transformation, much of the general public has shortened their attention span from twelve minutes to five minutes [9]. The drastic reduction of audience attention span can be attributed to the usage of smart devices, which wreak havoc with the users' attention spans and throw their concentration levels off [10]. With consideration to digital advertisements, it takes the average person seven seconds to decide if they endorse a company brand [3]. Thus, it becomes crucial for digital marketing campaigns to cater to these shorter attention spans and convert potential consumers quickly and effectively. The most successful digital marketing campaigns are the ones that consider the meticulous details of a digital post or advertisement and capture an individual's short attention span [4]. These details include esthetic factors such as color themes, picture and model staging, text formats, and more [8]. Roughly, 73% of the assessment of a digital advertisement or post is based on colors alone [8]. For example, the usage of colors can benefit companies in standing out from their competitors and influence the attitudes and purchasing decisions of consumers [7]. Colors, picture staging, and text formatting have the ability to strongly compel and trigger an individual's innate psychological and physiological predisposition to capture their attention and empower action [7]. In order to attract attention and standout in a highly saturated field of advertisements, companies have also resorted to using clever wordplay and a shortened text or no-text-at-all to capture the "less is more" approach. To captivate the modern attention span, brands shifted to shorten messages without .

As mentioned above, there have been many studies done on the importance of color, gradients, and shades in a marketing campaign; however, currently, there is not many research report providing a thorough study regarding how additional esthetic variables, including font and staging, affect specifically digital marketing. This paper will examine a survey conducted on participants to aim to understand how esthetic variables and demographic variables affect engagement with the company. These esthetic variables include color, font, and staging, while the demographic variables include gender, age, and employment. For this study, the specific research question is as follows: What variables, inclusive of esthetic and demographic, are the most important ones for a company to consider when developing a digital marketing campaign?

54.3 Research Method

This study will use random, willing participants to obtain data results. This data will be run through IBM SPSS statistics software, and relevant statistical tests will be applied; conclusions will then be drawn from the results of these statistical tests.

54.3.1 Experimental Participants

In this study, 254 participants were recruited from Amazon's Mechanical Turk for a study with the description of "Color in Advertising", where they were paid 25 cents. Participation was limited to participants located in the United States with a past approval rating in the Mechanical Turk system above 70%. The majority of participants (40.7%) were aged between 35 and 44 years old. Participants were recruited via Amazon Mechanical Turk. Most participants were male (57.4%) and married (77.8%).

54.3.2 Selection of Advertising Photos

The advertising photos were selected from a wide-array of popular lifestyle, fashion, beauty, and food brands that are either recognizable on a global or have a cult-like followings. Advertising photos selection was made based on the following criteria: (1) attention-grabbing imagery; (2) bold colors, texts, fonts, and sizing; and (3) emotional effectiveness of the advertisement.

54.3.3 Design of Survey Questions

An online experiment was conducted to collect data from June 7, 2022, until August 8, 2022. The independent variable was the selection of advertising photos (see Figs. 54.1, 54.2 and 54.3) shown to the participants. The dependent variable consisted of 32 questions (see Table 54.1). The dependent variables were divided into two sections: advertising photos and demographics. The first section, advertising photos, assessed participant's approval or disapproval of a company's advertising photo. An example question from this section is placing a subway advertising photo and asking the participant to rate how likely they would react to the statement: "I would eat at this restaurant". The next section examined the demographics of participants. An example question from this section was: "What best describes your employment status over the past three months"? The majority of questions were answered via a 5-point Likert scale, ranging from "strongly disagree" to "strongly agree".

Fig. 54.1 Advertising photos from the Figs. 54.1A–L in relation to the corresponding survey questions from Q1 to Q12 shown in the Table 54.1

54.4 Results

The survey in this research addresses two main themes and variables: esthetics and demographics. The following results are from performing the SPSS statistical tests.

54.4.1 Esthetics

Analysis revealed for Question 1, 2, 5, and 6, the most popular answers were somewhat agreed, and the least popular answer was somewhat disagreed. In regard to the frequency's percentages, Question 1, the popular answer (46.3%) and least popular (2.7%), Question 2, the popular answer (44.4%) and least popular (1.9%), Question

Fig. 54.2 Advertising photos from the Figs. 54.2A–L in relation to the corresponding survey questions from Q13 to Q24 shown in the Table 54.1

Fig. 54.3 Advertising photos from the Figs. 54.3A–H in relation to the corresponding survey questions from Q25 to Q32 shown in the Table 54.1

5, the popular answer (51.9%) and least popular (7.4%), and Question 7, the popular answer (46.3%) and least popular (5.6%).

The mean score for consumer liking of advertisements with only words was 3.35 (SD = 1.38). A one-sample t test indicated that this mean value was non-significantly different from the midpoint (neutral value of 3.0), $t\ (53) = 1.88, p = 0.066$, indicating that consumers neither significantly liked nor disliked the advertisement with only words.

A paired-samples t test was conducted to determine if potential consumers wanted to purchase more from a Lacoste advertisement with red colors or a Lacoste ad with blue colors. The results indicated the mean desire to purchase from a Lacoste advertisement with red colors (M = 3.89, SD = 1.09) was significantly greater

Table 54.1 Designed 32 survey questions (5-point Likert scale analysis. 1 = strongly disagree, 5 = strongly agree) in relation to the corresponding advertising photos, as shown in Figs. 54.1, 54.2 and 54.3 in this study

Photos	Designed survey questions in relation to the corresponding advertising photos
Figure 54.1A	Q1: This brand seems trustworthy. 5-point scale (1 = strongly disagree)
Figure 54.1B	Q2: I believe what this brand is saying. 5-point scale (1 = strongly disagree)
Figure 54.1C	Q3: I believe this brand cares about the world. 5-point scale (1 = strongly disagree)
Figure 54.1D	Q4: I believe this company cares about social causes. 5-point scale
Figure 54.1E	Q5: I believe this company has expertise. 5-point scale (1 = strongly disagree)
Figure 54.1F	Q6: I trust this company. 5-point scale (1 = strongly disagree, 5 = strongly agree)
Figure 54.1G	Q7: I believe this company has high-quality products. 5-point scale (1 = strongly disagree)
Figure 54.1H	Q8: I want to purchase this item. 5-point scale (1 = strongly disagree, 5 = strongly agree)
Figure 54.1I	Q9: I want this purchase from this company. 5-point scale (1 = strongly disagree)
Figure 54.1J	Q10: I want to buy this brand. 5-point scale (1 = strongly disagree, 5 = strongly agree)
Figure 54.1K	Q11: I want to buy this clothing. 5-point scale (1 = strongly disagree, 5 = strongly agree)
Figure 54.1L	Q12: I believe this brand cares about people. 5-point scale (1 = strongly disagree)
Figure 54.2A	Q13: I believe this product is of high quality. 5-point scale (1 = strongly disagree)
Figure 54.2B	Q14: I believe this product is of high quality. 5-point scale (1 = strongly disagree)
Figure 54.2C	Q15: I want to buy from this company. 5-point scale (1 = strongly disagree)
Figure 54.2D	Q16: I want to purchase from this company. 5-point scale (1 = strongly disagree)
Figure 54.2E	Q17: I want to purchase from this company. 5-point scale (1 = strongly disagree)
Figure 54.2F	Q18: This company has a fun personality. 5-point scale (1 = strongly disagree)
Figure 54.2G	Q19: This company feels like it sells high-quality product. 5-point scale
Figure 54.2H	Q20: I want to own products from this company. 5-point scale (1 = strongly disagree)
Figure 54.2I	Q21: I believe this brand creates high-quality products. 5-point scale
Figure 54.2J	Q22: I want to dine at this restaurant. 5-point scale (1 = strongly disagree)
Figure 54.2K	Q23: I think this restaurant would have good food. 5-point scale (1 = strongly disagree)
Figure 54.2L	Q24: I would eat at this restaurant. 5-point scale (1 = strongly disagree, 5 = strongly agree)

(continued)

Table 54.1 (continued)

Photos	Designed survey questions in relation to the corresponding advertising photos
Figure 54.3A	Q25: I want to visit this restaurant. 5-point scale (1 = strongly disagree, 5 = strongly agree)
Figure 54.3B	Q26: I want to purchase this product. 5-point scale (1 = strongly disagree)
Figure 54.3C	Q27: I want to eat this product. 5-point scale (1 = strongly disagree, 5 = strongly agree)
Figure 54.3D	Q28: I want to eat at this restaurant. 5-point scale (1 = strongly disagree)
Figure 54.3E	Q29: I want to try this food. 5-point scale (1 = strongly disagree, 5 = strongly agree)
Figure 54.3F	Q30: I would try this product. 5-point scale (1 = strongly disagree, 5 = strongly agree)
Figure 54.3G	Q31: I want to try this food. 5-point scale (1 = strongly disagree, 5 = strongly agree)
Figure 54.3H	Q32: I would purchase this product. 5-point scale (1 = strongly disagree)

than the mean desire to purchase from a Lacoste ad with blue colors (M = 3.76, SD = 1.24), t (53) = 0.87, $p < 0.001$. A paired-samples t test was conducted to determine if potential consumers wanted to purchase more in terms of high quality from a brightly colored advertisement or darker colored advertisement. The results indicated the mean desire to purchase from a brightly colored advertisement (M = 3.80, SD = 1.14) was significantly lesser than the mean desire to purchase from a darker colored advertisement (M = 4.07, SD = 1.15), t (53) = −2.13, $p = 0.038$.

54.4.2 *Demographics*

An independent samples t test was conducted to evaluate if gender affected perceived enjoyment of an advertisement featuring women. The test was non-significant, t (52) = 0.94, i = 0.35, although there was a tendency for males (M = 3.94, SD = 1.29) to perceive more enjoyment of an advertisement featuring women than females (M = 3.61, SD = 1.23).

A one-way analysis of variance was conducted to evaluate if desire to eat at a restaurant differed depending on employment status over the last three months. The independent variable, the employment status, included four options: full-time worker (M = 4.17, SD = 0.892), part-time worker (M = 3.80, SD = 1.64), unemployed (M = 1, SD = N/A), and stay-at-home parent (M = 1, SD = N/A). The dependent variable was the desire to eat at a restaurant measured on a 5-point scale. The ANOVA was significant, F (3) = 6.866, $p < 0.001$. Post-hoc tests utilizing Bonferroni are not able to be performed since one group has fewer than two cases.

A chi-square test was used to determine whether females or males are more or less likely to purchase the tea product in Question 31 and the fries in Question 32.

There was a tendency for males to be more likely to make a purchase. However, this difference was statistically non-significant for the tea product, χ^2 (4) = 2.26, p = 0.689, and non-significant for the fries, χ^2 (4) = 2.89, p = 0.575.

54.5 Discussion

Through understanding participant's tendencies for digital marketing advertisements, companies can use this analysis to improve their digital marketing campaigns through esthetics and targeting the right demographics. By analyzing the colors, font, and staging esthetics of the digital advertisements, it seems people are drawn to bold and bright colors when looking at advertisements. When comparing a red Lacoste advertisement versus a blue Lacoste advertisement in the paired-samples *t* tests, participants felt more compelled to purchase from the red colored advertisement. Likewise, between the luxury advertisements, with one brightly yellow colored advertisement and the other being a dark purple colored advertisement, participants were more likely to choose the bright yellow advertisement. Thus, a practical implication for companies to implement would be the use of bright and bold colors to capture the potential consumer's attention if they want to convert a purchase or seem higher quality. In terms of text on the advertisements, there was no significant preference for the amount of words on an advertisement, as demonstrated by the one-sample *t* test; colors and staging are more significant variables to consider.

With regards to demographics, there was no significant result presented between females and males viewing a digital advertisement; though there was a tendency for males to enjoy an advertisement more and purchase more, as shown through the chi-square and independent samples *t* test. Most participants were employed, and the ANOVA result was significant when it came to understanding the relationship between the desire to eat out and employment. The participants who were employed were more likely to agree when a question asked if they wanted to eat at a restaurant due to an advertisement. A practical implication of this for companies would be to place heavier emphasis on the esthetics in terms of color and staging and not emphasize the gender impact, as there is not much of a distinction. In addition, companies can also choose to place their digital advertisements or posts in places where they know employees frequent, such as LinkedIn, a social networking platform designed for working professionals.

If this study was to be repeated, there would be adjustments made. The answers participants gave would be more narrowed down—take out the "*somewhat agree*" and "*somewhat disagree*" option. As depicted by the frequency tests, the majority of individuals picked "*somewhat agree*", which in hindsight seems like a safe, objective answer. However, that did not provide proper analysis for drawing important conclusions. In addition, there would be more follow up questions regarding advertisements with only text, as there were only two questions regarding that topic, and there was insufficient data to draw a proper conclusion.

54.6 Conclusion

There are multiple conclusions to be drawn for companies to apply in their digital marketing campaigns and branding. A practical implication for companies to implement would be the use of bright and bold colors to capture the potential consumer's attention if they want to convert a purchase or seem higher quality. To go further, these bright and bold colors seem to draw in more attention in less time when it came to the participant's choosing the advertisement that was more compelling. There has been research on color psychology that demonstrates bright and bold colors leave users feeling more energetic and eliciting more positive than negative responses [1]. This supports the conclusion of this paper that participants respond favorably to these brightly colored marketing campaigns, and a company should expand their color palette in these areas. In terms of text on the advertisements, there was no significant preference for the number of words on an advertisement, as demonstrated by the one-sample *t* test; colors and staging are more significant variables to consider. Perhaps something to consider is keeping the amount of text minimal in comparison with the other color and staging esthetics of a marketing campaign. There have been studies that have shown when there is a lot of text and potential consumers lose interest in the product or service, which seems to be indicated by this study too [8].

Another practical implication based on the results is for companies would be to place heavier emphasis on the esthetics in terms of color and staging and not emphasize the gender impact, as there is not much of a distinction. In addition, companies can also choose to place their digital advertisements or posts in places where they know employees frequent, such as LinkedIn, a social networking platform designed for working professionals.

54.7 Future Research

In the opinion of the author, future area of research should include investigating more questions about the text implications since there did not seem to be a definitive answer about whether more or less text would affect and impact a marketing campaign advertisement. In addition, it would be interesting and useful to understand which colors specifically stood out to participants—even though they responded well to bright colors, does that mean all bright colors are equal in eliciting a positive response? The author plans to conduct more tests to see how imagery and texts' ability to captivate attention varies across different industries. Lastly, this study could benefit from testing the advertisements on various platforms to see if an additional variable would affect the results.

Acknowledgements The author would like to thank Professor Cynthia Martinez from Annenberg School for Communication and Journalism at University of Southern California for her advice for this research.

References

1. Briggs, S.: The science of attention: How to capture and hold the attention of easily distracted students. InformED, 28 June 2014. https://www.opencolleges.edu.au/informed/features/30-tricks-for-capturing-students-attention/
2. Chun, W., Fisher, A., Keenan, T.: New media, old media. In: A History and Theory Reader, 2nd edn. Published by Routledge Taylor & Francis Group (2016). https://www.routledge.com/New-Media-Old-Media-A-History-and-Theory-Reader/Chun-Fisher-Keenan/p/book/9781138021105
3. Gibbons, S.: You and your business have 7 seconds to make a first impression: Here's how to succeed. Forbes Newsletter, 19 June 2018. https://www.forbes.com/sites/serenitygibbons/2018/06/19/you-have-7-seconds-to-make-a-first-impression-heres-how-to-succeed/?sh=51a17ee656c2
4. Hunjet, A., Vuk, S.: The psychological impact of colors in marketing. Int. J. Vallis Aurea **3**(2) (2017). https://hrcak.srce.hr/192238
5. Khan, F., Siddiqui, K.: The importance of digital marketing: an exploratory study to find the perception and effectiveness of digital marketing amongst the marketing professionals in Pakistan. J. Inf. Syst. Oper. Manage. **7**(2) (2013). https://ideas.repec.org/a/rau/journl/v7y2013i2p221-228.html
6. Kingsnorth, S.: Digital Marketing Strategy: An Online Integrated Approach to Online Marketing, 3rd edn. Published by Kogan Page (2022). https://www.barnesandnoble.com/w/digital-marketing-strategy-simon-kingsnorth/1123181023
7. Singh, N., Srivastava, S.: Impact of colors on the psychology of marketing: a comprehensive overview. Sage J. **36**(2) (2011). https://doi.org/10.1177/0258042X1103600206
8. Singh, S.: Impact of color on marketing. Manage. Decis. **44**(6) (2006). https://www.deepdyve.com/lp/emerald-publishing/impact-of-color-on-marketing-eDaeeWlTsr
9. The Economic Times: Here's how technology affects our lives. The Economic Times tech, 14 Jan 2018. https://economictimes.indiatimes.com/tech/internet/heres-how-technology-affects-our-life/technology-addiction/slideshow/62497145.cms
10. The Guardian Labs: Is technology short-changing our attention spans? The Guardian Labs News, 22 Apr 2021. https://www.theguardian.com/sbs-on-demand--are-you-addicted-to-technology/2021/apr/23/is-technology-short-changing-our-attention-spans
11. Tiago, M., Verissimo, J.: Digital marketing and social media: Why bother? Bus. Horiz. **57**(6), 703–708 (2014). https://www.sciencedirect.com/science/article/abs/pii/S0007681314000949
12. Winterer, S.: What is a social media campaign? Digital Logic (2021). https://www.digitallogic.co/blog/what-is-a-social-media-campaign/
13. Wu, X., Zhang, F., Yu, Z.: Brand spillover as marketing strategy. Manage. Sci. **68**(7), 4755–5555 (2021). https://doi.org/10.1287/mnsc.2021.4165

Author Index

J. L. Reis et al. (eds.), *Marketing and Smart Technologies*, Smart Innovation, Systems and Technologies 337,
https://doi.org/10.1007/978-981-19-9099-1

Printed in the United States
by Baker & Taylor Publisher Services